W9-CHR-682

ANNUAL REVIEW OF FLUID MECHANICS

ANNUAL REVIEW OF FLUID MECHANICS

VOLUME 25, 1993

JOHN L. LUMLEY, *Co-Editor*
Cornell University

MILTON VAN DYKE, *Co-Editor*
Stanford University

HELEN L. REED, *Associate Editor*
Arizona State University

ANNUAL REVIEWS INC 4139 EL CAMINO WAY P.O. BOX 10139 PALO ALTO, CALIFORNIA 94303-0897

ANNUAL REVIEWS INC.
Palo Alto, California, USA

International Standard Serial Number: 0066-4189
International Standard Book Number: 0-8243-0725-9
Library of Congress Catalog Card Number: 74-80866

∞ The paper used in this publication meets the minimum requirements of American National Standard for Information Sciences—Permanence of Paper for Printed Library Materials, ANSI Z39.48-1984.

TYPESET BY BPCC-AUP GLASGOW LTD., SCOTLAND
PRINTED AND BOUND IN THE UNITED STATES OF AMERICA

PREFACE

Each year the Editorial Committee of the *Annual Review of Fluid Mechanics* meets just before Thanksgiving to select the topics for, and the authors who will be invited to contribute to, the volume which will be published some twenty-six months later. There are usually about ten of us in attendance, and each of us brings to the meeting a list of perhaps a dozen suggested topics, all of which may have several possible authors. We spend the day selling, browbeating, and negotiating to reduce the 120 initial topics to a short list of perhaps 25. Every year we leave in the meeting room 80% of the ideas that seemed so wonderful when we arrived. Of course, some of these are recycled the following year, if they still seem attractive.

We try to balance many things in our final selection, among them: theory and application; mathematics, computation, and experiment; the various branches of engineering and physics; trendiness and tradition; and we try to extend this balance over several years, like a blender in a winery.

The most encouraging aspect of our selection process is the apparently infinite renewability of the resource which is fluid mechanics. It is as encouraging as watching a grapevine renew itself each spring. In our case, not only is there no shortage of new sprouts each year, but the vine is vigorous enough to survive our drastic pruning, so necessary to maintain quality.

We hope you enjoy the result of our labors in this vineyard!

THE EDITORS

SOME RELATED ARTICLES IN OTHER *ANNUAL REVIEWS*

From the *Annual Review of Astronomy and Astrophysics*, Volume 30 (1992):

Dust-Gas Interactions and the Infrared Emission from Hot Astrophysical Plasmas, Eli Dwek and Richard G. Arendt

Solar Flares and Coronal Mass Ejections, S. W. Kahler

Smoothed Particle Hydrodynamics, J. J. Monaghan

From the *Annual Review of Earth and Planetary Science*, Volume 20 (1992):

Hotspot Volcanism and Mantle Plumes, Norman H. Sleep

Giant Planet Magnetospheres, Fran Bagenal

The Character of the Field During Geomagnetic Reversals, Scott W. Bogue and Ronald T. Merrill

Mixing in the Mantle, Louise H. Kellogg

From the *Annual Review of Energy and the Environment*, Volume 16 (1991):

Fluidized-Bed Combustion Technology, E. Stratos Tavoulareas

From the *Annual Review of Materials Science*, Volume 22 (1992):

Transient Liquid Phase Bonding, W. D. MacDonald and T. W. Eager

From the *Annual Review of Physical Chemistry*, Volume 42 (1991):

Dynamics of Suspended Colloidal Spheres, R. B. Jones and P. N. Pusey

Stratospheric Ozone Depletion, F. Sherwood Rowland

Annual Review of Fluid Mechanics
Volume 25, 1993

CONTENTS

(*continued*)

ANNUAL REVIEWS INC. is a nonprofit scientific publisher established to promote the advancement of the sciences. Beginning in 1932 with the *Annual Review of Biochemistry*, the Company has pursued as its principal function the publication of high quality, reasonably priced *Annual Review* volumes. The volumes are organized by Editors and Editorial Committees who invite qualified authors to contribute critical articles reviewing significant developments within each major discipline. The Editor-in-Chief invites those interested in serving as future Editorial Committee members to communicate directly with him. Annual Reviews Inc. is administered by a Board of Directors, whose members serve without compensation.

ANNUAL REVIEWS OF
Anthropology
Astronomy and Astrophysics
Biochemistry
Biophysics and Biomolecular Structure
Cell Biology
Computer Science
Earth and Planetary Sciences
Ecology and Systematics
Energy and the Environment
Entomology
Fluid Mechanics
Genetics
Immunology
Materials Science
Medicine
Microbiology
Neuroscience
Nuclear and Particle Science
Nutrition
Pharmacology and Toxicology
Physical Chemistry
Physiology
Phytopathology
Plant Physiology and Plant Molecular Biology
Psychology
Public Health
Sociology

SPECIAL PUBLICATIONS

Excitement and Fascination of Science, Vols 1, 2, and 3

Intelligence and Affectivity, by Jean Piaget

For the convenience of readers, a detachable order form/envelope is bound into the back of this volume.

Jean Léonard Marie Poiseuille (1797–1869). From a photographic portrait that appeared with the article by Brillouin (1930); oil-painted enhancement by SPS.

Annu. Rev. Fluid Mech. 1993. 25 : 1–19

THE HISTORY OF POISEUILLE'S LAW

Salvatore P. Sutera

Department of Mechanical Engineering, Washington University, St. Louis, Missouri 63130-4899

Richard Skalak

Department of Applied Mechanics and Engineering Sciences, University of California, San Diego, La Jolla, California 92093-0412

1. BIOGRAPHICAL HIGHLIGHTS AND MYSTERIES

Jean Léonard Marie Poiseuille entered the Ecole Polytechnique at the age of 18 in the fall of 1815. His residence there ended April 13, 1816 when the entire Ecole was disbanded for political reasons. He did not go back when it reopened but switched to the study of medicine instead. During his months at Ecole Polytechnique Poiseuille took courses from Cauchy, Ampère, Hachette, Arago, Petit, and Thénard. Brillouin (1930) attributes Poiseuille's extraordinary sense of experimental precision to the influence of his physics professor, the brilliant but short-lived (1791–1820) Alexis Petit, who along with P. L. Dulong discovered in 1819 that the molar specific heat of all solids tends to a constant at high temperature (Dulong-Petit rule). During his doctoral research on *The force of the aortic heart* (Poiseuille 1828), Poiseuille invented the U-tube mercury manometer (called the hemodynamometer) and used it to measure pressures in the arteries of horses and dogs. A recording version of the manometer, named the Poiseuille-Ludwig hemodynamometer, was used in medical schools until the 1960s and to this day blood pressures are reported in mm Hg due to Poiseuille's invention.

Between 1828 and 1868 Poiseuille published 15 articles ranging from

0066–4189/93/0115–0001$02.00

brief communications to the French Academy of Sciences to extensive monographs. A complete list of Poiseuille's publications is given under the Literature Cited section (from Pappenheimer 1978). It is remarkable that these few experimental papers have made the name of Poiseuille familiar in a variety of fields including engineering, physics, medicine, and biology. Following completion of his doctoral dissertation on the heart and pulse waves, Poiseuille turned his attention to hemodynamics in microcirculation. His observations of the mesenteric microcirculation of the frog (Poiseuille 1835) revealed that blood flow in the arterioles and venules features a plasma layer at the vessel wall in which there are few red cells, that "plasma-skimming" occurs at vessel bifurcations, and that white cells tend to adhere to the vessel wall. The realization that uncontrolled in vivo studies would not permit a clear formulation of the laws governing blood flow in microcirculation led him to undertake his careful and extensive studies of the flow of liquids in small diameter glass capillaries.

These studies presumably began sometime in the 1830s since in 1838 he gave a preliminary oral report on the effects of pressure and of tube length to the Société Philomatique (Poiseuille 1838). Then, in 1839, Poiseuille deposited with the French Academy of Sciences a sealed packet containing the results of his studies on the flow of water through glass tubes and the effects of pressure drop, tube length, tube diameter, and temperature. The purpose of this procedure was to establish priority. During the academic year 1840–1841 he made three oral communications (*Mémoires lus*) to the Academy of Sciences. Excerpts of these were subsequently published in the Academy's *Comptes Rendus* (Poiseuille 1840a,b; 1841). In January 1841 Poiseuille deposited another sealed packet of experimental results dealing with the flow of a variety of liquids through glass capillaries. Some of these results were communicated to the Academy in 1843 (Poiseuille 1843).

The results and conclusions presented by Poiseuille in 1840–1841 were considered sufficiently important that the Academy appointed an elite Special Commission to investigate their validity. This Commission, consisting of members Arago, Babinet, Piobert, and Régnault, met in 1842 and with Poiseuille repeated some of his experiments using his apparatus. In the course of this review, the Commission prevailed upon Poiseuille to do some new preliminary experiments using mercury and ethyl ether. The Commission reported back to the Academy on December 26, 1842 recommending that Poiseuille's work be approved and included in its entirety in *Mémoires des Savants Etrangers*, a publication of the Academy of Sciences. It actually appeared in the *Mémoires Presentés par Divers Savants à l'Académie Royale des Sciences de l'Institut de France* in 1846, seven full years after he delivered his first sealed packet to the Academy.

A complete English translation of this paper is available (in Bingham 1940). The Commission's report was published in the *Annales de Chimie et Physique* (Régnault et al 1843). Poiseuille's final contribution to the subject of liquid flow in narrow tubes appeared in September 1847. That paper presented measurements for (*i*) dilute aqueous salt solutions, (*ii*) aqueous solutions of bases, (*iii*) aqueous solutions of acids, (*iv*) mineral waters, (*v*) teas, (*vi*) wines and spirits, (*vii*) extracts of plants and roots, (*viii*) bovine serum and acidic solutions thereof, and (*ix*) a mixed group of ethers, alcohols, and solutions of ammonia. In each group flow times were compared to that of distilled water under the same conditions. It appears that these studies were motivated by Poiseuille's interest in the possible facilitation of capillary blood flow through medication.

There is no record of where Poiseuille did his work or how it was supported financially. His apparatus was elaborate and certainly required the services of an expert glassblower. The experiments were time-consuming (the calibration of a single capillary tube took as long as twelve hours) so he probably had technical assistance. Brillouin (1930) suggests the possibility that the well established physiologist Magendie provided space and necessary resources at La Salpêtrière Hospital in Paris. Pappenheimer (1978) suggests that a wealthy father-in-law may have made it possible for Poiseuille to dedicate himself to research. Apparently Poiseuille practiced medicine for a while because he was listed in a Paris directory of physicians dated 1845, but other evidence indicates that he did not practice medicine after 1844.

Original biographical information on Poiseuille's life is scarce. Brillouin (1930), Joly (1968), and Pappenheimer (1978) summarize most of the known information. Joly's biographical note, which was delivered prior to the presentation of the first Poiseuille medal to Robin Fåhraeus in 1966, is an especially eloquent testimony to the many facets of this scientist and his accomplishments. Joly points out that during his lifetime Poiseuille was only modestly recognized. In 1835, the Academy of Sciences awarded him half of the prize for experimental physiology (value unmentioned); in 1845, he won the prize for medicine and surgery (worth 700 francs), and in 1860, he received an honorable mention, again from the Academy of Sciences. Although Poiseuille was an elected member of the Paris Academy of Medicine, his numerous attempts to win election to the Academy of Sciences in the 1840s, 1850s, and 1860s were never successful.

Another mysterious aspect of Poiseuille's life concerns his circumstances and employment in later life. In 1858, he filed an application for a position in the Paris public school system. In 1860, Dr. Poiseuille went to work as Inspector of School Sanitation in the Seine district. Poiseuille, born on April 22, 1799, died in Paris, the city of his birth on December 26, 1869.

2. POISEUILLE'S EXPERIMENTS

Poiseuille set out to find a functional relationship among four variables: the volumetric efflux rate of distilled water from a tube Q, the driving pressure differential P, the tube length L, and the tube diameter D. The diameters of his glass tubes ranged from 0.015 to 0.6 mm, encompassing vessel sizes found in most microcirculatory systems but not quite that of human capillaries ($\sim$5 to 10 microns). Initially he planned to maintain a constant temperature of 10°C; subsequently he examined the influence of temperature from 0 to 45°C, still using distilled water as the test liquid. He later extended his studies to a great variety of other liquids (Poiseuille 1847).

Figure 1 is a drawing of the frontal view of Poiseuille's experimental apparatus. It is estimated that the apparatus stood between $2\frac{1}{2}$ and 3 meters tall. The heart of the system is the small capillary viscometer labeled *c–e–d* near the center of the figure just below the spindle-shaped bulb *M*. Because it is immersed in water inside a glass cylinder the viscometer is shown in dotted outline. A hand-operated pump (*h*, surrounded by a water jacket *X–Y*) was used to charge the vertical reservoir on the left with air and simultaneously to raise either a water column in the tall manometer *i–i* or a mercury column in the short manometer *i′–i′*. During pressurization the valve *R* leading to the viscometer was closed. Once the desired pressure, as indicated by one of the manometers, was reached, the pump discharge valve *R′* was closed, valve *R* was opened, and flow was driven through the test capillary *d*.

A close-up enlargement of the viscometer assembly is shown in Figure 2. The pointed bottom of the bulb *M* served to trap dust particles—which tended to settle out either from the air or the liquid used to clean the glassware—preventing them from falling into the capillary branch (*b″–c–e–d*). Poiseuille found it necessary to filter his distilled water repeatedly, sometimes as many as 20 times, to be rid of foreign particles. The entire test capillary was situated under water in a glass cylinder (*C–D–F–E*) which was surrounded by a water bath (*G–H–I–K*).

The underwater efflux of the capillary tubes was necessitated when Poiseuille discovered that he could not achieve reproducible results when the minuscule liquid flows (some as low as 0.10 cc in several hours) exited in air against the erratic resistance of surface tension. This problem was eliminated by underwater efflux, but required that flow be measured upstream by timing the passage of a liquid meniscus between two lines, *C* and *E* (Figure 3), which delimited a known volume of the spherical bulb *A–B*. The second smaller bulb *G* in Figure 3 provided the entrance to the horizontal test capillary *D* which was fused to *G* so as to provide an abrupt entrance. This was crucial to the accurate definition of tube length.

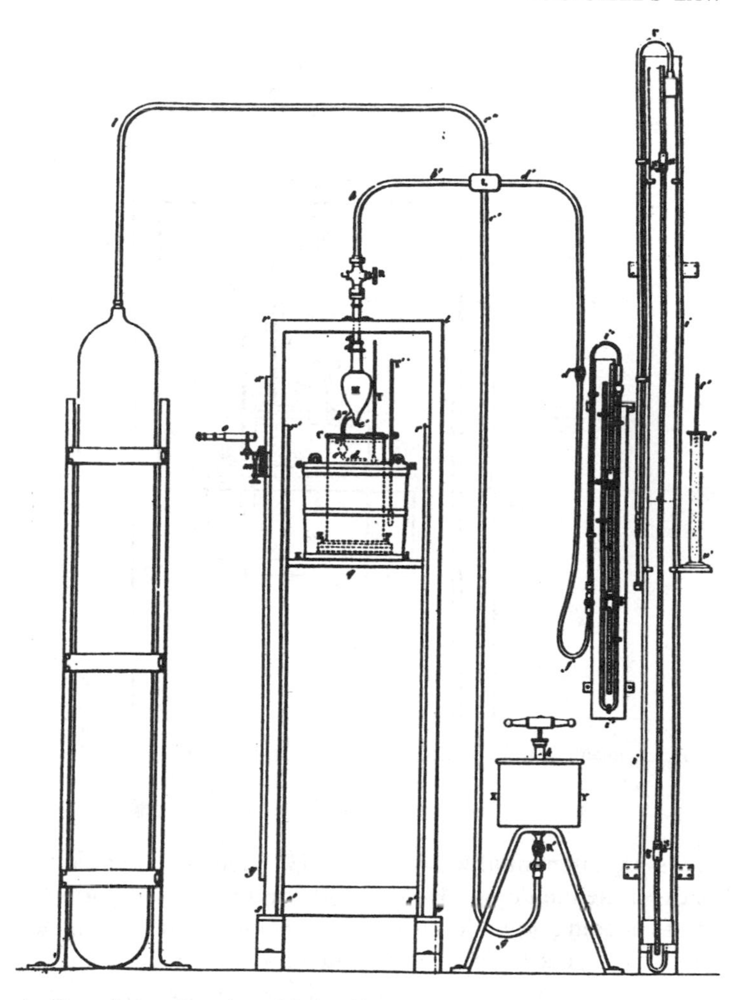

Figure 1 Frontal elevation view of Poiseuille's apparatus. Photocopy of one segment of a ten part fold-out plate published with Poiseuille's summary paper (1846).

Poiseuille extended his study of the influence of pressure up to about eight atmospheres. (At a pressure of 10 atmospheres, one of the bulbs *M* exploded.) For pressures above one atmosphere the spherical bulbs were replaced by a cylindrical vessel shown in Figure 4. These cylinders and the attached test capillaries (labeled *K–D* in Figure 4) were tested in air. In these cases, the effluxes were large enough so that surface tension was not a problem.

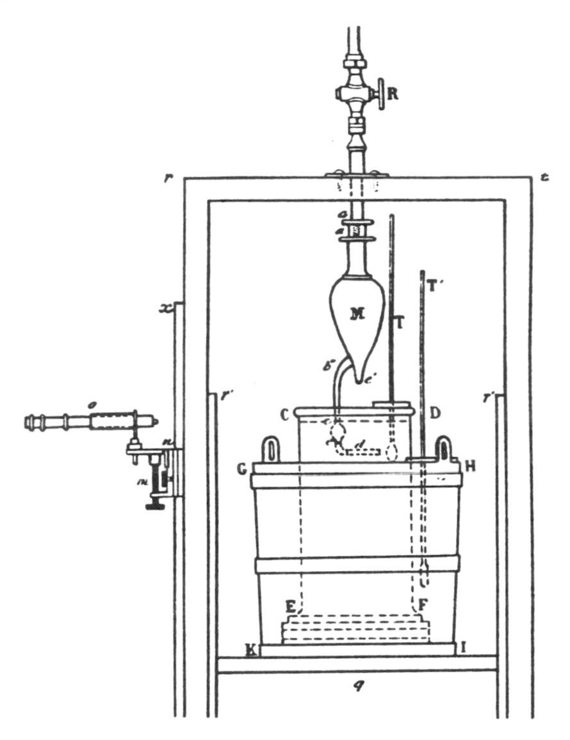

Figure 2 Enlargement of the viscometer assembly, from Figure 1.

The pressure differential was the primary independent variable in Poiseuille's experimental design. However, the head declined during outflow because of changing liquid levels in the manometer, the viscometer bulb, and the receiver vessel. Following contemporary understanding among hydraulic engineers, Poiseuille used the arithmetic average of the initial and final heads for P in his data analysis. He even performed an auxiliary experiment (one of several) to test the accuracy of this assumption. Bingham pointed out in his critique (1940) that the arithmetic mean is not rigorously the correct average to use but that, given the dimensions of his viscometer bulbs and the total heads applied, Poiseuille probably avoided any appreciable errors from this approximation. Poiseuille was meticulous in making second-order corrections for (*a*) the difference in the atmospheric pressures acting on the water in the open manometer leg S' and the free surface of the receiver vessel, (*b*) the weights of unequal air columns confined within the pressurized legs of the apparatus, and (*c*) capillarity in

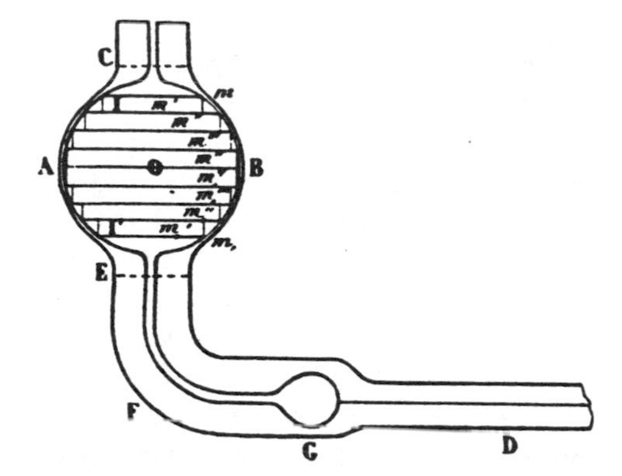

Figure 3 Detailed drawing of the spherical viscometer bulb (O) with attached test capillary (D) (from Poiseuille 1846). The horizontal lines m, m', ..., m_1 constructed inside the bulb were used by Poiscuille to argue that the elevation of the midplane (AOB) could be used to determine the average pressure under which the bulb volume was discharged.

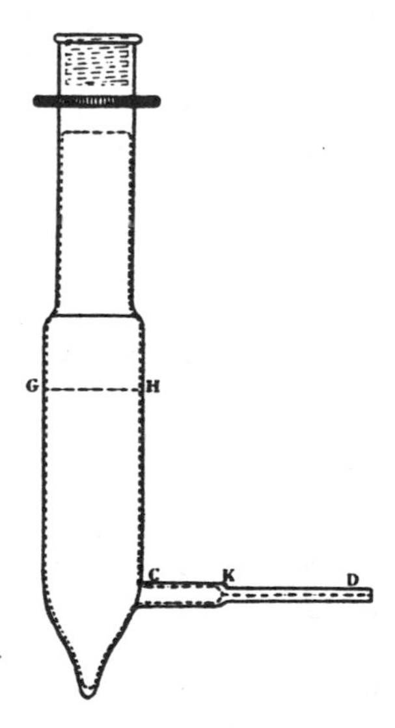

Figure 4 Cylindrical-conical viscometer bulb employed in high pressure experiments. KD is the capillary. From Poiseuille (1846).

the viscometer bulb. In one sample calculation Poiseuille showed that the correction due to different air column weights amounted to about 0.15%.

Correction for the capillarity in the spherical bulbs was more problematic because the area of the air-water interface varied continuously during flow. Poiseuille solved this by running auxiliary trials on graduated cylindrical bulbs wherein the capillarity was constant. First a capillary tube connected to a spherical bulb was tested. Then this tube was detached from the spherical bulb and reattached to a graduated cylindrical bulb. By carefully timing the outflow of a volume equal to that of the spherical bulb through the cylindrical bulb at the same average pressure and temperature, Poiseuille was able to figure the net capillarity correction of the original spherical bulb. This process was repeated "a great number of times" for each and every bulb used in the experiments. Again, by numerical example, Poiseuille also showed the capillarity correction to be of the order of three parts in 2000.

From a great number of glass tubes which he examined, Poiseuille selected a few which appeared to be fairly cylindrical along their length. This first screening was done by measuring the length of a thread of mercury a few centimeters long at different positions along the length of the tube. The cross section of a tube was then examined by cutting a small perpendicular section 2 to 3 mm long from one end, and grinding and polishing its faces until its thickness was reduced to about 0.1 mm. This thin annular disk was then placed between two plates of glass along with some Canada balsam and heated. The heating caused the balsam to flow into the small bore. This sandwich was then examined under the objective of a horizontal *Amici* microscope. Owing to the thinness of the annular disk, problems due to reflection, refraction, and diffraction were eliminated and the image of its bore was distinct and clear. By means of an illuminated chamber, a *camera lucida*, fitted to the microscope, an image of the bore was projected at a known magnification on the horizontal table of the microscope and its maximum and minimum diameters were measured with dividers and a millimeter scale. By this technique Poiseuille specified his tube diameters in millimeters, nominally from 0.015 to 0.6, to four and sometimes five places, i.e. to tenths or hundredths of a micron! One can question the significance of the fourth and fifth digits in these measured diameters, however, given that the original measurements of the magnified projected images were made by dividers and a millimeter scale and could be read perhaps to within $\frac{1}{2}$ part in 10 to 300 mm.

The lengths of the glass tubes were measured, after both ends were ground smooth, by means of a beam-compass equipped with a vernier scale. This tool (which was borrowed from the physical laboratory of the Collège de France courtesy of Monsieur Savart) could be read to within

1/20 to 1/40 mm. The series of seven tubes used in the "length study" ranged from 6.77 to 100.5 mm long.

Poiseuille recorded efflux times to the nearest quarter of a second but did not identify the particular timepiece (chronometer) used.

In most of the experiments dealing with the influence of pressure, tube diameter, and length, Poiseuille maintained the temperature of the bath surrounding the receiver vessel at 10°C. The temperature was indicated by the thermometer T (Figure 2) situated in the receiver with its bulb at the same level as the test capillary. This thermometer had divisions of fifths of a degree Celsius. Poiseuille's papers say nothing about how temperature was controlled. In two subsets of his experiments on the effect of pressure in which the driving heads were high, Poiseuille used tubes that were too long to fit in the receiving vessel. Hence these tests were performed in air at ambient temperatures varying from about 20°C all the way down to 7°C! (Apparently the laboratory was unheated.)

Poiseuille first studied the effect of pressure on flow. He began with a tube referred to as A, which was 100.5 mm long, determined its maximum and minimum internal diameters at each end (in this instance, exit end: 0.1395 mm, 0.1415 mm; entrance end: 0.1405 mm, 0.1430 mm), and fused it to the bulb G (Figure 3). Pressures of 385.870, 739.114, and 773.443 mm Hg at 10°C were established in succession and the corresponding flow times of 13.34085 cm^3 (the bulb volume at 10°C) of distilled water measured. These were 3505.75, 1830.75, and 1750.00 s, respectively. Next, successive portions of the end of the tube were cut off to provide test lengths of 51.1, 25.55, 15.75, 9.55, 6.775, and about 1 mm. The same procedure was followed with tubes B, C, D, E, F, G, H, I, and K with nominal internal diameters of 0.11, 0.085, 0.045, 0.03, 0.65, 0.63, 0.01, 0.09, and 0.13 mm, respectively. The lowest pressure applied was 74.29 mm of water and the highest was over 6000 mm Hg (about 8 atmospheres).

Poiseuille summarized his findings at this stage by the equation $Q = KP$, where the coefficient K was a function, to be determined, of tube length, diameter, and temperature. To investigate the influence of tube length Poiseuille took from his previous experiments on the A series of tubes all the data from those runs where the pressure was close to 775 mm Hg. Then, using his "law of pressures" he adjusted the measured flow times to correspond to a standard P of exactly 775 mm Hg. He was then able to show that the flow time was proportional to tube length (the "law of lengths") in a majority of his experiments. At this point Poiseuille could state that $K = K'/L$ and, therefore, $Q = K'P/L$, where K' was a function of tube diameter and temperature.

To determine the effect of tube diameter on flow Poiseuille (1847) stated that "we have measured the volumes of liquid flowing through tubes of different diameters under the same pressure, at the same temperature, in the same time, the tubes having the same length; and we have compared the efflux, taking the diameters of the tubes into account." In fact, Poiseuille used the data he already had in hand, interpolating as necessary and applying the "laws of pressure and length", to arrive at a set of volume-diameter data standardized to $P = 775$ mm Hg, $L = 25$ mm, $\delta t = 500$ s and $T = 10$°C. The volume of the bulb used in each experiment was accurately determined by weighing the mercury contained between the lines C and E (Figure 3) to the nearest 0.5 mg. Since these weighings were carried out at ambient temperature, the calculated bulb volumes were corrected for the thermal expansion of glass to find the correct volume at the standard temperature of 10°C.

To assign a diameter to one of his noncircular, noncylindrical tubes, Poiseuille first calculated a geometrical average diameter for each end. This was defined as the diameter of the circle having the same area as an ellipse with the maximum and minimum diameters of the tube section. The arithmetic average of the geometrical means at the two ends was taken as the average diameter of the tube.

Following the above scheme, Poiseuille analyzed the data of seven of his previous experiments from which he was able to discern that the efflux volumes (in 500 s) varied directly as the fourth power of the average diameter. He would now claim that

$$Q = K''PD^4/L, \tag{1}$$

K'' being simply a function of temperature and the type of liquid flowing. For 10°C his data yielded an average value of $K'' = 2495.224$ for distilled water expressed in mixed units of (mg/s)/(mm Hg) mm^3.

In his final series of experiments, Poiseuille explored the influence of temperature from a few tenths of a degree C to 45°C. He used four of his original tubes (before truncating them): A, C, D′, and E. In each case he corrected both tube diameter and bulb volume for thermal expansion or contraction relative to the reference state of 10°C. Recognizing that the dependence of K'' on the temperature T was nonlinear, he elected to seek a polynomial fit of the form $K'' = K_1(1+AT+A'T^2+A''T^3+\ldots)$ and found for distilled water:

$$K'' = 1836.7\,(1+0.033679T+0.00022099\,T^2), \tag{2}$$

where T is in °C.

Poiseuille recognized what are now called entrance effects, but did not come to precise conclusions. In his first series of experiments on the pressure effect beginning with tube A, Poiseuille found that the results obtained from shorter tubes deviated from the proportionality $Q = KP$.

He relegated these experiments to a "Second Series of Experiments" and excluded their data from his subsequent analyses. Poiseuille concluded that the "pressure law" would hold only if tube length exceeded a certain limit and that this limit depended on the tube diameter. He saw that the smaller the diameter, the smaller the limiting or minimum length. Beyond this observation Poiseuille had no explanation for the "Second Series." In one case, referring to the tube that was about 1 mm long, he opined that the "movement of fluid molecules" through the tube was not rectilinear. He recalled his observation of blood flow in a small diameter (0.15 mm), lateral branch from the mesenteric artery of a living frog. The "blood globules" could be seen to move along linear trajectories only if the artery was longer than about 2 mm.

The aberrant experiments (Second Series) encompassed a fairly wide range of Reynolds numbers, from close to 1 to 2600, but Poiseuille did not consider the relative roles of inertial and viscous forces in the development of tube flow. However, he expressed the belief that the pressure-flow proportionality would hold in capillary blood vessels longer than about 300 microns.

3. DERIVATION OF POISEUILLE'S LAW

Strictly speaking, Poiseuille's law as written by Poiseuille is Equation (1) above. The equation which is more usually referred to as Poiseuille's law was not derived by Poiseuille. The more usual form is:

$$Q = \pi D^4 P/128\mu L. \tag{3}$$

The difference between Equation (3) and Poiseuille's Equation (1) is simply that in Equation (3) Poiseuille's constant K'' is replaced by $\pi/128\mu$ where μ is the viscosity of the fluid. Although viscosity had been defined by Navier (1823) no mention of viscosity per se was made by Poiseuille. However, he clearly recognized that K'' was a function of temperature and the flowing liquid. Poiseuille's determinations of K'' for water were so accurate that the viscosity derived from K'' agrees with accepted values within 0.1% (Bingham 1922).

The first derivation of Equation (3) from the Navier-Stokes equations is usually attributed to Eduard Hagenbach (1833–1910), a physicist of Basel. Hagenbach's 1860 paper is reprinted in a book edited by L. Schiller (1933) who states in an appendix that at about the same time that Hagenbach's paper appeared, another derivation of Poiseuille's law was published by H. Jacobson (1860) based on lectures of Franz Neumann, a physicist of Königsberg. Neumann's own treatise did not appear until some years later (Neumann 1883). Bingham (1922) points out that derivations of

Poiseuille's law were also published by H. Helmholtz (1860), J. Stephan (1862), and E. Mathieu (1863).

Sir George Gabriel Stokes (1813–1903) of Cambridge University apparently solved the problem of Poiseuille flow as an application of the Navier-Stokes equations which he derived in the same paper in 1845. However, he did not publish the result because he was unsure of the boundary condition of zero velocity at the tube wall. He writes: "But having calculated, according to the conditions which I have mentioned, the discharge of long straight circular pipes and rectangular canals, and compared the resulting formulae with some of the experiments of Bossut and Du Buat, I found that the formulae did not at all agree with experiment." [Charles Bossut (1730–1814), Pierre Louis Georges Du Buat (1734–1809)]. Stokes was apparently unaware of Poiseuille's work at this time. Later in the same article (Stokes 1845), he discusses the flow in canals and points out the similarity to pipe flow under gravity at constant pressure. For the case of a circular pipe he writes: "In this case the solution is extremely easy" and gives the solution:

$$w = \frac{g\rho \sin\alpha}{4\mu}(\mathrm{a}^2 - r^2) + U. \tag{4}$$

Here w is the axial velocity, a and α are the radius and inclination of the pipe and U is the velocity of the fluid at the wall, which Stokes still leaves open. By 1851, Stokes felt quite sure of the no-slip condition for a viscous fluid at a rigid wall as he explicitly discusses it and uses it in his famous paper in which he derives Stokes law of drag on a sphere at low Reynolds number (Stokes 1851). But he did not remark further on pipe flow.

The naming of Equation (3) as Poiseuille's law is due to Hagenbach (1860) who, after giving the derivation, generously suggested calling it Poiseuille's law: "wir werden daher die obige Formel die POISEUILLE'sche Formel nennen." Jacobson (1860) also calls Equation (3) Poiseuille's law.

Hagenbach (1860) indicates a footnote that explains that Navier (1823) had arrived at a different equation, namely $Q = CPD^3/L$ where C is a constant [Claude Louis Marie Henri Navier (1785–1836)]. It is interesting to note that Thomas Young (1773–1829) tried to summarize existing pressure-drop formulae for flow of liquids in tubes in his Croonian Lecture of 1809 which was aimed at studying various aspects of blood flow, including wave propagation, in living organisms. He also quotes data of Bossut and Du Buat. His equations also give a dependence of Q approximately proportional to D^3. This was apparently a widespread opinion and explains why Bingham (1940) remarks on Poiseuille's work: "It was not a simple thing to go exactly counter to all of the established data and proposed formulas of the hydraulicians. It made it necessary to use the utmost possible precision."

An aspect of Poiseuille's law that is not explicitly covered in Poiseuille's work is the effect of gravity if the capillary is inclined. For this case Poiseuille's law may be written

$$Q = \frac{\pi D^4}{128\mu}\left(\frac{P}{L} + \rho X\right), \tag{5}$$

where ρ is the density of the fluid and X is the component of body force per unit mass in the direction of flow. All of Poiseuille's tests were carried out on horizontal tubes.

Another poorly documented aspect of Poiseuille flow history concerns who first solved and named the unidirectional flow between parallel plates commonly called two-dimensional Poiseuille flow. The form of Poiseuille's law that is the counterpart to Equation (5) in this case is:

$$q = \frac{H^4}{12\mu}\left(\frac{P}{L} + \rho X\right), \tag{6}$$

where q is defined as the discharge rate in a width H of the flow and H is the spacing of the plates. Poiseuille never mentioned flow between parallel plates, but such flows were well known to Stokes (1898) and were probably derived earlier.

4. HAGEN'S EXPERIMENTS

In 1839, a German hydraulic engineer, Gotthilf Heinrich Ludwig Hagen (1797–1884) of Berlin, published a paper on the flow of water in cylindrical tubes. His results were similar to those of Poiseuille, but less extensive and less accurate. However, they included some entrance effects and observations of the differences between laminar and turbulent flows. In the notation used above, Hagen's expression for the driving pressure difference was assumed to be of the form

$$P = \frac{1}{D^4}(ALQ + BQ^2) \tag{7}$$

where A and B are constants. Hagen found A to be dependent on temperature and expressed it in the form

$$A = a - bT + cT^2 \tag{8}$$

where a, b, and c are experimental constants. Hagen appreciated that the

Q^2 term in (7) was associated with generating the kinetic energy of the fluid and the term linear in Q was a fluid friction resistance. It is readily seen that for sufficiently small values of Q, the Q^2 term in Equation (7) should be negligible. Then solving Equation (7) for Q gives the same form as proposed by Poiseuille. Prandtl & Tietjens (1934) have converted Hagen's measurements of the coefficient A in Equations (7) and (8) to derive a plot of a friction factor vs the Reynolds number, $R_N = DV/\nu$, where V is the mean velocity and ν is the kinematic viscosity. Hagen's data fall very close to the theoretical line $f = 64/R_N$ (where f is the usual pipe friction factor) for a range of Reynolds numbers from about 70 to 1000. The coefficient of viscosity of water extracted from Hagen's data is also shown to agree closely with accepted values. In view of the fact that Hagen's results were quite accurate and preceded publication of Poiseuille's main papers in 1840 and 1841, Prandtl & Tietjens suggest that the laminar flow law should be called the Hagen-Poiseuille law as advocated by Ostwald (1925). It seems, however, that the majority opinion, as expressed by common usage, has settled on calling it Poiseuille's law. There are some points of rationale that can be raised in favor of this decision. It appears that Poiseuille and Hagen worked quite independently and were doing their experiments at about the same time. Their papers do not cross-reference each other's work, but Hagen in 1869 published an article pointing out that his 1839 paper preceded Poiseuille's work (Hagen 1869). Poiseuille's first paper is dated 1838, although his main results were not published until 1840 and 1841.

Hagen's tests were on three brass tubes of diameters 0.255, 0.401, and 0.591 cm and lengths of 47.4, 109, and 105 cm, respectively. In seeking the dependence of the pressure drop on tube diameter, he used a least squares fit to determine the appropriate exponent of the diameter and reported a value of -4.12—but suggested that since the possible errors in the measurements were not exactly known, a value of -4.0 be adopted. In Poiseuille's work, several more different diameters were used and the exponent -4.0 was more definitively established. Bingham (1940) concludes,

> It does not appear that entire historical justice can be done in a name and the coupling of several names together is cumbersome and unnecessary. The greatest importance must be attached to the fact that Poiseuille's paper brought conviction, whereas without it the rheological writings of all the others might have long remained unknown or never have been written.

5. EXTENSIONS AND USES OF POISEUILLE'S LAW

Historically, one of the interesting uses of Poiseuille's careful experiments was to provide evidence as to the correct boundary condition for a viscous flow at a solid boundary. Lamb (1932) remarks on the occurrence of the factor D^4 in the formula for discharge rate through a tube: "This last result is of importance as furnishing a conclusive proof that there is in these experiments no appreciable slippage of the fluid in contact with the wall." Lamb goes on to show that if there were slippage at the wall, there would be a correction to Poiseuille's law, which, in fact, Poiseuille's experiments show to be zero within measurable accuracy.

Deeley & Parr (1913) proposed naming the C.G.S. unit of viscosity the "Poise" in honor of Poiseuille. We quote from their paper on the viscosity of glacier ice:

> It would be a distinct advantage to have a name for the unit of viscosity expressed in C.G.S. units, and we would suggest that the word Poise be used for this; for it is to Poiseuille that we owe the experimental demonstration that when a liquid flows through a capillary tube of considerable length, at constant temperature, the viscosity is constant at all rates of shear, provided that the flow is not turbulent. In the case of a soft solid (plastic substance) the so-called viscosity is not the same for all rates of shear: whereas the viscosity of a liquid is a physical constant and should be named.

This usage is then found in the standards literature as early as 1918 (Perry 1955) and in later literature on weights, measures, and units (CGPM 1948, Mechtly 1964).

As a practical matter, the capillary viscometer is a simplified version of Poiseuille's test equipment, and its use is based on Poiseuille's law with interpretation based on Hagenbach's derivation, Equation (3).

It is interesting to take stock of progress made toward Poiseuille's original goal of understanding the laws of pressure distribution in a living circulation of blood. A great deal has been learned about the properties of blood cells and the flow of blood in the 20th century. When all is said and done, Poiseuille's law is a good approximation for blood flow provided the appropriate value for the apparent viscosity is used. Therein lies the rub. Red blood cells are very flexible and at low shear rates they aggregate into stacks called rouleaux. At higher shear stresses, disaggregation and deformation of the cells leads to decreasing viscosity (see Chien et al 1984, for example). Nevertheless, when bulk viscosity measurements are made (in large tubes) and compared to the apparent viscosity derived from tube flows in smaller tubes at the same hematocrit (cell concentration) and mean shear rate, the results agree quite closely for diameters down to about 29 μm (Barbee & Cokelet (1971).

Blood cells tend to move away from blood vessel walls [as observed by Poiseuille (1835) and many others] in small diameter vessels. This leads to a reduction in apparent viscosity as the diameter decreases—known as the Fåhraeus-Lindqvist (1931) effect. However, it has been shown that this result is largely due to the reduction in hematocrit which results from the centralization of the blood cells (Cokelet 1987). At sufficiently small diameters ($D < 8$ mm) there is a reverse Fåhraeus-Lindqvist effect, namely, the apparent viscosity increases with decreasing capillary diameter because the blood cells fill most of the lumen (Skalak 1990); so Poiseuille's law no longer holds.

As a direct proof of the extent of applicability of Poiseuille's law to the in vivo flow of blood, a summary of measurements due to Lipowsky et al (1978) is shown in Figure 5. A good correlation of the resistance per unit length of vessel is obtained with the exponent of vessel radius close to 4.0. Surely Poiseuille would have been glad to see this!

Poiseuille's law is one of the few equations derived from applied mechanics that is well known in the present medical community. It has been used to model other biological flows, besides blood flow. Pappenheimer (1978) explains how he used it to discuss flow through the endothelial layer

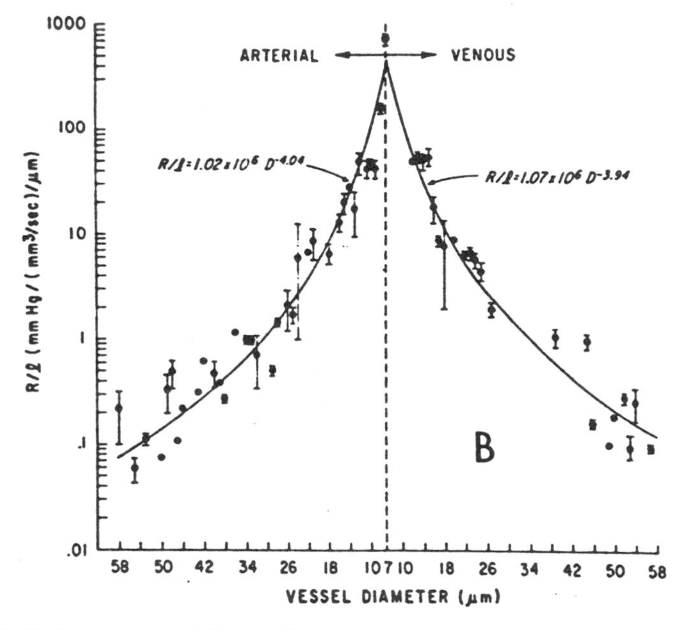

Figure 5 Resistance per unit length of vessel (R/L) where $R = \Delta P/Q$. The resistance R to blood flow is computed from simultaneous measurements of flow Q and pressure drop ΔP,in single unbranched vessels of mesentery. The solid curves are power law regressions of the form $R/L = aD^{m}$. (From Lipowsky et al 1978, by permission).

of blood vessels to approximate the size of pores or other channels that must exist to account for measured fluid transport.

Similarly, flow through porous media (Batchelor 1967, p. 233) and filters (Skalak et al 1987) have been modeled by defining equivalent Poiseuille flows. Like Stokes drag for sedimenting particles, Poiseuille's law allows an approximate length scale to be defined characterizing the geometry of laminar flows where the geometry is, in fact, much more complex.

Extensions of Poiseuille's law are legion, depending on which definition of an extension is adopted. A broad definition might be classes of flows which, in some limit of the range of the parameters involved, reduce again to Poiseuille's law. Thus, Poiseuille's law is one example of exact solutions of the Navier-Stokes equations (see Wang 1991 for a comprehensive discussion).

A case of interest to blood flow is the exact solution of the sinusoidally oscillatory rectilinear flow of a Newtonian fluid in a circular tube. This solution has also been published independently several times (McDonald 1974), but is known in the blood flow literature through the work of Womersley (1955) and McDonald (1974). At sufficiently low dimensionless frequency [$\alpha = a(\omega/\nu)^{1/2}$ where a is the tube radius and ω is the frequency], the oscillatory flow velocity profile and pressure gradient approach Poiseuille flow.

Another interesting extension is the so-called Hele-Shaw flows between parallel plates (Hele-Shaw 1898). The theory of these flows was given by Stokes (1898) who showed that the parabolic Poiseuille velocity profile was obtained in each component of velocity parallel to the bounding plates. He also pointed out the analogy of Hele-Shaw flows to two-dimensional inviscid flow.

Extensions and uses of Poiseuille flow listed above are just a few cases which readily come to mind. The authors apologize for neglect of the many additional cases and categories not covered in which the name of Poiseuille is involved. However, one of the most recent such references indicates how far afield Poiseuille's influence has extended (Chamkha 1991). Surely, Poiseuille would be surprised to see his name in the title: "Series solution for unsteady hydromagnetic Poiseuille two-phase flow." It shows how long and far the influence of Poiseuille has been manifest.

ACKNOWLEDGMENTS

The preparation of this article was supported in part by NIH Grants HL 12839 and HL 43026 from the National Heart, Lung, and Blood Institute.

Literature Cited

Barbee, J. H., Cokelet, G. R. 1971. Prediction of blood flow in tubes with diameters as small as 29 μm. *Microvas. Res.* 3: 17–21

Batchelor, G. K. 1967. *An Introduction to Fluid Dynamics*. Cambridge: Cambridge Univ. Press. 615 pp.

Bingham, E. C. 1922. *Fluidity and Plasticity*. New York: McGraw-Hill. 440 pp.

Bingham, E. C., ed. 1940. *Rheological Memoirs*, Vol. 1, 1: 1–101. Lancaster, PA: Lancaster Press. [Translation of Poiseuille (1846) by W. H. Herschel with critical notes by E. C. Bingham]

Brillouin, M. 1930. Jean Léonard Marie Poiseuille. *J. Rheol.* 1: 345–48

CGPM 1948. Comptes rendus des séances de la neuvième Conférence Générale des Poids et Mesures (CGPM). Gauthier-Villars et Cie (Paris), Resolution 7

Chamkha, R. J. 1991. Series solution for unsteady hydromagnetic Poiseuille two-phase flow. *Engineering Science Preprint* 28. Baltimore: Soc. Engr. Sci. 10 pp.

Chien, S., Usami, S., Skalak, R. 1984. Blood flow in small tubes. In *Handbook of Physiology*. The Cardiovascular System, Sec. 2, Vol. IV, Part 1, ed. E. M. Renkin, C. C. Michel, pp. 217–49. Bethesda, MD: Am. Physiol. Soc.

Cokelet, G. R. 1987. Rheology and tube flow of blood. In *Handbook of Bioengineering*, ed. R. Skalak, S. Chien, pp. 14.1–14.17. New York: McGraw-Hill

Deeley, R. M., Parr, P. H. 1913. On the viscosity of glacier ice. *Philos. Mag. VI*, XXVI: 85–111

Fåhraeus, R., Lindqvist, T. 1931. The viscosity of the blood in narrow capillary tubes. *Am. J. Physiol.* 96: 562–68

Hagen, G. H. L. 1839. Uber die Bewegung des Wassers in engen cylindrischen Röhren. *Poggendorf's Annalen der Physik und Chemie* 46: 423–42. (Reprinted, 1933, see Schiller, below)

Hagen, G. H. L. 1869. Bewegung des Wassers in cylindrischen, nahe horizontalen Leitungen. *Math. Abhand. Konig. Akad. Wissenchaften Berlin*, pp. 1–26

Hagenbach, E. 1860. Uber die Bestimmung der Zähigkeit einer Flüssigkeit durch den Ausfluss aus Röhren. *Poggendorf's Annalen der Physik und Chemie* 108: 385–426. (Reprinted, 1933, see Schiller, below)

Hele-Shaw, H. 1898. Flow of liquids in thin films. *Nature* 58: 34–36

Helmholtz, H. 1860. In *Scientific Papers of Hermann von Helmholtz*, Vol. 1. Leipzig: Johann Ambrosius Barth (published in 1892)

Jacobson, H. 1860. Beiträge zur Haemodynamik. *Arch. Anat. Physiol.* 80: 80–113

Joly, M. 1968. Notice biographique sur J. L. M. Poiseuille. *Hemorheology: Proc. First Int. Conf.*, Reykyavik, July 1966, pp. 29–31. Oxford, UK: Pergamon

Lamb, H. 1932. *Hydrodynamics*. New York: Dover

Lipowsky, H., Kovalcheck, S., Zweifach, B. W. 1978. The distribution of blood rheological parameters in the microcirculation of cat mesentery. *Circulation Res.* 43: 738–49

Mathieu, E. 1863. Sur le mouvement des liquides dans les tubes de très petit diamètre. *C. R. Acad. Sci. Paris* 57: 320–324

McDonald, D. A. 1974. *Blood Flow in Arteries*. Baltimore: Williams & Wilkens. 496 pp. 2nd ed.

Mechtly, E. A. 1964. The international system of units. Physical constants and conversion factors. Washington, DC: NASA SP-7012

Navier, C. L. M. H. 1823. Mémoire sur les lois du mouvement des fluides. *Mém. Acad. R. Sci.* 6: 389–441

Neumann, F. 1883. Einleitung in die theoretische Physik. *Herausgegeben von C. Pape*. Leipzig: Teubner. 291 pp.

Ostwald, F. W. 1925. Über die Geschwindigkeits-Funktion der Viscosität in dispersen Systemen. *Kolloid Z.* 36: 99

Pappenheimer, J. R. 1978. Contributions to microvascular research of Jean Léonard Marie Poiseuille. In *Handbook of Physiology*—The Cardiovascular System IV, Chap. 1, pp. 1–10

Perry, J. 1955. *The Story of Standards*, p. 218. New York: Funk & Wagnalls

Poiseuille, J. L. M. 1828. Recherches sur la force du coeur aortique. Paris: Didot le Jeune, Dissertation

Poiseuille, J. L. M. 1830. Recherches sur les causes du mouvement du sang dans les veines. *J. Physiol. Exp. Pathol.* 10: 277–95

Poiseuille, J. L. M. 1835. Recherches sur les causes du mouvement du sang dans les vaisseaux capillaires. *C. R. Acad. Sci.* 6: 554–60. Also appeared in *Mémoires des Savants Etrangers*. Paris: Acad. Sci., 1841, vol. VII: 105–75

Poiseuille, J. L. M. 1838. *Ecoulement des Liquides: Société Philomatique de Paris. Extraits des Procès-Verbaux des Séances Pendant l'Année 1838*. Paris: René et Cie, pp.1–3, 77–81

Poiseuille, J. L. M. 1840a. Recherches expérimentales sur le mouvement des liquides dans les tubes de très petits diamètres; I.

Influence de la pression sur la quantité de liquide qui traverse les tubes de très petits diamètres. *C. R. Acad. Sci.* 11: 961–67

Poiseuille, J. L. M. 1840b. Recherches expérimentales sur le mouvement des liquides dans les tubes de très petits diamètres; II. Influence de la longueur sur la quantité de liquide qui traverse les tubes de très petits diamètres; III. Influence du diamètre sur la quantité de liquide qui traverse les tubes de très petits diamètres. *C. R. Acad. Sci.* 11: 1041–48

Poiseuille, J. L. M. 1841. Recherches expérimentales sur le mouvement des liquides dans les tubes de très petits diamètres; IV. Influence de la temperature sur la quantité de liquide qui traverse les tubes de très petits diamètres. *C. R. Acad. Sci.* 12: 112–15

Poiseuille, J. L. M. 1843. Écoulement des liquides de nature différente dans les tubes de verre de très-petits diamètres. *C. R. Acad. Sci.* 16: 61–63. (Section 1 of "Recherches sur l'écoulement des liquides, considéré dans les capillaires vivants". *C. R. Acad. Sci.* 16: 60–72)

Poiseuille, J. L. M. 1845. Ventilation des navires. *C. R. Acad. Sci.* 21: 1427–32

Poiseuille, J. L. M. 1846. Recherches expérimentales sur le mouvement des liquides dans les tubes de très-petits diamètres. In *Mémoires presentés par divers savants à l'Académie Royale des Sciences de l'Institut de France*, IX: 433–544

Poiseuille, J. L. M. 1847. Sur le mouvement des liquides de nature différente dans les tubes de très petits diamètres. *Ann. Chim. Phys.*, 3rd series, XXI: 76–110 (plus Plate II)

Poiseuille, J. L. M. 1855. Recherches sur la respiration. *C. R. Acad. Sci.* 41: 1072–76

Poiseuille, J. L. M. 1858. De l'existence du glycose dans l'organisme animal. *C. R. Acad. Sci.* 46: 565–68, 677–79

Poiseuille, J. L. M. 1859. Recherches sur l'urée. *C. R. Acad. Sci.* 49: 164–67

Poiseuille, J. L. M. 1860. Sur la pression du sang dans le système artériel. *C. R. Acad. Sci.* 51: 238–42, (and 66: 886–90, 1868)

Prandtl, L., Tietjens, O. G. 1934. *Applied Hydro- and Aeromechanics*. New York: McGraw-Hill. 311 pp.

Régnault et al 1843. Rapport fait à l'Académie des Sciences, le 26 décembre 1842, au nom d'une Commission composée de MM Arago, Babinet, Piobert, Régnault (rapporteur) sur un Mémoire de M. le docteur Poiseuille. *Ann. Chim. Phys.* VII: 50–74

Schiller, L., ed. 1933. *Drei Klassiker der Strömungslehre: Hagen, Poiseuille, Hagenbach*. Leipzig: Akad. Verlagsgesellschaft. 97 pp.

Skalak, R., Soslowsky, L. , Schmalzer, E., Impelluso, T., Chien, S. 1987. Theory of filtration of mixed blood cell suspensions. *Biorheology* 24: 35–52

Skalak, R. 1990. Capillary flow: Past, present and future. *Biorheology* 27: 277–93

Stephan J. 1862. Uber die Bewegung flüssiger Körper. Wien: Sitzunsber. (2A) 46: 8–32 and 46: 495–521

Stokes, G. G. 1845. On the theories of the internal friction of fluids in motion, and of the equilibrium and motion of elastic solids. *Trans. Cambridge Phil. Soc.* 8: 287–341. (Reprinted, 1966, in *Mathematical and Physical Papers by G. G. Stokes*,Vol. 1: 75–129. New York: Johnson Reprint Corp.)

Stokes, G. G. 1851. On the effect of the internal friction of fluids on the motion of pendulums. *Trans. Cambridge Phil. Soc.* 9: 8–149. (Reprinted, 1966, in *Mathematical and Physical Papers by G. G. Stokes*, Vol. 3: 1–141. New York: Johnson Reprint Corp.)

Stokes, G. G. 1898. Mathematical proof of the identity of the stream lines obtained by means of a viscous film with those of a perfect fluid moving in two dimensions. *Report of the British Association*, 1989, pp. 143–47. (Reprinted, 1966, in *Mathematical and Physical Papers by G. G. Stokes*, Vol. 5: 278 82. New York: Johnson Reprint Corp.)

Wang, C. Y. 1991. Exact solutions of the steady-state Navier-Stokes equations. *Annu. Rev. Fluid Mech.* 23: 159–77

Womersey, J. R. 1955. Oscillatory motion of a viscous liquid in a thin-walled elastic tube. I. The linear approximation for long waves. *Philos. Mag.* 46: 199–221

Young, T. 1809. The Croonian Lecture. On the functions of the heart and arteries. *Phil. Trans. R. Soc. London* 98: 1–31 and 164–86

Annu. Rev. Fluid Mech. 1993. 25 : 21–53

THE STRUCTURE AND STABILITY OF LAMINAR FLAMES

John Buckmaster

Department of Aeronautical and Astronautical Engineering, University of Illinois, 104 South Mathews Avenue, Urbana, Illinois 61801

KEY WORDS: combustion modeling, microgravity combustion, acoustic-flame interaction, radiation, heterogeneous mixtures

1. INTRODUCTION

The study of laminar flames is at the heart of combustion theory, both laminar and turbulent, and this is the third discussion of the subject within these volumes. Williams's article in 1971 coincided with the start of a sustained examination of the topic in the Western World by fluid dynamicists and mathematicians trained in the use of modern tools of nonlinear mathematics, particularly asymptotic methods. Some of the first fruits of this work were summarized by Sivashinsky in 1983. It is not the purpose here to systematically describe the progress that has been made in the 10 years hence, although some of the discussion will necessarily be historical. Rather, the focus will be on a number of topics that are *au courant*, have fresh ingredients, and are not yet mature.

It is commonplace in combustion theory to adopt a simple one-step kinetic model characterized by an Arrhenius' temperature dependence. For premixed flames this might have the form $Y \rightarrow$ *products* at a rate

$$\Omega = D_1 Y \exp(-E/RT), \tag{1}$$

where Y is the mixture fraction, T the temperature, E/R an activation temperature, and D_1 a constant. For diffusion flames, $X + Y \rightarrow$ *products* at a rate

$$\Omega = D_1 XY \exp(-E/RT), \tag{2}$$

0066–4189/93/0115–0021$02.00

where $X(Y)$ is the oxygen (fuel) mass fraction. Asymptotic treatments are possible in the limit $E/RT_{\text{ref}} \to \infty$ (activation energy asymptotics) where T_{ref} is a reference temperature, and are the key to Sivashinsky's discussion (1983) and most of the work that has been done in the past 20 years. There are new developments in which reaction is handled quite differently, and these are briefly described below, but activation energy asymptotics remains a powerful and valuable tool. Its application to the linear stability of the plane premixed flame—a classical problem—provides a starting point for our discussion.

Stability of the Plane Flame

A simple model for unconfined premixed combustion is described by the equations

$$\rho C_{\text{p}} DT/Dt = \lambda\nabla^2 T + Q\Omega, \quad \rho\, DY/Dt = \rho \mathsf{D}\nabla^2 Y - \Omega, \quad \rho T = \text{constant}, \tag{3a,b,c}$$

together with the usual flow equations. Here ρ is density, C_{p} specific heat, λ the heat conduction coefficient, $\rho\mathsf{D}$ the mass diffusion coefficient, and Q the mixture heat content. When the activation energy is large, Ω is only significant in a thin region, a flame-sheet, and its limiting form is that of a Dirac delta-function with a strength $\sim \exp(-E/2RT_*)$ where T_* is the flame-temperature, i.e. the temperature at the sheet. This approximation has been used in one of two ways: as an ingredient of a formal asymptotic treatment in which the solution is expressed in terms of gauge functions that depend on E/RT_{ref}; or as a model—the *delta-function model*—in which E/RT_{ref} is subsequently treated as a finite parameter. An important example of the latter is the discussion of thermite combustion—solids that burn to form solids—by Matkowsky & Sivashinsky (1978). Unless defined otherwise, when the reaction term is replaced by a delta-function in our discussion it will be part of an asymptotic treatment.

Equations (3) have a solution that corresponds to a stationary plane flame characterized by a mass flux M. If the flame-sheet is located at $x = 0$, the cold fresh gas (subscript 1) at $x \to -\infty$, and the hot burnt gas (subscript 2) at $x \to +\infty$, the solution is:

$$\underline{s < 0} \quad T = T_1 + (QY_1/C_{\text{p}})\exp(s), \quad Y = Y_1(1 - \exp(\text{Le}\, s));$$

$$\underline{s > 0} \quad T = T_1 + (QY_1/C_{\text{p}}) \equiv T_{\text{a}}, \quad Y = 0; \tag{4}$$

$$s \equiv (MC_{\text{p}}/\lambda)x, \quad \text{Le} \equiv \lambda/\rho\mathsf{D}C_{\text{p}},$$

$$M \equiv \rho_1 W_{\text{ad}} = (RT_{\text{a}}^2/EQY_1)(2D_1\lambda C_{\text{p}}\text{Le})^{1/2}\exp(-E/2RT_{\text{a}}).$$

Note that in the domain $s < 0$ (the preheat zone)

$$T+(QY/C_p) = T_a+(QY_1/C_p)\cdot[\exp(s)-\exp(\mathrm{Le}\, s)], \tag{5}$$

and if the Lewis number is equal to 1 this is independent of s, corresponding to similarity between T and QY/C_p.

Nonsimilarity is a general source of instability in premixed flames in the limit $E/RT_a \to \infty$ (T_a, the *adiabatic flame-temperature*, is a suitable reference temperature) no matter how it is generated. Here it is appropriate to use the rubric *Turing instabilities* (Turing 1953) since the nonsimilarity arises because Le is different from 1.

Construction of the solution (4) is an elementary exercise, but the stability problem is more challenging since the linearized equations have nonconstant coefficients. A numerical treatment is described in Jackson & Kapila (1986), but a common strategy is to adopt the *constant-density model*: The density is set equal to a constant, the equation of state (Charles's law) is discarded, and the velocity is a prescribed solenoidal field. The stability problem for the plane flame is then straightforward.

If the prescribed velocity field is $(W_{ad}, 0)$ where W_{ad} is the *adiabatic flame speed*, and the infinitesimal flame-sheet displacement is

$$s = \phi(t, y) \equiv \varepsilon\, e^{\alpha t+iky}, \quad \varepsilon \to 0, \tag{6}$$

then the stability diagram Figure 1 can be constructed.

There are two regions of instability. The one on the left ($\mathrm{Le} < 1$) corresponds to *cellular flames*. This is a nonplanar instability ($\alpha \to 0$ as $k \to 0$). The right-hand region ($\mathrm{Le} > 1$) defines a traveling wave or pulsating instability that can be planar.

In the neighborhood of P, long wavelength disturbances grow very slowly so that weak nonlinearities can be incorporated into the analysis by means of a bifurcation analysis. In this way the Kuramoto-Sivashinsky (K-S) equation can be derived for the displacement ϕ, and when corrugations in the z-direction are also admitted this is

$$\phi_t+\tfrac{1}{2}(\nabla\phi)^2+\nabla^2\phi+4\nabla^4\phi = 0 \tag{7}$$

(valid to the left of P). Numerical simulations show that the flame-sheet adopts an irregular, unsteady, cellular configuration, and physical flames in mixtures with $\mathrm{Le} < 1$ can display similar behavior.

The K-S equation is the point of departure for two classes of analysis: (*a*) The chaotic behavior of its solutions has attracted the dynamical-systems community (e.g. Nicolaenko et al 1985), and (*b*) variations have been derived and/or proposed to account for a variety of physical ingredients—hydrodynamic effects, curvature, etc—e.g. Frankel & Sivashinsky (1991).

The K-S equation and its relatives provide a framework in which non-

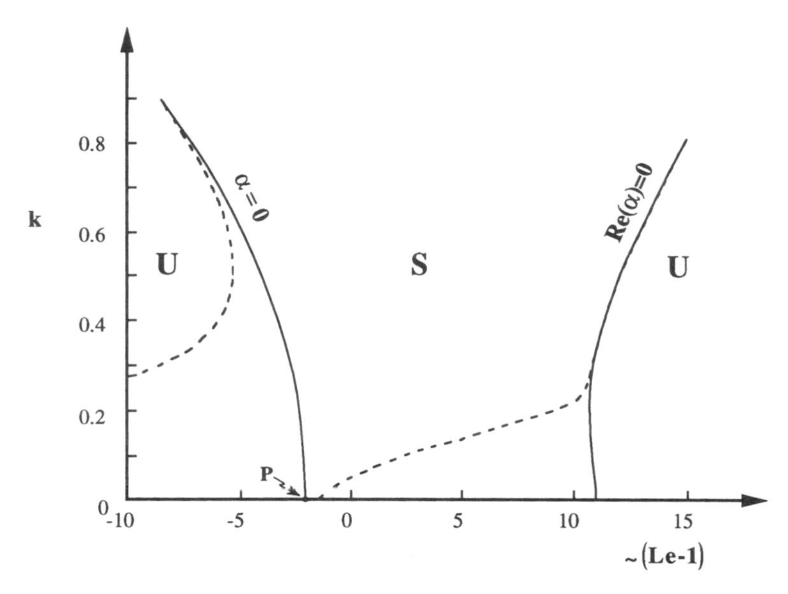

Figure 1 Stability diagram for a one-dimensional flame (wave number/scaled Lewis number plane). Stability boundaries are drawn as solid lines for the unbounded flame and as broken lines for the burner-attached flame. S ≡ stable, U ≡ unstable.

linear flame dynamics can be discussed. Nonlinear treatments are also possible when other bifurcations are examined. Thus rich dynamical behavior is associated with bifurcations at the right stability boundary when the wave numbers are restricted to discrete values and mode interactions can occur. Much of this work is concerned with the closely related problem of solids for which Le $= \infty$ and the bifurcation parameter is the activation energy rather than Le; a review of some of these aspects can be found in Margolis (1991). The subject has been energetically exploited by Matkowsky and his coworkers, much of whose work can be found in the *SIAM Journal on Applied Mathematics*, e.g. Bayliss & Matkowsky (1990). It is a subject complex in detail best left for summary by its authors.

The Turing instabilities identified in Figure 1 arise from the nonsimilarity generated when Le differs from 1. Nonsimilarity can also occur because of boundary conditions, and this can lead to instability even if Le = 1. An example is afforded by a flame supported by a porous-plug burner for which the boundary conditions are:

$$\underline{s=0} \quad T = T_1, \quad Y - \text{Le}^{-1}\, dY/ds = J_1, \quad \underline{s \to \infty} \quad T \text{ bounded}, \quad Y \to 0, \tag{8}$$

where J_1 is the mass-flux-fraction of combustible mixture in the supply. Typical stability boundaries generated for this problem are shown as broken lines in Figure 1 (Buckmaster 1983).

2. HETEROGENEOUS MIXTURES

Nonsimilarity also arises when the fuel is generated by gasification or evaporation of solid particles or liquid drops. A simple model for particle mixtures can be constructed when the loading is small, and has the ingredients:

$$\partial \rho_s/\partial t + \nabla \cdot (\rho_s \mathbf{u}_s) = -f(\rho_s, T_s),$$

$$\rho\, DY/Dt = \rho \mathrm{D} \nabla^2 Y + f(\rho_s, T_s) - \Omega. \tag{9a,b}$$

Here ρ_s is the particle density (solid mass per unit volume of gas/solid mixture) and small loading means $\rho_s/\rho \ll 1$. In the model the only physical role of the particles is to generate the fuel Y (by means of the pyrolysis function f) which then mixes with the supply gas and burns to generate heat. Formally the model corresponds to the limit $\rho_s/\rho \to 0$, $Q \to \infty$, $Q\rho_s/\rho$ fixed. Provided the oxidizer in the supply is in excess, Ω can be represented by (1) with Y vanishing behind the reaction zone, assumed to be a thin sheet located in a region where pyrolysis is complete.

In general there will be slip (thermal and kinematic) between the two phases so that $\mathbf{u}_s$ and T_s will differ from $\mathbf{u}$ and T:

$$D_s \mathbf{u}_s/Dt \equiv \partial \mathbf{u}_s/\partial t + (\mathbf{u}_s \cdot \nabla)\mathbf{u}_s = -(\mathbf{u}_s - \mathbf{u})/\tau_u, \tag{10a}$$

$$D_s T_s/Dt = -(T_s - T)/\tau_T, \tag{10b}$$

where τ_u and τ_T are slip times. The system is closed by the usual gas-phase equations together with appropriate boundary conditions.

Whether or not this model is a realistic representation of a specific mixture depends on the particle size, volatility, ignition temperature of the product gases, flame-thickness, etc. This is a complicated issue which we choose not to address here.

A representative stationary solution generated numerically and using the flame-sheet limit is shown in Figure 2. The chosen parameters are $\mathrm{Le} = 1$, $T_2/T_1 = 7$, $\tau_u/\tau_r = 1$, and $\tau_T/\tau_r = 1$ where τ_r is the flame-residence time $\lambda/(\rho C_p W^2)$ and W is the flame-speed. The pyrolysis function is

$$f = (\rho_{s_1}/\tau_r)(\rho_s/\rho_{s_1})^{2/3} F(T_s/T_1), \quad F(\psi) \equiv 5(\psi - 1)\exp[10 - 20/\psi]. \tag{11}$$

Note that Y vanishes in the supply gas.

To examine the one-dimensional stability of this solution we use the

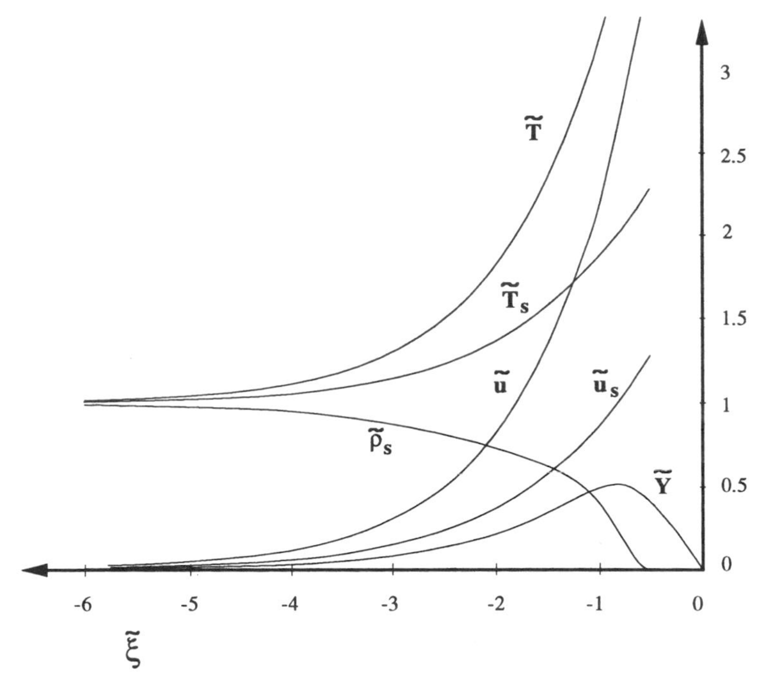

Figure 2 Steady solution for a particle/gas flame. These are nondimensional scaled variables. $\tilde{\xi}$ is the nondimensional distance from the flame-sheet. $\tilde{\mathbf{u}} \sim (\mathbf{u}-\mathbf{u}_1)$, $\tilde{\mathbf{u}}_s \sim (\mathbf{u}_s-\mathbf{u}_1)$.

distance from the flame-sheet as a coordinate and examine the behavior of small disturbances, i.e.

$$\xi = x - x_f(t), \quad x_f = \delta(t) \equiv \varepsilon e^{i\omega t}, \quad \varepsilon \to 0. \tag{12}$$

At the same time we adopt the delta-function model, rather than a formal asymptotic approach, for reasons discussed below.

A statement of global mass conservation can be deduced by integrating the continuity equation from $\xi = -\infty$ to $\xi = 0+$. When perturbed this is

$$-\rho_1\delta_t = \rho'_* u_* + \rho_*(u'_* - \delta_t) + \int_{-\infty}^{0} d\xi\, \rho'_t, \tag{13}$$

where primed quantities are perturbations, unprimed are steady-state, and the subscript * refers to the flame-sheet. u is the velocity in the laboratory frame. Now the perturbation ρ' may be written as

$$\rho' = \delta_t \phi(\xi, \omega), \tag{14}$$

where ϕ is the solution to a boundary value problem on the interval

$(-\infty, 0)$ generated by perturbing the field equations. When the frequency ω is specified, ϕ can be calculated numerically. Equation (13) is then a linear homogeneous relation between δ_t and u'_* with coefficients that depend upon ω.

A second relation of this kind can be deduced by examining global energy conservation, which in unperturbed form is

$$Q\int_{-\infty}^{0} d\xi(\rho_s+\rho Y)_t - Q\rho_{s_1}(u_{s_1}-\delta_t)+C_p\rho_1 T_1(u_*-u_1) = \lambda T_\xi(0+). \tag{15}$$

Setting the determinant of the pair consisting of Equation 13 and the perturbed form of Equation 15 equal to zero then defines an eigenvalue problem for the frequency ω that can easily be solved numerically. Again, instability can occur even if Le = 1. The list $\{(120, 0.14\pm i0.56)$, $(100, 0.002\pm i0.60)$, $(90, -0.005\pm i0.61)$, $(80, -0.12\pm i0.62)\}$ is of pairs $(E/RT_1, i\omega\tau_r)$ obtained for the same choices as in Figure 2 except that $\tau_u/\tau_r = \tau_T/\tau_r = 0.3$. E/R is the (finite) activation temperature that characterizes the delta-function model, and $i\omega\tau_r$ is the eigenvalue with $\mathrm{Re}(i\omega\tau_r) > 0$ corresponding to instability. There is a critical activation energy in the interval (90, 100) [E/RT_2 in the realistic range (12.8, 14.3)] for which there is neutral stability. Note that a formal asymptotic treatment valid when $E/RT_2 \to \infty$ would simply predict instability and the details presented here would be lost—a situation similar to that in Matkowsky & Sivashinsky (1978).

This type of problem is one that has received little attention (the results described above are from unpublished work of the author) and undoubtedly there is scope for additional modeling and analysis. Pulsations have been observed in flames propagating through lycopodium/air mixtures contained in long tubes (Berlad et al 1990)—an example, perhaps, of the instability identified here, albeit modified by acoustic waves generated within the confined space by the periodic flame motion. But, as we shall see, there are at least two other explanations, one of which is discussed in the next section.

The lycopodium/air experiments were conducted in a microgravity environment (*mugen*) essential to prevent particle settling. It is vital that theoretical work be tested against experiment, and increased access to mugens is likely to lead to more experiments using heterogeneous mixtures, and better understanding of this important subject.

3. ACOUSTIC INSTABILITIES

Gas Mixtures

It has been known for many years that a flame propagating through a mixture-filled tube can excite an acoustic instability, but the phenomenon is not well understood. The broad strokes are clear—Rayleigh's famous criterion must be satisfied so that energy is fed to the acoustic field, i.e.

$$\int_{\text{cycle}} \delta p \cdot \delta q \, dt > 0, \tag{16}$$

where δp is the pressure increment and δq is the incremental heat added to the flow. However, the details of how this comes about are not obvious.

Consider a tube containing a combustible mixture with a closed end located at $x = 0$ and an open end at $x = L$. The mixture is ignited at the open end and the flame travels to the left towards the closed end. The flame which—for the sake of analysis—we assume is flat, has a thickness several orders of magnitude smaller than L so that the flow-field can be analyzed in two parts: the flame-structure and the acoustic regions on each side. Analysis of the flame-structure generates connection conditions for the acoustic problem.

The acoustic disturbances have magnitudes:

$$(T'/T, \rho'/\rho, p'/\rho c^2, p'/\rho W^2) \sim (M_a, M_a, M_a, M_a^{-1})(u'/W), \tag{17}$$

where W is the flame-speed, $M_a = W/c$, and c is a representative sound speed. For small M_a (typically $\sim 10^{-3}$) the temperature and density perturbations in the cold gas have a negligible effect on the flame-structure. The discussion of the pressure is a little more complicated.

The momentum equation only admits spatial variations within the flame of order (ρW^2), and the equation of state is only affected by variations of order (ρc^2). The acoustically generated pressure is large on the first scale, small on the second, so that to leading order p' depends only on t within the flame, yet ρT is constant. The condition

$$[p']_-^+ = 0 \tag{18}$$

(the square-bracket denotes a jump across the flame) is one of the two connection conditions required for the calculation of the acoustic field.

It is apparent that the only acoustic fluctuation that affects the flame is u'. But in the combustion limit ($M_a \to 0$) this only alters the flame position and not its structure. If the velocity on the cold side of the flame is $u_1'(t)$, the temperature, density, velocity, and concentrations have the form

$$\psi = \psi\left(x + Wt - \int^t u_1' \, dt\right), \tag{19}$$

e.g.

$$T = T_1 + (QY_1/C_p)\exp\left[x + Wt - \int^t u_1'\,dt\right]$$

in the preheat zone described by (4). So the velocity jump generated by the steady flame is not altered by the acoustic fluctuations, and the second connection condition is

$$[u']_-^+ = 0. \tag{20}$$

Most significantly there are no fluctuations in the heat output and the Rayleigh criterion is not satisfied.

With the closed-end condition $u' = 0$ at $x = 0$, the open-end condition $p' = 0$ at $x = L$, and the flame at $x = h(0 < h < L)$ separating cold gas from hot gas, the frequency relation for perturbations $\sim e^{i\omega t}$ is

$$(\rho_2 c_2/\rho_1 c_1)\tan(\omega h/c_1)\tan[\omega(L-h)/c_2] = 1. \tag{21}$$

This defines neutrally stable modes.

If the observed instability is to be explained, clearly the physics implicit within the above discussion must be modified. Physical flames are not planar and it has been proposed that fluctuations in the flame area and/or interaction with the boundary layer at the tube wall can generate a fluctuating heat output. Incorporating these kinds of ingredients into a convincing mathematical treatment that can be tested against experiment is not easy however. Clavin et al (1990) have proposed that the instability is generated by Mach number dependent terms neglected in the derivation of (21). They argue that although the growth rate is small on the acoustic time scale L/c, it has a long time $(M^{-1}L/c)$ in which to act. This can only be convincing if an accurate accounting is made of the losses and, again, if the results are tested against experiment.

Heterogeneous Mixtures

The situation can be less complicated when the flame is sustained by heterogeneous mixtures of the kind discussed in Section 2. When the slip between the two phases is significant ($\tau_u \sim L/c$) the invariance expressed by (19) is broken and the acoustic field generates fluctuations in the mixture flux fraction with corresponding fluctuations in the heat release.

Clavin & Sun (1992) have carried out a detailed analysis for air/drop mixtures with parameters carefully chosen so that T and Y are nearly similar. This permits an asymptotic treatment closely related to the analysis which, for gas mixtures, generates the stability boundaries of Figure 1 [a NEF analysis, see Buckmaster & Ludford (1982)]. Near-similarity requires near-equidiffusion [$\mathrm{Le}-1 \sim (RT_2/E)$], small slip ($\tau_u c/L \sim RT_2/E$), and the completion of evaporation/gasification at a temperature close to the cold-boundary value T_1. These conditions will not always be met. But then the analysis of Section 2 (Equations 13–15) is readily adapted to the problem.

Here the flow ahead of the flame is not constant—there are fluctuations u'_1, u'_{s_1}. Thus the perturbed global conservation equations, with the integrals now extended to $\xi = +\infty$, contain terms proportional to u'_1, u'_2, and δ_t. These can be solved to yield a result of the form

$$u'_2/u'_1 = J(\omega), \tag{22}$$

replacing (20). J then appears as a multiplicative factor on the left of the frequency relation (21).

Only preliminary results are available at the present time (Buckmaster & Clavin 1992). When $h \uparrow L$, $\tan[\omega(L-h)/c_2] \to 0$ so that $J(\omega) \to \infty$ or $\tan(\omega h/c_1) \to \infty$. In the first case ω is a root corresponding to the intrinsic instability discussed earlier, and this will be modified as h is decreased. In the second case $\omega \to \omega_0 = (c_1/L)(n+1/2)\pi$, $n = 0, 1, 2, \ldots$ (the natural modes for the cold-gas-filled tube) and these also will be modified by decreasing h; when $\text{Im}[J(\omega_0)] > 0$ there is instability for small $(L-h)$. Some sample calculations are shown in Figure 3.

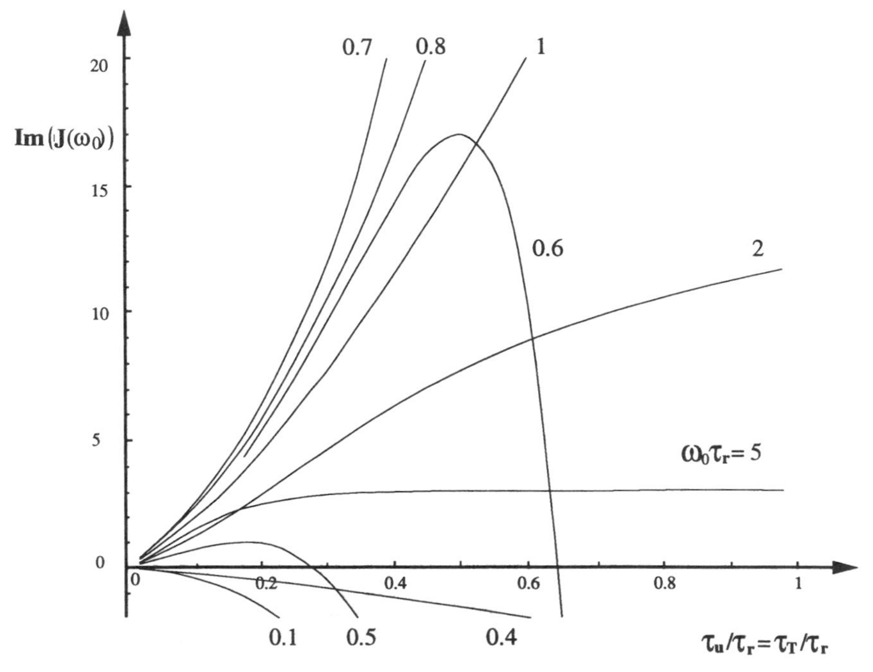

Figure 3 Stability results for particle/gas flames when $h \sim L$. Parameters other than τ_u, τ_T are the same as for Figure 2. Each curve is calculated for a fixed value of $\omega_0\tau_r$, and there is instability when $\text{Im}(J) > 0$.

More extensive results have been obtained by Clavin & Sun (1992) for the small-slip (near-similarity) model. An interesting conclusion is that for small values of $\omega\tau_r$ the fundamental ($n = 0$) is stable for any h, but there are values of h for which the other modes are unstable. Issues such as this have not yet been examined for the order $O(1)$ slip problem. Data from the lycopodium/air experiments (Berlad et al 1990) are consistent with excitation of the fundamental (1/4-wave) mode.

4. FLAME BALLS AND RELATED PHENOMENA

A happy byproduct of NASA's enthusiasm for space-station *Freedom* is institutionalized support for microgravity combustion experiments. These have been, and will continue to be, a rich playing field for modelers. In this section we discuss some phenomena that have their roots in the left stability boundary (Le < 1) of Figure 1, and are clear only in a mugen where buoyancy-induced motion is small. They are important on two accounts: (*a*) They are associated with near-limit mixtures, mixtures of such weak reactivity that they barely support combustion; and (*b*) they are linked to radiation—a physical ingredient worthy of further study.

As we have already noted in connection with Figure 1, flames of small Lewis number are subject to a cellular instability. In the case of lean hydrogen/air mixtures, for example, or methane mixtures containing a heavy diluent such as SF_6, the effective Lewis number is less than 0.5 (as small as 0.1 in some cases) and the instability is a strong one. Point ignition leads to a flame that rapidly breaks up as the cells separate. In some circumstances the fragments close upon themselves to form stationary spherical structures called *flame-balls* (Ronney 1990); see Figure 4.

The mass-averaged (fluid) velocity must vanish everywhere in the symmetric stationary combustion field, so that mixture reaches the reaction zone from the far-field through diffusive transport alone, and this is also the mechanism by which heat and combustion products escape to the far-field.

A simple mathematical model for flame-balls, first discussed by Zeldovich (see Zeldovich et al 1985), is described by Equations (3) with the material derivatives discarded and Ω equivalent to the flame-sheet model,

$$\Omega = B\exp(-E/2RT_*)\delta(r-r_*), \quad E/RT_* \to \infty. \tag{23}$$

Boundary conditions and constraints are

$$\underline{r < r_*} \quad Y = 0 \quad \underline{r \to \infty} \quad T \to T_1, Y \to Y_1. \tag{24}$$

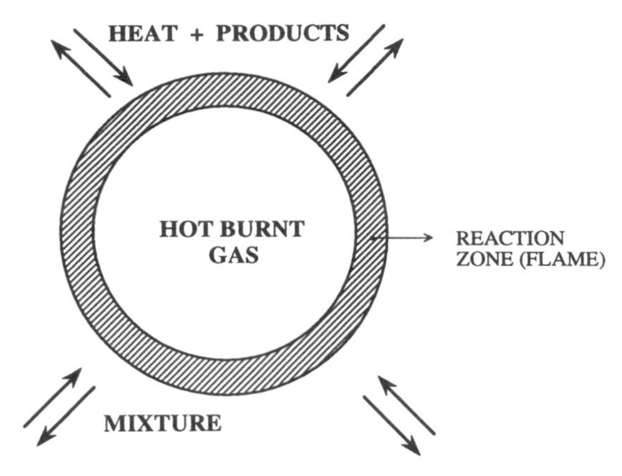

Figure 4 Stationary flame ball. The arrows represent diffusional fluxes—outward for heat and combustion products, inward for unburnt mixture.

An examination of the energy balance shows that the temperature in the burnt gas is

$$T_b = T_1 + (\rho \mathrm{D} Q/\lambda) Y_1 = T_1 + (Q/C_p \,\mathrm{Le}) Y_1, \tag{25}$$

which is significantly larger than the adiabatic flame-temperature T_a [cf (4)] when Le is small. Flame-balls are observed in mixtures for which T_a is less than the ignition temperature (as low as 600 K, for example).

The stationary solution is

$$\underline{r < r_*} \quad T = T_b, \quad Y = 0$$

$$\underline{r > r_*} \quad T = T_1 + (\rho \mathrm{D} Q Y_1/\lambda)(r_*/r), \quad Y = Y_1(1 - r_*/r); \tag{26}$$

$$r_* = R_z \equiv (\rho \mathrm{D} Y_1/B) \exp(E/2RT_b),$$

where R_z is the *Zeldovich radius.*

The one-dimensional stability of this solution can be examined by reinstating the material derivatives and adding the continuity equation. A simple approach, which does not alter the conclusion, is to adopt the constant-density model. The linear analysis, which is straightforward, shows that the Zeldovich solution (26) is unstable. Additional physical mechanisms must be incorporated into the mathematical model if the stable flame-balls of experiment are to be explained.

Consider a flame-ball that is a distance $O(ER_z/RT_b)$ from a cold wall (Joulin & Buckmaster 1990, unpublished). Boundary conditions at the wall are

$$T = T_1, \quad \partial Y/\partial n = 0. \tag{27}$$

On the scale $r = O(R_z)$—the near-field—the presence of the wall merely perturbs the combustion field by an $O(RT_b/E)$ amount. Without making any assumptions about the nature of this perturbation, other than that spherical symmetry is preserved to sufficient accuracy, solution of the near-field equations yields the formulas

$$1 = (r_*/R_z)\exp(C/2T_b); \tag{28}$$

$$\underline{r > r_*} \quad T = T_1 + (QY_1\rho D/\lambda)\cdot(r_*/r) + \cdots, \quad Y = Y_1 - Y_1(r_*/r) + \cdots; \tag{29a,b}$$

$$T + (Q\rho D/\lambda)Y = T_b + \beta^{-1}C + \cdots, \quad \beta = (E/RT_b); \tag{29c,d}$$

where C is a constant to be determined by examination of the far-field on the scale $r = O(\beta R_z)$. There,

$$T = T_1 + \beta^{-1}t + \cdots, \quad Y = Y_1 + \beta^{-1}y + \cdots, \tag{30}$$

where t and y are harmonic functions with poles at $r = 0$ to match with Equations 29a,b; they are readily determined using the method of images. Matching with (29c) then determines C,

$$C = -QY_1\rho Dr_*/\lambda h, \tag{31}$$

where βh is the distance separating the flame-ball from the wall. Combining (28) and (31) yields

$$1 = (r_*/R_z)\exp[-(r_*/R_z)\cdot(QY_1\rho DR_z/2T_b\lambda h)]. \tag{32}$$

This solution is sketched in Figure 5. The combustion field loses heat to the wall and the magnitude of this loss increases with decreasing h; quenching occurs when h is small enough. For heat losses less than the quenching value there are two solution branches, the lower one of which approaches the Zeldovich solution ($r_* \to R_z$) as $h \to \infty$. The entire lower branch is unstable to one-dimensional disturbances, as is a portion of the upper branch near the quenching point. Of crucial significance, however, is that a significant portion of the upper branch corresponds to stable solutions.

The evidence suggests that, in general, heat losses can stabilize flame-balls. An example that has been analyzed in detail (Buckmaster & Joulin 1991), with full accounting of the fluid mechanics ($\rho \neq$ constant) is that of a ball located in a weak nonuniform flow, a shear or straining flow. The flame is a source of heat, and convection will enhance the loss by an amount proportional to $U_c^{1/2}$ where U_c is a speed characterizing the flow. At the same time the flame is a sink of mixture, and convection enhances this inward flux by an amount proportional to $(U_c\,\mathrm{Le})^{1/2}$. If the Lewis

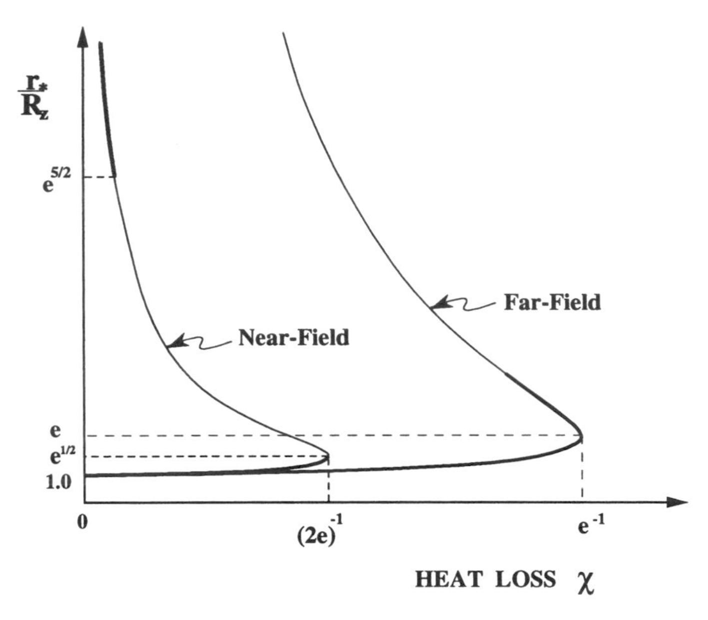

Figure 5 Flame-ball radius vs heat loss. For far-field losses $1 = (r_*/R_z)\exp(-\chi r_*/R_z)$, cf Equation (32); for near-field losses $1 = (r_*/R_z)\exp[-\chi(r_*/R_z)^2]$. The thick (thin) line denotes unstable (stable) solutions.

number were unity there would be no net effect on the energy flux, but for Le < 1 there is a net extraction and the result is identical to (32) when the heat-loss parameter is replaced by one proportional to $U_c^{1/2}(1-\text{Le}^{1/2})$.

The stabilizing agencies described so far are extrinsic in nature. Radiation is an intrinsic agency that also can be stabilizing and is undoubtedly of more significance to the experimental record. The Zeldovich model is then modified by the addition of the term $(-\nabla \cdot \mathbf{q})$ to the right side of the energy equation, where $\mathbf{q}$ is the radiative energy flux. A simple phenomenological model appropriate in the optically thin limit ($R_z \ll$ radiation absorption length) is defined by writing

$$\nabla \cdot \mathbf{q} = f(T), \quad f(T_1) = 0, \tag{33}$$

e.g.

$$f(T) \propto (T^4 - T_1^4). \tag{34}$$

The nature of the solutions depends on the relative magnitude of f, and there are two distinguished choices (Buckmaster et al 1990). For one, only losses in the far-field $r = O(\beta R_z)$ play a role, and the response is as in Figure 5 with $f'(T_1)$ playing the role of $1/h$. For the other, f must be cut off at some $O(R_z)$ radius, with losses only in the near-field. Calculations have only been done for a cutoff radius of r_* and then the response has the form

$$1 = (r_*/R_z)\exp[-\chi(r_*/R_z)^2], \tag{35}$$

where χ is proportional to $f(T_b)$ (the near-field curve shown in Figure 5).

The distinction between near- and far-field radiation losses is a creature of the asymptotics and of the assumed functional form of the losses. If the chief radiating species is a reaction product with mass fraction Y_p, e.g. H_2O, (33) is replaced by

$$\nabla \cdot \mathbf{q} = Y_p f(T), \quad f(T_1) = 0 \tag{36}$$

and Y_p vanishes like r^{-1} as $r \to \infty$. In this case there is a seamless contribution from both the near- and far-fields. This model is a more realistic reflection of the true physics for hydrogen/air mixtures than the choice (33), but at this time there are no analytical results to report.

Clearly there is need for an exact numerical treatment of flame-balls, and calculations of the steady structure are currently underway for full hydrogen/air chemistry, detailed transport modeling including an accounting of thermal diffusion for the light species (H, H_2), and a realistic model of radiation from H_2O (Buckmaster & Smooke 1992). Solutions constructed for different mixture strengths reveal a lean limit of about 3.4% H_2 by volume corresponding to the turning or quenching point in Figure 5. This is close to the observed limit (Whaling et al 1992).

Three-Dimensional Deformations

Our discussion of stability has so far been confined to one-dimensional perturbations. It is necessary also to examine three-dimensional perturbations, and these can be discussed by representing the corrugations of the flame-sheet as a linear combination of surface harmonics

$$\mathrm{P}_n^m(\cos\theta)\, e^{im\phi}\, e^{\alpha t}, \quad 0 \le m \le n, n > 1, \tag{37}$$

where θ and ϕ are the polar and azimuthal angles. Instability has only been identified when there are significant near-field losses. For the solution (35) there is instability if $(r_*/R_z) > e^{n+1/2}$, and the stability boundary is defined by $n = 2$ (Figure 5).

A transcritical bifurcation analysis can be carried out (Buckmaster et al 1992) that is valid in the neighborhood of this neutral stability point and this identifies the expected first observable manifestation of the instability. The ball deforms into an unsteady constant-volume prolate spheroid.

The response defined by Figure 5 (near-field losses) has certain implications insofar as experiment is concerned. The radiation parameter χ is controlled by the mixture strength since reducing (increasing) the reactivity increases (decreases) the importance of losses relative to chemical heat production. Thus for mixtures of low reactivity quenching will occur and flame-balls cannot exist, in agreement with observations. An increase in mixture reactivity will take us to the stability boundary and the generation of prolate spheroids, again in agreement with observations (Whaling et al 1992).

For certain mixtures rich unsteady three-dimensional behavior is observed, much of it associated with g-jitter—small fluctuations in the gravitational field which generate a nonuniform flow-field. Perhaps the most striking phenomenon of this nature is the generation of a *flame-string* (Whaling et al 1992) in which a ball is deformed into a long thread of flame of length twenty times or so the diameter. A flame-string is unstable: Sometimes it displays a "sausage" instability and collapses into a line of discrete balls; sometimes it displays a peristaltic instability with discrete balls pinched off the end towards which the wave propagates.

The modeling of flame-strings presents difficulties. It is not clear that there is an underlying stationary structure which can be the foundation of a stability analysis, and, if there is, it is not obvious what it should be. If one carries out a *flame-cylinder* analysis analogous to that for the flame-ball there are difficulties with the stationary solution since the source solution is now $\ln r$ rather than r^{-1} so that boundary conditions at infinity cannot be satisfied. If this problem is ignored, a stationary solution can be constructed provided T_* and r_* are specified. Symmetric perturbations of the form $e^{\alpha t + ikx} f(r)$ can then be constructed, where x is distance measured along the cylinder axis. These vanish at infinity and the eigenvalue problem is well defined, with the results shown in Figure 6. Attempts have been made to resolve the difficulty with the unperturbed solution by incorporating a velocity field, with only a quantitative effect on the results (Buckmaster 1992). Instability only occurs for k less than some critical value k_c, so that the instability can only be observed for sufficiently long strings. It is observed that the prolate spheroids discussed above neck and split when they become long enough, and this might be a manifestation of the same instability.

5. RADIATION EFFECTS

Radiation appears to play an important role in the behavior of flame-balls. This is true also for *self-extinguishing flames*—unsteady spherical flames generated by point ignition in near-limit mixtures which extinguish (because of radiative cooling) after reaching a radius of 2–7 cm (Ronney 1985). In these various systems radiation is relatively important because

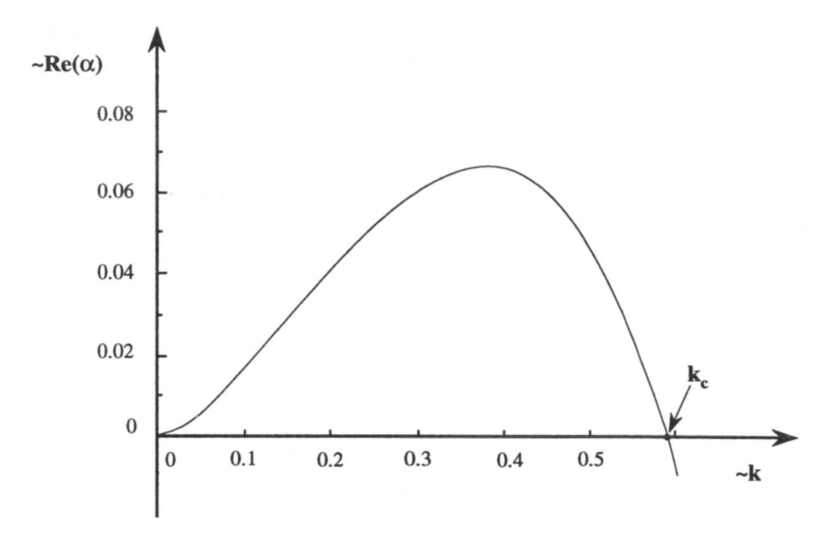

Figure 6 Growth rate vs wavenumber for flame-string perturbations.

of the low reactivity of the mixtures. For non-near-limit gas mixtures radiation tends to play only a passive role—a byproduct of combustion which does not affect it. This is less true in flames containing particles. Assuming black-body radiation, a six-micron diameter particle at 2000 K has a radiative power $\sim 10^{-4}$ W; so that for a density of 40 per mm^3, the power from 1 cm^3 of mixture is ~ 4 W. Compare this with the 1/3 W of power from 1 cm^3 of air at 2000 K and 1 bar, containing 10% H_2O.

It is difficult to study particle/gas mixtures at 1g because of particle settling. As access to mugens increases it is likely that there will be increased work in the area and radiation effects will be observed against which theory can be tested.

A discussion of the equations governing particle/fluid mixtures may be found in Marble (1970) and key ingredients are identified in (9)–(10). We are concerned here with inert particles that do not gasify. The simplest situation arises when the particles are so small that slip between the phases can be neglected. The governing equations are then those of a single fluid. Nonstandard ingredients that we wish to incorporate include the radiation term $\nabla \cdot \mathbf{q}$ in the energy equation together with appropriate equations governing the radiative transport. The latter are quite complicated so that it is useful, following much of the work of Joulin (see below), to adopt a differential approximation which has its roots in the astrophysics literature (see Chandrasekhar 1960). For example, for a nonscattering gray plane-parallel gas in thermodynamic equilibrium, variations in the radiation intensity are described by the equation

$$L\cos\theta\, dI/dx = -I + B, \quad B = \sigma T^4/\pi, \tag{38}$$

where x measures the distance into the gas, θ is the polar angle between the pencil of radiation and the x-axis, L is the absorption length, and σ is the Stefan-Boltzmann constant. The energy flux in the x-direction generated by pencils pointing in the positive x-direction $(0 \le \theta < \pi/2)$ is

$$q_+ = 2\pi \int_0^{\pi/2} I\cos\theta\sin\theta\, d\theta. \tag{39}$$

Thus

$$L\, dq_+/dx = -2\pi \int_0^{\pi/2} d\theta\, I\sin\theta + 2\pi B, \tag{40}$$

which can be approximated by

$$L\, dq_+/dx = -aq_+ + 2\pi B, \tag{41}$$

where the constant a accounts for the different weighting of the I integrals in (39) and (40).

Similarly, for pencils pointing in the negative x-direction

$$-L\, dq_-/dx = -aq_- + 2\pi B. \tag{42}$$

Equations (41) and (42) may be combined to generate an equation for the net energy flux $q = q_+ - q_-$:

$$L\frac{d}{dx}\left[L\frac{dq}{dx}\right] = a^2 q + 4L\frac{d}{dx}(\sigma T^4). \tag{43}$$

This is sufficient for the unbounded problems that we shall consider. Note that L is at the control of the experimentalist through variations of the particle loading and/or particle size. Note also that in the optically thin limit $(L \to \infty)$, $q_\pm$ are constants to leading order

$$q_\pm = (q_\pm)_0 + O(L^{-1}), \tag{44}$$

so that

$$dq/dx = 4\sigma T^4/L - (a/L)(q_{+_0} + q_{-_0}) + \cdots, \tag{45}$$

where the first term corresponds to emission and the second to absorption of the background radiation.

The analysis of a steady plane deflagration subject to the loss (45) is a familiar one (e.g. Williams 1985, p. 271), and leads to the result

$$(W/W_{\rm ad})^2 \ln (W/W_{\rm ad})^2 + \Psi = 0, \tag{46}$$

where Ψ represents the integrated heat losses. Quenching occurs for $\Psi > e^{-1}$, and $(W/W_{\rm ad}) = e^{-1/2}$ at the quenching limit.

In contrast, when q is defined by (43) there are no losses for an unbounded system, only energy redistribution, since q vanishes far ahead of and far behind the flame. This leads not to quenching but to enhancement of the flame-speed, as a simple analysis reveals (Joulin & Deshaies 1986).

The one-dimensional steady flame is described by (3) (with $D/Dt \rightarrow u\,d/dx$) with $(-dq/dx)$ added to the right side of the energy equation where q satisfies (43). The following conditions hold:

$$\underline{x \rightarrow -\infty} \quad T \rightarrow T_1, Y \rightarrow Y_1 \quad \underline{x \rightarrow +\infty} \quad T \rightarrow T_a, Y \rightarrow 0. \tag{47}$$

The goal is to determine u_1, the flame-speed, and the associated flame structure. Figure 7 shows the flame structure for a certain range of parameter values.

Two realistic assumptions are the key to a simple analysis: (*a*) L, the radiation-absorption length, is much larger than λ/MC_p, the conduction length, and (*b*) the radiative flux $q(\sim\sigma T_a^4)$ is small compared to the convective energy flux MC_pT_a (small Boltzmann number). The problem for q is then an outer problem in which the conduction/reaction structure appears as a discontinuity. On each side of this, q is defined by a convection/radiation balance with no emission (the σT^4 term is neglected) so that

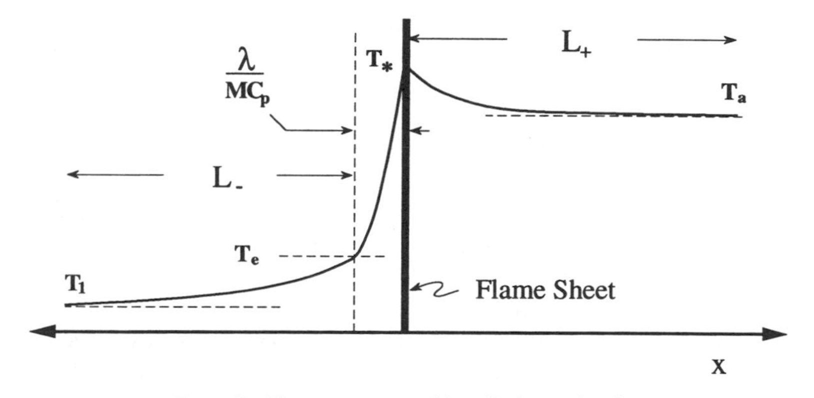

Figure 7 Flame structure with radiative preheating.

$$MC_pT+q = MC_pT_1 \quad \text{in } x < 0, \tag{48a}$$

and

$$q = C\exp(\mp ax/L\pm), \quad x \gtrless 0, \quad C = \text{const.}, \tag{48b}$$

where we have used the fact that q is continuous. On the other hand dq/dx is discontinuous, since from (43)

$$[L\,dq/dx]_-^+ = 4\sigma[T^4]_-^+ = 4\sigma(T_a^4 - T_1^4), \tag{49}$$

and this determines C. From (48) we then deduce the following expression for the temperature at the edge of the preheat zone:

$$(T_e - T_1) = 2\sigma(T_a^4 - T_1^4)/(aMC_p). \tag{50}$$

This is precisely the increment in flame-temperature $(T_* - T_a)$ and since, when the activation energy is large,

$$(W/W_{ad}) = (M/M_{ad}) = \exp[E(T_* - T_a)/2RT_a^2] \tag{51}$$

we have

$$(M/M_{ad})\ln(M/M_{ad}) = (E/RT_a)\sigma(T_a^4 - T_1^4)/(M_{ad}C_pT_aa), \tag{52}$$

a result that is independent of L. The right side of (52) is the product of a large activation energy and a small Boltzmann number, and can be $O(1)$.

The redistribution of energy responsible for the result (52) (so that $M > M_{ad}$) can be brought about in the absence of radiation by providing (through mechanical means) a highly effective conductive path from the hot to the cold gas, generating what is known as an *excess-enthalpy flame* (Weinberg 1974).

Slip and Flame Structure

The analysis described above does not account for the effects of slip between the two phases, and it is useful to examine this issue first in the absence of radiation. Accounting for significant velocity slip is difficult within an analytical framework (see Equation 10a) and some kind of extra-rational approximation is required. In Joulin (1987) velocity slip is neglected (i.e. $u = u_s$) and temperature slip is retained although these must be comparable in a fluid with Prandtl number of 1 or so. A slightly different approach is to adopt the constant density model ($\rho = \rho_s/K = \rho_1$); zero velocity slip then becomes a by-product ($u = u_s = u_1$). This is comparable to an Oseen approximation.

We assume that the loading K is small but has a non-negligible effect on the energy balance, so that the term $-\rho_s C_s(T - T_s)/\tau_T$ must be added to the right side of Equation 3a which is to be solved along with (3b) and

(10b). If $\dot{K}$ is $O(\beta^{-1})$ where $\beta = E/RT_a \gg 1$, this perturbation term has an $O(1)$ impact on the flame-speed. The strategy for dealing with such perturbations is well-established. To leading order T is described by the formulas (4) and these can be used to calculate T_s to leading order from (10b), and in this way the perturbation term is explicitly determined in terms of the unknown mass flux M. It is then a straightforward matter to calculate the change in flame-temperature generated by this term, namely

$$(T_* - T_a) = -K(C_s/C_p)(T_a - T_1)[1+2\tilde{\tau}(M/M_{ad})^2][1+\tilde{\tau}(M/M_{ad})^2]^{-2}$$
$$\tilde{\tau} = \tau_T/\tau_r, \quad \tau_r = \rho_1\lambda/(M_{ad}^2 C_p), \tag{53}$$

where τ_r is the residence time in the flame. Equation (51) then yields the conclusion

$$H \equiv \ln\left(\frac{M}{M_{ad}}\right) + \frac{\tilde{K}[1+2\tilde{\tau}(M/M_{ad})^2]}{[1+\tilde{\tau}(M/M_{ad})^2]^2} = 0,$$
$$\tilde{K} = EKC_s(T_a - T_1)/(2RT_a^2 C_p). \tag{54}$$

When radiative preheating is accounted for, the term

$$(E/2RT_a^2)(T_e - T_1) \equiv (E/RT_a)\sigma(T_a^4 - T_1^4)/(aMC_pT_a) \equiv \tilde{B}(M_{ad}/M) \tag{55}$$

is added to the right side of (54), cf (52) (Joulin 1987).

Results are shown in Figure 8. For small values of the slip (small

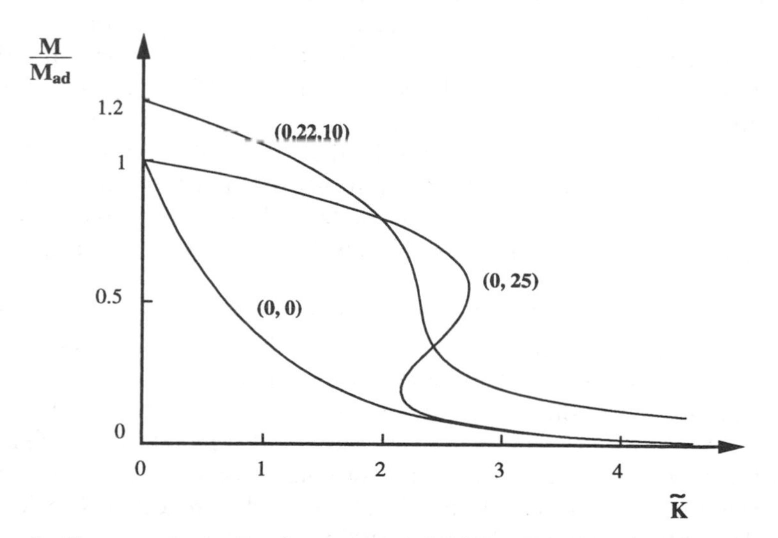

Figure 8 Flame-speed vs loading for a particle-laden flame. The curves are labeled with the pair of values $(\tilde{B}, \tilde{\tau})$.

particles) an increase in loading merely decreases the effective reactivity of the mixture, and the flame-speed falls monotonically. But for large particles, the response is not single valued and an increase in loading can lead to a discontinuous jump to a weak-burning branch. A nonvanishing Boltzmann number pushes the curves upward so that $M > M_{ad}$ for weak loading. This shift is important when stability is examined.

Stability

To discuss the stability problem we adopt a coordinate system attached to the flame so that the mass flux M is now time-dependent. One of the natural time scales is a slow time defined by the radiation length L. On this scale the outer problem [cf (48a)] is

$$C_p(\rho_1 \partial T/\partial t + M \partial T/\partial x) = -dq/dx,$$

$$T(-\infty, t) = T_1, \quad T(x, 0) = T_0(x), \tag{56a,b,c}$$

where dq/dx is given by (48b) and (49). With M and the initial temperature T_0 specified this determines T_e—the temperature at the outer edge of the diffusive preheat zone. The latter is quasi-steady so that from (54) and (55)

$$H = (E/2RT_a^2)(T_e - T_1). \tag{57}$$

This formula, together with (56) describes the slow variations in $M(t)$ (Joulin & Cambray 1987).

A second important time scale is the fast-time defined by the diffusive length $\lambda/M_{ad}C_p$. On this scale T_e is fixed and the stability is controlled by H. A sufficient condition for fast-time instability is $\partial H/\partial M < 0$ and this is necessary if modes of the kind described in Section 2 are not present (an instability linked to H is expected in view of the multiplicity of solutions; see Figure 8). This defines a curve in the $M - \tilde{K}$ plane (Figure 8) and radiative preheating tends to push the steady solution curve upwards into the unstable region. For some parameter values there is but a single stationary solution which is unstable on the fast time. What then is the flame behavior?

Consider the line A–C in Figure 9 (*top panel*); the corresponding variations in H are sketched in the lower part of the figure. We suppose that at some time M has the value defined by the point A. M is driven towards the stationary solution on the slow-time scale and passes through E to the point B where the fast-time takes over and M is driven to C. It then relaxes on the slow-time back towards the stationary solution, but when D is reached there is a fast transition to E and the cycle is repeated. The details are described in Joulin & Cambray (1987).

Earlier, it was noted that oscillations have been observed in flames

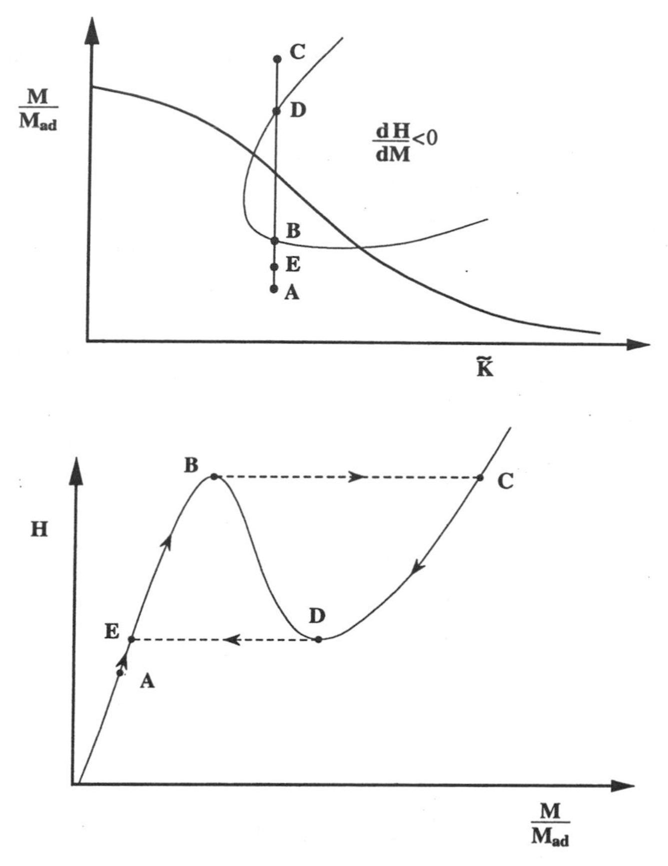

Figure 9 (*Top*) Steady solution curve which penetrates the fast-time instability domain for some loading values. (*Bottom*) Variations of H with M, $\tilde{K}$ fixed, corresponding to the line A–C shown above.

supported by lycopodium/air mixtures (Berlad et al 1990). In these pages we have identified three distinct physical mechanisms that could be responsible: acoustic instabilities driven by slip between the two phases; an instability of the diffusive structure associated with nonsimilarity of the T, Y fields (nonuniformity in the enthalpy); and a structural instability (relaxation oscillation) in which radiative preheating plays a vital role.

6. THE IODATE OXIDATION OF ARSENOUS ACID AND "LIQUID FLAME FRONTS"

On a scale large compared to λ/MC_p a premixed flame appears as a hydrodynamic discontinuity separating cold unburnt gas from hot burnt gas. If this front travels with speed W_{ad} relative to the cold gas which has velocity $\mathbf{v}$, the flame motion is described by the *kinematic flame equation*

$$\partial G/\partial t + \mathbf{v} \cdot \nabla G = W_{ad} |\nabla G|, \tag{58}$$

where the flame is located at $G = 0$. This can be used to examine the response to a turbulent velocity field. The goal is to calculate the average speed—turbulent flame speed—with which the distorted front propagates. There have been both numerical studies and analytical treatments—the latter using renormalization group (RNG) theory (Yakhot 1988). Two assumptions underlie this work: Flame-stretch and curvature do not alter the flame speed, which remains equal to W_{ad}; and the velocity induced by the flame-front distortions is neglected. It is not known what effect these assumptions have on the accuracy of the results.

In the absence of a density jump across the front there is no induced velocity and the first assumption is thereby isolated. A constant density flame is not part of physical reality, but there are *liquids* which support reactive fronts that travel at well defined speeds, and within these the density is sensibly constant. Experimental results for these "liquid flames" provide a valuable test bed for theory (Buckley et al 1991).

Consider an aqueous solution of iodate (IO_3^-) and arsenous acid (H_3AsO_3) for which the arsenous acid is in excess. This can support a reactive front controlled by the Dushman reaction and the Roebuck reaction (Hanna et al 1982):

$$IO_3^- + 5I^- + 6H^+ \rightarrow 3I_2 + 3H_2O, \tag{59A}$$

$$H_3AsO_3 + I_2 + H_2O \rightarrow 2I^- + H_3AsO_4 + 2H^+. \tag{59B}$$

These define an autocatalytic process in which I^- is required for the consumption of the iodate, but is only available as a product which diffuses forward through the front. The role played here by I^- is similar to that which the temperature plays in thermal models of gas flames such as (1). The reaction rates are

$$R_A = -\tfrac{1}{5}d[I^-]/dt = (k_1 + k_2[I^-])[IO_3^-][H^+]^2[I^-],$$

$$R_B = -d[I_2]/dt = k_3[I_2][H_3AsO_3]/[I^-][H^+], \tag{60}$$

where the H^+ concentration is effectively fixed by buffering the solution; k_1, k_2, and k_3 are known constants.

The process (B) is much faster than (A) which is, accordingly, rate defining: The rate at which I_2 is produced by (A) is equal to the rate at which I_2 is consumed by (B), i.e.

$$3R_A - R_B \approx 0. \tag{61}$$

Put slightly differently $d[I_2]/dt$ is equal to the left side of (61) but is much smaller than either term, hence the approximation. This is a *steady-state approximation* of the kind which plays an important role in reduced chemistry models for gas flames (see below). A discussion of this approximation as a formal asymptotic limit may be found in Peters (1991). When formulas (60) are used, Equation (61) is an algebraic relation between the concentrations.

Conservation of iodine atoms implies, for a homogeneous process,

$$[IO_3^-]+[I^-] = [IO_3^-]_0, \tag{62}$$

which is the initial concentration, and this will be true also for a spatially dependent problem if the diffusion coefficients for each species are equal. Then the reaction rate R_A depends only on the unknown concentration $[I^-] \equiv C$ and for a reactive front traveling with speed u the structure is defined by the boundary-value problem

$$u\,dC/dx = \mathsf{D}\,d^2C/dx^2+(k_1+k_2C)[H^+]^2C([IO_3^-]_0-C)$$

$$\underline{x\to-\infty} \quad C\to 0, \quad \underline{x\to+\infty} \quad C\to[IO_3^-]_0. \tag{63}$$

The solution is (Hanna et al 1982)

$$C = (IO_3^-)_0(1+e^{-k(x-x_0)})^{-1},$$

$$k = [IO_3^-]_0[H^+](k_2/2\mathsf{D})^{1/2}, \quad u = (k_1/k)[IO_3^-]_0[H^+]^2+\mathsf{D}k. \tag{64}$$

Ronney (see Buckley 1991) has embarked on an experimental program in which these fronts are distorted by nonuniform flows and the average speed of propagation is measured. Concurrently, theoretical investigations are underway.

7. APPROXIMATE KINETIC MECHANISMS AND ASYMPTOTIC APPROXIMATIONS

In the preceding section we saw how two reactions can be reduced to one by adopting a steady-state approximation. A similar reduction is also possible for more complex kinetic schemes (Peters 1985). Consider, for example, the combustion of a methane/air mixture. A minimum kinetic scheme consists of 25 reactions for the 15 species H, O_2, OH, O, H_2, H_2O, HO_2, CO, CO_2, CH_4, CH_3, CH_2O, HCO, CH_3O, and H_2O_2. If we assume that the species OH, O, HO_2, CH_3, CH_2O, HCO, CH_3O, and H_2O_2 are in steady-state (cf I_2 of the previous section) then the 25 reactions can be reduced to a set of 4 reversible reactions

$$CH_4+2H+H_2O \rightleftarrows CO+4H_2 \tag{65a}$$

$$CO + H_2O \rightleftarrows CO_2 + H_2 \tag{65b}$$

$$H + H + M \rightleftarrows H_2 + M \tag{65c}$$

$$O_2 + 3H_2 \rightleftarrows 2H + 2H_2O, \tag{65d}$$

where the forward and backward rates for each of these global reactions can be expressed in terms of the fundamental rates of the initial 25 reactions. These rates depend on the concentrations of the steady-state species which are determined by algebraic relations, cf (61).

A two-step reversible mechanism can be constructed by assuming further that H is in steady state, and the water-gas shift (65b) is in partial equilibrium.

A great many assumptions are made in this process, but numerical calculations show that reduced mechanisms can be quite accurate, both for premixed flames and for diffusion flames.

Although the subject is a young one, a significant amount of work has been done by a mere handful of pioneers, and the book edited by Smooke (1991) provides a useful introduction. Reduced mechanisms have been used in two ways: as an ingredient of numerical simulations which are thereby simplified; and as the foundation for approximate analytical treatments. The value to numerical simulations arises, in principle, because of the reduced burden on computational resources, particularly in situations where these are necessarily under stress (unsteady, multi-dimensional problems). In practice there can be difficulties. Newton-iteration convergence can be handicapped by the algebraic constraints. Also, the reduced reaction rates have singularities (removable) where certain concentrations vanish (e.g. the rate for (65a) contains the factor $[CH_4][H][H_2O]/[H_2]$) and these can be a problem during time integrations (Rogg 1991). Strategies are needed for dealing with these difficulties if the approach is to fulfill its promise.

More clearly linked to the theme of this article is the analytical work founded on reduced kinetic schemes. The challenges here are significant. The algebraic relations must be simplified, preferably in a rational fashion. Much of the reasoning is *ex post facto*, informed by numerical solutions. These reveal, for example, that for methane/air flames, under certain conditions, the reaction

$$H_2 + OH \rightleftarrows H_2O + H \tag{66}$$

is near equilibrium; thus a simple approximation for the OH concentration can be written down.

Numerical solutions also reveal key characteristics of the flame-structure. Consider, for example, a stoichiometric methane/air mixture. The temperature rises monotonically and rapidly from the cold boundary value T_1 (300 K) to about 1800 K (T_0). This occurs over a distance which we shall take to be 1 unit. Beyond T_0 the temperature continues to increase, but at a much slower rate, from 1800 K to about 2100 K (T_2) over a distance ~ 2 units. The methane concentration $[CH_4]$ falls monotonically over a distance ~ 1, becoming small in the neighborhood of $T = T_0$, and the oxygen concentration $[O_2]$ behaves similarly. Near the cold boundary the balance is a convection-diffusion one (preheat zone) but CH_4/O_2 consumption becomes more important as the temperature rises, and displays a sharp maximum in the neighborhood of $T = T_0$. The distance δ over which consumption is significant is ~ 0.2.

Consumption of methane and oxygen generates minor species together with CH_3, H, OH, O, CO_2, H_2O, CO, and H_2. Of these, CO and H_2 are consumed downstream of $T = T_0$ over a distance $(\varepsilon) \sim 0.4$. (These various distances are estimated by "eye-balling" the calculated profiles, and so are quite subjective.) The two reaction regions identified here—CH_4/O_2 consumption, followed by CO/H_2 oxidation—correspond to the steps in the two-step mechanism (Figure 10).

The key to analysis in this case is the assumption that ε and δ can be treated as small parameters, and $1 \gg \varepsilon \gg \delta$. This is a bold move, but the results are better than one might expect from the estimates above. δ is proportional to the ratio of the rates k_1 and k_{11}

$$H + O_2 \xrightarrow{k_1} OH + O, \quad CH_4 + H \xrightarrow{k_{11}} CH_3 + H_2 \tag{67}$$

evaluated at T_0, hence the rubric *ratio-ratio asymptotics* (Williams 1991). The other parameter, ε, is proportional to $(T_2 - T_0)/(T_2 - T_1)$.

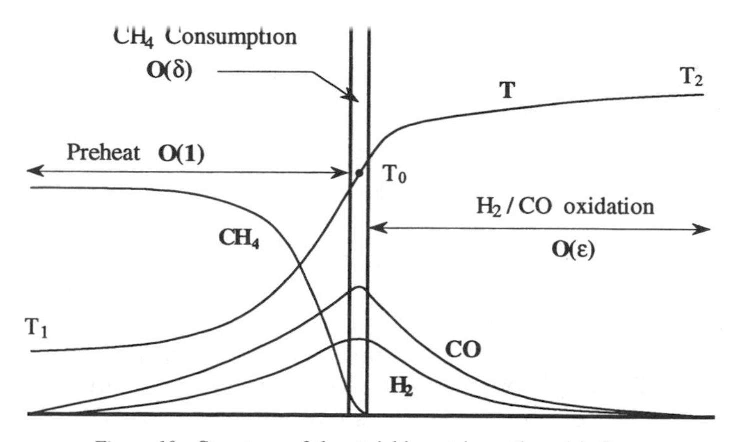

Figure 10 Structure of the stoichiometric methane/air flame.

Bold moves can have substantial disadvantages. Here each mixture and flame type requires its own ad hoc approximations involving fixed parameters that are not unambiguously small. Under these circumstances it is difficult to have a priori confidence in the results, so that they lose independent predictive power. This is quite different from activation energy asymptotics in which a single universal model is used which is of proven robustness. Perhaps the most important thing that will emerge from the reduced chemistry work is a like paradigm, and there are efforts being made in that direction.

A characteristic of reduced chemistry systems is that they predict absolute *kinetic flammability limits*—mixture strengths beyond which flame propagation is not sustainable even in the absence of losses. There is every reason to believe that this is a creature of the approximation, not of the full kinetics on which the approximation is based. Giovangigli & Smooke (1991) have constructed numerical solutions for premixed methane/air and hydrogen/air flames on (necessarily) finite computational domains and have shown that limits are linked exclusively to the length of the domain. For example, on a domain of 1 m length the lean limit equivalence ratio for hydrogen/air mixtures is 0.197 with a flame-speed of 0.195 mm/s, and for a 10 m domain the limit is 0.153 with a speed of 0.018 mm/s. (Compare these with the radiation induced flame-ball lean limit of 0.085, defined in Section 4.) In each case the limit is associated with an interaction of the flame with the downstream boundary. The slow thick flames associated with weakly reactive mixtures are very sensitive to losses and the experimental limits are a consequence of these losses. The "limits" predicted by reduced kinetic schemes are, in fact, mixture strengths at which the sensitivity becomes large, and so correspond, roughly, to practical flammability limits.

8. TRIBRACHIAL (3-ARMED) OR TRIPLE FLAMES

We conclude our discussion with a brief description of an important flame configuration that can arise when the mixture strength is nonuniform in space. Imagine a premixed flame sustained by a mixture moving in the x-direction in which the fuel and oxidizer concentrations vary linearly with y. Suppose the mixture is in stoichiometric proportion on the x-axis, fuel-lean in $y > 0$, and fuel-rich in $y < 0$. Then the burnt gas will contain hot unburnt oxidizer in $y > 0$, hot unburnt fuel in $y < 0$, and these excess reactants will diffuse towards each other to generate a diffusion flame that trails the two branches of the premixed flame. We shall call this configuration a τ-flame—both a short-hand term and an icon.

τ-flames have been observed in nonuniform mixtures, but their importance arises in nonpremixed combustion. A flame-sheet approximation is often a good one for diffusion flames, and the sheet separates the fuel from the oxidizer. Holes can be created in this sheet, corresponding to local extinction, when a local effective Damkohler number (D_2) becomes too small (Williams 1985, p. 76).

When all Lewis numbers equal unity, D_2 is proportional to the inverse of the scalar dissipation rate and the latter is continuously varying in a turbulent flame according to fluctuations in the size and strength of the turbulent eddies. Thus holes can be created, "heal," and be recreated under the influence of these fluctuations. During a time and at a place in which there is a hole in the sheet, oxidizer and fuel can pass from one side of the sheet to the other, creating a nonuniform mixture. Healing occurs by propagation of a τ-flame at the edge of the hole. There are two propagation issues here: inward propagation of a τ-flame to seal the hole, and the outward propagation of the perimeter as the hole expands.

Let us suppose that chemical reaction is modeled by the simple process (2), that the Lewis numbers of X and Y are both unity, and that the stoichiometric proportions are 1:1. Consider a plane laminar flame in which the fresh mixture is defined by

$$T = T_1, \quad X = X_1 = X_0 + \beta^{-1}X', \quad Y = Y_1 = X_0 + \beta^{-1}Y'$$

$$\beta = E/RT_{a_0} \gg 1, \quad T_{a_0} = T_1 + QX_0/C_p. \tag{68}$$

T_{a_0} is the adiabatic flame temperature when the perturbations X' and Y' are both zero.

It is not difficult, using activation-energy asymptotics, to show that the adiabatic flame speed is given by the formula

$$M = M_0 \exp(X'Q/2C_pT_{a_0})[1 + (Y' - X')Q/2C_pT_{a_0}]^{1/2}, \tag{69}$$

when $Y' \geq X'$, with the interchange $X' \rightleftarrows Y'$ when $X' \geq Y'$.

Suppose now that the perturbations X' and Y' vary linearly with y, the coordinate transverse to the flow. To preserve symmetry we make the choice

$$X' = \alpha y/l, \quad Y' = -\alpha y/l, \quad l \equiv \lambda/M_0C_p, \tag{70}$$

where α is a positive nondimensional concentration gradient.

Now the flame is not planar—it must curve so that the normal component of the incoming flow is equal in magnitude to the local flame speed. Using the constant-density model leaves the flow unaffected by the flame and simplifies the discussion.

We restrict attention to weak concentration gradients ($\alpha \to 0$) so that Equation (69) and its mate reduce to

$$M = M_0[1-(Q^2\alpha^2y^2)/(4l^2C_p^2T_{a_0}^2)] \tag{71}$$

and the flame-sheet has the shape

$$x = Q\alpha y^2/(2\sqrt{2}lC_pT_{a_0}). \tag{72}$$

In the scenario so far the flame is assumed stationary in a flow moving with mass flux M_0, but this cannot, in fact, be the case. As a consequence of the flame curvature the flame experiences stretch—a pseudo-Lagrangian area increase here generated by a nonvanishing velocity component tangential to the sheet—and a stretched flame does not travel at the adiabatic flame speed. The stretch, equal to $(M_0Q\alpha)/(\sqrt{2}lC_p\rho T_{a_0})$, is small, and there is a universal relationship between small stretch and the flame-speed perturbation that it generates. This correction may be calculated for a flame in a uniform mixture ($X' = Y' = 0$) since the two small effects (of concentration gradient and stretch) are, to first order, additive. In this way, we deduce that the τ-flame travels with a speed corresponding to a mass flux

$$M_0[1-Q\alpha/\sqrt{2}C_pT_a]. \tag{73}$$

The concentration gradient generates the τ-structure and, at the same time, slows the flame (Dold 1989).

To account for larger values of α the problem (which is a free-boundary problem when reaction is confined to thin sheets) must be solved numerically. This has been done by Hartley & Dold (1991). They show that the flame speed decreases monotonically with increasing α, and asymptotes to zero as $\alpha \to \infty$.

The preceding discussion is not directly relevant to turbulent diffusion flames for these are subject to strain which not only generates the holes, but affects the rate at which they open or close. Dold et al (1991) have described a simple configuration in which these effects can be examined. Consider a planar counter-flow of fuel and oxidizer supporting a diffusion flame-sheet. The gases move in the x-y plane with the sheet located at $x = x_0$. This is a classical configuration used to study diffusion flames both numerically and experimentally. If the rate of strain is too large quenching will occur.

We now discard the part of the sheet lying in $z > 0$ so that, instantaneously, the remainder occupies the region $x = x_0$, $z < 0$. Now the sheet is subject not only to strain but to edge effects, and the edge will advance or retreat according to the prevailing conditions. When the strain is small a τ structure will develop and the edge will advance into the region $z > 0$. But for large strain—close to the quenching value for the uncut configuration—the edge retreats, again as a τ-structure. Details and numerical results are described by Dold et al (1991).

9. CONCLUDING REMARKS

We have examined a number of different topics here, but they help to identify three themes that will play an important role in laminar-flame theory in the coming years.

1. Microgravity experiments. Access to drop-towers, the space-shuttle, and—in due course, perhaps—space station *Freedom*, encourages the development of experiments well designed to isolate the fundamental physics of combustion. Modeling will play an important role in this endeavor.
2. Kinetic modeling. Despite assertions by some to the contrary, activation-energy asymptotics is not dead. It has its limitations however, and the development of alternative tools is an important goal. At the present time reduced kinetic schemes require great effort and ingenuity merely for the calculation of steady structures. What are needed are robust models simple enough to permit dynamical calculations.
3. Turbulence modeling. The use of laminar flame concepts is an important strategy in turbulent-flame modeling, and will continue to be so.

ACKNOWLEDGMENTS

This article could not have been written without the long-term support of my combustion research by the Air Force Office of Scientific Research, Dr. A. Nachman, monitor. I am also grateful for support by NASA's microgravity combustion program, Dr. Howard Ross, monitor.

Literature Cited

Bayliss, A., Matkowsky, B. J. 1990. Two routes to chaos in solid fuel combustion. *SIAM J. Appl. Math.* 50: 436–59

Berlad, A. L., Ross, H., Facca, L., Tangirala, V. 1990. Particle cloud flames in acoustic fields. *Combust. Flame* 82: 448–50

Buckley, S. G. 1991. *Liquid-phase autocatalytic reactions: a useful paradigm for turbulent combustion studies.* Unpubl. undergrad. rep., Dept. Mech. & Aerospace Eng., Princeton Univ.

Buckley, S. G., Shy, S. S., Ronney, P. D. 1991. *Experimental simulation of premixed turbulent combustion using a liquid-phase autocatalytic reaction.* Presented at Spring Tech. Meet., Combust. Inst., Western States Sect., March 1991, Boulder

Buckmaster, J. D. 1983. Stability of the porous plug burner flame. *SIAM J. Appl. Math.* 43: 1335–49

Buckmaster, J. D. 1992. A flame-string model and its stability. *Combust. Sci. Technol.* In press

Buckmaster, J. D., Ludford, G. S. S. 1982. *Theory of Laminar Flames.* New York: Cambridge Univ. Press

Buckmaster, J., Clavin, P. 1992. An acoustic-instability theory for particle-cloud flames. *24th Symp. (Int.) on Combust. The Combust. Inst., Pittsburgh.* To appear

Buckmaster, J., Gessman, R., Ronney, P. 1992. The three-dimensional dynamics of flame-balls. *24th Symp. (Int.) on Combust. The Combust. Inst., Pittsburgh.* To appear

Buckmaster, J., Joulin, G. 1991. Flame balls stabilized by suspension in fluid with a steady linear ambient velocity distribution. *J. Fluid Mech.* 227: 407–27

Buckmaster, J., Joulin, G., Ronney, P. 1990. The structure and stability of non-adiabatic flame balls. *Combust. Flame* 79: 381–92

Buckmaster, J., Smooke, M. D. 1992. *Analytical and numerical modeling of flame-balls in hydrogen-air mixtures.* Unpubl. Rep., Univ. Ill.

Chandrasekhar, S. 1960. *Radiative Transfer.* NY: Dover

Clavin, P., Pelce, P., He, L. 1990. One-dimensional vibrating instability of planar flames propagating in tubes. *J. Fluid Mech.* 216: 299–322

Clavin, P., Sun, J. 1992. Theory of acoustic instabilities of planar flames propagating in sprays or particle-laden gases. *Combust. Sci. Technol.* 78: 265–88

Dold, J. W. 1989. Flame propagation in a nonuniform mixture: analysis of a slowly varying triple flame. *Combust. Flame* 76: 71–88

Dold, J. W., Hartley, L. J., Green, D. 1991. Dynamics of laminar triple-flamelet structures in non-premixed turbulent combustion. In *Dynamical Issues in Combustion Theory, IMA Vol. Math. Appl.*, ed. P. C. Fife, A. Linan, F. Williams, 35: 83–105. New York: Springer-Verlag

Frankel, M. L., Sivashinsky, G. I. 1991. Surface evolution equations modelling intrinsic dynamics of premixed flames. In *Fluid Dynamical Aspects of Combustion Theory, Pitman Res. Notes Math.*, ed. M. Onofri, A. Tesei, 223: 30–42. Harlow, UK: Longman

Giovangigli, V., Smooke, M. D. 1991. Application of continuation methods to plane premixed laminar flames. *Rep. ME-100-91.* Yale Univ.

Hanna, A., Saul, A., Showalter, K. 1982. Detailed studies of propagation fronts in the iodate oxidation of arsenous acid. *J. Am. Chem. Soc.* 104: 3838–44

Hartley, L. J., Dold, J. W. 1991. Flame propagation in a nonuniform mixture: analysis of a propagating triple-flame. *Combust. Sci. Technol.* 80: 23–46

Jackson, T. L., Kapila, A. K. 1986. Effect of thermal expansion on the stability of plane, freely propagating flames. Part II: Incorporation of gravity and heat loss. *Combust. Sci. Technol.* 49: 305–17

Joulin, G. 1987. Temperature-lags and radiative transfer in particle-laden gaseous flames. Part 1: Steady planar fronts. *Combust. Sci. Technol.* 52: 377–95

Joulin, G., Cambray, P. 1987. Temperature-lags and radiative transfer in particle-laden gaseous flames. Part 2: Unsteady propagations. *Combust. Sci. Technol.* 52: 397–412

Joulin, G., Deshaies, B. 1986. On radiation-affected flame propagation in gaseous mixtures seeded with inert particles. *Combust. Sci. Technol.* 47: 299–315

Marble, F. E. 1970. Dynamics of dusty gases. *Annu. Rev. Fluid Mech.* 2: 397–446

Margolis, S. B. 1991. The transition to non-steady deflagration in gasless combustion. *Prog. Energy Combust. Sci.* 7: 135–62

Matkowsky, B. J., Sivashinsky, G. I. 1978. Propagation of a pulsating reaction front in solid fuel combustion. *SIAM J. Appl. Math.* 35: 465–78

Nicolaenko, B., Scheurer, B., Temam, R. 1985. Some global dynamical properties of the Kuramoto-Sivashinsky equation: nonlinear stability and attractors. *Physica* 16D: 155–83

Peters, N. 1985. Numerical and asymptotic analysis of systematically reduced reaction schemes for hydrocarbon flames. *Numerical Simulation of Combustion Phenomena, Lect. Notes Phys.*, ed. R. Glowinski, B. Larrouturou, R. Teman, p. 90. New York: Springer-Verlag

Peters, N. 1991. See Smooke 1991, p. 48

Rogg, B. 1991. See Smooke 1991, p. 159

Ronney, P. 1985. Effect of gravity on laminar premixed gas combustion. II. Ignition and extinction phenomena. *Combust. Flame* 62: 121–33

Ronney, P. 1990. Near limit flame structures at low Lewis number. *Combust. Flame* 82: 1–14

Sivashinsky, G. I. 1983. Instabilities, pattern formation, and turbulence in flames. *Annu. Rev. Fluid Mech.* 15: 179–99

Smooke, M. D., ed. 1991. *Reduced Kinetic Mechanisms and Asymptotic Approximations for Methane-Air Flames, Lect. Notes Phys.* 384. Berlin: Springer-Verlag

Turing, A. M. 1953. The chemical basis of morphogenesis. *Philos. Trans. R. Soc. London Ser. B* 237: 37–72

Weinberg, F. 1974. The first half-million years of combustion research and today's burning problems. *15th Symp. (Int.) on Combust., The Combust. Inst., Pittsburgh*, pp. 1–17

Whaling, K. N., Abud-Madrid, A., Ronney, P. D. 1992. *Structure and stability of near-*

limit flames with low Lewis number. Unpubl. Rep., Univ. Princeton
Williams, F. A. 1971. Theory of combustion in laminar flows. *Annu. Rev. Fluid Mech.* 3: 171–88
Williams, F. 1985. *Combustion Theory*. Menlo Park: Benjamin/Cummings. p. 271
Williams, F. 1991. See Smooke 1991, p. 68
Yakhot, V. 1988. Propagation velocity of premixed turbulent flames. *Combust. Sci. Technol.* 60: 191–214
Zeldovich, Ya. B., Barenblatt, G. I., Librovich, V. B., Makhviladze, G. M. 1985. *The Mathematical Theory of Combustion and Explosions*, p. 327. New York: Consultants Bureau

Annu. Rev. Fluid Mech. 1993. 25 : 55–97

RESONANT INTERACTIONS AMONG SURFACE WATER WAVES

J. L. Hammack

Departments of Geosciences and Mathematics, The Pennsylvania State University, University Park, Pennsylvania 16802

D. M. Henderson

Department of Mathematics, The Pennsylvania State University, University Park, Pennsylvania 16802

KEY WORDS: nonlinear dynamics, gravity-capillary waves, dispersive waves

INTRODUCTION

Thirty years have passed since Phillips (1960) pioneered a view of weak, nonlinear interactions among gravity waves on the surface of deep water. His view emphasized the special role of "resonant" interactions, so-called because they have the mathematical form of a resonantly forced, linear oscillator. Unlike weak nonresonant interactions, weak resonant interactions can cause significant energy transfer among wavetrains and profoundly affect wavefield evolution. Early skepticism[1] of Phillips' view was dispelled unequivocally by experiments presented in the companion papers of Longuet-Higgins & Smith (1966) and McGoldrick et al (1966). This experimental confirmation of RIT (resonant interaction theory) and RIT's general applicability to a variety of phenomena (Benney 1962) made it one of the principle catalysts for the rapid expansion in the understanding of nonlinear wave phenomena that has occurred during the last thirty years.

[1] See the discussion comments in Section 4, pages 163–200, of *Ocean Wave Spectra Proceedings of a Conference* (1963) held in Easton, Maryland, May 1–4, 1961.

0066–4189/93/0115–0055$02.00

Ten years have passed since Phillips (1981a) presented a retrospective of resonant interaction theory which noted its maturity and suggested that it might be approaching its limits of usefulness. The maturity of RIT then and its older age now are evidenced by the plethora of literature which include review articles (Phillips 1967, 1974, 1981a; Yuen & Lake 1980, 1982) and comprehensive descriptions in monographs (Phillips 1977, LeBlond & Mysak 1978, West 1981, Ablowitz & Segur 1981, Craik 1985). Because of this maturity and the preponderance of readily available survey material this review is narrowly focused. In particular, we do not review coupling between surface and internal waves, internal waves, planetary waves, trapped waves, or waves in shallow water. Our objective is to review resonant interaction theories and experiments for waves on the surface of a deep layer of water. RIT is more than a framework for unifying wave phenomena—RIT is predictive. In fact, it provides one of the few analytically tractable models (the three-wave equations) of weak interactions among wavetrains with arbitrary wavelengths and directions of propagation. Yet, quantitative comparisons between RIT and RIE (resonant interaction experiments) are rarer than generally believed. Moreover, the success of RIT in quantitatively predicting the outcome of experiments is less established than generally believed. The theme of this review is that although RIT is mature, RIE is not, and perhaps its maturing will provide the basis for further theoretical developments.

An outline of the review is as follows. First, we review RIT in a variety of dynamical settings with an emphasis on the underlying approximations. We examine the elementary interactions of three- and four-wave resonances as well as deterministic and stochastic models for wavefields comprising either a broad or narrow spectrum of waves interacting in multiple and coupled sets. We also review some special settings in which higher-order resonances have been examined. Second, we review RIE in a variety of dynamical settings with an emphasis on those aspects that are predicted. We begin with a review of experimental investigations on the applicability of Kelvin's dispersion relation. We then review experiments involving gravity, gravity-capillary, and capillary wavetrains. We conclude by briefly summarizing results from comparisons between RIT and RIE and by noting the particularly glaring absence of controlled experiments in some applications.

RESONANT INTERACTION THEORY

In order to emphasize the basic notions of RIT and the generality of its application, we briefly outline its development for an arbitrary, nonlinear, energy conserving dynamical system. This outline, which closely parallels

Segur (1984) and Craik (1985), leads to necessary kinematical conditions for the existence of resonance, which underlie all applications of RIT. We then review a variety of dynamical analyses for surface waves in deep water that lie within the framework of RIT.

Basic Notions and Kinematical Considerations

Consider a nonlinear, energy-conserving dynamical system represented by

$$N(\phi) = 0, \tag{1}$$

in which N is a nonlinear operator, $\phi(\mathbf{x}, t)$ is a solution of (1), $\mathbf{x} = (x, y, z)$ is a position vector, and t is time. Suppose $\phi = 0$ is a (neutrally) stable equilibrium solution of (1), e.g. the quiescent surface for water waves. Infinitesimal deformations from this equilibrium state are found by linearizing (1) to obtain

$$L(\phi_1) = 0, \tag{2}$$

in which L is a linear operator and ϕ_1 is a solution of (2). If L has constant coefficients as it does for water waves, ϕ_1 has the form:

$$\phi_1 \sim \exp[i(\mathbf{k}\cdot\mathbf{x} - \omega t)], \tag{3}$$

in which $\mathbf{k} = (k_i) = (l, m, n)$ is a wavenumber vector (or simply wavevector) and ω is the wave frequency. Substituting (3) into (2) yields a dispersion relation for the linear problem:

$$\omega = W(\mathbf{k}). \tag{4a}$$

We require that ω exists in the sense that for fixed $\mathbf{k}$ there is a countable set of possible values for ω and that $W(\mathbf{k})$ is real-valued for real-valued $\mathbf{k}$. We also require that waves are nontrivially dispersive (Whitham 1974, p. 365), i.e.

$$\det\left(\frac{\partial^2 W(\mathbf{k})}{\partial k_i \partial k_j}\right) \neq 0, \qquad (i, j = 1, 2, 3). \tag{4b}$$

(Hence, in the context of water waves we avoid shallow water.) Now return to (1) and seek small-but-finite deformations from the equilibrium state by means of a formal power series:

$$\phi(\mathbf{x}, t; \varepsilon) = \sum_{r=1}^{\infty} \varepsilon^r \phi_r(\mathbf{x}, t), \qquad 0 < \varepsilon \ll 1. \tag{5}$$

At $O(\varepsilon)$ of (5), ϕ_1 satisfies (2) and can be represented by a superposition of S linear wavetrains, i.e.

$$\phi_1 = \sum_{s=1}^{S} (A_s e^{i\theta_s} + A_s^* e^{-i\theta_s}), \tag{6a}$$

in which A_s is the complex amplitude, A_s^* is its complex conjugate, and

$$\theta_s = \mathbf{k}_s \cdot \mathbf{x} - \omega_s t, \tag{6b}$$

is the wave phase. At this order, the S wavetrains interact linearly. The first nonlinear interactions appear at $O(\varepsilon^2)$ for which

$$L(\phi_2) = Q(\phi_1), \tag{7}$$

where Q is a quadratic operator. For definiteness, suppose we examine the nonlinear interactions between two wavetrains ($S = 2$); then Q yields terms like:

$$Q(\phi_1) = A_1 A_1^* + A_1^2 e^{i2\theta_1} + A_1^{*2} e^{-i2\theta_1} + A_1 A_2 e^{i(\theta_1+\theta_2)} + A_1 A_2^* e^{i(\theta_1-\theta_2)} + \cdots \tag{8}$$

The first three terms in (8) represent self-interactions of wavetrain $s = 1$; the first term corresponds to its mean flow while the second and third terms correspond to its second harmonic. The other terms shown in (8), and their complex conjugates, correspond to sum and difference wavetrains formed by nonlinear interactions between wavetrains $s = 1$ and $s = 2$. Supposing these quadratic interactions do not vanish identically we can examine the terms of (8) to see if any satisfy the dispersion relation of (4), i.e. we define

$$\theta_1 \pm \theta_2 = \theta_3 \quad \text{or} \quad (\mathbf{k}_1 \pm \mathbf{k}_2) \cdot \mathbf{x} - (\omega_1 \pm \omega_2)t = \pm \mathbf{k}_3 \cdot \mathbf{x} - (\pm \omega_3)t. \tag{9}$$

[The second-harmonic terms in (8) are a special case of (9) with $\theta_1 = \theta_2$.] If $(\mathbf{k}_3, \omega_3)$ satisfy (4), then (7) is the equation of a resonantly forced, linear oscillator. Hence, ϕ_2 will grow linearly in time so that the underlying assumption of small deformations in (5) is violated when $\varepsilon t = O(1)$. When the dispersion relation satisfies $W(\mathbf{k}) = -W(-\mathbf{k})$, as in water waves, we can write (9) in the form

$$\mathbf{k}_1 = \mathbf{k}_2 \pm \mathbf{k}_3, \tag{10a}$$

$$\omega_1 = \omega_2 \pm \omega_3. \tag{10b}$$

Equations (10) are necessary conditions for resonant three-wave interactions. The special case of $\theta_1 = \theta_2$ in (9) is termed second-harmonic (or internal) resonance. Phillips (1960) showed that resonant triads are not possible for deep-water gravity waves. McGoldrick (1965) showed that both second-harmonic and triadic resonances are possible for deep-water gravity-capillary waves. Second-harmonic resonance was originally noted by Harrison (1909) and investigated by Wilton (1915). Known as Wilton ripples, second-harmonic resonance occurs for a wavetrain with wavenumber $k_0 = (\varrho g/2T)^{1/2}$, in which ϱ is the mass density, T is the surface tension, and g is the gravitational force per unit mass. [Triadic resonances also occur in shallow water where waves are nondispersive at leading order so that (10) are satisfied somewhat trivially.]

Proceeding to $O(\varepsilon^3)$, in which cubic nonlinear interactions occur on the right-hand side of (7), we allow three wavetrains ($S = 3$) in (6a). Then, assuming that triadic resonances do not occur, considerations similar to those above lead to necessary conditions for four-wave resonances:

$$\mathbf{k}_1 + \mathbf{k}_2 = \mathbf{k}_3 + \mathbf{k}_4, \tag{11a}$$

$$\omega_1 + \omega_2 = \omega_3 + \omega_4, \tag{11b}$$

in which we have made use of a dynamical result of Hasselmann (1962) to eliminate the usual $\pm$ signs. If Equations (11) are satisfied, the underlying assumption of small deformations in (5) is violated when $\varepsilon^2 t = O(1)$. Equations (11) include the special case of third-harmonic (internal) resonance as well as the degenerate case with $\omega_1 = \omega_2 = \omega_3 = \omega_4$ and $|\mathbf{k}_1| = |\mathbf{k}_2| = |\mathbf{k}_3| = |\mathbf{k}_4|$, i.e. the wavevectors in (11a) form a rhombus. Since a single wavetrain satisfies the latter degenerate case with collinear wavevectors, quartic resonances—unlike triadic resonances—are possible for any system. Phillips (1960) first showed that nondegenerate quartic resonances are the first to occur for gravity waves on deep water.

The above procedure can be continued for resonant quintets, sextets, etc; however, RIT in this general form is typically not exploited beyond triadic and quartic resonances. The reasons are twofold. Philosophically, it is generally assumed that the lowest-order resonant interactions that occur will dominate wavefield evolution; hence, the inevitable existence of resonant quartets mitigates interest in higher-order effects. In practice, the labor involved in the dynamical calculations of higher-order resonances is daunting. Nevertheless, the perturbative approach of RIT has been extended to higher-order resonances in some special cases which we discuss subsequently.

Dynamical Considerations

When the kinematical conditions of (10) or (11) are satisfied, a separate dynamical analysis is required to follow long-time evolution of the wavefield. Dynamical analyses have branched in several directions depending on the number of resonant wave sets possible for the underlying (linear) wavefield of (6a).

ELEMENTARY INTERACTIONS In the simplest cases, which we term elementary resonant interactions, a single triad or a single quartet of waves is studied. Two common methods of analysis are the method of multiple scales (e.g. Benney 1962, McGoldrick 1965) and variational techniques (e.g. Simmons 1969). Both methods lead to a set of coupled, nonlinear partial differential equations for the complex amplitudes of the interacting wavetrains. These equations are similar for all dynamical systems, differing only in real-valued interaction coefficients, which depend on the linear properties of the underlying wavetrains. The equations for a single resonant triad (Ball 1964, Bretherton 1964, McGoldrick 1965, Benney & Newell 1967, Simmons 1969) can be written in the general form

$$(\partial_t + \mathbf{U}_s \cdot \nabla) A_s = i\gamma_s A^*_{s+1} A^*_{s+2}, \tag{12}$$

in which s is interpreted modulo 3, γ_s are the interaction coefficients, and $\mathbf{U}_s = \partial W(\mathbf{k}_s)/\partial \mathbf{k}$ are the respective (constant) group velocities. The interaction coefficients for deep-water gravity-capillary waves are given by McGoldrick (1965) and in their most perspicuous form by Simmons (1969) as

$$\gamma_s = -\frac{Jk_s}{4\omega_s}, \text{ (no sum)} \qquad J = \sum_{j=1}^{3} \omega_j \omega_{j+1}(1 + \mathbf{e}_j \cdot \mathbf{e}_{j+1}), \tag{13a,b}$$

in which $\mathbf{e}_j = \mathbf{k}_j/k_j$ and j is interpreted modulo 3. The three-wave equations of (12) have a rich structure (see Craik 1985 for a comprehensive discussion). They are an infinite-dimensional Hamiltonian system which is completely integrable (i.e. the Hamilton-Jacobi equation is separable), and they can be solved exactly for a wide class of initial data by the inverse scattering transform (see Kaup 1981 and the references cited there). In the absence of spatial gradients, they have exact solutions in terms of Jacobi elliptic functions (Ball 1964, Bretherton 1964), which exhibit periodic exchanges of energy among the three waves as well as the phenomenon of recurrence. There are a number of other exact solutions for special cases (e.g. see Craik 1985). Resonant triads occur on a time scale $t \sim t_0/\varepsilon$, in which t_0 is a characteristic waveperiod.

The equations for a single resonant quartet (Benney 1962, Bretherton 1964, Benney & Newell 1967) can be written in the compact form

$$(\partial_t + \mathbf{U}_s \cdot \nabla) A_s = iA_s \sum_{j=1}^{4} \gamma_{sj} |A_j|^2 + i\Gamma\omega_s A^*_{s+\delta} A_{s+2\delta} A_{s+3\delta}, \tag{14}$$

in which s is interpreted modulo 4; $\delta = +1$ for s odd; $\delta = -1$ for s even; and γ_{sj} and Γ are real-valued interaction coefficients. The interaction coefficients in the matrix (γ_{sj}) with $s, j = 1, 2, 3, 4$ are discussed by Phillips (1977, p. 85) for deep-water gravity waves; their form is given by Longuet-Higgins & Phillips (1962). These coefficients account for the nonlinear dispersion of the waves with the diagonal terms corresponding to self interactions (Stokes 1847) and the off-diagonal terms corresponding to mutual interactions between wave pairs. The coefficient Γ accounts for energy exchange among the four waves; it is a complicated function of the frequencies and wavevector configuration of the underlying waves. An explicit expression for Γ in the general case has not been presented; however, Benney (1962) hints at its complexity and Longuet-Higgins (1962) gives an explicit result when (11) is satisfied with two coincident wavetrains (also see McGoldrick et al 1966). The four-wave equations of (14) have many interesting properties (see Craik 1985 for a review). They are not known to be integrable (Ablowitz & Segur 1981, p. 312), but exact solutions in terms of Jacobi elliptic functions exist when there are no spatial gradients (Bretherton 1964, Boyd & Turner 1978). Hence, as in the three-wave equations, periodic exchanges of energy among the wavetrains are possible. Resonant quartets occur on a time scale $t \sim t_0/\varepsilon^2$.

The elementary interactions described by the three- and four-wave equations are the backbone of RIT; their predictions and experimental validation are the primary basis for RIT's general acceptance. Yet, in many practical applications these equations are not sufficiently general, since a system that admits one resonant set of waves often admits many resonant sets simultaneously, as is the case for water waves.

MULTIPLE RESONANT INTERACTIONS When the underlying (linear) wavefield of (6a) comprises many waves—e.g. a discrete spectrum of S waves or a continuous spectrum—multiple and coupled resonances are often possible. A distinction is usually made between broad spectra and narrow spectra, since the latter enable some useful simplifications. In addition, the existence of multiple and coupled resonant sets among both broad and narrow spectra allows a choice between deterministic or stochastic descriptions. For broad spectra, stochastic descriptions dominate owing to the escalating complexity of deterministic descriptions and the absence of known initial conditions for systems like ocean waves. It should be noted also that the simplifying analyses[2] that result from adopting stochastic descriptions are purchased with assumptions on the nature of the underlying randomness; results can differ markedly with different assumptions. [Reviews of stochastic descriptions of water waves can also be found in Phillips (1977), West (1981), and Yuen & Lake (1982).]

[2] Simplification is obscured by the burdensome symbolism in many studies. We note the importance of symbolism in mathematics, e.g. see Whitehead (1958, p. 39) who cites the classical example of arithmetic-progress stalled until the arabic symbolism replaced the roman symbolism.

Broad-spectrum interactions Ablowitz & Haberman (1975a,b) used a deterministic approach with S large to study a general system with multi-triad resonances. They used the methods of Ablowitz et al (1974) to obtain a set of S integrable, nonlinear partial differential equations. These are coupled through M interlocked triads in which $M \geq (S-1)/2$. Ablowitz & Haberman (1975b) also considered the case of multiquartet interactions. The special case ($M = 2$, $S = 5$) of two resonant triads with one wavetrain in common was examined in the context of plasma waves by Wilhelmsson & Pavlenko (1973) [see also Weiland & Wilhelmsson (1977, p. 121)], who obtained exact solutions in terms of Jacobi elliptic functions—similar to the results for elementary interactions.

Stochastic descriptions of nonlinearly interacting water waves were pioneered by Hasselmann (1962, 1963a,b, 1966, 1967a) who recognized the special role of four-wave resonances for surface gravity waves and asserted the random-phase approximation or equivalently, asserted that wavefields are spatially homogeneous (and Gaussian). The random-phase assertion (RPA) is wrong for the multitriad or multiquartet descriptions of Ablowitz & Haberman (1975a,b). Whether it is right for resonant interactions among a broad spectrum of gravity or gravity-capillary water waves cannot be answered unequivocally. The RPA is plausible, and it appears useful in a practical sense for studying plasma waves (Davidson 1972, Section 13). In essence, the RPA relegates nonlinearity in four-wave resonances to acting on the slow time scale $t \sim t_0/\varepsilon^4$ whereas randomness acts on the faster time scale $t \sim t_0/\varepsilon^2$ (e.g. see Yuen & Lake 1982); hence, randomness dominates nonlinearity. The diminished nonlinearity allows energy transfers among resonant waves; however, the resonant set contains passive waves, which grow to a steady state, rather than active waves, which exchange energy periodically as in elementary resonant interactions. Thus, energy transfers are irreversible. For a discrete broad spectrum of underlying waves, Hasselmann's stochastic model yields an evolution (transport) equation for the spectral density of wave action per unit mass $N_1 := N(\mathbf{k}_1, t)$ with the form

$$\frac{\partial N_1}{\partial t} = \int_{-\infty}^{\infty}\int_{-\infty}^{\infty}\int_{-\infty}^{\infty} Q(\mathbf{k}_1, \mathbf{k}_2, \mathbf{k}_3, \mathbf{k}_4)\,[N_3N_4(N_1+N_2)-N_1N_2(N_3+N_4)]$$
$$\times\, \delta(\mathbf{k}_1+\mathbf{k}_2-\mathbf{k}_3-\mathbf{k}_4)\delta(\omega_1+\omega_2-\omega_3-\omega_4)\,d\mathbf{k}_2\,d\mathbf{k}_3\,d\mathbf{k}_4, \quad (15)$$

in which $Q(\mathbf{k}_1, \mathbf{k}_2, \mathbf{k}_3, \mathbf{k}_4)$ is a complicated interaction coefficient (see Webb 1978, for a succinct listing) and the Dirac delta functions select contributions only from resonant interactions. Not only is the content of (15) difficult to see, it is difficult to calculate. Rough calculations using (15) and (JONSWAP) oceanic data (Sell & Hasselmann 1972, Hasselmann et al 1973 as referenced by Phillips 1977, p. 139) indicated that energy flows toward a broad spectral peak, but the results were inconclusive (Phillips 1977, p. 138). More well-founded calculations using (15) and a broad spectrum by Webb (1978), Masuda (1981, 1986), Hasselmann & Hasselmann (1985), and Hasselmann et al (1985) also showed, among other things, that energy flows toward a broad spectral peak from high wavenumbers. More recent calculations by Resio & Perrie (1991) affirmed this result and examined energy fluxes among spectral regions for a variety of spectral parameters.

Valenzuela & Laing (1972), Holliday (1977), and van Gastel (1987a) used a stochastic model similar to Hasselmann's to study a broad discrete spectrum of gravity-capillary waves, in which resonant triads occur. Van Gastel (1987a) gives an especially perspicuous derivation of the transport equations, which she writes in terms of the variance density G according to

$$\frac{\partial G_1}{\partial t} = \frac{k_1}{16\pi\omega_1} \int_{-\infty}^{\infty} \int_{-\infty}^{\infty} J^2(\mathbf{k}_1, \mathbf{k}_2, \mathbf{k}_3) \left[\frac{k_1}{\omega_1} G_2 G_3 - \frac{k_2}{\omega_2} G_3 G_1 - \frac{k_3}{\omega_3} G_1 G_2 \right] \times \delta(\mathbf{k}_1 - \mathbf{k}_2 - \mathbf{k}_3) \delta(\omega_1 - \omega_2 - \omega_3)\, d\mathbf{k}_2\, d\mathbf{k}_3. \quad (16)$$

[The relation between the variance density G in (16) and wave action density per unit mass N in (15) is discussed by van Gastel (1987b).] The interaction coefficients J^2 that appear in the stochastic model of (16) are proportional to the square of those in the deterministic model of (12) and (13) for an elementary resonant triad. A similar revealing connection between the interaction coefficients for gravity waves in the stochastic model of (15) with those of the deterministic elementary resonant quartet of (14) has not been found. Van Gastel (1987a) also used calculations to show that nonlinear interactions caused energy to flow away from a broad spectral peak toward wavenumbers larger than that of Wilton ripples [$k_0 = (\varrho g/2T)^{1/2}$]. In her stochastic model the interactions within a resonant triad were independent of other triads. The time scale of randomness was $t \sim t_0/\varepsilon$, whereas the time scale of nonlinearity was $t \sim t_0/\varepsilon^2$.

The effects of spatial inhomogeneities for a broad spectrum of waves were investigated by Willebrand (1975) who estimated that they are small in deep water, but the arguments are subtle. Watson & West (1975) developed a stochastic model that included specific external mechanisms to account for spatial inhomogeneities ab initio. They showed that non-resonant interactions between these mechanisms and the wavefield were important. A general stochastic model for a broad continuous spectrum of waves which did not use the RPA was presented by Benney & Saffman (1966) and Benney & Newell (1969). Their analyses made essential use of the smoothness of the underlying wave spectrum; hence, their results cannot be compared directly with the discrete spectral results, in which the spectrum is a sum of Dirac delta functions. Segur (1984) suggested that this smoothness amounts to enhancing linear dispersion relative to nonlinear coupling so that their model applies in a different regime from those of elementary resonant interactions and Hasselmann's model. Segur (1984) developed a stochastic model—which did not use the RPA—for coupled, resonant triads of interacting, localized wave packets. His derivation closely paralleled Boltzmann's derivation for a dilute gas of interacting molecules (kinetic theory), but important differences arose in consequence of fundamental differences between interacting molecules and interacting wave packets. In particular, two interacting wave packets generate a third ab initio so that the essential assumption of a dilute system was eventually violated. Segur's final evolution equation for the probability density of finding a particular wave packet at a particular time and location in phase space differed considerably from Boltzmann's equation for dilute gases. Segur suggested that his model probably does not have stable equilibrium solutions, which play a central role in the kinetic theory of gases.

Narrow-spectrum interactions When the wavefield has a narrow spectrum and four-wave resonances are the first to occur, expansions about the dominant wavetrain reduce both deterministic and stochastic descriptions to simpler forms. [In terms of (11a), the wavevectors of a narrow spectrum are nearly collinear and nearly equal in magnitudes.] Benney & Roskes (1969), Davey & Stewartson (1974), Djordjevic & Redekopp (1977), and Ablowitz & Segur (1979; 1981, p. 317) exploited the spectrum's narrowness deterministically. They obtained (dimensionless) evolution equations in terms of the complex amplitude A with the form

$$iA_t + \zeta A_{xx} + \mu A_{yy} = \chi |A|^2 A + \chi_1 \Phi_x A, \tag{17a}$$

$$\beta \Phi_{xx} + \Phi_{yy} = -\beta_1 (|A|^2)_x, \tag{17b}$$

in which the coefficients ζ, μ, χ, χ_1, β, and β_1 are functions of the dominant wavetrain (see the aforementioned references or Perlin & Hammack 1991), Φ is a velocity potential for the mean flow that is induced by the wavefield, and the spatial coordinates are referenced to a frame moving with the wavetrain at its group velocity. [We also require that $(kh)^2 \gg \varepsilon$ to avoid shallow water.] Equations (17), which are often termed the Davey-Stewartson equations, assume that the dominant wavetrain propagates mainly in the x-direction, i.e. its wavevector is, say, $\mathbf{k} = (l_0, 0)$. These equations have many interesting variations. For example, in the deep-water limit ($kh \to \infty$) the mean flow disappears yielding

$$iA_t + \zeta A_{xx} + \mu A_{yy} = \chi |A|^2 A, \tag{18}$$

which was first derived by Zakharov (1968) and, according to Ablowitz & Segur (1981, p. 322), is probably not solvable by the inverse scattering transform. Equation (18) can also be derived directly from the four-wave equations (Phillips 1981a). The mean flow and amplitude equations in (17) decouple when either the transverse (y) variations are neglected to obtain

$$iA_t + \zeta A_{xx} = \chi |A|^2 A, \tag{19}$$

or when the longitudinal (x) variations are neglected to obtain

$$iA_t + \mu A_{yy} = \chi |A|^2 A. \tag{20}$$

Both (19) and (20) are nonlinear Schroedinger (NLS) equations, first derived for longitudinal modulations by Benney & Newell (1967), Zakharov (1968), and Hasimoto & Ono (1972). [A comprehensive review of NLS equations in the context of water waves is given by Peregrine (1983).] These equations are completely integrable and solvable by the inverse scattering transform (Zakharov & Shabat 1972). In particular, (19) predicts the Benjamin & Feir (B-F, 1967) instability of a wavetrain to longitudinal, modulational (sideband) perturbations and also models the instability's long-time evolution. The B-F instability occurs when $\zeta\psi < 0$; the unstable wavenumbers lie in longitudinal sidebands $l_0 \pm \delta l$, in which $\delta l = |A_0|(-2\psi/\zeta)^{1/2}$ and $|A_0|$ is the initial amplitude of the underlying wavetrain. Comprehensive reviews of (18) and (19), the B-F instability, and its long-time behavior, as well as experimental and numerical results are presented by Yuen & Lake (1980, 1982), and we discuss them in more detail subsequently. In a little noticed paper, Phillips (1967) showed that the B-F instability could be obtained directly from the four-wave equations for gravity waves. Importantly, his results showed that the B-F instability of gravity waves is possible for longitudinal and transverse sideband modulation, i.e. for oblique wavetrain perturbations.

Equation (20) is a nonlinear Schroedinger equation for a wavetrain that propagates in the x-direction but is modulated in the transverse y-direction; it is commonly used in wave diffraction studies (e.g. Zakharov & Shabat 1972). Perlin & Hammack (1991) used (20) to study transverse sideband instabilities of gravity-capillary wavetrains. When $\mu\chi < 0$ in (20), a wavetrain with wavevector $\mathbf{k} = (l_0, 0)$ is unstable to wavevectors $\mathbf{k} = (l_0 \pm \delta m)$, in which $\delta m = |A_0|(-2\chi/\mu)^{1/2}$. According to this result, all wavetrains with wavenumbers $k > k_0 = (\varrho g/2T)^{1/2}$, i.e. wavenumbers greater than that of Wilton ripples, have unstable sidebands of transverse wavenumbers while those with $k < k_0$, which includes gravity waves, do not. Moreover, Perlin & Hammack (1991) showed that, according to (19) and (20), the growth rates of the most unstable longitudinal and transverse modes are equal. A more general result for gravity waves was obtained by Hayes (1973), Davey & Stewartson (1974), Alber (1978), and Martin & Yuen (1980, repeated in Yuen & Lake 1980, 1982) who performed a linear stability analysis directly on (17), (18), or their equivalent. They found an unbounded band of oblique wavetrain perturbations with both transverse and longitudinal wavenumbers that destabilized the wavetrain. [This result is consistent with that of (20) when the longitudinal perturbation wavenumber vanishes.] However, as noted by Martin & Yuen (1980), this unboundedness introduced perturbations which violated the narrow-spectrum approximation that underlies (17) and (18). This violation led Crawford et al (1981a) to abandon (17) and its variations in favor of a less restrictive model equation.

Crawford et al (1981a) investigated the Zakharov integral equation (Zakharov 1968):

$$\frac{\partial B_1}{\partial t} = \int_{-\infty}^{\infty}\int_{-\infty}^{\infty}\int_{-\infty}^{\infty} T(\mathbf{k}_1, \mathbf{k}_2, \mathbf{k}_3, \mathbf{k}_4) B_2^* B_3 B_4 \delta(\mathbf{k}_1 + \mathbf{k}_2 - \mathbf{k}_3 - \mathbf{k}_4) \times e^{i\Delta\omega}\, d\mathbf{k}_2\, d\mathbf{k}_3\, d\mathbf{k}_4, \quad (21)$$

in which $B(\mathbf{k}, t)$ is a complex envelope spectral function, T is an interaction coefficient, and $\Delta\omega = \omega_1 + \omega_2 - \omega_3 - \omega_4 = O(\varepsilon^2)$ accounts for slight detuning of the resonance conditions. Strictly speaking, (21) is not restricted to a narrow spectrum, rather it requires that the spectrum have dominant components that lie within a bandwidth of $O(\varepsilon^2)$ around the resonant curves of (11). When the resonance conditions of (11) are satisfied exactly, the four-wave equations are recovered. The B-F instability can be studied utilizing the detuning in (21). Results show that gravity waves are unstable to oblique wave perturbations and that the band of unstable wavenumbers is bounded. Importantly, the most unstable perturbation occurs when the transverse wavenumbers vanish so that the classical longitudinal B-F instability is dominant. A comprehensive review of (21) and further results are given in Yuen & Lake (1980, 1982).

Stochastic descriptions for narrow spectra received a major impetus from Longuet-Higgins (1976) who used Equation (18) as a starting point, assumed a discrete spectrum of waves with uncorrelated phases (i.e. the random phase assertion), and averaged and summed over time to obtain Hasselmann's Equation (15) with an explicit expression for the interaction coefficient, $Q(k, k, k, k) = 4\pi k^6$, in which k is the magnitude of the dominant wavevector. Fox (1976) used this result and (JONSWAP) oceanic data to perform computations with (15). His results differed markedly from the earlier computations by Sell & Hasselmann (1972); he found that energy flows away from, instead of toward, the spectral peak. Dungey & Hui (1979) improved Longuet-Higgins (1976) result by accounting for the finite spectral width of the narrow spectrum and showed that the flow of energy away from the spectral peak decreases as a consequence of its finite width. Masuda (1981, 1986) clarified these varying results using (15) without approximating the interaction coefficient Q for spectral shape. His well-founded computations showed that although energy does flow away from the peak in a broad spectrum, it flows toward a peak in a narrow spectrum. Moreover, Masuda found that in a spectrum with both low- and high-frequency peaks, the energy exchanges occurred such that the high-frequency peak was reduced and the low-frequency peak was intensified. Resio & Perrie (1991) also examined the effects of spectrum peakedness on energy fluxes within the spectrum. All of these results for gravity waves were consistent with van Gastel's (1987a) results for gravity-capillary waves.

Yuen & Lake (1982) started with the Zakharov integral Equation (21) and derived a stochastic model which did not use the RPA. The resulting wavefield comprised modulations (spatial inhomogeneity) with a time scale $t \sim t_0/\varepsilon^2$ while nonlinear interactions (energy exchange) occurred on a time scale $t \sim t_0/\varepsilon^4$ as in (15). Crawford et al (1980) used this stochastic model to test the stability of a narrow, homogeneous spectrum to inhomogeneous disturbances. They found a band of perturbation wavenumbers to which the narrow homogeneous spectrum was unstable; the bandwidth depended inversely on the strength of randomness. As the randomness approached zero they recovered the B-F instability; as randomness increased, the B-F instability diminished and finally disappeared so that the spectrum was stable. Alber (1978) obtained similar results beginning with (17). Janssen (1983b) studied the long-time behavior of an unstable sideband mode for varying degrees of randomness. He found that the sideband's spectral amplitude initially overshot and then oscillated in a damped manner to its long-time value. Both the overshoot and damped oscillations disappeared as the width of the homogeneous spectrum increased.

HIGHER-ORDER RESONANT INTERACTIONS Dysthe (1979) extended the nonlinear Schroedinger equation (NLS) for gravity waves to $O(\varepsilon^4)$ and found that the mean flow, which is a degenerate wave, varies as a consequence to wavetrain modulations. The higher-order NLS equation predicted a large but bounded band of unstable oblique wave perturbations, and, similar to the results of Crawford et al (1981a) for the Zakharov equation, it predicted that the most unstable perturbations for gravity waves are collinear. Calculations using this model for collinear perturbations by Lo & Mei (1985) showed good agreement with Keller's experimental data, which is (only) reported there. Further calculations by Lo & Mei (1987) using oblique perturbations showed that the higher-order NLS did not lead to a basic violation of its narrow-band approximation as (18) did. Janssen (1983a) used Dysthe's higher-order NLS equation to study a random field of gravity waves with the random phase approximation and recovered the results of Dungey & Hui (1979).

Hogan (1985, 1986) extended the Zakharov equation to $O(\varepsilon^4)$ for a narrow spectrum of gravity-capillary waves. Stiassnie (1984) showed that the higher-order NLS equation could be obtained as a special case of the Zakharov equation. Stiassnie & Shemer (1984) extended the Zakharov equation for gravity waves to include quintet interactions. Their calculations for wavetrain stability were qualitatively similar to more exact results of McLean (1982a,b) who used direct computations of the unapproximated (Euler) equations. Stiassnie & Shemer (1987) used the higher-order Zakharov equation to study the coupled evolution of resonant quartets and resonant quintets, which lead to the instability of deep-water gravity wavetrains.

SUMMARY COMMENTS ON RIT In summary, we emphasize several aspects of RIT whose importance will be clearer in the subsequent comparisons of theoretical and experimental results. First, RIT is a perturbative description which supposes wave-wave interactions are weak. Hence, RIT depends crucially on the existence of a single small parameter ε, regardless of the number of wavetrains involved. Second, RIT neglects nonresonant interactions, which are those excluded by kinematical considerations, i.e. by Equations (10) and (11), and those that satisfy kinematical conditions but are excluded by dynamical considerations. There is experimental evidence to suggest that caution be used before neglecting dynamically nonresonant interactions (Perlin et al 1990). Third, most of the discussion herein has assumed that the wavetrains satisfy the resonance conditions exactly. Precise tuning is not necessary and detuning is easily incorporated into RIT (e.g. see Craik 1985) as it was in (21). Fourth, the RIT presented herein neglects water viscosity, although the weak viscous effects that are typical of water waves may be significant, especially over the long time scales associated with the weak interactions of RIT. Weak viscosity accounts for three important effects in experiments: It attenuates wavetrain amplitudes, it detunes resonances, and it imposes a minimum ε in order for the growth of inviscid instabilities to occur. All of these effects are easily incorporated into RIT. For example, Miles (1984a,b) noted that the effects of weak (linear) viscosity are included in nonlinear evolution equations by letting $\partial/\partial t \to \partial/\partial t+\beta$, in which β is a viscous damping rate which usually varies among wavetrains. A comprehensive discussion of resonant interactions in systems that do not conserve energy can be found in Craik (1985, 1986); also see Craik & Moroz (1988) and Murakami (1987).

RESONANT INTERACTION EXPERIMENTS

In the previous section we reviewed some of the theoretical literature in which the special role of resonant interactions provided the framework for analysis. In this section we review the experimental literature in which there are measurements that allow us to test the predictions of RIT, either qualitatively or quantitatively. We begin by reviewing experiments that address the applicability of Kelvin's linear dispersion relation for water waves on a free surface. This dispersion relation is crucial to all applications of RIT, and it has an interesting history of experimental verification which illustrates the difficulties of measuring wavevectors on a two-dimensional water surface. Then we review resonant interaction experiments for gravity waves, gravity-capillary waves, and capillary waves.

Wave Dispersion on a Water Surface

DISPERSION RELATION AND WAVE CLASSIFICATION Kelvin[3] (1871) obtained the dispersion relation for infinitesimal wavetrains ($\varepsilon \to 0$) on the free surface of an inviscid water layer with constant density ϱ, constant surface tension T, quiescent depth h, and gravitational force per unit mass g, i.e.

$$c_2 = \frac{\omega^2}{k^2} = \frac{g}{k}(1+\tau)\tanh kh. \tag{22}$$

Here c is defined as the wave celerity and $\tau = Tk^2/\varrho g$ measures the relative importance of surface tension and gravitation, and resembles a reciprocal Bond number. To avoid shallow water we take $kh \gg 1$ so that the relevant perturbation parameter is the wave steepness $\varepsilon = ak$. Wilton (1915) showed that there is a countably infinite family of internal resonances among surface waves corresponding to wavetrains with $\tau = 1/n$, $n = 2, 3, \ldots$. With the values $T = 73$ dyn/cm, $\varrho = 1$ gm/cm^3, and $g = 980$ cm/s^2, second-harmonic resonance occurs at $\tau = 1/2$ for a wavetrain with $f = \omega/2\pi = 9.8$ Hz, third-harmonic resonance occurs at $\tau = 1/3$ for a wavetrain with $f = 8.4$ Hz, and so on. It is convenient to use (22) and the theoretical results for resonant triads and quartets (modulational instabilities) reviewed in

[3] Kelvin did not consider the effects of finite depth, which we have included in (22), but he did include the effects of air density, which we have neglected.

the previous section to define the following (distinct) frequency ranges for classifying surface waves. We define gravity wavetrains as those with $\tau < 0.155$, which corresponds to $f < 6.4$ Hz. Wavetrains in this frequency range are modulationally unstable with collinear wave perturbations being dominant (Djordjevic & Redekopp 1977, Crawford et al 1981a). (Note that the group velocity of water waves is minimum for $f = 6.4$ Hz.) We define capillary wavetrains as those with $\tau > 2.0$, which corresponds to $f > 19.6$ Hz; resonant triads are possible in this frequency range (Simmons 1969). (Note that $f = 19.6$ Hz is the second harmonic of Wilton ripples.) We define gravity-capillary waves (or ripples) as those with $6.4 < f < 19.6$ Hz. Waves in this frequency band are predicted to have three types of weakly nonlinear behavior. Wavetrains with $6.4 < f < 9.8$ Hz are stable to modulational instabilities (Hogan 1985, Djordjevic & Redekopp 1977); however, five members of the family of Wilton ripples ($n = 2$–6) are embedded in this band. Wavetrains with $f > 9.8$ Hz are modulationally unstable to collinear and oblique wavetrain perturbations (Perlin & Hammack 1991). All of these resonances and wave classifications are summarized in Figure 1.

DISPERSION RELATION EXPERIMENTS Fundamental to the application of RIT is the accuracy of the linear dispersion relation; yet, experimental verification of (22) is deceivingly difficult. Von Matthiessen (1889) first conducted experiments to test the validity of Kelvin's Equation (22) using a porcelain basin and a variety of liquids. He used two tuning forks with attached dippers to excite two wavetrains with circular wavecrests; the frequency range was 8.4–1024 Hz for experiments using distilled water. Von Matthiessen used two procedures to measure wavelengths. First, he counted the number of wave crests in the standing wavefield between the two tuning forks, which were separated by a known distance. Second, he measured wavelengths of the progressing waves to either side of the tuning forks by examining reflections of wavefield images on the polished tuning fork. (Presumably, he used an intermittent light source to render the images stationary, but the details are unclear to us.) Von Matthiessen used a standard value (73 dyn/cm) for surface tension in calculating predicted wavelengths with (22). Errors between predicted and measured wavelengths ranged between 0–12% with a mean error of 3.6%. We note that von Matthiessen's lowest measurement frequency (8.4 Hz) corresponds to the third-harmonic resonance. He observed that more crests occurred than expected for this wavetrain (only), undoubtably resulting from internal resonance, which was unknown at his time.

Rayleigh (1890) indirectly tested the validity of (22) using more refined experiments in his efforts to measure surface tension on clean and greasy

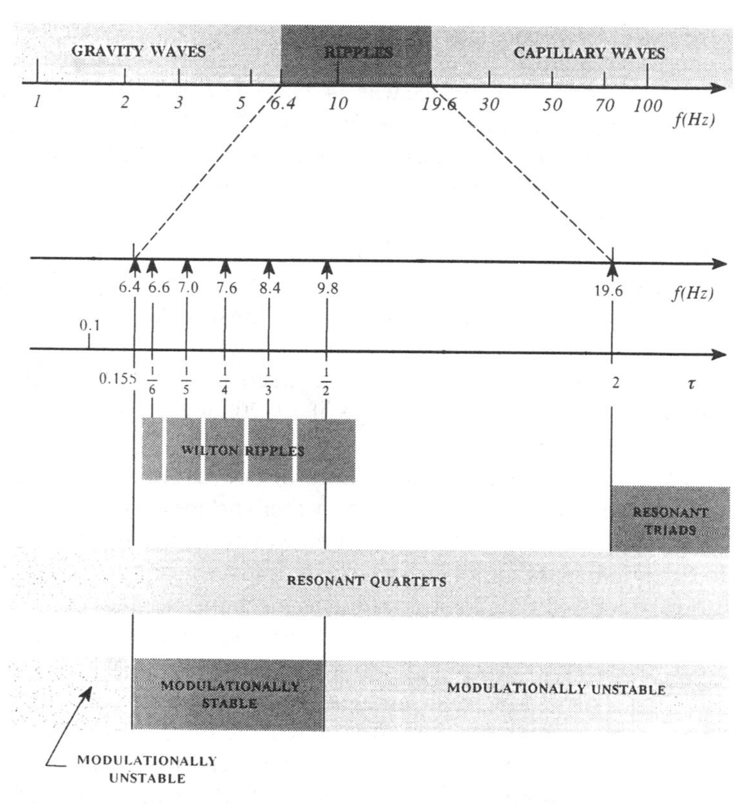

Figure 1 Surface wave classification in terms of frequency f and dimensionless surface tension τ based on weakly nonlinear wavetrain instabilities.

water surfaces. He used a 12 in × 12 in basin filled with either tap or distilled water, and cleaned the surface using an expandable brass hoop which pushed impurities aside. Like von Matthiessen, Rayleigh used tuning forks to generate waves, but unlike von Matthiessen, he used a 2.5 in glass plate as a paddle so as to generate plane waves. His wave amplitudes were so small that they could not be seen with the unaided eye. Rayleigh observed them using an optical arrangement comprised of a light source (a small gas flame) located in the focal plane of a large (6 in diameter) lens, which was located just above and parallel to the water surface. An image was formed on a screen by light after double passage through the lens; this image was rendered stationary by making the light source intermittent with a plate in front of the light which vibrated isoperiodically with the tuning fork used to generate waves. Rayleigh calibrated image lengths by viewing a bar of known dimension situated in close proximity to the water surface. He estimated that his measurements were in error by less than 1%. Using wave frequencies of 42.12 and 124.9 Hz, Rayleigh found that the mean tension of a clear surface was 74.2 and 73.6 dyn/cm, respectively, which indicated a discrepancy of about 1%. This "method of ripples" which Rayleigh used for measuring surface tension (at the suggestion of P. G. Tait) remains a standard method today (Freundlich 1922, p. 15; Adamson 1990, p. 40).

It is testimony to the difficulty of such experiments, and perhaps Rayleigh's influence, that further controlled experiments to verify (22) apparently were not reported until 1955 by Dobroklonskii. According to Scott (1981), Dobroklonskii used careful procedures which included continual renewal of the water surface in order to keep it clean. Dobroklonskii examined waves in the frequency range of 12–200 Hz and obtained good agreement with (22).

Davies & Vose (1965) examined the accuracy of (22) for capillary waves on water surfaces that were clean and on water surfaces that were covered with monomolecular films. They generated waves with frequencies of 50–920 Hz in a small glass tank (the size was not given) using a glass paddle driven horizontally by an electromagnetic mechanism; frequencies were carefully monitored. They cleaned the glass with chromic acid followed by washings with phosphoric acid, tap water, and distilled water. They cleaned water surfaces by spreading talc which was subsequently vacuumed. Surface tension was continually monitored with a du Nouy tensiometer. Like Rayleigh, Davies & Vose measured wavelengths using an optical arrangement and stroboscopic light source. Their experiments showed Kelvin's equation to be accurate to within 1.5% for clean surfaces over the entire frequency range. Unfortunately, they encountered (unspecified) experimental difficulties when measuring wavelengths on surfaces with insoluble films which precluded definitive results. Nevertheless, they described Kelvin's equation as approximately valid when films exist. (We note that the dispersion relation depends weakly on the viscoelastic properties of a surface film, e.g. see Levich 1962, p. 613 or Cini & Lombardini 1978.)

Scott (1981) examined surface waves with frequencies of 2–10 Hz using the extraordinary measures of Davies & Vose (1965) in a larger 9.5 cm × 100 cm basin in which the water depth was 2.9 cm. Scott used wetted, ground glass sidewalls, which were cleaned similarly to the procedures of Davies & Vose (1965), in order to maintain a static contact angle of zero and minimize edge effects. He used doubly distilled water, and vacuumed the water surface before each experiment. Scott generated waves at one end of the channel using a vertically oscillating wedge whose frequency was carefully monitored. He used a nonintrusive proximity gauge to measure passing waves, and measured wavelengths with an accuracy of $\pm 0.4\%$ by traversing the wavefield at known distances and number of wavelengths. The mean deviation between measured and predicted wavelengths in Scott's experiments was 1.3%; this deviation arose mainly from the higher-frequency waves.

Henderson & Lee (1986) conducted experiments in a 91 cm $\times$ 30 cm wave channel; the water depth was 2.54 cm. They generated wavetrains in the range of 6–20 Hz using a vertical plate hinged at the bottom and pushed horizontally at the top by an electromagnetic servomechanism. They measured wavelengths by traversing the wavefield with an in situ gauge at known distances and number of wavelengths. They used doubly distilled water and special cleaning procedures but did not employ the extreme measures of either Davies & Vose (1965) or Scott (1981) to clean the water surface. (Undoubtably, films were present on their surfaces.) Nevertheless, Henderson & Lee obtained excellent agreement between measured and predicted wavelengths using $T = 73.0$ dyn/cm in (22) when the water was filtered of particles with nominal sizes larger than 0.2 μm. When the water was not filtered of particles, a value of $T = 54.0$ dyn/cm in (22) gave similar agreement between predicted and measured data. The special role of particulate contamination does not appear to have been noticed previously, although it might be responsible for Rayleigh's observation that on occasion distilled water proved less satisfactory than tap water. (Typically, the distilling process does not remove small particles.)

The experimental difficulties in measuring frequencies and wavelengths for simple wavetrains are greatly increased when wavefields comprise many wavelengths and directions of propagation. Yet, such measurements are crucial in distinguishing between waves that arise from resonant interactions, which satisfy (22), and those that arise from nonresonant interactions, which don't satisfy (22). This measurement difficulty may account for the curious and potentially damning (as regards the application of RIT) results for wind wavefields obtained by numerous investigators beginning with Ramamonjiarisoa & Coantic (1976) who used two, in situ, wave gauges and filtered space-time correlations to obtain $c(k)$. [See Yuen & Lake (1982) for a discussion and a list of references; also see Ramamonjiarisoa & Mollo-Christensen (1979, 1981) and discussions by Huang (1981), Komen (1980), and Papadimitrakis (1986).] Measurements indicated that the phase speeds of waves with frequencies greater than that of the dominant wavetrain were constant—departing significantly from the predictions of (22). Crawford et al (1981b) attributed these departures to nonlinearity and spectral bandwidth. Huang & Tung (1977) and Longuet-Higgins (1977) attributed them to directional effects. Phillips (1981b) argued that these departures were due primarily to convection by the orbital velocities of the dominant (long) waves, and were thus a manifestation of the measurement technique. Plant & Wright (1979) used Doppler radar measurements of a wavefield, which avoid some of the inherent problems in two-probe correlation techniques, and found that the high-frequency waves dispersed according to (22). Gotwols & Irani (1980) used optical remote sensing measurements of the ocean surface and found excellent agreement between (22) and measured phase speeds. Barrick (1986) examined the role of (22) in radar measurements and concluded that (22) and Phillips' (1981b) appraisal were correct. Melville (1983) used measurements with two in situ gauges and an envelope-detection data analysis to find that (22), suitably modified by the Stokes' (1847) correction for large wave steepnesses, was accurate for strongly modulated wavetrains.

Gravity-Wave Experiments

The first comparisons between the predictions of resonant interaction theory and gravity-wave measurements were reported in companion papers by Longuet-Higgins & Smith (1966) and McGoldrick et al (1966). Both experiments were designed according to a suggestion by Longuet-Higgins (1962) in which (11) has the form $\mathbf{k}_1+\mathbf{k}_2 = 2\mathbf{k}_3$ and $\omega_1+\omega_2 = 2\omega_3$. Both Longuet-Higgins & Smith and McGoldrick et al generated waves *2* and *3* with two orthogonal wavemakers along adjacent basin walls; beaches were located along opposing walls to minimize reflections. They obtained continuous time signals of the water-surface displacements with in situ wave gauges at different distances from the wavemakers. They processed these signals with a sharply tuned, band-pass filter with variable center frequency so as to obtain amplitude-frequency spectra. They examined the spectral amplitudes at different gauges in order to obtain the spatial growth/decay rates of nascent/generated waves. According to the four-wave equations (14), wave *1* should be generated ab initio by resonant energy transfers from waves *2* and *3*; theoretical predictions of its growth rate which accounted for detuning were obtained using the four-wave equations. Their measured spectra contained large peak amplitudes at frequencies of the generated waves, and small peak amplitudes at frequencies of superharmonics and many sum-and-difference frequencies that included $\omega_1 = 2\omega_3-\omega_2$. In each spectrum, the spectral amplitude at ω_1 exceeded that of superharmonic and other sum-and-difference frequencies, suggesting that it resulted from a resonant rather than nonresonant interaction. More convincingly, comparison of spectral amplitudes at different gauge sites showed that, unlike all other spectral peak amplitudes, the amplitudes at ω_1 grew linearly with distance as predicted by RIT. Moreover, the measured growth rates were 20% higher than those predicted by RIT, which did not account for viscous attenuation. These experimental results firmly established the special role of resonant interactions.

The discovery of the modulational instability of a Stokes gravity wavetrain evolved from theoretical studies by Lighthill (1965) and Whitham (1967) and culminated with the definitive study by Benjamin & Feir (1967), who were strongly motivated by Feir's laboratory experiments. [Zakharov (1968) also discovered this instability theoretically, but his terse paper was unnoticed in the West until sometime later.] Unfortunately, Benjamin & Feir did not give a detailed report of their experiments as they originally intended (Benjamin & Feir 1967), although Benjamin (1967) did give a brief report. Benjamin & Feir generated wavetrains at one end of a long tank using a mechanically operated paddle whose motion was either regular or modulated to generate collinear wavetrains at the predicted most-unstable sideband frequencies. (The experiments were conducted in two facilities.) They used in situ wave gauges along the channel to obtain continuous time signals for the vertical motion of the water surface. They obtained modulational periods and sideband growth rates directly from wave-gauge records and from spectral analyses. Their measured modulational periods agreed well with those predicted. Their measured growth rates agreed qualitatively with those predicted; however, even when the effects of viscous damping were considered, predicted growth rates were nearly twice those measured. Lake & Yuen (1977) reexamined the data of Benjamin (1967) and attributed the large discrepancies to inappropriate values for the steepnesses of generated waves, which did not have the proper shape of Stokes waves. Lake & Yuen (1977) corrected the growth rates measured by Benjamin and obtained good agreement with predictions. Longuet-Higgins (1978b) used the original data of Benjamin and found better agreement with his theoretical results that accounted for finite-amplitude effects on growth rates. Regardless of refinements to measurements or theoretical predictions, Benjamin's & Feir's pioneering experiments showed unequivocally that deep-water, gravity wavetrains are unstable to modulations that result from collinear, resonant-quartet interactions with waves of nearly the same (sideband) frequencies.

In spite of the excitement and interest caused by the discovery of the Benjamin-Feir instability, ten years passed before additional laboratory experiments on the instability and long-time evolution of a deep-water gravity wavetrain were reported by Lake et al (1977). Lake et al were motivated in part by interest in the nonlinear Schroedinger equation and its exact solution for localized initial data by the inverse scattering transform. [Related experiments on deep-water wave packets were reported by Feir (1967) and Yuen & Lake (1975).] They generated wavetrains with frequencies of 1–5 Hz using a wavemaker, driven by a programmable servomechanism, located at one end of a 40 ft tank which contained a beach for absorbing wave energy at the opposite end. They used a linear array of in situ wave gauges along the wave channel to measure water surface motion. The continuous time signals from these gauges were recorded on FM tape for subsequent playback to an analog power-spectrum analyzer. Most of the experimental wavetrains had steepnesses in the range $0.10 \leq \varepsilon \leq 0.35$; wave breaking occurred for the steepest wavetrains. As expected, the wavetrains were unstable to sideband frequencies, and the measured, initial spatial growth rates agreed well with predictions. By exploiting the control available in these experiments, Lake et al were able to generate wavetrains with the same amount of sideband amplification that they measured at the gauge site farthest from the wavemaker. In this manner they effectively observed wavetrain evolution over much longer distances than their tank length. These long-time measurements showed preferential growth of the lower (frequency) sideband; this asymmetry is not predicted by either the Benjamin-Feir analysis or the nonlinear Schroedinger equation. Lake et al also found that the growth of modulations eventually ceased, followed by a period of demodulation and near recurrence of the original wavetrain. However, at recurrence, most of the energy resided in a wavetrain at the lower sideband frequency. An unequivocal explanation for this frequency downshift has not been established, but it seems to be related to strongly nonlinear effects such as wave breaking. (For example, see Melville 1982, 1983; Su et al 1982; Janssen 1983a; Hatori & Toba 1983; Trulsen & Dysthe 1990; Hara & Mei 1991.)

Melville (1983) conducted experiments in a wave channel that was 28 m $\times$ 50 cm and filled with water to a depth of 60 cm. He generated 2-Hz wavetrains with large steepnesses ($0.23 \leq \varepsilon \leq 0.29$) using a paddle driven by a hydraulic servomechanism. A beach, which began 16 m from the paddle, was used to dissipate wave energy. Melville measured the motion of the water surface using two in situ wave gauges located 8 cm apart and obtained discrete time signals at frequencies of 100- and 200-Hz. He used a Hilbert transform (envelope detection) technique to process these signals which, unlike spectral techniques, rendered temporal information on frequency, wavenumber, and phase-speed modulation. Melville found that the dispersion relation (22) was accurate for gravity waves when modified to account for weak nonlinearity (Stokes 1847). He also found that very rapid variations in wave phase occurred near minima in the wavetrain's envelope; these phase jumps were similar to the "crest pairing" described by Ramamonjiarisoa & Mollo-Christensen (1979). (Coalescence of wave crests is a possible explanation for the frequency downshift mentioned above.) Melville (1982) examined large-amplitude wavetrains and found two distinct regions of behavior. For $\varepsilon \leq 0.29$ wavetrain evolution remained (sensibly) two-dimensional and the Benjamin-Feir instability was dominant; asymmetric sideband growth occurred subsequent to wave breaking. For $\varepsilon \geq 0.31$, the wavetrains became three-dimensional and the Benjamin-Feir instability was no longer dominant. Melville's experiments and his nontraditional data analysis provided new insight into the richness of the Benjamin-Feir instability whose qualitative features persisted for wave amplitudes well outside the putative range of validity of four-wave interaction theory.

Su et al (1982), Su (1982), and Su & Green (1984) reported experiments on deep-water gravity waves performed in two large experimental facilities: a 137.2 m × 3.7 m × 3.7 m tank and a 340 m × 100 m × 1 m (outdoor) basin. They used mechanical, plunger-type wavemakers, one that spanned the 3.7 m tank width and one that spanned 15.8 m of the 100 m basin width. Su et al (1982) and Su (1982) generated wavetrains with frequencies of 1.05–1.55 Hz and large steepnesses of $0.16 \leq \varepsilon \leq 0.34$ which led to breaking. They observed a fascinating sequence of instabilities in these large-amplitude, long-crested deep-water wavetrains which were measured with in situ wave gauges and photographic records. They interpreted their measurements within the framework of normal-mode instabilities calculated by Longuet-Higgins (1978a,b) and McLean (1982a) and within the framework of wavetrain bifurcations calculated by Meiron et al (1982). These calculations were based on the unapproximated (Euler) equations; hence, their results are not directly within the purview of RIT; nevertheless, the qualitative features of McLean's calculations were obtained by Stiassnie & Shemer (1984, 1987), who used a perturbative approach to extend the Zakharov equation to include resonant quintets. RIT supposes a time-scale separation between resonant quartets and quintets which was not evident in the measurements of Su and coworkers for large-amplitude wavetrains. Instead, they observed that a two-dimensional wavetrain near the wavemaker evolved quickly into a three-dimensional wavetrain of spilling breakers followed by another transition to a modulated two-dimensional wavetrain with a lower steepness and frequency. Although these results are only tenuously interpreted in terms of RIT, it is noteworthy that Su and coworkers found that large-amplitude bifurcated wavetrains exhibited the Benjamin-Feir instability.

Su & Green (1984) conducted experiments on deep-water gravity wavetrains with a single frequency of 1.23 Hz and steepnesses of $0.09 \leq \varepsilon \leq 0.20$ using the tank described above and a linear array of ten in situ wave gauges along the tank. They observed the following pattern of wavetrain evolution. The Benjamin-Feir instability [resonant quartets or the "Type I" instability of McLean (1982a,b)] developed gradually at first, but then grew more rapidly causing intense modulations of the wavetrain. When the spectral amplitude of the lower sideband was about one-half that of the original wavetrain, the modulations produced distinct wave packets whose amplitudes were as much as 50% greater than those of the original wavetrains. [Bonmarin & Ramamonjiarisoa (1985) found a similar fast-growing modulation leading to significant amplification of wave amplitudes in their experimental studies of breaking waves.] At this stage three-dimensional, resonant quintets (the "Type II" instability of McLean) grew rapidly, then subsided leaving a two-dimensional wavetrain that exhibited modest modulations and had a frequency that was down-shifted from the original sideband growth. Su & Green suggested that the Type I instability triggered the growth of Type II instabilities. They also noted that the coexistence and interactions between these two types of instabilities depended crucially on the tank's width exceeding twice the wavelength of the generated wavetrain. Stiassnie & Shemer (1987) examined the coupled behavior between Type I and Type II instabilities observed in Su's & Green's experiments using their extended Zakharov equation. Their theoretical predictions were qualitatively similar to the measurements of Su & Green; however, Stiassnie & Shemer disagreed with the causality of Type II instabilities by Type I instabilities. Instead, their results suggested that whenever the level of Type I instabilities was substantially greater than that of Type II, the Type II instability was suppressed.

Wu et al (1979) conducted laboratory experiments to investigate the applicability of Hasselmann's stochastic model of wave-wave interactions. They used a wind-wave tank comprising a closed channel, which was 37.7 m long, 2 m high, and 1 m wide; a fan that drew air over a layer of water 1 m deep; and five in situ wave gauges located at 3 m intervals along the 23 m long, glass-walled test section of the tank. They also measured pressure fluctuations in the air channel using crystal pressure transducers placed near each wave gauge. They performed three experiments at wind speeds of 7.1, 8.0, and 8.9 m/s, and obtained continuous time signals from wave gauges and pressure transducers which were amplified, low-pass filtered, and digitized to obtain 40-Hz discrete time signals. They used ten-minute records and partitioned them to form 50 records with which they calculated one-dimensional energy spectra and cross-spectra between the fluctuating pressure and wave heights. These spectra were then ensemble averaged at each gauge site. They used the averaged energy spectra to find the net energy change at each spectral frequency (2–9 Hz) between two gauge sites. The net energy change comprised three processes: energy transferred from the wind to the waves, energy transferred among waves by nonlinear interactions, and energy dissipated by wave breaking and viscous effects. In order to investigate energy transfer by resonant interactions, it is necessary to isolate each of these three processes, which, as Wu et al note, is extremely difficult to do experimentally. To this end they adopted parametric models for each process. In particular, they replaced the nonlinear wave-wave interactions in Hasselmann's model (14) by a model due to Barnett (1968), in which the measured energy spectra were used as input. Their adopted models for energy dissipation and energy transfer from the wind used the measured energy spectra and cross-spectra as inputs, respectively. They computed theoretical predictions for energy transfer by nonlinear wave-wave interactions using Barnett's model. They computed "experimental" (sic) values for energy transfer by nonlinear wave-wave interactions by subtracting predicted values for energy transfer from the wind and energy dissipation from the measured net energy change. Their comparisons between theoretical and experimental values at four measurement sites showed satisfactory agreement for low and intermediate frequencies, but unsatisfactory agreement at higher frequencies.

Gravity-Capillary and Capillary Wave Experiments

Simmons (1969) derived the governing equations for a degenerate resonant triad of gravity-capillary waves corresponding to second-harmonic resonance, i.e. in Equation (10) $\mathbf{k}_1 = 2\mathbf{k}_0$, $\omega_1 = 2\omega_0$, and $(\mathbf{k}_0, \omega_0)$ correspond to Wilton ripples with $f_0 = 9.8$ Hz. McGoldrick (1970a) elaborated on Simmons' analysis and showed that the permanent-form solution obtained by Wilton (1915) is not expected to occur in a viscous fluid owing to its special choice of initial conditions. McGoldrick (1970b) performed experiments on second-harmonic resonance in which Wilton ripples were generated by a wedge-shaped paddle which was partially immersed in a water surface and oscillated vertically by an electromagnetic servomechanism. The paddle, which had a glass-covered front face, spanned the 61 cm width of a 3 m-long basin. McGoldrick used an in situ wave gauge, capable of detecting wave amplitudes as small as 10^{-3} mm and made measurements at 1 cm intervals away from the paddle. As in previous experiments (McGoldrick et al 1966), he processed the continuous time signals from the wave gauge using a sharply tuned, band-pass filter which enabled him to measure the spectral amplitudes at the wavetrain's fundamental and superharmonic frequencies. McGoldrick used tap water in the experiments and found to his consternation that a surface film was present, requiring him to lower the surface tension value in (22) by about 30% from its clean-surface value. In fact, this lowering of surface tension could not be detected by a du Nouy tensiometer, so McGoldrick used Kelvin's Equation (22) as Rayleigh (1890) had done to measure T. This film also led to enhanced viscous attenuation of the waves, which were extinguished before they reached the end of the basin. Nevertheless, McGoldrick's experiments clearly showed the strength of second-harmonic resonance; he notes: "... the interaction process is so dramatic that it can be seen by eye!" The measured data showed significant growth of the superharmonic amplitude at the expense of the fundamental prior to both waves being extinguished by viscosity. In addition, the relative phase between the two waves remained near its predicted value of $\pi/2$. These striking results occurred in spite of the exceedingly small wave steepnesses ($\varepsilon < 0.05$) used in the experiments. Henderson & Hammack (1987) also performed experiments on Wilton ripples and its second harmonic. When they generated a wavetrain of Wilton ripples, it rapidly transferred a significant portion of its energy to its 19.6-Hz superharmonic, and there was a proliferation of higher-frequency superharmonics in the measured spectra. When they generated a 19.6-Hz wavetrain, it slowly transferred an insignificant portion of its energy to the 9.8-Hz subharmonic; Benjamin-Feir instabilities were dominant. Perlin & Hammack (1991) used remote sensing techniques (described in more detail below) to measure the two-dimensional wavenumber spectrum when a 9.8-Hz wavetrain was generated in a channel. They observed the proliferation of spectral peaks at superharmonic wavenumbers and significant growth of transverse side-band wavenumbers for the 9.8- and 19.6-Hz wavetrains. The absence of sideband growth in frequency spectra indicated that the presence of side-band growth in the wavenumber spectra resulted from rhombus-quartet interactions.

McGoldrick (1972) revisited the second-harmonic resonance for gravity-capillary waves and extended the analysis to the third-harmonic resonance. The analytical features of these two cases were then generalized to fourth-, fifth-, and higher-harmonic resonances. McGoldrick performed experiments on these higher-order internal resonances using apparatus and techniques similar to those of McGoldrick (1970b) except that the water surface was renewed every 20 min by flushing it over a weir so that the surface tension was the clean-surface value. He measured amplitude response curves in the vicinity of the predicted resonant frequencies for third-, fourth-, and sixth-harmonic[4] resonances, and observed strong responses in the vicinity of all of these higher-order resonances. As is common in externally forced systems, the maximum responses occurred at different frequencies (slightly lower) from those predicted by (22). McGoldrick's

[4] McGoldrick's Figure 8 and his corresponding discussion refer to sixth-harmonic resonance throughout; however, the period of 0.144 s labeled in his Figure 8 more closely corresponds to fifth-harmonic resonance.

results also showed that resonance persisted even when there was substantial detuning, so much in fact that the response curves for third- and fourth-harmonic resonance nearly joined. Third-harmonic resonance was excited down to 8.06 Hz while fourth-harmonic resonance was excited up to 7.94 Hz. Perlin & Hammack (1991) explored the small intervening frequency band, and found that third-harmonic resonance occurred for an 8.00-Hz wavetrain. Therefore, internal resonance occurs everywhere within the frequency band between the (discrete) theoretical frequencies of third- and fourth-harmonic resonance. (This behavior is indicated in Figure 1 by the shaded bands about each internal-resonance.) Moreover, both McGoldrick (1972) and Perlin & Hammack (1991) observed that these internal resonances were easily excited by waves with surprisingly small steepnesses. McGoldrick excited sixth-harmonic resonances using a wavetrain with $\varepsilon = 0.15$—although the nonlinear terms responsible for this interaction in the theory are $O(\varepsilon^6)$. The finite width and joining of the internal-resonance response curves and their excitation by exceedingly small wave steepnesses make the stability of wavetrains in the frequency band of 6.4–9.8 Hz to modulations (Hogan 1985) of little practical consequence. These experimental results are testimony to the robustness of internal resonances, but are somewhat disturbing from the perturbative point of view of RIT.

Banerjee & Korpel (1982) generated capillary wavetrains in the frequency range of 30–100 Hz by horizontally oscillating a vertical plate immersed just below the surface of tap water which nearly filled a 30 cm × 30 cm × 5 cm basin. They used an electromagnetic mechanism to drive the paddle, which did not span the basin width. They measured the temporal motion of the water surface with an in situ gauge and its spatial motion by setting the basin on the Fresnel lens of an overhead projector which was modified with a Schlieren system and strobe lamp. This apparatus illuminated the wavefield, forming images with a contrast in light intensities related to the local water depth. The strobe light rendered the images stationary so that photographs could be made. Banerjee & Korpel found that subharmonic wavetrains occurred in all of their experiments when the paddle stroke exceeded a threshold value. These subharmonic waves formed a standing-wave pattern which was readily observed at the wavemaker and to its side. They attributed these results to a resonant triad between the generated wavetrain and two oblique, subharmonic wavetrains. Henderson & Hammack (1987) performed similar experiments, but did not observe the subharmonic behavior reported by Banerjee & Korpel. Based on Hogan's (1984) reinterpretation of Banerjee's & Korpel's data, which indicated that strongly nonlinear waves were being generated, and their own experiments, Henderson & Hammack (1987) cited a list of circumstantial evidence indicating that subharmonic cross waves—which are a strongly nonlinear, trapped wave phenomenon and outside the purview of weakly nonlinear RIT—were generated by Banerjee & Korpel.

Banerjee et al (1983) used the experimental apparatus described above to examine the weakly nonlinear phenomenon of focusing, which occurs in (17) and is discussed in the context of water waves by Ablowitz & Segur (1979) and Peregrine (1983). According to (17), a localized packet of short-crested waves with frequencies greater than that of Wilton ripples will evolve a singularity in finite time as a consequence of focusing. The phenomenon of focusing is easily understood using the nonlinear dispersion relations given by Wilton (1915), i.e.

$$c(k \approx k_0; \varepsilon) = 3\left(\frac{gT}{2\varrho}\right)^{1/4}\left(1 \mp \frac{1}{4}\varepsilon\right), \tag{23}$$

and using the analysis of Pierson & Fife (1961) for wavetrains with wavenumbers near $k = k_0$. In (23) the upper $(-)$ sign is chosen for waves with $k > k_0$ and the lower $(+)$ sign is chosen for waves with $k < k_0$. When a short-crested wavetrain is generated with $k > k_0$, large-amplitude regions of crests propagate more slowly than low-amplitude regions so that the crests turn on themselves, i.e. they focus when $\partial c/\partial\varepsilon < 0$. When a short-crested wavetrain is generated with $k < k_0$, the wave crests bend in opposite directions, i.e. they defocus when $\partial c/\partial\varepsilon > 0$. Banerjee et al (1983) used a wave paddle with a length of 3 cm to generate localized packets of short-crested wavetrains with frequencies of 20, 30, and 50 Hz; focusing was expected for all of these wavetrains. They observed focusing for the 30- and 50-Hz wavetrains, but not for the 20-Hz wavetrain. They gave no explanation for the absence of focusing by the 20-Hz wavetrain.

Perlin & Hammack (1991) performed experiments on gravity-capillary and capillary wavetrains with frequencies of 8.0–25.0 Hz and moderate steepnesses ($\varepsilon < 0.3$). They conducted experiments in a wave channel measuring 91 cm × 30.5 cm with a water depth of 4.9 cm. They generated wavetrains using a wedge-shaped paddle which spanned the channel width and was oscillated vertically in the water surface by an electromagnetic servomechanism. They cleaned all of the materials contacting the water with ethyl alcohol before each experiment and used doubly distilled water which was filtered of organic material, minerals, and particles with nominal sizes greater than 0.2 μm. They measured water surface motion at different positions along the channel using an in situ wave gauge. Continuous time signals were band-pass filtered, amplified, and sampled at 250 Hz to obtain discrete time signals. They used these data to compute amplitude-frequency spectra with a frequency resolution of 0.015 Hz. In addition, they measured two-dimensional wavenumber spectra of wavefields using a high-speed imaging system with an imager looking down on the water surface from above the channel. The imager comprised a 128×128 pixel array that measured discrete-space gray-levels in a 22 cm $\times$ 22 cm surface area, which began 23 cm from the wavemaker; the wavenumber resolution was 0.284 rad/cm. An image calibration showed that the correlation coefficient between these gray levels and measurements of a nearby in situ wave gauge was 0.89. All of the frequency and wavenumber spectra for wavetrains in the frequency range 9.8–19.6 Hz, in which resonant quartets are the first to occur, showed the following results: (*a*) Frequency spectra showed distinct spectral peaks at frequencies of the generated wavetrain and its superharmonics with no indication of sideband growth. (*b*) Wavenumber spectra showed spectral amplitudes spreading outward from the peak of the generated wavetrain in a circular arc with a radius equal to the magnitude of the generated wavetrain's wavevector. In other words, rhombus-quartet sideband instabilities were dominant. (The latter result is apparent in Figure 3, which is taken from Perlin et al (1990) and is discussed below.) Perlin & Hammack (1991) compared their measurements to predictions of the NLS Equations (19) and (20) for longitudinal and transverse sideband growth, respectively. There was some evidence of longitudinal sideband growth in the wavenumber spectra, but much less than predicted by (19). On the other hand, there was much more growth of transverse sideband amplitudes (at the frequency of the generated wavetrain) than predicted by (20). This dominance of sideband rhombus-quartet instabilities was not predicted by the numerical calculations of Zhang & Melville (1987) using the unapproximated (Euler) equations. Perlin & Hammack also found that sideband rhombus-quartet instabilities remained dominant for wavetrains with $f > 19.6$ Hz where resonant triads are possible—when the phenomenon of selective amplification (Henderson & Hammack 1987, Perlin et al 1990) does not occur.

Henderson & Hammack (1987) reported experiments on the stability of capillary wavetrains with frequencies of 22–46 Hz where resonant triad interactions are expected to occur. They used the experimental facilities and procedures described in Perlin & Hammack (1991). Their experiments were designed specifically to test the stability of wavetrains for which resonant triads are possible, as predicted by Simmons (1969) for an elementary resonant triad of capillary waves. Simmons showed that a generated wavetrain, say $(\mathbf{k}_1, \omega_1)$, can form resonant triads with two waves (summed) from a continuum of waves in a lower-frequency (closed) band B_l and with two waves (differenced) from a higher-frequency (open-ended) band B_h. Simmons also showed that waves in B_h could not amplify when their

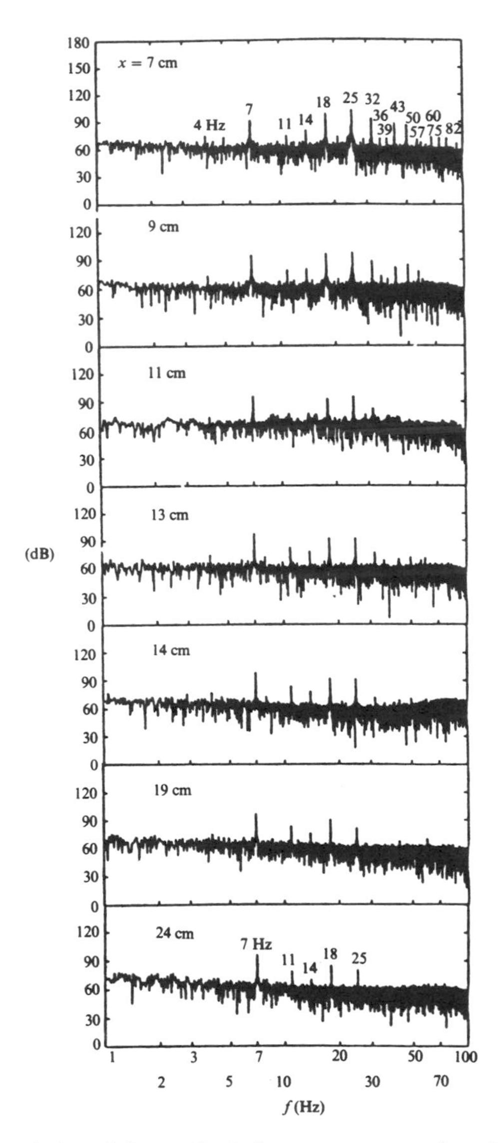

Figure 2 Spatial evolution of the amplitude-frequency spectrum for a 25-Hz wavetrain in the presence of a 57-Hz background wavetrain; $h = 5$ cm, $T = 73$ dyn/cm (from Perlin et al 1990; courtesy of Cambridge University Press).

amplitudes were infinitesimal relative to that of the generated wavetrain. [This result is generally true for resonant triads according to a theorem by Hasselmann (1967b).] Waves in B_l can amplify; however, since the interaction coefficients of (13) vary smoothly across B_l, Simmons conjectured that selective amplification is unlikely. Instead, all of the background waves in B_l are expected to amplify and destabilize the generated wavetrain. Henderson's & Hammack's experiments showed that wavetrains with $f > 19.6$ Hz were unstable to background waves with frequencies in B_l; however, in most experiments a single triad with two waves from B_l was selectively amplified. In some experiments, the spectral amplitude(s) of the wave(s) in B_l grew, decayed, and grew again during propagation down the channel, i.e. energy cycled, as the exact solutions of the three-wave equations suggest. Henderson & Hammack did not measure wavenumber spectra in their experiments; however, this cyclic behavior, which cannot occur in nonresonant interactions, led them to conjecture that the waves in B_l formed a resonant triad with the generated wavetrain. In their experiments the selected triad varied with the frequency of the generated wavetrain. However, the same triad amplified repeatably, regardless of the generated wavetrain's amplitude, which was increased from its viscous threshold value until subharmonic cross waves evolved at the wavemaker paddle. They gave no explanation for either the presence of selective amplification in some experiments or its absence in others.

A surprising and serendipitous answer to the riddle of selective amplification described above was found by Perlin et al (1990). When a new computer system was installed in the laboratory, Perlin et al conducted the experiments of Henderson & Hammack (1987), and found *no* triads selectively amplified. When they repeated the experiments with the older computer system, selective amplification returned. They traced the different results to the computers' analog output systems, which provided command signals to the wavemaker. Both devices were comparable in specification; however, the inadvertent noise level at the electrical power frequency of 60 Hz was about 1/5 lower in the newer device. This small difference, which was represented by signal-to-noise ratios of about 100 in the older device and 200 in the newer device, was sufficient to alter radically the outcome of an experiment. Perlin et al (1990) conjectured the following explanation of and algorithm for predicting selective amplification of two wavetrains in B_l by a wavetrain f_1 of moderate amplitude and an infinitesimal background wavetrain f_0 in B_h. First, both f_1 and f_0 become directionally unstable by sideband rhombus-quartet instabilities described by Perlin & Hammack (1991). This instability is crucial since f_1 and f_0 are initially collinear and generally cannot satisfy (10a) with any other wavetrain. The directional instability allows wavetrains to amplify with the proper wavevector configurations to satisfy (10a). Thereby the generated wavetrains form a resonant triad with the difference wavetrain $(f_0 - f_1)$. The difference wavetrain cannot amplify in accordance with Hasselmann's (1967b) theorem; however, Perlin et al used numerical calculations of the three-wave equations to show that $(f_0 - f_1)$ does exchange energy with f_0. Second, $(f_0 - f_1)$ becomes directionally unstable and a new resonant triad forms with frequencies $(f_1, f_0 - f_1, 2f_1 - f_0)$. Again, the new difference wavetrain cannot amplify but it does exchange energy with the previous difference wavetrain. This process continues until a difference wavetrain with a frequency in B_l, say f_2, is excited and becomes directionally unstable; since f_2 is in B_l, it can and does amplify. Thereby a resonant triad with frequencies $(f_1, f_2, f_1 - f_2)$ is seeded by f_0 and is selectively amplified. This algorithm can be continued to determine if more than one triad with two waves in B_l can be selectively amplified. The algorithm comprises exceedingly weak nonlinear interactions that are normally ignored; yet, when they occurred, they determined the outcome of an experiment. Perlin et al used this algorithm and correctly predicted the presence and absence of selective amplification in all of the experiments in Henderson & Hammack (1987). Since the 60-Hz noise level of the newer computer system was below the threshold to cause selective amplification, Perlin et al (1990) added artificial noise at discrete frequencies in B_h and tested their algorithm further. Results from one of their experiments is shown in Figure 2, which presents amplitude-frequency spectra computed from the discrete time signals (sampled at 250 Hz) of a wave gauge at various distances (x) from the wavemaker. They programmed the wavemaker to generate a 25-Hz wavetrain with steepness $\varepsilon \approx 0.20$ and a 57-Hz wavetrain with steepness $\varepsilon \approx 0.04$. According to their selection algorithm, two triads with waves in B_l, which has the range of about 5.0–20.0 Hz, were possible: The first triad to occur comprised wavetrains with frequencies of 25, 7, and 18 Hz; the second triad comprised wavetrains with frequencies of 25, 11, and 14 Hz. At the first station of measurement in Figure 2 ($x = 7$ cm), which is seven wavelengths of the 25-Hz wavetrain from the wavemaker, superharmonics and sum-and-difference wavetrains proliferate; the 57-Hz wavetrain is barely discernible. At the last measurement station ($x = 24$ cm), five distinct wavetrains dominate, and they correspond to those expected in the two, coupled resonant triads. The spectral amplitudes of the 7- and 18-Hz wavetrain at $x = 24$ cm are larger than those of the 25-Hz wavetrain. The spectral amplitudes of the 25-Hz wavetrain gradually diminish during propagation while those of the 7-Hz wavetrain, which is already large by $x = 7$ cm, remain about constant. The spectral amplitudes of the 18-Hz wavetrain, which is already large by $x = 7$ cm, decrease slightly, then increase slightly, and then decrease again during propagation. The spectral amplitudes of the 11- and 14-Hz wavetrains grow, diminish, disappear, reemerge, grow, and diminish again during propagation. This behavior strongly suggests that two, coupled triads are evolving in the wave channel. Perlin et al also measured two-dimensional wavenumber spectra using the apparatus described in Perlin & Hammack (1991). The result for one image is shown in Figure 3, which shows ten equally spaced contour levels of the spectral amplitudes in the positive quadrant of wavenumber space. The wavenumber axes (l, m) are rotated so that the wavevector of the 25-Hz wavetrain generated by the wavemaker bisects this quadrant. The circles drawn in Figure 3 correspond to the isotropic dispersion relation (22) at frequencies of 25, 18, 14, 11, and 7 Hz. Wave energy is concentrated in arcs at wavenumbers corresponding to the aforementioned wave frequencies; hence, they are free waves resulting from resonant interactions. Moreover, the directional spread of the wave energy about these circular arcs is a manifestation of sideband rhombus-quartet instabilities—necessary in order for the wavetrains to have the proper wavevector configurations satisfying (10a).

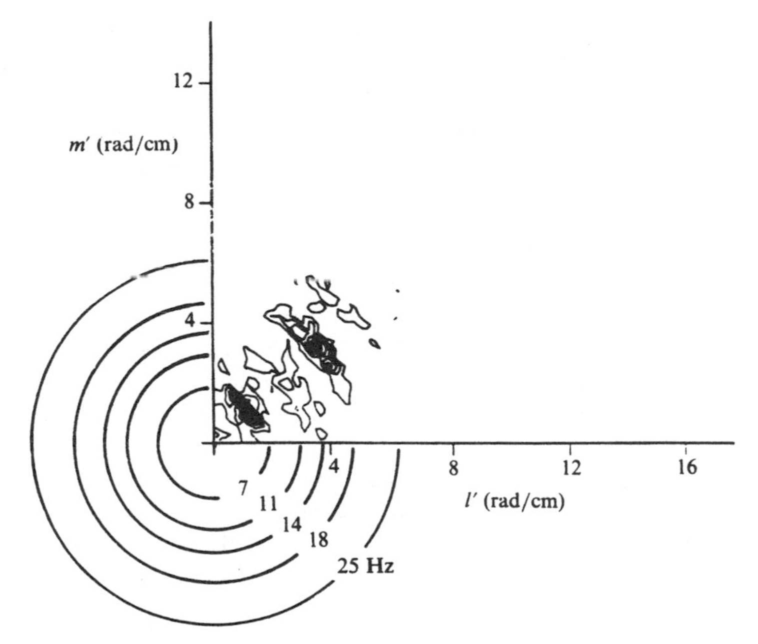

Figure 3 Contour map of amplitude-wavevector spectrum for a 25-Hz wavetrain in the presence of a 57-Hz background wavetrain; $h = 5$ cm, $T = 73$ dyn/cm (from Perlin et al 1990; courtesy of Cambridge University Press).

The sensitivity of triadic resonances among capillary wavetrains to discrete components in ubiquitous background noise led Perlin et al (1990) to an experimental investigation in which they added artificial "white" noise at the wavemaker to the dominant wavetrain and, in some experiments, a discrete noise component which seeded selective amplification. They characterized the broad-spectrum background noise by a signal-to-noise ratio (SNR) defined as

$$\mathrm{SNR} = \frac{\text{rms of wavetrain signal}}{\text{rms of white noise signal}},$$

in which rms is the root-mean-square of deviations in the command signal voltages from their mean value. (The relationship between voltage and stroke of the wavemaker was nearly linear.) They found that selective amplification was reduced when the broad-spectrum background noise was amplified; it disappeared completely when the SNR of the discrete noise component was about 1/10. They then examined the effect of (only) broad-spectrum random noise on the evolution of a dominant wavetrain by conducting experiments in the following manner. First, they obtained 16 discrete time series of wave gauge data and computed frequency spectra for the natural background noise in the tank when the wavemaker was powered, but not moving. These 16 frequency spectra were ensemble averaged to obtain a representative spectrum for the natural background noise, $\langle\mathfrak{B}\rangle$, in the tank. Then they conducted 16 experiments using white-noise command signals to the wavemaker. Wave gauge data were obtained and analyzed for each experiment and an ensemble-averaged frequency spectrum for the random waves in the presence of the tank's natural background noise, $\langle\mathfrak{R}+\mathfrak{B}\rangle$, was found. Using only a 25-Hz wavetrain, an additional 16 experiments were performed to find an ensemble-averaged frequency spectrum for the dominant wavetrain in the presence of the tank's natural background noise, $\langle\mathfrak{T}+\mathfrak{B}\rangle$. Then they added the dominant-wavetrain signal (SNR = 10) to the 16 command signals of white noise and conducted 16 more experiments to find an ensemble-averaged frequency spectrum for the dominant wavetrain with random waves and the tank's background noise, $\langle\mathfrak{T}+\mathfrak{R}+\mathfrak{B}\rangle$. They combined these four ensembled-averaged spectra linearly to obtain the average frequency spectrum for the random waves alone, i.e. $\langle\mathfrak{R}\rangle = \langle\mathfrak{R}+\mathfrak{B}\rangle-\langle\mathfrak{B}\rangle$ and the dominant wavetrain alone, i.e. $\langle\mathfrak{T}\rangle = \langle\mathfrak{T}+\mathfrak{B}\rangle-\langle\mathfrak{B}\rangle$. To display the effects of nonlinear interactions they calculated

$$\langle\mathfrak{N}\rangle := \langle\mathfrak{T}+\mathfrak{R}+\mathfrak{B}\rangle-\langle\mathfrak{T}\rangle-\langle\mathfrak{R}\rangle-\langle\mathfrak{B}\rangle; \tag{24}$$

hence, $\langle\mathfrak{N}\rangle$ is null if nonlinear interactions are insignificant. (They tacitly assumed that the natural background noise in the tank was sufficiently small to interact linearly with the random and dominant wavetrains.) They performed this series of experiments at three gauge sites downstream of the wavemaker. Their main results are presented in Figure 4 where the ensemble-averaged spectra $\langle \mathfrak{N} \rangle$ are shown at three downstream measurement stations. Note that $\langle \mathfrak{N} \rangle$ is not null, indicating that significant non-

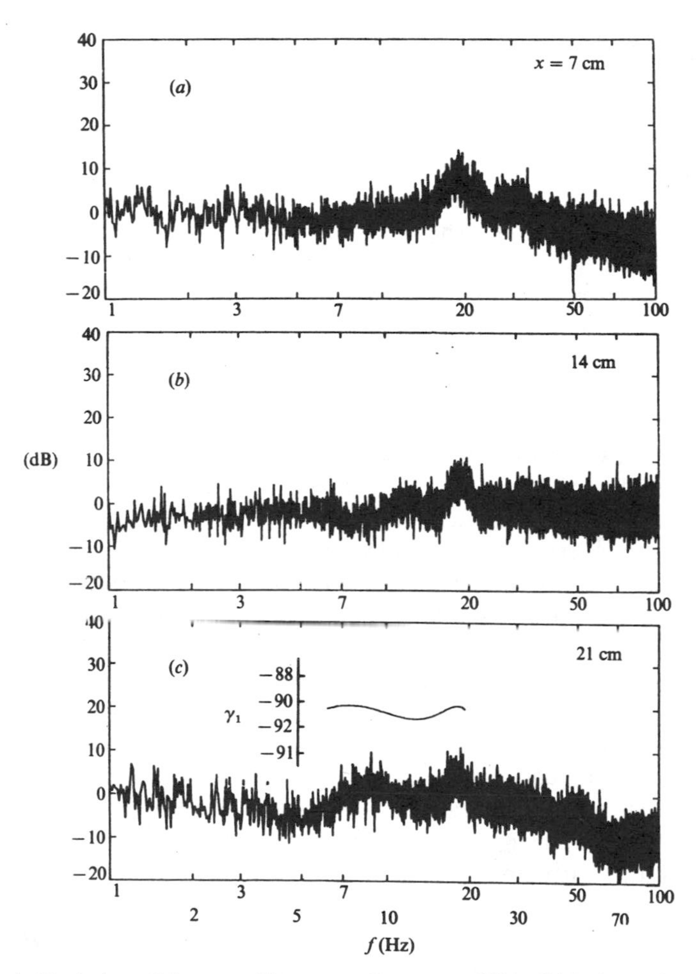

Figure 4 Evolution of the ensemble-averaged spectrum $\langle \mathfrak{N} \rangle$, which shows the effects of nonlinear wave-wave interactions, down the channel; $h = 5$ cm, $T = 73$ dyn/cm. Inset shows the theoretical interaction coefficients γ_1 for triadic interactions across B_1 (from Perlin et al 1990; courtesy of Cambridge University Press).

linear interactions occurred. At $x = 7$ cm, spectral amplitudes between 16 and 22 Hz increased with a maximum amplification at 19 Hz. Note that this frequency band is in and near the low-frequency continuum B_l (5.0–20.0 Hz) of the 25-Hz wavetrain. Spectral amplitudes in the high-frequency continuum B_h (≥ 30.0 Hz) decreased. At $x = 14$ cm spectral amplitudes between 1 and 7 Hz decreased while those between 7 and 22 Hz increased in a bimodal manner with maxima near 12 and 19 Hz and a minimum near 15 Hz. Beyond 22 Hz, spectral amplitudes at $x = 14$ cm show no amplification or attenuation; hence, the waves in B_h have actually gained energy while propagating from $x = 7$ cm. At $x = 21$ cm spectral amplitudes from 1–6 Hz remained lowered while amplitudes in B_h decreased again. The spectral amplitudes in B_l remain amplified with the bimodal distribution observed at the previous measurement station. This spectral distribution is similar to the form of the dynamical interaction coefficient γ_l in (13) over B_l for the 25-Hz wavetrain, which is shown superposed in Figure 4. Perlin et al concluded that, when the background noise spectrum is broad-banded, all of the wavetrains in the low-frequency continuum are amplified by a dominant wavetrain in accordance with predictions of RIT for elementary resonant triads.

CONCLUDING REMARKS

We have reviewed a variety of theoretical investigations (RIT) that were founded on Phillips' (1960) view that resonant wave-wave interactions play a special role. These theoretical investigations branched in two directions. Deterministic investigations included studies of wavefields comprised of resonant interactions among a single triad or a single quartet of waves, a small number of coupled triads or quartets, or a narrow spectrum of waves. Stochastic investigations included studies of wavefields comprised of resonant interactions among either a broad or a narrow spectrum of waves. We have also reviewed a variety of experimental investigations (RIE) that enabled some aspects of RIT to be tested. These experimental investigations branched in four directions. There are numerous experimental investigations that studied the validity of Kelvin's linear dispersion and the stability of a single wavetrain to artificial and naturally occurring background perturbations. We have found only two experimental investigations that have studied the interactions among more than one wavetrain of comparable (initial) amplitudes and only one experimental investigation that addressed the relative roles of nonlinearity and randomness for a broad spectrum of waves.

Many meticulous laboratory experiments showed that Kelvin's dispersion relation for infinitesimal waves (22), which is crucial to all applications of resonant interaction theory, is accurate over a wide range of frequencies, even for wavetrains with moderate steepnesses. Good quantitative agreement between theoretical and experimental investigations of wavetrain stability was found for the celebrated Benjamin-Feir (B-F) instability, which is a degenerate resonant four-wave interaction between two wavetrains (each counted twice) that are collinear or nearly collinear and have nearly the same frequencies and wavevector magnitudes. Good qualitative agreement persisted for wavetrains with near-breaking steepnesses in experiments that used narrow (relative to wavelength) channels, which inhibited oblique B-F instabilities. Results differed dramatically between gravity and gravity-capillary wavetrains in experiments that used wide channels, which allowed for the possibility of oblique instabilities. For gravity waves with initially moderate steepnesses, evolution comprised a sequence of two- and three-dimensional instabilities, which included the collinear B-F instability. However, these instabilities occurred too rapidly for a straightforward application of RIT, which supposes a rank ordering of disparate time scales for resonant triads, quartets, quintets, etc. For gravity-capillary and capillary wavetrains with initially moderate steepnesses, evolution was dominated by degenerate rhombus-quartet instabilities; this dominance is not predicted by existing RIT. In the absence of the phenomenon of selective amplification, rhombus-quartet instabilities were also dominant for capillary wavetrains, even though resonant-triad interactions are possible.

Selective amplification is an experimentally observed phenomenon in which a discrete background wave with exceedingly small steepness seeds the destabilization of a capillary wavetrain by one or more (coupled) resonant triads. Seeding appears to occur through a sequence of weak, dynamically nonresonant interactions that are normally neglected; yet, when they occur, they determine the outcome of an experiment. Whether selection occurs depends on the nature of the ubiquitous high-frequency background noise—present in any water-wave system. If the high-frequency spectrum of background noise is narrow-banded so that it contains a discrete component, selection occurs. If the spectrum is broad-banded, selection does not occur. In the latter case, experiments suggest that the entire continuum of dynamically possible low-frequency triads evolve according to RIT for single resonant triads.

The important role of degenerate resonances was also shown by the internal resonances known as Wilton ripples. These resonances, which were observed to $O(\varepsilon^6)$, are excited by wavetrains with exceedingly small steepnesses and with significant frequency detuning. In fact, these internal resonances effectively fill the band of modulationally stable wavetrains (6.4–9.8 Hz). Internal resonant interactions, like the sequence of interactions that lead to selective amplification, evolve quickly so that the rank ordering of evolution time scales supposed by RIT is violated.

Experimental investigations have focused on the degenerate resonances that play an especially important role for studying the stability of a single wavetrain, almost to the exclusion of studies on nondegenerate resonances, i.e. resonant interactions among wavetrains which satisfy (10) and (11) with arbitrary frequencies and wavevector configurations. The absence of controlled experiments to test RIT in this general, deterministic setting is striking; in fact, we found only one controlled experiment in which two different wavetrains were generated by two wavemakers and their interaction studied—and this was the first experiment reported in the companion papers by Longuet-Higgins & Smith (1966) and McGoldrick et al (1966) which established the special role of resonant interactions. The absence of controlled experiments to test RIT in its general stochastic settings is even more striking. We found only one experimental investigation that attempted to establish the validity of stochastic models that use the random phase approximation, which requires randomness to dominate nonlinear wave-wave interactions. That experiment used wind to generate nearly collinear wavetrains in a laboratory channel. While wind generation resembles the oceanic application of the stochastic models, it does not allow definitive conclusions about the relative roles of nonlinearity and randomness because of poorly understood contributions from wind and wave-breaking processes.

Phillips (1981a) noted that perhaps the simple ideas of resonant interaction theory have reached their natural limits and that further progress would depend on new mathematics, new physics, and new intuition. Perhaps this new intuition and new physics may be found in the richness of behavior observed in resonant interaction experiments.

Acknowledgments

JLH is grateful for financial support by the Office of Naval Research under contract N00014-89-J-1335 and the Army Research Office under contract DAAL03-89-K-0150.

Literature Cited

Ablowitz, M. J., Haberman, R. 1975a. Nonlinear evolution equations—two and three dimensions. *Phys. Rev. Lett.* 35: 1185–88

Ablowitz, M. J., Haberman, R. 1975b. Resonantly coupled nonlinear evolution equations. *J. Math. Phys.* 16: 2301–5

Ablowitz, M. J., Kaup, D. J., Newell, A. C., Segur, H. 1974. The inverse scattering transform-Fourier analysis for nonlinear problems. *Stud. Appl. Math.* 53: 249–315

Ablowitz, M. J., Segur, H. 1979. On the evolution of packets of water waves. *J. Fluid Mech.* 92: 691–715

Ablowitz, M. J., Segur, H. 1981. *Solitons and the Inverse Scattering Transform*. Philadelphia: S.I.A.M.

Adamson, A. W. 1990. *Physical Chemistry of Surfaces*. New York: Wiley. 5th ed.

Alber, I. E. 1978. The effects of randomness on the stability of two-dimensional surface wavetrains. *Proc. R. Soc. London Ser. A* 363: 525–46

Ball, F. K. 1964. Energy transfer between external and internal gravity waves. *J. Fluid Mech*. 19: 465–78

Banerjee, P. P., Korpel, A. 1982. Subharmonic generation by resonant three-wave interaction of deep-water capillary waves. *Phys. Fluids* 25: 1938–43

Banerjee, P. P., Korpel, A., Lonngren, K. E. 1983. Self-refraction of nonlinear capillary-gravity waves. *Phys. Fluids* 26: 2393–98

Barnett, T. P. 1968. On the generation, dissipation, and prediction of ocean wind waves. *J. Geophys. Res*. 73: 513–29

Barrick, D. E. 1986. The role of the gravity-wave dispersion relation in HF radar measurements of the sea surface. *IEEE J. Ocean. Eng*. 11: 286–92

Benjamin, T. B. 1967. Instability of periodic wavetrains in nonlinear dispersive systems. *Proc. R. Soc. London Ser. A* 299: 59–75

Benjamin, T. B., Feir, J. E. 1967. The disintegration of wavetrains on deep water. *J. Fluid Mech*. 27: 417–30

Benney, D. J. 1962. Non-linear gravity wave interactions. *J. Fluid Mech*. 14: 577–84

Benney, D. J., Newell, A. C. 1967. The propagation of nonlinear wave envelopes. *J. Math. Phys*. 46: 133–39

Benney, D. J., Newell, A. C. 1969. Random wave closures. *Stud. Appl. Math*. 48: 29–53

Benney, D. J., Roskes, G. J. 1969. Wave instabilities. *Stud. Appl. Math*. 48: 377–85

Benney, D. J., Saffman, P. G. 1966. Nonlinear interactions of random waves in a dispersive medium. *Proc. R. Soc. London Ser. A* 289: 301–20

Bonmarin, P., Ramamonjiarisoa, A. 1985. Deformation to breaking of deep water gravity waves. *Exp. Fluids* 3: 11–16

Boyd, T. J. M., Turner, J. G. 1978. Three- and four-wave interactions in plasmas. *J. Math. Phys*. 19: 1403–13

Bretherton, F. P. 1964. Resonant interactions between waves. The case of discrete oscillations. *J. Fluid Mech*. 20: 457–79

Cini, R., Lombardini, P. P. 1978. Damping effect of monolayers on surface wave motion in a liquid. *J. Colloid Interface Sci*. 65: 387–89

Craik, A. D. D. 1985. *Wave Interactions and Fluid Flows*. Cambridge: Cambridge Univ. Press

Craik, A. D. D. 1986. Exact solutions of non-conservative equations for three-wave and second-harmonic resonance. *Proc. R. Soc. London Ser. A* 406: 1–12

Craik, A. D. D., Moroz, I. M. 1988. Temporal evolution of interacting waves in non-conservative systems: some exact solutions. *Wave Motion* 10: 443–52

Crawford, D. R., Saffman, P. G., Yuen, H. C. 1980. Evolution of a random inhomogeneous field of nonlinear deep-water gravity waves. *Wave Motion* 2: 1–16

Crawford, D. R., Lake, B. M., Saffman, P. G., Yuen, H. C. 1981a. Stability of weakly nonlinear deep-water waves in two and three dimensions. *J. Fluid Mech*. 105: 177–91

Crawford, D. R., Lake, B. M., Saffman, P. G., Yuen, H. C. 1981b. Effects of nonlinearity and spectral bandwidth on the dispersion relation and component phase speeds of surface gravity waves. *J. Fluid Mech*. 112: 1–32

Davey, A., Stewartson, K. 1974. On three-dimensional packets of surface waves. *Proc. R. Soc. London Ser. A* 338: 101–10

Davidson, R. C. 1972. *Methods in Nonlinear Plasma Theory*. New York: Academic

Davies, J. T., Vose, R. W. 1965. On the damping of capillary waves by surface films. *Proc. R. Soc. London Ser. A* 286: 218–34

Djordjevic, V. D., Redekopp, L. G. 1977. On two-dimensional packets of capillary-gravity waves. *J. Fluid Mech*. 79: 703–14

Dobroklonskii, S. V. 1955. The damping of capillary-gravity waves on the surface of clean water. *Tr. M.G.I. Akad. Nauk SSSR* 6: 43–57 (In Russian)

Dungey, J. C., Hui, W. H. 1979. Nonlinear energy transfer in a narrow gravity-wave spectrum. *Proc. R. Soc. London Ser. A* 368: 239–65

Dysthe, K. B. 1979. Note on a modification to the nonlinear Schroedinger equation for application to deep water waves. *Proc. R. Soc. London Ser. A* 369: 105–14

Feir, J. E. 1967. Discussion: some results from wave pulse experiments. *Proc. R. Soc. London Ser. A* 299: 54–58

Fox, M. J. H. 1976. On the nonlinear transfer of energy in the peak of a gravity-wave spectrum. II. *Proc. R. Soc. London Ser. A* 348: 467–83

Freundlich, H. 1922. *Colloid and Capillary Chemistry*. Transl. H. S. Hatfield. New York: Dutton

Gotwols, B. L., Irani, G. B. 1980. Optical determination of the phase velocity of short gravity waves. *J. Geophys. Res*. 85: 3964–70

Hara, T., Mei, C. C. 1991. Frequency downshift in narrowbanded surface waves under the influence of wind. *J. Fluid Mech.* 230: 429–77

Harrison, W. J. 1909. The influence of viscosity and capillarity on waves of finite amplitude. *Proc. London Math. Soc.* 7: 107–21

Hasimoto, H., Ono, H. 1972. Nonlinear modulation of gravity waves. *J. Phys. Soc. Jpn.* 33: 805–11

Hasselmann, K. 1962. On the non-linear energy transfer in a gravity-wave spectrum. Part 1. General theory. *J. Fluid Mech.* 12: 481–500

Hasselmann, K. 1963a. On the non-linear energy transfer in a gravity-wave spectrum. Part 2. Conservation theorems; wave-particle analogy; irreversibility. *J. Fluid Mech.* 15: 273–81

Hasselmann, K. 1963b. On the non-linear energy transfer in a gravity-wave spectrum. Part 3. Evaluation of the energy flux and swell-sea interaction for a Neumann spectrum. *J. Fluid Mech.* 15: 385–98

Hasselmann, K. 1966. Feynman diagrams and interaction rules of wave-wave scattering processes. *Rev. Geophys.* 4: 1–32

Hasselmann, K. 1967a. Nonlinear interactions treated by the methods of theoretical physics (with application to the generation of waves by wind). *Proc. R. Soc. London Ser. A* 299: 77–100

Hasselmann, K. 1967b. A criterion for nonlinear wave stability. *J. Fluid Mech.* 30: 737–39

Hasselmann, K., Barnett, T. P., Bouws, E., Carlson, H., Cartwright, D. E., et al. 1973. Measurements of wind-wave growth and swell decay during the Joint North Sea Wave Project (JONSWAP). *Dtsch. Hydrogr. Z.* A8, No. 12. 95 pp.

Hasselmann, S., Hasselmann, K. 1985. Computations and parameterizations of the nonlinear energy transfer in a gravity-wave spectrum. Part I: A new method for efficient computations of the exact nonlinear transfer integral. *J. Phys. Oceanogr.* 15: 1369–77

Hasselmann, S., Hasselmann, K., Allender, J. H., Barnett, T. P. 1985. Computations and parameterizations of the nonlinear energy transfer in a gravity-wave spectrum. Part II: Parameterizations of the nonlinear energy transfer for application in wave models. *J. Phys. Oceanogr.* 15: 1378–91

Hatori, M., Toba, Y. 1983. Transition of mechanically generated regular waves to wind waves under the action of wind. *J. Fluid Mech.* 130: 397–409

Hayes, W. D. 1973. Group velocity and nonlinear dispersive wave propagation. *Proc. R. Soc. London Ser. A* 332: 199–221

Henderson, D. M., Hammack, J. L. 1987. Experiments on ripple instabilities. Part 1. Resonant triads. *J. Fluid Mech.* 184: 15–41

Henderson, D. M., Lee, R. C. 1986. Laboratory generation and propagation of ripples. *Phys. Fluids* 29: 619–24

Hogan, S. J. 1984. Subharmonic generation of deep-water capillary waves. *Phys. Fluids* 27: 42–45

Hogan, S. J. 1985. The fourth-order evolution equation for deep-water gravity-capillary waves. *Proc. R. Soc. London Ser. A* 402: 359–72

Hogan, S. J. 1986. The potential form of the fourth-order evolution equation for deep-water gravity-capillary waves. *Phys. Fluids* 29: 3479–80

Holliday, D. 1977. On nonlinear interactions in a spectrum of inviscid gravity-capillary surface waves. *J. Fluid Mech.* 83: 737–49

Huang, N. E. 1981. Comment on "Modulation characteristics of sea surface waves" by A. Ramamonjiarisoa and E. Mollo-Christensen. *J. Geophys. Res.* 86: 2073–75

Huang, N. E., Tung, C.-C. 1977. The influence of the directional energy distribution on the nonlinear dispersion relation in a random gravity wave field. *J. Phys. Oceanogr.* 7: 403–14

Janssen, P. A. E. M. 1983a. On a fourth-order envelope equation for deep-water waves. *J. Fluid Mech.* 126: 1–11

Janssen, P. A. E. M. 1983b. Long-time behaviour of a random inhomogeneous field of weakly nonlinear surface gravity waves. *J. Fluid Mech.* 133: 113–32

Kaup, D. J. 1981. The solution of the general initial value problem for the full three dimensional three-wave resonant interaction. *Physica* 3D: 374–95

Kelvin, Lord (Thompson, W.) 1871. Part IV. (Letter to Professor Tait, of date August 23, 1871.) and Part V. Waves under motive power of gravity and cohesion jointly, without wind. *Phil. Mag.* XLII: 370–77

Komen, G. J. 1980. Spatial correlations in wind-generated water waves. *J. Geophys. Res.* 85: 3311–14

Lake, B. M., Yuen, H. C. 1977. A note on some nonlinear water-wave experiments and the comparison of data with theory. *J. Fluid Mech.* 83: 75–81

Lake, B. M., Yuen, H. C., Rungaldier, H., Ferguson, W. E. 1977. Nonlinear deep-water waves: theory and experiment. Part 2. Evolution of a continuous wave train. *J. Fluid Mech.* 83: 49–74

LeBlond, P. H., Mysak, L. A. 1978. *Waves in the Ocean*. New York: Elsevier Oceanogr. Ser. Vol. 20

Levich, V. G. 1962. *Physicochemical Hydrodynamics*. Transl. Scripta Technica, Inc. Englewood Cliffs: Prentice-Hall (From Russian)

Lighthill, M. J. 1965. Contributions to the theory of waves in non-linear dispersive systems. *J. Inst. Math. Its Appl.* 1: 269–306

Lo, E., Mei, C. C. 1985. A numerical study of water-wave modulation based on a higher-order nonlinear Schroedinger equation. *J. Fluid Mech.* 150: 395–416

Lo, E. Y., Mei, C. C. 1987. Slow evolution of nonlinear deep water waves in two horizontal directions: a numerical study. *Wave Motion* 9: 245–59

Longuet-Higgins, M. S. 1962. Resonant interactions between two trains of gravity waves. *J. Fluid Mech.* 12: 321–32

Longuet-Higgins, M. S. 1976. On the nonlinear transfer of energy in the peak of a gravity-wave spectrum: a simplified model. *Proc. R. Soc. London Ser. A* 347: 311–28

Longuet-Higgins, M. S. 1977. Some effects of finite steepness on the generation of waves by wind. In *A Voyage of Discovery*, ed. M. Angel, pp. 393–403. New York: Pergamon

Longuet-Higgins, M. S. 1978a. The instabilities of gravity waves of finite amplitude in deep water I. Superharmonics. *Proc. R. Soc. London Ser. A* 360: 471–88

Longuet-Higgins, M. S. 1978b. The instabilities of gravity waves of finite amplitude in deep water II. Subharmonics. *Proc. R. Soc. London Ser. A* 360: 489–505

Longuet-Higgins, M. S., Phillips, O. M. 1962. Phase velocity effects in tertiary wave interactions. *J. Fluid Mech.* 12: 333–36

Longuet-Higgins, M. S., Smith, N. D. 1966. An experiment on third-order resonant wave interactions. *J. Fluid Mech.* 25: 417–35

Martin, D. U., Yuen, H. C. 1980. Quasi-recurring energy leakage in the two-space-dimensional nonlinear Schroedinger equation. *Phys. Fluids* 23: 881–83

Masuda, A. 1981. Nonlinear energy transfer between wind waves. *J. Phys. Oceanogr.* 10: 2082–93

Masuda, A. 1986. Nonlinear energy transfer between random gravity waves. Some computational results and their interpretation. In *Wave Dynamics and Radio Probing of the Ocean Surface*, ed. O. M. Phillips, K. Hasselmann, pp. 41–57. New York: Plenum

McGoldrick, L. F. 1965. Resonant interactions among capillary-gravity waves. *J. Fluid Mech.* 21: 305–31

McGoldrick, L. F. 1970a. On Wilton's ripples: a special case of resonant interactions. *J. Fluid Mech.* 42: 193–200

McGoldrick, L. F. 1970b. An experiment on second-order capillary gravity resonant wave interactions. *J. Fluid Mech.* 40: 251–71

McGoldrick, L. F. 1972. On the rippling of small waves: a harmonic nonlinear nearly resonant interaction. *J. Fluid Mech.* 52: 725–51

McGoldrick, L. F., Phillips, O. M., Huang, N. E., Hodgson, T. H. 1966. Measurements of third-order resonant wave interactions. *J. Fluid Mech.* 25: 437–56

McLean, J. W. 1982a. Instabilities of finite-amplitude water waves. *J. Fluid Mech.* 114: 315–30

McLean, J. W. 1982b. Instabilities of finite-amplitude gravity waves on water of finite depth. *J. Fluid Mech.* 114: 331–41

Meiron, D. I., Saffman, P. G., Yuen, H. C. 1982. Calculation of steady three-dimensional deep-water waves. *J. Fluid Mech.* 124: 109–21

Melville, W. K. 1982. The instability and breaking of deep-water waves. *J. Fluid Mech.* 115: 165–85

Melville, W. K. 1983. Wave modulation and breakdown. *J. Fluid Mech.* 128: 489–506

Miles, J. W. 1984a. Nonlinear Faraday resonance. *J. Fluid Mech.* 146: 285–302

Miles, J. W. 1984b. On damped resonant interactions. *J. Phys. Oceanogr.* 14: 1677–78

Murakami, Y. 1987. Damped four-wave resonant interaction with external forcing. *Wave Motion* 9: 393–400

Ocean Wave Spectra Proceedings of a Conference, 1963. Englewood Cliffs: Prentice-Hall

Papadimitrakis, Y. A. 1986. On the structure of artificially generated water wave trains. *J. Geophys. Res.* 91: 14,237–49

Peregrine, D. H. 1983. Water waves, nonlinear Schroedinger equations and their solutions. *J. Austr. Math. Soc. Ser. B* 25: 16–43

Perlin, M., Hammack, J. 1991. Experiments on ripple instabilities. Part 3. Resonant quartets of the Benjamin-Feir type. *J. Fluid Mech.* 229: 229–68

Perlin, M., Henderson, D., Hammack, J. 1990. Experiments on ripple instabilities. Part 2. Selective amplification of resonant triads. *J. Fluid Mech.* 219: 51–80

Phillips, O. M. 1960. On the dynamics of unsteady gravity waves of finite amplitude. Part 1. The elementary interactions. *J. Fluid Mech.* 9: 193–217

Phillips, O. M. 1967. Theoretical and experimental studies of gravity wave interactions. *Proc. R. Soc. London Ser. A* 299: 104–19

Phillips, O. M. 1974. Nonlinear dispersive waves. *Annu. Rev. Fluid Mech.* 6: 93–110
Phillips, O. M. 1977. *The Dynamics of the Upper Ocean.* Cambridge: Cambridge Univ. Press. 2nd ed.
Phillips, O. M. 1981a. Wave interactions—the evolution of an idea. *J. Fluid Mech.* 106: 215–27
Phillips, O. M. 1981b. The dispersion of short wavelets in the presence of a dominant long wave. *J. Fluid Mech.* 107: 465–85
Pierson, W. J., Fife, P. 1961. Some nonlinear properties of long-crested periodic waves with lengths near 2.44 centimeters. *J. Geophys. Res.* 66: 163–79
Plant, W. J., Wright, J. W. 1979. Spectral decomposition of short gravity wave systems. *J. Phys. Oceanogr.* 9: 621–24
Ramamonjiarisoa, A., Coantic, M. 1976. Loi experimental de dispersion des vagues produites par le vent sur une faible longueur d'action. *C. R. Acad. Sci. Paris Ser. B* 282: 111–13
Ramamonjiarisoa, A., Mollo-Christensen, E. 1979. Modulation characteristics of sea surface waves. *J. Geophys. Res.* 84: 7769–75
Ramamonjiarisoa, A., Mollo-Christensen, E. 1981. Reply. *J. Geophys. Res.* 86: 2076–77
Rayleigh, Lord 1890. On the tension of water surfaces, clean and contaminated, investigated by the method of ripples. *Phil. Mag.* 30: 386–400
Resio, D., Perrie, W. 1991. A numerical study of nonlinear energy fluxes due to wave-wave interactions. Part 1. Methodology and basic results. *J. Fluid Mech.* 223: 603–29
Scott, J. C. 1981. The propagation of capillary-gravity waves on a clean water surface. *J. Fluid Mech.* 108: 127–31
Segur, H. 1984. Toward a new kinetic theory for resonant triads. *Contemp. Math.* 28: 281–313
Sell, W., Hasselmann, K. 1972. Computations of nonlinear energy transfer for JONSWAP and empirical wind wave spectra. *Rep. Inst. Geophys.*, Univ. Hamburg
Simmons, W. F. 1969. A variational method for weak resonant wave interactions. *Proc. R. Soc. London Ser. A* 309: 551–75
Stiassnie, M. 1984. Note on the modified nonlinear Schroedinger equation for deep water waves. *Wave Motion* 6: 431–33
Stiassnie, M., Shemer, L. 1984. On modifications of the Zakharov equation for surface gravity waves. *J. Fluid Mech.* 143: 47–67
Stiassnie, M., Shemer, L. 1987. Energy computations for evolution of class I and II instabilities of Stokes waves. *J. Fluid Mech.* 174: 299–312
Stokes, G. G. 1847. On the theory of oscillatory waves. *Cambridge Philos. Soc. Trans.* 8: 441–55
Su, M. Y. 1982. Three-dimensional deep-water waves. Part 1. Experimental measurement of skew and symmetric wave patterns. *J. Fluid Mech.* 124: 73–108
Su, M. Y., Bergin, M., Marler, P., Myrick, R. 1982. Experiments on nonlinear instabilities and evolution of steep gravity-wave trains. *J. Fluid Mech.* 124: 45–72
Su, M. Y., Green, A. W. 1984. Coupled two- and three-dimensional instabilities of surface gravity waves. *Phys. Fluids* 27: 2595–97
Trulsen, K., Dysthe, K. B. 1990. Frequency down-shift through self modulation and breaking. In *Water Wave Kinematics.* 178: 561–72. Dordrecht: Kluwer (Applied Sci., Ser. E)
Valenzuela, G. R., Laing, M. B. 1972. Nonlinear energy transfer in gravity-capillary wave spectra, with applications. *J. Fluid Mech.* 54: 507–20
van Gastel, K. 1987a. Nonlinear interactions of gravity-capillary waves: Lagrangian theory and effects on the spectrum. *J. Fluid Mech.* 182: 499–523
van Gastel, K. 1987b. Imaging by *X* band radar of subsurface features: a nonlinear phenomenon. *J. Geophys. Res.* 92: 11,857–65
von Matthiessen, L. 1889. Experimentelle Untersuchungen ueber das Thomson'sche Gesetz der Wellenbewegung auf Fluessigkeitein unter der Wirkung der Schwere und Cohaesion. *Ann. Phys.* 38: 118–30
Watson, K. M., West, B. J. 1975. A transport-equation description of nonlinear ocean surface wave interactions. *J. Fluid Mech.* 70: 815–26
Webb, D. J. 1978. Non-linear transfers between sea waves. *Deep-Sea Res.* 25: 279–98
Weiland, J., Wilhelmsson, H. 1977. *Coherent Non-linear Interaction of Waves in Plasmas*, 88. New York: Pergamon. (Int. Ser. Natural Philos.)
West, B. J. 1981. *Deep Water Gravity Waves*, 146. New York: Springer-Verlag
Whitehead, A. N. 1958. *An Introduction to Mathematics.* New York: Oxford Univ. Press
Whitham, G. B. 1967. Non-linear dispersion of water waves. *J. Fluid Mech.* 27: 399–412
Whitham, G. B. 1974. *Linear and Nonlinear Waves.* New York: Wiley-Interscience
Wilhelmsson, H., Pavlenko, V. P. 1973. Five wave interaction—a possibility for enhancement of optical or microwave radiation by nonlinear coupling to explosively unstable plasma waves. *Phys. Scr.* 7: 213–16

Willebrand, J. 1975. Energy transport in a nonlinear and inhomogeneous random gravity wave field. *J. Fluid Mech.* 70: 113–26

Wilton, J. R. 1915. On ripples. *Phil. Mag.* 29: 688–700

Wu, H.-Y., Hsu, E.-Y., Street, R. L. 1979. Experimental study of nonlinear wave-wave interaction and white-cap dissipation of wind-generated waves. *Dyn. Atmos. Oceans* 3: 55–78

Yuen, H. C., Lake, B. M. 1975. Nonlinear deep water waves: theory and experiment. *Phys. Fluids* 18: 956–60

Yuen, H. C., Lake, B. M. 1980. Instabilities of waves on deep water. *Annu. Rev. Fluid Mech.* 12: 303–34

Yuen, H. C., Lake, B. M. 1982. Nonlinear dynamics of deep-water gravity waves. *Adv. Appl. Mech.* 22: 67–229

Zakharov, V. E. 1968. Stability of periodic waves of finite amplitude on the surface of a deep fluid. *J. Appl. Mech. Tech. Phys.* 2: 190–94. (From Russian)

Zakharov, V. E., Shabat, A. B. 1972. Exact theory of two-dimensional self-focusing and one-dimensional self-modulation of waves in nonlinear media. *Sov. Phys. JETP* 34: 62–69

Zhang, J., Melville, W. K. 1987. Three-dimensional instabilities of nonlinear gravity-capillary waves. *J. Fluid Mech.* 174: 187–208

Annu. Rev. Fluid Mech. 1993. 25 : 99–114

FLOW-INDUCED VIBRATIONS IN ARRAYS OF CYLINDERS

Peter M. Moretti

School of Mechanical and Aerospace Engineering, Oklahoma State University, Stillwater, Oklahoma 74078-0545

KEY WORDS: fluid/structure interaction, flow stability, tube arrays, tube bundles

INTRODUCTION

There is extreme variety in flow-induced vibrations; they are observed in many diverse situations ranging from reeds in musical instruments to fluttering panels in supersonic airplanes. In order to keep the discussion within bounds, we will focus on one versatile set of simple geometries: arrays of circular cylinders in cross flow. By changing cylinder arrangement and the parameters of spacing, damping, and mass ratio, several different basic types of self-sustaining oscillations can be produced. Because of the industrial importance of tube bundles in heat exchangers, a wealth of observations and analyses has accumulated in the literature (Naudascher & Rockwell 1980).

We limit ourselves to discussing cross flow because of the fundamental differences between parallel-flow instabilities and the separated-flow mechanisms occurring in cross flow. In parallel flow in or outside of tubes, the cylindrical walls form continuous flow boundaries which curve when there is a vibrational deflection. The dynamic equation for a single tube is a beam equation with "gyroscopic equation" terms added:

$$(M+m)\frac{\partial^2 y}{\partial t^2}+2mU\frac{\partial^2 y}{\partial t\,\partial x}+mU^2\frac{\partial^2 y}{\partial x^2}+\mathrm{EI}\frac{\partial^4 y}{\partial x^4}=0, \tag{1}$$

where U is a velocity of the fluid; M is the mass (per unit length) of the beam; and m is the equivalent added mass due to the fluid—not necessarily the same coefficient in each term, as explained by Chang et al (1991) in connection with a different but related problem. For an array of cylinders, the concept of equivalent added mass must be extended to include mass

0066–4189/93/0115–0099$02.00

coupling between the parallel beams. The mixed-derivative "Coriolis" term makes analysis of gyroscopic equations somewhat difficult, but it is clear that the "centrifugal" term resembles that of a compression load on a column. Therefore we expect to find a critical value of the fluid velocity beyond which the tube buckles. At even higher velocities, there can be traveling wave solutions (Paidoussis 1979).

On the other hand, in cross flow, the flow boundaries are discontinuous and the flow is forced to separate and reattach. The instabilities in boundary and shear layers, and the sensitivity of the separation point to pressure gradients, give rise to intermittent and/or asymmetrical processes described by names like "vortex shedding" and "Coanda flow." Understanding the mutual interactions between these fluid phenomena and structural vibrations is a challenge.

We can classify these fluid/solid interactions into three categories:

CATEGORY 1: Solid bodies in a flow are buffeted by randomly fluctuating pressures due to the turbulence in the approaching fluid (Chen 1991). The turbulence field is essentially an external forcing function, and the action is largely one-way, from fluid to solid. In the absence of a source of large eddies or fluctuating flow, the vibrations rarely build up to large amplitudes.

CATEGORY 2: Bluff bodies shed vortices in their wakes and experience corresponding reaction forces from the fluid. The vortex streets can be observed in the fluid even if the bodies do not move, but both the frequency and the magnitude of the forces can be modified by oscillation of the solid bodies (Cheng & Moretti 1991). Therefore the interaction is not strictly one-way, although the fluid dynamics dominate the process, as discussed below in the section on Vortex Shedding.

CATEGORY 3: Beyond a critical velocity, we observe autonomous vibrations which involve a fully two-way fluid/solid interaction. If the tubes do not vibrate, the flow does not produce the alternating forces, and vice versa. The nature of the interaction is discussed in the section on Flutter.

In complex situations, these distinctions can become blurred: One tube's wake vortices may become another tube's turbulence field. Nevertheless, these categories form a useful framework for sorting out diverse observations.

VORTEX SHEDDING

Single Cylinders

The classic reference case for Category 2 phenomena is vortex shedding from a single circular cylinder in cross flow. Over a wide range of Reynolds numbers, flow visualization shows a "street" of alternating vortices in the far wake (Van Dyke 1982, Nakayama et al 1988), in approximately the arrangement shown to be stable by von Karman. The frequency of alternation can be expressed by the dimensionless Strouhal number:

$$S \equiv \frac{f_{vs}D}{U}, \tag{2}$$

where f_{vs} is the vortex-shedding frequency; D is the diameter of the cylinder; and U is a velocity of the fluid—the free-stream velocity in the case of a solitary cylinder. Theoretically, S is a function of the Reynolds number; however, the Strouhal number remains almost constant near a value of 0.2 over a wide range of Reynolds numbers, from less than 1,000 to more than 100,000. The significance of this observation is that the shedding frequency is controlled not by the local parameters of the boundary layer on the cylinder, but by the geometry of the wake. Even at much higher, super-critical Reynolds numbers, with turbulent boundary layers on the cylinder, similar Strouhal numbers are sometimes seen. Apparently the pressure and velocity fields from the departing vortices influence either the separation of the boundary layers and/or the stability of the shear layers, which act as amplifiers for the recurring disturbances and thus maintain the vortex street. This view is supported by the observation that cancellation of the acoustic field by "anti-sound" can suppress the vortex street (Ffowcs Williams & Zhao 1989). It is easy to conceive of slight pressure gradients and velocity fluctuations affecting the boundary-layer profiles and thus the flow separation from a circular cylinder. However, we also have to explain how vortex streets are generated by, for example, triangular cylinders with the separation points fixed at the corners. Presumably shear layer instability plays an important role.

When we look at the near wake of a circular cylinder in more detail, we note that the shear layers of the just-separated flow generate numerous small eddies—the Bloor-Gerrard vortices (Bloor 1964). Their frequency of occurrence varies with the square root of the Reynolds number, and therefore is a function of the local flow conditions. Their role in wake development is ambiguous. On the one hand, some of them may move ahead while others fall behind as they all move downstream, to coalesce and join into the larger pattern of the von Karman vortex street. Such a coalescence of vorticity may be a mechanism by which the process can move from the local-parameter scale to the periodic-wake scale. On the other hand, they may act mainly to thicken the shear layer and dissipate into wake turbulence (Unal & Rockwell 1984).

These and other mechanisms involved in the formation of the undulating wake have been studied from various perspectives (Gerrard 1966, Berger & Wille 1972, Bearman 1984). No simple answers have come forth; our understanding of the feedback loop—vortex wake, pressure/velocity fields, near-field flow development, and vortex generation back into the wake—is still incomplete.

Synchronization processes give us clues to some of the processes in the control loop. Vortex shedding generates an acoustic field normal to the flow direction and to the tubes, which is usually modeled as a dipole source related to fluctuating lift (Blevins 1984). Conversely, acoustic fields interact with the vortex shedding (Tadrist et al 1991). Blevins (1985) confirmed that a strong transverse acoustic field will synchronize the vortex shedding and keep it in phase over the entire length of a cylinder; he also showed that it can lock the vortex shedding into an external excitation even when the excitation frequency is not exactly the unstimulated vortex-shedding frequency. He attributed these effects to the velocity fields of the acoustic waves.

Forced oscillation of the cylinder normal to the flow direction has a similar influence on vortex shedding. The vortex shedding locks in with the forced vibration over a range of excitation frequencies. However, experiments (Cheng & Moretti 1991) yield two surprises: First, the frequency range over which the vortex shedding locks in with the forced vibration increases with forcing amplitude up to a point, but narrows down again for amplitudes greater than one-fourth diameter; secondly, at the outer edges of the lock-in range, two frequencies of vortex shedding exist at the same time—one is the same as the unstimulated vortex-shedding frequency, the other is the forcing frequency. This last observation creates some conceptual difficulties: In general, the superposition of two different vortex streets should not be stable, but apparently both coexist for a distance of several diameters before instability disorganizes and reorganizes the vortex patterns.

Classical aerodynamic theory relates lift in two-dimensional flow to vorticity in the flow field. The magnitude of the lift is proportional to a vortex flow field centered within the body. Any increment of change in the lift force is accompanied by an increment of change in the vortex strength bound to the cylinder, which must be balanced by an opposite vortex being shed into the fluid. Therefore, when we see vortices being shed, we expect to find fluctuating lift forces on the cylinder, and vice versa. This process permits us to close the vicious circle for self-sustaining oscillations of flexibly mounted cylinders: Vortex shedding causes alternating lift forces which deflect the flexible cylinder, and the resulting vibrational deflections lock the vortex shedding in with the vibration. If the fluid forces are small and the structural damping low, the periodic response will take place at the natural frequency of the flexible element, and the unstimulated vortex-shedding frequency will have to be close to that natural frequency before it can lock in.

The forced-vibration data cited above (Cheng & Moretti 1991) give us some indications of the conditions under which sustained free vibrations can take place. First, the response must be sufficiently large in amplitude for the vortex shedding to lock in with the vibration. The larger the difference between unstimulated vortex shedding and structural natural frequency, the closer this response amplitude must be to one-fourth of a diameter. Since the resonant amplitude will be proportional to the density of the fluid and inversely proportional to damping, the lock-in band will be wider for a lower mass-and-damping parameter, defined as

$$d_{\mathrm{m}} = \frac{D\delta}{\rho D^2}, \tag{3}$$

where M is the mass-per-unit-length of the cylinder, ρ the density of the fluid, D the diameter of the cylinder, and δ the logarithmic decrement, a dimensionless measure of the structural damping. For very large values of this parameter the response becomes very small even right at resonance—so small that it disappears among the background noise of Category 1 vibrations. This explains the results of Scruton (1963), who obtained the range of velocities for which synchronization of the vortex shedding with the natural frequency is obtained. This is shown in Figure 1. The ordinate of normalized or "reduced" dimensionless velocity is

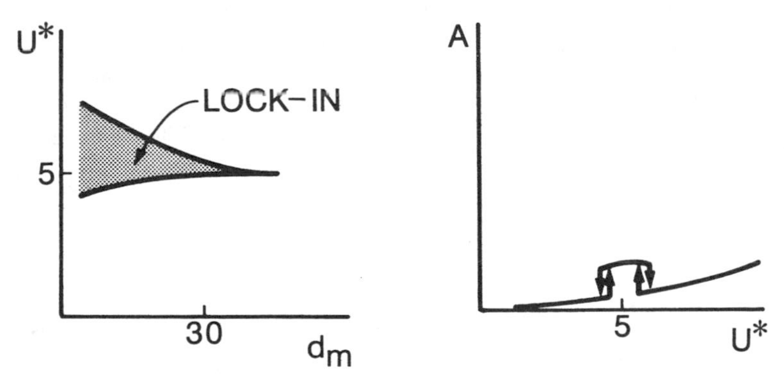

Figure 1 (*Left*) Region of locked-in vortex shedding and structural vibration for a single elastic cylinder in cross-flow. The abscissa is the mass-and-damping parameter from Equation 3; the ordinate is the reduced velocity from Equation 4.

Figure 2 (*Right*) Amplitude response of a single elastic cylinder with varying reduced velocity, showing jump phenomenon. Amplitude scale depends on system and flow parameters.

$$U^* \equiv \frac{U}{f_{str}D}, \tag{4}$$

where f_{str} is the structural frequency. When the structural frequency and the vortex-shedding frequency are the same, the reduced velocity is the inverse of the Strouhal number; therefore, the vortex-shedding response of a solitary cylinder occurs when the reduced velocity is near 5.

Secondly, since a minimum amplitude is required to start the lock-in process, we expect that there is a hysteresis and jump phenomenon: When a system first approaches critical conditions, lock-in occurs late; but once lock-in has started, it tends to maintain itself (Figure 2). Also, since the ability of vibrations to control vortex shedding falls off again at large amplitudes, free vibrations due to vortex shedding tend to be limited in amplitude, leading to a flat-topped response curve rather than a sharply peaking resonance.

Since single cylinders have been so extensively studied, they form a useful reference case for comparison with the flow-induced vibrations in rows and arrays of tubes. The similarities and also the differences help us understand the processes at work.

Tube Rows

When we arrange a row of tubes across the flow, we expect each tube to behave like a solitary cylinder, possibly with the added feature of each wake synchronizing itself with the motions and wakes of adjacent tubes. However, careful observation shows a more complex interaction: The flow across a closely spaced row of tubes no longer passes straight through them, but deflects into remarkable new patterns.

If the pitch-to-diameter (P/D) ratio is between 1.5 and 2.2, some of the wakes will be wider, others will narrow down, in an alternating pattern. Another way to describe this is to say that the jets of flow between the wakes deflect towards each other in pairs, meeting behind the narrow wakes (Figure 3).

In other words, we can regard a tube row not merely as a row of bluff objects, but alternatively also as a row of diffusers. Straight flow in abrupt two-dimensional diffusers tends to be unstable; the two stable states have the flow attached to one wall or the other. Similarly, the tube row has two stable flow states: one with the narrow wakes behind all the even-numbered tubes, and the other with the narrow wakes behind all the odd-numbered tubes instead. At smaller values of P/D—equivalent to more abrupt diffusers—the deflected jets angle more sharply.

This phenomenon is related to the Coanda effect. It is, like the von Karman vortex street, a wake-controlled phenomenon; but, unlike the von

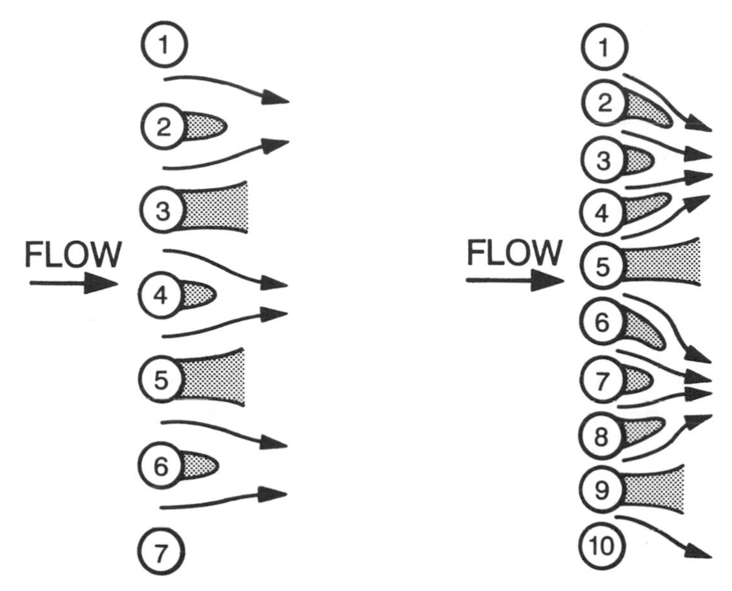

Figure 3 (*Left*) Simple jet pairing after a tube row with P/D value between 1.5 and 2.2.
Figure 4 (*Right*) Complex jet pairing after a tube row with P/D value less than 1.5.

Karman vortex street, it tends to maintain a continuing deflection rather than an alternation. The two processes of steady (Coanda) and alternating (von Karman) deflections do not appear to compete with each other (a vortex street can be seen in each wake) but the vortex-shedding frequency is lower in the wider wakes. We might say that each vortex street seeks stability in its own wake, rather than to lock in with the wider or narrower adjacent wakes. Thus, at P/D values between 1.5 and 2, we observe two different vortex-shedding frequencies being generated in the same row. It has been speculated that the lower-frequency vortex street, associated with a wider and much longer wake, is stronger than the other, but so far the actual evidence has been ambiguous.

When we look at P/D values of less than 1.5, the Coanda instability phenomenon becomes so much more severe that it acquires an entirely new character. This is an example of how quantitative changes can lead to a qualitatively new regime. We have published photographs of colored dye streaks through a tube row in a water tunnel (Moretti 1986) that show such extreme flow angles that we initially doubted the symmetry of the test section. However, subsequent experiments in a wind tunnel showed the same effect (Moretti & Cheng 1987). At small P/D ratios, the jets do not simply join in pairs, but the combined jets pair again into groups of four jets (Figure 4). Rather than only two stable states, there are now four flow patterns which can persist. Occasionally, in response to some random perturbation, the flow may switch from one pattern to another (Cheng & Moretti 1988). Observations in other facilities confirm the existence of multi-stable flows (Zdravkovich & Stonebanks 1988).

The complexity of these flow fields accounts for the observation of more than two vortex-shedding frequencies behind tube rows if the P/D ratio is small. The lowest frequency will generally be lower than the frequency for a solitary cylinder, although the value of the Strouhal number may be higher than 0.2 because Strouhal number in tube rows is usually given on the basis of mean velocity in the gap between the cylinders rather than in the approaching and departing fluid. This gives us an explanation of the effect of transverse P/D on vortex shedding; the effect of longitudinal (flow-direction) P/D is more difficult to understand.

Tube Arrays

It is hard to imagine that vortex shedding can even exist within a closely spaced tube array, but there are two observations that suggest that it does. First, if there is an acoustic resonance, then gas flows through tube bundles generate tones at certain Strouhal numbers. These Strouhal numbers are a function of array geometry, but they are independent of Reynolds number over significant ranges, paralleling the behavior of solitary cylinders. Secondly, if there is a structural resonance, then liquid flows through flexible tube arrays generate vibrations at similar Strouhal numbers. Therefore, the analogy with single-cylinder vortex shedding is useful: We can control vibrations by avoiding resonance. All we need is a catalog of Strouhal numbers as a function of the geometry, which is defined by P/D ratios and either in-line or staggered arrangement.

Intuition leads us to suspect that the longitudinal (flow-direction) P/D ought to be the crucial parameter for tube columns or arrays, since it relates to the distance an eddy has to travel from interaction to interaction with solid boundaries. Available data for long columns of tubes are scanty. For arrays, correlation to longitudinal P/D has not been consistently successful; rather, it has been found necessary to plot experimental Strouhal numbers against both transverse and longitudinal P/D, either as a family of curves or as a contour map (Moretti 1974).

An obstacle to a simple theory explaining the dependence of Strouhal number on geometry is that multi-stable flow patterns are observed even within bundles (Cheng & Moretti 1989). The flow is even more complicated than that observed after tube rows, and has a great many stable states. As a result, we expect to see complex, possibly chaotic systems; not surprisingly, several values of the Strouhal number can take effect (Ishigai & Nishikawa

1975, Ziada et al 1988). For some common equilateral-triangle and square arrangements with minimum P/D greater than 1.8, accumulated experience has led to a consensus on what the most worrisome Strouhal number is; for other cases, comparison of different sources shows substantial disagreements (Chenoweth 1977).

FLUTTER

Jet Switching

In many cases, vortex-shedding does not cause any structural problems, for example in low-density gas flows, where the mass-and-damping parameter of Equation 3 is too large to permit lock-in, or the amplitudes are too low to be of interest. Nevertheless, if the velocity through an array is sufficiently increased, there will be violent flutter, evidently arising from some Category 3 phenomenon.

The first possible mechanism to be proposed was jet switching due to drag-direction tube motion. As we have described above, a tube row with a P/D between 1.5 and 2.2 will have wide and narrow wakes behind alternate tubes. As a result, alternate tubes have higher and lower drag forces acting on them. Roberts (1966) showed that a stream-wise motion of alternate tubes could interact with the wakes and jets, causing them to switch from one state (narrow wakes behind even-numbered tubes) to the other (narrow wakes behind odd-numbered tubes). He showed that a feedback loop could be constructed from the changes in drag, the response of the tubes, and the switching of jets. He measured the critical velocity needed to sustain oscillation, and showed that the nondimensional or "reduced" critical velocity (Equation 4) is proportional to the square root of the mass-and-damping parameter (Equation 3). He did not wish to extrapolate this relationship to low mass and damping parameter values. On the basis of theoretical considerations, but without experimental evidence, he argued that jet switching could not drive self-sustaining oscillations at low velocities. He believed that there is a minimum value of the critical velocity that takes effect at low values of the mass-and-damping parameter, where a square-root relationship would otherwise lead to even lower values.

Purely flow-direction vibrations like Roberts' are relatively rare; most destructive vibrations have a strong lift-direction component. It appears that other, lateral-motion instabilities usually occur first, at lower critical velocities than Roberts' switching mechanism. For P/D values over 1.5 these lateral motions cannot be explained by jet switching, because between P/D ratios of 1.5 to 2.2 the simple jet pairings are symmetrical behind each tube and generate no lift forces. However, if we look at systems with smaller P/D ratios, we find that the wakes are no longer symmetrical behind every tube; every group of four tubes has one very wide, high-drag wake and two wakes curving away from it on either side (Figure 4). This curvature is associated with lift forces (Moretti 1986). Therefore the multi-stable flows in tightly packed bundles open the door to more complex jet-switching patterns.

Fluid-Elastic Whirling

Connors (1970) conducted free-vibration experiments in a short tube row with a P/D of 1.41, and observed elliptical motions of individual tubes, with alternate tubes moving predominantly in the lift direction, and the others in the drag direction. In other tests, he carefully measured forces on firmly held tubes in a short tube row as a function of particular relative displacements of three of the tubes. For static displacements similar to those found during free vibration, he found not only the switching forces associated with multi-stable flows, but also continuous variations in lift and drag. He subtracted the contribution of switching forces, on the basis of Roberts' argument that these would not be effective at low velocity. The remaining continuously varying forces were sufficient to drive complex patterns of whirling. Like Roberts (1966), Connors found the critical velocity to be proportional to the square root of the mass-and-damping parameter.

The question remains whether jet switching contributes to the fluid-elastic whirling process. Flow visualizations on free vibrations do not permit us to distinguish between deflections caused by wake instability which then lead to lift forces, and other kinds of feedback instabilities which cause lift forces that then lead to deflected wakes. Of course, from a practical viewpoint, it is not necessary to know the source of the instability as long as the critical velocity is known as a function of the mass-and-damping parameter. But from a scientific viewpoint the job is not finished until we can predict the critical velocity at previously untested tube arrangements and arbitrary transverse and longitudinal P/D ratios. Therefore the search for the nature of the underlying instability continues. Hara (1987) carried out experiments on a short tube row with one forced and two free cylinders. He concluded that the whirling forces and jet switching were closely tied to the shifting of the separation points on the cylinder surfaces.

The interesting thing about Connors' model is that it is in a certain sense a two-degree-of-freedom model. It requires mobility from more than one tube. In a simplified description, tubes in a long row move in a pattern that repeats in groups of four: tubes 1, 5, 9, etc move downstream while tubes 3, 7, 11, etc move upstream; this constitutes one degree of freedom.

Ninety degrees out of phase with this motion, tubes 2, 6, 10, etc move to one side while tubes 4, 8, 12, etc move to the other side; this constitutes another degree of freedom. The flow extracts energy from the fluid because the drag of a cylinder depends on the spacing from adjacent cylinders, and the lift on another cylinder depends on the alignment of its neighbors in the row. The two degrees of freedom are coupled by the hydrodynamic mass coupling provided by the inertia of the fluid. The analogy here is to flutter in an airplane wing, where one degree of freedom is twisting, and the other is deflection in the direction of lift, and the two degrees of freedom are structurally coupled because the center of mass is behind the center of twist; the combined out-of-phase action of deflection and twist extracts energy from the flow due to the dependence of lift on angle of attack.

The pioneering studies of Roberts and Connors dealt with uniform rows of tubes. Blevins (1974) extended the approach based on quasi-static fluid forces and argued that the proportionality of the critical velocity to the square root of the mass-and-damping parameter is widely applicable even in structurally nonuniform rows; one merely needs to know the proportionality constant appropriate to P/D and other given parameters. Since then, a number of models have been proposed specifically for either in-line or staggered tube arrays. Chen (1984), Paidoussis (1987), and Blevins (1990) summarize several of them; they are generally semi-empirical expressions which are fitted to data on a plot of critical velocity versus the mass-and-damping parameter.

Single-Tube Instability

Southworth & Zdravkovich (1975) showed that even a single flexible cylinder in a rigid array can sustain flow-induced vibrations. Lever & Weaver (1986a, 1986b) developed a model based on a single flexible cylinder. One way in which single-tube galloping can be explained is the postulation of a time lag in the redistribution of the fluid flow around the moving tube when there is a lateral displacement of that tube. The moving tube squeezes down the flow channel on one side of it; the velocity temporarily increases and the pressure, by Bernoulli's law, temporarily decreases; this reduced pressure can be translated into a phase-leading force or a negative damping.

It should be noted here that there can be several instability mechanisms, each with different numbers of tubes involved and different relative motions of the tubes. Only the one with the lowest critical velocity—the one to go into action first—is important to us. Intuitively, we believe that an array that permits many tubes to move should become unstable at a lower velocity than one constrained to let only one tube move. Blevins

(1990) reports that an unconstrained in-line array can sustain vibrations at a lower velocity than one in which the first and last tube rows are held rigidly. However, Lever & Rzentkowsky (1988) show that the difference between one and seven flexible tubes is not so much the threshold at which increasing velocity starts the oscillations as it is the amplitude of the motions, as well as the continued sustained vibrations when the flow is decreased. We recall that the jet-switching mechanism of Roberts (1966) is a "hard" instability, one that needs an initial perturbation to leave the unmoved equilibrium state. We see here a hint that one kind of instability mechanism might start the oscillations, while another might amplify, modify, and sustain them.

Unified Theories

S.S. Chen (1988) has approached the problem from the viewpoint of structural equations with fluid-dynamic coefficients which are not quasi-static. The linearized equation for small responses of one mode of one element is:

$$(M+m)\frac{\partial^2 y}{\partial t^2}+(C+c)\frac{\partial y}{\partial t}+(K+k)y = 0, \tag{5}$$

where the capitalized coefficients represent structural properties, and the lower-case letters fluid-induced coefficients. We note that oscillatory solutions can be obtained in several ways: a sufficiently negative coefficient c on the velocity term; a phase shift in the coefficient k of the displacement term; and/or coupling terms to another element. These coupling terms can include an equivalent mass coefficient m which covers not only the conventional hydrodynamic mass (Moretti & Lowery 1976), but also hydrodynamic mass coupling between elements; a fluid damping c which includes cylinder-to-cylinder interaction-dependent fluid damping; and a fluid force k which includes relative-displacement-dependent forces arising in the fluid flow.

Tanaka & Takahara (1981) and Tanaka & Ohta (1982) conducted experiments in which unsteady forces were measured. The resulting stability curves show two mechanisms that can explain flow-induced vibrations. One is a velocity mechanism driven by the fluid damping coefficients; it leads to a critical velocity threshold like the one we identify with fluid-elastic whirling (Connors 1970), as shown in the upper part of Figure 5. The other is a stiffness mechanism that is driven by relative-displacement coefficients and leads to a zone of instability that resembles Figure 1, as shown in the lower part of Figure 5. Sweeping through a range of velocities one might see a response as in Figure 6, where the amplitude of the first

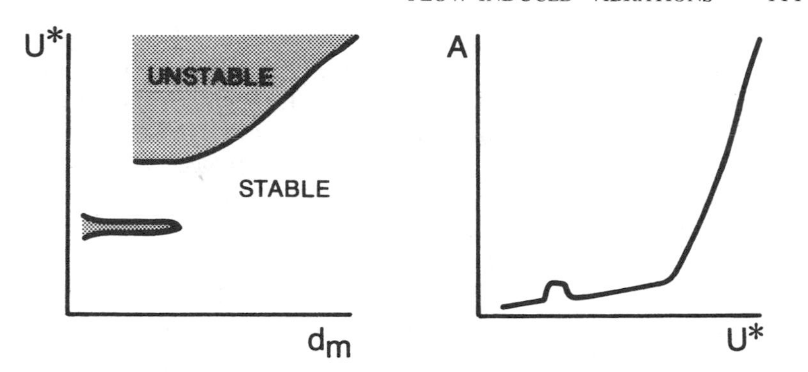

Figure 5 (*Left*) Stability regions for a flexible tube row or array; compare with Figure 1 for a single cylinder. Coordinate scales depend on tube arrangement and P/D value.
Figure 6 (*Right*) Amplitude response of a flexible tube row or array as a function of reduced velocity. Coordinate scales depend on system and flow parameters.

hump, associated with the stiffness mechanism, may be insignificant for gas flows but damagingly large for liquid flows.

The kind of instability that leads to this first hump is indistinguishable from vortex-shedding excitation. We cannot tell whether a wake instability causes the vortex shedding and tube response, or whether another instability is active and leads to displacements which cause the vortex shedding. In other words, aerodynamic theory requires that vortex shedding and changes in lift occur together—it does not tell us which is the cause and which the effect. But from a fluid-dynamics viewpoint, we want to know the source of the instability and the mechanism by which it is amplified. Looking back to the clues we have in the classic von Karman and Coanda phenomena, the answers should be in the stability of boundary layers, flow separations, and shear layers; in the travel of eddies downstream; and in the travel of pressure perturbations upstream.

The other kind of flow-induced vibration, which leads to the sharp increase in amplitude beyond a critical threshold velocity, can be attributed to many different conceptual mechanisms: velocity-mechanism instability, fluid-elastic whirling, phase shifting, or jet switching. Various models all plot the log of normalized critical velocity against the log of the mass-and-damping parameter, but the slope of the curve given by different sources varies from nil to 0.5, as summarized by Pettigrew & Taylor (1991).

New Approaches

In view of the difficulty of seeing the details of the flow structures around and between vibrating cylinders, it appears promising to use numerical simulations to demonstrate essential features of the flow. Insights into vortex formation at a single cylinder have been obtained by Braza et al (1986). Also, the mechanism of Category 1 excitation has been demonstrated by numerical large-eddy simulation (Hassan et al 1990). However, studies of flow through arrays do not necessarily show the known destabilizing forces (Singh et al 1989). For forced oscillations, disagreements between simulation and experiment show up at large amplitudes (Blevins & Chilukuri 1989).

The obvious nonlinearities demonstrated by the jump phenomena and amplitude limits (Figure 1), together with the irregularity of observed system motions, have spawned an interest in applying chaos analysis (Corless 1988).

Other interest focuses on the exceptions to the overall patterns of flow-induced vibrations: the experiments which do show some influence of Reynolds number or of free-stream turbulence, and the occasional observation of stream-wise vibration. It is to be hoped that these unusual cases can give some insight into the underlying physics.

APPLICATION

The state-of-the-art in predicting flow-induced vibrations and designing vibration-resistant heat exchangers is very uneven. On the side of the structural parameters, natural frequencies can be estimated easily and accurately (Nguyen et al 1984), but in realistic configurations the almost equally important structural damping factors vary unpredictably (Lowery & Moretti 1978). On the side of the fluid mechanics, complex computational approaches are being attempted (Kim 1987), but the experimental data on such basic information as the lift coefficients from vortex shedding are wildly inconsistent (Blevins 1990). Putting both sides together, Chen (1987) presents sophisticated analyses of the entire fluid/structural system, but has to depend on limited information on fluid-dynamic coefficients.

Industry handles the limitations by using design procedures with big safety factors. But to improve our predictive abilities, it will be necessary to continue basic research on the underlying phenomena, particularly the near-wake fluid dynamic processes.

Acknowledgments

We are grateful to Jia-Qi Cai at Oklahoma State University; Minter Cheng at Feng Chia University, Taiwan; Qing-De Nie (C. D. Nieh) at Tianjin

University, PRC; and David A. Steininger at the Electric Power Research Institute (EPRI) for discussions which clarified the issues in this review.

Literature Cited

Au-Yang, M. K., Hara, F., eds. 1991. *Flow-Induced Vibration and Wear—1991*, PVP-Vol. 206. New York: ASME. 147 pp.

Bearman, P. W. 1984. Vortex shedding from oscillating bluff bodies. *Annu. Rev. Fluid Mech.* 16: 195–222

Berger, E., Wille, R. 1972. Periodic flow phenomena. *Annu. Rev. Fluid Mech.* 4: 313–40

Blevins, R. D. 1974. Fluid elastic whirling of a tube row. *J. Pressure Vessel Technol.* 98: 263–67

Blevins, R. D. 1984. Review of sound induced by vortex shedding from cylinders. *J. Sound Vib.* 92(4): 455–70

Blevins. R. D. 1985. The effect of sound on vortex shedding from a cylinder. *J. Fluid Mech.* 161: 217–37

Blevins, R. D. 1990. *Flow-Induced Vibration.* New York: Van Nostrand Reinhold. 2nd ed.

Blevins, R. D., Chilukuri, R. 1989. Flow over vibrating cylinders. In *Some Unanswered Questions in Fluid Mechanics*, ASME Paper 89-WA/FE-5, ed. L. M. Trefethen, R. L. Panton, p. 16. New York: ASME

Bloor, M. S. 1964. The transition to turbulence in the wake of a circular cylinder. *J. Fluid Mech.* 19: 290–304

Braza, M., Chassaing, P., Ha Minh, H. 1986. Numerical study and physical analysis of the pressure and velocity fields in the near wake of a circular cylinder. *J. Fluid Mech.* 165: 79–130

Chang, Y. B., Fox, S. J., Lilley, D. G., Moretti, P. M. 1991. Aerodynamics of moving belts, tapes, and webs. In *Machinery Dynamics and Element Vibrations*, DE-Vol. 36, ed. T. C. Huang, N. C. Perkins, K. W. Wang, E. G. Lovell, C. C. Chao et al, pp. 33–40. New York: ASME

Chen, S. S. 1984. Guidelines for the instability flow velocity of tube arrays in cross flow. *J. Sound Vib.* 93: 439–55

Chen, S. S. 1987. *Flow-Induced Vibrations of Cylindrical Structures.* New York: Hemisphere

Chen, S. S. 1988. Some issues concerning fluid-elastic instability of a group of circular cylinders in crossflow. See Paidoussis et al 1988a, pp. 1–24

Chen, S. S. 1991. Flow-induced vibration of an array of cylinders; Part I. *Shock Vib. Dig.* 23(12): 3–9

Cheng, M., Moretti, P. M. 1988. Experimental study of the flow field downstream of a single tube row. *Exp. Thermal Fluid Sci.* 1(1): 69–74

Cheng, M., Moretti, P. M. 1989. Flow instabilities in tube bundles. In *Flow-Induced Vibrations—1989*, PVP-Vol. 154, ed. M. K. Au-Yang, S. S. Chen, S. Kaneko, R. Chilukuri, pp. 11–15. New York: ASME

Cheng, M., Moretti, P. M. 1991. Lock-in phenomena on a single cylinder with forced transverse vibration. See Au-Yang & Hara 1991, pp. 129–33

Chenoweth, J. M., ed. 1977. *Flow-Induced Tube Vibrations in Shell-and-Tube Heat Exchangers*, Final Rep. ERDA Div. Conserv. Res. Technol. Contract No. EY-76-C-03-1273

Connors, H. J. 1970. Fluidelastic vibration of tube arrays excited by cross flow. In *Flow-Induced Vibration in Heat Exchangers*, ed. D. D. Reiff, pp. 42–56. New York: ASME

Corless, R. M. 1988. Chaos in a flow-induced vibration model. See Reischman et al 1988, pp. 77–85

Ffowcs Williams, J. E., Zhao, B. C. 1989. The active control of vortex shedding. *J. Fluids Struc.* 3: 115–22

Gerrard, J. H. 1966. The mechanics of the formation region of vortices behind bluff bodies. *J. Fluid Mech.* 25: 401–13

Hara, F. 1987. Unsteady fluid dynamic forces acting on a single row of cylinders vibrating in a cross flow. In *Flow-Induced Vibrations 1987*, PVP-Vol. 122, ed. M. K. Au-Yang, S. S. Chen, pp. 51–58. New York: ASME

Hassan, Y. A., Bagwell, T. G., Steininger, D. A. 1990. Large eddy simulation of fluctuating forces on a square pitched tube array and comparison with experiment. In *Flow-Induced Vibration 1990*, PVP-Vol. 189, ed. S. S. Chen, K. Fujita, M. K. Au-Yang, pp. 45–49. New York: ASME

Ishigai, S., Nishikawa, E. 1975. Experimental study of the structure of gas flow in tube banks with tube axes normal to flow—Part 2: On the structure of gas flow in single-column, single-row, and double-row tube banks. *Bull. JSME* 18(119): 528–35

Kim, S.-W. 1987. A Critical Evaluation of Various Methods for the Analysis of

Flow-Solid Interaction in a Nest of Thin Cylinders Subjected To Cross Flows. *NASA Contractor Rep. CR-178996; Natl. Tech. Inf. Serv. (NTIS) N87-18781/NSP*

Lever, J. H., Rzentkowsky, G. 1988. An investigation into the post-stable behavior of a tube array in cross-flow. See Paidoussis et al 1988a, pp. 95–110

Lever, J. H., Weaver, D. S. 1986a. On the stability of heat exchanger tube bundles—Part 1: Modified theoretical model. *J. Sound Vib.* 107(3): 375–92

Lever, J. H., Weaver, D. S. 1986b. On the stability of heat exchanger tube bundles—Part 2: Numerical results and comparison with experiments. *J. Sound Vib.* 107(3): 393–410

Lowery, R. L., Moretti, P. M. 1978. Natural frequencies and damping of tubes on multiple supports. In *Heat Transfer Research and Applications*, AIChE Symp. Ser. vol. 74 no. 174, ed. J. C. Chen, pp. 1–5. New York: AIChE

Moretti, P. M. 1974. A critical review of the literature and research on flow-induced vibrations in heat exchangers. In *Heat Transfer Research and Design*, AIChE Symp. Ser. vol. 70 no. 138, ed. D. Gidaspow, pp. 185–89. New York: AIChE

Moretti, P. M. 1986. Caught in a cross flow—the paradox of flow-induced vibrations. *Mech. Eng.* 108(12): 56–61; errata in 109(2): 2

Moretti, P. M., Cheng, M. 1987. Instability of flow through tube rows. *J. Fluids Eng.* 109(2): 197–98

Moretti, P. M., Lowery, R. L. 1976. Hydrodynamic inertia coefficients of a tube surrounded by rigid tubes. *J. Pressure Vessel Technol.* 98J(3): 190–93

Nakayama, Y., Japan Society of Mechanical Engineers, eds. 1988. *Visualized Flow*. Oxford: Pergamon. 137 pp.

Naudascher, E., Rockwell, D., eds. 1980. *Practical Experiences with Flow-Induced Vibrations*. Berlin: Springer-Verlag

Nguyen, D. C., Lester, T., Good, J. K., Lowery, R. L., Moretti, P. M. 1984. Lowest natural frequencies of multiply supported U-tubes. *J. Pressure Vessel Technol.* 106(4): 414–16

Paidoussis, M. P. 1979. The dynamics of clusters of flexible cylinders in axial flow: theory and experiments. *J. Sound Vib.* 65(3): 391–417

Paidoussis, M. P. 1987. Flow-induced instabilities of cylindrical structures. *Appl. Mech. Rev.* 40: 163–75

Paidoussis, M. P., Chen, S. S., Bernstein, M. D., eds. 1988a. *Flow-Induced Vibration and Noise in Cylinder Arrays*, 1988 Int. Symp. on Flow-Induced Vibration and Noise, Vol. 3. New York: ASME. 277 pp.

Paidoussis, M. P., Griffin, O. M., Dalton, C., eds. 1988b. *Flow-Induced Vibration in Cylindrical Structures: Solitary Cylinders and Arrays in Cross-Flow*, 1988 Int. Symp. on Flow-Induced Vibration and Noise, Vol. 1. New York: ASME. 142 pp.

Pettigrew, M. J., Taylor, C. E. 1991. Fluidelastic instability of heat exchanger tube bundles: review and design recommendations. *J. Pressure Vessel Technol.* 113(2): 242–56

Reischman, M. M., Paidoussis, M. P., Hansen, R. J., eds. 1988. *Nonlinear Interaction Effects and Chaotic Motions*, 1988 Int. Symp. on Flow-Induced Vibration and Noise, Vol. 7. New York: ASME. 181 pp.

Roberts, B. W. 1966. Low frequency, aeroelastic vibrations in a cascade of circular cylinders. *Mech. Eng. Sci. Monogr. No.* 4, pp. 1–29

Scruton, C. 1963. On the wind excited oscillations of stacks, towers, and masts. *Proc. Conf. Wind Effects on Buildings and Structures*, Pap. 16, pp. 798–832. England: Natl. Phys. Lab.

Singh, P., Caussignac, Ph., Fortes, A., Joseph, D. D., Lundgren, T. 1989. Stability of periodic arrays of cylinders across the stream by direct simulation. *J. Fluid Mech.* 205: 553–71

Southworth, P. J., Zdravkovich, M. M. 1975. Cross flow induced vibrations of finite tube banks in in-line arrangement. *J. Mech. Eng. Sci.* 17: 190–98

Tadrist, H., Martin, R., Tadrist, L., Seguin, P. 1991. Experimental investigation of the aeroacoustic phenomenon around a circular cylinder. *J. Sound Vib.* 146(2): 223–41

Tanaka, H., Takahara, S. 1981. Flow-induced vibration of tube array in cross flow. *J. Sound Vib.* 77(1): 19–67

Tanaka, H., Ohta, K. 1982. Flow-induced vibration of tube arrays with various pitch-to-diameter ratios. *J. Pressure Vessel Technol.* 104: 168–74

Unal, M. F., Rockwell, D. 1984. The role of shear layer stability in vortex shedding from cylinders. *Phys. Fluids* 27(11): 2598–99

Van Dyke, M., ed. 1982. *An Album of Fluid Motion*. Stanford, Calif.: Parabolic. 176 pp.

Zdravkovich, M. M., Stonebanks, K. L. 1988. Intrinsically non-uniform and metastable flow in and behind tube arrays. See Paidoussis et al 1988b, pp. 61–73

Ziada, S., Oengoren, A., Buhlmann, E. T. 1988. On acoustical resonance in tube arrays: Part I—Experiments. See Paidoussis et al 1988a, pp. 219–44

Annu. Rev. Fluid Mech. 1993. 25: 115–49

AERODYNAMICS OF HORIZONTAL-AXIS WIND TURBINES[1]

A. C. Hansen

Mechanical Engineering Department, University of Utah

C. P. Butterfield

National Renewable Energy Laboratory, Golden, Colorado

KEY WORDS: dynamic inflow, dynamic stall, airfoil, wind energy

INTRODUCTION

Though wind turbines and windmills have been used for centuries, the application of rotor aerodynamics technology to improve reliability and reduce costs of wind-generated energy has only been pursued in earnest for the past 15 years. This paper provides an overview of recent research and development pertaining to the aerodynamics of the horizontal-axis wind turbine rotor. But before discussing aerodynamics in some detail, we provide a brief overview of the size and nature of the wind energy industry and the types of rotors commonly in use today.

After a hasty development, spawned by concerns about energy supplies in the 1970s and the resulting tax credits, three major wind parks in California have recovered from growing pains and are now generating energy at 8 cents per kilowatt hour (kWh) (Rashkin & Nguyen 1988). The California Energy Commission (CEC) reports more than 15,000 turbines generated 2.5 billion kWh in California in 1990, and preliminary estimates indicate 2.7 billion kWh were generated in 1991. This is enough energy to supply all the residential needs of a city the size of San Francisco. Wind turbines in Denmark generated 0.7 billion kWh in 1991, or 2.4% of

[1] The US government has the right to retain a nonexclusive, royalty-free license in and to any copyright covering this paper.

Denmark's total electrical consumption. The majority of these turbines are horizontal-axis wind turbines (HAWTs) with a minority (3%) made up of vertical-axis wind turbines (VAWTs) called Darrieus rotors. More than 60% of the installed capacity operating today was installed after the tax credits expired in 1985.

Most of the HAWTs and VAWTs range in rotor size from 15 m (50 ft) to 30 m (100 ft) in diameter and produce from 80 kW to 500 kW peak power. The largest turbine ever built was developed by NASA with funding from the U.S. Department of Energy (DOE). This 98 m (320 ft) diameter, two-blade turbine uses a teetered rotor to generate a maximum power of five megawatts in 13 meter per second winds. This turbine is currently operated by Hawaiian Electric Renewable Systems (HERS) on the island of Oahu. New turbines will generate energy at 5 cents per kWh by 1995 and 3 cents per kWh by the year 2000 according to the American Wind Energy Association (AWEA 1991).

Many developers in the USA, who are less restricted than Europeans by availability of suitable land, believe 300 kW to 500 kW turbines are the most cost effective to produce and maintain (Holley et al 1987). This conclusion is supported by the development and operation experience gained from the large MOD–2s (Lynette et al 1989).

Rotor Configuration Trends

Some trends in turbine configuration have developed over the past fifteen years, though no single configuration which is clearly superior has emerged. HAWTs have been more widely used than VAWTs. As mentioned above, only about 3% of the turbines installed in California are VAWTs. [This paper will not discuss the aerodynamics of VAWTs, though many aspects of HAWT aerodynamics are pertinent to the VAWT as well. More information on the VAWT can be found in a recent survey paper by Touryan et al (1987).]

HAWT rotors are generally classified according to the rotor orientation (upwind or downwind of the tower), blade articulation (rigid or teetering), and number of blades (generally two or three blades). Downwind turbines were favored initially in the USA, but the trend has been toward greater use of upwind rotors with a current split between upwind (55%) and downwind (45%) configurations.

Downwind orientation allows blades to deflect away from the tower when thrust loading increases. Coning can also be easily introduced to decrease mean blade loads by balancing aerodynamic loads with centrifugal loads. Figure 1 shows typical upwind and downwind configurations along with definitions for blade coning and yaw orientation.

Free yaw, or passive orientation with the wind direction, is also possible

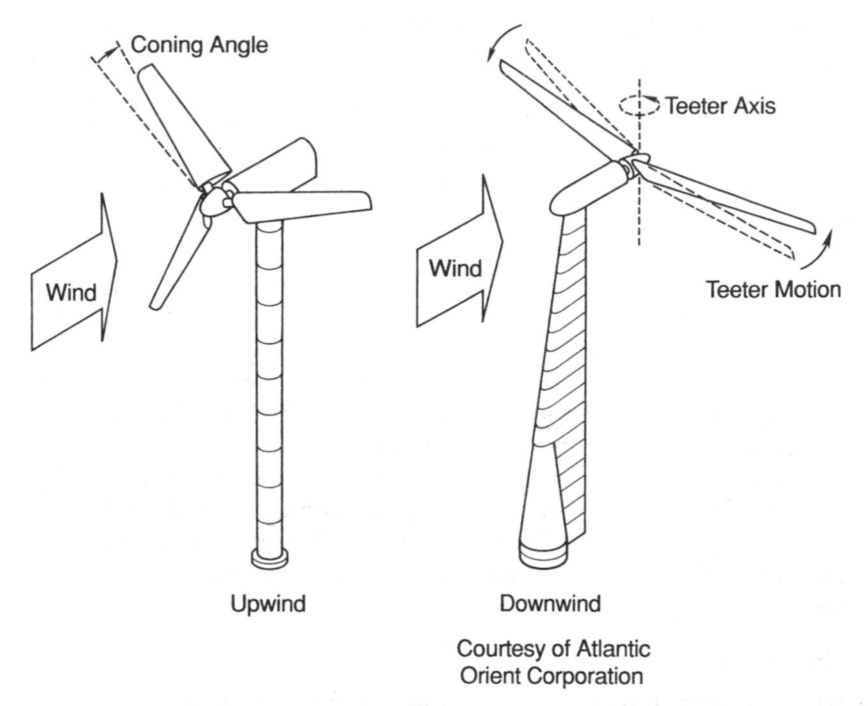

Figure 1 Schematic of the two common rotor configurations. Upwind, rigid hub, three-bladed turbine is common Danish configuration. Downwind, teetered, two-bladed turbine leads to light weight, large HAWT design.

with downwind configurations but yaw must be actively controlled with upwind configurations. Free-yaw systems rely on rotor thrust loads and blade moments to orient the turbine. Net yaw moments for rigid rotors are sensitive to inflow asymmetry caused by turbulence, wind shear, and vertical wind. These are in addition to the moments caused by changes in wind direction which are commonly, though often incorrectly, considered the dominant cause of yaw loads.

Some early downwind turbine designs developed a reputation for generating subaudible noise as the blades passed through the tower shadow (tower wake). Most downwind turbines operating today have greater tower clearances and lower tip speeds which result in negligible infrasound emissions (Kelley & McKenna 1985).

Blade Articulation

Several different rotor blade articulations have been tested. Only two have survived as lasting trends: the three-blade, rigid rotor and the two-blade, teetering rotor. The rigid, three-blade rotor attaches the blade to a hub using a stiff cantilevered joint. The first bending mode natural frequency of such a blade is typically greater than twice the rotor rotation speed (two cycles per revolution of the rotor, denoted as 2p). Cyclic loads on rigid blades are generally higher than on teetering blades of the same diameter (Hock et al 1988). McNerney et al (1990) describe a 33 m, 300 kW turbine currently under development which reflects a mature version of this configuration.

Teetered, two-blade rotors use relatively stiff blades rigidly connected to a hub, but the hub is attached to the main drive shaft through a hinge called a teeter hinge. This type of rotor is commonly used in tail rotors and some main rotors on helicopters. Two-blade rotors usually require teeter hinges or flexible root connections to reduce dynamic loading which results from nonaxisymmetric mass moments of inertia. In normal operation the cyclic loads on the teetering rotor are low, but there is risk of teeter-stop bumping ("mast bumping" in helicopter terminology) that can greatly increase dynamic loads in unusual situations. Lynnette (1991) describes a 26 m (86 ft), 300 kW, teetered turbine currently under development.

Number of Blades

Most two-blade rotors operating today use teetering hinges but all three-blade rotors use rigid root connections. For small turbines (smaller than 50 ft) rigid, three-blade rotors are inexpensive and simple and yield the lowest system cost. As the turbines become larger, blade weight (and hence cost) increases in proportion to the third power of the rotor diameter while power output increases as the square of the diameter (Holley et al 1987). This makes it cost effective to reduce the number of blades to two and add the complexity of a teeter hinge or flexbeams to reduce blade loads. In the mid-scale rotor size (15 to 30 m) it is difficult to determine whether three rigidly mounted blades or two teetered blades is more cost effective. In many cases the choice between two- and three-blade rotors has been driven by designers' lack of experience and the potential risk of high development cost rather than technical and economic merit. Currently only 10% of the installed turbines installed have two blades, yet approximately 60% of all new designs being considered in the U.S. are two-blade, teetered rotors.

Control Strategies

Horizontal-axis turbines use several different types of aerodynamic control to achieve peak power and optimum performance control. Nearly all turbines use an induction or synchronous generator interconnected with the utility grid. These generators maintain a constant rotor speed during normal operation, so aerodynamic control is needed only to limit and optimize power output. More than 60% of the turbines installed use fixed-pitch blades to take advantage of the simplicity and reliability that results from fewer moving parts. Passive power regulation is achieved by allowing the airfoils to stall. As wind speed increases stall progresses outboard along the span of the blade causing decreased lift and increased drag. This balances the increasing available wind power against increasing drag to achieve a nearly constant power output.

One disadvantage of stall-controlled rotors is that they must withstand steadily increasing thrust loads with increasing wind speed because drag loads continue to increase as the blade stalls. Another disadvantage is the difficulty of predicting aerodynamic loads in deep stall.

The remaining turbines use either full span or partial span blade pitch control. Peak power is controlled by adjusting blade pitch angle to progressively lower angles of attack to control increasing wind loading. Pitch control offers the advantage of more positive power control, decreasing thrust loads as blades pitch toward feather in high winds and low parked rotor loads while the turbine is not operating in extreme winds. One disadvantage of pitch control is the lack of peak power control during turbulent wind conditions. Power excursions can exceed twice the rated power levels before the pitch-control system can respond.

The final selection between pitch control and stall control is often driven by the life cycle cost of the pitch-control system and its maintenance compared to the additional cost of the slightly stronger stall-control blade which can withstand the thrust loads.

Additional Aerodynamic Controls

In addition to partial-span and full-span pitch control, several types of aerodynamic brakes have been used for stall-controlled rotors. Pitchable tips and pivoting tip vanes have been used with reasonable success. Jamieson et al (1991) and Snyder et al (1987) describe a variety of other devices which have been tested but not implemented in production turbines.

Variable speed control has been considered as a means of improving the aerodynamic efficiency of the rotor and reducing dynamic loads. This type of control results in the rotor speed changing to maintain a constant ratio between blade tip speed and wind speed ("tip speed ratio"). When an optimum tip speed ratio is chosen the rotor can operate near peak efficiency at all wind speeds up to rated power. Variable speed operation offers a maximum of 20% to 25% additional annual energy capture (Hock & Tu 1986). As the cost of variable speed drives drop, this control option will become more attractive to designers.

The general discussion presented above is intended to inform the reader of the current state of the wind energy industry and trends which have emerged over the past fifteen years. Additional information can be found in a recent book which provides a good introduction to most aspects of wind turbine technology (Freris 1990). The following sections will focus on one of the most challenging areas of wind turbine design: steady and unsteady HAWT aerodynamics. Blade planform optimization and airfoil selection and design will be discussed to highlight unusual requirements of the HAWT airfoil. Rotational effects which become important after stall will be summarized. An overview of rotor performance analyses and airload models for structural dynamics predictions will also be provided. Finally, the current understanding of the existence and importance of unsteady aerodynamic effects and atmospheric turbulence models will be presented.

AIRFOIL CHARACTERISTICS

Modern HAWT blades have been designed using airfoil "families." The blade tip is designed using a thin airfoil, for high lift to drag ratio, and the root region is designed using a thick version of the same airfoil for structural efficiency. Typical Reynolds numbers are in the range between 500,000 and 2 million. A catalog of airfoil data for low Reynolds numbers which was compiled by Miley (1982) has been used by many designers.

Aviation airfoils such as the NACA 44xx and NACA 230xx were popular airfoils because they had high maximum lift coefficients (C_{Lmax}), low pitching moment, and low minimum drag (C_{Dmin}). Early wind turbine designers felt that minor differences in airfoil performance characteristics were far less important than optimizing blade twist and taper. For this reason little attention was paid to the task of airfoil selection. Airfoils which were in common use by the helicopter industry were chosen because the helicopter was viewed as a similar application.

After 1983 turbine designers became aware of airfoils such as the NASA LS(1) MOD (McGhee et al 1979). This airfoil was chosen by U.S. and British designers for its reduced sensitivity to leading edge roughness. Danish wind turbine designers began to use the NACA 63(2)-xx instead of the NACA 44xx for the same reason.

The experience gained from operating the standard aviation class airfoils highlighted the shortcomings of these airfoils for wind turbine applications. Stall-controlled HAWTs commonly produced too much power in high winds which caused generator damage. Viterna & Corrigan (1981) describe poor prediction of peak power on a NASA test turbine. Designers began to realize that a better understanding of airfoil stall performance was critically important. Stall-control turbines were operating with some part of the blade in deep stall for more than 50% of the life of the machine. Peak power and peak blade loads all occurred while the turbine was operating with most of the blade stalled, and predicted loads were only 50% to 70% of measured loads. Deep stall will be discussed in greater detail in a later section. Leading edge roughness also affected rotor performance. When the blades accumulate insects, smog, and dirt along the leading edge, power output can drop to 40% of its clean value, resulting in a significant loss of energy. Several researchers have reported the effects of insect accumulation on rotor performance for two different turbines (Clark & Davis 1991, Eggleston 1991, Musial et al 1990, Yekutieli & Clark 1987).

In general, clean wind turbine blade performance was not high enough in the 8–12 m/s wind speed range (the most frequently occurring wind speeds), and too high for the infrequent wind speeds above 20 m/s. Excess power levels caused by high wind storms would cause generator damage or electrical breaker activation. When the storms passed, the turbines would not be available for optimum energy producing winds. Blade roughness caused large energy losses and was a maintenance liability. Designers realized that airfoil selection criteria had to change to achieve reliable performance.

Airfoil Selection Criteria

Fueled by the disappointing stall-control turbine performance, researchers changed the blade optimization approach. Prior to 1985 blades were optimized primarily by adjusting blade twist and taper to achieve near optimum wake induced velocity distributions along the blade. This optimization approach is described by Glauert (1935) and later modified by Wilson & Lissaman (1974). This approach, however, did not address the practical problem of limiting peak power through airfoil stall. Nor did it include the possibility of variable airfoil performance characteristics to tailor rotor performance characteristics.

Stall-controlled blades needed airfoils with restrained maximum lift coefficients in their tip region. This would reduce the peak power generated for a specified rotor diameter. Consequently the rotor diameter could be increased to improve performance at frequently occurring medium wind speeds without increasing the power and drive train capacity. The rotor swept area and annual energy production can be increased by 20–30% as a result of this change in design philosophy.

In 1985 Tangler's and Somer's (Tangler 1987) new airfoil families allowed Jackson (Jackson & Migliore 1987) to add variable airfoil performance characteristics to the blade geometry optimization matrix. Jackson used airfoils with low C_{Lmax} in the tip region and reduced the tip chord to control rotor peak power. He used higher C_{Lmax} airfoils and increased chord distributions in the blade mid-span to increase rotor performance in the medium wind speed range. Finally twist distribution was used to trade off improved low wind performance for constant power stall characteristics in high winds. Garrad, Morgan, and Hill (Hill & Garrad 1989, Morgan & Garrad 1988) also quantified the benefits of decreasing C_{Lmax} from the root region to the tip region.

Feathering Rotors

Specifications for pitch-controlled rotors are different. Airfoils for pitch-controlled rotors should have high maximum lift coefficients so stall is avoided at high lift. This also results in lower chord length and lower blade cost. The pitch-control system can be used to adjust the angle of attack to maximize lift to drag ratio for all wind speeds up to maximum power. Thus airfoil performance can be optimized for a small region of the lift curve. Roughness insensitivity is important for pitch-controlled rotors since soiling can degrade the maximum lift coefficient below the design point of the airfoil. Glauert's optimization approach for achieving optimum twist and taper can be used in pitch-controlled blade designs with fewer compromises than the stall-controlled blade since tailoring of the power curve to achieve flat stall characteristics is not necessary.

Stall characteristics for the pitch-controlled rotor have been considered unimportant but operating experience has shown stalling can reduce power excursions common on pitch-control rotors operating in turbulent winds.

New Airfoil Designs

New airfoil design codes became available in the mid 1980s which suddenly opened the door to airfoils designed specifically for HAWTs. A blade design could now have an optimum airfoil designed for every spanwise location. Eppler (1990) developed the code that was first used for this purpose by wind turbine engineers.

The "Eppler Code" is a combination of three theories: a conformal-mapping technique, a panel technique, and a boundary layer analysis. These are combined to optimize boundary layer characteristics and airfoil shapes that achieve specified performance criteria.

Other codes have been used for wind turbine airfoil designs. Lock & Williams (1987) developed a viscous-inviscid interaction analysis which Hill (Hill & Garrad 1989) believes is more accurate than the Eppler code when predicting airfoil performance with separated flows. Drela (1991) also describes a computational fluid dynamics analysis. However the Eppler code is currently the popular code since its intent is to determine the optimum shape of an airfoil designed to achieve a specified behavior rather than determine the behavior of a specified airfoil geometry. Recently

Dini (Dini et al 1991) has improved the accuracy of the Eppler code by including a laminar separation bubble model which helps the code accurately predict minimum drag.

Using the "Eppler Code" Somers (Tangler 1987) developed three "special purpose families" of airfoils for three different classes of turbines. These airfoils have been tested on 7.5 m long blades and shown to increase annual energy production by 15–25% without increasing peak power.

Static Stall

As mentioned in previous sections, designers have not been able to predict high wind speed performance accurately on stall-controlled turbines. Viterna (Viterna & Corrigan 1981) was the first to suggest that airfoil stall characteristics were causing the errors. Rotor performance codes used wind tunnel data to describe the post-stall airfoil lift and drag properties. Viterna speculated that these properties were modified by spanwise flow and pressure gradients. He developed an empirical correction, based on blade aspect ratio, which improved the predictions of peak power. This correction remains the most widely used stall performance model in the US, even though it has little foundation in the basic physical mechanisms of stall. More recently a number of researchers have conducted tests that show how rotating blade airfoil performance differs from wind tunnel measurements on a nonrotating blade (Butterfield 1989b, Butterfield et al 1990, Hales 1991, Madsen 1990, Ronsten et al 1989, Wood 1991). All these tests have confirmed Viterna's speculation of rotating blade stall characteristics.

Figure 2 shows normal force coefficient measurements for several spanwise locations compared to the same airfoil's performance in the wind tunnel. The performance compares surprisingly well at pre-stall angles of attack. However the airfoil on the rotating blade does not experience the same drop in lift during stall that is seen in the wind tunnel data. The curves also show that C_{Lmax} increases by approximately 10% at the 47% and 63% spanwise stations. The inboard station at 30% span shows the normal force continuously increasing without any signs of stall drop-off. Normally one would expect C_{Lmax} to drop with Reynolds number at inboard stations as described by Jacobs (Jacobs & Sherman 1937). The test data show the opposite trend.

The differences observed in the normal and chordwise forces would not seem to explain the large discrepancies in measured and predicted peak power on operating turbines. But if these normal and tangent components are resolved into local rotor thrust and torque coefficients the discrepancies between the rotating blade and wind tunnel data are more dramatic. Figure 3 shows such curves for the 63% station again. The rotating airfoil torque

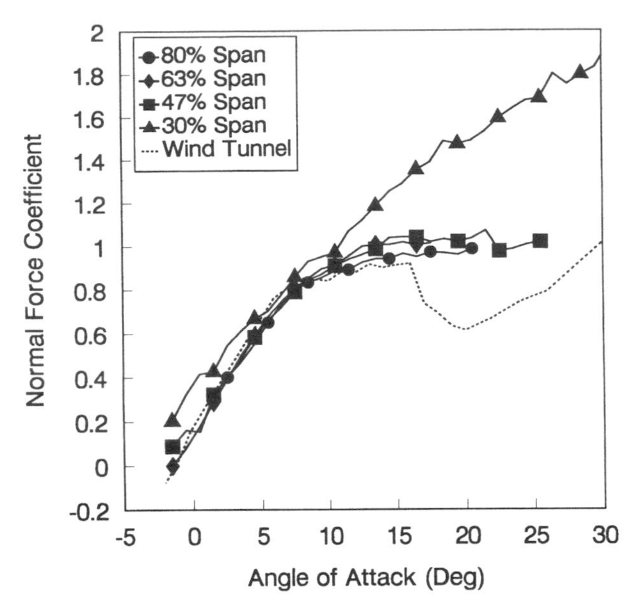

Figure 2 Normal force coefficients measured on a rotating blade and in a wind tunnel. This figure shows spanwise variation in performance for an S809 airfoil on a constant 1/2 m chord, zero twist, 5 m blade, rotating at 72 RPM (Combined Experiment Rotor—CER).

coefficient is 3 times higher than the wind tunnel data in the 16–20° angle of attack range. This discrepancy does explain the performance prediction errors which wind turbine designers have commonly experienced.

Pressure distributions were compared to gain insight into the cause of this discrepancy. Figure 4 shows a comparison of rotating blade (RB) pressure distribution data with wind tunnel (WT) pressure distribution data for high angle of attack (AOA). From these curves it is clear that the RB airfoil has a suction peak near the leading edge while the WT airfoil does not. The suction peak implies that flow remains attached at the leading edge while the WT airfoil experiences leading edge flow separation. The leading edge suction peaks cause the observed increase in normal and tangent forces. This may be related to spanwise flow in the laminar separation bubble which is located at the leading edge during stall. Liquid crystal surface stress measurements have positively identified laminar separation bubbles on the rotating blade. These are common for low Reynolds numbers. Bubble bursting affects the loss of leading edge suction peak and hence causes the precipitous drop in lift at stall. If spanwise flow in this bubble stabilizes the bubble then it may explain why the rotating blade airfoils do not stall with a precipitous drop-off.

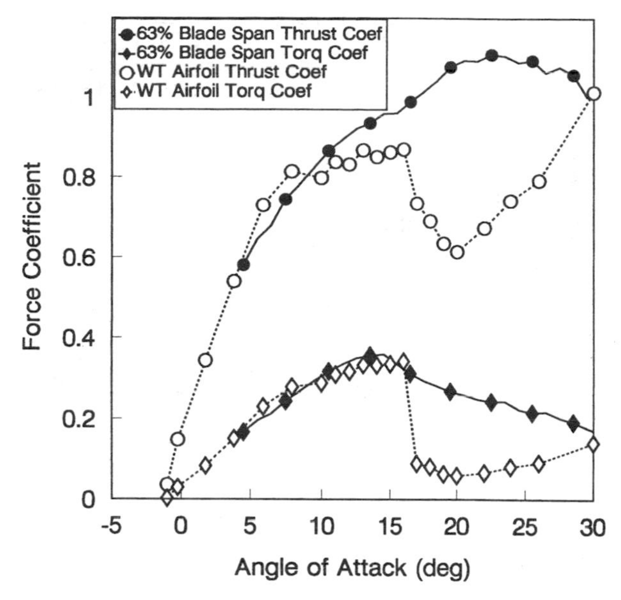

Figure 3 Thrust and torque coefficients measured on a rotating blade and in a wind tunnel. This figure shows rotating blade effects on blade element thrust and torque coefficients for a 10 m rotor diameter, three-blade turbine (CER).

It has long been known that rotation of a blade delays the occurrence of static stall. Himmelskamp first noted the effect on fan blades (Schlichting 1978) and a number of investigators examined the importance of delayed static stall to helicopters in forward flight (e.g. McCroskey & Yaggy 1968, Velkoff et al 1971, Young & Williams 1972). In these investigations an outward radial (spanwise) flow generated by centrifugal pumping was postulated or predicted. The spanwise flow results in a Coriolis acceleration which acts as a favorable chordwise pressure gradient. This pressure gradient delays the boundary layer separation and increases the maximum lift coefficient.

More recently the problem has been studied as it pertains to wind turbines. It is very important to the stall-controlled wind turbine designer because power regulation depends upon stall and detailed understanding of stall behavior is required for accurate prediction of peak rotor power. Sorensen included centrifugal and Coriolis terms in a viscous-inviscid matching solution and predicted a 30% increase in the maximum lift due to the spanwise flow and additional Coriolis "pressure gradient" (Sorensen 1986). Snel also attributed the delayed stall to the spanwise flow in the separated flow zone of the blade (Snel 1991).

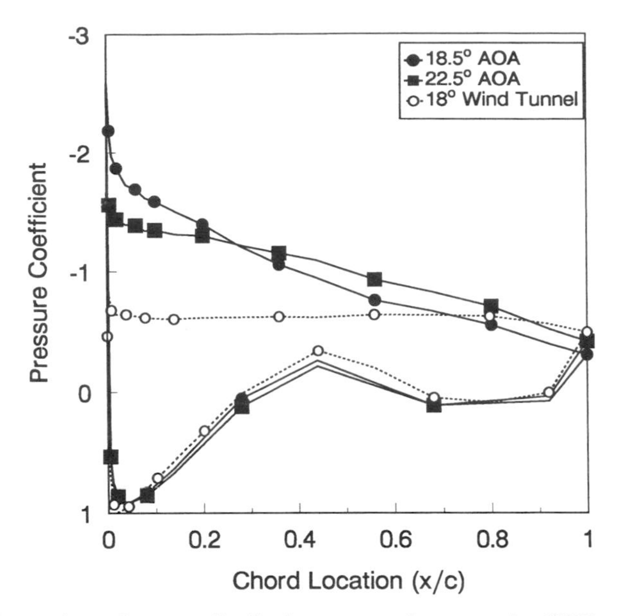

Figure 4 Comparison of pressure distributions measured on a rotating (CER at 63% span) and nonrotating blade. Every other pressure tap location is shown by a center symbol. Wind tunnel data show zero gradient on suction side of airfoil at stall compared to delayed stall and gradient which persists to high angles of attack on the rotating blade.

In a recent paper Eggers & Digumarthi (1992) present a method for scaling the rotational effects from 2-D wind tunnel data. This method offers a simple means of making the necessary corrections in a BEM analysis and thus retains the computational efficiency needed for engineering design. The authors develop power-series solutions to the Navier-Stokes equations. The solutions show that spanwise flow in the separation bubble creates a constant chordwise pressure gradient (via the Coriolis acceleration) on the low pressure side of the airfoil. This is in contrast to the nearly constant pressure in the separation zone of a nonrotating airfoil. The high pressure side (and hence the trailing edge pressure) is not affected by rotation. This assumption is verified by examination of the Combined Experiment data (see Figure 2). It is then possible to estimate the area of the triangular pressure distribution on the leeward surface and the resulting increment in normal and tangential aerodynamic coefficients. The method is applied to the Combined Experiment rotor with the result that power and blade root bending moment are predicted within the scatter of the test data over the entire range of wind speeds tested. If this method continues to be confirmed by additional validation studies, it will offer a simple and reliable means of accounting for rotational effects.

Wood (1991) has presented an alternative explanation for the delay of static stall. He uses a panel method to calculate pressure distributions on a rotating blade and shows that trailing vorticity near the root affects the leading edge suction. He predicts a decrease in the leading edge suction and therefore a decreased adverse pressure gradient and delayed separation. He further states that the aforementioned effects of centrifugal pumping and Coriolis acceleration in the boundary layer cannot be sufficient to give the required change in pressure distribution. However, he does not seem to consider the importance of the spanwise flow inside the separation zone which Eggers and Snel find of first-order importance (Eggers & Digumarthi 1992, Snel 1991). Wood concludes that local solidity is the dominant parameter and verifies his results using data from a ducted rotor with a solidity of 21% and the blade root at $r/R = 0.32$. He shows that for a 50% reduction in solidity the effect is considerably less important. Typical rotor solidities for wind turbines are 5–10%, while the root cut-out is generally much less than 30% of the span. So it remains to be determined if the effects Wood describes will be of significance to wind turbine analysis.

ROTOR PERFORMANCE ANALYSIS

Rotor performance analysis is concerned with the estimation of mean power output of the rotor and aerodynamic loads acting upon the blades and rotor. Estimation of dynamic structural loads will be discussed in the next section. Performance analysis is one of the first and most critical steps in designing a rotor and has, as a result, received a great deal of attention from a number of researchers. All wind turbine designers known to the authors employ a blade element/momentum (BEM) method for performance analysis. Two factors contribute to the popularity of BEM methods. First, they have been proven accurate for a wide variety of rotors and flow conditions. Second, they are simple to learn and use and are readily implemented on virtually any desktop computer.

A number of researchers are investigating use of various vortex trajectory calculations for performance analysis. These methods offer some advantages over BEM techniques but also have some serious drawbacks. They are not commonly used by designers today, but the potential advantages of the methods promise increased use as computer power increases and experience with the models shows which are the most accurate.

Simple actuator disk theory has also been applied to wind turbines to provide insight to the basic mechanisms and limitations of rotors. Though this method cannot be used to determine optimal blade planform or airfoil characteristics, it does lead to the conclusion that the theoretical maximum rotor efficiency (power coefficient) is 59.3% in steady-state conditions (Wilson & Lissaman 1974). The induced velocity equals 1/3 the approaching wind speed when this optimal efficiency is obtained.

The sections that follow discuss BEM and vortex wake methods with emphasis on the advantages and shortcomings of each technique.

Blade Element/Momentum Methods

Blade element/momentum calculations are performed today much the same way they were first introduced to wind turbine designers in 1974 (Wilson & Lissaman 1974). The basic method assumes the blade can be analyzed as a number of independent elements (or strips). The induced velocity at each element is found by performing a momentum balance for an annular control volume containing the blade element in question and the air bounded by the stream surfaces extending upwind and downwind of the element. The aerodynamic forces on the element are calculated using two-dimensional lift and drag coefficients at the geometric angle of attack of the blade element relative to the local flow velocity (including induced velocity effects).

In the full implementations of the method, iteration is required to determine the induced velocity consistent with the aerodynamic forces. The calculations can be performed much more quickly by assuming that all angles are small and the lift coefficient is proportional to the angle of attack. This permits derivation of an explicit relationship between the induced velocity and global rotor parameters, thus avoiding the need for time-consuming iteration. However, many situations of interest to the designer (such as performance in deep stall) cannot be calculated accurately with the linearized equations (Van Grol et al 1991a,b). Furthermore, computer capabilities have improved to the point that iteration poses little extra difficulty to the performance analyst.

The basic BEM technique has a number of limitations which are frequently encountered in wind turbine applications. Many of these limitations have been overcome using empirical relations derived from helicopter, propeller, and wind turbine experience. Glauert (1948) determined that the simple momentum balance is not valid for high induced velocities (exceeding approximately 40% of the free-stream speed). Thus, most BEM methods include Glauert's empirical relation for induced velocities for high disk loading conditions.

All wind turbines operate with some portion of the blade stalled for a portion of the time. As mentioned earlier, many designs depend upon aerodynamic stall to limit rotor power in high winds. This means that inboard blade sections routinely encounter very high angles of attack and some rotors have outboard stations that occasionally operate in deep stall.

Section lift and drag data at high angles of attack are not available for many airfoils of interest to wind turbine analysts. Viterna & Corrigan (1981) developed a semi-empirical relation to extrapolate section lift and drag data to angles of attack up to 90°. The method extrapolates from known airfoil characteristics below stall, to flat-plate characteristics (at the appropriate aspect ratio) at high angles. This technique is now widely used and has resulted in improved correlation between measured and predicted performance.

Blade tip and hub losses are generally accounted for in BEM methods using the Prandtl relationship (Hibbs & Radkey 1981). This method is simple and fast and also results in improved accuracy in the predictions. Many approaches have been used to implement the Prandtl tip loss model, but the differences appear to have little effect on the rotor performance predictions (Van Grol et al 1991a,b).

In spite of the assumptions and oversimplifications made in BEM theory, the method often predicts rotor performance with acceptable accuracy. Numerous authors have investigated the method in comparisons with tests in the natural wind and wind tunnels (Tangler 1982; Van Grol et al 1991a,b). Van Grol et al explore the differences in a number of models currently used in Europe in the most comprehensive study to date of BEM application to wind turbines. In conditions below deep stall and at low yaw error the method is accurate provided airfoil characteristics are known for the appropriate Reynolds number and surface roughness. In the cases examined by Van Grol, the power and annual energy yield were predicted with an uncertainty of $\pm 8\%$. The greatest difficulty in obtaining accurate predictions below stall is determination of the appropriate airfoil section characteristics.

In high winds, when much of a fixed-pitch blade is stalled, the method generally predicts power output considerably lower than the measured value. Figure 5 shows a comparison of predicted and measured rotor power for a stalling rotor (Musial et al 1990). This failure to predict the maximum rotor power output has caused many difficulties and is the subject of ongoing research. One known cause of the increase in maximum power is delayed static stall at the inboard sections (discussed in the previous section). Rasmussen (Rasmussen et al 1988) used measured power and blade flap moment data to reconstruct the airfoil polars for the blade under test. The resulting airfoil characteristics were quite different from 2-D data and qualitatively similar to the rotating-blade polars presented earlier. When a systematic method is proven for determining the rotating, 3-D characteristics of blade sections (hopefully from 2-D data) then the BEM method should be capable of accurately predicting peak power output of a rotor as well as power in lower winds.

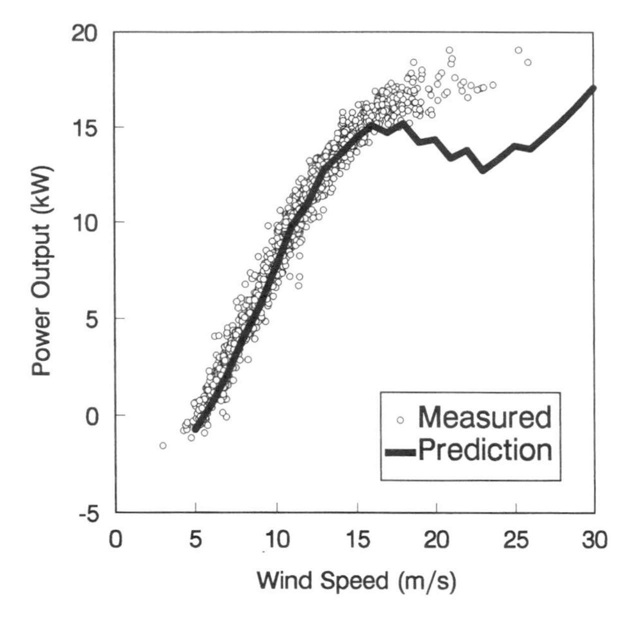

Figure 5 Comparison of predicted and measured rotor power (CER). Predictions agree with measurements at low and moderate wind speeds. Measurements exceed predictions at high wind speeds due to delayed stall.

Vortex Wake Methods

The blade element/momentum method offers the advantage of ease of understanding and use as well as minimal computation requirements. However, there are a number of situations where it is not reasonable to expect BEM methods to offer the greatest accuracy. Situations of interest to the designer include yawed flow and unsteady aerodynamics (particularly dynamic inflow effects). Though methods are available for incorporating these complex effects into a BEM analysis, it is desirable to seek analytical methods which are fundamentally better suited to complex flows. One obvious alternative is detailed calculation of the induced velocity field by determining the distribution of vorticity in the wake. Vortex wake methods are widely used for performance and loads analysis by helicopter designers. Thus, it is reasonable to examine their potential for wind turbine rotors.

The greatest obstacle to widespread use of vortex wake methods is the computation burden. No programs are currently available that calculate the details of an unsteady, three-dimensional, free wake in a reasonable time, regardless of computer platform. Thus the problem is one of finding a balance between model simplification (and limitation) and computation time. To date, HAWT designers have found no method which they prefer over BEM techniques. For this reason, the discussion of vortex wake methods will be brief and is included because of the future potential rather than current use.

Prescribed wake techniques have been applied using both lifting line and lifting surface formulations. Kocurek adapted a helicopter hovering lifting surface method for use with wind turbine rotors (Kocurek 1987). He concludes the method is most useful for studying the sensitivity of performance predictions to the variety of model parameters. He also notes the importance of the stall model—which has already been pointed out as a major uncertainty in any performance or loads analysis. The method has not been widely used or reported.

A progress report on the development of nonlinear lifting line and lifting surface models is given by Gould & Fiddes (1991). They conclude that for one example rotor the power output and thrust loading were insensitive to the shape of the prescribed wake. A wake expansion of 35% produced a change in thrust of 1% and a change of 4% in power prediction when compared with a cylindrical wake. No comment is made on the sensitivity of the spanwise load distribution to the wake shape. This work is being extended to examine yawed flow and three-dimensional boundary layer effects. These are the areas where the method might be expected to be superior to the less expensive BEM techniques in use today.

Free-wake methods avoid the difficulty of prescribing a wake geometry but introduce more computation expense. Early work in the U.S. produced a free wake model to analyze the NASA Mod–0A turbine (Afjeh & Keith 1986). More recently a free wake method was compared with a prescribed wake and the BEM method and found to produce little difference in the predicted power output of the rotor (Simoes & Graham 1991).

An alternative to the vortex wake methods is use of an asymptotic acceleration potential. Some work has been conducted at the Delft University of Technology and reported by van Bussel (1991). The method offers the advantage of reduced computing cost but is unproven in all but the simplest of flows.

DYNAMIC LOADS ANALYSIS

Estimation of dynamic loads is the most important and most difficult task facing the user of wind turbine aerodynamics models. The dynamic structural loads which a rotor will experience play the major role in determining the lifetime of the rotor. Obviously, aerodynamic loads are a major source of loading and must be well understood before the structural response can be accurately determined. The stochastic nature of the wind, flexibility of the wind turbine structure, a need to minimize the weight of the structure, and the unsteady, three-dimensional character of the flow all combine to make accurate prediction of all loading scenarios impossible. It is not surprising that a number of methods have been devised for estimating loads and each of the methods has its strengths and weaknesses. The models differ primarily in the degree of simplification in the aerodynamics analysis and in the number and types of structural degrees of freedom which are considered. They can also be categorized into time-domain and frequency-domain models. No universally accepted method has been identified, nor is it likely this will happen for several years.

There is no doubt that this area is receiving the most attention from the research community and turbine designers. It is probably the analytical discipline with the greatest potential for turbine reliability improvement and cost reduction.

Simplified Techniques

The simplest techniques for aeroelastic analysis of a rotor are often useful for gaining insight to dominant mechanisms and rotor parameters (Eggleston & Stoddard 1987). Unfortunately, they are seldom accurate enough for estimating design loads. The methods lead to analytic solutions of blade equations of motion and thereby lend themselves readily to preliminary design tradeoffs and use in education. Steady, linear aerodynamics assumptions (linear lift curve, small angle approximations, zero drag, and linear blade twist and taper) are employed to make the aerodynamic analysis tractable. Generally only one or two structural degrees of freedom (DOFs) are considered. Blade flap (or teeter) motion (with a rigid blade and equivalent root hinge/spring) and yaw are two commonly used DOFs. Coupled modes of vibration and tower motions are generally neglected.

FLAP Code for Loads Predictions

The FLAP code was developed at Oregon State University and is now maintained, distributed, and validated by the National Renewable Energy Laboratory (NREL)(Wright et al 1987, 1991). FLAP is the public-domain code most often used by designers in the U.S. It is also similar to a number of codes employed in Europe. FLAP is used to estimate blade load distributions on a constant-speed rotor in steady or turbulent winds. It uses a linearized BEM aerodynamic model with a frozen wake. It can include up to four flap modes of vibration (including teeter motion, if desired). No other degrees of freedom are modeled, though yaw angle and yaw rate can be specified inputs. FLAP operates in the time domain and can use wind inputs from a turbulence model, a data file, or constant values.

As is seen in Figure 6, the method is capable of predicting the blade flap moments with reasonable accuracy (Wright & Butterfield 1992). The spectral content is well represented both qualitatively and quantitatively over most frequencies. However, discrepancies can be seen. For example at 0.6 Hz the predicted power spectral density (PSD) is 1.3 times the measured value and the peak at 6 Hz (the first edgewise frequency) is not predicted within two orders of magnitude. The effect of these differences upon the fatigue life of the turbine has not been determined. Also, it is known that some turbines have other structural degrees of freedom which cannot be ignored. Tower motion is important in some instances, and free-yaw motion cannot be predicted by FLAP.

Furthermore, there are situations where FLAP is known to be inaccurate. These include high wind, where much of a blade is stalled; yawed flow, where skewed wake effects are important; and highly unsteady conditions (such as rapid pitch changes, yawed flow with resulting dynamic stall on the blades, or variable rotor speed operation).

For these reasons NREL is now developing a software system which will be capable of analyzing an arbitrary number of structural modes of

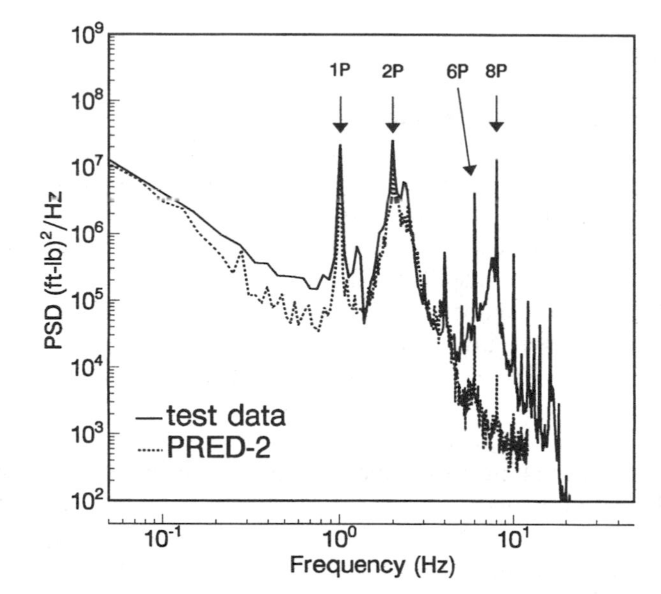

Figure 6 Power spectral densities of measured and predicted blade root flap moments for a two-blade, teetering, 24 m diameter rotor operating in 16 m/s wind.

vibration. The method will be implemented using the commercially available ADAMS® code for the dynamics analysis. Aerodynamic forcing functions will be implemented in subroutines written specifically for wind turbine analysis. The subroutines will use the BEM method with modifications to account for effects of unsteady aerodynamics and delayed static stall. This new software will offer the analyst a wider choice of degrees of freedom, variable speed rotor simulation (including start-up and shut-down), and improved unsteady aerodynamics capability. Of course, the price for this capability will be increased computation time and cost.

Other Codes

A number of other methods have been developed for estimation of dynamic loads on a rotor (for example, see Wilson et al 1991, Snel & Lindenburg 1990, Bierbooms 1990a, Kretz & Rasmussen 1990). The codes differ in the number and types of structural modes of vibration which are included as degrees of freedom. Accurate analysis of some rotors requires consideration of tower motions or coupling of the flap motions of the blades. Other turbines are sufficiently stiff so that only simple blade flap motion need be considered.

Some of these loads analyses operate in the frequency domain while others operate in the time domain. Of course, the frequency-domain models require linear systems and therefore employ the small angle assumptions for angle of attack and the approximate, linear relationship between the induced velocity and the rotor parameters. These models offer advantages of rapid execution, direct input of the wind fluctuation spectrum (which is often readily available), and direct output of loads spectra for use in fatigue life estimation. Their disadvantages are that they are unable to model effects of: static or dynamic stall, large amplitude motions—which are often the worst-case design loads, worst-case gust loads (specified in the time domain), and rotor start-up or shut-down.

The time-domain models avoid these difficulties with the frequency-domain methods but have other limitations. To generate a spectrum of fatigue loads the designer must run a large number of cases or long simulations with wind time series synthesized from measurements or turbulence models. The time-domain calculations are more computationally intensive, requiring longer run times or more expensive computer systems. It is likely that both methods will continue to be used in the design process. Frequency-domain methods can be used to generate high-cycle fatigue spectra while time-domain models can be used to explore transients, worst-case loads, effects of stall, and other nonlinearities.

All of the engineering design codes depend upon some form of the BEM method for calculating aerodynamic loads on the blades, but there are a number of differences in the detailed implementation. This is particularly true regarding the assumptions employed when analyzing unsteady conditions. Many codes do not include dynamic stall effects, while others include dynamic stall to improve predictions in high winds and yawed flow. The implementation of dynamic stall in the models is discussed in a later section.

Many of the analyses assume the wake is "frozen" and dependent only upon the mean inflow conditions. Others assume the wake is in quasi-steady equilibrium with all changes in blade pitch or motion and inflow velocities. In fact, the truth lies between these two extremes and the importance of the dynamics of the inflow is still a subject of investigation. Bierbooms (1990a,b) assumes the wake behaves as a first order system with a time constant which is twice the rotor diameter divided by the free stream wind speed. Dynamic inflow theory of Pitt & Peters (1981) has also been applied to the calculation of inertial lag in the response of a wake to sudden changes in wind speed or pitch angle (Swift 1981). Dynamic inflow theory is discussed further in a later section.

Yaw Dynamics and Aerodynamics

Whether under active or passive (free) yaw control, all HAWTs operate some fraction of the time with the wind direction misaligned with the axis of rotation of the rotor (yaw error). It is common for a rotor to encounter yaw angles of $\pm 30^\circ$ for durations of several minutes. Active yaw-controlled rotors typically respond to a time-averaged wind direction with averaging times of several minutes. Instantaneous yaw misalignment of $\pm 30^\circ$ is frequently observed. Some free-yaw rotors—which nominally operate with the rotor downwind of the tower—have been observed operating upwind of the tower for extended periods while others persistently operate at a small yaw error (5–10°). Others experience high yaw rates which result in fatigue failures induced by gyroscopic moments. Historically, yaw related problems have been the second-leading cause of turbine downtime in California wind parks (Lynette 1988).

All of the performance and loads calculation methods discussed above analyze the yawed rotor by calculating the relative velocity components of the blade element from the total wind vector and applying BEM theory. The component of ambient wind that lies in the plane of rotation is not altered by the presence of the rotor. Effects of spanwise flow and skewing of the rotor wake are neglected. However, a number of researchers have shown that BEM theory is not suitable for accurate estimation of the wake or the yaw moments on the rotor (de Vries 1985, Hansen, Butterfield et al 1990, Hansen & Cui 1989, Swift 1981). The skewing of the wake alters the distribution of induced velocity across the disk and invalidates the basic assumption of BEM theory that each blade element operates independent of all others.

Fortunately methods are available for modifying BEM theory and achieving acceptable accuracy with yawed flow predictions. Swift was the first to suggest use of the Pitt & Peters dynamic inflow theory for wind turbine analysis (Gaonkar & Peters 1986, Pitt & Peters 1981, Swift 1981). Subsequent comparisons of predictions with wind tunnel and field test data by Hansen and coworkers showed that a quasi-steady implementation of this method yields substantial improvements in the prediction of yaw moments (Hansen 1992).

Accurate prediction of yaw and pitch moments of a rigid-hub rotor has also been shown to be contingent upon modeling of dynamic stall of the airfoil (Hansen 1992). Yawed flow, tower shadow, or wind shear cause cyclic variations in angle of attack with a frequency of 1p. The resulting stall hysteresis creates an asymmetry in the blade loading with respect to a vertical axis through the hub. This asymmetry has a large effect on the yaw moments, with increases of 50% common when dynamic stall effects are included in the calculations. Yaw moments are particularly sensitive to dynamic stall because they result from small differences in large flap moments at the roots of each of the blades. Dynamic stall on wind turbine blades will be discussed in greater detail in a later section of this paper.

Yaw dynamics research has also pointed out the need for consideration of all components of turbulence approaching the rotor disk. The authors believe that full understanding of all the net rotor loads will not be possible until vertical and lateral turbulent fluctuations as well as the longitudinal fluctuations are included in turbulence models for wind turbines.

Summary of Dynamic Loads Estimation

In summary, many methods have been developed to estimate dynamic structural loads resulting from aerodynamic forces on the wind turbine rotor. There are many situations where the dynamic models perform well and yield credible information for the designer. The requirements for accurate modeling are: (*a*) the appropriate degrees of freedom must be considered (this varies from one turbine to the next), (*b*) the static properties of the rotating airfoil sections must be known at the appropriate Reynolds number and surface roughness, (*c*) the wind inputs must be specified in detail (with temporal and spatial resolution) and, (*d*) unsteady aerodynamic effects must be negligible (or accurately represented). None of these requirements can be taken for granted, particularly when designing a new turbine. Hence research is continuing to develop less restrictive models and to generate wind and airfoil databases which are more suitable for wind turbine applications.

UNSTEADY AERODYNAMICS

Wind turbines operate at all times in an unsteady aerodynamic environment. The blade element forces vary in time and space as a result of ambient turbulence, persistent shear in the ambient wind, blade vibratory motions, control inputs, and skewed flow. The analysis of HAWT blade loads is subdivided into two major areas: dynamic stall and dynamic inflow. Dynamic stall refers to the unsteady aerodynamic effects in the immediate vicinity of the blade. Airfoil characteristics have been shown to be quite different from steady-state wind tunnel test results when the angle of attack varies rapidly. McCroskey (1981) presents an excellent review of dynamic stall. Dynamic inflow refers to lagging in the response of the induced velocity field of a rotor following rapid changes in the rotor operating state. The mass of the air in the wake of the rotor makes it impossible for the wake to respond instantaneously to a change in rotor loading such as might be observed after a change in blade pitch angle.

Dynamic Stall Experiments

Prior to 1988 dynamic stall and unsteady aerodynamic effects were not included in HAWT performance and load analyses. Hibbs (1985) analytically examined the effect of dynamic stall on HAWT performance and concluded that dynamic stall could be ignored when predicting performance. He speculated that dynamic loads could be affected by dynamic stall but did not quantify the potential effects.

In 1988 Butterfield (1989a) was able to quantify both the existence of dynamic stall and its effect on rotor loads by measuring pressure distributions on a 10 m HAWT. Dynamic stall was shown to occur under a variety of inflow conditions, including turbulence, tower shadow, and yawed flow. Figure 7 shows typical rotating blade dynamic stall measurements during yawed operation, compared to static, wind-tunnel test data. The blade azimuth positions are marked on the hysteresis curve to illustrate the difference between normal forces on opposite sides of the rotor (90° and 270°). As mentioned earlier, the existence of hysteresis and its phasing relative to azimuth position significantly increases yaw loads. Figure 8 shows the tangential force coefficients for the same test conditions.

Dynamic stall formation can be detected through close examination of pressure distribution time sequences. Figure 9 shows one such pressure-time sequence. Suction surface pressures are plotted for one rotor revolution for 30° yawed operation. The vortex development is implied by

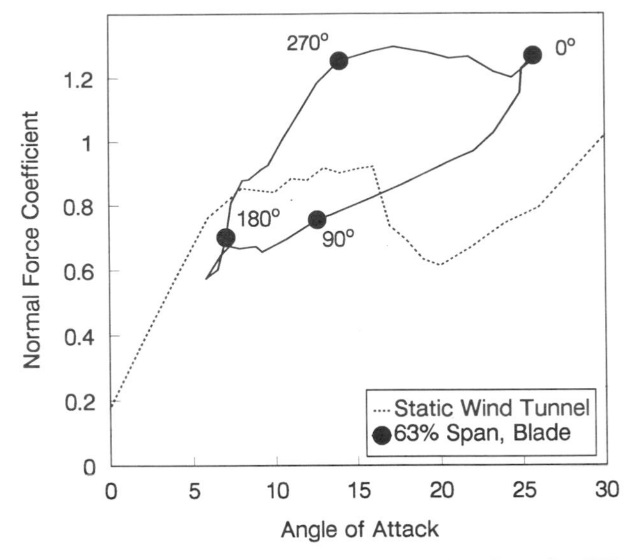

Figure 7 Normal force coefficients measured during dynamic stall on the CER operating in 30° yawed flow and 14 m/s wind. Rotor azimuth positions highlight the differential normal loading from one side of the rotor (90°) to the opposite side (270°). The top of the rotor is oriented at 0°. This difference increases yaw loads.

the leading edge suction peak rise. At 180° azimuth position the suction peak drops and pressure maxima can be seen moving towards the trailing edge as the blade progresses in time.

Dynamic Stall Models in Current Use

At least three models are currently incorporated in dynamics analyses to represent dynamic stall. All are semi-empirical in nature and require some a priori knowledge of the airfoil characteristics. They differ primarily in the degree of fundamental physical basis of the model (as opposed to curve-fitting wind tunnel test data). A method developed by Gormont (1973) was first applied to wind turbines in the analysis of a Darrieus rotor (Berg 1983). The same method was later applied to analysis of the yaw dynamics of HAWTs (Hansen 1992). The Gormont method determines an effective angle of attack based upon the actual angle, the time rate of change of the angle, and two constants determined by experiment. Values for the constants are suggested for the case when dynamic stall data are not available for the airfoil and conditions of interest. When data for the S809 airfoil were obtained in full-scale field testing, it was found that constants were different from those suggested and in fact varied along the span of the blade. Nonetheless, constants were determined which

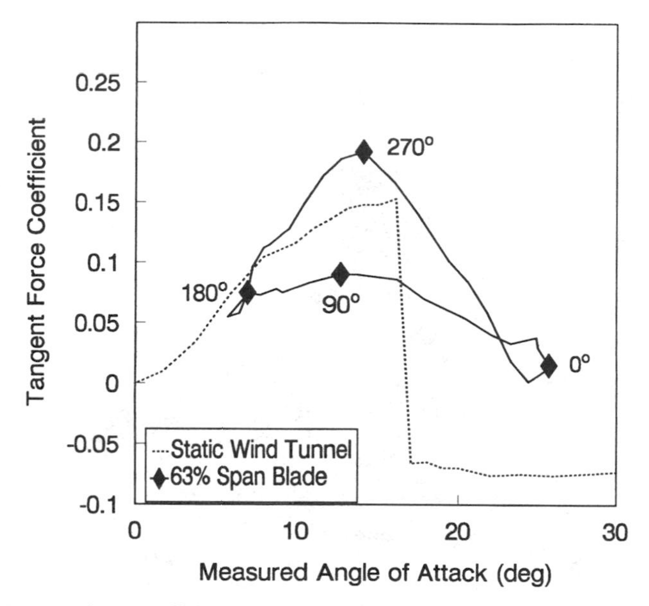

Figure 8 Tangent force coefficients measured during dynamic stall on the CER operating in 30° yawed flow and 14 m/s wind. Rotor azimuth positions highlight the differential inplane loading from one side of the rotor (90°) to the opposite side (270°).

adequately represented the hysteresis observed at the outboard stations and improvement was noted in the accuracy of the flap and yaw moments predicted when the Gormont model was used.

More recently, Beddoes & Leishman have proposed a model which is more deeply founded in the physical mechanisms active in dynamic stall. The method still relies upon empirical constants, but there are indications that the constants will apply to a broad variety of airfoils and reduced frequencies (Leishman & Beddoes 1989). The method is implemented as a series of indicial functions or responses to step changes in angle of attack. As such it is readily incorporated into time-domain models with algorithms that are relatively insensitive to time step and not dependent upon accurate estimation of the time rate of change of angle of attack. This approach is currently used in Europe and is in the process of being programmed into ADAMS® code aerodynamics routines at NREL.

A method developed by ONERA (Petot 1983, Tran & Petot 1982) has also been applied to wind turbine analysis (Bierbooms 1990a). In this method linear, ordinary differential equations with constant coefficients are developed to describe the linear and nonlinear components of the lift, drag, and moment coefficients. Again, empirical constants are required in

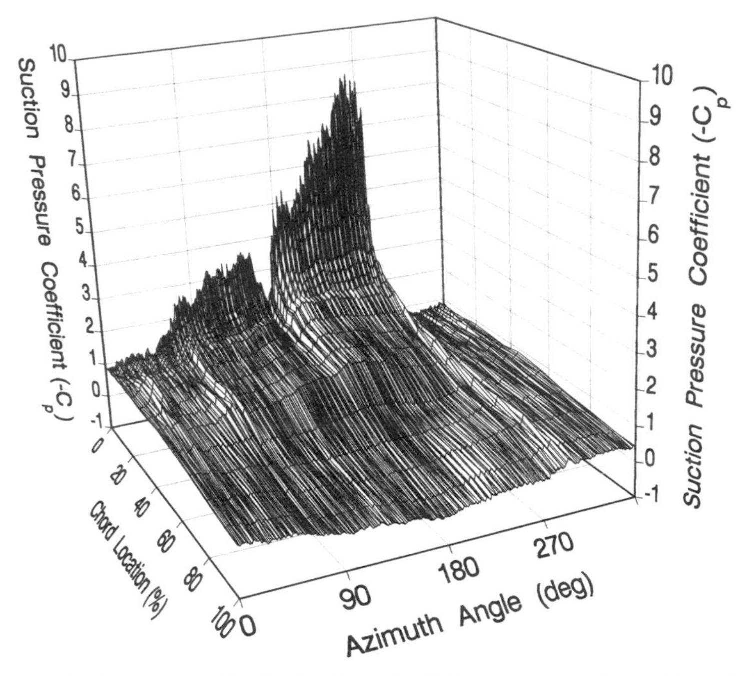

Figure 9 Suction pressure distribution from the 63% span, averaged over 25 revolutions of the CER during 30° yawed operation and 15 m/s wind. This figure shows the average variation of suction pressures throughout one revolution. The tower shadow can be seen as a dip in the pressures at 180°. The increase in leading edge suction peak followed by a sudden collapse and pressure convection implies the shedding of a vortex.

the model. Ideally they should be obtained by testing the airfoil in question. However, values are suggested for use when dynamic stall data are not available for the particular airfoil. Validation studies for application of the ONERA method to wind turbines are incomplete and somewhat inconclusive. One study compared the ONERA and Gormont models and concluded that the ONERA method provided a more accurate representation of the hysteresis on a 2-D, nonrotating airfoil (Bierbooms 1991). Another found the ONERA model gave poorer results than the Gormont method for a NACA 4418 airfoil on a 8.5 m test turbine and superior results for a 64-series airfoil on a 500 kW Windmaster rotor (Yseznasni et al 1991).

At this point it is clear that dynamic stall must be considered in some critical loading situations for wind turbine rotors. Research is continuing to explore the most accurate and practical method to implement this aspect of unsteady aerodynamics.

Dynamic Inflow

Dynamic inflow has only recently been the subject of investigation in wind turbine design. The CEC Joule project is sponsoring a joint investigation of dynamic inflow modeling by six European organizations. As a part of that project, Snel & Schepers (1991) presented a paper which clearly establishes by both testing and analysis that the induction lag associated with rapid changes in blade pitch or wind speed can have a large effect on system loads. Tests of the Danish Tjaereborg 2 MW turbine showed 50% overshoot in rotor torque and blade flap moment when the pitch was quickly changed 2 degrees.

Wake equilibrium assumptions appear to affect the analysis of the teetering rotor, even in situations one would not generally consider highly dynamic. Figure 10 shows the predicted teeter amplitude of an ESI–80 rotor under two different conditions. The calculations were performed using the YawDyn computer code described above and by Hansen (1992). In the first, the wake is assumed to be in equilibrium with the forces on the rotor caused by all motions of the rotor. Specifically, the additional

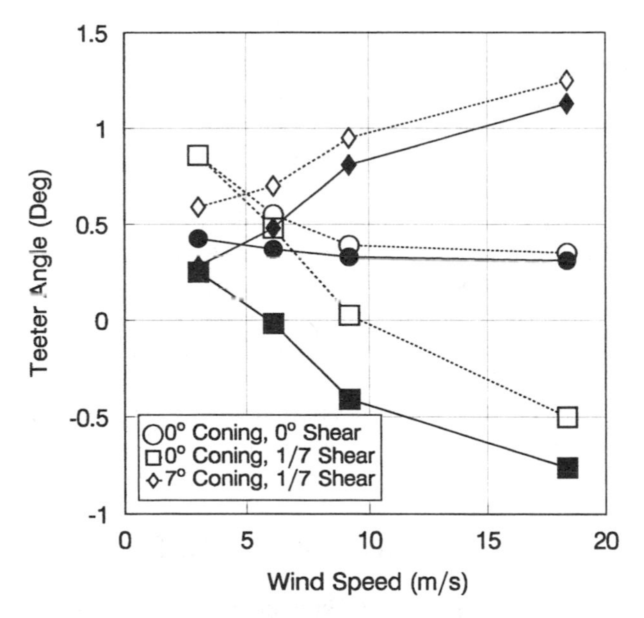

Figure 10 ESI-80 (modified) teeter amplitudes predicted by YawDyn code. Wind speed, vertical power-law shear coefficient, and rotor coning angle as shown on graph. No yaw angle, horizontal shear, or vertical component of wind. Lift curve slope = 5.2, drag coefficient = 0, no static or dynamic stall model. The frozen wake (*solid symbols*) and equilibrium wake (*open symbols*) assumptions are both shown.

force on the blade that is caused by the blade motion (the aerodynamic damping force) is assumed to influence the induced velocities. In the second curve (the frozen wake curve) the wake is frozen at the strength calculated in the absence of teeter velocities. It is clear that in situations where the flap velocities can be of the order of the wind velocity, the assumption regarding the equilibrium of the wake will have a large effect on the estimated motions and loads.

Two general approaches are being taken to analyze dynamic inflow. The first is suitable for use with the BEM method and the second is more computationally intensive. In the simpler approach, a linear, first-order model is assumed for the induced velocity. A time constant for the induction lag is determined in a variety of ways. Bierbooms (1990a) uses a time constant which is twice the time scale for the rotor disk, $2D/V$. Other time-constant methods are described by Snel & Schepers (1991). The dynamic inflow theory of Pitt & Peters provides one method for determining the time constants and has the additional advantage of incorporating skewed wake effects (Gaonkar & Peters 1986, Pitt & Peters 1981). This is the same method described earlier for quasi-steady analysis of yawed rotors. The time constants in this method asymptotically approach actuator disk values for flow normal to the rotor (zero yaw) and results from detailed vortex trajectory calculations for flow parallel to the rotor plane.

The second approach to analyzing dynamic inflow uses a vortex wake calculation. This method is inherently more satisfying since it enables one to directly calculate the motion of the vortex trajectories without needing to make assumptions regarding the time constant of the flow. In principle these methods can also treat dynamic stall directly, since the location and motion of all shed and trailing vorticity is determined at every time step. However, other assumptions, such as convection velocities and/or wake shape, are required to make the analysis tractable with current computers. At this time there is not sufficient data to determine whether the more complex methods are more accurate.

For the next several years it is likely that engineering analysis programs will be based upon time constants determined either by experiment or by exercising the vortex wake computations.

EFFECTS OF TURBULENCE ON ROTOR LOADS

Two approaches have generally been taken to estimate the effects of atmospheric turbulence upon the structure of a wind turbine. The first and more common approach is a discrete gust method. Loads are calculated in the time domain as the rotor is subjected to a specified time history of wind speed, direction, and distribution across the rotor disk. Selection of gust characteristics is often done on an ad hoc basis using wind data from field measurements at turbine sites or from handbooks such as that compiled by Frost et al (1978).

The second method calculates the response to stochastic wind with specified power spectral density and coherence functions. Calculations are done either in the frequency domain or in the time domain using time series synthesized from the specified spectral characteristics (Dragt 1984, Madsen & McNerney 1989). The vast quantity of wind characteristics data which is available from boundary-layer meteorology experiments is a major advantage of these techniques. But little is known about the applicability of the existing database to locations with good wind energy potential or to winds which will cause extreme loads on the rotors. Another advantage will justify use of statistical wind descriptions for the foreseeable future: The accumulation of fatigue cycles during "normal" operation of a turbine is easily predicted using this approach. Normal operation is difficult to define and varies from one turbine to the next. But in this context it means operation in the linear range below stall and at constant rotor speed. It excludes stalled conditions, start-up or shut-down, extreme gusts, or other events that cause extreme load cycles.

At first glance it may appear that stochastic wind inputs would have little effect on the rotor loads. Typical rotor rotation and natural frequencies are greater than 0.5 Hz. The energy content of the atmospheric turbulence at this frequency is orders of magnitude below that at lower frequencies. But full-scale rotor testing and analysis has shown that "rotational sampling" of the turbulence by a blade provides significant rotor excitation at frequencies that are integer multiples of the rotor speed (Connell 1982, George & Connell 1984). Turbulence length scales are of the same order of magnitude as a large HAWT diameter. Thus a rotating blade passing through a spatially varying wind field is excited at the frequency of passage. This shifts energy to higher frequencies and causes more fatigue accumulation than was originally anticipated.

The realization that rotational sampling increases the importance of turbulence led to development of a number of models to simulate stochastic loads on the HAWT rotor. Some models generate a wind record that would be experienced by a single point on a rotating blade. Others generate a full wind field which varies spatially in three dimensions. A good review of four of these models is given by Walker et al (1989). They conclude that the methods which give the more accurate wind simulation are, unfortunately, also more difficult to interface with loads calculations. The method of Veers (1988) was found to give acceptable results in a form that can be readily adapted for loads calculations. One problem with Veers' method was the computation time required to generate a wind field. Winkelaar (1991) developed an alternative method to implement the Veers model and found a significant savings in computer time. The Veers model was used to generate the blade responses shown in Figure 6 above. It is clear that inclusion of the longitudinal component of turbulence in the loads analysis improves the accuracy of the predictions.

All of the turbulence models currently in use for loads analysis consider only the longitudinal component of planetary boundary layer turbulence. Kelly & Wright (1991) reported measurements upwind of, and within, a wind park in the San Gorgonio Pass of California. They found that the turbulence characteristics are altered by the wind park with the most notable change being a decrease in integral length scale to more closely coincide with the diameter of the rotors in the park. In this situation the Veers model predicts peak gusts which were often only 50% of the observed peak gusts and the energy content at the rotor frequency is underestimated as well. This shift in integral scale and inability to estimate peak winds has disturbing implications for the modeling of wind park rotor loads.

The inability of the models to generate three components of wind fluctuations is greater cause for concern. Some rotor responses are highly sensitive to lateral or vertical components of wind fluctuations. As mentioned earlier, yaw moments of a rigid rotor are more sensitive to vertical wind than to any other wind disturbance with similar probability of occurrence. The teeter response of a rotor is also sensitive to in-plane components of wind disturbance. Recently, detailed wind measurements have been reported from wind turbine sites that are known to cause higher rotor failure rates or have greater maintenance requirements (Wendell et al 1991). These data were processed to show the effects of rotational sampling. Some observations by the authors are worth noting in this paper because of the implications for rotor design. The wind experienced by a single rotating blade element has accelerations of 25–50 m/s^2. Wind direction shifts of 90° over a period of 10–20 seconds were not uncommon. Vertical wind components of 10 m/s were frequently observed at one site. This wind component is of the order of the longitudinal wind and more than sufficient to induce dynamic stall effects on the blades.

One other finding of Wendell's work may be the most significant in terms of current practice. Wendell states, "Comparison of the time series depicting these quantities at two sites showed that the turbulence intensity (the commonly used descriptor of turbulence) did not adequately characterize the turbulence at these sites." The desire to have simple indicators of the severity of a site has led to the use of turbulence intensity as the most common (often only) descriptor. But the turbulence intensity conveys no information regarding the spatial variations of wind speed, the occurrence or severity of lateral or vertical wind in complex terrain, or the frequency content of the energy spectrum.

Much work remains to be done in finding descriptors and statistics of wind characteristics of greatest importance to wind turbine structural loading. Estimation of extreme loads remains particularly difficult. To solve this problem methods will have to be devised to present the wind characteristics in such a way that turbine manufacturers or certification agencies can understand the risks associated with placing turbines in various sites.

CONCLUSIONS

Important strides have been made over the past 15 years in the understanding of HAWT aerodynamics. Turbine performance and reliability improvements have resulted from experience gained in field operations and better understanding of the technology. This means design methods are now available that reduce the risk of introducing a new system to the marketplace.

Power output of a rotor can be estimated within $\pm 8\%$ below stall. Less accurate estimates for a stalled rotor have been attributed to marked changes in the stall characteristics of a rotating blade. It appears that radial flow in the separation zone of a stalled airfoil induces a spanwise pressure gradient and maintains leading edge suction and elevated lift forces. Research is continuing to validate this hypothesis and develop methods for predicting lift and drag characteristics of rotating blades.

New airfoils have been designed specifically for use on stall-controlled wind turbines. As a result, annual energy yields have increased 15–25% without incurring higher peak power levels or structural loads. In addition, realization of the critical importance of leading-edge soiling has led to improved airfoil selection criteria.

Dynamic loads can be predicted accurately for normal operation of turbines whose structural dynamics and airfoil characteristics are well understood. Prediction of extreme loads and off-design conditions remains a more formidable challenge. There is compelling experimental and analytical evidence that dynamic stall and dynamic inflow effects can increase some operating loads 50–100%. Methods are available for incorporating these effects into blade element/momentum analyses. But the most accurate and efficient means have not been determined, nor is it certain that modifications of blade element theory will be sufficient.

Understanding the importance of rotational sampling of atmospheric turbulence by a rotor blade has clarified the role of turbulence in fatigue design. Models are now available for generating time-series of longitudinal wind speed for use in loads calculations. However, the importance of lateral and vertical velocity fluctuations has been explored only recently. Research is continuing to seek the best methods for determining, and using, extreme wind statistics pertinent to HAWTs and the wide variety of sites in which they are located.

ACKNOWLEDGMENTS

Many colleagues in laboratories, corporations, and universities in the U.S. and Europe have assisted greatly in the preparation of this paper through open discussions and sharing of publications and data. Their support and the support of the U.S. Department of Energy are gratefully acknowledged.

Literature Cited

Afjeh, A. A., Keith, T. G. 1986. A simplified free wake method for horizontal axis wind turbine performance prediction. *Trans. ASME, J. Fluids Eng.* 108: 400–6

AWEA 1991. *Wind Energy—A Resource for the* 1990*s and Beyond.* Washington, DC: Am. Wind Energy Assoc.

Berg, D. E. 1983. Recent improvements to the VDART3 VAWT code. *Wind/Solar Energy Conf., Kansas City, MO*

Bierbooms, W. 1990a. A dynamic model of a flexible rotor: part I, description of the mathematical model. *Inst. for Wind Energy, Delft Univ. Technol., IW* 89.034*R*

Bierbooms, W. 1990b. A dynamic model of a flexible rotor: part II, description of the computer program. *Inst. for Wind Energy, Delft Univ. Technol., IW* 89.034*R*

Bierbooms, W. 1991. A comparison between unsteady aerodynamic models. *Proc. Eur. Wind Energy Conf., EWEC '91, Amsterdam*, pp. 13–17

Butterfield, C. P. 1989a. Aerodynamic pressure and flow-visualization measurement from a rotating wind turbine blade. *Eighth ASME Wind Energy Symp., Houston, TX*, pp. 245–256

Butterfield, C. P. 1989b. Three-dimensional airfoil performance measurements on a rotating wing. *Proc. Eur. Wind Energy Conf., Glasgow, Scotland*

Butterfield, C. P., Simms, D. A., Musial, W. P., Scott, G. N. 1990. Spanwise aerodynamic loads on a rotating wind turbine blade. *Sol. Energy Res. Inst., SERI/TP–*257–3983 (Also appears in *Proc. of the Windpower '90 Conf., Am. Wind Energy Assoc., Washington, DC, September* 1990)

Clark, R. N., Davis, R. G. 1991. Performance changes caused by rotor blade surface debris. *Windpower '91, Palm Spring, CA*

Connell, J. R. 1982. The spectrum of wind speed fluctuations encountered by a rotating blade of a WECS. *Sol. Energy* 29(5): 363–75

de Vries, O. 1985. Comment on the yaw stability of a horizontal-axis wind turbine at small angles of yaw. *Wind Eng.* 9(1): 42–49

Dini, P., Selig, M. D., Maughmer, M. D. 1991. A simplified method for separated boundary layers. *AIAA Pap.* 91–3285

Dragt, J. B. 1984. The spectra of wind speed fluctuations met by a rotating blade, and resulting load fluctuations. *Proc. Eur. Wind Energy Conf., Hamburg*

Drela, M. 1991. Improvements in low Reynolds number airoil flow predictions with ISES and XFOIL. *MIT CFL Rep.* 91–5

Eggers, A. J., Digumarthi, R. V. 1992. Approximate scaling of rotational effects of mean aerodynamic moments and power generated by the combined experiment rotor blades operating in deep-stalled flow. *Eleventh ASME Wind Energy Symp., Houston, TX*

Eggleston, D. M. 1991. Wind turbine bug roughness sampling and power degradation. *Windpower '91, Palm Springs, CA*

Eggleston, D. M., Stoddard, F. S. 1987. *Wind Turbine Engineering Design.* New York: Van Nostrand Reinhold

Eppler, R. 1990. *Airfoil Design and Data.* Berlin/New York: Springer-Verlag

Freris, L. L. 1990. *Wind Energy Conversion Systems.* Englewood Cliffs, NJ: Prentice Hall

Frost, W., Long, B. H., Turner, R. E. 1978. Engineering handbook on the atmospheric environmental guidelines for use in wind turbine generator development. *NASA Tech. Pap.* 1359

Gaonkar, G. H., Peters, D. A. 1986. Effectiveness of current dynamic-inflow models in hover and forward flight. *J. Am. Helicopter Soc.* 31(2): 47–57

George, R. L., Connell, J. R. 1984. Rotationally sampled wind characteristics and correlations with MOD–0A wind turbine response. *Pac. Northwest Lab., PNL*–5238

Glauert, H. 1935. *Aerodynamic Theory*. Berlin: Julius Springer

Glauert, H. 1948. *The Elements of Aero Foil and Airscrew Theory*. Cambridge: Cambridge Univ. Press

Gormont, R. E. 1973. A mathematical model of unsteady aerodynamics and radial flow for application to helicopter rotors. *US Army Air Mobility Res. Dev. Lab., USAAMRDL Technical Rep.* 76–67

Gould, J., Fiddes, S. P. 1991. Computational methods for the performance predictions of HAWTS. *Proc. Eur. Wind Energy Conf., EWEC '91, Amsterdam*, pp. 29–33

Hales, R. L. 1991. Dynamic stall on horizontal-axis wind turbines. *Proc. Eur. Wind Energy Conf., EWEC '91, Amsterdam*, pp. 34 39

Hansen, A. C. 1992. Yaw dynamics of horizontal axis wind turbines: final report. *Natl. Renewable Energy Lab., NREL Tech. Rep.* 442–4822

Hansen, A. C., Butterfield, C. P., Cui, X. 1990. Yaw loads and motions of a horizontal axis wind turbine. *J. Sol. Energy Eng.* 112(4): 310–14

Hansen, A. C., Cui, X. 1989. Analyses and observations of wind turbine yaw dynamics. *J. Sol. Energy Eng.* 111(4): 367–71

Hibbs, B. D. 1985. HAWT performance with dynamic stall. *AeroVironment Inc. Rep.*

Hibbs, B., Radkey, R. L. 1981. Small wind energy conversion systems rotor performance model comparison study. *Rockwell Int., Rocky Flats Plant, RFP*–4074/13470/36331/81–0

Hill, D. C., Garrad, A. D. 1989. Design of airfoils for wind turbine use. *Eur. Wind Energy Conf., Glasgow, Scotland*

Hock, S., Tu, P. K. C. 1986. Variable-speed operation of wind turbines: impact on energy capture and economics. *Am. Sol. Energy Soc. Annu. Meet., Boulder, CO*

Hock, S. M., Thresher, R. W., Wright, A. D. 1988. A comparison of results from dynamic-response field tests. *Sol. Energy Res. Inst., SERI/TP*–217–3423

Holley, W. E., Aitkenhead, W., McNerney, G., Rogers, E. 1987. Estimating a wind turbine rotor diameter to minimize the installed cost of energy. *Sixth ASME Wind Energy Symp., Dallas, TX*

Jackson, K. L., Migliore, P. G. 1987. Design of wind turbine blades employing advanced airfoils. *Windpower '87, San Francisco, CA*

Jacobs, E. N., Sherman, A. 1937. Airfoil section characteristics as affected by variations of the Reynolds number. *NACA Rep.* 586

Jamieson, P., Bowles, A., Derrick, A., Leithead, W., Rogers, M. 1991. Innovative concepts for aerodynamic control of wind turbine rotors. *Proc. Eur. Wind Energy Conf., EWEC '91, Amsterdam*, pp. 819–823

Kelley, N. D., McKenna, H. E. 1985. Acoustic noise associated with the MOD–1 wind turbine: its source, impact and control. *Sol. Energy Res. Inst., SERI/TP*–635–116

Kelly, N. D., Wright, A. D. 1991. A comparison of predicted and observed turbulent wind fields present in natural and internal wind park environments. *Windpower '91, Palm Springs, CA*

Kocurek, D. 1987. Lifting surface performance analysis for horizontal axis wind turbines. *SERI/STR*–217–3163

Kretz, A., Rasmussen, F. 1990. An aeroelastic model with applications to a two-bladed wind turbine. *Riso Natl. Lab., Riso-M*–2884

Leishman, J. G., Beddoes, T. S. 1989. A semi-empirical model for dynamic stall. *J. Am. Helicopter Soc.* 34(3): 3–17

Lock, R. C., Williams, B. R. 1987. Viscous-inviscid interactions in external aerodynamics. *Prog. Aerospace Sci.* 24: 51–71

Lynette, R. 1988. California wind farms operational data collection and analysis. *Sol. Energy Res. Inst. Rep.*

Lynette, R. 1991. Advanced 275 kW wind turbine. *Windpower '91, Palm Springs, CA*

Lynette, R., Conover, K., Young, J. 1989. Experiences with commercial wind turbine design. *Electr. Power Res. Inst., Final Rep., Project* 1590–12, *EPRI GS*–6245, Vol. 1

Madsen, H. A. 1990. Measured airfoil characteristics of three blade segments on a 19m HAWT rotor. *Recent Developments in the Aerodynamics of Wind Turbines, Proc. BWEA Workshop, Univ. Nottingham*

Madsen, P. H., McNerney, G. M. 1989. Frequency domain modelling of free yaw response of wind turbines to wind turbulence. *Eighth ASME Wind Energy Symp., Houston, TX*

McCroskey, W. J. 1981. The phenomenon of dynamic stall. *NASA-TM*–81264

McCroskey, W. J., Yaggy, P. F. 1968. Laminar boundary layers on helicopter rotors in forward flight. *AIAA J.* 6(10): 1919–26

McGhee, R. J., Beasley, W. D., Whitcomb, R. T. 1979. NASA low- and medium-speed airfoil development. *NASA TM*78709

McNerney, G. M., Erdman, W., Steeley, W. J., DeMeo, E. A. 1990. The EPRI-utility-USW advanced wind turbine program—1990 update. *Windpower '90, Washington, DC*

Miley, S. J. 1982. A catalog of low Reynolds number airfoil data for wind turbine applications. *Rockwell Int., Rocky Flats Plant, RFP*–3387

Morgan, C. A., Garrad, A. D. 1988. The design of optimum rotors for HAWTs. *Brit. Wind Energy Assoc. Wind Energy Conf., London*

Musial, W. D., Butterfield, C. P., Jenks, M. D. 1990. A comparison of two- and three-dimensional S809 airfoil properties for rough and smooth HAWT rotor operation. *Ninth ASME Wind Energy Symp., New Orleans, LA*

Petot, D. 1983. Progress in the semi-empirical prediction of the aerodynamic forces due to large amplitude oscillations of an airfoil in attached or separated flow. *Ninth Eur. Rotorcraft Forum, Stresa, Italy*

Pitt, D. M., Peters, D. A. 1981. Theoretical predictions of dynamic inflow derivatives. *Vertica* 5(1): 21–34

Rashkin, S., Nguyen, T. H. P. 1985–1989. Results from the wind project performance reporting system. 1985–1988 *Annu. Rep. Calif. Energy Commission*

Rasmussen, F., Peresen, S. M., Larsen, G. 1988. Investigations of aerodynamics, structural dynamics and fatigue on Danwin 180 kW. *Riso Natl, Lab., Denmark, Riso-M*–2727

Ronsten, G., Dahlberg, J. A., Meijer, S. 1989. Pressure measurements on a 5.35 m HAWT in CARDC 12 × 16 m wind tunnel compared to theoretical pressure distributions. *Eur. Wind Energy Conf., Glasgow, Scotland*

Schlichting, H. 1978. *Boundary-Layer Theory*. New York: McGraw-Hill

Simoes, F. J., Graham, J. M. R. 1991. Application of a free vortex wake model to a horizontal axis wind turbine. *Proc. Eur. Wind Energy Conf., EWEC '91, Amsterdam*, pp. 46–50

Snel, H. 1991. Scaling laws for the boundary layer flow on rotating wind turbine blades. *CEC Aerodyn. Specialists Conf.*

Snel, H., Lindenburg, C. 1990. Aeroelastic rotor system code for horizontal axis wind turbines: PHATAS-II. *Eur. Wind Energy Conf., Madrid*

Snel, H., Schepers, J. G. 1991. Engineering models for dynamic inflow phenomena. *Proc. Eur. Wind Energy Conf., EWEC '91, Amsterdam*

Snyder, M. H., Wentz, W. H., Cao, H. 1987. Additional reflection plane tests of control devices on a NACA 23024 airfoil. *Wichita State Univ., WER*–26

Sorensen, J. N. 1986. Three-level, viscous-inviscid interaction technique for the prediction of separated flow past rotating wings. PhD dissertation. Tech. Univ. Denmark

Swift, A. H. P. 1981. The effects of yawed flow on wind turbine rotors. PhD dissertation. Washington Univ.

Tangler, J. L. 1982. Comparison of wind turbine performance prediction and measurement. *J. Sol. Energy Eng.* 104: 84–88

Tangler, J. L. 1987. Status of the special-purpose airfoil families. *Sol. Energy Res. Inst., SERI/TP*–217–3264

Touryan, K. J., Strickland, J. H., Berg, D. E. 1987. Electric power from vertical-axis wind turbines. *J. Propuls.* 3(6): 481–493

Tran, C. T., Petot, D. 1982. Semi-empirical model for the dynamic stall of airfoils in view of the application to the calculation of responses of a helicopter blade in forward flight. *Vertica* 6: 219–39

van Bussel, G. J. W. 1991. The use of the asymptotic acceleration potential method for horizontal axis wind turbine aerodynamics. *Proc. Eur. Wind Energy Conf., EWEC '91, Amsterdam*, pp. 18–23

Van Grol, H. J., Snel, H., Schepers, J. G. 1991a. Wind turbine benchmark exercise on mechanical loads: a state of the art report, Volume I (Part A). *Netherlands Energy Res. Found. ECN, ECN-C*—91–030

Van Grol, H. J., Snel, H., Schepers, J. G. 1991b. Wind turbine benchmark exercise on mechanical loads: a state of the art report, Volume I (Part B). *Netherlands Energy Res. Found. ECN, ECN-C*—91–031

Veers, P. S. 1988. Three-dimensional wind simulation. *Sandia Natl. Lab., SAND*88–0152

Velkoff, H. R., Blaser, D. A., Jones, K. M. 1971. Boundary-layer discontinuity on a helicopter rotor blade in hovering. *J. Aircraft* 8(2): 101–7

Viterna, L. A., Corrigan, R. D. 1981. Fixed-pitch rotor performance of large HAWTs. *DOE/NASA Workshop on Large HAWTs, Cleveland, OH*

Walker, S. N., Weber, T. L., Wilson, R. E. 1989. A comparison of wind turbulence simulation models for stochastic loads analysis for horizontal-axis wind turbines. *Sol. Energy Res. Inst., SERI/STR*–217–3463

Wendell, L. L., Morris, V. R., Tomich, S. D., Gower, G. L. 1991. Turbulence

characterization for wind energy development. *Windpower '91, Palm Springs, CA*
Wilson, R. E., Lissaman, P. B. S. 1974. Applied aerodynamics of wind power machines. *Oregon State Univ., NTIS PB* 238594
Wilson, R. E., Weber, T. L., Walker, S. N. 1991. Prediction of fatigue loads on a free-yaw HAWT. *Tenth ASME Wind Energy Symp., Houston, TX*
Winkelaar, D. 1991. Fast three-dimensional wind simulation and the prediction of stochastic blade loads. *Tenth ASME Wind Energy Symp., Houston, TX*
Wood, D. H. 1991. A three-dimensional analysis of stall-delay on a horizontal-axis wind turbine. *J. Wind Eng. Ind. Aerodyn.* 37: 1–14
Wright, A. D., Buhl, M. L. Thresher, R. W. 1987. FLAP Code development and validation. *SERI, TR*–217–3125
Wright, A. D., Butterfield, C. P. 1992. The NREL teetering hub rotor code: final results and conclusions. *Eleventh ASME Wind Energy Symp., Houston, TX*
Wright, A. D., Thresher, R. W., Butterfield, C. P. 1991. Status of a teetered rotor code development and validation project. *Tenth ASME Wind Energy Symp., Houston, TX*
Yekutieli, O., Clark, R. N. 1987. Influence of blade surface roughness on the performance of wind turbines. *Sixth ASME Wind Energy Symp., Dallas, TX*
Young, W. H., Williams, J. C. 1972. Boundary-layer separation on rotating blades in forward flight. *AIAA J.* 10(12): 1613–19
Yseznasni, A., Derdelinckx, R., Hirsch, C. 1991. Influence of dynamic stall in the aerodynamic study of HAWT. *Proc. Eur. Wind Energy Conf., EWEC '91, Amsterdam*, pp. 56–62

Annu. Rev. Fluid Mech. 1993. 25 : 151–181

UP-TO-DATE GASDYNAMIC MODELS OF HYPERSONIC AERODYNAMICS AND HEAT TRANSFER WITH REAL GAS PROPERTIES

G. A. Tirsky

Laboratory of Physical and Chemical Gasdynamics,
Institute of Mechanics at Moscow State University,
Michurinsky pr. 1, 119899, Moscow, Russia

KEY WORDS: thermal and chemical nonequilibrium, catalytic reactions, excited particles, electron temperature, vibrational temperature, vibration-dissociation coupling

INTRODUCTION

Hypersonic aerodynamics and heat transfer problems have received renewed interest during the past decade (Anderson 1984). This revival has been due to the design and creation of space vehicles that move along glide reentry trajectories in the upper layers of the Earth's atmosphere ($H = 50$–100 km) at hypersonic velocities ($V_\infty < 7.8$ km s^{-1}; i.e. Space Shuttle, Buran), and the development of space vehicles with reentry trajectories in the upper atmospheric layers (at altitudes $H = 70$–100 km) which use aerodynamic braking during recovery from geosynchronous orbit ($V_\infty = 7$–11 km s^{-1}) at nominal reentry velocities $V_\infty = 10$ km s^{-1} (Walberg 1983). Typical trajectories of various spacecraft and the regions of different physical and chemical processes in the shock layer near such vehicles are shown in Figure 1.

Unlike the aerodynamic and heat transfer characteristics of conventional aircraft, these problems are characterized at reentry (or ascending) trajectories by a wide range of Reynolds numbers ($100 < Re_\infty$), large

0066–4189/93/0115–0151$02.00

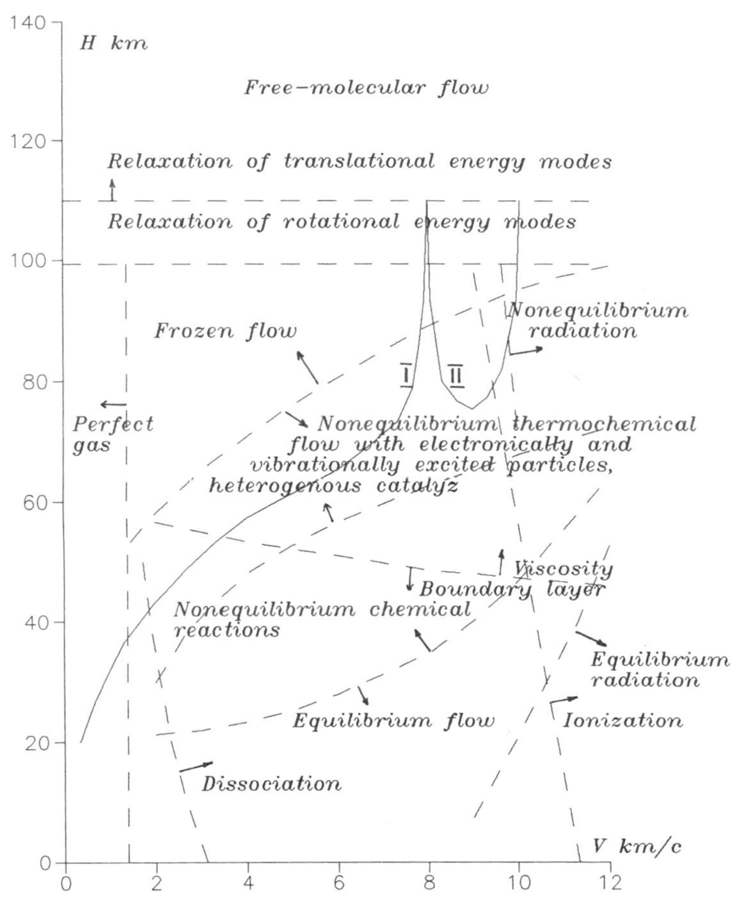

Figure 1 Regions of different physical and chemical processes in the shock layer and typical trajectories of various spacecraft. I—Space Shuttle, Buran trajectories; II—aeroassisted orbital transfer vehicles.

Mach numbers ($M_\infty < 30$), and therefore by high temperatures immediately behind the bow shock wave (up to several tens of thousands of degrees before the equilibrium temperature sets in). The stagnation pressure along the reentry trajectory behind the shock varies from 10^{-4}–10^{-3} atm (the beginning of the continuum flow regime) up to hundreds of atmospheres in the sea-level layers. The low and moderate Reynolds numbers at high altitude ($H > 50$–60 km) make it necessary to turn to more universal gasdynamical models than the Prandtl one, where second order [$\mathcal{O}(\mathrm{Re}^{-1/2})$] and higher order terms are disregarded. The asymptotic theory of second order boundary layers—which has been developed since the 1960s (Van

Dyke 1962)—becomes labor-consuming when nonequilibrium physico-chemical processes are involved. A two-layer model of a thin viscous shock layer (TVSL) was proposed at the beginning of the 1960s by Cheng (1961). This may be reduced to solution of the boundary-layer equations in the entire region between the body and the unknown detached bow shock wave, which is assumed to lie equidistant from the surface of the body, and where generalized Rankine-Hugoniot conditions are imposed (Cheng 1961, Magomedov 1970). This model was widely used because of its mathematical simplicity (equations of parobolic type, solved by marching methods) and the possibility of taking into account the various physical-chemical processes occurring throughout the shock layer (see surveys by Gershbein et al 1985, Peigin & Tirsky 1988).

However the TVSL model has an essential shortcoming, which arises from the appearance of the zero-pressure point at the surface of the body (about 60° on the sphere), beyond which the solution can't be extended. This model's low accuracy in the downstream region, where the shock layer thickness is no longer small, leads to underestimated values of the pressure, viscous drag, and heat transfer rates (Tirsky & Utjuzhnikov 1989).

During the 1960s and early 1970s, new approaches in the mathematical modeling of supersonic and hypersonic aerodynamics and heat transfer appeared. These were, as a rule, connected with the development of more universal, but mathematically more complicated, gasdynamical models of mixed type. Beginning with the foundation laid by Tolstykh (1966), two models were proposed:

1. simplified (parabolized) Navier-Stokes equations (PNS), where the second derivatives along the marching direction—which are responsible for perturbations upstream from molecular transfer processes (viscosity, heat conductivity, and diffusion)—are absent;
2. a full equation model for the viscous shock layer (VSL), which is similar to the PNS but differs from it by the absence of a viscous term in the normal momentum equation and which is coupled with the boundary conditions at the unknown shock wave in the form of generalized Rankine-Hugoniot relations (Davis 1970, Tirsky 1975).

The PNS and VSL models provide better results for the flow field than the asymptotic theory of high order boundary layers as regards either expenditure of computing time or accuracy (Davis & Rubin 1980). As mentioned above, high gas temperature in the shock layer (up to 50,000 K) is the essential feature of hypersonic flow past a body, which at large altitude ($H > 60$ km) leads to nonequilibrium dissociation-recombination and ionization-neutralization processes. At a translational temperature $T > 8000$ K, these processes proceed against a background of relaxation of internal degrees of freedom (rotational, vibrational, and electronic). Behind the bow shock wave or at the outer boundary of the merged shock layer, lagging (compression flow) of the vibrational temperature of the molecules and molecular ions occurs. This lagging affects (decreases) the reaction rates, resulting in a so-called vibration-dissociation coupling (CVD) and, at higher temperature, electron-ionization coupling (CEI). Conversely, relaxation of internal degrees of freedom depends on dissociation-recombination processes [vibration-dissociation-vibration coupling (CVDV)] and the electron temperature relaxation depends on the ionization-neutralization process [electron-ionization-electron coupling (CEIE)], i.e. the system has negative feedback. These interactions lead to a considerable temperature rise in the shock layer, thereby slowing down the electron temperature relaxation and reducing the number of internal degrees of freedom by one. This results in an increase of convective heat flux to the body.

Since homogenous recombination and neutralization occur slowly at high altitudes, exothermic heterogenous processes at the body surface become crucial to the magnitude of the convective heat flux. These depend on the catalytic properties of the surface. The maximum input of convective heat flux to the body caused by heterogenous recombination of dissociated air is as much as 3–4 times that due to usual heat conductivity (Goulard 1958). When ionization takes place this influence is even greater.

The generation of vibrationally and electronically excited metastable molecules at the surface is an important feature of heterogenous atom generation, which causes a decrease of heat flux to the wall (Berkut et al 1985a, 1986a,b,c, 1987, 1988; Doroshenko et al 1988, 1989, 1990a,b). At higher pressures, electronically and vibrationally excited particles are generated in the gas phase recombination; this also lowers the heat flux. In this case the vibrational distribution function of the molecules may be essentially non-Boltzmann. At a low wall temperature ($T_w = 300$ K), an inverted population is possible in the near surface layer. And, finally, at high altitude, concentration of some components or electrons in the shock layer may exceed equilibrium values, causing intensive light emission from the nonequilibrium region of the shock layer. In some cases it is more intense than emission from the equilibrium region. This is the so-called enhancement radiation (Teare et al 1962; Allen et al 1963; Park 1983, 1984a; Candler & Park 1988; Sasoh et al 1990). Unlike during the 1960s and 1970s, when some of these processes were investigated separately and when the structure of only strong shock waves was considered, current studies attempt to account for all of these processes simultaneously, incorporating viscosity, heat conductivity, and diffusion by solving equations either in the VSL, PNS, or full Navier-Stokes models.

This review focuses on the influence of multicomponent diffusion, thermodynamic and chemical nonequilibrium processes, heterogenous reactions, generation of vibrationally and electronically excited particles by recombination, and on resistance and heat transfer in hypersonic flow past bodies at low and moderate Reynolds numbers.

1. MASS, MOMENTUM, AND ENERGY EQUATIONS FOR GAS MIXTURES

Hydrodynamic equations for multicomponent, multitemperature, chemically active, relaxing and emitting gas mixtures, which consist of both neutral and charged particles and where external electromagnetic fields are absent, are the mathematical foundation for the study of hypersonic aerodynamics and heat transfer when dissipative and physico-chemical processes are present. These equations may be obtained either phenomenologically or from kinetic models. The phenomenological approach—which is based on applying the general laws of classical mechanics and thermodynamics to the interfering and mutually penetrating continuums [first introduced by Maxwell (1868) and Stefan (1871)]—may be found in courses on the thermodynamics of irreversible processes.

The second approach—the kinetic one—is based on deducing the hydrodynamic equations from the Boltzmann kinetic equations and their appropriate moments. The two approaches yield equations with identical structure, although the phenomenological approach is a more general one, since it is not coupled with any particular model or with the simplifications (binary collisions, correlation independency, molecular chaos, etc.) used in the kinetic approach.

For a gas mixture of N species these equations are (see Gogosov & Polyansky 1976 for a review of the derivation):

$$\frac{\partial \rho_i}{\partial t}+\nabla\cdot(\rho_i\mathbf{v}+\mathbf{J}_i)=\omega_i \qquad (i=1,\ldots,N) \tag{1.1}$$

$$\frac{\partial \rho_i\mathbf{v}_i}{\partial t}+\nabla\cdot(\rho_i\mathbf{v}\mathbf{v}+\mathbf{v}\mathbf{J}_i+\mathbf{J}_i\mathbf{v})-\nabla\mathscr{P}_i=\rho_i\mathbf{F}_{gi}+\rho_i\mathbf{F}_{qi}+\rho_i\mathbf{F}_{\tau i}+\omega_i\mathbf{v} \tag{1.2}$$

$$\frac{\partial}{\partial t}\left[\rho_i\left(\frac{v^2}{2}+e_i+\mathbf{v}\cdot\mathbf{V}_i\right]+\nabla\cdot\left[\rho_i\left(\frac{v^2}{2}+e_i\right)\mathbf{v}-\mathscr{P}_i\cdot\mathbf{v}+\mathbf{J}_{qi}+\mathbf{J}_i\frac{v^2}{2}\right.\right.$$
$$\left.+(\mathbf{J}_i\cdot\mathbf{v})\mathbf{v}-\mathbf{J}_{Ri}\right]=\rho_i(\mathbf{F}_{gi}+\mathbf{F}_{qi})\cdot\mathbf{v}+\rho_i\mathbf{F}_{\tau i}\cdot\mathbf{v}+\omega_i\frac{v^2}{2}+Q_i \tag{1.3}$$

$$p_i=n_ikT_i \tag{1.4}$$

where

$$c_i = \rho_i/\rho, \quad \rho = \sum_{k=1}^{N} \rho_k, \quad \rho_i = n_i m_i, \quad \mathbf{V}_i = \mathbf{v}_i - \mathbf{v}, \quad \mathbf{J}_i = \rho_i \mathbf{V}_i$$

$$\mathbf{F}_{\tau i} = \sum_{k=1}^{N} \nu_{ik}(\mathbf{v}_k - \mathbf{v}_i), \quad \mathbf{F}_{qi} = \frac{q_i}{m_i}\left[\mathbf{E} + \frac{1}{c}(\mathbf{v}_i \times \mathbf{H})\right], \quad \mathscr{P}_i = -p_i \mathscr{I} + \mathscr{T}_i.$$

Here, n_i is the i-th species number density, m_i the particle mass, ρ_i the density, c_i the mass fraction, $\mathbf{v}_i$ the statistical mean velocity, $\mathbf{V}_i$ the diffusive velocity, $\mathbf{J}_i$ the mass diffusive flux, ω_i the mass source due to chemical reactions, $\mathbf{F}_{gi}$ the nonelectromagnetic force, $\mathbf{F}_{qi}$ the electromagnetic force, $\mathbf{F}_{\tau i}$ the drag force on the i-th particle by all others, e_i the specific internal energy (including the energy of the formation), $\mathbf{J}_{Ri}$ the specific radiation flux, Q_i the energy exchange term between the species, $\mathscr{P}_i$ the partial tension tensor, p_i the partial pressure, $\mathscr{T}_i$ the partial viscous tension tensor, T_i the temperature, q_i the charge, ν_{ik} the effective frequency of collision between particles of the i-th continuum and the k-th one; $\mathbf{E}$ and $\mathbf{H}$ are electric and magnetic fields, ρ is the mass density of the mixture, $\mathbf{v}$ the mean mass velocity, c the speed of light, k is Boltzmann's constant, $\mathscr{I}$ is a unit tensor, and $\nabla \cdot$ is the divergence operator.

If we multiply Equation 1.2 by $\mathbf{v}$ and subtract this from Equation 1.3, the equation for the heat flux into the i-th continuum yields

$$\frac{d(\rho_i e_i)}{dt} + \rho_i e_i \nabla \cdot \mathbf{v} + \nabla \cdot (\mathbf{J}_{qi} - \mathbf{J}_{Ri}) + \mathbf{J}_i \cdot \frac{d\mathbf{v}}{dt}$$
$$- \mathscr{P}_i : \nabla \mathbf{v} = (\mathbf{F}_{gi} + \mathbf{F}_{qi}) \cdot \mathbf{J}_i + Q_i \qquad (i = 1, \ldots, N). \quad (1.5)$$

The hydrodynamic equations for the mixture as a whole may be obtained by summing Equations 1.1–1.4 in consecutive order and taking into account total mass conservation in all possible reactions, momentum conservation in all possible collisions, and conservation of the total exchange energy:

$$\frac{\partial \rho}{\partial t} + \nabla \cdot (\rho \mathbf{v}) = 0 \qquad (1.6)$$

$$\frac{\partial \rho \mathbf{v}}{\partial t} + \nabla \cdot (\rho \mathbf{v}\mathbf{v} - \mathscr{T}) + \nabla p = \sum_{k=1}^{N} \rho_k (\mathbf{F}_{gk} + \mathbf{F}_{ek}) \qquad (1.7)$$

$$\frac{\partial}{\partial t}\left[\rho\left(\frac{v^2}{2} + e\right)\right] + \nabla \cdot \left[\rho\left(\frac{v^2}{2} + e\right)\mathbf{v} + p \cdot \mathbf{v}\right.$$
$$\left. - \mathscr{T} \cdot \mathbf{v} + \mathbf{J}_q - \mathbf{J}_R\right] = \sum_{k=1}^{N} \rho_k \mathbf{F}_k \cdot \mathbf{v}_k \quad (1.8)$$

$$p = \sum_{i=1}^{N} n_i k T_i \tag{1.9}$$

where the pressure, viscous tension tensor, and specific internal energy are given by

$$p = \sum_{k=1}^{N} p_k, \qquad \mathcal{T} = \sum_{k=1}^{N} \mathcal{T}_k, \qquad \rho e = \sum_{k=1}^{N} \rho_k e_k$$

and the total specific heat flux, total specific radiative heat flux, and the external mass force, $\mathbf{F}_i$, which acts on the i-th continuum are

$$\mathbf{J}_\mathrm{q} = \sum_{k=1}^{N} \mathbf{J}_{\mathrm{q}k}, \qquad \mathbf{J}_\mathrm{R} = \sum_{k=1}^{N} \mathbf{J}_{\mathrm{R}k}, \qquad \mathbf{F}_i = \mathbf{F}_{\mathrm{g}i} + \mathbf{F}_{\mathrm{q}i}.$$

To complete the adduced set of equations for the gas mixture explicit expressions are needed for the internal energy of the i-th component e_i, for the $\mathbf{J}_i$, $\mathbf{J}_\mathrm{q}$, and $\mathbf{J}_\mathrm{R}$ fluxes, and for the tension tensor. Expressions for the source terms ω_i and Q_i also need to be deduced.

2. INTERNAL ENERGY OF THE SPECIES

If we suppose a local equilibrium (Maxwell) energy distribution function for the translational degrees of freedom of the particles (molecules, atoms, and ions) and a local equilibrium (Boltzmann) distribution function for their internal degrees of freedom, the expression for the internal energy of the i-th constituent is

$$e_i = e_i^{(\mathrm{T})}(T_i) + e_i^{(\mathrm{R})}(T_i^{(\mathrm{R})}) + e_i^{(\mathrm{V})}(T_i^{(\mathrm{V})}) + e_i^{(\mathrm{E})}(T_i^{(\mathrm{E})}) + h_{i0}. \tag{2.1}$$

Here the first four terms represent the specific energies of the translational (T), rotational (R), vibrational (V), and electronic (E) degrees of freedom at their appropriate temperatures; h_{i0} is the formation energy (specific reaction heat at absolute zero) of the particle of the i-th component.

For hypersonic flow at an altitude $H < 100$ km, for simplicity we assume that the translational and rotational degrees of freedom of heavy particles are in equilibrium, i.e. $T_i^{(\mathrm{R})} = T_i = T$. Expressions for $e_i^{(\mathrm{V})}$ and $e_i^{(\mathrm{E})}$ may be found in reference books. Within the framework of the harmonic oscillator model

$$e_i^{(\mathrm{V})} = \frac{k\theta_{\mathrm{V}i}}{m}(\exp t_i - 1)^{-1}, \qquad \theta_{\mathrm{V}i} = \frac{h\nu_i}{k}, \qquad t_i = \frac{\theta_{\mathrm{V}i}}{T_i^{(\mathrm{V})}} \tag{2.2}$$

where $\theta_{\mathrm{V}i}$ is the characteristic vibrational temperature and ν_i the oscillation frequency.

Electronic levels are generally excited by free electrons, whose energy is characterized by the electron temperature T_e. An equilibrium (Boltzmann) energy distribution function with temperature T_e is usually assumed, i.e. $T_i^{(E)} = T_e$. This assumption is close to being valid for the lower electron energy levels. The population of the highly excited levels deviates from the equilibrium distribution, but their input to the total energy at temperature $T_e < 10{,}000$ K is negligible. Until complete thermodynamic equilibrium behind the shock wave sets in, the vibrational temperature $T_i^{(V)}$ of the molecules and molecular ions and the electron temperature T_e, which is equal to $T_i^{(E)}$, relax to the translational temperature T. Thus, if relaxation of internal degrees of freedom with quasi-equilibrium temperatures $T_i^{(V)}$, $T_i^{(E)} = T_e$ is included in the model, then we also need the appropriate energy balance equations for the vibrational degree of freedom and the equation for the energy of the free electrons T_e.

3. ENERGY BALANCE EQUATION FOR INTERNAL DEGREES OF FREEDOM

If we multiply the quantity balance equation for the i-th species, which is in quantum state α, by the value of the energy of this state $\varepsilon_{i\alpha}$ and sum over all α we get (Galkin & Kogan 1969, Tirsky 1987)

$$\rho \frac{d(c_i e_i^{(\mathrm{in})})}{dt} + \nabla \cdot \mathbf{q}_i^{(\mathrm{in})} = Q_i^{(\mathrm{in})} \quad (i = 1, \ldots, N), \quad (\mathrm{in}) = (\mathrm{R}), (\mathrm{V}), (\mathrm{E}) \tag{3.1}$$

where

$$e_i^{(\mathrm{in})} = \langle \varepsilon_{i\alpha} \rangle / m_i, \quad n_i \langle \varepsilon_{i\alpha} \rangle = \sum_\alpha n_{i\alpha} \varepsilon_{i\alpha}, \quad n_i = \sum_\alpha n_{i\alpha}$$

$$\mathbf{q}_i^{(\mathrm{in})} = \sum_\alpha n_{i\alpha} \varepsilon_{i\alpha} \mathbf{V}_{i\alpha}, \quad \mathbf{V}_{i\alpha} = \mathbf{v}_{i\alpha} - \mathbf{v}, \quad Q_i^{(\mathrm{in})} = \sum_\alpha \varepsilon_{i\alpha} \dot{n}_{i\alpha}.$$

Here $\mathbf{q}_i^{(\mathrm{in})}$ is the specific energy flux in the i-th species internal degrees of freedom and $e_i^{(\mathrm{in})}$ is the mean specific energy of the internal degrees of freedom of the i-th constituent. The $Q_i^{(\mathrm{in})}$ are the exchange source terms. To complete the set of Equations (3.1), expressions for the fluxes $\mathbf{q}_i^{(\mathrm{in})}$ and sources $Q_i^{(\mathrm{in})}$ are required.

4. ELECTRON ENERGY BALANCE EQUATION

Because of the mass difference between electrons and heavy particles, there is a low kinetic energy exchange rate in electron-heavy particle collisions. Thus, the electron temperature T_e in the shock layer immediately behind the bow shock is lower than the heavy particle temperature T. So an equation for T_e or for the energy of the free electrons is necessary: $e_e = 3/2(k/m_e)T_e + h_{eo}$. If we rewrite Equation (1.5) using the electron temperature T_e, we obtain

$$\frac{3}{2}\frac{d(n_e T_e)}{dt} + \frac{5}{2}n_e T_e \nabla \cdot \mathbf{v} + \nabla \cdot (\mathbf{J}_{qe} - h_{eo}\mathbf{J}_e - \mathbf{J}_{Re}) + \mathbf{J}_i \cdot \frac{d\mathbf{v}}{dt} - \mathscr{T}_e : \nabla \mathbf{v} = (\mathbf{F}_{ge} + \mathbf{F}_{qe}) \cdot \mathbf{J}_e + Q_e - h_{eo}\dot{\omega}_e. \quad (4.1)$$

5. TRANSFER COEFFICIENTS AND EQUATIONS

The explicit expressions for the diffusion flux $\mathbf{J}_i$, total heat flux $\mathbf{J}_q$, internal energy fluxes $\mathbf{q}_i^{(in)}$, radiation fluxes $\mathbf{J}_{Ri}$, and the partial component of the tension tensor $\mathscr{P}_i$ for the mixture and for its particular species—the so-called transfer equations—are crucial to deriving a complete system of equations for the multicomponent and multitemperature gas mixture. At this stage one must apply either (*a*) the principles of thermodynamics of irreversible processes together with Onsager's principle of kinetic coefficient symmetry or (*b*) the kinetic approach which uses, for example, the Chapman-Enskog method for the solution to the Boltzmann equations. Either approach yields an absolutely identical structure for the transfer equations (Kolesnichenko & Tirsky 1976). For neutral atomic gas mixtures the transfer coefficients are obtained via Sonine polynomials (Hirschfelder et al 1954) in a restricted model using low order terms for the disturbed distribution function calculation. As a rule, the error does not exceed 0.3% for the viscosity coefficient and 0.5% for the heat conductivity coefficients, and is up to 10% for the thermal diffusion coefficients, provided the temperature doesn't exceed the dissociation temperature (2000–4000 K).

For partly ionized gases higher order approximations are necessary (Devoto 1966). For example, a third order approximation is required for the heat conduction coefficients of air (Vasil'evsky et al 1986). The standard procedure in the Chapman-Enskog method yields transfer equations in the form of expressions for the fluxes across the boundaries via gradients of hydrodynamic variables. In this case, complicated expressions are obtained, in the form of ratios of $(N\xi+1)$- and $(N\xi)$-order determinants, for the multicomponent diffusion and thermal diffusion coefficients $D_{ij}(\xi)$ and $D_i^T(\xi)$, and for the heat conduction coefficients $\lambda(\xi)$. (ξ is the number of terms retained in the Sonine polynomial expansion of the disturbed distribution function.) Transfer equations in a such complicated form are rarely used for any practical tasks. Tirsky (1974) and Kolesnikov & Tirsky

(1982) deduced exact transfer equations with simpler transfer coefficients in any approximation:

$$\mathbf{d}_i = \sum_{k=1}^{N} \frac{x_i x_k}{\mathscr{D}_{ik}(1) f_{ik}(\xi)} (\mathbf{V}_k - \mathbf{V}_i) - k_{\mathrm{T}i}(\xi) \nabla \ln T \qquad (i = 1, \ldots, N) \tag{5.1}$$

$$\mathbf{J}_{\mathrm{q}} = -\lambda(\xi)\nabla T + nkT \sum_{j=1}^{N} (k_{\mathrm{T}i}/\rho_j)\mathbf{J}_j + \sum_{k=1}^{N} h_k \mathbf{J}_k \tag{5.2}$$

where

$$\mathbf{d}_i = \nabla x_i + (x_i - c_i)\nabla \ln p - (c_i/p)\left(\rho \mathbf{F}_i - \sum_{k=1}^{N} \rho_k \mathbf{F}_k\right)$$

$$x_i = n_i/n, \qquad n = \sum_{k=1}^{N} n_k, \qquad m = \sum_{k=1}^{N} m_k x_k.$$

Here correction functions $f_{ik}(\xi)$, thermal diffusion ratio $k_{\mathrm{T}i}$, and heat conduction coefficient $\lambda(\xi)$ are expressed in terms of the determinants of the rank $N(\xi - 1)$. The determinant components are just full integrals, which depend on the collision integrals of each species pair and on the mixture composition. $\mathscr{D}_{ik}(1)$ is the first order binary diffusion coefficient.

Transfer equations in the form of (5.1) and (5.2) have also been derived phenomenologically using the methods of thermodynamics of irreversible processes (Kolesnichenko & Tirsky 1976, Pilyugin & Tirsky 1989). These equations are widely applied to solve various viscous multicomponent gas flow problems, especially when numerical methods of high-order approximation are used and when the fluxes $\mathbf{J}_i (i = 1, \ldots, N)$ and $\mathbf{J}_{\mathrm{q}}$ are considered to be the unknown functions (Kovalev & Suslov 1981, Peigin 1987).

6. INTERNAL DEGREES OF FREEDOM ENERGY FLUXES

Allowing the relaxation of internal degrees of freedom doesn't affect the particles' interaction potentials which depend on their molecular mass (Eucken approximation). Galkin & Kogan (1969) and Tirsky (1987) obtained the following expression for the specific energy flux of the particles' internal motion:

$$\mathbf{q}_i^{(\mathrm{in})} = \sum_{\alpha} n_{i\alpha}\varepsilon_{i\alpha}\mathbf{V}_{i\alpha} = e_i^{(\mathrm{in})}\mathbf{J}_i - \rho_i \mathscr{D}_i \nabla e_i^{(\mathrm{in})} = e_i^{(\mathrm{in})}\mathbf{J}_i + \rho \mathscr{D}_i e_i^{(\mathrm{in})} \nabla c_i - \rho \mathscr{D}_i \nabla (c_i e_i^{(\mathrm{in})}),$$

$$\mathscr{D}_i^{-1} = \sum_{k=1}^{N} x_k \mathscr{D}_{ik}^{-1}. \tag{6.1}$$

To account for the transfer Equations (5.1) for a mixture of electroneutral species ($\mathbf{F}_i = \mathbf{F}_g$), it is more convenient to express the first of Equations (6.1) as (Tirsky 1987):

$$\mathbf{q}_i^{(\text{in})} = -\rho\mathscr{D}_i\nabla(c_i e_i^{(\text{in})}) + c_i e_i^{(\text{in})}\Bigg[(m/\Delta_i)\sum_{k=1}^{N}(\Delta_{ik}^{(0)}/m_k)\mathbf{J}_k - \rho\mathscr{D}_i\alpha_{\text{T}i}^{(0)}\nabla\ln T - \rho\mathscr{D}_i\alpha_{\text{p}}^{(0)}\nabla\ln p\Bigg] \tag{6.2}$$

where

$$\Delta_{ik}^{(0)} = \Delta_{ik} + \sum_{s=1}^{N} x_s\left(\frac{m_k}{m} - \frac{m_s}{m}\right)\Delta_{ks}, \qquad \Delta_{ik}^{-1} \doteq n\mathscr{D}_{ik}$$

$$\Delta_i = \sum_{k=1}^{N} x_k\Delta_{ik}, \quad \alpha_{\text{T}i}^{(0)} = \alpha_{\text{T}i} - \sum_{s=1}^{N} c_s\alpha_{\text{T}s}, \quad k_{\text{T}i} = x_i\alpha_{\text{T}i}$$

$$\alpha_{\text{p}i}^{(0)} = \alpha_{\text{p}i} - \sum_{s=1}^{N} c_s\alpha_{\text{p}s}, \quad k_{\text{p}i} = x_i\alpha_{\text{p}i} = x_i - c_i.$$

Here $\alpha_{\text{T}i}$ and $\alpha_{\text{P}i}$ are the i-th species thermal and barodiffusion factors. Notice that $c_i e_i^{(\text{in})}$ is the energy of the internal degrees of freedom of the i-th species per unit mass of mixture.

If the internal quantum states have Boltzmann energy distributions with temperatures $T_i^{(\text{in})}$, then

$$\mathbf{q}_i^{(\text{in})} = e_i^{(\text{in})}\mathbf{J}_i - \lambda_i^{(\text{in})}\nabla T_i^{(\text{in})}, \qquad \lambda^{(\text{in})} = \rho\mathscr{D}_i c_i c_{vi}^{(\text{in})}, \qquad c_{vi}^{(\text{in})} = \frac{de_i^{(\text{in})}}{dT_i^{(\text{in})}} \tag{6.3}$$

where $\lambda_i^{(\text{in})}$ is the i-th species heat conduction coefficient of the energy deposited in the internal degrees of freedom, which depends on the temperature $T_i^{(\text{in})}$ [(in) = (R), (V), or (E)]. Notice that if one adds expression (6.1) to the expression for the total heat flux (5.2), written for structureless particles, then the total heat flux for a gas mixture with internal degrees of freedom appears:

$$\mathbf{J}_\text{q} = -\lambda\nabla T + \sum_{k=1}^{N} h_k^{(\text{T})}\mathbf{J}_k - \sum_{k=1}^{N}\rho_k\mathscr{D}_k\nabla e_k^{(\text{in})} \tag{6.4}$$

where

$$h_i^{(T)} = h_i + \alpha_{Ti} kT/m_i, \qquad h_i = e_i + kT/m_i.$$

Here e_i is described by (2.1). In the case when expression (6.3) is valid

$$\mathbf{J}_q = -\lambda \nabla T - \sum_{k=1}^{N} \lambda_k^{(in)} \nabla T_k^{(in)} + \sum_{k=1}^{N} h_k^{(T)} \mathbf{J}_k. \tag{6.5}$$

A similar expression may be obtained for a mixture that contains charged particles (partly ionized gas), if we exclude the electric field in relations (5.1) (which accounts for the charged particle motion, by the mixture quasineutrality condition). Since the magnetic field **H** has a relativistically small effect it may be dismissed. Notice that only expressions with $\Delta_{ik}^{(0)}$ alter (6.2) (Tirsky 1978, Benilov & Tirsky 1979).

7. DISSOCIATION AND IONIZATION KINETICS FOR A ONE-TEMPERATURE GAS MIXTURE

If the translational temperature behind the bow shock wave is less than 8000–10,000 K, characteristic times of the physico-chemical processes in the air conform to the following hierarchy (Stupochenko et al 1967):

$$\tau_{TT} < \tau_{RT} \ll \tau_{VT} \ll \tau_D < \tau_I \tag{7.1}$$

where τ_{TT}, τ_{RT}, and τ_{VT} are the translational, rotational, and vibrational energy relaxation times, and τ_D and τ_I are the characteristic dissociation and ionization times. In this case the region of relaxation behind the bow shock wave has a specific structure which consists of sequential relaxation zones. Since dissociation and ionization proceed against a background of energy quasi-equilibrium between all degrees of freedom, a common gas temperature may be involved.

Now let's consider the expressions for the source terms from Equations (1.1) and (1.7). If it is established that R reactions proceed in the N-component mixture

$$\sum_{i=1}^{N} \nu'_{ir} A_i \underset{k_r^-}{\overset{k_r^+}{\rightleftharpoons}} \sum_{i=1}^{N} \nu''_{ir} A_i \qquad (r = 1, \ldots, R) \tag{7.2}$$

then expressions for the ω_i may be obtained. We must invoke the principles of mass action and particle balance:

$$\omega_i = m_i \dot{n}_i = m_i n \sum_{r=1}^{R} (\nu''_{ir} - \nu'_{ir}) e_r / \tau_r \tag{7.3}$$

where

$$\tau_r^{-1} = (p/kT)^{\nu''_r - 1} k_r^- = n^{\nu''_r - 1} k_r^-,$$

$$e_r = K_{pr} p^{\nu'_r - \nu''_r} \prod_{k=1}^{N} x_k^{\nu_{kr}} - \prod_{k=1}^{N} x_k^{\nu''_{kr}}$$

$$\nu'_r = \sum_{k=1}^{N} \nu'_{kr}, \quad \nu''_r = \sum_{k=1}^{N} \nu''_{kr}, \quad K_{pr} = \kappa_r (kT)^{\nu''_r - \nu'_r}, \quad K_r = k_r^+ / k_r^- .$$

Here ν'_{ir} and ν''_{ir} are the stoichiometric coefficients, A_i the symbols of chemical species, k_r^+ and k_r^- the rate constants of direct and back reactions, τ_r the "chemical" time of the back reactions, and e_r the values characterizing the reaction deviations from chemical equilibrium.

Phenomenological kinetics is actually based on the approach described above. It uses expressions for the rate constants k_r^+ and k_r^- that are taken from experiment by processing the data within the framework of classical collision theory. At the present time much literature on generalized data on the dissociation and ionization of air in conditions relevant to flight vehicles reentering the Earth's atmosphere has been published (Krivonosova et al 1987, Lin & Teare 1963, Kang & Dunn 1972, Blottner 1969, Bortner 1969, Park 1984b). An analysis of these data (Tirsky et al 1990) shows that the dissociation constants of O_2 vary over 0.5 up to 2 orders, whereas those for N_2 constants vary over 1–1.5 orders. The largest variation (up to 3 orders) is observed for the NO dissociation rate constant (the mass concentration of NO in the dissociated air, as a rule, doesn't exceed 0.05). For the exchange ($O+N_2 \rightleftarrows N+NO$, $O+NO \rightleftarrows N+O_2$) the reaction constants vary by not more than two orders. The ionization constants have significantly larger variation (Gurovich et al 1990).

A heat flux calculation along the Buran reentry trajectory, using all six kinetic data sets mentioned above, has an error up to 30% (Tirsky et al 1990). The corresponding error in the equilibrium wall temperature at the front stagnation point reaches 60–80 K when $k_{wi} = 0$ for a noncatalytic wall.

8. THERMODYNAMIC AND CHEMICAL NONEQUILIBRIUM REGIMES OF THE HYPERSONIC FLOW PAST A BODY

At air temperatures in the shock layer of about 8000–10,000 K, the time τ_{VT} is equal to the times τ_D and τ_I (Stupochenko et al 1967). In this case dissociation starts and proceeds under conditions when the vibrational relaxation is incomplete. This leads to so-called vibration-dissociation coupling. Dissociation occurs on all possible vibrational levels, not only from the upper ones. It occurs at rather low temperature in accordance with the "ladder" dissociation model, in which the molecule is raised "upstairs" in the system of vibrational levels to be situated in the energy gap kT near the dissociation threshold. A subsequent collision results in dissociation. Since the dissociation constant depends on the vibrational population distribution function—which in Boltzmann equilibrium depends on the vibrational temperature—then, eventually, the dissociation constant depends on not only the translational temperature, but the vibrational one $T^{(V)}$ also, i.e.

$$k_D(T, T^{(V)}) = k_D(T)V(T, T^{(V)})$$

$$V(T, T^{(V)}) \leq 1 \quad \text{for} \quad T^{(V)} \leq T, V(T, T) \equiv 1 \tag{8.1}$$

where $k_D(T) = k_r(T)$ is the equilibrium (one-temperature) dissociation constant discussed in Section 7.

The multiplier $V(T, T^{(V)})$ was introduced by Hammerling et al (1959), Losev & Generalov (1961), Treanore & Marrone (1962), and others to explain the luminescence and relaxation zone structure behind strong shock waves. Recently Park (1988) processed the experimental data for O_2 and N_2 dissociation constants (obtained in shock tubes during the 1960s and 1970s) within the framework of the two-temperature model. He supposed that the rate constants k_D were functions of the geometrical mean of the translational and vibrational temperatures. This suggestion has a unified set of dissociation constants for N_2 and O_2 in collisions with N_2, N, O_2, and O and covers a large amount of the experimental findings with an error within 1.5 in magnitude. Density and concentration depend weakly on this error because of the negative feedback that exists in the vibration-dissociation coupling (see Section 9).

There are many nonequilibrium kinetic models for air described in the literature. For example, the formula suggested by Marrone & Treanor (1963) is

$$V = \frac{Q(T)Q(T_U)}{Q(T^{(V)})Q(-U)}$$

where Q are the statistical sums, $T_U^{-1} = (T^{(V)-1} - T^{-1} - U^{-1})$, and U is the characteristic "negative temperature" of the probability distribution function of the vibrational states. The latter appears in the initial expression for the probability of a reaction from the vibrational level with energy ε_α:

$$p \sim \exp[-(D - \varepsilon_\alpha)/U]$$

where D is the dissociation energy. U is a model parameter. At $U = \infty$ equal probabilities of reactions from all vibrational states occur; when U decreases the reaction probability rises as the level number increases. Other expressions for V were deduced by Kuksenko & Losev (1969), Smekhov

(1982), Kuznetsov (1965, 1971), Wray et al (1970), and Smekhov & Losev (1979). The influence of nonequilibrium kinetics on heat transfer for hypersonic flow is described in the next section.

9. EXCHANGE TERMS IN THE RATE EQUATIONS FOR INTERNAL DEGREES OF FREEDOM: THE ROLE OF VIBRATION-DISSOCIATION-VIBRATION COUPLING (CVDV)

The energy balance Equation (3.1) for internal degrees of freedom contains an exchange energy source term $Q_i^{(\mathrm{in})}$ which in the general case for vibrational degrees of freedom has the following form:

$$Q_i^{(\mathrm{V})} = Q(T, V_i)+Q(e, V_i)+Q(R, V_i)+Q(V, V_i)+Q(E, V_i)+Q(C, V_i). \tag{9.1}$$

Here the first term on the right-hand-side is the i-th constituent vibrational exchange rate with the translational degrees of freedom of the heavy particles per unit time, the second term is the exchange rate with the translational degrees of the freedom of free electrons, the third the exchange rate with the rotational degrees of freedom, the fourth the exchange rate with the vibrational degrees of freedom of the other species, the fifth the exchange rate with the electron degrees of freedom, and the last item is coupled with the appearance (or disappearance) of vibrational energy in the chemical reactions. When considering reentry into the Earth's atmosphere, the first, fourth, and last terms in (9.1) make the crucial contribution in the case of weak ionization. For these terms, Tirsky & Shcherbak (1988) obtained the following expressions for dissociated air covering six standard reactions (M is an arbitrary particle):

$$\mathrm{O_2+M} \underset{k_{1M}^{-}}{\overset{k_{1M}^{+}}{\rightleftarrows}} \mathrm{O+O+M}, \qquad \mathrm{O+N_2} \underset{k_{4M}^{-}}{\overset{k_{4M}^{+}}{\rightleftarrows}} \mathrm{N+NO},$$

$$\mathrm{N_2+M} \underset{k_{2M}^{-}}{\overset{k_{2M}^{+}}{\rightleftarrows}} \mathrm{N+N+M}, \qquad \mathrm{O+NO} \underset{k_{5M}^{-}}{\overset{k_{5M}^{+}}{\rightleftarrows}} \mathrm{N+O_2},$$

$$\mathrm{NO+M} \underset{k_{3M}^{-}}{\overset{k_{3M}^{+}}{\rightleftarrows}} \mathrm{N+O+M}, \qquad \mathrm{O_2+N_2} \underset{k_{6M}^{-}}{\overset{k_{6M}^{+}}{\rightleftarrows}} \mathrm{NO+NO}.$$

In terms of the harmonic oscillator model:

$$Q(T, V_i) = \rho[e_i(T)-e_i(T_i^{(\mathrm{V})})]\tau^{-1}(T, V_i)$$

$$Q(V,V_i)=\rho\sum_{l=\text{mol.}}\frac{1}{t_l\tau_{il}U_l(T)}[(U_i+\gamma_i h\nu_i)U_l e^{t_l}-(U_l+\gamma_l h\nu_l)U_i e^{t_i}]$$

$$Q(C,V_i)\equiv Q(C,V_{O_2})=-n\tau_1^{-1}[\varepsilon_D(O_2)K_{p1}p^{-1}V(T,T_{O_2}^{(V)})x_{O_2}-\varepsilon_R(O_2)x_O^2]$$
$$+n\tau_5^{-1}(\kappa_5 x_O x_{NO}-x_{NO}x_{O_2})-n\tau_6^{-1}(x_{O_2}x_{N_2}-x_{NO})$$

where

$$\tau^{-1}(T,V_i)=(1-e^{-t_i})\sum_{k=1}^{N}Z_{ik}(P_{10})_i^{(k)},$$

$$\tau_{il}^{-1}=\tau_{il}^{-1}(V,V^1)=Z_{il}(Q_{10}^{01})_i^{(l)}e^{-t_i}=Z_{il}Q_i^{(l)},$$

$$Q_i^{(l)}=(Q_{10}^{01})_i^{(l)}e^{-t_i}=(Q_{01}^{10})_i^{(l)}e^{-t_l},$$

$$U_i(T)=c_ikT/m_i,\quad \gamma_i=c_i/m_i,\quad t_i=h\nu_i/kT=\theta_{Vi}/T,$$

$$\varepsilon_{Di}=h\nu_i[\exp(t_{U_i})-1]^{-1}-d_ih\nu_i[\exp(d_it_{U_i})-1]^{-1},$$

$$t_{U_i}=(h\nu_i/k)(T_i^{(V)-1}-T^{-1}-U_i^{-1}),\quad \varepsilon_{Ri}=\varepsilon_{Di}(T_i^{(V)}=T).$$

Here $\tau(T,V_i)$ is the i-th species vibration-translation relaxation time, Z_{ik} is the frequency of binary collisions, $(P_{10})_i^{(k)}$ is the first vibrational level deactivation probability in the i-j species collisions, τ_{ij} is the i-j vibrational relaxation time, $(Q_{10}^{01})_i^{(l)}$ the probability of a transition from the first excited level to the ground molecular state for a particle of the i-th component by i-l species collisions (when transition from the ground to the first level of the molecule of l-th component occurs), $\varepsilon_D(O_2)$ is the alteration of the mean vibrational energy per dissociated particle (per act of dissociation), $\varepsilon_R(O_2)$ just the same but for recombination, $(d-1)$ is the number of levels in the constrained harmonic oscillator model, and τ_i is the back reaction time (see 7.3). The transition probability data are reviewed by Millikan & White (1963) and Dmitrieva (1987).

At high temperature ($T>10{,}000$ K) vibration relaxation times taken from Millikan & White (1963) are underestimated. This causes a decrease in the detachment of the shock wave on which the emission maximum is achieved (Park 1984a). Park (1984a) updated the vibration relaxation model by including an effective collision cross-section for the calculation of $\tau(T,V_i)$ at high temperatures.

Thus, two-temperature kinetics (8.1) and the influence of back reactions via the term $Q(c,V_i)$ in (9.1) result in vibration-dissociation-vibration coupling (CVDV). Ladnova (1964, 1969) using the TVSL model and Tirsky (1989), Zhluktov & Tirsky (1990), and Tirsky & Shcherbak (1990) using the total viscous shock layer model showed that along both the Space Shuttle and Buran descending path CVDV increases the heat flux up to 25% and the equilibrium wall temperature up to 100 K. But uncertainties remain in the probabilities of dissociation from different vibrational levels, so no final conclusions about the influence of CVDV on heat transfer can yet be made, though its influence may be essential.

10. HETEROGENEOUS RECOMBINATION AND DEACTIVATION OF INTERNAL DEGREES OF FREEDOM

Before stating this problem, expressions for the convective heat flux to the wall (6.5) need to be considered.

Let L denote the number of independent (basic) components, in particular chemical elements and electrons. Then, chemical symbols for the other components (reaction products) $A_i (i = L+1, \ldots, N)$ may be expressed via the basic constituents B_j in the following form without loss of generality:

$$A_i = \sum_{j=1}^{L} \nu_{ij} B_j - q_i \qquad (i = L+1, \ldots, N) \tag{10.1}$$

where ν_{ij} are the stoichiometric matrix elements and q_i is the reaction heat per gram of product A_i. According to (10.1) the expressions for the product enthalpies h $(i = L+1, \ldots, N)$ are:

$$h_i = \sum_{j=1}^{L} \nu_{ij} h_j - q_i \qquad (i = L+1, \ldots, N). \tag{10.2}$$

If we define the element concentrations c_j^* and the element diffusive fluxes $\mathbf{J}_j^*$, then according to (10.1)

$$c_j^* = c_j + \sum_{k=L+1}^{N} \nu_{kj} \frac{m_j}{m_k} c_k, \qquad \mathbf{J}_j^* = \mathbf{J}_j + \sum_{k=L+1}^{N} \nu_{kj} \frac{m_j}{m_k} \mathbf{J}_k. \tag{10.3}$$

After excluding the $h_i (i = L+1, \ldots, N)$ from (6.5) using (10.2) and $\mathbf{J}_j (i = 1, \ldots, L)$ using (10.3), Equation (6.5) transforms to

$$\mathbf{J}_\mathrm{q} = -\lambda(\xi)\nabla T - \sum_{i=1}^{N} \lambda_i^{(\mathrm{in})} \nabla T_i^{(\mathrm{in})} + \sum_{j=1}^{L} h_j^{(\mathrm{T})} \mathbf{J}_j^* - \sum_{k=L+1}^{N} q_k^{(\mathrm{T})} \mathbf{J}_k \tag{10.4}$$

where

$$h_j^{(\mathrm{T})} = h_j + kT\alpha_{\mathrm{T}j}/m_j \qquad (j = 1, \ldots, L)$$

$$q_i^{(\mathrm{T})} = q_i - kT\left(\alpha_{\mathrm{T}i} - \sum_{j=1}^{L} \nu_{ij}\alpha_{\mathrm{T}j}\right) m_i^{-1}.$$

It follows from this, that if we use the generalized specific enthalpies $h_j^{(\mathrm{T})}$

instead of the basic constituent ones $h_j (i = L+1, \ldots, N)$, and the generalized reaction heats $q_i^{(T)}$ instead of the reaction heats q_i $(i = L+1, \ldots, N)$, and take into account the diffusive thermal effect $(\alpha_{Ti} \neq 0)$, the expression for the total heat flux (10.4) coincides in its form with the expression for $\mathbf{J}_q$ without accounting for the diffusive thermal effect. If we substitute (10.4) into (1.8), the energy equation yields

$$\mathbf{J}_q|_w = \varepsilon\sigma T_w^4 \tag{10.5}$$

where ε is a measure of the surface blackness and σ is the Stefan-Boltzmann constant. This equation may be used to find the gas temperature with the boundary condition of heat balance at the wall. The magnitude of the heat flux (10.4), and thus of the radiation equilibrium temperature of the wall (10.5), depend essentially on the boundary conditions for the species concentrations at the wall. The boundary conditions at the thermochemically stable surface include L conditions for the elemental diffusive fluxes at the wall

$$\mathbf{J}_{jw}^* = 0 \qquad (j = 1, \ldots, L) \tag{10.6}$$

and the mass balance equations for the reaction products

$$\mathbf{J}_{iw} = \rho_w k_{wi} c_{iw} \qquad (i = L+1, \ldots, N) \tag{10.7}$$

where k_{wi} is the effective catalytic constant ($k_{wi} = 0$ for an uncatalytic wall, $k_{wi} = \infty$ for a perfect catalytic wall). From (10.4) and (10.1) an expression for the heat flux to the wall may be deduced (for simplicity, relaxation is considered to be already completed at the wall), i.e.

$$\mathbf{J}_{qw} = -\lambda \frac{\partial T}{\partial y}\bigg|_{y=0} - \sum_{k=L+1}^{N} q_k^{(T)} \mathbf{J}_k|_{y=0}. \tag{10.8}$$

Thus at $k_{wi} = 0$ we have $\mathbf{J}_{qw} = -\lambda \partial T/\partial y|_{y=0}$. If $k_{wi} = \infty$ [and $c_{iw} = 0$ from (10.7)], the heat flux increases by a factor of 3–4 via the last term in (10.8). When the Space Shuttle or Buran vehicles descend, the heat transfer due to diffusion contributes up to 50% of the maximum total heat flux (at an altitude of about 70 km) (Shcherbak 1987, Tirsky et al 1990).

Thus, the catalytic constants are of great importance for a valid quantative determination of the heat flux to the wall. These constants are determined, as a rule, from experiments. Experimental research aimed at determining the k_{wi} of advanced heat-protective tiles (constructed on a base of SiO_2 with low catalytic constants $k_{wi} \sim 1\ \mathrm{ms}^{-1}$) was carried out with an electrodeless MW-plasmatrone by Gordeev et al (1983, 1985), Kolesnikow & Yukushin (1989), with an arc plasmatrone by Zhestkov & Knivel (1987), Scott (1980), Stewart et al (1988), and with a shock tube by

Berkut et al (1985a,b, 1986d,e). These experiments only covered wall temperatures up to 1000 K at a total pressure of more than 10^{-2} atm. Thus the need to derive theoretical expressions for k_{wi} arises.

Tirsky et al (1985) and Kovalev & Suslov (1987) developed a phenomenological model for catalytic reactions, which accounts for physical and chemical absorption, the interaction between the impinging atoms and adatoms (adsorbed atoms), and between the adatoms themselves. A model of the Langmuir layer with ideal adsorption was applied.

Structural expressions for k_{wi} deduced in this way are expressed as functions of temperature, pressure, and species concentration. This allows us to correctly interpret the experiments described in the literature which were carried out in a narrow temperature interval, and to determine the heat fluxes directed toward vehicles moving in the upper atmospheric layers at hypersonic velocity, which change speed and flight altitude over a wide range. Those investigations were carried out and further developed by Kuznetsov et al (1987).

In the case of a relaxing gas flow, the problem appears of setting boundary conditions for the energy (temperature) of the internal degrees of freedom. Let's consider, as an example, the case of vibrational degrees of freedom. If we proceed from the assumption that the vibrational energy relaxes when particles impinge the wall and appears or disappears in the dissociation processes, the following boundary condition may be obtained (Kaleshko & Lunkin 1970, Tirsky & Shcherbak 1988):

$$\left(e_i^{(V)}\mathbf{J}_i-\rho\mathscr{D}_i c_i\frac{\partial e_i^{(V)}}{\partial y}\right)_{w} = \rho_w k_w^{(V)}[e_i(T_w)-e_{iw}]$$
$$-\rho_w k_w m_i^{-1}[\varepsilon_{iDw}K_{pi}p^{-1}V(T,T_i^{(V)})x_{m_i}-\varepsilon_{iRw}x_{Ai}^2]_w$$

where $k_{wi}^{(V)}$ is a constant of heterogenous vibrational deactivation for the i-th species, A_i and M_i refer to atoms and molecules, and ε_{iD} and ε_{iR} are variations of the mean vibrational energy in a single act of dissociation or recombination. Presently, there is difficulty in finding an exact numerical determination of the $k_{wi}^{(V)}$ constants for air. So, two extremes are analyzed: $k_{wi}^{(V)} = 0$ (noncatalytic wall with respect to deactivation of internal degrees of freedom) and $k_{wi}^{(V)} = \infty$ (perfect catalytic wall). These extremes result in a maximum deviation in the heat flux value of 15%, when the glide reentry trajectory into the Earth's atmosphere is considered (Tirsky & Shcherbak 1990).

Notice that the catalytic properties of the surface significantly affect the separation of diffusive elements caused by the mutual influence of multicomponent diffusion and the selective catalytic action of the body on the nitrogen and oxygen atoms. The maximal deviation of the oxygen concentration along the glide reentry trajectory is about 20% (Kovalev & Suslov 1988). The profile of the electron density n_e across the shock layer depends weakly on the catalytic properties of the surface.

11. INTERACTION BETWEEN GAS-PHASE EXCHANGE AND HETEROGENOUS REACTIONS OF RECOMBINED ATOMS IN DISSOCIATED AIR

This interaction affects convective heat flux to a body when it undergoes aerobraking in the upper atmospheric layers (Rosner & Cibrian 1974, Agaphonov & Nicol'sky 1980) or when gasdynamical experiments are carried out aimed at determining the catalytic properties of the materials in the flow of dissociated air past models (Voronkin & Zalogin 1980). When dissociated air flows around most of the heat-protective coverings in practical use, the heterogenous oxygen atom recombination rate is usually faster than that for nitrogen atoms [$k_w(O) > k_w(N)$]. Owing to this a noticeable amount of the O_2 molecules is generated near the body surface which causes a gas-phase exchange reaction cycle:

$$O_2 + N \rightarrow NO + O + 32.4 \text{ kcal/mol}$$

$$NO + N \rightarrow N_2 + O + 76.4 \text{ kcal/mol}. \tag{11.1}$$

The final result is the generation of one nitrogen molecule and two oxygen atoms and, since these reactions are exothermic, an increase in heat flux to the surface of the body.

12. GENERATION OF EXCITED PARTICLES IN THE FLOW AND AT THE SURFACE OF THE BODY

Since relaxation of internal degrees of freedom in rarefied air at high altitude ($H > 60$ km) is slow, electronically excited particles can be generated. For example, it was found that a considerable amount of electronically excited O_2 molecules in states $a^1\Delta_g$, $b^1\Sigma_g^+$, and $A^3\Sigma_u^+$ is generated by the heterogenous recombination of oxygen atoms, which appear in the gas phase because of their desorption from the surface (Melin & Madix 1971, Black & Slander 1981). The generation of nitrogen molecules in the $A^3\Sigma_u^+$ and $B^3\Pi_g$ metastable states by heterogenous atom recombination was reported by Harteck et al (1960), Reeves et al (1960), and Schub (1983).

Berkut et al (1985a, 1988) developed a kinetic scheme of heterogenous and homogenous oxygen recombination, including the generation of electronically excited oxygen molecules whose energy is less than the dissociation threshold. (This situation corresponds to the low-altitude part of multipurpose vehicle trajectories, when their speed isn't sufficient to provide conditions for nitrogen dissociation). It has the following form:

$$\mathrm{O}+\mathrm{O}(S) \rightarrow \mathrm{O}_2+S$$

$$\mathrm{O}+\mathrm{O}(S) \rightarrow \mathrm{O}_2^*(a, b, A)+S$$

$$\mathrm{O}_2^*(a, b, A)+M \rightarrow \mathrm{O}_2^*(a, b)+M$$

$$\mathrm{O}_2^*(a, b, A)+M \rightarrow \mathrm{O}_2+M$$

$$\mathrm{O}_2^*(A)+\mathrm{O} \rightarrow \mathrm{O}^*(S)+\mathrm{O}_2$$

$$\mathrm{O}^*(S)+\mathrm{O}_2^*(a) \rightarrow \mathrm{O}_2^*(A)+\mathrm{O}$$

$$\mathrm{O}^*(S)+M \rightarrow \mathrm{O}+M$$

$$\mathrm{O}_2^*(a, b, A)+S \rightarrow \mathrm{O}_2^*(a, b)+S$$

$$\mathrm{O}_2^*(a, b, A) \rightarrow \mathrm{O}_2+S$$

$$\mathrm{O}^*(S)+S \rightarrow \mathrm{O}+S$$

where $\mathrm{O}_2^*(a, b, A)$ denotes the $a^1\Delta_g$, $b^1\Sigma_g^+$, and $A^3\Sigma_u^+$ electronic forms of oxygen, S is the surface of the body, and M is any particle in the gas phase.

In the framework of this reaction scheme, equations for the boundary layer surrounding the stagnation point were solved for the following regimes: $p < 10^{-3}$–10^{-2}, $T_e = 3500$–4500 K, $\beta = dU_e/dx|_{x=0} = 10^4$–$10^5$ s^{-1}. It appears that the energy carried away by the excited particles may decrease the heat flux to the surface by 10–20%. This decrease in heat flux is generally preceded by the ejection from the surface of electronically excited particles in metastable states [singlet oxygen $\mathrm{O}_2^*(a)$]. This circumstance must be taken into account when the probability of heterogenous recombination is measured via gasdynamical methods (Berkut et al 1986e).

At low pressure, electronically excited particles are generally generated at the surface of the body. As was shown by Polack et al (1976) and Brennen & McIntyre (1982), electronically excited N_2 molecules ($A^3\Sigma_u^+$, $B^3\Pi_g$) may be generated at some surfaces by heterogenous nitrogen recombination. In turn, electronically excited nitrogen atoms $\mathrm{N}(^2P)$ are generated in the gas-phase quenching of $\mathrm{N}_2(A^3\Sigma_u^+)$ molecules by nitrogen atoms (Polack et al 1976). The following scheme for generating $\mathrm{N}_2^*(A, B)$

molecules and $N^*(P)$ atoms and for their subsequent quenching was suggested by Berkut et al (1986b, 1987, 1988):

$$N + N(S) \rightarrow N_2 + S$$

$$N + N(S) \rightarrow N_2^*(A, B) + S$$

$$N_2^*(A) + N \rightarrow N_2 + N^*(P)$$

$$N_2^*(A) + N_2^*(A) \rightarrow N_2 + N_2^*(B)$$

$$N_2^*(A) + M \rightarrow N_2 + M$$

$$N_2^*(B) + M \rightarrow N_2(A) + M$$

$$N^*(P) + M \rightarrow N + M$$

$$N_2^*(A, B), N^*(P) + S \rightarrow N_2, N + S$$

$$N_2^*(B) \rightarrow N_2^*(A) + h\nu$$

where $N_2^*(A, B)$ and $N^*(P)$ are the $A^3\Sigma_u^+$, $B^3\Pi_g$, or $N(^2P)$ electronically excited states of the nitrogen molecules or atoms, respectively. The equations for the boundary layer, coupled with the indicated set of non-equilibrium reactions, were solved for conditions of gasdynamical experiments in a shock tube. These experiments were aimed at determining the probability of heterogenous recombination, and the nature of conditions for a descending Space Shuttle. At the stagnation point, the pressure was assumed to be less than 10 atm, $T_e = 5500$–6500 K, and $\beta = dU_e/dx|_{x=0} = 10^4$–$10^5\ s^{-1}$.

It was found that the excited particles may carry away up to 40% of the energy (Berkut et al 1986b, 1987, 1988). The electronically-excited particles in the metastable states $N_2(A)$ carry away most of this energy, but vibrationally-excited N_2 molecules may also have some contribution. Indeed, when the pressure is raised gas-phase reactions intensify. Doroshenko et al (1988, 1989, 1990a,b) developed a kinetic scheme for nitrogen atom recombination in the gas phase, which accounts for generation and recombination of either electronically- or vibrationally-excited particles. Two recombination channels with the same yield were considered:

$$N(^4S) + N(^4S) + M \rightarrow N_2(X^1\Sigma_g^+, v = 50) + M$$

$$N(^4S) + N(^4S) + M \rightarrow N_2(X^1\Sigma_g^+, v = 25) + M.$$

The kinetics of the vibration energy transfer within the nitrogen molecule include as many as 67 vibrational levels. In order to solve the appropriate set of gasdynamical and kinetic equations in the region of the stagnation point within the framework of boundary layer theory, the evolution numerical scheme was developed. The calculations showed, that for typical conditions at the outer edge of the boundary layer ($p = 0.1$–1 atm, $T_e = 6000$–8000 K), the energy carried away by vibrationally-excited nitrogen may reach 20–40% of the total heat flux. This is in agreement with experimental findings. Notice that the energy is generally carried away by molecules with excited vibrational levels $v \lesssim 10$–15. The vibrational distribution of N_2 molecules is essentially a non-Boltzmann one, and at low temperatures of the body surface ($T_w \sim 300$ K) an inverted population is possible in the near surface layer.

13. INFLUENCE OF ELECTRONICALLY EXCITED PARTICLES ON GAS-PHASE KINETICS

Berkut et al (1986c) demonstrated that nonequilibrium excited particles affect the exchange reaction cycle (11.1) mostly at the near surface layer. The presence of a large amount of singlet oxygen, which is generated at the surface, inhibits the exchange reaction cycle, since the first reaction has lower efficiency (by about two orders of magnitude) in comparison with the others. This phenomenon is of practical importance since it decreases the number of nitrogen atoms, replacing them by oxygen atoms along the vehicle. This, in turn, affects the heat transfer, reducing it by 10–15%.

14. LOCAL CHEMICAL EQUILIBRIUM FLOWS OF GAS MIXTURES WITH DIFFERENT SPECIES DIFFUSIVE PROPERTIES

This section is included in this survey because an approximate method for chemical equilibrium flows, which accounts for dissipative processes, is widely used. This approach is valid only when the binary diffusion coefficients are equal and thermal and barodiffusion are neglected. Indeed, in the case of dissociated air, the two latter processes contribute little to the effective transfer coefficients, but the difference in diffusion coefficients is noticeable. It is especially important for flows over ablating or burning surfaces when the gas mixture contains species with different masses. The conditions at which chemical equilibrium flow forms (i.e. when the maximal "chemical" time of back reactions become much less than the hydrodynamic time) appear when the vehicle altitude comes down to $H = 40$–50 km at Mach number $M_\infty > 6$. As a rule, under these conditions, the boundary layer forms with an outer inviscid flow surrounding it. For this reason chemical equilibrium flows are often considered within the framework of solutions to the Prandtl or Euler equations. This approach uses equations for a homogenous (one-component) gas, but the effective heat conduction coefficient and effective heat capacity are employed. These depend on the pressure and temperature and are defined at constant elemental concentration. For a rigorous analysis, elemental diffusion should be included in the model, since the different diffusive properties of the constituents cause their separation (Tirsky 1969, 1978). Now let's consider the main features of a rigorous formulation of the problem of chemical equilibrium flow of a multicomponent gas mixture with different diffusive properties of each species.

There exists a voluminous literature devoted to transfer phenomena in chemical (ionization) equilibrium gas mixtures. Nernst (1926) was the first to consider heat transfer in a gas at rest taking into account the additional diffusive transfer of "chemical" energy. Later on, this approach for binary gas mixtures at rest in which only dissociation occurs was developed by Von Haase (1953), Von Meixner (1953), and Hirschfelder (1957). Physical substantiation of this process comes from reaction products diffusing from the regions of higher temperature (where they are generated) to regions of lower temperature, where back reactions occur and where heat is released, thereby increasing the effective heat conduction coefficients.

For a multicomponent gas mixture at rest ($\nabla p = 0$) with any number of fast dissociation reactions, an effective heat conduction coefficient was obtained by Batler & Brokaw (1957) and Brokaw (1960) and for equilibrium partially ionized mixtures by Krinberg (1965) and Luchina (1975). In all these papers only the effective heat conductance coefficient for the media at rest was deduced, i.e. molecular heat transfer only, but not molecular mass transfer by elemental diffusion. The latter always occurs in multicomponent mixtures of constituents with different diffusive properties provided temperature gradients exists (even when thermal diffusion is absent) or a pressure gradient exists. The diffusion of elements yields an additional term in the heat flux; the pressure gradient yields an additional term in the energy equation. The diffusion of elements in equilibrium flows also causes an increase in the number of cross effective transfer coefficients, resulting in a diffusion flow of certain particular elements that depends on the gradients of the other element concentrations (Tirsky 1969, Suslov et al 1971, Vasil'evsky & Tirsky 1991).

In the case when the maximal "chemical" reaction time is much less than the hydrodynamic one ($t_{\text{hyd}} \sim L/V$), the convective diffusion equations for the reaction products degenerate into conditions of local chemical equilibrium [Guldberg-Vaage conditions for chemical reactions and Saha conditions for ionization, i.e. $e_r = 0$ in (7.3)]. The equations $e_r = 0$ ($r = L+1, \ldots, N$) may be considered as first integrals of the Navier-Stokes equations and used for simplifying the remaining differential equations.

As a result, the total convective heat flux (10.4) for chemical equilibrium flows yields (Tirsky 1978):

$$\mathbf{J}_q = -\lambda_{\text{eff}}\nabla T - \sum_{j=1}^{L} b_j^* \mathbf{J}_j^* \tag{14.1}$$

where

$$\lambda_{\text{eff}} = \lambda + \lambda_R, \qquad b_j^* = b_j(q_i) - h_j^{(T)}.$$

Here λ_{eff} is the effective heat conduction coefficient, equal to the sum of the molecular heat conduction coefficient—which characterizes collisional heat transfer between internal and translational degrees of freedom, and the heat conduction coefficient λ_R—which characterizes the energy transferred by chemical reactions via the reaction products diffusion. b_j is a known function of the reaction heats, species diffusive properties, and mixture composition. For an open system of a liquid particle in a flow the enthalpy change is

$$dh = c_{\text{peff}}\, dT - a(v, q_i)/\rho\, dp - \sum_{j=1}^{L} a_j^*\, dc_j^* \tag{14.2}$$

where

$$c_{\text{peff}} = c_p + c_R, \qquad a_j^* = a_j(q_i) - h_j.$$

Here c_p is the specific heat at constant pressure, c_R is the additional specific heat coupled to the heat absorption in the reactions, and $a(v, q_i)$ and $a_j(q_i)$ are known functions of the reaction heats and chemical composition (Tirsky 1978). Using (14.2) the heat flux (14.1) is

$$\mathbf{J}_q = -\frac{\mu}{\sigma_{\text{eff}}}\left[\nabla h + a(v, q_i)\frac{\nabla p}{\rho} + \sum_{j=1}^{L}\left(a_j^*\nabla c_j^* + \frac{\sigma_{\text{eff}}}{\mu} b_j^* \mathbf{J}_j^*\right)\right] \tag{14.3}$$

where the effective Prandtl number is

$$\sigma_{\text{eff}} = \frac{\mu c_{\text{peff}}}{\lambda_{\text{eff}}}.$$

If we substitute (14.3) into the energy Equation (1.8), the energy equation for chemical equilibrium flow yields:

$$\rho\frac{dH}{dt} - \frac{\partial p}{\partial t} = \nabla\cdot\left\{\frac{\mu}{\sigma_{\text{eff}}}\left[\nabla H + \frac{\sigma_{\text{eff}}}{\mu}\mathcal{T}\cdot\mathbf{v} - \nabla\left(\frac{\mathbf{v}^2}{2}\right)\right.\right.$$
$$\left.\left. + a(v, q_i)\frac{\nabla p}{\rho} + \sum_{j=1}^{L}\left(a_j^*\nabla c_j^* + \frac{\sigma_{\text{eff}}}{\mu} b_j^* \mathbf{J}_j^*\right)\right] + \mathbf{J}_R\right\}. \tag{14.4}$$

The concentrations c_j^* and the elemental diffusion fluxes (10.3) should be found from the equations for elemental diffusion:

$$\rho \frac{dc_j^*}{dt} + \nabla \cdot \mathbf{J}_j^* = 0 \qquad (j = 1, \ldots, L) \tag{14.5}$$

which can be supplemented by the transfer equations (5.1), which in the case of chemical equilibrium flow transform into (Vasil'evsky et al 1986)

$$\nabla c_j^* + \left(K_{\mathrm{T}j}^* - \frac{m_j}{m} \delta_j^{(\mathrm{e})} \right) \nabla \ln T + K_{\mathrm{p}j}^* \nabla \ln p = -\frac{S_j^{(\mathrm{e})}}{\mu} \mathbf{J}_j^* + \frac{m_j S_j^{(\mathrm{e})}}{\mu} \sum_{l=1}^{L} \frac{\alpha_{jl}^{(\mathrm{e})}}{m_l} \mathbf{J}_l^* \tag{14.6}$$

where $S_j^{(\mathrm{e})}$ is the effective Schmidt number. For the model mixture, when the coefficients of the binary diffusion are equal ($\mathscr{D}_{ij} = D$), $\delta_j^{(\mathrm{e})} = 0$. Then the right-hand-side part of (14.6) becomes proportional to the diffusion flux of each element, i.e. to $\mathbf{J}_j^*$. Then, if thermal and barodiffusion are neglected, the Fick law is obtained from (14.6) for the elemental diffusion: $\mathbf{J}_j^* = -\rho D \nabla c_j^*$. In this case, for boundary conditions

$$c_j^*(\infty) = c_{j\infty}^*, \qquad \mathbf{J}_{j\mathrm{w}}^* = 0 \qquad (j = 1, \ldots, L) \tag{14.7}$$

the solution is: $c_j^* = c_{j\infty}^*$, $\mathbf{J}_j^* = 0 (j = 1, \ldots, L)$. The energy equation (14.4) becomes similar to the one for a homogeneous liquid with an effective Prandtl number. This approximate approach is used in most cases. If one does not use the simplifying assumption $\mathscr{D}_{ij} = D$, the problem becomes more complicated and the effect of elemental separation in the flow yields c_j^* which is not equal to $c_{j\infty}^*$ in the freestream.

Element diffusion models become simpler for two-element mixtures. This assumption is valid for many of the planetary atmospheres in the solar system: Earth (O_2, N_2), Venus and Mars (CO_2), and Jupiter and Saturn (H_2, He).

Since the mixture is quasineutral, $c_{\mathrm{e}}^* = 0$ for the third element (electrons). When electrical current to the body is absent $\mathbf{J}_{\mathrm{ew}}^* = 0$ (Benilov & Tirsky 1979). In this case, for the chemical equilibrium flows relations (14.6) also take on the form of Fick's law (j and l denote two elements):

$$\mathbf{J}_j^* = -\mathbf{J}_l^* = -\frac{\mu}{S_j^{(\mathrm{e})}} \left[\nabla c_j^* + K_{\mathrm{p}j}^* \nabla \ln p + \left(K_{\mathrm{T}j}^* - \frac{m_j}{m} \delta_j^{(\mathrm{e})} \right) \nabla \ln T \right]. \tag{14.8}$$

It is important to emphasize that the coefficient $(m_j/m)\delta_j^{(\mathrm{e})}$ is of $\mathscr{O}(1)$ in the case of different diffusion coefficients $\mathscr{D}_{ij}$ and the effect of elemental separation has finite values in this case. But the influence of it on heat transfer in air is rather small ($\approx$ 5–8%) (Vasil'evsky et al 1986, Vasil'evsky & Tirsky 1991).

15. CONCLUSIONS

Practically all atomic-molecular and ionic processes become important in the shock layer near a body when it moves at high altitude in the atmospheres of the Earth or other planets at hypersonic velocity. The rates of these reactions at temperatures $T > 8000$–10,000 K are of the same order as the residence time of fluid particles in the shock layer. During the 1960s and 1970s some of these processes were investigated separately for strong shock waves. The main feature of recent real gas hypersonic flow investigations is that they take into account all of these processes simultaneously, considering viscosity, heat conduction, and diffusion in the framework of PNS, TVSL, or the complete Navier-Stokes equations. There are at least two problems remaining. The first is the absence of reliable quantitative data about many kinetic constants at high temperature (two-temperature dissociation and ionization constants, constants of exchange reactions, TV, VV′—relaxation constants, etc). Secondly, hypersonic flows over real configurations, when all the abovementioned physical-chemical processes are accounted for, represent a substantial calculation problem from the point of view of either the development of new and more effective algorithms or of restricted speeds of contemporary computers.

Literature Cited

Agafonov, V. P., Nikol'sky, V. S. 1980. The interaction of homogenous and surface reactions at strong dissociated air flow in the boundary layer. *Uch. Zap. TsAGI* 2: 46–53

Allen, R. A., Rose, P. H., Camm, J. C. 1963. Nonequilibrium and equilibrium radiation at super-satellite re-entry velocities. *IAS Pap. No. 63-77*

Anderson, J. D. 1984. A survey of modern research in hypersonic aerodynamics. *AIAA Pap. No. 84-1578*

Batler, J. N., Brokaw, R. S. 1957. Thermal conductivity of gas mixtures in chemical equilibrium. *J. Chem. Phys.* 26: 1470–75

Benilov, M. S., Tirsky, G. A. 1979. To the calculation of electric effects in ionized gas flow around electroconductive body. *Prikl. Mat. Mekh.* 43: 286–304

Berkut, V. D., Kudryavtsev, N. N., Novikov, S. S. 1985a. The influence of the formation electronically excited O_2 molecules by heterogeneous atoms recombination on the heat transfer to blunted bodies surface and plate in hypersonic flow. In *Fizicheskaya Gazodinamika.* Minsk. Inst. Heat and Mass Transfer Akad. Nauk BSSR. pp. 74–92

Berkut, V. D., Kovtun, V. V., Kudryavtsev, N. N., Novikov, S. S. 1985b. Method of determination probability of heterogeneous atom recombination induced by interaction of supersonic flow with surfaces. *Khim. Fiz.* 4: 673–83

Berkut, V. D., Kudryavtsev, N. N., Novikov, S. S. 1986a. The influence of electronically excited molecules formation in the heterogeneous atoms recombination on the heat transfer to the surface of flow around a body. *Khim. Fiz.* 5: 95–105

Berkut, V. D., Kudryavtsev, N. N., Novikov, S. S. 1986b. The influence of the formation electronically excited nitrogen molecules at heterogeneous atoms recombination on the heat transfer to the surface. *Khim. Vys. Energ.* 4: 374–80

Berkut, V. D., Kudryavtsev, N. N., Novikov, S. S. 1986c. The influence of electronically excited molecules formated on surface on chemical reaction kinetics in supersonic flow around a body. *8th*

Vsesojuzny Simp. Goreniyu Izlucheniyu. pp. 106–10
Berkut, V. D., Kovtun, V. V., Kudryavtsev, N. N., Novikov, S. S., Sharovatov, A. I. 1986d. Determination of momentary values of heterogeneous atom recombination probability in shock-tube experiments. Minsk. Inst. Heat and Mass Transfer Akad. Nauk BSSR. *Preprint No. 12*
Berkut, V. D., Kovtun, V. V., Kudryavtsev, N. N. 1986e. Thermophysical properties of surfaces induced by chemical energy accommodation of supersonic flow of dissociated gas. In *Obz. Teplofiz. Svoystvam Veschestv*. Akad. Nauk SSSR. Inst. Visokikh Temperature 2(58): 3–135
Berkut, V. D., Kudryavtsev, N. N., Novikov, S. S. 1987. The heat transfer to a surface blown by dissociated air with formation of electronically excited molecules in heterogeneous recombination reactions. *Teplofiz. Vys. Temp.* 25: 340–48
Berkut, V. D., Kudryavtsev, N. N., Novikov, S. S., Smetanin, V. V. 1988. The influence of the heterogeneous formation of electronically excited molecules on the heat transfer in hypersonic air flow around a body. Inst. Problem Mekh. Akad. Nauk SSSR. *Preprint No. 347*
Black, G., Slanger, T. G. 1981. Production of $O_2(a^1\Delta_g)$ by oxygen atom recombination on pyrex surface. *J. Chem. Phys.* 74: 6517–19
Blottner, F. G. 1969. Viscous shock layer at the stagnation point with nonequilibrium air chemistry. *AIAA J.* 7: 2281–88
Bortner, M. H. 1969. A review of rate constants of selected reaction of interest in reentry flow in the atmosphere. *NBS Tech. Note 484*
Brennan, W., McIntyre. 1982. Vibrational relaxation and electronic mutation of metastable nitrogen molecules generated by nitrogen atom recombination on cobalt and nickel. *Chem. Phys. Lett.* 90: 457–60
Brokaw, R. S. 1960. Thermal conductivity of gas mixtures in chemical equilibrium. 2. *J. Chem. Phys.* 32: 936–39
Candler, C., Park, C. 1988. The computation of radiation from nonequilibrium hypersonic flows. *AIAA Pap. No. 88-2678*
Cheng, H. K. 1961. Hypersonic shock-layer theory of the stagnation region at low Reynolds number. *Proc. Heat Transfer Fluid Mech. Inst.* Stanford Press. pp. 161–75
Davis, R. T. 1970. Numerical solution of the hypersonic viscous shock-layer equation. *AIAA J.* 8: 843–51
Davis, R. T., Rubin, S. G. 1980. Non-Navier-Stokes viscous flow computations. *Comput. Fluids* 8: 101–31
Devoto, R. S. 1966. Transport properties of ionized monatomic gases. *Phys. Fluids* 9: 1230–40
Dmitrieva, I. K. 1987. Analysis and estimation of the agreement of data on times and rate constants of vibrational relaxation of nitrogen and oxygen molecules. Minsk. Inst. of Heat and Mass Transfer. Akad. Nauk BSSR. *Preprint No. 11*. 32 pp.
Doroshenko, V. M., Kudryavtsev, N. N., Novikov, V. V., Smetanin, V. V. 1988. The influence of formation of vibrationally excited molecules of nitrogen on the heat transfer at recombination of the atoms in boundary layer. *Dokl. Akad. Nauk SSSR* 301: 1131–35
Doroshenko, V. M., Kudryavtsev, N. N., Smetanin, V. V. 1989. The heat transfer at supersonic dissociated nitrogen flow around a body with consideration of vibrationally excited particles in boundary layer and on the body surface. *Mat. Modelirovan.* 12: 13–21
Doroshenko, V. M., Kudryavtsev, N. N., Smetanin, V. V. 1990a. Macroscopic model of vibrational relaxation in heat transfer problems in supersonic flow around bodies. *Teplofiz. Vys. Temp.* 5: 952–59
Doroshenko, V. M., Kudryavtsev, N. N., Smetanin, V. V. 1990b. Vibrational nonequilibrium in the dissociated nitrogen flow past a blunt body. USSR Acad. Sci., Inst. High Temp. Moscow. *Preprint No. 1–294*
Galkin, V. S., Kogan, M. N. 1969. About nonequilibrium multiatomic gas flow equation in Eucken approximation. In *Problemy Gidrodinamiki i Mekh. Sploshnoi Sredy*, ed. M. A. Lavrentev, pp. 119–28. Moscow: Nauka
Gershbein, E. A., Peigin, S. V., Tirsky, G. A. 1985. Supersonic flow around bodies at low and moderate Reynolds number. *Itogi Nauki Tekh. VINITI. Ser. Mech. Fluids Gases* 19: 3–85
Gogosov, V. V., Polyansky, V. A. 1976. Electrodynamics: problems and applications, governing equations, discontinuous solutions. *Itogi Nauki Tekh. VINITI. Ser. Mech. Fluids Gases* 10: 5–85
Gordeev, A. N., Kolesnikov, A. F., Yakushin, M. I. 1983. The investigation of heat transfer using models in subsonic jet streams of induction plasmotron. *Izv. Akad. Nauk SSSR, Mekh. Zhidk. Gaza* 6: 129–35
Gordeev, A. N., Kolesnikov, A. F., Yakushin, M. I. 1985. The influence of catalytic activity of surface on non-

equilibrium heat transfer in subsonic stream of dissociated nitrogen. *Izv. Akad. Nauk SSSR, Mekh. Zhidk. Gaza* 3: 166–72

Goulard, R. 1958. On catalytic recombination rates in hypersonic stagnation heat transfer. *Jet Propul.* 28: 737–45

Gurovich, I. M., Tirsky, G. A., Shcherbak, V. G. 1990. Comparison of chemical reactions data systems for ionized hypersonic air blunt body flow. Moscow Inst. Mekh. Moscow Univ. *Preprint No. 4015*

Hammerling, P., Teare, J. D., Kivel, B. 1959. Theory of radiation from luminous waves in nitrogen. *Phys. Fluids* 2: 422–26

Harteck, P., Reeves, R. R., Mannella, G. 1960. Surface-catalyzed atom recombination that produce excited molecules. *Can. J. Chem.* 38: 1648–51

Hirschfelder, J. O. 1957. Heat transfer in chemical reaction mixtures. *J. Chem. Phys.* 26: 108–12

Hirschfelder, J. O., Curtis, C. F., Bird, R. B. 1954. *Molecular Theory of Gases and Liquids*. New York: Wiley

Kaleshko, S. B., Lunkin, Yu. P. 1970. Laminar boundary layer on plate with arbitrary catalytic properties in vibrationally and dissociationally relaxating gas. *Tr. Leningr. Politekh. Inst.* 313: 5–12

Kang, S. W., Dunn, M. G. 1972. Hypersonic viscous shock layer with chemical nonequilibrium for spherically blunted cones. *AIAA J.* 10: 1361–62

Kolesnichenko, A. V., Tirsky, G. A. 1976. Stefan-Maxwell relations and heat transfer in the non-ideal multicomponent continuous media. *Chislen. Meto-dy Mekh. Sploshnoy Sredy* 7: 106–21

Kolesnikov, A. F., Tirsky, G. A. 1982. Hydrodynamics equations of partially ionized multicomponent gas mixtures with higher approximation transport coefficients consideration. In *Molecular Gasodynamics*, ed. V. V. Struminsky, pp. 20–44. Moscow: Nauka

Kolesnikov, A. F., Yakushin, M. I. 1989. On determination of effective probabilities of heterogeneous atom recombination by heat fluxes to a surface in dissociated air flow. *Mat. Modelirovan.* 3: 44–60

Kovalev, V. L., Suslov, O. N. 1981. The high order approximation finite difference method for integration of chemically nonequilibrium multicomponent viscous shock layer equations. In *Giperzvukovie Prostran. Techeniya pri Nalichii Fiziko-Khimich. Prevraschenij*, ed. G. A. Tirsky. Inst. Mekh., Moscow Univ. pp. 113–37

Kovalev, V. L., Suslov, O. N. 1987. The model of interaction of partially ionized air with catalytic surface. In *Issledovaniya po Giperzvuk. Aerodinam. i Teplo-obmenu s Uchyotom Neravnov. Khimich. Reaktzy*, ed. G. A. Tirsky. Inst. Mekh., Moscow Univ. pp. 58–69

Kovalev, V. L., Suslov, O. N. 1988. The effect of diffusion elements separation on catalytic surface. *Izv. Akad. Nauk SSSR, Mekh. Zhidk. Gaza* 4: 115–21

Krinberg, I. A. 1965. The influence of ionization reactions on plasma heat conductivity. *Teplofiz. Vys. Temp.* 3: 20–31

Krivonosova, O. E., Losev, S. A., Nalivayko, V. P., Mukoseev, J. K., Shatalov, O. P. 1987. Recommended data on chemical reaction rate constant of molecules, consisting of N-O atoms. In *Khimiya Plazmy*, ed. B. M. Smirnov. Moscow. 14: 3–21

Kuksenko, P. V., Losev, S. A. 1969. The excitation and deactivation of two-atomic molecules at collisions at high temperature gas. *Dokl. Akad. Nauk SSSR* 185: 69–72

Kuznetsov, N. M. 1965. The interaction of vibrational relaxation and two-atomic molecules dissociation processes. *Dokl. Akad. Nauk. SSSR* 164: 1097–1100

Kuznetsov, N. M. 1971. The molecular dissociating kinetics in the molecular gas. *Theoretich. Exp. Khim.* 7: 22–33

Kuznetsov, V. M., Kuznetsov, M. M., Kolesnikov, A. F., Yakuchin, M. I. 1987. Theoretical and experimental problems of heterogeneous catalysis on surfaces in dissociated gas flow. In *Modelirovanie v Mekhanike*. Novosibirsk Inst. Teor. i Prikl. Mekh. Sib. otdelenie Akad. Nauk SSSR 3: 83–104

Ladnova, L. A. 1964. The laminar boundary layer on the plate with consideration of thermodynamical and chemical non-equilibrium. *Vestn. Leningr. Univ. Ser. Mat. Mekh.* 19: 114–28

Ladnova, L. A. 1969. The nonequilibrium viscous shock layer with arbitrary catalytic activity of the surface. *Vestn. Leningr. Univ. Ser. Mat. Mekh.* 13: 106–12

Lin, S. C., Teare, J. D. 1963. Rate of ionization behind shock waves in air 2. Theoretical interpretation. *Phys. Fluids* 6: 355–75

Losev, S. A., Generalov, N. A. 1961. To the investigation of excitation of vibration and dissociation of oxygen molecules at high temperature. *Dokl. Akad. Nauk SSSR* 141: 1072–75

Luchina, A. A. 1975. The influence of chemical reactions on heat transfer and diffusive fluxes in plasma. *Inzh. Fiz. Zh.* 29: 7–14

Magomedov, K. M. 1970. The hypersonic viscous gas flow past blunt bodies. *Izv. Akad. Nauk USSR, Mekh. Zhidk. Gaza* 2: 45–56

Marrone, P. V., Treanore, C. E. 1963.

Chemical relaxation with preferential dissociation from excited vibrational levels. *Phys. Fluids* 6: 1215–21

Maxwell, J. C. 1868. On the dynamical theory of gases. *Phil. Mag. J. Sci. Ser. 4* 35: 185–217

Melin, G. A., Madix, R. J. 1971. Energy accommodation during oxygen atom recombination on metal surfaces. *Trans. Faraday Soc.* 67: 198–206

Millikan, R. C., White, D. R. 1963. Systematics of vibrational relaxation. *J. Chem. Phys.* 39: 3209–13

Nernst, W. F. M. 1926. *Theoretishe Chemie vom Standpunkte der Avogadro schein Regel und der Thermodynamik*. Stuttgart: Enke

Park, C. 1983. Radiation enhancement by nonequilibrium in earth atmosphere. *AIAA Pap. No. 83-0410*

Park, C. 1984a. Calculation of nonequilibrium radiation in AOTV flight regimes. *AIAA Pap. No. 84-0306*

Park, C. 1984b. Problems of rate chemistry in the flight regimes of aeroassisted orbital transfer vehicles. *AIAA Pap. No. 84-1730*

Park, C. 1988. Two-temperature interpretation of dissociation rate date for N_2 and O_2. *AIAA Pap. No. 88-0458*

Peigin, S. V. 1987. The high-order approximation numerical method for solving two-dimensional boundary layer problems. *Zh. Vych. Mat. Mat. Fiz.* 16: 118–33

Peigin, S. V., Tirsky, G. A. 1988. Three-dimensional problems of super and hypersonic viscous gas flows past bodies. *Itogi Nauki Tekh. VINITI. Ser. Mech. Fluids Gases* 22: 62–177

Pilyugin, N. N., Tirsky, G. A. 1989. *Dynamics of Ionized Radiating Gas*. Moscow: Moscow Univ. Publish. 309 pp.

Polack, L. S., Slovetsky, D. I., Todesayte, R. D. 1976. De-excitation rate coefficients metastable particles $N_2(A^3\Sigma_u^+, v = 1, 2)$ by nitrogen atoms and molecules. *Khim. Vys. Energ.* 1: 54–70

Reeves, R. R., Mannella, G., Harteck, P. 1960. Formation of excited NO and N_2 by wall catalysis. *J. Chem. Phys.* 32: 946–47

Rosner, D. E., Cibrian, R. 1974. Nonequilibrium stagnation region aerodynamic heating of hypersonic glide vehicles. *AIAA Pap. No. 74-0755*

Sasoh, A., Chang, X., Fujiwara, T. 1990. Equilibrium and nonequilibrium radiation heat transfer over a reentry blunt body. *AIAA Pap. No. 90-2113*

Schub, B. R. 1983. *Heterogeneous relaxation of internal energy of molecules and nonequilibrium processes on the body surfaces*. PhD thesis. Inst. Khemich. Fiziki Akad. Nauk. SSSR. 284 pp.

Scott, C. P. 1980. Catalytic recombination of nitrogen and oxygen on high-temperature reusable surface insulation. *AIAA Pap. No. 80-1477*

Shcherbak, V. G. 1987. The numerical investigation of the nonequilibrium flow structure in the hypersonic three-dimensional flow around blunt bodies. *Izv. Akad. Nauk. SSSR, Mekh. Zhidk. Gaza* 5: 143–50

Smekhov, G. D. 1982. Application of adiabatic principle to calculation of two-atomic molecule dissociation constant rate. In *Neravnovesnye Techeniya Gaza i Optimalny Formy tel v Giperzvukov. Potoke*, ed. G. G. Tchorny, pp. 30–38. Moscow: Inst. Mekh. Moscow Univ.

Smekhov, G. D., Losev, S. A. 1979. The influence of vibrational-rotational excitation on two-atomic molecules dissociation. *Teor. Eksp. Khim.* 15: 492–567

Stefan, J. 1871. Uber das Gleichgewicht und die Bewegung insbesondere die Diffusion von Gasgemengen. *Akad. Wissenschaft Abheitlung II. Heft 1. s.* 63–124

Stewart, D. A., Henline, W. D., Kolodziej, P., Rincha, E. M. 1988. Effect of surface catalysis on heating to ceramic coated thermal protection systems for transatmospheric vehicles. *AIAA Pap. No. 88-2706*

Stupochenko, Ye. V., Losev, S. A., Osipov, A. I. 1967. *Relaxation in Shock Waves*. New York: Springer-Verlag

Suslov, O. N., Tirsky, G. A., Shchennikov, V. V. 1971. Description of chemically equilibrium multicomponent ionized mixtures using Navier-Stokes and Prandtl equations. *Prikl. Mekh. Tekh. Fiz.* 1: 73–89

Teare, J. D., Georgiev, S., Allen, R. A. 1962. Radiation from the nonequilibrium shock-front. In *Hypersonic Flow Research*, ed. F. R. Riddell, pp. 281–317. New York: Academic

Tirsky, G. A. 1969. The successive approximation method for integration of equation of multicomponent boundary layer with chemical reactions. Moscow Inst. Mekh. Moscow Univ. *Preprint No. 1016*

Tirsky, G. A. 1974. Equations of motion of partially ionized multicomponent gas mixtures in normal Cauchy form with exact transport coefficients. In *Tr. Inst. Mekh. Moscow Univ.*, ed. S. S. Grigoryan, 32: 6–22

Tirsky, G. A. 1975. To the theory of the hypersonic viscous chemically active gas flow past plane and axisymmetric blunt bodies with gas injection. In *Tr. Inst. Mekh. Moscow Univ.*, ed. G. A. Tirsky, 39: 5–28

Tirsky, G. A. 1978. Hydrodynamic description of chemically nonequilibrium partially ionized non-ideal gas mixtures. In *Nekotorye Voprosy Mekh. Sploschnoi Sredy*, ed. G. G. Tchorny, pp. 114–43. Moscow: Inst. Mekh., Moscow Univ.

Tirsky, G. A., Alpherov, V. I., Kovalev, V. L., Suslov, O. N. 1985. Viscous gas flow around a body at nonequilibrium homogeneous and heterogeneous reactions regime. In *Mekh. Neodnorodnykh System*. Novosibirsk Inst. Teor. Prikl. Mekh. Sib. otdelenie Akad. Nauk SSSR. pp. 255–80

Tirsky, G. A. 1987. Semiphenomenological deducing of the hydrodynamics equations of the multi-atomic gas mixtures with excited internal degrees of freedom. In *Mekhanika. Sovremennye Problemy*, ed. G. G. Tchorny, pp. 79–86. Moscow: Inst. Mech. Moscow Univ.

Tirsky, G. A., Shcherbak, V. G. 1988. The chemically and thermodynamically nonequilibrium flow around a body at small and moderate Reynolds numbers. Moscow Inst. Mekh., Moscow Univ. *Preprint No. 3645*

Tirsky, G. A. 1989. The thermodynamically non-equilibrium effects in hypersonic viscous flow around a body. In *Modeli Mekh. Neodnorodnykh Sistem*. Novosibirsk, Inst. Teor. Prikl. Mekh. Sib. otdel. Akad. Nauk SSSR. pp. 66–92

Tirsky, G. A., Utyuzhnikov, S. V. 1989. Comparison of thin and full viscous shock layer models in the problem of the supersonic viscous flows past blunted cones. *Prikl. Mat. Mekh.* 53: 963–69

Tirsky, A. G., Shcherbak, V. G. 1990. The influence of vibrational relaxation on the viscous chemically non-equilibrium air flow around a body. *Izv. Akad. Nauk SSSR, Mekh. Zhidk. Gaza* 1: 151–57

Tirsky, A. G., Shchelin, V. S. Shcherbak, V. G. 1990. The influence of uncertainty in chemical reaction on convective heat transfer. *Izv. Akad. Nauk SSSR, Mekh. Zhidk. Gaza* 6: 146–51

Tolstykh, A. I. 1966. About the numerical solution of the supersonic viscous flow problems past blunt bodies. *Zh. Vych. Mat. Mat. Fiz.* 6: 113–20

Treanore, C. E., Marrone, P. V. 1962. Effect of dissociation on the rate vibrational relaxation. *Phys. Fluids* 5: 1022–26

Van Dyke, M. D. 1962. Higher approximations in boundary layer theory. Part 1. General analysis. *J. Fluid Mech.* 17: 161–77, 481–95

Vasil'evsky, S. A., Tirsky, G. A. 1991. The effect of element diffusion in chemical equilibrium flows of multicomponent gases. In *Sovrem. Gasodinamich. i Fiziko-Khimich. Modeli Giperzvuk. Aerodinamiki i Teploobmena*, ed. L. I. Sedov, pp. 195–230. Moscow Inst. Mekh. Moscow Univ.

Vasil'evsky, S. A., Sokolova, I. A., Tirsky, G. A. 1986. The determination and calculation of effective transport coefficients for chemically equilibrium flows of partially ionized and dissociated gas mixtures. *Zh. Prikl. Mekh. Tehn. Fiz.* 1: 68–79

Von Meixner, J. 1953. Zur Theorie der Wärmeleitfähigkeit reagierender fluider Mischungen. *Z. Naturforsch.* 7a: 57–64

Von Haase, R. 1953. Zur thermodynamik der irreversibler prozess. *Z. Naturforsch.* 8a: 71–80

Voronkin, V. G., Zalogin, G. N. 1980. On mechanism of atomic nitrogen recombination near catalytic surface in dissociated air flow. *Izv. Akad. Nauk SSSR, Mekh. Zhidk. Gaza* 3: 156–58

Walberg, G. D. 1983. Aeroassisted orbital transfer—window opens a mission. *Astronaut. Aeronaut.* 21(11): 36–43

Wray, K. L., Feldman, E. V., Lewis, P. F. 1970. Shock tube study of the effects of vibrational energy of N_2 on the kinetics of the $O + N_2 \rightarrow NO + N$ reactions. *J. Chem. Phys.* 53: 4131–36

Zhestkov, B. E., Knivel, A. Ya. 1987. Experimental investigation of heterogeneous recombination. *Tr. TsAGI Vyp.* 2111: 215–27

Zhluktov, S. V., Tirsky, G. A. 1990. The influence of vibrational-dissociational interaction on the heat transfer and drag in hypersonic flow around a body. *Izv. Akad. Nauk SSSR, Mekh. Zhidk. Gaza* 1: 158–67

Annu. Rev. Fluid Mech. 1993. 25:183–214

COMPUTATIONAL METHODS FOR THE AERODYNAMIC DESIGN OF AIRCRAFT COMPONENTS

Th. E. Labrujère and J. W. Slooff

Aerodynamics Division, National Aerospace Laboratory NLR,
A. Fokkerweg 2, 1059 CM Amsterdam, The Netherlands

KEY WORDS: inverse design, optimal control, multi-point design, design constraints

INTRODUCTION

The present article reviews state-of-the-art computational aerodynamic design methods. The review is limited to methods aimed directly at the determination of geometries for which certain specified aerodynamic properties can be obtained, with or without constraints on the geometry. Cut-and-try methodologies, which utilize analysis methods only, are not considered. The review is further limited to methods which are considered representative of different approaches and to methods illustrating the latest developments. Also, airfoil and wing design methods are emphasized because of the present authors' background. Additional material may be found in the reviews by Slooff (1983), Sobieczky (1989), Meauzé (1989), and Dulikravich (1990).

In general, the development of computational design methods aims at reducing man-in-the-loop activities (i. e. increasing the level of automation) during the design process. Although automation may reduce design processing time as well as the dependence of the result on the expertise of the designer, its success depends heavily on reliability and accuracy of the computational methods and on how well the designer has set his goals.

The history of computational design method development clearly

0066–4189/93/0115–0183$02.00

reflects this dualism. It shows continuous efforts to acquire easy-to-use methods, which unfortunately sometimes happen to produce undesirable results. An example of such a result (Volpe 1989) is the airfoil designed to be shockfree, but with so-called hanging (secondary) shocks in the flow field. These cause inefficient behavior even at the design condition and drag increase due to boundary layer separation at off-design conditions.

The first computational methods for airfoil design arose from treatment of the inverse problem. This involves determining the shape of an airfoil such that on its contour an a priori prescribed pressure distribution exists at the flow condition considered. Here, the basic idea is that the designer can formulate the design requirements in terms of a target pressure distribution. Methods of this type are generally referred to as inverse design methods. The formulation of a well-posed inverse problem is not at all trivial, as has already been demonstrated by Betz (1934) and by Mangler (1938) for incompressible flow. Incorporation of inverse methods in practical designs has led to additional user requirements with respect to control over the geometry. As a consequence, the problem is often complicated by the introduction of constraints with respect to the geometry. Furthermore, in attempts to extend the range of applicability of inverse methods, increasingly complicated flow equations (full potential, Euler, Navier-Stokes) are being used. Both factors have led to a considerable increase in the effort to develop inverse design methods.

Hicks et al (1976) introduced an alternative to inverse design methods by formulating the concept of direct numerical optimization. Design methods based on this concept are formed by coupling aerodynamic analysis methods with numerical minimization schemes. The user specifies the design requirements in terms of a cost function, which takes into account any constraints. In this way existing analysis codes can be used directly for design purposes, without the need to solve the corresponding, often complicated inverse problem. Furthermore, improvements made to the analysis code become directly available for design as well. Another advantage of this type of method is its flexibility with respect to the selection of design objectives. Unfortunately, a major disadvantage of direct numerical optimization is the large amount of computing time needed for each iteration step. As a consequence, the development of design methods which follow this kind of approach shows various attempts at decreasing computing time—e.g. by introducing so-called aerofunction shapes to represent the geometry, the number of design variables can be reduced (Aidala et al 1983). Other authors, e.g. Rizk (1989), left the black box idea and mixed the minimization scheme with the analysis code to convert it into a design code.

Because of its potential flexibility with respect to the formulation of design objectives, design by optimization draws more and more attention. As an alternative to the developments mentioned in the preceding paragraph, it attempts to increase efficiency by using gradient search techniques and determining the gradients in a computationally cheap way. Most effective in this respect is probably the method of Newton iteration (Drela 1986), but the development of methods based on this concept is rather laborious. Other investigations concern the application of the calculus of variations, often referred to as optimal control (Pironneau 1983). In this approach the gradients needed for determining the search direction are calculated by solving an adjoint problem, which is usually similar to the corresponding analysis problem. Compared with direct numerical optimization it seems an effective method: Given an estimate of the geometry to be determined, the computational effort for one geometrical correction is of the same order of magnitude as for each computation required to evaluate the cost function using the analysis method.

The greater part of the present article is devoted to the single-point design problem. This problem involves determining an aerodynamic shape with specified characteristics at one single design condition. For practical aircraft design, however, it is not sufficient to consider only one design condition. That is why, gradually, methods are being developed to deal with the multi-point design problem, i.e. the optimization of an aerodynamic shape wherein an a priori weighted compromise is achieved between the required characteristics for different design conditions.

The majority of single-point design methods involve treating the inverse problem. Since the solution of the inverse problem may lead to shapes that are impractical from the designer's point of view, additional constraints on the geometry may be required. These constraints may be introduced by formulating mixed direct-inverse problems, in which one part of the geometry is kept fixed and treated just as in the direct analysis problem, while the other part is designed. Alternatively, the inverse problem may be reformulated as a minimization problem taking geometric constraints into account, and fulfilling design requirements approximately. From a practical engineering point of view, indirect methods for solving the inverse problem, such as hodograph and fictitious gas methods, are not attractive because they lack control over both geometry and aerodynamic characteristics. Therefore, such types of inverse methods are not discussed here.

Methods following the direct numerical optimization approach allow, in principle, a wider design philosophy than methods for treating the inverse design problem. They might aim at direct realization of certain aerodynamic goals—such as low drag or high lift—without depending on the designer's knowledge of the detailed aerodynamic characteristics of a given shape or pressure distribution. In principle, direct numerical optimization methods are equally applicable to both single-point and multi-point design problems. A paper by Jameson (1988) on the application of the calculus of variations to inverse design problems has drawn attention to possible advantages of applying this approach to optimization problems.

To define an inverse problem requires one to first specify a pressure (or velocity) distribution on the geometry to be determined. In the direct numerical optimization approach, one tries to avoid this. In practice, however, definition of a cost function in terms of quantities such as drag or drag-to-lift ratio does not seem to be feasible. Thus, optimization problems are often also formulated in terms of target pressure distributions. As a consequence, we will pay some attention to specifying the target pressure distribution as a special optimization problem.

INVERSE DESIGN

Existence of Solutions

The first method, applicable to the inverse design of airfoils in incompressible flow, devised by Betz (1934) and reconsidered by Mangler (1938), was based on conformal mapping of the airfoil onto a circle. It was shown that three conditions had to be satisfied by the prescribed pressure distribution in order to ensure the existence of a closed airfoil which could generate that pressure distribution at a prescribed onset flow condition. These constraints are given by the integral relations:

$$\int_0^{2\pi} \log\left|\frac{q_0(\omega)}{q_\infty}\right| \left\{\begin{matrix} 1 \\ \cos\omega \\ \sin\omega \end{matrix}\right\} d\omega = 0,$$

where $q_0(\omega)$ is the tangential velocity on the airfoil surface derived from the prescribed pressure distribution, q_∞ is the freestream speed, and ω is the polar angle in the circle's plane.

The first constraint expresses the regularity condition establishing a unique relationship between the prescribed velocity and the freestream speed. The other two constraints are derived from the requirement of the airfoil contour to be closed.

Later, on the verge of the computer era, attention was again drawn to these constraints by Lighthill (1945) and by Timman (1951). Since then, the necessity of taking these constraints into account when developing an inverse airfoil design method has been the subject of much discussion. The possible existence of similar constraints for other types of flow has also been studied. Woods (1952) was able to formulate constraints for compressible subcritical flow using the Von Karman-Tsien gas model. But, so far, explicit formulation of similar constraints for more general types of flow has not appeared. Nevertheless, it is usually assumed that similar constraints exist for all inverse airfoil design problems.

Early conformal mapping methods demonstrated yet another consequence of arbitrarily prescribing the pressure distribution—i.e. the appearance of self-intersecting geometries ("crossing-over," fish tail airfoils) as a solution to the problem. So, even if all consistency constraints are satisfied by the target pressure distribution, the result may still not have any practical value. Some authors have drawn the conclusion that in this case respecification of the target pressure distribution is inevitable. Other authors devised methods that incorporated geometric constraints in an attempt to reduce the class of admissible solutions to realistic airfoils. In this way uncertainty in the correct pressure distribution behavior near the forward stagnation point can be removed by prescribing, either exactly or approximately, a part of the leading edge region. The trailing edge thickness may be introduced as another geometric constraint. In this way the inverse problem may be recast in a mixed direct-inverse problem, where one part of the geometry is prescribed and the other part is designed. Or, alternatively, the inverse problem is reformulated as a least squares minimization problem, where the prescribed pressure distribution is satisfied approximately. Then, the constraints on the geometry are taken into account either exactly by adding constraint terms to the least squares cost function using Lagrange multipliers or in a approximate least squares way. Unfortunately, the existence of a (unique) solution of the minimization problem has never been proven.

As an alternative to applying geometric constraints, one can attempt to achieve well-posedness by introducing free parameters in the prescribed pressure distribution. These parameters are determined as part of the solution, such that the constraints on the pressure distribution will be automatically satisfied; the specific choice for the adjustable free parameters determines implicitly the class of admissible solutions. Volpe & Melnik (1981) have shown that the regularity constraint associated with the relation between the freestream speed and the prescribed pressure distribution may be satisfied by introducing the freestream speed as a free parameter while maintaining a specified location of the forward stagnation point. Drela (1986) chose to fix the freestream speed, but left the location of the forward stagnation point in physical space unspecified. The constraints associated with trailing edge closure are sometimes assumed to be implicitly fulfilled. As an alternative certain functions with free parameters may be added to the prescribed pressure distribution, so that it can be adjusted to comply with the required trailing edge thickness.

Apart from the need for constraints related to the well-posedness of the inverse problem and constraints needed to prevent nonphysical solutions, constraints may be required for more practical reasons, e.g. from the point of view of the structural engineer. But also, constraints on the geometry as well as on the aerodynamic characteristics may be required to avoid undesirable off-design behavior.

The latter situation may be illustrated by the case of a transonic shock-free airfoil design, which has been mentioned by several authors e.g. Volpe (1989). The airfoil of Figure 1, intended to be shockfree by prescribing a pressure distribution with a smooth recompression along the contour, and designed by solving a well-posed inverse problem, exhibits a so-called hanging (secondary) shock in the flow field (see Figure 2). As a result the drag coefficient is large, even at the design point. According to Sobieczky (1989), the occurrence of this secondary shock is associated with the concave part of the upper side of the airfoil. Such a result might be avoided by putting a constraint on the curvature. Another remedy has also been suggested: specifying the target pressure distribution so that its point of inflection lies in the locally subsonic region and not in the locally supersonic region (see Figure 3) (R. D. Cedar, unpublished observations).

With respect to the three-dimensional inverse problem for wing design, the situation is still quite unclear. Even for incompressible flow, the requirements for a well-posed inverse problem have not yet been formulated. Without the application of constraints, the general design problem with an arbitrarily prescribed pressure distribution seems to be ill-posed; see

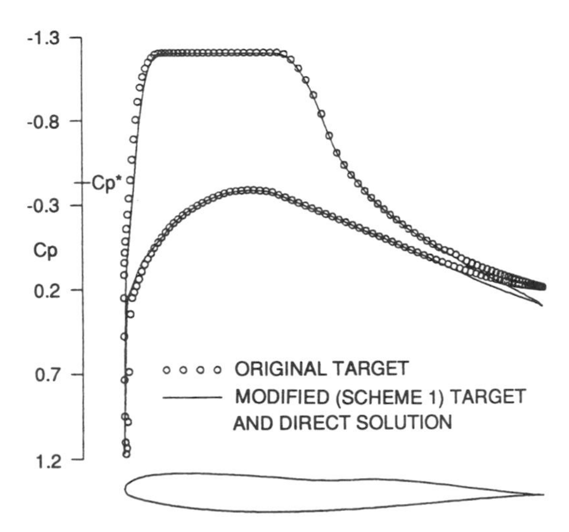

Figure 1 Design of a "shock-free" airfoil by prescribing a smooth recompression, $M_{\infty} = 0.8$, $\alpha = 0°$, $C_l = 0.4801$, $C_d = 0.0232$ (Volpe 1989).

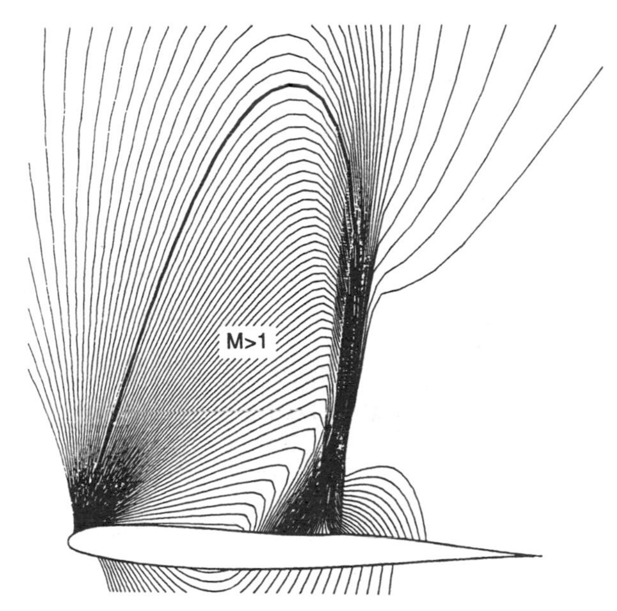

Figure 2 Design point isomachs for the "shock-free" airfoil of Figure 1 ,$M_\infty = 0.8$, $\alpha = 0°$; contours shown at 0.01 intervals beginning with M = 0.810 (Volpe 1989).

Slooff (1983). Takanashi (1984) reported an example of inverse wing design exhibiting a root section instability. Ratcliff & Carlson (1989) presented an example of spanwise oscillations in the wing geometry obtained by means of their inverse wing design method. Neither of these phenomena has been explained satisfactorily. Several authors have found a way out of these problems by either applying explicit constraints on the geometry or by considering a mixed direct-inverse problem instead of a fully-inverse problem.

Coupled-Solution Methods

Inverse methods are sometimes classified as being either iterative or noniterative. Here, the term noniterative is rather confusing. It is used in the sense that the geometry is determined directly by solving a boundary value problem (see Figure 4), thus avoiding the application of an iterative procedure with successive updates of the geometry. However, the inverse boundary value problem is nonlinear in essence and its solution requires an iterative process. The inverse method of Drela & Giles (1987) is an example of a noniterative method in the above sense, but it is sometimes referred to as a typical example of a design method applying Newton iteration.

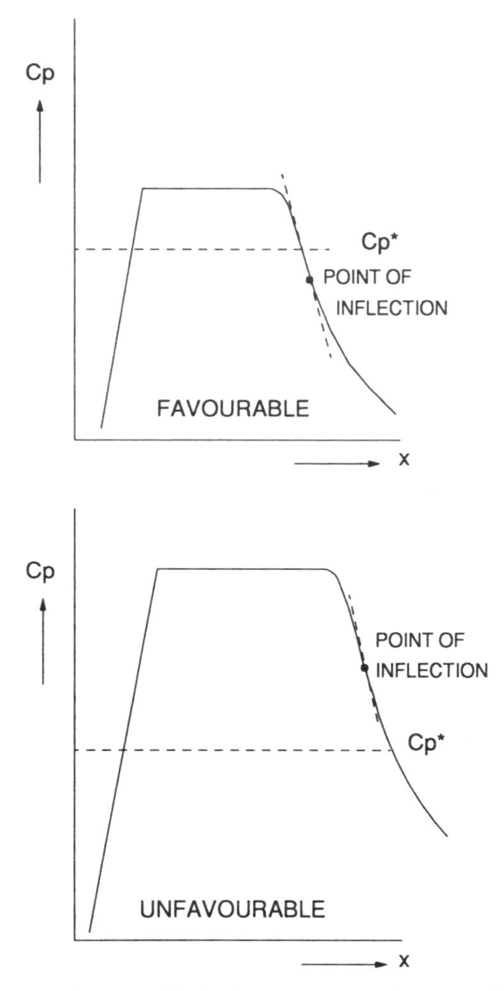

Figure 3 Suggested location of inflection point in target pressure distribution.

In the so-called noniterative methods the flow variables as well as the unknown geometric parameters (either explicitly or implicitly) are considered as one set of unknowns and as such are tightly coupled. Therefore, in the following, this type of method will be referred to as a "coupled-solution method." These methods often apply a mapping technique wherein a computational domain with fixed boundaries is obtained. Then both the flow quantities and the mapping variables have to be solved from the reformulated boundary value problem. The geometry is obtained either

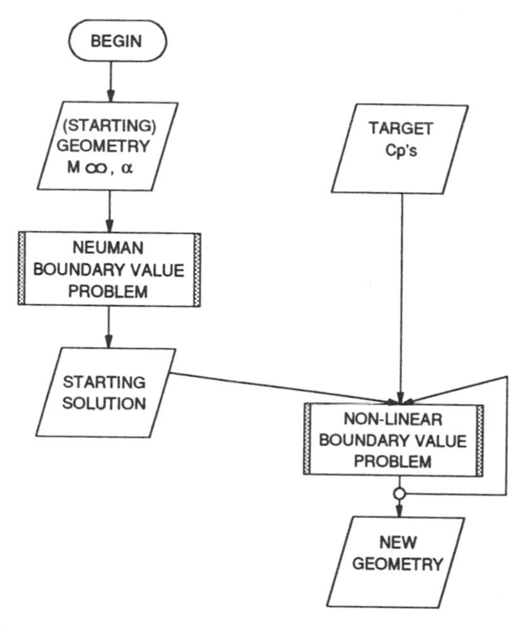

Figure 4 Flow chart of coupled solution method for the inverse design problem.

directly as part of the solution or afterwards from the inverse mapping. The potential/stream function method of Dedoussis et al (1992) is a typical example of this type of approach.

Examples of coupled-solution methods are the panel methods formulated for 2-D subsonic potential flow by Ormsbee & Chen (1972), Bristow (1976), and Labrujère (1978). A method based on transonic small perturbation theory has been formulated by Shankar (1980). These older methods have already been reviewed by Slooff (1983). Here, attention will be given to more recent developments.

BARRON & AN (1991) When full potential flow problems with exact boundary conditions are considered, one usually chooses to apply numerical grid generation techniques in order to obtain body-fitted coordinates. When design problems are considered with free boundaries, it will then be necessary to adapt the grid to geometry modifications during the solution procedure. The grid generation and adaption process may be avoided if the problem is formulated in so-called streamwise coordinates. Barron (1990) applied the Von Mises transformation to the 2-D inverse design problem for incompressible potential flow. Applying this transformation, the Cartesian coordinates (x,y) are replaced by the Von Mises coordinates

(x,ψ) where ψ is the stream function. The velocity components are then given by

$$u = \psi_y = 1/y_\psi, v = -\psi_x = y_x/y_\psi,$$

and the governing flow equation transforms to

$$y_\psi^2 y_{xx} - 2y_x y_\psi y_{x\psi} + (1+y_x^2) y_{\psi\psi} = 0.$$

For symmetric flow, the flow domain transforms to the rectangular, fixed boundary domain depicted in Figure 5. By prescribing the target pressure distribution as a function of x, the boundary conditions for y become:

$$y = \psi \quad \text{in the far field,}$$

$$y = 0 \quad \text{on } \psi = 0, -\infty < x \leq x_{le} \quad \text{and} \quad x_{te} \leq x < \infty,$$

$$[1 - C_p(x)]\, y_\psi^2 - y_x^2 = 1 \quad \text{on } \psi = 0,\ x_{le} < x < x_{te}.$$

In this way the problem of solving a linear equation in a domain with partly unknown curved boundaries is transformed into a problem of solving a nonlinear equation on a rectangular domain with fixed boundaries. By solving the latter problem for the function $y(x,\psi)$ which defines the coordinate transformation, the shape of the airfoil is directly determined as part of the solution. Barron & An (1991) extended application of this method to transonic flow, applying the Von Mises transformation to the Euler equations.

The method is attractive in that both the analysis and design problems can be treated in a similar way requiring about the same computational effort. The treatment of the mixed analysis-design problem should also be feasible. However, the matter of constraints on the prescribed pressure distribution has not been addressed and, so far, only results for symmetric reconstruction test cases (where the pressure distribution generated by a given airfoil is prescribed as the target) have been shown. Most probably, application of the present method will meet with difficulties when a design case with arbitrarily prescribed pressure distribution is to be treated.

DEDOUSSIS ET AL (1992) A similar type of approach is followed by Dedoussis et al (1992) in developing a full potential, subsonic, inverse method for the design of 3-D internal flow configurations. A body-fitted coordinate transformation is obtained by introducing the potential function ϕ and two orthogonal stream functions ψ and η as independent variables. By specifying the velocity distribution (pressure distribution) on the walls as well as on the inlet and outlet sections of the channel, a boundary value problem on the (ϕ,ψ,η) space is defined.

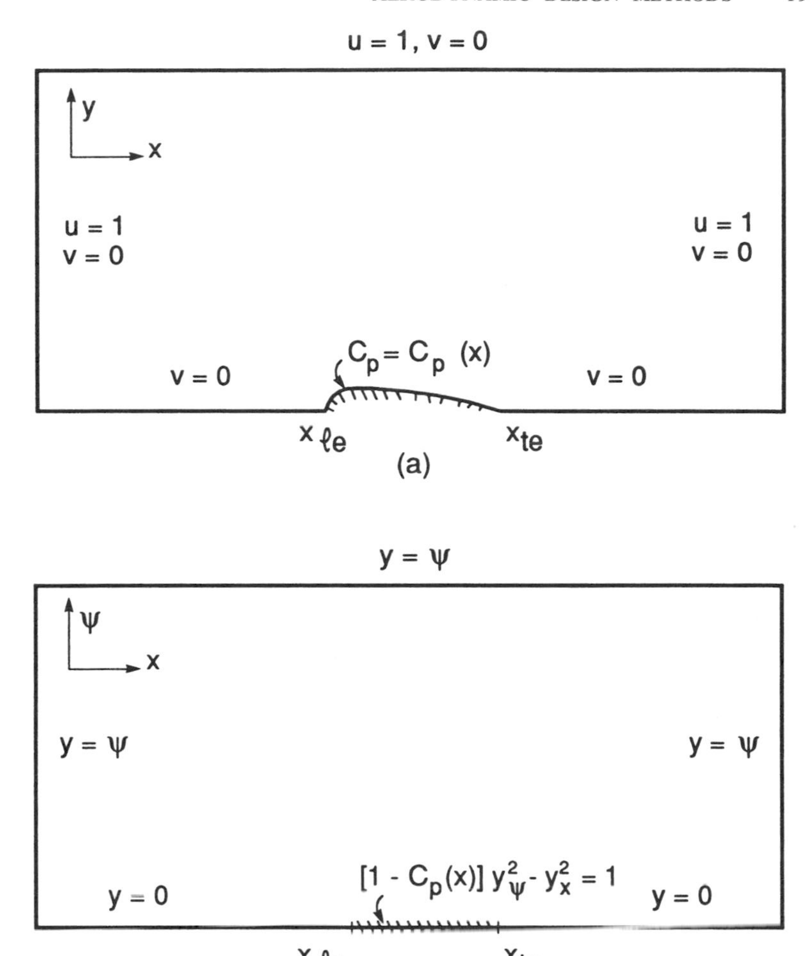

Figure 5 (*a*) Physical domain and boundary conditions; (*b*) computational domain and boundary conditions (Barron & An 1991).

In the case of axisymmetric flow, the governing equations in the meridional η = constant plane, for the velocity V and for the cross section of the elementary stream tube t,

$$V[(\ln V)_{\phi\phi}+(\ln \rho)_{\phi\phi}]+\rho t(\ln V)_{\psi\psi}+1/2V[(\ln V)_{\phi}^{2}-(\ln t)_{\phi}^{2}-(\ln \rho)_{\phi}^{2}]$$
$$-\rho t(\ln V)_{\psi}[(\ln V)_{\psi}-(\ln t)_{\psi}]=0,$$

$$(\ln t)_{\phi\phi} - (\ln \rho)_{\phi}(\ln t)_{\phi}(\ln t)_{\phi} + \rho t/V[(\ln V)_{\psi\psi} + (\ln V)_{\psi}(\ln \rho)_{\psi}] = 0,$$

are integrated in a rectangular domain. On the solid wall a target velocity distribution is specified as $V(\phi)$, and at the inlet and outlet the velocity is taken to be uniform. The potential ϕ is related to the arclength s via the relation $d\phi = Vds$, so that $V(s)$ may be specified as well. Together with the usual relation between density and velocity for a perfect gas, these equations form a closed system of nonlinear equations which is solved for V, t, and ρ. Afterwards, the geometry of the channel is determined by integrating along the streamlines ψ = constant.

So far, the method has only been applied to reconstruction test cases. It would be interesting to see applications to real design problems.

DRELA & GILES (1987) The 2-D design and analysis method of Drela & Giles (1987) is based on the Euler equations in conservation form and takes boundary layer effects into account. For discretization a grid is used in which one set of coordinate lines is formed by the streamlines. Figure 6 shows the definition of a conservation control cell. Because there is no convection across the streamlines, the continuity and energy equations can be replaced by a constant mass flux and stagnation enthalpy condition for each streamtube. Instead of the standard Euler variables—e. g. density ρ, pressure p, and velocities u and v, as well as both node coordinates (x,y)—only the density ρ and the normal position n of the grid nodes have to be considered as variables. The streamline grid is determined as part of the solution, which implies that the design and analysis mode of the method differ only in the specific form of the boundary condition on the airfoil surface.

In the full-inverse mode of the method, the regularity constraint on the pressure distribution is satisfied implicitly by leaving the exact position of the forward stagnation point unspecified, later to be determined as part of the solution. In an attempt to render the inverse problem well-posed for a prescribed trailing-edge thickness, two free parameters are introduced in the target pressure distribution by means of two auxiliary shape functions. These parameters appear in the prescribed pressure boundary conditions and are added as unknowns to the set of unknown flow variables. In the case of a mixed direct-inverse application (i.e. where part of the geometry is prescribed), free parameters are introduced in a similar way in order to allow the imposition of geometrical continuity conditions.

Optionally, the method can be applied by taking boundary layer effects into account. This is established via the displacement surface concept. To this end the Von Karman integral momentum equation, the kinetic energy shape parameter equation, and a dissipation lag equation are introduced at each airfoil surface node in order to govern the boundary layer variables,

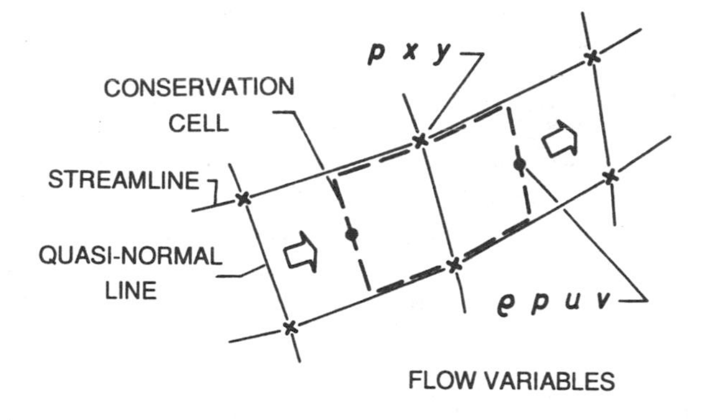

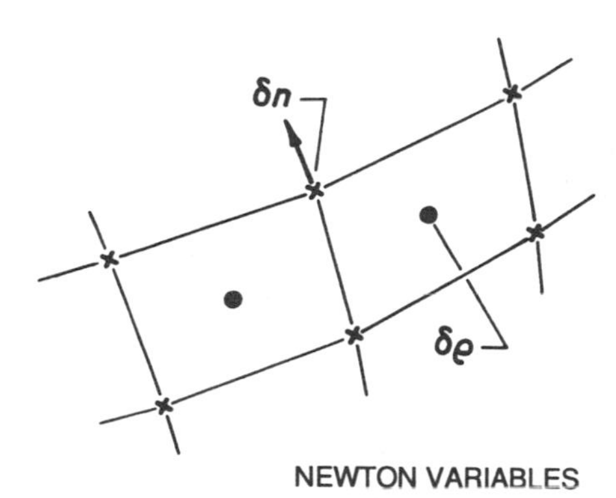

Figure 6 Euler grid node and variable locations (Drela & Giles 1987).

i.e. displacement thickness, momentum thickness, and shear stress coefficient. These equations are added to the discrete Euler equations so that a fully coupled viscous/inviscid nonlinear system is obtained.

In both analysis and inverse cases the complete nonlinear system is solved by means of a global Newton-Raphson method for the set of unknowns formed by density and normal grid position in each grid node, pressure distribution parameters, and boundary layer variables.

Decoupled-Solution Methods

A second class of inverse methods is formed by the iterative decoupled-solution methods in which the flow variables and geometric parameters are decoupled in the solution process. There are three types of methods:

Dirichlet methods, Neumann or residual-correction methods, and variational methods. All methods start with an initial guess of the geometry to be determined. First, in each subsequent iteration step, a boundary value problem is solved for a given estimate of the geometry. With the Dirichlet method this boundary value problem is of Dirichlet type. With the Neumann or residual-correction methods and with the variational methods, this boundary value problem is of Neumann type. Then, a correction to the geometry is derived from the solution of this boundary value problem (see Figure 7). In the majority of these methods one tries to reduce the computational effort for the geometry correction as much as possible.

A large variety of decoupled-solution methods has been developed in the past decade. Nearly all 3-D design methods are of this type. The idea of decoupling the flow and geometry solutions in inverse design is in most cases inspired by the desire to take maximum advantage of the fact that

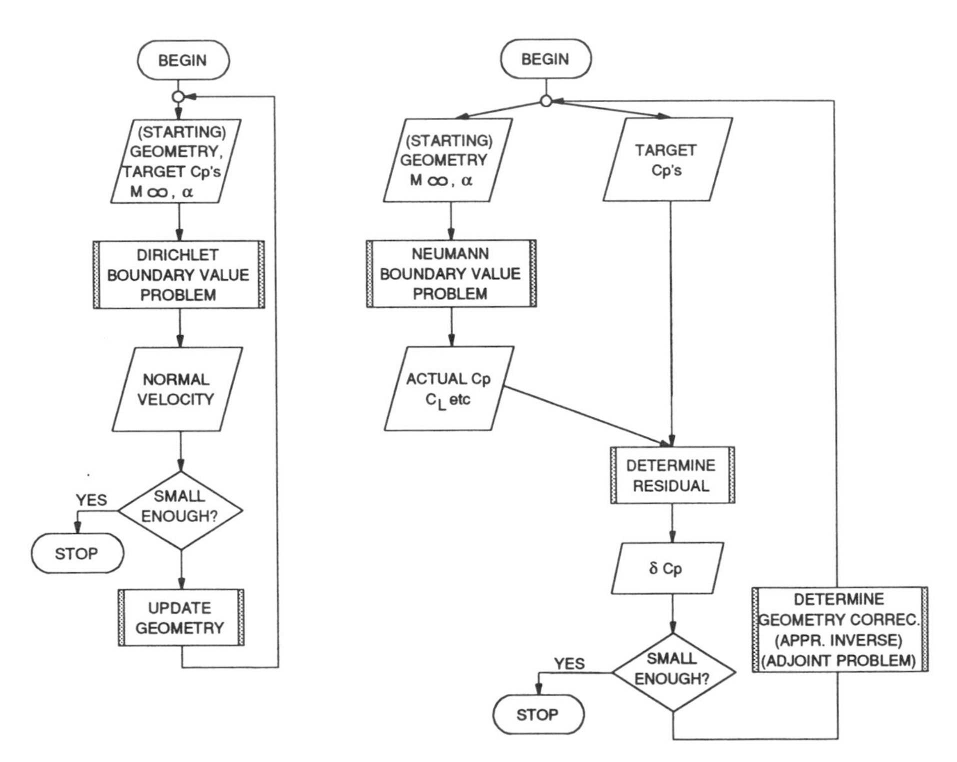

Figure 7 Flow charts of decoupled-solution methods for the inverse design problem; (*left*) Dirichlet type, (*right*) Neumann and variational type.

analysis methods have been developed for many applications in different flow regimes and for sometimes complex configurations. Another advantage of decoupled-solution methods is the fact that, in general, geometric constraints can be implemented much more easily in a separate geometry update procedure than in a complete system of equations for flow as well as geometry variables.

DIRICHLET METHODS Solving a Dirichlet problem, for which the boundary condition of prescribed tangential velocity is derived from the target pressure distribution, leads to a flow field with nonzero normal velocity on the boundary. Aimed at removal of this transpiration, a geometry update is determined by applying either the transpiration model based on mass flux conservation or the streamline model based on alignment with the streamlines. In the majority of Dirichlet decoupled-solution methods, existing flow solvers have been modified in order to accept Dirichlet (pressure type) boundary conditions in addition to the usual Neumann (flow tangency) boundary condition. Typical examples of a Dirichlet decoupled-solution method are that of Henne (1980) for transonic wing design, where the transpiration model is used for determination of a geometry update, and that of Volpe (1989) for transonic airfoil design, where the streamline model is used for determining geometry updates.

Gally & Carlson (1987) presented an extension of an earlier method developed for orthogonal grids to a body-fitted nonorthogonal curvilinear grid for the mixed direct-inverse transonic design problem. The method is based on the finite volume full potential method of Jameson & Caughey (1977) in which the boundary condition of flow tangency applied in analysis is replaced by a specification of the perturbation potential in the inverse design regions. The value of the perturbation potential is derived from the prescribed target pressure making use of previous estimates of the flow variables. During the iteration procedure the actual geometry used is updated periodically by aligning it to the streamlines. By excluding the wing leading edge region from the inverse design regions, the problem of how to apply the regularity condition at the leading edge is circumvented. Prescription of the trailing edge thickness is made possible by means of a relofting process in which a displacement thickness is added to the airfoil contour. This displacement is distributed such that it is zero at the leading edge and compensates for a possibly calculated deviation of the trailing edge. As a direct consequence of this procedure, deviations from the target have to be accepted as a result; there are no further means of control.

Dirichlet methods based on panel method technology have been developed by Fornasier (1989) and Kubrynski (1991). Despite the limitation with respect to the description of real flow, panel methods are still widely applied because of their capability in treating complex configurations.

In the panel method of Fornasier (1989), surface distributions of sources as well as doublets are utilized. This offers the opportunity to relate the local source strength to the normal component of the freestream velocity and to relate the tangential derivative of the doublet strength to the tangential velocity. As such, the source distribution provides direct information on the geometry, and the doublet distribution provides direct information on the flow field. Direct and inverse problems lead to the same type of equations and as a consequence a mixed direct-inverse problem can be treated as well. In regions with given geometry the source strength is predetermined, whereas in regions with prescribed velocity the doublet strength is predetermined using the current guess of the geometry. Application of the boundary condition of zero internal perturbation potential leads to a linear system of equations from which the remaining unknown singularity strengths are determined. The new source distribution is then used to update the geometry.

The panel method of Kubrynski (1991) has features similar to those of Fornasier's method (1989). Here, also, source as well as doublet distributions are used, although the definition of the singularity strengths is different. The source strengths are related to the mass flow through the body surface (zero in the analysis case) and the doublet strengths are related to the velocity potential. The boundary condition of zero internal potential is applied to derive an integral equation for the doublet strengths. An inverse or mixed direct-inverse problem is solved by the following iteration process. For a given guess of the geometry, the doublet distribution is determined for a source distribution of zero strength. Then a geometry correction is determined, whose aim is to minimize the difference between the approximated actual pressure and the target pressure. The geometry is not actually updated, but geometry modifications are modelled by means of the transpiration concept which is used to determine a local source strength. The source distribution thus determined gives rise to an incremental doublet distribution associated with a change in the approximated actual pressure distribution. After minimizing the differences between the approximated actual pressure and the target pressure, the shape of the configuration is updated and the whole process is repeated until satisfactory convergence has been obtained. An interesting feature of this method is the fact that the pressure distribution may be prescribed on one part of the configuration and that a different part of the configuration may be reshaped in attempts to realize that pressure distribution. This is of particular interest for fuselage-wing, pylon-wing, pylon-nacelle, and other interference problems. Figure 8 shows an example of pylon-nacelle junction design.

NEUMANN OR RESIDUAL-CORRECTION METHODS Solving the Neumann problem for a given estimate of the geometry to be determined leads to a pressure distribution along the contour, which deviates from the target pressure distribution. In the methods based on the residual-correction approach, the key problem is to relate the calculated differences between the actual pressure distribution on the current estimate of the geometry and the target pressure distribution (the residual) to required changes in the geometry. Obviously, the art in developing a residual-correction method is to find an optimum between the computational effort for determining the required geometry correction and the number of iterations needed to obtain a converged solution. This geometry correction may be estimated by means of a simple correction rule, making use of relations between geometry changes and pressure differences known from linearized flow theory. In other Neumann methods the geometry correction is determined

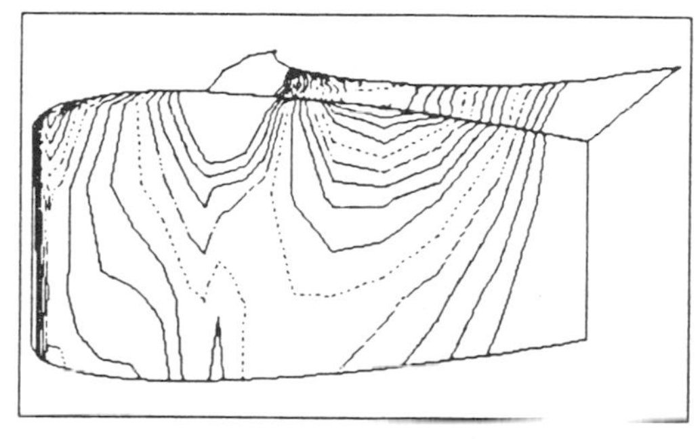

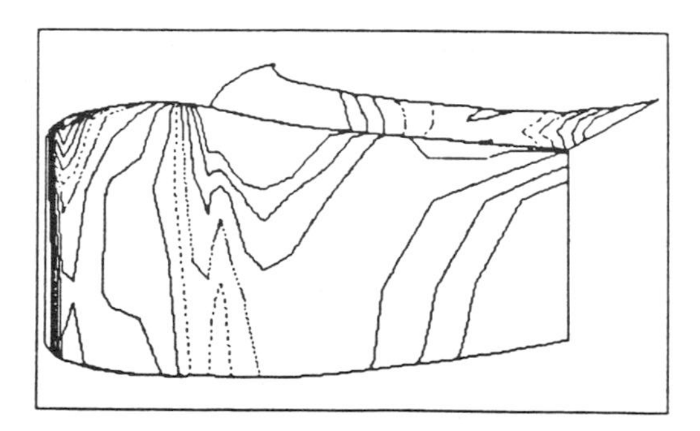

Figure 8 Contour of nacelle and pylon, before (*top*) and after (*bottom*) design process (Kubrynski 1991).

by applying a coupled solution method to an approximate inverse problem, which is derived from the actual inverse problem—e.g. by applying similarity rules or by linearizing the flow equations. In the latter case, the gain in computational effort is due to the reduced complexity of the approximate inverse problem as compared to the actual inverse problem. The Neumann decoupled-solution methods try to utilize the analysis methods for the solution of the Neumann problem as a black-box.

In 1974, Barger & Brooks presented a streamline curvature method in which they utilized the possibility of relating a local change in surface curvature to a change in local velocity. Since then, quite a number of methods have been developed following that concept. Subsequent refinements and modifications made the concept applicable to design problems based on the full potential equation (e.g. Campbell & Smith 1987), the Euler equations (e.g. Bell & Cedar 1991), and the Navier-Stokes equations (e.g. Malone et al 1989).

The method of Bell & Cedar (1991) is an extension of the method of Campbell & Smith (1987) for application to engine-nacelle redesign. Special care is taken to preserve the essence of the original cross-sectional shape of the nacelle. Greff et al (1991) described a 2-D airfoil design code for viscous-transonic flow. The approximate inverse problem is defined using a modified Von Karman-Tsien rule for the derivation of an equivalent subsonic target from the calculated differences between the transonic pressure distribution on the current estimate of the geometry and the target pressure distribution. The approximate inverse problem is solved by means of an inverse panel method. Takanashi (1984) presented a method for transonic wing design using for geometry correction an integral equation method to solve an approximate inverse problem on the basis of transonic small disturbance theory. Brandsma & Fray (1989) presented a method for transonic wing design utilizing linearized compressible flow theory for the definition of an approximate inverse problem. The constraints introduced on the geometry lead to a least squares minimization problem which is solved with the aid of linearized panel method technology.

So far, the method of Takanashi (1984) seems to be the most widely applied residual-correction method. It has been coupled with analysis methods on the basis of Euler equations as well as Navier-Stokes equations. It has been applied to 2-D as well as 3-D and to transonic as well as supersonic design problems (Fujii & Takanashi 1991). Hua & Zhang (1990) have modified this method by replacing the numerical integrations applied in the integral equation method by analytical integrations, thus reducing computing time. They also added a smoothing technique in order to smooth the curvature of the designed geometry. The approach of

Takanashi (1984) has also been followed by Zhu et al (1991) for transonic airfoil design; they introduced a modification for taking into account the regularity condition at the forward stagnation point.

VARIATIONAL METHODS Application of the calculus of variations (optimal control theory) to the solution of the inverse design problem leads to the formulation of two strongly related flow problems: (*a*)the Neumann problem of flow analysis for a given geometry and (*b*) the so-called adjoint problem in which the residual differences between current and target pressure distribution determine the boundary condition. Usually this adjoint problem is of the same type as the corresponding analysis problem, which implies that a solution method may be readily derived from an available analysis method. One attempts to determine the geometry correction as accurately as possible using the solution to the adjoint problem for determining a search direction for the geometry update. Application of this type of geometry correction method leads to an increase in computation time when compared with simpler types of geometry corrections. It is, however, expected to be more robust. It might also have a positive effect on the speed of convergence of the whole process. The variational approach has been applied by Bristeau et al (1985) to flow analysis problems. Pironneau (1983) gave an extensive survey of possible applications to optimum shape design for systems described by elliptic flow equations.

The concept of the variational approach may be best explained with the aid of a simple inverse airfoil design problem. To this end, consider the nonlifting incompressible potential flow around a symmetric airfoil where the leading and trailing edge stagnation points are fixed. Assume the tangential velocity on the airfoil contour to be prescribed as a function of the chordwise coordinate x and the airfoil contour to be represented by $y(x)$. Then the inverse design problem may be formulated as the minimization of the functional $F(y)$:

$$F(y) = \oint_{\Gamma} [\phi_s(y) - V_s]^2 \, ds.$$

Here s is the arclength of the contour, ϕ_s is the actual tangential velocity, and V_s is the target velocity.

Considering incompressible potential flow around a given airfoil, the equations

$$\Delta\phi = 0 \qquad \text{in } \Omega,$$

$$\frac{\partial \phi}{\partial n} = 0 \qquad \text{on } \Gamma,$$

$$\frac{\partial \phi}{\partial n} = \mathbf{q}_\infty \cdot \mathbf{n}_\infty \qquad \text{on } \Gamma_\infty,$$

determine the velocity potential ϕ in the flow domain Ω apart from a constant which may be determined by prescribing the potential at some point. The flow domain Ω is bounded at the inner side by the airfoil contour Γ and at the other side at infinity by Γ_∞. Application of the calculus of variations to the minimization problem leads to the definition of an adjoint problem. For a given airfoil contour and given velocity potential ϕ in the flow domain, this adjoint problem amounts to the determination of the co-state variable w, apart from a constant, from the following equations:

$$\Delta w = 0 \qquad \text{in } \Omega,$$

$$\frac{\partial w}{\partial n} = -2\frac{\partial}{\partial s}(\phi_s - V_s) \qquad \text{on } \Gamma,$$

$$\frac{\partial w}{\partial n} = 0 \qquad \text{on } \Gamma_\infty.$$

With the aid of the solution to this problem, the first variation of the functional F can be determined from

$$\delta F = \oint_\Gamma h(x)\delta y(x)\, dx,$$

with

$$h(x) = -\nabla w \nabla \phi + \frac{d}{dx}[\phi_s - V_s]^2 \frac{dy}{ds}.$$

For two successive estimates of the airfoil contour y^i and y^{i+1} the difference between the associated values of the functional F is to first order approximated by

$$F^{i+1} - F^i \approx \delta F.$$

Thus, choosing $y^{i+1} = y^i + \delta y$ with $\delta y(x) = -\varepsilon h(x)$ and $\varepsilon > 0$, such that $\delta F < 0$, a reduction of F is ensured. After modifying the airfoil contour in this way, ϕ, w, and δy are calculated again. The whole process is repeated until a minimum is reached.

Application of the variational approach to the inverse design of airfoils in subsonic potential flow has been pioneered by Angrand (1980). Beux & Dervieux (1991a) treated the case of inverse design for internal subsonic flow governed by the Euler equations. An important issue with respect to implementation of the variational approach has been discussed by Frank & Shubin (1990). They have shown that in order to obtain a convergent process, it may be necessary to discretize the analysis problem first and apply the calculus of variations to the discretized problem, instead of applying the variational approach directly to the continuous problem.

DESIGN BY OPTIMIZATION

Direct Numerical Optimization

The concept of direct numerical optimization is illustrated by Figure 9. In principle it allows the minimization of any aerodynamic cost function. Implementation of a concept like this is feasible only if sufficient computer resources are available. In 3-D wing design especially, the number of design variables is so large that practical application of the concept seems to be remote—even some 15 years after the publication of the idea by Hicks et al (1976).

Nevertheless, several authors have considered the idea worthwhile for further investigation. There are three major aspects in direct numerical optimization worth considering when attempting to make the approach more feasible. First of all, the objective function should be chosen such that it closely reflects the designers' requirements, bearing in mind, however, the possibilities offered by the analysis codes, in particular with respect to the accuracy of their solution. Secondly, a dominant role with respect to computing time is played by the number of design variables; therefore, several attempts have been made to reduce this number by choosing appropriate shape functions for geometrical representation. A third important factor is the optimization algorithm applied to determine the design variables. It should be efficient, fast, and robust. It should be able to treat a reasonable, not too small number of variables and allow for nonlinear constraints.

Following the black box idea, it seems reasonable to express the objective function in terms of global aerodynamic characteristics such as lift-to-drag ratio. In 2-D problems this seems to be feasible. In 3-D, however, such objective functions seem to be less appropriate. Even if an analysis method is available for accurate prediction of the global quantity considered, the computational effort for its calculation may be prohibitive. Also, maintaining global characteristics as design criteria necessarily inhibits direct control over local flow characteristics. This may lead to additional aerodynamic constraints or else to undesirable pressure distributions. As a consequence there is a tendency to rely on objective functions in terms of pressure distributions . Thus formulated, design by optimization seems to offer nothing more than the inverse design technology treated in the

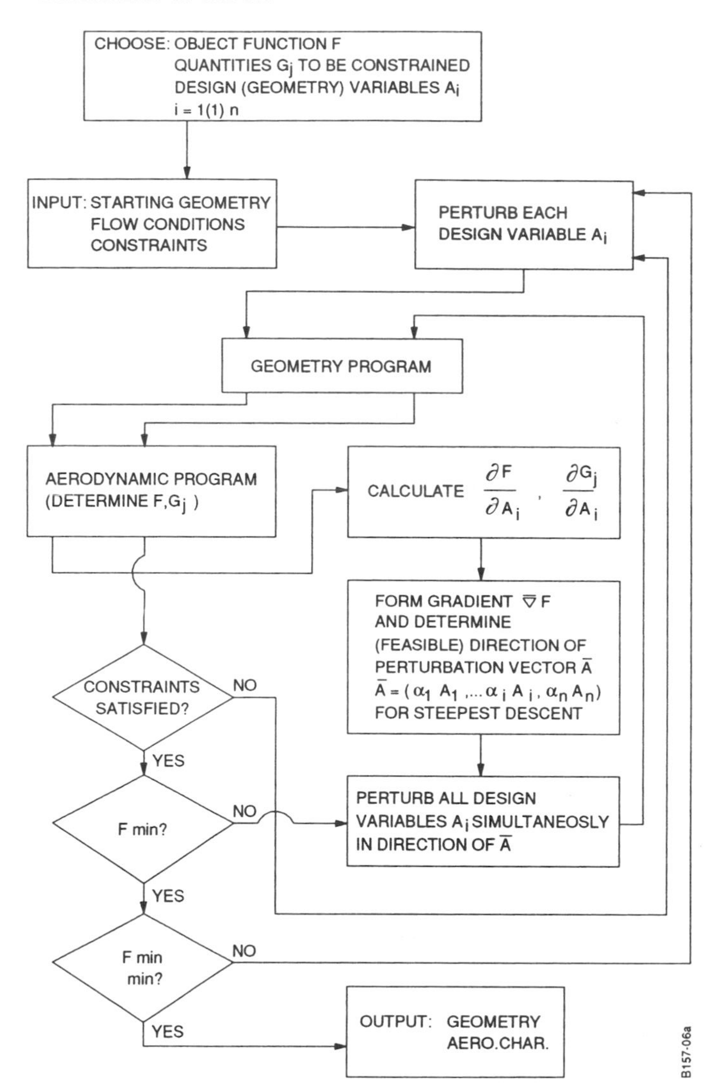

Figure 9 Flow chart of design by direct numerical optimization.

previous section. However, there are at least three clear advantages: 1. greater possibilities of applying geometric constraints, 2. multi-point optimization would seem to be more feasible, and 3. better possibilities for multi-disciplinary design applications.

The use of a discrete set of points for representing an airfoil or wing contour seems virtually out of the question because of the large number of design variables involved. Therefore the contour is represented by means of a limited number of shape functions. These shape functions can be of purely analytical nature, but it is probably more efficient to use, for instance, aerofunction shapes of the kind described by Aidala et al (1983). By means of these shape functions, geometry modifications are related directly to changes in aerodynamic characteristics, e.g. a particular behavior of the pressure distribution. The shape functions themselves result from solving inverse (re)design problems in which a specific pressure distribution is prescribed. The same concept has been applied by Destarac & Reneaux (1990) for airfoil and wing design problems as well as for minimizing wing/engine interference effects. An apparent drawback of this type of shape function is that their effect is associated with a specific design condition (Mach number) and, moreover, depends on the initial geometry applied in the inverse calculation. Another approach is to select appropriate existing airfoils and build an airfoil library from which by linear combination (resulting from the optimization) a new airfoil may be obtained. This approach has of course the same drawback. Low-speed high-lift airfoils are of a considerably different nature than high-speed low-drag airfoils. Nevertheless, an effective combination of both approaches has been applied by Reneaux & Thibert (1985) for airfoil design.

Yet another idea for reducing the computational effort has been described by Beux & Dervieux (1991b) who introduced the concept of hierarchical parametrization. It is assumed that the geometry can be described by means of a parametric representation with different sets of a different number of parameters, and that it will be possible to derive the coarser sets from finer ones by appropriate interpolation procedures. It is demonstrated that the convergence of the optimization process is considerably increased by applying alternately coarser and finer levels of representation, as in a multigrid process.

The original idea of the numerical optimization technique was to treat the analysis code as a black box for evaluation of the objective function and to use an optimizer for determination of geometry modifications. Investigations have been performed in attempts to increase the efficiency of the optimizers, e.g. by Cosentino & Holst (1986). However, being faced with the fact that many of the analysis problems are nonlinear and are solved iteratively, it is not surprising that the idea came up to mix the outer (optimization) iteration and the inner (analysis) iteration steps, as e.g. described by Rizk (1989). Though the idea has been pioneered for only a few design variables, it seems to be promising. Nevertheless its usefulness remains to be demonstrated for problems involving a larger, more practical, number of design variables.

Variational Approach

The variational approach, described above as a method for treating inverse design problems, may also be followed for the solution of optimization problems. Here, the key problem is the formulation of the adjoint problem from which the search direction for optimization will be determined.

So far, only a few papers have appeared concerning the variational approach as a potential means for aerodynamic design by optimization. Cabuk et al (1991) applied the variational approach to the problem of optimizing a diffuser wherein a maximum pressure rise is provided. The method is based on the incompressible Navier-Stokes equations. For the analysis problem the boundary condition of no slip is imposed on the solid wall. At the entrance and exit a Dirichlet boundary condition is applied by specifying the streamwise velocity and assuming the transverse velocity to be zero. In that case it is shown that an adjoint problem may be formulated in which the governing equations are similar but not identical to the Navier-Stokes equations and which may be interpreted as a direct problem with slightly different boundary conditions (derived from analysis of a given estimate of the geometry). The numerical algorithm for the solution of the adjoint problem is similar to the algorithm for the analysis problem.

MULTI-POINT DESIGN

Though the majority of optimization methods mentioned above are equally well applicable to multi-point design problems, the present article has, up to this point, dealt with single-point design problems, i.e. inverse design or optimization for one single design condition. Drela (1990) presented a very convincing example of the usefulness of computational multi-point design of an airfoil aimed at drag reduction. The following results were obtained by means of an extension of a previous code (Drela & Giles 1987) by adding an optimization mode using shape functions for geometry representation.

Figure 10 shows the C_l–C_d polars for four different airfoils. The solid line represents the polar of a given airfoil LA203A, for which redesign calculations were performed. One single-point redesign calculation has been performed for a design condition related to a lift-coefficient of $C_l = 1.08$. A second single-point redesign has been performed at $C_l = 1.5$. Comparison with the polar of the original airfoil clearly shows a considerable reduction of the drag at the design points. However, it is clear that the drag reduction is realized only in the vicinity of the design points and that the single-point improved airfoils can be considered inferior to

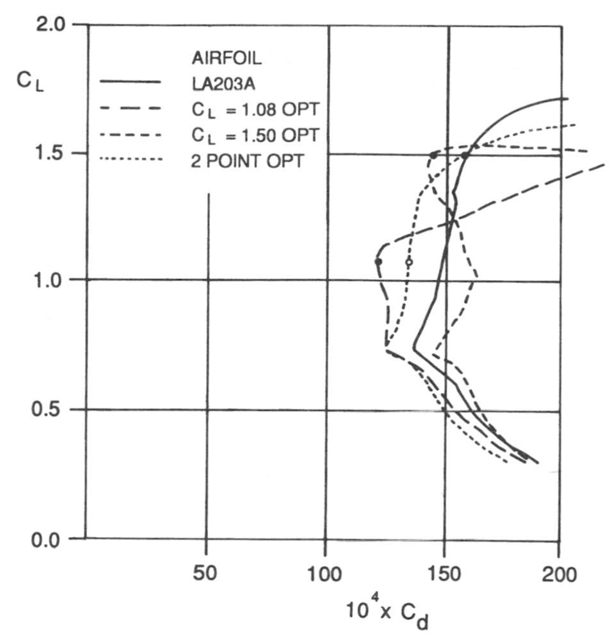

Figure 10 Calculated polars for original LA203A, single-point, and two-point optimized airfoils (Drela 1990).

the original airfoil in an overall sense. The fourth polar, referred to as "2 point opt," belongs to an airfoil which has been obtained by means of a two-point optimization using a weighted sum of the two C_d values at the two C_l design points as the objective function. A larger weight was placed on the $C_l = 1.5$ point because the polar of the other single-point design indicates a considerable loss in C_{lmax}. The two-point optimization polar shows a far more attractive overall behavior of the airfoil with a considerable overall reduction of drag at the cost of a significant, though perhaps acceptably small, loss in C_{lmax}. Ultimately the value of such a design depends of course on the choice of the design goal, but the merits of multi-point design are clearly demonstrated by this example.

Potentially, the methods developed for direct numerical optimization are applicable to multi-point design problems by simply extending the objective function. This has been demonstrated by Reneaux & Allongue (1989) for the problem of rotor blade design where, because of the combination of forward and rotating movement of the rotor, airfoils have to operate under largely different conditions at the same flight speed. However, application of direct optimization for this kind of problem is of course even more computationally costly than for single-point design.

Some authors (e.g. Selg & Maughmer 1991, Kubrynski 1991) have taken a different point of view to multi-point design. In order to meet requirements for different operational conditions, they simply divide the geometry to be determined into parts which are assigned separately to each of the different design conditions. In this way they are able to adapt inverse methods to multi-point design. But of course they could have applied their methods equally well to a number of single-point mixed direct-inverse problems, each time designing a different part of the geometry and fixing the part of the geometry that should not be changed.

In the authors' opinion a really practical multi-point design method is not yet available. Work is currently in progress at NLR which explores the possibilities offered by the variational approach. This approach seems to be as equally well suited for optimization purposes as the direct numerical optimization approach, at least in cases where the objective function is formulated in terms of target pressure distributions. It has the advantage that the representation of the geometry is not restricted to the use of shape functions; it offers the same potential as inverse methods. As with the direct numerical optimization approach, constraints can easily be implemented.

PRESSURE DISTRIBUTION OPTIMIZATION

As mentioned earlier, many design methods are based on minimization of an objective function formulated in terms of prescribed (target) pressure distributions. This leaves the user with the problem of translating his design goals into properly defined pressure distributions exhibiting the required aerodynamic characteristics.

Though skilful designers are capable of producing successful designs, the design efficiency can be improved by providing the designer with tools for target pressure specification. For this purpose two codes have been developed at NLR. One code, developed by Van den Dam (1989), aims at optimizing spanwise load distributions for minimum induced, and viscous drag by taking into account aerodynamic, flight-mechanical, and structural constraints. It is based on lifting-line approximations using the conservation laws of momentum to determine the induced drag and simple, semi-empirical rules for calculating the sectional viscous drag in terms of section lift, pitching moment coefficient, and airfoil thickness. Propeller slipstream interaction with the lifting surfaces may be considered as long as one assumes that each propeller sheds a helical vortex sheet not influenced by the presence of the wing and confined to a cylindrical stream tube parallel to the freestream direction. The velocity distribution inside the slipstream is assumed to be known. Through variational calculus, a set of optimality equations is derived from the object function augmented with constraint terms using Lagrange multipliers. Application of appropriate discretization then leads to a system of linear equations for the bound circulation (span loading) and Lagrange multipliers. For example, Figure 11 shows the optimal spanwise circulation distribution which is determined after taking into account a propeller induced velocity distribution. Clearly the optimal distribution differs greatly from the "clean wing" (wing without propeller) distribution. Application of this distribution would restore much of the loss associated with the slipstream swirl.

The other code aims at optimizing chordwise sectional pressure distributions, subject to constraints on e.g. lift, pitching moment, and airfoil thickness. This code can be considered as an interactive optimization system for the solution of optimization problems defined by the user with respect to its object function, design variables, and constraints. It has been applied by Van Egmond (1989) for selecting appropriate target pressure distributions for transonic and subsonic flow. His investigations resulted in the selection of a number of relatively simple pressure distribution shape functions leading to a pressure distribution representation as depicted schematically in Figure 12. This representation involves a limited number of design variables in the form of coefficients and exponents. As an example of the practical applicability of the code, results are shown for a case study which uses the above representation. Drag was determined by means of boundary layer calculations based on Thwaites method for laminar flow and Green's lag-entrainment method for turbulent flow. The example is a

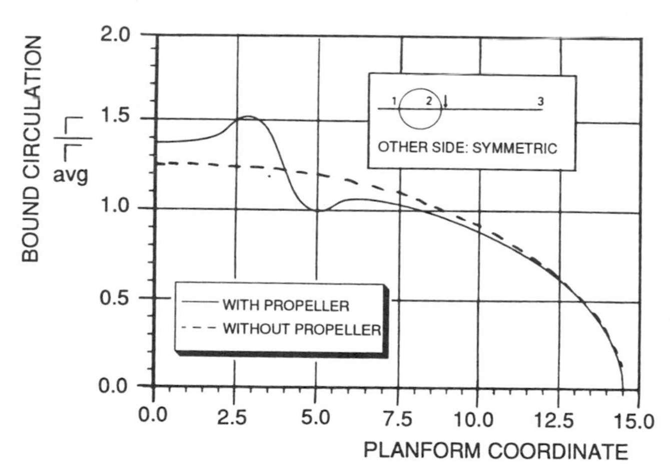

Figure 11 Optimal bound circulation distribution for a wing with two up-inboard rotating propellers (Van den Dam 1989).

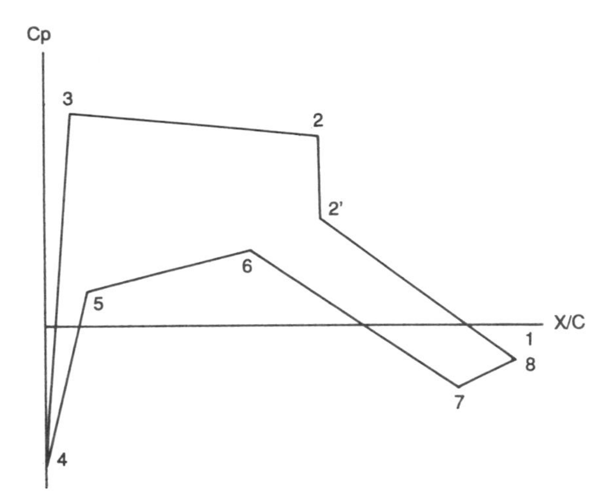

Figure 12 Schematic representation of pressure distributions (Van Egmond 1989). The numbers refer to characteristic parameters in the optimization process.

demonstration of the code's capability for designing high-lift airfoils. The intention was to maximize lift by changing only the upper surface pressure distribution for a fixed (arbitrarily chosen) lower surface pressure distribution under the additional constraint that the flow had to remain attached and subsonic everywhere on the airfoil. By constraining the shape function coefficients to produce a rooftop pressure distribution with a Stratford type pressure recovery, the result shown in Figure 13 (*solid line*) was obtained. This result compares favorably with Liebeck's solution for high lift as presented by Smith (1974). Application of the code with the upper surface pressure distribution entirely free, led to a solution with a slightly higher lift coefficient (*dashed line* in Figure 13). Application of NLR's inverse airfoil design system led to the corresponding geometries. Comparison shows that the second pressure distribution leads to an airfoil shape with a less extreme curvature distribution.

The idea of pressure distribution optimization prior to application of an inverse design method has also been followed by Lekoudis et al (1986). They developed an inverse boundary layer method based on prescription of the skin friction as target. Application of the method results in a pressure distribution which may be used as input for an inverse design method.

CONCLUSIONS

Summarizing the state of the art in computational methods for airfoil and wing design, it may be concluded that versatile methods are available

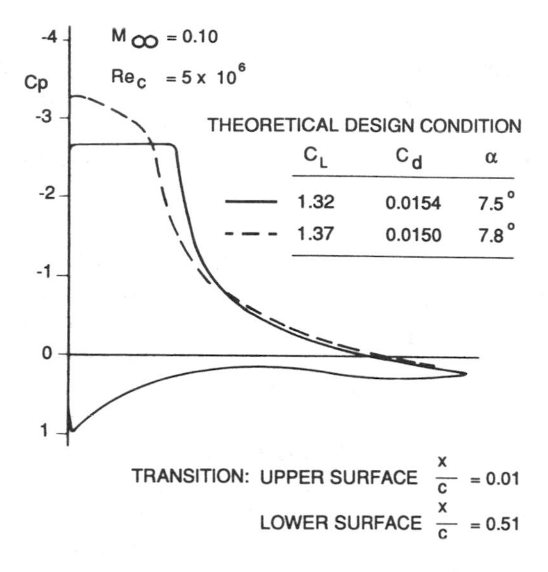

Figure 13 Pressure distribution optimization for a high-lift airfoil (Van Egmond 1989).

nowadays for the solution of the full inverse and mixed direct-inverse problem of 2-D airfoil design. Various methods exist for different types of flow, ranging from incompressible potential flow to compressible Navier-Stokes flow. Though explicit formulation of the conditions for well-posedness of the inverse problem in other than incompressible flow has not (yet) appeared to be possible, ways have been suggested to make the problem well-posed implicitly.

When discussing methods for inverse design, distinction has been made between coupled- and decoupled-solution methods, the difference being whether or not the flow variables and the parameters representing the geometry are considered as one set of unknowns and as such are tightly coupled or are decoupled in the solution process. In itself this distinction is not of practical significance when choosing a method for application to a particular design problem. Although the coupled-solution methods appear to be faster and more robust, their development is much more laborious. Also, it is far from easy to extend the domain of applicability of a coupled-solution method; most decoupled-solution methods offer more flexibility in this respect. Therefore, it is not surprising that the number of coupled-solution methods is rather limited.

The majority of methods for inverse design of 3-D wings seem to be of the residual-correction type. The main reasons for this include: the possibility to take full advantage of the existence of analysis methods, which are implemented as black boxes, and the possibility to combine different analysis methods with the same correction procedure in order to solve the inverse problem for flows of different complexity. Successful applications for 3-D wing design have been reported, but so far a definitive answer to the question of well-posedness of the 3-D inverse wing design problem has not been given.

For practical applications there is a need to further develop 2-D inverse methods since few existing methods take geometric constraints (apart from trailing edge thickness) into account. Even the powerful method of Drela (1990) in full-inverse mode does not allow for such constraints; difficulties at blunt leading edges near the stagnation point had to be circumvented by applying a mixed-inverse mode in which the leading edge is kept fixed. So far, only a few methods have been developed where geometric constraints have been implemented; see e.g. Labrujère (1978), Ribaut & Martin (1986), Brandsma & Fray (1989), and Kubrynski (1991).

Implementation of geometric and other constraints in direct numerical optimization methods is relatively easy. However, the inherent limitation with respect to geometric representation as well as the computational effort involved still makes this approach unattractive, especially for 3-D wing design. Nevertheless, further investigation of improvements of this approach, such as the hierarchical parametrization concept of Beux & Dervieux (1991b) or the Rizk (1989) optimization, seems to be worthwhile.

The application of the calculus of variations to the development of alternatives for the solution of optimization problems seems to offer perspectives, especially for problems where the objective function is formulated in terms of prescribed pressure distributions. It does not lead to limitations in geometry representation and allows for the implementation of geometric and other constraints.

Given the fact that the majority of design methods are based on prescribed pressure distributions, the development of tools for target pressure distribution selection seems to be mandatory, especially for 3-D wing design. Finally, it may be remarked that up until now the application of computational design methods still requires a lot of expertise of the designer, not only in setting design goals but also in handling the methods as design tools.

Literature Cited

AGARD 1989. *Proc. Conf. Comput. Methods Aerodyn. Des. (Inverse) and Optim., Loen, Norway*, AGARD CP–463

AGARD 1990. Special course on inverse methods for airfoil design for aeronautical and turbomachinery applications. *AGARD Rep. No.*780

Aidala, P. V., Davis, W. H. Jr., Mason, W. H. 1983. Smart aerodynamic optimization. *AIAA Pap.* 83–1863

Angrand, F. 1980. Méthode numeriques pour des problèmes de conception optimale en aerodynamique. Thèse de 3ème cycle. L'université Pierre et Marie Curie, Paris. 89 pp.

Barger, R. L., Brooks, C. W. 1974. A streamline curvature method for design of supercritical and subcritical airfoils. *NASA TN D*–7770

Barron, R. M. 1990. A non-iterative technique for design of aerofoils in incompressible potential flow. *Commun. Appl. Numer. Methods* 6: 557–64

Barron, R. M., An, C.-F. 1991. Analysis and design of transonic air-foils using streamwise coordinates. See Dulikravich 1991, pp. 359–70

Bell, R. A., Cedar, R. D. 1991. An inverse method for the aerodynamic design of three-dimensional aircraft engine nacelles. See Dulikravich 1991, pp.405–17

Betz, A. 1934. Änderung eines Profils zur Erziehlung einer vorgegebenen Änderung der Druckverteilung. *Luftfahrtforschung* 11: 158–64

Beux, F., Dervieux, A. 1991a. Exact-gradient shape optimization of a 2D Euler flow. *INRIA contr. Brite/Euram proj* 1082, 12-*month rep. part* 1

Beux, F., Dervieux, A. 1991b. A hierarchical approach for shape optimization. *INRIA contr. Brite/Euram proj.* 1082, 12-*month rep. part* 2.

Brandsma, F. J., Fray, J. M. J. 1989. A system for transonic wing design with geometric constraints based on an inverse method. *NLR TP* 89179 *L*

Bristeau, M. O., Pironneau, O., Glowinski, R., Périaux, J., Perrier, P., et al 1985. On the numerical solution of nonlinear problems in fluid dynamics by least squares and finite element methods (II). Application to transonic flow simulations. *Comput. Methods Appl. Mech. Eng.* 51: 363–94

Bristow, D. R. 1976. A new surface singularity method for multi-element airfoil analysis and design. *AIAA Pap.*76–20

Cabuk, H., Sung, C.-H., Modi, V. 1991. Adjoint operator approach to shape design for internal incompressible flows. See Dulikravich 1991, pp. 391–404

Campbell, R. L., Smith, L. A. 1987. A hybrid algorithm for transonic airfoil and wing design. *AIAA Pap.*87–2552

Cosentino, G. B., Holst, T. L. 1986. Numerical optimization design of advanced transonic wing configurations. *J. Aircraft* 233: 192–99

Dedoussis, V., Chaviaropoulos, P., Papailiou, K. D. 1992. A fully 3-D inverse method applied to the design of axisymmetric ducts. *Proc. TURBO EXPO, 37th Int. Gas Turbine Aeroengine Congr. Expo., Cologne.* In press

Destarac, D., Reneaux, J. 1990. Transport aircraft aerodynamic improvement by numerical optimization. *Int. Counc. Aeronaut. Sci. Pap.* 90–6.7.4

Drela, M. 1986. Two-dimensional transonic aerodynamic design and analysis using the Euler equations. *MIT Gas Turbine & Plasma Dyn. Lab. Rep. No.* 187

Drela, M. 1990. Viscous and inviscid inverse schemes using Newton's method. See AGARD 1990, Pap. 9

Drela, M., Giles, M. B. 1987. ISES : A two-dimensional viscous aerodynamic design and analysis code. *AIAA Pap.* 87–0424

Dulikravich, G. S. 1990. Aerodynamic shape design. See AGARD 1990, Pap. 1

Dulikravich, G. S., ed. 1991. *Proc. 3rd Int. Conf. Inverse Des. Concepts and Optim. in Eng. Sci. ICIDES III*

Fornasier, L. 1989. An iterative procedure for the design of pressure-specified three-dimensional configurations at subsonic and supersonic speeds by means of a higher-order panel method. See AGARD 1989, Pap. 6

Frank, P. D., Shubin, G. R. 1990. A comparison of optimization-based approaches for a model computational aerodynamics design problem. *BCS-ECA-TR*–136

Fujii, K., Takanashi, S. 1991. Aerodynamic aircraft design methods and their notable applications—Survey of the activity in Japan. See Dulikravich 1991, pp. 31–44

Gally, T. A., Carlson, L. A. 1987. Inviscid transonic wing design using inverse methods in curvilinear coordinates. *AIAA Pap.* 87–2551

Greff, E., Forbrich, D., Schwarten, H. 1991. Application of direct inverse analogy method (DIVA) and viscous design optimization techniques. See Dulikravich 1991, pp. 307–24

Henne, P. A. 1980. An inverse transonic wing design method. *AIAA Pap.* 80–0330

Hicks, R. M., Vanderplaats, G. N., Murman, E. M., King, R. R. 1976. Airfoil section drag reduction at transonic speeds by numerical optimization. *Soc. Automot. Eng. Pap.* 760477

Hua, J., Zhang, Z. Y. 1990. Transonic wing design for transport aircraft. *Int. Counc. Aeronaut. Sci. Pap.* 90–3.7.4, *ICAS Congr.*

Jameson, A. 1988. Aerodynamic design via control theory. *NASA CR*–181749, *ICASE Rep. No.*88–64

Jameson, A., Caughey, D. A. 1977. A finite volume method for transonic potential flow calculations. *Proc. AIAA 3rd Comput. Fluid Dyn. Conf., Albuquerque*, pp.35–54

Kubrynski, K. 1991. Design of 3-dimensional complex airplane configurations with specified pressure distribution via optimization. See Dulikravich 1991, pp.263–80

Labrujère, Th. E. 1978. Multi-element airfoil design by optimization. *NLR MP* 78023 *U*

Lekoudis, S. G., Sankar, N. L., Malone, J. B. 1986. The application of inverse boundary layer methods to the three-dimensional viscous design problem. *Commun. Appl. Numer. Methods* 2: 57–61

Lighthill, M. J. 1945. A new method of two-dimensional aerodynamic design. *Aeronaut. Res. Counc. Rep. Memo.* 2112

Malone, J. B., Narramore, J. C., Sankar, L. N. 1989. An efficient airfoil design method using the Navier-Stokes equations. See AGARD 1989, Pap. 5

Mangler, W. 1938. Die Berechnung eines Tragflügelprofils mit vorgeschriebener Druckverteilung. *Jahrb. Deutsche Luftfahrt forsch.* 1938, pp. I46–I53

Meauzé, G. 1989. Overview of blading design methods. *AGARD Lect. Ser. No.* 167

Ormsbee, A. I., Chen, A. W. 1972. Multiple element airfoils optimized for maximum lift coefficient. *AIAA J.* 10: 1620–24

Pironneau, O. 1983. Optimal shape design for elliptic systems. In *Springer Series in Computational Physics*, ed. H. Cabannes, M. Holt, H. B. Keller, J. Killeen, S. A. Orszag. Heidelberg: Springer-Verlag. 192 pp.

Ratcliff, R. R., Carlson, L. A. 1989. A direct-inverse transonic wing-design method in curvilinear coordinates including viscous-interaction. *AIAA Pap.* 89–2204

Reneaux, J., Thibert, J. J. 1985. The use of numerical optimization for airfoil design. *AIAA Pap.* 85–5026

Reneaux, J., Allongue, M. 1989. Defenitions de profils et des pales d'helicoptere par optimisation numerique. See AGARD 1989, Pap. 19

Ribaut, M., Martin, D. 1986. A quasi three-dimensional inverse design method using source and vortex integral equations. *Commun. Appl. Numer. Methods* 2: 63–72

Rizk, M. H. 1989. Aerodynamic optimization by simultaneously updating flow variables and design parameters. See AGARD 1989, Pap. 15

Selg, M. S., Maughmer, M. D. 1991. A multipoint inverse airfoil design method based on conformal mapping. *AIAA Pap.* 91–0069

Shankar, V. 1980. Computational transonic inverse procedure for wing design with automatic trailing edge closure. *AIAA Pap.* 80–1390

Slooff, J. W. 1983. Computational methods for subsonic and transonic aerodynamic design. *NLR MP* 83006 *U*, *AGARD Rep. No.* 712

Smith, A. M. O. 1974. High lift aerodynamics. *AIAA Pap.* 74–939

Sobieczky, H. 1989. Progress in inverse design and optimization in aerodynamics. See AGARD 1989, Pap. 1

Takanashi, S. 1984. An iterative procedure for three-dimensional transonic wing design by the integral equation method. *AIAA Pap.* 84–2155

Timman, R. 1951. The direct and the inverse problem of aerofoil theory. A method to obtain numerical solutions. *Natl. Aeronaut. Res. Inst. Rep. F.*16

Van den Dam, R. F. 1989. Constrained spanload optimization for minimum drag of multi-lifting-surface configurations. See AGARD 1989, Pap. 16

Van Egmond, J. A. 1989. Numerical optimization of target pressure distributions for subsonic and transonic airfoil design. See AGARD 1989, Pap. 17

Volpe, G. 1989. Inverse design of airfoil contours: Constraints, numerical method and applications. See AGARD 1989, Pap. 4

Volpe, G., Melnik, R. E. 1981. The role of constraints in the inverse design problem for transonic airfoils. *AIAA Pap.* 81–1233

Woods, L. C. 1952. Aerofoil design in two-dimensional subsonic compressible flow. *Aeronaut. Res. Counc. Rep. Memo.* 2845

Zhu, Z., Xia, Z., Wu, L. 1991. An inverse method with regularity condition for transonic airfoil design. See Dulikravich 1991, pp. 541–49

Annu. Rev. Fluid Mech. 1993. 25: 215–40

SURFACE WAVES AND COASTAL DYNAMICS

Chiang C. Mei

Dept. of Civil Engineering, Massachusetts Institute of Technology, Cambridge, Massachusetts 02139

Philip L.-F. Liu

School of Civil and Environmental Engineering, Cornell University, Ithaca, New York 14853

KEY WORDS: shoreline processes, coastal engineering

INTRODUCTION

Unlike in the open sea, water motion in the coastal region is distinguished by the vital role of the solid boundaries. If surface waves are generated by distant storms, they must first propagate across the continental shelf towards the coast. Upon entering shoaling waters, they are either refracted by varying depth or current, or diffracted around abrupt bathymetric features such as submarine ridges or valleys, losing part of their energy back to the deep sea. Waves continuing their shoreward march give up some energy by dissipation near the bottom. Nevertheless each crest becomes steeper and mightier, and makes its final display of power by breaking and splashing on the shoreline, and sending countless sand particles afloat.

This review covers theoretical advances made during the past decade for predicting or understanding wind-induced waves and associated coastal processes. We first discuss the offshore propagation over many wavelengths. An effective approximation which permits practical calculations of both refraction and diffraction over a mildly sloping bottom will be mentioned. For still longer distances nonlinear effects accumulate to become important; recent progress in the parabolic approximation for the

0066–4189/93/0115–0215$02.00

refraction and diffraction of short waves is also reviewed. We then turn our attention to the nearshore region and survey the infragravity waves which are caused by nonlinear interactions of short waves within a narrow frequency band. In particular we discuss the effects of a gradual or sudden change of depth or a steady current on the modulation of short waves and the generation of long infragravity waves. This is followed by a discussion of infragravity waves which resonate in harbors and the strong resonance of edge waves on beaches. The final topics are concerned with the causes and effects of bars and ripples on the seabed.

Many other topics on coastal hydrodynamics that have been surveyed in *Annual Review of Fluid Mechanics* and elsewhere are not treated here. These are: harbor seiching (Miles 1974, Mei 1983), breaking waves on beaches (Peregrine 1983), tsunamis (Voit 1987, Kajiura & Shuto 1990), surf-zone dynamics and circulation (Battjes 1988, Battjes et al 1990), and wave-current interaction in bottom boundary layers (Grant & Madsen 1986, Sleath 1990).

DIFFRACTION AND REFRACTION OVER A LARGE REGION

Linear Approximation for a Mildly Sloping Bottom

For planning a proper defense for beaches and harbors against the perpetual onslaught of sea waves, engineers need reliable estimates of the wave climate in the vicinity of the project site. Since long records of direct measurement are rarely available, rational extrapolation must usually be made from wave data or wind conditions far offshore. Therefore the theoretical prediction of wave propagation from a distant boundary is the first step in engineering design. In principle mathematical tools for the three-dimensional computation of diffraction and refraction are available for infinitesimal waves (Mei 1978), primarily for the immediate neighborhood within a few wavelengths of a platform. The coastal engineer must, however, consider slowly varying topography and wave propagation over hundreds or more wavelengths. Efforts for reducing the computations are necessary, and have been sought by reducing the dimension of the computational problem. The forerunner of this kind of effort is the *ray approximation* for infinitesimal waves over depths that vary gently over horizontal distances much longer than the local wavelength. In this approximation one first finds the rays by a geometrical optics approximation, and then calculates the spatial variation of the wave envelope along the rays. Numerical discretization can be done in steps not necessarily small compared to a typical wavelength. However such an approximation fails near the caustic or the focal regions, where neighboring wave rays intersect;

diffraction and possibly nonlinearity are important. While ad hoc methods for local remedies are available, it is not always convenient to implement them in practice.

Within the linear wave theory framework, an improvement was first suggested by Berkhoff (1972) who proposed a two-dimensional theory which can deal with large regions of refraction and diffraction. Heuristically over a slowly varying topography evanescent modes are not important except in the immediate neighborhood of a three-dimensional obstacle. For a monochromatic wave with frequency ω and free-surface displacement η, it is reasonable to represent the velocity potential, which formally represents the propagating mode only, by

$$\phi = \frac{-ig\eta}{\omega} \frac{\cosh k(z+h)}{\cosh kh} e^{-i\omega t}, \tag{1}$$

where $k(x, y)$ and $h(x, y)$ vary slowly in x and y according to the usual linear dispersion relation

$$\omega^2 = gk \tanh kh \tag{2}$$

and g is the acceleration of gravity. By a perturbation argument Smith & Spinks (1975) have shown that the free surface displacement η must satisfy the following partial differential equation:

$$\nabla \cdot (CC_g \nabla \eta) + \frac{\omega^2}{g} \eta = 0, \tag{3}$$

where C and C_g are the local phase and group velocities of a plane progressive wave. This elliptic equation is asymptotically valid for sufficiently small $\mu (= |\nabla h|/kh$ to leading order) and is known as the *mild-slope equation*. An indication of its versatility can be seen in two limits. For long waves in shallow water the limit of (3) at $kh \ll 1$ is known to be valid even if $\mu = \mathcal{O}(1)$ as long as linearization is acceptable. On the other hand if the depth is constant but kh is arbitrary, (3) reduces to the Helmholtz equation where k satisfies (2). Both limits can be used to calculate diffraction legitimately. Thus (3) should be a good interpolation for all kh. Recently Miles (1991) rederived Equation (3) via a variational principle and showed that it conserves not only wave action, but also wave energy.

In applying (3) to field problems its efficiency lies between the fully three-dimensional schemes and the one-dimensional ray theory, and practical methods based on finite elements for linearized long waves (Chen & Mei 1974) have been developed (e.g. Houston 1981, Tsay & Liu 1983). If there is an obstable whose length scale is comparable to the wavelength, this equation can in principle be used for the far field where evanescent modes are of diminishing importance. In the neighborhood of the obstacle, the problem is strictly three dimensional and many numerical techniques exist for its solution (Mei 1978). The entire problem can then be completed by matched asymptotics.

The mild-slope equation was extended first by Booij (1981) to include the effects of varying currents. After some modifications and corrections on Booij's original derivation, the general mild-slope equation can be written (Kirby 1983, 1984; Liu 1990) as

$$\frac{D^2\eta}{Dt^2}+(\nabla\cdot\mathbf{U})\frac{D\eta}{Dt}-\nabla\cdot(CC_g\nabla\eta)+(\sigma^2-k^2CC_g)\eta=0, \tag{4}$$

where σ is the intrinsic frequency of the wave-current system,

$$\sigma=\omega-\mathbf{k}\cdot\mathbf{U}$$

$$\sigma^2=gk\tanh kh$$

$$\nabla\times\mathbf{k}=0. \tag{5}$$

The total time derivative appearing in (3) is defined as

$$\frac{D}{Dt}=-i\omega+\mathbf{U}\cdot\nabla. \tag{6}$$

The coefficients in the mild-slope equation (4) are determined from the ray theory described by (5). Difficulties in defining the direction of the wavenumber vector $\mathbf{k}$ arise in the regions where the ray theory becomes invalid, such as near caustics. Kostense et al (1986) developed an iterative finite-element scheme by first ignoring the current effects in determining the wavenumber vector. They then updated the $\mathbf{k}$ vector from the solutions of the mild-slope equation.

Thus far there has been no allowance for nonlinearity.

Linear Parabolic Approximation

For essentially forward-propagation problems, the so-called *parabolic approximation* expands the validity of ray theory and is much more efficient than the mild-slope equation. The basic idea was originated for electromagnetic waves by Leontovich & Fock (1944), and can be readily illustrated for a pure diffraction problem, i.e. the edge of the shadow zone behind a long breakwater. Consider a semi-infinite breakwater lying along the positive y axis which is struck by a plane wave of normal incidence originating from $x\sim-\infty$. The ray passing the tip of the breakwater separates the domain $x>0$ into an illuminated zone ($y<0$) and a shadow zone ($y > 0$). According to the ray approximation, waves in the illuminated zone are progressive with amplitude A equal to that of the incident wave, i.e. A_o. But A must suddenly drop to zero on the shadow side of the ray. This discontinuity is of course a shortcoming of the ray approximation. An effective remedy is to insert a boundary layer along the edge ray within which the progressive-wave amplitude is modulated slowly in the transverse direction and even more slowly in the direction of wave propagation. In this way an approximate equation is obtained for the wave envelope:

$$-2ik\frac{\partial A}{\partial x} + \frac{\partial^2 A}{\partial y^2} = 0. \tag{7}$$

Equation (7) is a Schrödinger equation where the coordinate x is time-like; it is effective in a parabolic domain $\mathcal{O}(ky/\sqrt{kx}) = 1$.

The same idea can be easily extended for a curved ray and hence for progressive waves that are simultaneously diffracted and refracted. Within the linearized context, Liu & Mei (1976) employed the concept to find the wave field in the neighborhood of a breakwater on a beach with parallel (one-dimensional) depth contours. Their solutions have been compared to laboratory data with excellent agreement (Liu 1982, Tsay & Liu 1982).

For two-dimensional bathymetry the first extension was given by Radder (1979). Stable numerical techniques for the heat equation, such as the Crank-Nicolson method, can be easily applied. Numerical discretization can be coarser than the local wavelength since the governing equation describes the envelope. One of the important applications is to calculate the wave field in the vicinity of caustics, where ray theory fails. Lozano & Liu (1980) compared numerical solutions near cusped caustics generated by a topographical lens with laboratory data (Whalin 1971). Although there is qualitative agreement, nonlinearity—significant very near the focal region—is not well predicted by linear theory. The nonlinear effects were later considered by Liu & Tsay (1984).

Although the parabolic approximation has been used primarily for forward propagation, extensions have been made by Liu & Tsay (1983) who deduced a pair of parabolic equations coupling both forward and backward wave fields. These coupled equations are then solved numerically by an iterative scheme. For scattering by a submerged breakwater, agreement between the solutions from the mild-slope equation and from the parabolic approximation is very good.

As indicated before, the parabolic equation (7) is valid in a narrow boundary layer within which waves propagate primarily in the longitudinal direction. This restriction for small angles can be relaxed by adding higher-order derivatives, i.e.

$$-2ik\frac{\partial A}{\partial x}+\frac{\partial^2 A}{\partial y^2}-\frac{1}{4k^2}\frac{\partial^4 A}{\partial y^4}=0. \tag{8}$$

Other similar higher-order approximations have been discussed by Kirby (1986c) and Liu (1990). Most recently, Dalrymple & Kirby (1988) and Dalrymple et al (1988, 1989) developed a model for very wide-angle diffraction. They decomposed the entire wave field into an angular spectrum by applying a Fourier transformation with respect to the alongshore coordinate to the mild-slope equation. Therefore, the model is limited to problems that are periodic in the alongshore direction.

The parabolic approximation has also been applied to the wave-current interaction system (Booij 1981, Liu 1983, Kirby 1983). Because of the difficulty in determining the wavenumber vector in the mild-slope equation (7), the application of the parabolic approximation has been limited. High quality laboratory data are needed to verify the approximate theories.

Nonlinear Parabolic Approximation

Quite apart from the study of diffraction it is known that the Benjamin-Feir instability of nonlinear Stokes waves due to colinear sideband disturbances is governed by the parabolic Schrödinger equation involving a third-order nonlinearity (Zakharov 1968). To study the effects of slightly oblique sidebands on unidirectional Stokes waves, Benney & Roskes (1969) have extended the nonlinear Schrödinger equation to two horizontal dimensions for finite and constant depth, and obtained two evolution equations coupling the short-wave envelope A and the potential $\bar{\phi}$ of the induced long wave (see also Davey & Stewartson 1974). These equations are valid if the characteristic scales are such that $\varepsilon^2\omega t \leq \mathcal{O}(1)$, $\varepsilon^2 kx \leq \mathcal{O}(1)$, and $\varepsilon ky = \mathcal{O}(1)$.

Now if the envelope is steady, i.e. $\partial/\partial t = 0$, this set of evolution equations can be combined to give

$$2ik\frac{\partial A}{\partial x}+\frac{\partial^2 A}{\partial y^2}+k^4\frac{C}{C_g}\frac{\cosh 4kh+8-2\tanh^2 kh}{8\sinh^4 kh}|A|^2A=0 \tag{9}$$

which is the nonlinear extension of (7). Equation (9) has been applied to nonlinear diffraction problems where the domain of interest satisfies the above scale assumptions. For example, it was used by Yue & Mei (1980) to study the head-sea incidence of steady Stokes waves on a slender island with vertical sides. This leads to an initial-boundary-value problem readily solved by the Crank-Nicolson scheme. For the case of a thin wedge, one side of which also describes the grazing incidence on a semi-infinite breakwater, it was found that the wave envelope exhibits the Mach-stem phenomenon, in that the wave crests are suddenly bent along a straight line inclined at an angle slightly larger than the angle between the incident wave and the breakwater. Between this line and the breakwater the wave crests are nearly perpendicular to the breakwater. Allowing for transient evolution, Yue (1980) has further calculated the diffraction of a plane Stokes wave in a canal whose width changes linearly from one constant to another smaller constant. Another application was made by Stamnes et al (1983) to predict the waves in the focal region. The practical motivation was to let plane incident waves pass over a series of underwater lenses, so that wave crests would become concentric circular arcs and converge to a point. Wave energy at the focal point can then be harnessed (Mehlum & Stamnes 1978). Steady nonlinear theory was found to compare quite well with extensive experimental data (Stamnes et al 1983). In particular, nonlinearity reduces the amplitude at the focal region and hence the prospects of energy conversion.

For horizontally varying bathymetry, extensions have been made by Liu & Tsay (1984), Kirby (1984, 1986b), and Kirby & Dalrymple (1983). The nonlinear parabolic wave equation for the second-order Stokes wave can be written as

$$2ik_o\frac{\partial A}{\partial x}+\frac{\partial^2 A}{\partial y^2}+\left[k^2-k_o^2-\frac{\nabla^2(CC_g)^{1/2}}{(CC_g)^{1/2}}+i\frac{\partial k_o}{\partial x}\right]A$$
$$-k^4\frac{C}{C_g}\frac{\cosh 4kh+8-2\tanh^2 kh}{8\sinh^4 kh}|A|^2A=0, \quad (10)$$

in which k_o is a reference wavenumber. This equation was used to calculate wave amplitudes in a focal region caused by a topographical lens (Whalin 1971). In the case in which the incident wave contains no higher harmonics and can be described by the small-amplitude wave theory, the second harmonic component becomes as large as the first harmonic component in the caustics. Liu & Yoon (1986) further applied the theory to examine the forward scattering of second-order Stokes waves by a depth discontinuity. When waves propagate from a shallow water region h_1 to a deeper region h_2, the line of depth discontinuity coincides with the linear caustics line when the incident angle is smaller than the critical angle

$$\theta_{cr}=\frac{\pi}{2}-\sin^{-1}\left(\frac{C_1}{C_2}\right), \quad (11)$$

where C_1 and C_2 are the phase speeds associated with water depth h_1 and h_2, respectively. The angle of incidence is defined as the angle between the direction of wave propagation and the line of depth discontinuity. Liu & Yoon demonstrated that stem waves, similar to those seen along the breakwater, develop along linear caustics. Nonlinearities become more significant for larger incident-wave amplitudes and smaller angles of incidence. For wide-angle diffraction of Stokes waves, angular-spectrum theory has been extended by Suh et al (1990).

In fairly shallow water, wave propagation can be described more adequately by Boussinesq equations. Liu et al (1985) derived evolution equations for spectral-wave components in a slowly varying two-dimensional domain. Their model equations are the extension of the K-P equation (Kadomtsev & Petviashvili 1970) to a varying depth. The original K-P equation extends the Korteweg–de Vries equation to include weak transverse modulation and is also a kind of parabolic approximation.

INFRAGRAVITY WAVES INDUCED BY SHORT WAVES

From wind swell records along beaches Munk (1949) and Tucker (1950) observed long waves of period between 1 and 5 minutes, and found strong correlation between the long-wave and the swell envelope. For storms of 1–2 days duration the long-wave height is roughly one-tenth of the swell amplitude. Munk attributed this correlation to the nonlinear interaction between waves and coined the term *surf beats* for these long waves. He suggested that they may be of relevance to harbor or bay oscillations. The nonlinear mechanism was first explained theoretically by Longuet-Higgins & Stewart (1962) through the idea of *radiation stresses*. Since surf beats are appreciable on the beach only, they have also been suspected of being edge waves trapped on the beach. Gallagher (1971) gave a theory in which groups of swells with a narrow frequency band generate long-period forcing at second order and resonate long-period edge waves on a beach. Recent field experiments give strong confirmation of this speculation (Holman 1981, Oltman-Shay & Guza 1987). Long waves in this range of periods are called *infragravity waves* in coastal engineering and oceanography.

For narrow-banded waves the radiation stress is just a depth-averaged and time-averaged momentum flux arising at the second order in wave steepness, similar to the Reynolds stress in turbulence. If the short wave consists of two nearly equal frequencies, the radiation stress forces long-period motion at the difference frequency of the short waves. In deep water the long wave travels at the group velocity of the short wave, and is a depression with its maximum amplitude beneath the peak of the group envelope. The long wave has been called the *set-down* wave, or the *forced* or *bound* long wave.

Sand (1982) has demonstrated in wave tanks that when generating wave groups by oscillating pistons, two kinds of long waves are radiated: a bound (set-down) wave propagating at the group velocity of the short wave, and another which travels at the long-wave speed of $\sqrt{gh}$. It turns out that these two types of waves must in general coexist whenever there is a lateral boundary or a change of depth. This means that bathymetry, man-made structures, or coastal land features can change the wavenumber spectrum. These changes have been studied for several representative cases, discussed below.

Short Waves over Varying Depth or in Currents

Molin (1982) considered groups of short waves incident from deep water to shoaling depth. A more consistent theory was given by Mei & Benmoussa (1984) by extending the ray approximation. Specifically, for a train of monochromatic waves with local amplitude A the period-average of the mean surface displacement ξ satisfies the long-wave equation

$$\frac{\partial}{\partial x_i} gh \frac{\partial \xi}{\partial x_i} - \frac{\partial^2 \xi}{\partial t^2} = \frac{1}{\rho} \frac{\partial^2 S_{ij}}{\partial x_i \partial x_j}, \tag{12}$$

where S_{ij} are the components of the radiation stress tensor of the progressive wave given by

$$S_{ij} = \frac{\rho g A^2}{2}\left[\left(\frac{C_g}{C} - \frac{1}{2}\right)\delta_{ij} + \frac{C_g}{C} k_i k_j\right]. \tag{13}$$

Now C_g, C, and k are slowly varying in x, y through the depth h. Wherever h is not constant the radiation stress also varies, forcing infragravity waves. So far only one-dimensional bathymetries with obliquely incident long-crested waves have been studied. At leading order, short waves are refracted according to the laws of geometrical optics. Unlike unidirectional waves in constant depth, the envelope propagates away from the short carrier waves upon entering the shoaling water. At second order the long wave consists of the bound long wave and free long waves. The free long waves are generated in the region of variable depth and radiate outward. Over a submarine ridge, the free long waves can be trapped, while the short waves pass through (Mei & Benmoussa 1984, see Liu 1989 for corrections).

Equations (12) have also been used as the basis for modeling surf beats as a second-order consequence of incident wave groups breaking on the beach. Symonds et al (1982) first suggested the importance of the variation of the breaker-line position. Recently Nakaza & Hino (1991) described a one-dimensional theory and experiments for normally incident, periodically modulated short waves breaking on a long reef of constant depth, connected to the deep sea by a sloping transition. On the reef the amplitude of the breaking wave is related to the local depth empirically—as is usual in the theory of longshore currents (i.e. currents that flow along the shore and are induced by obliquely-incident breaking waves). Both bound long waves and free long waves are of course present; the latter may be weakly resonated by the finite length of the reef.

A sudden change of depth refracts and scatters short waves. For a wide rectangular shelf the full second-order problem has been examined by Agnon & Mei (1988). They also studied transient long waves trapped on a long shelf. If a finite packet of short waves is incident obliquely on a shelf, the long waves trapped on the shelf can reverberate long after the short-wave packet has passed.

The generation of long waves due to the refraction of short waves over a one-dimensional shear current has been investigated by Liu et al (1990). Similar to the situation with depth variation, the short waves, the wave envelope, and the free long waves propagate in different directions over the shear current region. The bound long waves, however, always propagate with the wave envelope. The free long waves may radiate away from or be trapped in the shear current region, depending on the angle of incidence and the current velocity.

In both varying depth and current cases, caustics may exist in short waves as well as in the wave envelope. The treatment of caustics has not been included in long-wave analysis. Moreover, because the linear long-wave equation (12) is derived from the Stokes wave theory, it is inadequate for describing waves in fairly shallow water. More work is needed to extend present theoretical knowledge to include nonlinear effects in long-wave propagation.

Short Waves Diffracted by Harbors or Artificial Islands

Long-period oscillations inside a harbor may induce breaking of mooring lines, or make loading and unloading operations hazardous. For many harbors the horizontal dimensions and depth are such that the lowest few modes have periods longer than a few minutes. Direct excitation of these natural modes is possible by incident long waves of comparable periods, such as tsunamis. Such theories were begun by Miles & Munk (1961), and have been reviewed previously by Miles (1974) and Mei (1983). Extensions to include nonlinear effects have also been developed on the basis of Boussinesq equations (Rogers & Mei 1978, Lepelletier 1980). Now computational models based on these theories are also available.

Though the effects can be damaging, tsunamis are, however, rare events of concern only to certain harbors in the Pacific, Alaska, and South America. Many more harbors along the seacoasts or in the Great Lakes of North America are plagued by long-period oscillations even though the wave spectrum outside consists only of wind swell of periods of only a few seconds. This is clearly seen in the wave records at Barber's Point Harbor, Honolulu, taken by the U.S. Corps of Engineers and cited by Mei (1991). Since these long periods lie in the range of the natural frequency of most mooring systems, their presence must be understood and accounted for in the design of breakwaters.

Bowers (1977) was the first to link the theory of Longuet-Higgins & Stewart on the set-down of short waves to long-period harbor oscillations. He studied a semi-infinitely long channel with a narrow bay of finite length at the closed end. By sending narrow-banded incident waves from the wider channel, he found—theoretically and experimentally—both bound long waves (set-down) and free long waves of speed $\sqrt{gh}$. It is these free waves which are resonated in the narrow bay when the bandwidth of the incident short waves is properly chosen. Mei & Agnon (1989) extended this theory for a harbor set in from an open coast. For analytical convenience they examined a narrow bay of rectangular planform, with its longitudinal axis intersecting the straight coastline at a right angle. They further assumed that the incident wavelength is much less than the bay width which is in turn much less than the bay length L. At first order the incident waves were taken to be normally incident and sinusoidally modulated with the group period Ω which is much smaller than the frequency ω of the incident carrier wave. In addition to simple reflection from the coast and from the end of the bay, diffraction from the corners of the bay entrance is treated by the parabolic approximation. Short waves reflected from both the coast and the bay end generate bound long waves propagating with the associated wave groups. If $K_g = 2\Omega/C_g$ denotes the wavenumber of the bound long wave, the bay is resonated if $K_g L \simeq (2n+1)\pi$, when the bound long wave reflected from the end of the bay is out of phase with respect to the bound long wave reflected from the coastline by $\pi/2$. They also deduced the approximate factors of amplification at resonance peaks both for the narrow bay and for a two-dimensional harbor whose back side is parallel to the coastline.

Relaxing some of the restrictions on the harbor geometry imposed by Mei & Agnon (1989), Wu & Liu (1990) examined the long-wave excitation in a rectangular harbor protected by a pair of breakwaters parallel to the shoreline. In their study, the harbor mouth is assumed to be wider than the short wavelength and oblique incident wave groups are considered. Using an integral-equation method, the first-order wave field is solved exactly. Analytical expressions for both bound and free long waves inside and outside the harbor are obtained. From numerical examples, Wu & Liu showed that only the free long waves are resonated at the lowest mode;

the bound long waves may be ignored for all practical purposes. For higher modes both bound and free long waves are equally important. For engineering applications it is desirable to extend the theory to broad-banded incident waves and arbitrary depth, and to account empirically for boundary friction and dissipation due to the breaking of short waves on beaches.

The linearized theory of scattering by a vertical cylinder of circular planform is well known and is one of the few geometries solvable exactly. If the ratio $ka = \mathcal{O}(1)$ where k denotes the short-wave wavenumber and a the cylinder radius, then second-order diffraction is already a complex mathematical problem (Molin 1979). Zhou & Liu (1987) treated a large cylinder of radius a where $k^2Aa = \mathcal{O}(1)$. At first order the incident plane wave and all the partial waves corresponding to various angular modes have amplitudes slowly varying in horizontal directions. At second order the long waves are governed by an equation similar to (12). Now the bound long wave associated with the incident group of short waves acts as an incident long wave. In addition, there is local forcing due to slow variations of the radiation stresses. Thus long waves with characteristic speed $\sqrt{gh}$ are forced both from infinity and locally. From the asymptotic solution, Zhou & Liu calculated the wave field and wave forces. Based on their theory, Masuda et al (1989) extended the calculations to $ka = 50$ and found that on the shadow side, only long waves are appreciable while short waves are absent. This implies that a moored ship on the lee side may be freed from short-period heaving and rolling, but is still subject to the perils of slow-drift oscillations.

Strong Resonance of Trapped Waves

Although edge waves can be forced by local wind or earthquakes, on a plane beach of great length they cannot be linearly excited by incident waves from far offshore with the same frequency. Aside from the second-order mechanism discussed by Gallagher (1971), normally incident short waves can resonate long edge waves by a subharmonic mechanism predicted by Guza & Bowen (1976; see also Rockcliff 1978, Minzoni & Whitham 1977). The subharmonic mechanism is effective only when the incident wave is tuned within a relatively narrow frequency band, though the amplification ratio at resonance can be very large—$\mathcal{O}(ka^{1/2})$. Still, there must be some energy in the low-frequency part of the incident-wave spectrum for this mechanism to be effective.

Field experiments by Oltman-Shay & Guza (1987) strongly support the basic ideas behind Gallagher's theory in which drastic approximations for wave shoaling and breaking are made. In particular the resonant amplitude of the edge wave depends crucially on bottom friction which is hard to estimate quantitatively. Foda & Mei (1981) reexamined the mechanism for two kinds of trapped long waves—one on a submarine ridge and the other on a beach. They assumed the incident short waves to be weakly nonlinear and the bottom slope to be as small as the short-wave steepness. The long waves are allowed to attain a displacement amplitude comparable to that of the short waves and are periodic in the direction parallel to the depth contours (i.e. proportional to $\cos Ky$). The normally incident short waves are assumed to be sinusoidally modulated in time at the modulation frequency Ω and to include a second-order perturbation modulated in y in the form of $\cos Ky$ also. As in the subharmonic case of Guza & Bowen, long waves are radiated at the frequency 2Ω. The evolution equation for the complex amplitude D of the long trapped wave is found to have the following form:

$$\frac{\partial D}{\partial t}+(\gamma_o+\gamma_1 A_o^2)D+\gamma_2|D|^2D = -\gamma_3 A_o^2 b_o, \tag{14}$$

where γ_n ($n = 0, 1, 2, 3$) are coupling coefficients, with γ_2 corresponding to damping due to the radiation of long waves of the frequency 2Ω. In the evolution equation above, A_o is the incident short-wave amplitude in deep water, and b_o is the amplitude of its perturbation with longshore modulation. For a submarine ridge, all the coupling coefficients are obtained theoretically for an inviscid fluid. Frictional dissipation can be added to radiation damping without causing qualitative changes. Transient responses due to Gaussian envelopes for both A_o and b_o have been examined. Foda & Mei further modified the theory for a plane beach by adopting the empirical model that the short waves alone break along a line where the local amplitude reaches a certain fraction of the still-water depth. Evolution of the edge-wave amplitude due to long groups of incident waves with a Gaussian envelope has been studied theoretically. Unfortunately controlled experiments in the laboratory are not yet available.

Far Infragravity Waves

In their field study, Oltman-Shay et al (1989) observed that long-wave motions with frequencies in the range of 10^{-3}–10^{-2} Hz exist. These long-wave frequencies lie outside the traditional low-frequency limit of the infragravity band. Bowen & Holman (1989) demonstrated theoretically that these nearshore long waves are generated by the shear instability of longshore currents (see the review of Battjes et al 1990). They called these waves Far Infragravity (or FIG) waves. The mechanism of the shear instability depends on the conservation of potential vorticity with the background vorticity field supplied by the shear structure of the longshore current. For the case of constant depth, the governing equations reduce to the inviscid Orr-Sommerfeld, or Rayleigh equations. Using a simplified (triangular) longshore current profile and a constant depth, Bowen & Holman showed that wavelengths of shear waves are of the order of twice the longshore current width. The celerities are roughly equal to one third of the peak longshore velocity. The growth time is exponential with an *e*-folding time scale of about half a wave period. The simple theory has been compared to field data (Oltman-Shay et al 1989) with good qualitative agreement. Dodd & Thornton (1990) extended the simple theory to consider a more realistic sloping-beach profile. They also examined the energy transfer between the longshore current and shear waves and showed that the necessary condition for instabilities to grow is that the cross-shore gradient of the horizontal Reynolds stress must be nonzero.

Shear waves can have a significant impact on the nearshore mixing process, as they may provide an effective horizontal eddy viscosity ten times greater than that due to the incident gravity waves alone. Further theoretical studies and measurements for more complex current systems are certainly needed.

WAVE INTERACTION WITH LONGSHORE BARS

Submerged sandbars that are parallel to the shorelines of lakes and oceans have been studied by geologists and oceanographers for a long time. Their presence has been noted in the Great Lakes and open seacoasts (see Mei 1985 for references). Katoh (1984) showed that bars are found over more than half of the coastline surrounding Japan. Typically these multiple bars exist on very mild beaches with slopes of less than 5 per thousand. They number from 3 to 17 while their spacing may range from 10 m to 480 m, which generally increases offshore. How are the bars created? How do they change the wave climate?

Laboratory experiments by Bagnold (1946) show that near a seawall where all the incident waves are reflected to give rise to a standing-wave system, bars grow from an initially flat sandy bottom at a spacing equal to half of the incident wavelength. This spacing is precisely that between nodes or antinodes of the standing-wave system. It is known that time-averaged Reynolds stresses force circulation cells in the bottom boundary layer such that fluid particles near the bottom drift towards the antinodes while those near the top drift towards the nodes. Thus heavy sand particles would tend to accumulate under the antinodes, while light particles would tend to accumulate at the nodes. This process not only initiates sandbars at a spacing equal to half of the incident wavelength but also enhances sediment sorting (Carter et al 1973, Irie et al 1984). Carter et al (1973) also showed by a constant eddy-viscosity model that the reflection coefficient must be larger than 0.414 for cells of mean circulation to form. How then can there be such a strong reflection on a mild beach?

One possibility is that on a mild sandy beach, plunging breakers can dislodge sand particles and deposit them nearby, creating a bar near the first line of breaking. This so-called *break-point bar* had been established in the laboratory by Keulegan (1944) for the U.S. Army Corps of Engineers prior to the landing of Allied Forces at Normandy in WWII. Being an obstacle at very shallow depth, this bar can increase reflection towards the sea and start a partially standing wave—hence it encourages the formation of sandbars at half-wavelength spacings. This picture is supported by the laboratory experiments of Boczar-Karakiewicz et al (1981) and Irie et al (1984) who performed tests on a gently sloping sandy beach terminated near the beach by a steep seawall.

By passing waves over a train of periodically spaced rigid bars on the bottom of a wave tank, Heathershaw (1982) showed decisively that reflection can be very strong if the incident wavelength is twice the wavelength of the bars. This mechanism is called *Bragg resonance* in crystallography. With such a matching of phases, waves reflected from successive bars are in phase and therefore reinforce one another, resulting in strong total reflection. This mechanism suggests that strong reflection of waves is both the cause and effect of periodic sandbars. Experiments by Belzons et al (1991) further showed that bed waves with two wavelengths can cause Bragg resonance subharmonically. A theory on Bragg resonance was initiated by Davies & Heathershaw (1984) whose predictions are good away from resonance. Mei (1985) gave a uniformly valid theory near and at resonance to predict the amplitude of the reflected waves for intermediate depth $kh = \mathcal{O}(1)$. Theoretical extensions have been made by Kirby (1986a,b). Mei stressed the role of a threshold frequency $\Omega_o = \omega kD/2 \sinh 2kh$, where D is the bar amplitude. Let the wavenumbers of the bars and the waves be $2k$ and $k+K$, respectively, with $K \ll k$, and define the detuning frequency $\Omega = C_g K$. Then the envelopes of the incident and reflected waves over the bar region are dispersive, hence undulatory, with the distance of propagation if $\Omega > \Omega_o$, and they are nondispersive, hence monotonic, if $\Omega < \Omega_o$. The threshold has been verified experimentally by Hara & Mei (1987) for wave packets. Moreover, Hara & Mei have extended the theory to second order in wave slope and studied the infragravity waves induced by nonlinearity. If a packet of short waves is incident on a finite packet of bars, there are two kinds of long waves generated over the bars and radiated away from the zone of bars on both sides. The first is locked to the short-wave envelopes, and the second moves at the long-wave speed $\sqrt{gh}$. If the region of bars is sufficiently long, short waves can be entirely reflected while long waves are found on the transmission side. These features, which have been confirmed by experiments (Hara & Mei 1987), imply that bars can alter the wave climate in the nearshore zone.

In nature the factors affecting the bar-forming process can of course be much more complex. Wave records at Hakui Beach, Japan exhibit two spectral peaks: a high spectral peak at 10-second period, and another lower peak at 100 seconds. Katoh (1984) reasoned that while short waves mobilize the sediments into suspension, the mass transport in the long waves distributes the suspension to form sandbars. In wave-tank experiments with a sandy bottom, Boczar-Karakewicz et al (1981) noted that a few hours after waves start, bars form and reflection increases, in accordance with the Bragg-resonance/mass-transport theory. Afterwards, however, the standing wave may eventually break, generating turbulence, and the bars are wiped out. Waves become progressive again with a rather strong second harmonic. Long dunes then form with a wavelength comparable to the recurrence length of the second harmonic. Boczar-Karakewicz and her colleagues have proposed a numerical model coupling sand motion with the Boussinesq equations which accounts for both non-linearity and dispersion. Therefore the Bragg-resonance mechanism may be just one phase of a much more complex process (Benjamin et al 1987). Much remains to be done on the evolution of sand and waves when the latter are transformed from standing to progressive waves.

The Bragg-resonance mechanism also suggests a possible means for constructing coastal protection devices, such as an artificial bar field. With a proper design of heights and spacings, the bar field may significantly reduce the wave energy arriving at the surf zone. Any such installation would necessarily be of finite extent in the alongshore direction. Mei et al (1988) examined several practical aspects of Bragg scattering by a bar field, including angle of incidence, end effects of a bar field, and beach slope. They found that at the critical incidence angle, $\theta = \pi/4$, the bar field loses all effects on waves. If the detuning frequency is low, there is only a small neighborhood near the critical angle where reflection is minimal. As the detuning frequency increases, this neighborhood grows wider. Because of the bars it is likely that the resulting localized depression in the mean set-up behind the bar field would generate a nearshore current. To evaluate the effectiveness of the bar field in the presence of a current, Kirby (1988) showed that a current can shift resonant frequencies significantly and enhance wave reflection. Another possible application is in the design of a harbor entrance by either placing a bar field in the entrance channel or introducing undulations in the channel width (Liu 1987). For a doubly periodic seabed where $h(x, y)$ is periodic in both x and y, the Bragg-resonance theory has been given by Naciri & Mei (1988).

When sandbars are in very shallow water, the wave field is better described by the Boussinesq equations. Yoon & Liu (1987) studied the resonant reflection of a modulated wave train and cnoidal waves. For the modulated wave train nonlinearity reduces not only the maximum reflection but also the range of the detuning frequencies where reflection is significant. On the other hand, for cnoidal waves, nonlinearity causes stronger reflection of the first-harmonic component over the bar field than that predicted by linear theory.

OSCILLATORY FLOWS AND BED RIPPLES

Sand ripples are dynamically important in coastal waters because they roughen the sea bottom, and enhance frictional dissipation in and resistance to the water motion above. Bagnold (1946) was the first to clearly demonstrate that the existence of ripples is a consequence of water motion above. He spread a layer of sand over the horizontal bottom of a water tank, and oscillated the bottom horizontally beneath the otherwise still water. When the relative speed exceeded a certain threshold, most of the sand particles on top of the sand layer began to roll, and soon to form ripple crests aligned normal to the direction of oscillation. He called these *rolling-grain ripples*. The same phenomenon can also be seen on a monolayer of sand grains on the smooth bottom of a wave tank (Carter et al 1973, Kaneko & Honji 1979). If the speed of the oscillatory water over the ripples already formed is increased sufficiently, vortices start to form in the water behind the ripple crests, dislodging particles into suspension and further sharpening the ripple profile. Bagnold calls these *vortex ripples*. There are many ways vortex ripples can be initiated on an initially flat sand layer. One of the easiest is to stand a thin plate vertically across the width of a wave tank, with the sharp edge protruding above the sand. An oscillating flow above quickly induces vortices near the sharp edge, excavating sand particles on both sides and piling them against the plate. The outer edges of the excavated troughs provide new points of flow separation; another pair of troughs is formed with sand piling between them. In this way sand ripples spread away from the protruding edge and reshape the entire sandy bed. More extensive experiments on the spontaneous generation of vortex ripples have been reported by Sleath & Ellis (1978). The book by Sleath (1984) summarizes many interesting experiments on this topic.

Initiation of Rolling-Grain Ripples

To shed more light on the initiation of rolling-grain ripples from a flat bed, Kaneko & Honji (1979) examined a row of glass beads aligned initially with the direction of the oscillating flow. As soon as the beads start to roll back and forth, two neighboring particles first approach each other to form a pair. Later the axes of the pairs tend to rotate so as to become perpendicular to the flow. Gradually two neighboring pairs attract each other to form a new aggregate involving more beads. Eventually rows of aggregated beads are seen, which they called *particle waves*. The boundary layer in these tests was laminar with the Stokes-layer thickness comparable to the bead diameter. Prompted by the theory for two well-separated sedimenting particles (Vasseur & Cox 1976), Kaneko & Honji conjectured that hydrodynamic interaction of two adjacent particles should be important, and that two particles should attract each other if they are in tandem, and repel each other if they are side by side. This conjecture has been studied theoretically by Shibata & Mei (1990), who applied an Oseen approximation to study the inertia effect in the far field of a localized force resting on the bottom of a Stokes boundary layer. The force is used to simulate the presence of a small particle. Convective inertia, effective in the far field, induces a certain asymmetry in the velocity distribution. In particular if the force is in the negative x direction while the ambient flow is along the x axis, the x component of the velocity is asymmetric so as to exert a stronger pull on a particle far to its right then the push on a particle far to its left. In this way two particles along the direction of water motion should attract each other. On the other hand if the velocity field due to a localized force is along the $+y$ axis while the ambient oscillation is along the x axis, then the perturbed velocity is essentially in the $+y$ direction on the side of $x > 0$ but has a counterclockwise cell on the side of $x < 0$. Thus if there are three particles originally along the x axis, any transverse displacement of the mid particle would tend to rotate the two other particles counterclockwise. These results are consistent with Kaneko & Honji qualitatively, however the computed velocity is very weak and no direct experimental evidence is available. It would also be interesting to find a theory for the instability of a monolayer of evenly distributed particles rolling or sliding under an oscillatory boundary layer.

Vortices Over Ripples

Theoretical attempts have been made in two directions: (*a*) models for the transient development of oscillating vortices due to flow separation near fixed ripple crests and (*b*) models incorporating the transport of sediments to predict the ripple evolution. All models are so far based on Navier-Stokes equations with constant viscosity.

The typical wavelength of sand ripples is $\mathcal{O}(10\text{ cm})$ while that of the gravity waves is ten or a hundred times larger. As a first approximation one may take the length of the water wave to be infinite, so that the ambient flow is horizontally uniform but periodic in time. A number of theories have been advanced. Before discussing them, note that there are several length scales in the problem: $2\pi/k$ = ripple wavelength, a = ripple amplitude, A = orbital amplitude of ambient oscillations, and $\delta = \sqrt{2\nu/\omega}$ = Stokes boundary-layer thickness (where ν = kinematic viscosity and ω = oscillation frequency). From these, three length ratios can be formed: the ripple slope $\varepsilon = ka$, the Keulegan-Carpenter number $\alpha = A/a$, and the boundary-layer thickness $\sigma = k\delta$. Various alternatives can also be defined. For example the ratio $R = (A/\delta)^2 = (\alpha/\sigma)^2$ can be regarded as the Reynolds number.

We first discuss a potential-flow model by Longuet-Higgins (1981) who studied vortex generation over sharp-crested ripples. Flow separation is assumed to occur only at the ripple crests where a vortex sheet emanates. The sheet is approximated by a string of discrete vortices with the strength determined from the potential-flow velocity at a certain distance above the crest. During each period the cloud of vortices is replaced by a pair of point vortices which eventually rise far above the ripples. Viscosity is introduced by another heuristic hypothesis to help damp the vortices and facilitate the numerical computation. The model therefore involves several ad hoc heuristic criteria and it is difficult to ascertain the realm of its validity. It is not easy to modify it for smooth ripples or to use it to estimate the shear stress on the bed.

There are many laminar viscous models based on the Navier-Stokes equations. In particular several perturbation theories are available if one or more of the length ratios are small. For small Reynolds numbers, Lyne (1971) gave the first perturbation for $\varepsilon/\sigma = a/\delta \ll 1$, i.e. for ripples completely immersed in the laminar boundary layer. The Navier-Stokes equations are linearized at every order. He found at second order that stationary cells of mean circulation exist inside the Stokes boundary layer. These cells correspond to Eulerian streaming induced by Reynolds stresses; the direction of the streaming velocity along the ripple surface is always from the trough to the crest, therefore tending to transport the sediments away from the troughs and toward the ripple crests. Extensions have been made by Sleath (1974) who calculated the Lagrangian drift velocity and accounted for spatial nonuniformity due to the finite length of the surface wave. A higher order analysis has been given by Kaneko & Honji (1979) for small α but finite ε/δ; their theory has been confirmed by experiments with a water/glycerin mixture of high viscosity. A similar analysis has been reported by Matsunaga & Honji (1989). Vittori (1989) obtained results for $\alpha = \mathcal{O}(1)$ but included $\mathcal{O}(\varepsilon/\sigma)^2$ terms only.

It is well known that for small-amplitude oscillations around a large cylinder ($\varepsilon \ll 1$), the induced streaming velocity U_s does not vanish at the outer edge of the Stokes boundary layer. If the Reynolds number $R = U_s/k\nu$ is large, an outer boundary layer of thickness $\delta_s = \mathcal{O}(1/kR^{1/2})$ must exist so that U_s can become zero outside the thicker layer. Extending this idea Hara & Mei (1990a) carried out an asymptotic analysis for two cases: (*i*) small α but finite ε and (*ii*) small ε but finite α. In case (*i*) they found stationary circulation cells in the outer boundary layer, therefore *outside* the Stokes boundary layer. In case (*ii*) the oscillation amplitude was found to be comparable to the ripple wavelength, with cells convected to and fro by the ambient oscillations. Thus these are the vortices drifting high above the Stokes boundary layer.

A strictly numerical solution of the full Navier-Stokes equations as an initial boundary value problem in a semi-infinite strip is a difficult task. Earlier works by Sleath (1975) and Sato et al (1984) used finite differences and discretized both spatial coordinates to solve the stream function and the vorticity alternately at each time step. Rather large grids were used and the results appear crude. More accurate computations have been done by the pseudo-spectral method in which the solution is represented by a Fourier series in the coordinate parallel to the ripple surface (Shum 1988, Blondeaux & Vittori 1991a). For moderately high Reynolds numbers the computing cost is large; one must either limit the total time for each set of parameters to a few oscillation periods or limit the number of parameter sets. For moderately high Reynolds numbers, a vortex generated in the earlier period is found to drift with the ambient flow. As the ambient flow decelerates, a new vortex is initiated on the lee side of the crest. This wall vortex grows and rises above the bed, and is convected past the crest by the reversed flow in the second half period; it grows larger and leaves the wall near the end of the complete period, while another wall vortex is initiated on the opposite lee of the same crest. When the same ε and α are kept, a large R gives a more complex vortex structure. These features have also been found by Hara et al (1992) by assuming the motion to be quasi-steady, i.e. periodic in time. More interestingly, for sufficiently large Reynolds numbers, a periodic ambient flow leads to nonperiodic flow near the ripple after many cycles. This result, which is of course due to nonlinearity, has been noted also by Sleath (1975). It suggests the threshold of transition to turbulence. Blondeaux & Vittori (1991b) carried the computation for one set of parameters to 200 oscillation periods, and found that the boundary-layer flow eventually becomes chaotic through a cascade of period-doublings. Further numerical and laboratory work appears worthwhile to determine the properties of the chaotic state, which can yield theoretical information on other practical matters such as the friction factor and the rate of wave damping.

Evolution of Ripples

In a series of remarkable papers Blondeaux (1990) and Vittori & Blondeaux (1990) studied the formation of rolling-grain ripples theoretically and experimentally. They carried out a perturbation analysis based on the Navier-Stokes equations and invoked an empirical law of sediment transport which is a relation between the sediment-flux rate and the local flow. Such laws are available only from steady-flow experiments in a current flume. Ignoring the threshold stress in the empirical transport formula, Vittori & Blondeaux pursued a second-order analysis. The induced streaming is found to render a periodically disturbed sand surface unstable under certain conditions; the wavelength corresponding to the fastest growth is then taken to be the initial wavelength of a rolling-grain ripple. In their second paper they went on to the next order and deduced a Landau-Stuart equation for the nonlinear instability of the sand surface. The final ripple height is found as a function of the ripple length and other flow parameters. These predictions are in broad agreement with available experiments.

In nature ripples are not always long crested; sometimes they appear staggered in alternating rows in the form of layers of bricks. Laboratory observations have been reported by Bagnold (1946), Sleath & Ellis (1978), and Matsunaga & Honji (1980). Bagnold suggested that brick-like ripples occur when $kA < 1$ [or 3 according to Sleath (1984)]. There are two possible mechanisms for this phenomenon. One is purely hydrodynamical: centrifugal instability due to the finite curvature of the ripple form; the other is due to the instability of the movable sand. To support the first mechanism Matsunaga & Honji conducted experiments in a water flume with periodic rows of half-circular cylinders fixed on a flat bed. By oscillating the bed horizontally at a fixed frequency, they found that beyond a certain amplitude, smoke streaks from metallic compounds imbedded in the cylinders can be seen at periodic spacings along the crests of the cylinders, and the streak pattern is repeated after two ripple wavelengths. They also placed glass beads between the half cylinders and found them to drift and form a brick-like pattern. These evidences suggest that subharmonic hydrodynamic instability can initiate brick-like ripples in the sea. This type of instability has been studied before theoretically by Hall (1984) and experimentally by Honji (1981) and Sarpkaya (1986) for a single cylinder in oscillating flow. The physics is associated with the Taylor-Görtler instability due to the curvature of the boundary (i.e. centrifugal force). Hara & Mei (1990b) extended Hall's theory for gentle periodic ripples. The finite spacing between adjacent ripple crests now affects the instability characteristics. As the Keulegan-Carpenter number α increases beyond a certain threshold, periodic disturbances in the spanwise direction indeed occur subharmonically, leading to an induced streaming velocity pattern that is consistent with the brick pattern of ripple crests. The predictions are in order-of-magnitude accord with the observations of Matsunaga & Honji, despite the substantial difference in ripple shape and slope and the likelihood of vortices in the experiments. The predicted values are also in order-of-magnitude accord with Sleath's experiments, which were for sand ripples.

Vittori & Blondeaux (1992) have given yet another mechanism by extending their sand-layer instability theory from two to three dimensions. This theory appears to agree better quantitatively with experiments, but it does not explain the phenomenon observed by Matsunaga & Honji for rigid ripples. In nature both mechanisms are likely to be effective and complement each other, and should be combined with further improvements to account properly for vortex shedding and turbulence. As in existing theories for sand waves in steady currents, uncertainties remain because few advances have been made in the sand dynamics itself. These uncertainties of course are passed on to any sand-wave theory. More must be learned about the dynamics of sediments in water.

CONCLUDING REMARKS

In the past, investigations on the causes and effects of coastal hydrodynamics have been largely motivated by the needs for geophysical understanding, and in coastal defense and construction. The recent upsurge of interest in the preservation of the coastal environment, however, adds not only new impetus but also new dimensions to these studies. For example mixing due to waves and currents greatly enhances the transfer of chemical and biological elements near the seabed and the air-sea interface, and in the surf zone. In addition to improving our ability to predict quantitatively the motion of water itself and of noncohesive sediments, we also need to expand our knowledge of the diffusion and transport of very fine substances for which physico-chemical processes such as coagulation are an integral part. Reproduction and growth of marine life, being most vigorous in the coastal zone, calls for many refinements in fluid mechanics hence close collaboration between hydrodynamicists and marine biologists. Such a call has been eloquently promulgated by the biologist M. W. Denny in a fascinating new book (Denny 1988). The work of a coastal hydrodynamicist is never done.

ACKNOWLEDGMENTS

CCM wishes to acknowledge the support for many years of the Office of Naval Research (Fluid Dynamics and Ocean Engineering Programs), and the National Science Foundation (Fluid, Hydraulics and Particulate Systems, and Ocean Engineering Programs). PLFL wishes to acknowledge the support of the New York Sea Grant Institute, the Mathematical Sciences Institute (U.S. Army Research Office), and the National Science Foundation (Fluid, Hydraulics and Particulate Systems Program, and Environmental and Ocean Systems Program).

Literature Cited

Agnon, Y., Mei, C. C. 1988. Trapping and resonance of long shelf waves due to groups of short waves. *J. Fluid Mech.* 195: 201–22

Bagnold, R. A. 1946. Motion of waves in shallow water: interaction of waves and sand bottom. *Proc. R. Soc. London* 87: 1–15

Battjes, J. A. 1988. Surf-zone dynamics. *Annu. Rev. Fluid Mech.* 20: 257–93

Battjes, J. A., Sobey, R. J., Stive, M. J. F. 1990. Nearshore circulation. In *The Sea: Ocean Engineering Science*, Vol. 9, ed. B. LeMehaute, D. M. Hanes, pp.467–93. New York: Wiley

Belzons, M., Rey, V., Guazzelli, E. 1991. Subharmonic Bragg resonance for surface gravity waves. *Europhys. Letts* 16: 189–94

Benjamin, T. B., Boczar-Karakiewicz, B., Prichard, W. G. 1987. Reflection of water waves in a channel with corrugated bed. *J. Fluid Mech.* 18: 249–74

Benny, D. J., Roskes, G. J. 1969. Wave instabilities. *Stud. Appl. Math.* 48: 377–85

Berkhoff, J. C. W. 1972. Computation of combined refraction-diffraction. *Proc. 13th Int. Conf. on Coastal Eng.* ASCE, pp. 471–90

Blondeaux, P. 1990. Sand ripples under sea waves. Part 1. Ripple formation. *J. Fluid Mech.* 218: 1–18

Blondeaux, P., Vittori, G. 1991a. Vorticity dynamics in an oscillatory flow over a rippled bed. *J. Fluid Mech.* 226: 257–89

Blondeaux, P., Vittori, G. 1991b. A route to chaos in an oscillatory flow: Feingenbaum Scenario. *Phys. Fluids* 3: 2492–95

Boczar-Karakiewicz, B., Pablinska, B., Winieki, J. 1981. Formation of sand bars by surface waves. Laboratory experiments. *Pol. Akad. NAUK, Inst. Budouvictwa Wodnego Gdansk Rozpr. Hydrotech.* 43: 111–25

Booij, N. 1981. Gravity waves on water with non-uniform depth and current. *Rep. No.* 81–1, Delft Univ. of Technol.

Bowen, A. J., Holman, R. A. 1989. Shear instabilities of the mean longshore current. *J. Geophys. Res.* 95: 18,023–30

Bowers, E. C. 1977. Harbour resonance due to set-down beneath wave groups. *J. Fluid Mech.* 79: 71–92

Carter, T. G., Liu, P. L.-F., Mei, C. C. 1973. Mass transport by waves and offshore sand bedforms. *J. Waterways, Harbor and Coastal Eng. Div. ASCE* 99: 165–84

Chen, H. S., Mei, C. C. 1974. Oscillations and wave forces in an offshore harbor. Parsons Lab., *MIT Rep.* 190

Dalrymple, R. A., Kirby, J. T. 1988. Models for very wide-angle water waves and wave diffraction. *J. Fluid Mech.* 192: 33–50

Dalrymple, R. A., Suh, K. D., Kirby, J. T., Chae, J. W. 1989. Models for very wide-angle water waves and wave diffraction. Part 2. Irregular bathymetry. *J. Fluid Mech.* 201: 299–322

Davey. A., Stewartson, K. 1974. On three-dimensional packets of surface waves. *Proc. R. Soc. London Ser. A* 338: 101–10

Davies, A. G., Heathershaw, A. D. 1984. Surface wave propagation over sinusoidally varying topography. *J. Fluid Mech.* 144: 419–43

Denny, M. W. 1988. *Biology and the Mechanics of the Wave-Swept Environment*. Princeton: Princeton Univ. Press

Dodd, N., Thornton, E. B. 1990. Growth and energetics of shear waves in the nearshore. *J. Geophys. Res.* 95: 16,075–83

Foda, M. A., Mei, C. C. 1981. Nonlinear excitation of long trapped waves by a group of short swells. *J. Fluid Mech.* 111: 319–45

Gallagher, B. 1971. Generation of surf beat by non-linear wave interactions. *J. Fluid Mech.* 49: 1–20

Grant, W. D., Madsen, O. S. 1986. The continental-shelf bottom boundary layer. *Annu. Rev. Fluid Mech.* 18: 265–305

Guza, R. T., Bowen, A. J. 1976. Finite amplitude Stokes edge waves. *J. Mar. Res.* 34: 269–93

Hall, P. 1984. On the stability of the unsteady boundary layer on a cylinder oscillating transversely in a viscous fluid. *J. Fluid Mech.* 146: 347–67

Hara, T., Mei, C. C. 1987. Bragg scattering of surface waves by periodic bars: theory and experiment. *J. Fluid Mech.* 178: 221–41

Hara, T., Mei, C. C. 1990a. Oscillating flows over periodic ripples. *J. Fluid Mech.* 211: 183–209

Hara, T., Mei, C. C. 1990b. Centrifugal instability of an oscillatory flow over periodic ripples *J. Fluid Mech.* 217: 1–32

Hara, T., Mei, C. C., Shum, K. T. 1992. Oscillatory flow over periodic ripples of finite slope. *Phys. Fluids A* 4(7): 1–12

Heathershaw, A. D. 1982. Seabed-wave resonance and sand bar growth. *Nature* 296: 343–45

Holman, R. A. 1981. Infragravity energy in the surf zone. *J. Geophy. Res.* 86: 6442–50

Honji, H. 1981. Streaked flow around an oscillating circular cylinder. *J. Fluid Mech.* 107: 509–20

Houston, J. R. 1981. Combined refraction and diffraction of short waves using the finite element method. *Appl. Ocean Res.* 3: 163–70

Irie, I., Nadaoka, K., Kondo, T., Terasaki, T. 1984. Two dimensional seabed scour in front of breakwater by standing waves. *Rep. Port & Harbor Res. Inst.* 23: 3–52 (In Japanese)

Kadomtsev, B. B., Petviashvili, V. I. 1970. On the stability of solitary waves in weakly dispersive media. *Sov. Phys. Dokl.* 15: 539–41

Kajiura, K., Shuto, N. 1990. Tsunamis. In *The Sea*: *Ocean Engineering Science*, Vol. 9, ed. B. LeMehaute, D. M. Hanes, pp. 395–422. New York: Wiley

Kaneko, A., Honji, H. 1979. Double structures of steady streaming in the oscillatory flow over a wavy bed. *J. Fluid Mech.* 93: 727–36

Katoh, K. 1984. Multiple longshore bars formed by long period standing waves. *Rep. Port & Harbour Res. Inst.* 23(3): 3–31

Keulegan, G. J. 1944. An experimental study of submarine sandbars. *Tech. Rep. No.* 3 *Beach Erosion Board*, U. S. Army Corps Eng.

Kirby, J. T. 1983. *Propagation of weakly-nonlinear surface water waves in region with varying depth and currents.* PhD thesis. Univ. Delaware Dept. Civil Eng., Newark, Del.

Kirby, J. T. 1984. A note on linear surface wave-current interaction over slowly varying topography. *J. Geophys. Res.* 89: 745–47

Kirby, J. T. 1986a. A general wave equation for waves over rippled beds. *J. Fluid Mech.* 162: 171–86

Kirby, J. T. 1986b. On the gradual reflection of weakly nonlinear Stokes waves in regions with varying topography. *J. Fluid Mech.* 162: 171–86

Kirby, J. T. 1986c. Higher-order approximation in the parabolic equation method for water waves. *J. Geophys. Res.* 91: 933–52

Kirby, J. T. 1988. Current effects on resonant reflection of surface water waves by sandbars. *J. Fluid Mech.* 186: 501–20

Kirby, J. T., Dalrymple, R. A. 1983. A parabolic equation for the combined refraction-diffraction of Stokes waves by mildly varying topography. *J. Fluid Mech.* 136: 453–66

Kostense, J. K., Meijer, K. L., Dingemans, M. W., Mynett, A. E., van den Bosch, P. 1986. Wave energy dissipation in arbitrary shaped harbours of varying depth. *Proc. 20th Int. Conf. on Coastal Eng.* ASCE, pp. 2002–16

Leontovich, M., Fock, V. A. 1944. A method of solution of problems of electromagnetic wave propagation along the Earth's surface. *Izv. Akad. Nauk., USSR* 8: 16–22

Lepelletier, T. G. 1980. Tsunamis—Harbor oscil|lations induced by nonlinear transient long waves. *Rep. No. KH-R*-41, Keck Lab., Calif. Inst. Technol.

Liu, P. L.-F. 1982. Combined refraction and diffraction: comparison between theory and experiments. *J. Geophys. Res.* 87: 5723–30

Liu, P. L.-F. 1983. Wave-current interactions on a slowly varying topography. *J. Geophys. Res.* 88: 4421–26

Liu, P. L.-F. 1987. Resonant reflection of water waves in a long channel with corrugated boundaries. *J. Fluid Mech.* 179: 371–81

Liu, P. L.-F. 1989. A note on long waves induced by short-wave group over a shelf. *J. Fluid Mech.* 205: 163–70

Liu, P. L.-F. 1990. Wave transformation. In *The Sea*: *Ocean Engineering Science* Vol. 9, ed. B. LeMehaute, D. M. Hanes, pp. 27–63. New York: Wiley

Liu, P. L.-F., Mei, C. C. 1976. Water motion on a beach in the presence of a breakwater 1. Waves. *J. Geophys. Res.* 81: 3079–84

Liu, P. L.-F., Tsay, T. K. 1983. On weak reflection of water waves. *J. Fluid Mech.* 131: 59–71

Liu, P. L.-F., Tsay, T. K. 1984. Refraction-diffraction model for weakly nonlinear water waves. *J. Fluid Mech.* 141: 265–74

Liu, P. L.-F., Yoon, S. B. 1986. Stem waves along a depth discontinuity. *J. Geophys. Res.* 91: 3979–82

Liu, P. L.-F., Yoon, S. B., Kirby, J. T. 1985. Nonlinear refraction- diffraction of waves

in shallow water. *J. Fluid Mech.* 153: 185–201
Liu, P. L.-F., Dingeman, M. W., Kostenese, J. K. 1990. Long waves generation due to the refraction of short-wave groups over a shear current. *J. Phys. Oceanogr.* 20: 53–59
Longuet-Higgins, M. S. 1981. Oscillating flow over steep sand ripples. *J. Fluid Mech.* 107: 1–35
Longuet-Higgins, M. S., Stewart, R. W. 1962. Radiation stress and mass transport in gravity wave, with applications to "surf beats." *J. Fluid Mech.* 13: 481–504
Lozano, C. J., Liu, P. L.-F. 1980. Refraction-diffraction model for linear surface water waves. *J. Fluid Mech.* 101: 705–20
Lyne, W. H. 1971. Unsteady viscous flow over a wavy wall. *J. Fluid Mech.* 50: 30–48
Masuda, K., Mei, C. C., Nagai, T. 1989. A nonlinear diffraction problem of second order long waves by a huge circular cylinder. *Proc. of the Ocean Eng. Symp.*, Soc. of Naval Arch., Japan, pp. 363–70
Matsunaga, N., Honji, H. 1980. Formation mechanism of brick pattern ripples. *Rep. Res. Inst. Appl. Mech.* Kyushu Univ. 28: 27–38
Mehlum E., Stamnes, J. J. 1978. On the focusing of ocean swells and the significance in power production. *Central Inst. Ind. Res. Blindern Oslo, S. E. Rept.* 78–04 08–3
Mei, C. C. 1978. Numerical methods in water wave diffraction and radiation. *Annu. Rev. Fluid Mech.* 10: 393–416
Mei, C. C. 1983. *The Applied Dynamics of Ocean Surface Waves.* New York: Wiley-Interscience
Mei, C. C. 1985. Resonant reflection of surface water waves by periodic sandbars. *J. Fluid Mech.* 152: 315–35
Mei, C. C. 1991. Short wave excitation of long waves in a basin. In *Dynamics of Marine Vehicles and Structures in Waves*, ed. W. G. Price, P. Temarel, A. J. Keane, pp. 37–47. Amsterdam: Elsevier
Mei, C. C., Benmoussa, C. 1984. Long waves induced by short-wave groups over an uneven bottom. *J. Fluid Mech.* 139: 219–35
Mei, C. C., Agnon, Y. 1989. Long-period oscillations in a harbour induced by incident short waves. *J. Fluid Mech.* 208: 595–608
Mei, C. C., Hara T., Naciri, M. 1988. Note on Bragg scattering of water waves by parallel bars on the seabed. *J. Fluid Mech.* 186: 147–62
Miles, J. W. 1974. Harbor seiching. *Annu. Rev. Fluid Mech.* 6: 17–35
Miles, J. W. 1991. Variational approximations for gravity waves in water of variable depth. *J. Fluid Mech.* 232: 681–88
Miles, J. W., Munk, W. 1961. Harbor paradox. *J. Waterways & Harbor Div. ASCE* 87: 111–30
Minzoni, A. A., Whitham, G. B. 1977. On the excitation of edge wave on beaches. *J. Fluid Mech.* 79: 272–87
Molin, B. 1979. Second-order diffraction loads upon three-dimensional bodies. *Appl. Ocean Res.* 1: 197–202
Molin, B. 1982. On the generation of long-period second-order free-waves due to changes in the bottom profile. *Pap. Ship Res. Inst.*, Tokyo, Japan 68
Munk, W. H. 1949. Surf beat. *Trans. Am. Geophys. Union* 30: 849–54
Naciri, M., Mei, C. C. 1988. Bragg scattering of water waves by a doubly periodic seabed. *J. Fluid Mech.* 192: 51–74
Nakaza, E., Hino, M. 1991. Bore-like surf beat in a reef zone caused by wave groups of incident short period waves. *Fluid Dyn. Res.* 7: 89–100
Oltman-Shay, J., Guza, R. T. 1987. Infragravity edge waves observations on two California beaches. *J. Phy. Ocean* 17: 644–53
Oltman-Shay, J., Howd, P. A., Birkemeier, W. A. 1989. Shear instabilities of the mean longshore current. 2. Field observations. *J. Geophys. Res.* 94: 18,031–42
Peregrine, D. H. 1983. Breaking waves on beaches. *Annu. Rev. Fluid Mech.* 15: 149–78
Radder, A. C. 1979. On the parabolic equation method for water wave propagation. *J. Fluid Mech.* 95: 159–76
Rockcliff, N. 1978. Finite amplitude effects in free and forced edge waves. *Math. Proc. Cambridge Philos. Soc.* 83: 463–79
Rogers, S. R., Mei, C. C. 1978. Nonlinear resonant excitation of a long and narrow bay. *J. Fluid Mech.* 88: 161–80
Sand, S. E. 1982. Wave grouping described by bounded long waves. *Ocean Eng.* 9: 567–80
Sarpkaya, T. 1986. Force on a circular cylinder in viscous oscillatory flow at low Keulegan-Carpenter numbers. *J. Fluid Mech.* 165: 61–71
Sato, S., Mimura, N., Watanabe, A. 1984. Oscillatory boundary layer flow over ripple beds. *Proc. 19th Conf. on Coastal Eng.*, pp. 2293–309
Shibata, M., Mei, C. C. 1990. Inertia effects of a localized force distribution near a wall in a slow shear flow. *Phys. Fluids A* 2: 1094–104
Shum, K. T. 1988. *A numerical study of vortex dynamics over rigid ripples.* PhD thesis. MIT Dept. Civil Eng., Cambridge, Mass.

Sleath, J. F. A. 1974. Mass transport over a rough bed. *J. Mar. Res.* 32: 13–24

Sleath, J. F. A. 1975. A contribution to the study of vortex ripples. *J. Hydraul. Res.* 13: 315–28

Sleath, J. F. A. 1984. *Seabed Mechanics*. New York: Wiley

Sleath, J. F. A. 1990. Seabed boundary layers. In *The Sea: Ocean Engineering Science*, vol. 9, Part B, ed. B. LeMehaute, D. M. Hanes. New York: Wiley

Sleath, J. F. A., Ellis, A. C. 1978. Ripple geometry in oscillatory flow. *Univ. Cambridge, Dept. Eng. Rep. A/hydrau. TR2*

Smith, R., Sprinks, T. 1975. Scattering of surface waves by a conical island. *J. Fluid Mech.* 72: 373–84

Stamnes, J. J., Lovhaugen, O., Spjelkavik, B., Mei, C. C., Lo, E., Yue, D. K. P. 1983. Nonlinear focusing of surface waves by a lens-theory and experiment. *J. Fluid Mech.* 135: 71–94

Suh, K. D., Dalrymple, R. A., Kirby, J. T. 1990. An angular spectrum model for propagation of Stokes waves. *J. Fluid Mech.* 221: 205–32

Symonds, G., Huntley, D. A., Bowen, A. J. 1982. Two-dimensional surf beat: long wave generation by a time-varying breakpoint. *J. Geophys. Res.* 87: 492–98

Tucker, M. J. 1950. Surfbeats: sea waves of 1–5 minutes period. *Proc. R. Soc. London Ser. A* 202: 565–73

Tsay, T.-K., Liu, P. L.-F. 1982. Numerical solution of water wave refraction diffraction problems in the parabolic approximation. *J. Geophys. Res.* 87: 7932–40

Tsay, T.-K., Liu, P. L.-F. 1983. A finite element model for wave refraction and diffraction. *Appl. Ocean Res.* 5: 30–37

Vasseur, P., Cox, R. G. 1976. The lateral migration of a spherical particle in two dimensional shear flows. *J. Fluid Mech.* 78: 385–413

Vittori, G. 1989. Nonlinear viscous oscillatory flow over a small amplitude wavy wall. *J. Hydraul. Res.* 27: 267–80

Vittori, G., Blondeaux, P. 1990. Sand ripples under sea waves, Part 2. Finite amplitude development. *J. Fluid Mech.* 218: 19–39

Vittori, G., Blondeaux, P. 1992. Sand Ripples under sea waves, Part 3. Brick-pattern ripple formation. *J. Fluid Mech.* 239: 23–45

Voit, S. S. 1987. Tsunamis *Annu. Rev. Fluid Mech.* 19: 217–36

Whalin, R. W. 1971. The limit of applicability of linear wave refraction theory in a convergence zone. *Res. Rep. H–71–3, U. S. Army Corps Eng.*, Waterways Exp. Station, Vicksburg, Mississippi

Wu, J.-K., Liu, P. L.-F. 1990. Harbour excitations by incident wave groups. *J. Fluid Mech.* 217: 595–613

Yoon, S. B., Liu, P. L.-F. 1987. Resonant reflection of shallow-water waves due to corrugated boundaries. *J. Fluid Mech.* 180: 451–69

Yue, D. K. P. 1980. *Numerical study of Stokes wave diffraction at grazing incidence*. PhD thesis. MIT, Dept. Civil Eng., Cambridge, Mass.

Yue, D. K. P., Mei, C. C. 1980. Forward diffraction of Stokes' waves by a thin wedge. *J. Fluid Mech.* 99: 33–55

Zakharov, V. E. 1968. Stability of periodic waves of finite amplitude on surface of a deep fluid. *J. Appl. Mech. Tech. Phys.* 2: 190–94

Zhou, C., Liu, P. L.-F. 1987. Second-order low-frequency wave forces on a vertical cylinder. *J. Fluid Mech.* 175: 143–55

Annu. Rev. Fluid Mech. 1993. 25 : 241–89

VORTICES IN ROTATING FLUIDS

E. J. Hopfinger

Institut de Mécanique-LEGI, U.J.F. et CNRS, B.P.53,
38041 Grenoble Cédex, France

G. J. F. van Heijst

Department of Technical Physics, Eindhoven University of Technology,
P.O. Box 513, 5600 MB Eindhoven, The Netherlands

KEY WORDS: vortex structure, vortex merger, instability of vortices, baroclinic vortices, vorticity concentration

1. INTRODUCTION

The emergence of coherent vortex structures is a characteristic feature of quasi-geostrophic or two-dimensional turbulence and because of their relevance to large-scale geophysical flows, the dynamics of these structures has been studied increasingly over the past decade. In the oceans and in the atmospheres vortices or eddies are abundant. For this reason, the study of isolated vortices in rotating fluids—including their dynamics, instability properties, mutual interaction behavior, and effects due to bottom topography—is of fundamental interest for refined models of geostrophic turbulence and of the general oceanic and atmospheric circulations. Processes of heat transfer, dispersion of biochemical components, and the transport of other physical properties are closely connected with coherent structures. The dynamics of two-dimensional vortices in homogeneous fluids is also a key problem in free shear flows. With or without background rotation, the main question concerns the stability of these vortices to three-dimensional disturbances. In addition, two-dimensional vortices also play an important role in tokamak-confined plasmas as well as in astrophysical situations such as accretion discs of neutron stars.

During the past decade, considerable insight has been gained in the

0066–4189/93/0115–0241$02.00

general dynamics of two-dimensional vortices, both through analytical and numerical studies and through laboratory experiments. In particular, substantial progress has been made on the instability behavior and interaction properties of barotropic vortices. The instability of these vortices and the emergence of dipolar and tripolar vortices was demonstrated theoretically (Flierl 1988), numerically (Carton et al 1989), and also by experiments in a rotating fluid (van Heijst & Kloosterziel 1989, Kloosterziel & van Heijst 1991). In a strain field, two-dimensional vortices may have an elliptic shape. A model closely describing the properties of elliptic vortices and their stability was developed by Dritschel & Legras (1991). For a better understanding of the dynamics of evolving geostrophic turbulence, insight into the merging of two vortices is essential. The merging process has been studied in some detail numerically (Overman & Zabusky 1982, Dritschel 1985) and a criterion for merger has been shown to exist. Laboratory experiments on merging of barotropic vortices in a rotating fluid, conducted by Griffiths & Hopfinger (1987), confirmed the theoretical prediction for the case of anticyclonic vortices; in laboratory experiments with a free-surface rotating fluid, the curvature of the upper surface may cause the vortices to drift and make cyclones behave differently from anticyclones.

Vortices in rotating, stratified fluids are subject to subtle physical processes which are as yet not fully understood. The stability conditions (baroclinic instability) of such vortices depend on a number of parameters, which makes it difficult to derive a simple criterion. The interaction properties raise even more questions. Experiments by Griffths & Hopfinger (1987) indicate that merger of baroclinic vortices in a two-layer stratified fluid depends strongly on the radius of deformation compared with the vortex size. Contour dynamics calculations of the same problem by Polvani et al (1989), who represent the vortices by uniform potential vorticity patches, show, however, that merging is fairly insensitive to stratification. When the vortices are represented as relative vorticity patches a quite different result emerges (Verron et al 1990). A related problem is the vertical alignment of baroclinic vortices and finite area "heton" interactions (Polvani 1991, Griffiths & Hopfinger 1986). In what follows we discuss the genesis and stability of vortices in rotating fluids and the fundamental physical processes. Emphasis is given to isolated vortices and the interaction of pairs in barotropic and stratified fluids. The emergence of geostrophic, intense vortices in a rotating turbulent fluid is also discussed. The most striking vortices—giant eddies occurring in the atmospheres of the major planets—will only be mentioned briefly. For a more extensive treatment of their dynamical properties and the different models, the reader is referred to the review by Read (1992).

2. BAROTROPIC VORTICES

Vortical flow structures in which surfaces of constant pressure and constant density coincide are commonly referred to as barotropic vortices. In such vortices buoyancy forces are absent, which is evidently the case in a constant density fluid. In an incompressible fluid, rotating with constant angular velocity Ω about a vertical axis fixed in space, it is possible to generate two-dimensional vortices by virtue of the well known Taylor-Proudman theorem

$$f(\mathbf{k}\cdot\nabla)\mathbf{v} = 0, \tag{1}$$

where $f = 2\Omega$ is the Coriolis parameter, $\mathbf{k}$ the unit vector in the direction of the axis of rotation and $\mathbf{v}$ the relative fluid velocity. This theorem states that the flow is invariant in axial direction and hence is two-dimensional. The flow is called geostrophic when the Coriolis force is balanced by the horizontal pressure gradient force, i.e. $f\mathbf{k}\times\mathbf{v} = -\nabla p/\rho$, with p and ρ the (reduced) pressure and the density, respectively. For a geostrophic balance to exist in a real, steady or quasi-steady flow, it is necessary that inertia and viscous forces (in the absence of buoyancy forces) be negligibly small, as expressed by the Rossby number $\mathrm{Ro} = V/fL \ll 1$ and the Ekman number $\mathrm{E} = \nu/fH^2 \ll 1$, respectively, where V is a characteristic velocity, L a horizontal length scale, and H a vertical length scale. Barotropic vortices are said to be quasi-geostrophic when Ro and E are small but finite; the geostrophic balance exists at order zero only.

2.1 *Vortex Structure*

For the description of monopolar vortices in a rotating fluid it is convenient to introduce a polar coordinate system $\mathbf{r} = (r, \theta, z)$, with the z-axis parallel with the rotation axis, and express the flow relative to the rotating frame in the corresponding velocity components, i.e. $\mathbf{v} = (u, v, w)$. A purely azimuthal flow about the z-axis is expressed as $\mathbf{v} = [0, v(r), 0]$, and is governed by the radial component of the momentum equation

$$\frac{v^2}{r} + fv = -\frac{1}{\rho}\frac{\partial p}{\partial r}. \tag{2}$$

The azimuthal velocity of this vortical flow can be written as

$$v(r) = -\frac{1}{2}fr \pm \left[\left(\frac{1}{2}fr\right)^2 - \frac{r}{\rho}\frac{\partial p}{\partial r}\right]^{1/2} \tag{3}$$

and, depending on the sign of the pressure gradient, one can distinguish four different types of vortices (see Holton 1979). Because the two anomalous types are physically less relevant (they are likely to be unstable) we confine ourselves here to the regular types: the cyclonic vortex around a low-pressure center, and the anticyclonic vortex around a high-pressure core. The term "cyclonic" refers to a (positive) vortex with the same rotation sense as the system's background rotation, whereas "anticyclonic" refers to vortices with opposite rotation sense.

For two-dimensional flow it is convenient to describe the motion in terms of the vorticity equation, which then takes a scalar form. In the case of flow in a horizontal plane with velocity $\mathbf{v} = (u, v, 0)$ the vorticity $\boldsymbol{\omega} = \omega\mathbf{k}$ is given by:

$$\frac{\partial \omega}{\partial t} + J(\omega, \psi) = \nu \nabla^2 \omega \qquad (4)$$

where J is the Jacobian operator and ψ the stream function, defined according to $\mathbf{v} = -\mathbf{k} \times \nabla\psi$. By definition ψ satisfies $\nabla^2\psi = -\omega$. Note that ω represents here the absolute vorticity, which is equal to the sum of the relative vorticity and the planetary vorticity: $\omega = \omega_{\text{rel}} + f$. From the vorticity equation (4) it is inferred that steady inviscid flows are described by $J(\omega, \psi) = 0$, and this implies that $\omega = F(\psi)$, with F *any* integrable function. It is an open question as to which functional relationship F applies for specific steady flow structures. For example, for steady axisymmetric vortex flows *any* radial distributions of $\omega(r)$ and $\psi(r)$ form solutions of the inviscid vorticity equation. An important property of inviscid two-dimensional flows is the conservation of vorticity, as expressed by (4), with $\nu = 0$. In addition, it can be demonstrated that such flows possess more conservation properties (in fact an infinite number), the most important ones being conservation of kinetic energy and conservation of enstrophy (defined as $\frac{1}{2}\omega^2$). The simultaneous conservation of energy and enstrophy is responsible for the spectral flux of energy to smaller wave numbers, i.e. to larger scales of motion (see e.g. Pedlosky 1987)—a phenomenon commonly known as the *inverse energy cascade* in two-dimensional turbulence. Phenomenologically, this inverse cascade results in a self-organization of the evolving flow: Energy initially distributed over both larger and smaller eddies eventually becomes concentrated in larger coherent vortex structures, giving the flow an "ordered" appearance. It should be remarked that, on the other hand, the flow's enstrophy is characterized by a regular cascade towards the smaller scales of motion. The phenomenon of "ordering" in two-dimensional turbulence is nicely demonstrated in numerical simulations by e.g. McWilliams (1984) and Legras et al (1988). The most common vorticity structure that is found in such simulations is the *monopolar vortex*, consisting of a single set of closed streamlines with one common center. Although these vortices may show a definite tendency towards an axisymmetric shape, elliptical vortices are observed under certain conditions and these are also considered as being monopolar. It should be kept in mind that the term "monopolar" here refers to the existence of just a single center of the vortex motion, and not to the vorticity being single-signed in the entire vortex domain; in fact, most circularly symmetric vortices consist of a vorticity core of one sign enclosed by a band of oppositely-signed vorticity. Another vortex type observed to occur in evolving two-dimensional flows is the *dipolar vortex*, which can be concisely described as a linear arrangement of two closely packed patches of opposite vorticity. In contrast to the monopolar vortex, which contains a net angular momentum, the dipolar vortex is characterized by a nonzero linear momentum, as can be concluded from its steady translation in the direction defined by the dipole axis. When the dipole is completely symmetric the translation is directed along a straight line; in the case of an asymmetric dipole the vortex structure is observed to translate along a curved trajectory.

In one particular case, Legras et al (1988) found a *tripolar vortex* to arise spontaneously in the evolving two-dimensional flow. Such a coherent structure consists of a linear arrangement of three oppositely-signed vorticity patches: an elliptical core of, say, positive vorticity adjoined at its longer sides by two satellite vortices of negative vorticity. Similar to the monopolar vortex, the tripolar vortex contains net angular momentum, and the structure shows a steady rotation of its axis around the center of the core vortex. As will be discussed further on, both dipolar and tripolar vortices may also arise from unstable monoplar vortices, as was observed in numerical simulations and in laboratory experiments on vortices in rotating fluids. In the inviscid case, all these different types of vortices are described by the vorticity equation (4), with the viscous term being zero. When these vortices are quasi-steady (i.e. stationary as seen in a reference frame comoving with the vortex structure), this equation is reduced to $J(\omega, \psi) = 0$—at least when ω and ψ are corrected for the moving frame of reference. The fact that *any* relationship between ω and ψ provides a solution to this reduced equation implies a serious difficulty in the description of real vortices. For example, one may assume the vorticity being distributed in patches of constant ω-values, or according to a linear relation $\omega = F(\psi)$, but (although interesting solutions have thus been constructed which correctly describe some of the vortex characteristics observed in physical reality) a firm physical basis for these assumptions is lacking.

Some important clues to the solutions of this problem may be found, however, by considering the effect of a small amount of viscosity. Effectively this means that the viscous term in (4) is nonzero, but small. In this limit it may be expected that the inverse energy cascade is still active in bringing the kinetic energy to the large scales of motions, while the enstrophy still shows a spectral flux to larger wave numbers. Because viscous dissipation is most effective on the smallest length scales, it is thus anticipated that enstrophy shows some dissipation, while the energy is more or less conserved. This remarkable "selective decay" property of slightly viscous two-dimensional flows suggests that one should search for solutions that—for a given kinetic energy—contain a minimum of enstrophy. Some such attempts have indeed been (partially) successful and will be commented on later in this review.

LABORATORY EXPERIMENTS Cyclonic and anticyclonic barotropic vortices can be created in the laboratory in various ways. A convenient way of producing vortices is by applying the so-called *gravitational collapse* technique, for which purpose a bottomless thin-walled cylinder is placed in a rotating tank filled with a homogeneous fluid, with the fluid level inside the cylinder differing from that outside it. By withdrawing the cylinder vertically, a radial gravity-driven flow arises immediately, which is quickly deflected in azimuthal direction (as can be understood from conservation of angular momentum). After a period of typically π/Ω an equilibrium state is reached in which the motion is purely azimuthal, and the vortex formation is completed. Except for the motion in the thin Ekman layer at the bottom of the tank, the azimuthal flow in this state is governed by a balance of radial forces: the Coriolis force, the centrifugal force, and the pressure gradient force. In this way one can generate cyclonic vortices by making the level of fluid initially inside the cylinder lower than outside, whereas anticyclonic vortex flow should be obtained by making that level higher. For given values of Ω, the water depth H, and the diameter of the inner cylinder, the level difference ΔH obviously is the controlling parameter for the vortex intensity.

An alternative way of producing either cyclonic or anticyclonic vortices in a rotating fluid is by locally stirring the fluid with a thin rod. This is most conveniently done by briefly stirring the fluid confined in a bottomless cylinder placed in the tank, then allowing the stirring-induced motion to become organized in a purely azimuthal flow, and then releasing this vortex by quickly lifting the cylinder vertically. Although the withdrawal of the cylinder induces some turbulent motions locally, these usually decay very quickly, and a nearly two-dimensional vortex results. Another technique of creating barotropic vortices in a rotating fluid is by locally withdrawing some volume of fluid through a sink, e.g. by syphoning it through a thin tube. By conservation of angular momentum, this suction technique results in cyclonic vortices; within a few rotation periods after stopping the forcing, the vortex flow is close to two-dimensional (except, of course, in the thin bottom Ekman layer). In the reverse case, careful addition of fluid through a point source in the rotating fluid yields an anticyclonic vortex.

Although these techniques all give rise to vortex motions that are initially circularly symmetric (and which might eventually become unstable), the characteristics of the resulting vortices may be crucially different. This is illustrated by the measured radial distributions of the azimuthal velocity $v(r)$ of a suction vortex and a stirring vortex as presented in Figure 1. These velocity profiles were obtained by streak photography of passive

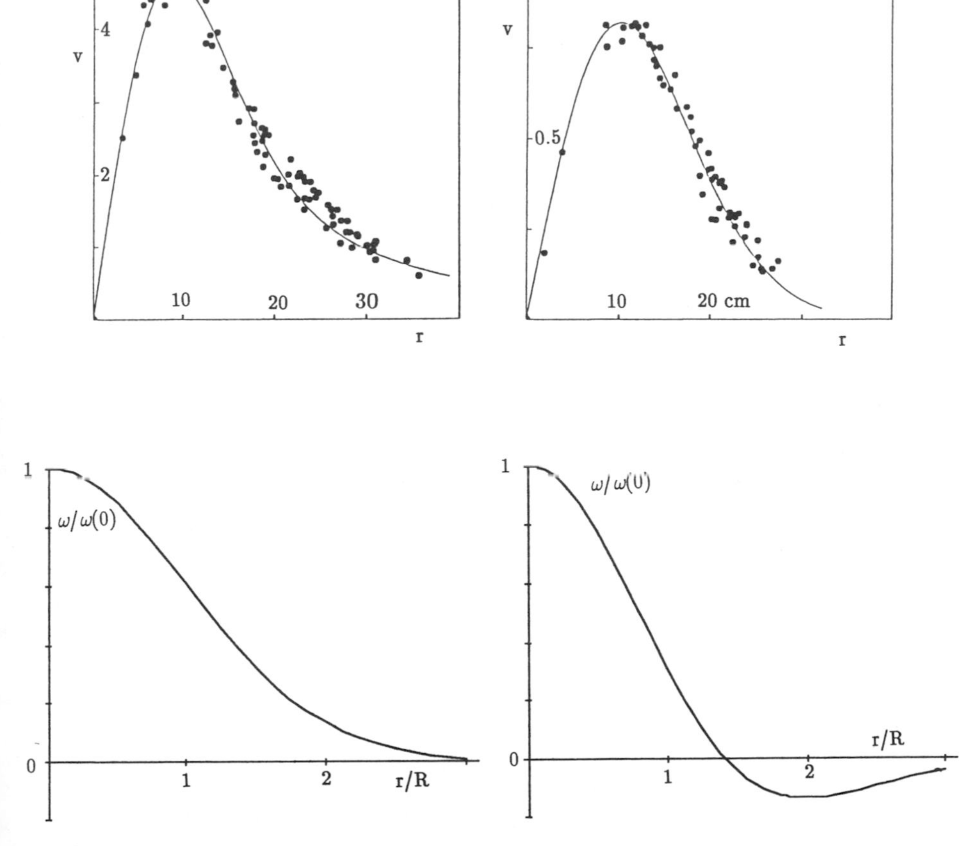

Figure 1 Typical radial distributions of the azimuthal velocity measured in a sink-produced vortex (*a*) and in a stirring-produced vortex (*b*), taken from Kloosterziel & van Heijst (1991). The corresponding vorticity profiles are shown schematically below.

tracer particles floating at the free upper surface of the fluid. It is observed that the profiles behave differently for larger radii: The distribution of the suction-induced vortex has a tail that behaves to good approximation as r^{-1}, whereas the velocity profile of the stirring-induced vortex drops off to zero much more quickly. In the same graph the corresponding vorticity profiles are given, which reveal an important difference in their dynamical structure: The suction-induced vortex has a single-signed vorticity, whereas the stirring-induced vortex consists of a core of positive vorticity, surrounded by a ring of negative vorticity. In fact, this latter vortex has no net vorticity: The area integral over ω yields zero, so that its circulation at larger distances, $\Gamma(r \to \infty)$, is also equal to zero, implying that this vortex is *isolated*. As will be addressed later in this paper, the instability properties of isolated and nonisolated vortices are essentially different. Notwithstanding these differences both vortices have a continuous vorticity distribution, with a maximum value at the vortex center.

MODELING APPROACHES The simplest models of a nonisolated vortex flow are the potential vortex (also referred to as the point or discrete vortex) and the Rankine vortex, but it is obvious that, although both give the correct behavior $v(r \to \infty) \sim r^{-1}$, these do not provide a good description of the vorticity distribution closer to the center of the vortex. If one allows for viscous diffusion of vorticity, the singularities or discontinuities in these vortices are removed. For example, the analytical solution of the viscously decaying potential vortex was obtained by Oseen (see Lamb 1932), and although its smooth velocity and vorticity profiles show some resemblance to the profiles of a real nonisolated vortex (see Figure 1*a*), they are strongly time-dependent. Of course, barotropic vortices in a rotating fluid show spin-down, but—as will be addressed in the next section—this is accomplished through the action of the bottom Ekman layer, not by radial diffusion of vorticity.

For practical purposes, one may consider a snap-shot of the Oseen vortex at some time t^*, for which one obtains

$$v(r) = \frac{\Gamma}{2\pi r}[1 - \exp(-r^2/\lambda^2)] \tag{5a}$$

$$\omega(r) = \frac{\Gamma}{\pi\lambda^2}\exp(-r^2/\lambda^2) \tag{5b}$$

which is known as the *Lamb vortex* (see Saffman & Baker 1979). The length scale λ is equal to $\lambda = 2\sqrt{\nu t^*}$, and it provides a measure of the radius of the nonzero vorticity core of the vortex. Although the Lamb vortex is not stationary for viscous flows, it is a desingularized potential vortex with continuous distributions of the flow quantities, and it has been used as such in a number of numerical simulations of inviscid 2D vortex flows. The expressions (5) for the azimuthal velocity and the vorticity of the Lamb vortex are identical to those of the *Burgers vortex*. This latter vortex, however, is an essentially three-dimensional flow structure in which the radial diffusion of vorticity is exactly balanced by radial advection of vorticity production by axial stretching of vortex tubes, in such a way that a steady flow structure results. More specifically, the radial and axial velocity components are given by $u = -\frac{1}{2}ar$ and $w = az$, respectively, where z is the axial coordinate and $a = 4\nu/\lambda^2$.

A relatively simple model describing isolated vortex structures with continuous vorticity and velocity distributions was used by Carton et al (1989) in a numerical instability study of monopolar vortices. The expressions for the nondimensional quantities $\tilde{\omega} = \omega R/V$ and $\tilde{v} = v/V$ (with R as a length scale and V a velocity scale) of this vortex are

$$\tilde{\omega}(\tilde{r}; q) = (1 - \tfrac{1}{2}q\tilde{r}^q)\exp(-\tilde{r}^q) \tag{6a}$$

$$\tilde{v}(\tilde{r}; q) = \tfrac{1}{2}\tilde{r}\exp(-\tilde{r}^q) \tag{6b}$$

with $\tilde{r} = r/R$, and q being the steepness parameter that controls the shape of the profiles. For the particular case $q = 2$ one obtains (while rescaling the radial coordinate by $\tilde{r} = \bar{r}/\sqrt{2}$) the so-called *Gaussian vortex*:

$$\tilde{\omega}(\bar{r}) = (1 - \tfrac{1}{2}\bar{r}^2)\exp(-\tfrac{1}{2}\bar{r}^2). \tag{7a}$$

This vortex bears the name Gaussian because of the shape of the stream function distribution, which is given by

$$\tilde{\psi}(\bar{r}) = \tfrac{1}{2}\exp(-\tfrac{1}{2}\bar{r}^2). \tag{7b}$$

Because of its relatively simple mathematical structure and its continuous properties, the Gaussian vortex is often used to fit observational velocity data of e.g. oceanic or laboratory vortices (see, for example, Simpson et al 1984).

By eliminating $\bar{r}$ from (7a) and (7b), one easily derives the ω, ψ relationship for this vortex:

$$\tilde{\omega} = F(\tilde{\psi}) = 2\tilde{\psi}[1 + \log(2\tilde{\psi})] \tag{8}$$

which is shown graphically in Figure 2. Also shown in this figure is an experimentally determined scatter plot of the ω, ψ relationship of a cyclonic stirring-produced vortex in a rotating fluid (van Heijst et al 1991). Apart from the different scalings, the similarity with that of the Gaussian vortex is obvious. It is seen from both plots that the function $F(\psi)$ is close to linear in the positive-vorticity core of the vortex, whereas it is strongly

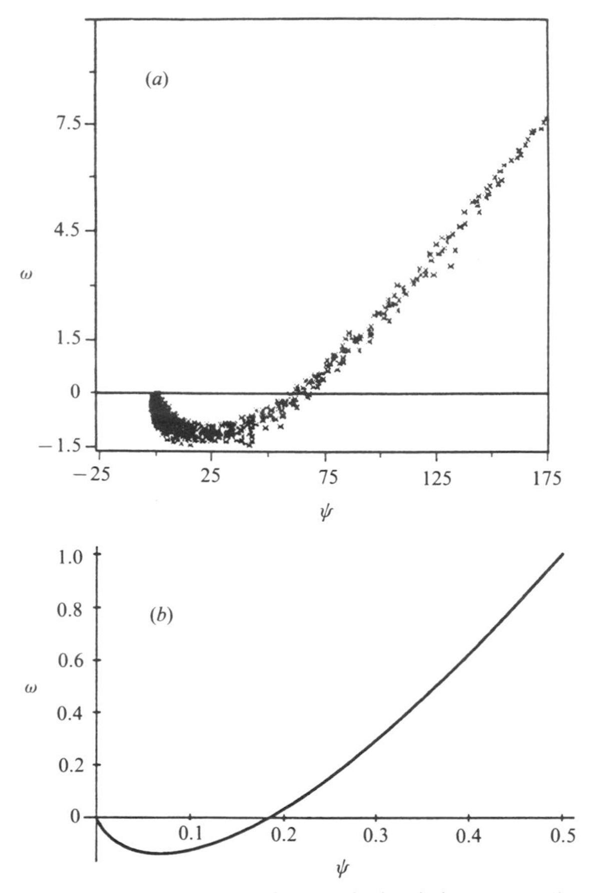

Figure 2 (*a*) The scatter plot measured for a cyclonic stirring vortex. (*b*) The theoretical (ω, ψ) relation for the Gaussian vortex, as given by (8). (After van Heijst et al 1991.)

nonlinear in the region of negative vorticity. As will be discussed in more detail in Section 2.3, this type of isolated vortex may be unstable, in the sense that the vorticity structure undergoes serious distortions, resulting in a substantial redistribution of the vorticity.

An interesting approach in the analysis of two-dimensional vortices can be made by considering the general dynamical properties of an evolving two-dimensional flow itself. Because in slightly viscous two-dimensional flows the enstrophy dissipates much faster than the kinetic energy (which becomes concentrated mainly in the larger scales of motions, i.e. in the vortices), it is physically not unreasonable to seek stationary vortex solutions that possess a *minimum enstrophy* for a given energy. Such a minimum-enstrophy analysis can be carried out by applying variational techniques, as was done by Leith (1984) for an axisymmetric isolated vortex. In this way he obtained a vortex solution on a circular disk $0 < r < R$ in the form of Bessel functions of the first kind

$$v(r) = A[J_1(\gamma r) - \frac{r}{R} J_1(\gamma R)] \tag{9a}$$

$$\omega(r) = A'[J_0(\gamma r) - J_0(\gamma R)], \tag{9b}$$

where A and A' are constants determined by the kinetic energy and the angular momentum contained in the vortex, and γ is an eigenvalue determined by the first zero of J_2, i.e. $J_2(\gamma R) = 0$, hence $\gamma R = 5.1356$. It can be verified that this particular vortex satisfies a linear relationship $\omega = F(\psi)$. In an analysis of coherent vortex solutions on a β-plane, Stern (1975) constructed a dipolar vortex solution (commonly referred to as "modon") which also satisfies minimum-enstrophy requirements, and which is also characterized by a linear relationship between the stream function and the (potential) vorticity.

Although the approach of seeking solutions that satisfy such requirements of a minimum enstrophy results in vortices that possess some resemblance with real (laboratory or ocean) vortices, it is an open question as to whether any evolving two-dimensional flow reaches the minimum-enstrophy state or whether the decay stops earlier (in the shape of another circularly symmetric vortex), having exhausted the eddy enstrophy needed for radial mixing and adjustment. Also, it could well be that some asymmetric vortex has a lower enstrophy than the symmetric vorticity structure. In this respect it is important to mention that in a numerical study of the minimum-enstrophy vortex given by (9), Leith (1984) found that, when subjected to a random perturbing flow field, the vortex exhibited a large-scale asymmetry which eventually led to the formation of a tripolar vortex with even lower enstrophy.

In another approach, the methods of statistical physics have been applied in investigating the behavior of large numbers of discrete vortices. In this approach a continuous vorticity distribution is modeled by a high-density cluster of potential vortices, which leads to a finite-dimensional Hamiltonian system. Joyce & Montgomery (1973) have studied this problem within the context of the dynamics of a two-dimensional guiding-center plasma, and derived a solution for the most probable density of such a discrete system. Essentially, this most-probable state analysis leads to a sinh-Poisson equation for the equilibrium vorticity:

$$\nabla^2\psi = -\omega = C \sinh(\beta\psi), \tag{10}$$

where C is a constant, and β is a Lagrangian multiplier (formally an inverse temperature). Recently, a different approach was taken by Robert & Sommeria (1991): Taking into account all the known constants of motion of the two-dimensional Euler equations, they defined the microscopic structure of a flow in terms of a probability function of vorticity in every point of the flow field. In a statistical sense, this probability distribution represents a local description of the small-scale fluctuations of the microscopic vorticity functions. Maximization of the entropy (for which a generalized form of Boltzmann's mixing entropy was derived) gives the equilibrium states of the flow. For the case of isolated vortex structures as such equilibrium states, Robert & Sommeria obtained a functional relationship (ω, ψ) very similar to the sinh-relationship (10).

Strong evidence for the validity of this sinh-Poisson relation was recently obtained by Montgomery et al (1992) in a high-resolution numerical study of a decaying turbulent viscous two-dimensional flow. After a few hundred eddy-turnover times it was observed that the flow relaxes to a state very close to the maximum entropy (or most probable) state described by (10), rather than to a state described by a linear (ω, ψ) relationship as predicted by the selective decay approach.

2.2 *Spin-Down of Barotropic Vortices*

The decay process of barotropic vortices in a rotating fluid is generally referred to as "spin-down," although strictly speaking the decay of an anticyclonic vortex should be termed "spin-up." In most practical laboratory situations the spin-down of vortices is established by an essentially three-dimensional effect, e.g. through the action of the thin viscous Ekman layer at the bottom of the tank. In addition, the vortex evolution may be considerably affected by the deformability of the free upper surface of the fluid. Laboratory experiments on a decaying cyclonic vortex produced by the suction technique (Kloosterziel & van Heijst 1992) have revealed that during the decay process the azimuthal velocity distributions $v(r)$ can be accurately fitted with curves representing members of a one-parameter family of self-similar solutions of the radial diffusion equation, as derived by Kloosterziel (1990a,b). These solutions take the following form

$$v(r) = VP_\alpha\left(\frac{r}{R}\right)\exp\left[-\frac{1}{2}(r/R)^2\right], \tag{11}$$

where V is the velocity scale and R the length scale as before, and P_α is a polynomial satisfying $P_\alpha(0) = 0$. The corresponding polynomial arising in the expression for the vorticity $\omega(r)$ associated with (11) represents the polynomial solution of the Laguerre equation (see Kloosterziel 1990a,b). Although the physical relevance of the radial diffusion is, under most practical circumstances, very limited (because it is completely dominated by effects due to the Ekman-layer action), the self-similar profiles offer a good diagnostic tool for the fitting of observed velocity profiles. The two extreme cases, $\alpha = 1$ and $\alpha = 3$, in the range of practical applicability correspond to the Lamb vortex (5) and the Gaussian vortex (7), respectively.

In their study of the decaying vortex, Kloosterziel & van Heijst (1992) measured velocity profiles $v(r)$ at subsequent stages of the decay process, and the evolution of the profile shape was conveniently observed by scaling the fitted curves onto each other, in such a way that they all have a maximum $v/V = 1$ at $\bar{r} = r/r_{max} = 1$. It was found that the velocity profile in the initial stage of the decay process is closely approximated by the Lamb profile given by (5a), corresponding to the member $\alpha = 1$ of Kloosterziel's (1990a,b) family of self-similar solutions (11). At subsequent stages, however, this sink-induced vortex shows a considerable steepening of the velocity profile at larger radii ($\bar{r} > 1$). Moreover, the evolving profiles tend to one particular curve, which was found to be close to the Gaussian profile (7), i.e. (11) with $\alpha = 3$. Other experiments by Kloosterziel & van Heijst (1992) on vortices created by the stirring and the collapse technique have revealed a similar tendency of steepening, i.e. a tendency of velocity profiles to reach a state that is in good approximation given by a Gaussian profile. Generally stated, this means that these vortices are evolving towards a barotropically more *unstable* state (see Gent & McWilliams 1986, Flierl 1988, Carton & McWilliams 1989), and, indeed, in a number of cases instabilities have been observed to arise eventually (Kloosterziel & van Heijst 1991), as will be discussed in Section 2.3.

The vortex evolution is a manifestation of a continuous redistribution of angular momentum in the vortex, which—although affected by the deformability of the free surface—is mainly caused by *viscous* processes. In principle, one can distinguish two different viscous processes: (*a*) radial diffusion of vorticity, resulting in changes in the vorticity distribution on a diffusive time-scale $T_d = L^2/\nu$ (where L is the vortex scale), and (*b*) vorticity changes due to a secondary circulation driven by the viscous Ekman layer at the tank bottom. The time-scale associated with this latter mechanism is $T_E = \Omega^{-1}E^{-1/2}$, with the Ekman number defined here as $E = \nu/\Omega H^2$. The relative importance of both viscous processes is given by the ratio of their time-scales, which is equal to $T_E/T_d = (H/L)^2 E^{1/2}$. In practical laboratory situations $H/L = O(1)$ and $E \ll 1$, such that the Ekman circulation is by far the dominating process in the vortex evolution. In fact, the Ekman circulation contributes to the evolution of the vortex interior in two ways. The suction (blowing) established by the bottom Ekman layer results in the stretching (shrinking) of vortex tubes in the interior of the vortex, and thus causes changes in the vorticity distribution. Secondly, the secondary circulation driven by the bottom Ekman layer results in an advection of angular momentum. This latter process is essentially nonlinear, and its importance is measured by the Rossby number of the vortex flow ($\mathrm{Ro} = V/Rf$). For small Rossby numbers ($\mathrm{Ro} \ll 1$) the Ekman suction imposed at the edge of the Ekman layer is locally linearly related to the vorticity in the interior of the vortex (see e.g. Greenspan 1968). For larger values of the Rossby number [$\mathrm{Ro} = O(1)$], however, this relation is essentially nonlinear, and besides, the Ekman suction velocity is no longer locally related to the vorticity in the vortex interior. Because it was found that a linear Ekman suction condition still holds for relatively large Rossby numbers ($\mathrm{Ro} \leq 0.6$; see Greenspan 1968), it is not unreasonable to use this condition in a study of relatively strong vortices, as did Kloosterziel (1990a,b) and Kloosterziel & van Heijst (1992). In their analysis of the vortex evolution they derived that the time-dependent azimuthal velocity $v(r, t)$ is governed by the inviscid form of Wedemeyer's (1964) equation, for which numerical solutions were presented. In a recent study, Maas (1992) showed that this problem can be solved analytically for an arbitrary initial profile. The solutions thus obtained show the characteristics of the evolving vortex flow as measured by Kloosterziel & van Heijst (1992), e.g. a clear outward shift of the position $r_{\max}$ of the maximum velocity magnitude, and also the steepening of the velocity profile for $r > r_{\max}$. Moreover, Maas's (1992) solution reveals that for certain initial conditions "breaking" may occur, for example, when the absolute vorticity of the initial profile is negative at some radius. This "breaking" phenomenon may be associated with instability of the vortex flow, and this aspect will be addressed in more detail in Section 2.3.

Another effect that may considerably influence the vortex evolution is due to the deformation of the free surface of the fluid, the importance of which is expressed by the Froude number $\mathrm{F} \equiv \Omega^2 L^2/g\bar{H}$, where $\bar{H}$ is the fluid depth in the nonrotating tank. In recent theoretical and experimental studies by Kloosterziel (1990b), O'Donnell & Linden (1991), and Kloosterziel & van Heijst (1992) it was found that, for linear spin-up of fluid in a flat-bottomed tank, the free surface has the effect of slowing down the adjustment process, and the time-scale increases by a factor $(1+\mathrm{F}/4)$.

2.3 *Instability of Barotropic Vortices*

Depending on their general vorticity structure, monopolar barotropic vortices may be unstable to small perturbations. As remarked in Section 2.1, many vortices have a shielded vorticity structure, with the vortex core surrounded by a ring of oppositely-signed vorticity (see Figure 1*b*). Such a vortex generally has much steeper vorticity and velocity gradients than the nonisolated vortex (see Figure 1*a*), and it is therefore expected to be more susceptible to barotropic instabilities. In an analytical study of two-dimensional vortex instability Flierl (1988) approximated the continuous vorticity distribution of the vortex by a circular region of uniform vorticity surrounded by a ring of uniform, but different, vorticity. For isolated vortices he found that such piecewise-constant vorticity structures are unstable, in particular, when the outer ring is relatively narrow. The most unstable mode has an azimuthal wavenumber $m = 2$, but higher modes are found when the outer ring is more narrow. Note that for an isolated vortex a narrower outer ring implies a larger vorticity difference between core and ring. Similar results were obtained by Gent & McWilliams (1986) in a numerical study of two-dimensional instabilities of isolated vortices with continuously distributed vorticity. In such two-dimensional studies the stability properties of vortices do not depend on the sign of the vortex, i.e. on whether they are cyclonic or anticyclonic. In a rotating fluid, however, the stability behavior of cyclonic and anticyclonic vortices has been observed to be essentially different, implying that three-dimensional effects are likely to play an important role in that case.

Recent laboratory experiments by Kloosterziel & van Heijst (1991) on unstable barotropic stirring-induced vortices have revealed that both cyclonic and anticyclonic vortices usually show a rapid transition to non-axisymmetric patterns: The cyclonic stirring vortices were observed to transform into a tripolar flow structure, whereas the anticyclonic vortices generally broke up into two dipolar structures. The scenario of tripole formation from an unstable cyclonic monopolar vortex is illustrated by Figure 3. For the purpose of flow visualization, the stirred fluid in the inner cylinder was dyed. The vortex is released by lifting the inner cylinder. Immediately after release the dye patch obtains a rather irregular appearance (Figure 3*a*), due to the local turbulent mixing at the circumference of the vortex associated with the lifting of the cylinder. After a few rotation periods, however, the flow reorganizes in an approximately axisymmetric vortex (Figure 3*b*). In the next stages one observes a growing asymmetry with azimuthal wavenumber $m = 2$ (Figure 3*c*), leading to the formation of a tripolar vortex (Figure 3*d*). Once such a tripole is formed, it is persistent, without showing any shape changes. The entire tripole structure rotates in a solid-body fashion in a cyclonic sense relative to the rotating system; the motion in its elliptical core is cyclonic, whereas anticyclonic relative flow occurs in the satellites. Laboratory experiments by van Heijst & Kloosterziel (1989) and van Heijst et al (1991) have shown that during the formation of the tripole the negative vorticity initially confined in a

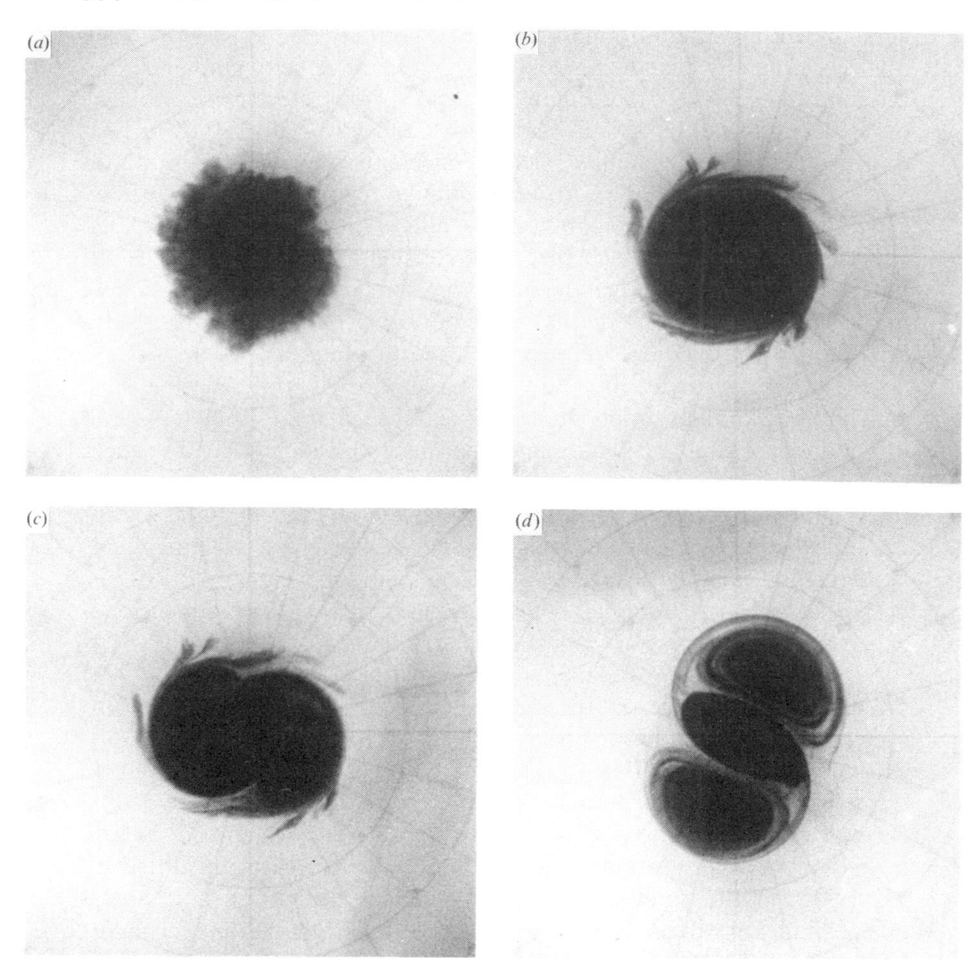

Figure 3 Photographs showing the evolution of an unstable cyclonic, stirring-induced barotropic vortex into a tripolar vortex structure. (After Kloosterziel & van Heijst 1991.)

narrow ring surrounding the core of the monopolar vortex is redistributed in the two satellite vortices. As remarked in Section 2.1, it was found in a numerical solution by Leith (1984) that under certain conditions minimum-enstrophy vortices may show a similar transition to a tripolar structure, which might suggest that the tripolar vortex is characterized by an even lower enstrophy than the initial monopolar vortex. It is not an easy task to prove this theoretically, because, owing to the tripole's complicated geometry, a satisfactory analytical solution for this coherent vortex structure is not yet available. In a high-resolution numerical simulation of forced two-dimensional turbulence (Legras et al 1988) the tripole was observed to arise spontaneously in a randomly generated flow field, most likely as the product of an unstable monopolar vortex.

Other investigators (Carton et al 1989, Orlandi & van Heijst 1992) have numerically studied the unstable monopolar vortex given by (6) when subjected to perturbations either with a prescribed azimuthal wavenumber or with a random character. These studies demonstrated that the steepness of the vorticity gradient in the initial vortex is a crucial factor for the evolving instability. It was confirmed that $m = 2$ is the most unstable mode. In particular, it was found by Orlandi & van Heijst (1992) that steeper vorticity profiles (corresponding to a narrow band of high-amplitude negative vorticity surrounding the positive-vorticity core) give rise to higher-order instability modes, which is in agreement with Flierl's results. In one particular case the vortex showed a rapid transition to a vortex structure consisting of a triangular core with three satellite vortices at its sides (mode $m = 3$). This vortex structure appeared to be very unstable, however, and it was observed to quickly split up into two dipolar vortices.

In contrast to the tripole formation from a cyclonic stirring-induced vortex, the laboratory experiments by Kloosterziel & van Heijst (1991) revealed that the anticyclonic vortex behaves differently, as can be observed in Figure 4. The first photograph (Figure 4*a*) was taken just after lifting the inner cylinder, and the irregular contour of the dyed region indicates turbulent mixing at the circumference of the vortex, similar to what is observed in the experiment of Figure 3. The elongated shape of the dye patch suggests the presence of a wavenumber $m = 2$ perturbation, and soon thereafter (Figure 4*b*) one indeed observes two cyclonic vortices emerging from the dyed region. These vortices move away from the central area, and anticyclonic vorticity is seen to concentrate into small eddies on one side of each cyclonic vortex (Figure 4*c*). Finally, two dipolar vortices are formed (Figure 4*d*), which perform looping excursions in a cyclonic direction—an effect due to their asymmetric vorticity structure. This scenario is very similar to that found by Flierl (1988; see Figure 14 in that paper) for the $m = 2$ instability of a piecewise-constant vorticity patch.

Careful observation of the flow immediately after the release of the anticyclonic vortex in the ambient solidly-rotating fluid has revealed that vigorous three-dimensional overturning motions arise in the body of the vortex (not only at its circumference, as in Figure 3*a*), suggesting the occurrence of centrifugal instability. Although the three-dimensional instability appears to dominate the flow at this stage, it is likely that simultaneously a nonaxisymmetric two-dimensional instability mode is present. In the next stage, the vertical motions are quickly damped, and the two-dimensional instability mode appears to survive, as can be observed on

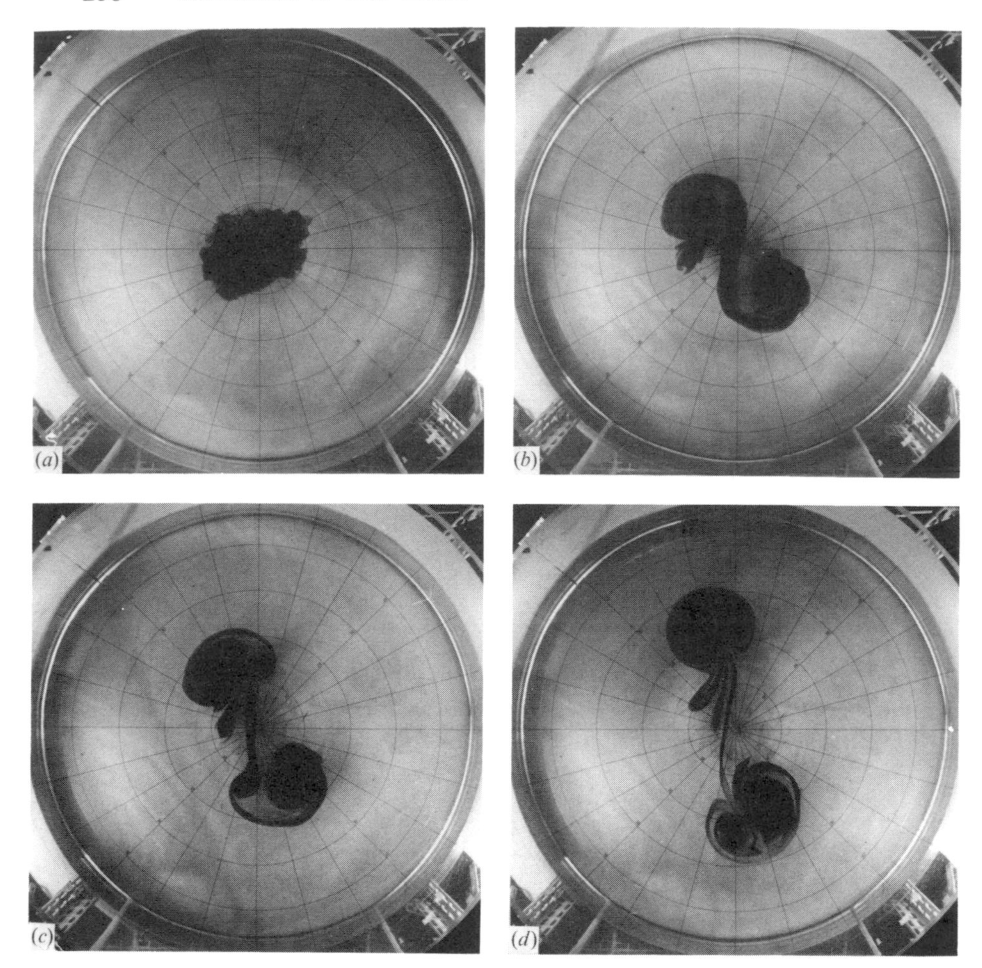

Figure 4 Photographs showing the evolution of an unstable anticyclonic, stirring-induced barotropic vortex, leading to the formation of two dipolar vortices. (After Kloosterziel & van Heijst 1991.)

the subsequent pictures of the flow evolution. This example shows that the observed instabilities may be a complicated mixture of both two-dimensional and three-dimensional, and both axisymmetric and non-axisymmetric modes. Although the $m = 2$ mode seems to be the dominant one, in a few experiments with anticyclonic stirring-induced vortices, Kloosterziel & van Heijst (1991) observed that the $m = 3$ mode grew to a finite amplitude, giving rise to a vortex with a triangular cyclonic core, accompanied by three anticyclonic satellite vortices. As in the numerical simulation by Orlandi & van Heijst (1992), this vortex turned out to be unstable: Two satellites quickly merged, momentarily yielding an unstable anticyclonic tripolar structure, which subsequently split up into two dipolar vortices.

In similar experiments on cyclonic vortices produced by the gravitational-collapse technique, it was found that under certain conditions these vortices also become unstable, but in a more gradual way (Kloosterziel & van Heijst 1991). For example, the vortices generally showed a gradual transition from a circularly symmetric flow pattern to an elliptic structure. In some cases, the flow pattern was observed to change back and forth between circular and elliptical, in a similar way as found in a theoretical study by Cushman-Roisin et al (1985). In other cases the amplitude of the instability was greater, giving rise to tripole formation as seen in the experiment of Figure 3. An interesting feature observed in the experiments is that in some cases the released vortex remains axisymmetric for a relatively long time, and then shows a rapid transition to a tripolar structure. This effect is attributed to the steepening effect of the Ekman circulation during the spin-down of the vortex (see Section 2.2), which tends to bring the vortex gradually into a more unstable state. In other cases the vortex could be observed to remain circularly symmetric throughout its lifetime; in such cases the vortex motion has entirely decayed before the steepening process could transform it into a barotropically-unstable state. For purely two-dimensional flows, Rayleigh's inflection-point theorem states that the necessary condition for instability is for the gradient of the vorticity to change sign somewhere. In this respect the amplitude and the sign (cyclonic or anticyclonic) of the vortex flow is of no importance. According to this theorem, all isolated vortices are unstable. However, laboratory experiments have revealed that shielded vortices may be stable, for the reason just described.

If one allows for three-dimensional motions (as e.g. occurring in the initial stage of the experiment shown in Figure 4), Rayleigh's circulation theorem states that a necessary and sufficient condition for a stationary inviscid swirling flow with azimuthal velocity $v(r)$ to be stable to axisymmetric disturbances is that the square of the circulation does not decrease anywhere in the flow:

$$\frac{d}{dr}(rv)^2 \geq 0. \tag{12}$$

On the other hand, the swirling flow is unstable if the circulation decreases at some radius, giving rise to *centrifugal instability* in the form of axisymmetric overturning motions in coaxial rings of fluid.

This circulation theorem by Rayleigh was extended by Kloosterziel & van Heijst (1991) to swirling (vortex) flow in a rotating fluid or—more generally—on an f-plane, yielding the following criterion for stability:

$$\frac{d}{dr}\left(rv+\frac{1}{2}fr^2\right)^2 \geq 0. \tag{13}$$

For the special case of a vortex centered on the axis of a rotating tank ($f = 2\Omega$), this extended criterion could also be obtained directly from Rayleigh's criterion (12) by replacing v by the absolute velocity $v + \Omega r$. Note, however, that the extended criterion applies also to off-center vortices in a rotating fluid. In nondimensional form, after introducing a velocity scale V and a length scale R, the extended criterion (13) can also be written as

$$\left.(\varepsilon\tilde{v}+\tilde{r})(\varepsilon\tilde{\omega}+2) \begin{array}{l} > 0 \rightarrow \text{stability} \\ < 0 \rightarrow \text{instability} \end{array}\right\} \tag{14}$$

with $\tilde{v} = v/V$, $\tilde{\omega} = \omega R/V$, $\tilde{r} = r/R$, and $\varepsilon = 2\,\mathrm{Ro} = V/R\Omega$. According to this result, a vortex is stable or marginally stable if the product of the absolute vorticity and the absolute velocity $\varepsilon\tilde{v}+\tilde{r}$ is positive or zero everywhere. In order to illustrate the implications of this instability criterion, Kloosterziel & van Heijst (1991) have presented graphs of the product $(\varepsilon\tilde{v}+\tilde{r})(\varepsilon\tilde{\omega}+2)$ as a function of $\tilde{r}$ for $\varepsilon = 0.5$, 2, and 5, for a Lamb-type vortex [see (5), which serves as a model of the sink-induced vortex],

$$\tilde{v}_{\text{sink}}(\tilde{r}) = \frac{1}{\tilde{r}}\left[1-\exp\left(-\frac{1}{2}\tilde{r}^2\right)\right] \tag{15a}$$

$$\tilde{\omega}_{\text{sink}}(\tilde{r}) = \exp\left(-\tfrac{1}{2}\tilde{r}^2\right), \tag{15b}$$

and for a Gaussian vortex [see (7), which serves as a model of the stirring-induced vortex],

$$\tilde{v}_{\text{stir}}(\tilde{r}) = \tfrac{1}{2}\tilde{r}\cdot\exp\left(-\tfrac{1}{2}\tilde{r}^2\right) \tag{16a}$$

$$\tilde{\omega}_{\text{stir}}(\tilde{r}) = \left(1-\tfrac{1}{2}\tilde{r}^2\right)\exp\left(-\tfrac{1}{2}\tilde{r}^2\right) \tag{16b}$$

both for the cyclonic and the anticyclonic case. Their results are reproduced here in Figure 5. The velocity profiles are scaled in such a way that $\tilde{v}_{\max} = 1$ at $\tilde{r}_{\max} = 1$, and accordingly the Rossby number is thus based on the position and the amplitude of the maximum velocity. It is seen that the cyclonic nonisolated (sink-induced) vortex is stable irrespective of the Rossby number (Figure 5*a*), whereas its anticyclonic counterpart becomes unstable if the Rossby number exceeds a critical value (Figure 5*b*). This

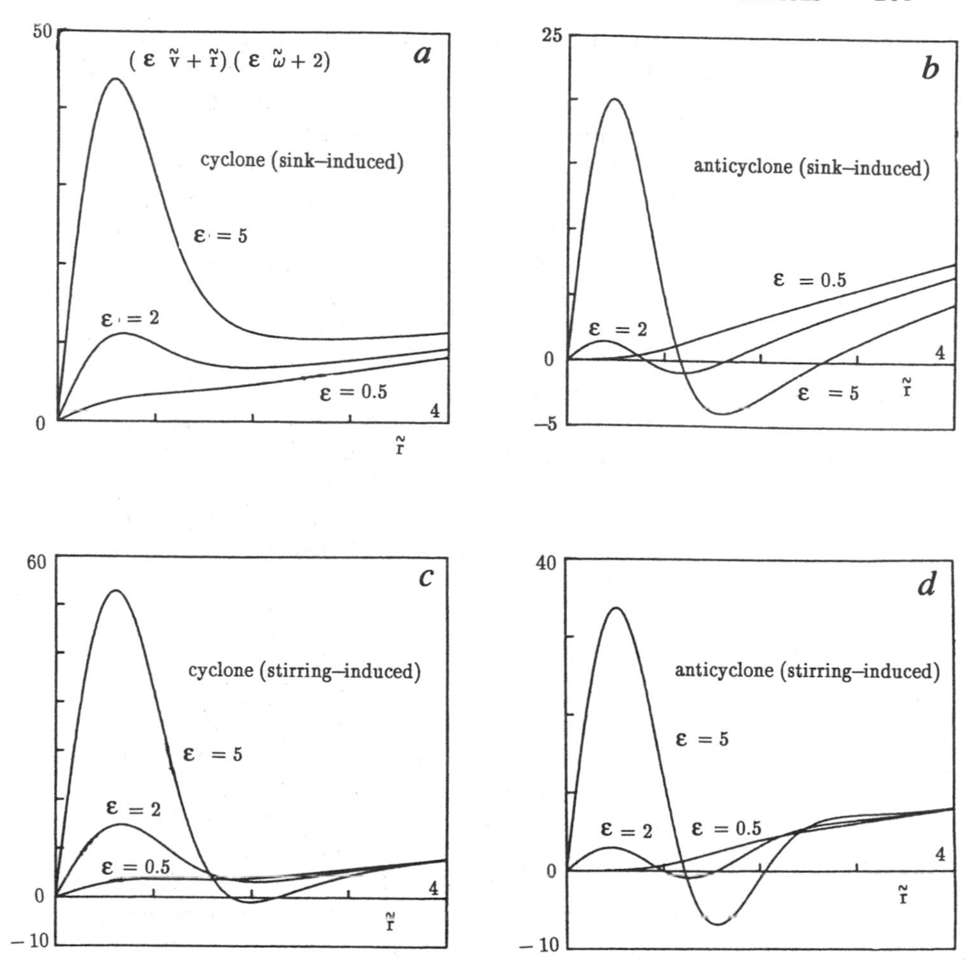

Figure 5 Graphical presentation of the product of the nondimensional absolute vorticity and velocity as a function of $\tilde{r}$ for $\varepsilon = 0.5$, 2, and 5 for (*a*) the cyclonic and (*b*) the anticyclonic non-isolated (sink-induced) vortex given by (15), and for (*c*) the cyclonic and (*d*) the anticyclonic isolated (stirring-induced) vortex given by (16). (After Kloosterziel & van Heijst 1991.)

critical value was found to be $\varepsilon_c \approx 0.57$. The isolated (stirring-induced) vortex can be unstable in both the cyclonic and the anticyclonic case (Figure 5*c*,*d*), e.g. when twice the Rossby number exceeds critical values of $\varepsilon_c \approx 4.5$ and 0.65, respectively. Generally stated, only very weak anticyclonic vortices are centrifugally stable, while only very strong cyclonic vortices are centrifugally unstable.

This result provides the explanation of the difference in instability behavior of the vortices shown in Figures 3 and 4: The cyclonic vortex remained centrifugally stable and showed a two-dimensional barotropic instability leading to tripole formation, whereas the anticyclonic vortex first became centrifugally unstable before splitting into two dipoles. In their experiments on anticyclonic vortices produced by the reversed sink-technique (by carefully injecting fluid into a rotating fluid), Griffiths & Hopfinger (1987) observed that these vortices remained stable. The Rossby number was in their case estimated to be approximately 0.15 (i.e. $\varepsilon = 0.3$), and this is indeed lower than the critical value of 0.57 cited above. In a recent study of vortex spin-down, cited in Section 2.2, Maas (1992) analyzed the possible occurrence of shock formation due to steepening velocity profiles, and he derived a *dynamical* breaking criterion: $\varepsilon\tilde{\omega}+2 > 0$. For the case of the absolute velocity being positive at all radii, this dynamical criterion is equivalent to Rayleigh's extended kinematical criterion (14) for the onset of centrifugal instability.

2.4 *Topography Effects*

It has been established in a number of studies (e.g. Adem 1956, McWilliams & Flierl 1979, Mied & Lindemann 1979) that monopolar vortices on a β-plane show a clearly recognizable translational motion. It is well-known that the β-effect on large-scale geophysical flows can be simulated in the laboratory by a suitably chosen bottom topography in a rotating fluid confined in a tank spinning at constant speed. In order to be able to appreciate the relevance of studying topography effects on vortices in rotating fluids to geophysical flows, we will first discuss some general features of the flow dynamics. Inviscid quasi-geostrophic flow (assumed to be confined to scales much smaller than the Rossby deformation radius) on a β-plane is governed by

$$\frac{\partial \omega}{\partial t} + \mathbf{v} \cdot \nabla(\omega + f_0 + \beta y) = 0. \tag{17}$$

The latitudinal variation of the Coriolis parameter is approximated here by $f = f_0 + \beta y$, with f_0 and β constants, and y the local northward Cartesian coordinate (the corresponding Cartesian coordinates x and z point locally eastward and vertically upward, respectively). By applying the relationship $\omega = -\nabla^2\psi$ (where ψ is the stream function as before) this equation can be written in the form

$$\frac{\partial}{\partial t}(\nabla^2\psi)+J(\nabla^2\psi,\psi)+\beta\frac{\partial\psi}{\partial x}=0, \tag{18}$$

which describes the local changes in the relative vorticity due to advection of relative vorticity (second term) and advection of planetary vorticity [i.e. $\mathbf{v}\cdot\nabla(f_0+\beta y)$, as represented by the third term].

As a first approach, solutions of this vorticity equation may be obtained for vortex flows with small relative velocities (small Rossby numbers), for which the Jacobian term is negligible. The resulting linear equation

$$\frac{\partial}{\partial t}(\nabla^2\psi)+\beta\frac{\partial\psi}{\partial x}=0 \tag{19}$$

is invariant under the transformation $\{\psi\}\rightarrow\{-\psi\}$ so that cyclonic vortices will show the same behavior as anticyclonic vortices. In one of the earliest studies of vortex motion on a β-plane Bjerknes & Holmboe (1944) analyzed the vortex motion governed by (19), and predicted a westward translation. In a later study, Flierl (1977) showed that the so-called Bessel eddies [cf (19)] travel westward with a translation speed that is essentially dependent on the spatial scale of motion (an effect previously noted by Tojo 1953), implying that the solution shows dispersion. Due to this effect the vortex disintegrates and the kinetic energy is radiated away by Rossby waves. For the case of an initially circular vortex in the absence of any background flow this Rossby-wave radiation wake is symmetric about the east-west axis—as can be understood from the invariance of (19) under the transformation $\{y\}\rightarrow\{-y\}$, Inspection of the *full* equation (18) reveals that this equation is invariant under the transformation $\{\psi,y\}\rightarrow\{-\psi,-y\}$, which implies that cyclonic and anticyclonic vortices with arbitrary (nonzero) Rossby number behave differently: When a cyclonic vortex would move northward, an anticyclonic vortex moves southward (and vice versa). It can be verified that (18) has *no* steadily translating monopolar vortex solutions, nor any other (quasi-)stationary monopole solutions. The vortex evolution according to (18) is therefore an extremely intricate process. The first steps in the analysis of this problem were made by Rossby (1948) and some years later by Adem (1956), who expanded the stream function ψ in a Taylor series in time t. The combination of the lowest-order effects then predicts an initial tendency of the isolated cyclonic vortex to propagate to the northwest. In view of the transformation properties mentioned before, an anticyclonic vortex will therefore show an initial tendency to move to the southwest. Although there has been some criticism of Adem's analysis, later numerical simulations of the full nonlinear equations (by e.g. McWilliams & Flierl 1979, Mied & Lindemann 1979, McWilliams, Gent & Norton 1986) have confirmed that this initial drift tendency is obeyed.

Moreover, the overall tendency for many turnover times of the cyclonic vortex is to move to the northwest, though along a somewhat complicated trajectory. In contrast to the linear evolution according to the Rossby wave equation (19), the nonlinear evolution is characterized by the energy remaining fairly localized, provided that the vortex flow is sufficiently nonlinear. As such, the vortex is a coherent nonlinear entity.

In the shallow-water approximation (see e.g. Pedlosky 1987) the inviscid relative flow in a rotating system preserves its potential vorticity:

$$\frac{D}{Dt}\left(\frac{\omega+f}{h}\right) = 0, \tag{20}$$

where $h(x, y, t)$ is the depth of the fluid column. This conservation property implies that the relative vorticity ω of a fluid column with constant depth h moving northward into regions of larger f-values (on the northern hemisphere) will decrease, and vice versa for motion in southward direction. On the other hand, in a system with constant f (as in a rotating tank of fluid) a suitable choice of the bottom topography may give the same dynamical features: The relative vorticity of a fluid column moving into shallower areas will decrease, whereas it increases when the column moves into deeper parts (under the assumption $|\omega| < |f|$). The local gradient of h thus defines the local compass directions, with the direction of decreasing depth corresponding to "north," and that of increasing depth to "south." For the flow in a rotating tank with a certain bottom topography given by $\eta(x, y)$, such that the fluid depth is given by $h(x, y) = H - \eta(x, y)$, where H is some constant reference depth, the term h^{-1} in (20) can be expanded in the small parameter $\delta = \eta/H$ (it is assumed that the bottom topography is shallow). Under the additional restriction that the Rossby number is small, the conservation equation (20) takes the following form:

$$\frac{D}{Dt}\{\omega + f + \eta f/H\} = 0. \tag{21}$$

For the special case of an inclined bottom with a constant slope s in the y-direction, $\eta = sy$, and with $f = f_0 = 2\Omega$ the equation is written exactly in the same form as (17), now with $\beta \equiv 2s\Omega/H$. A proper choice of s, H, and Ω obviously yields a flow configuration that is dynamically equivalent to that on a β-plane on a rotating sphere. This dynamical similarity has been exploited by a number of investigators (e.g. Firing & Beardsley 1976, Masuda et al 1989, Carnevale et al 1991b) in the experimental study of vortex motion on a β-plane. Their laboratory experiments have confirmed the general translational motion of cyclonic and anticyclonic vortices in northward and southward directions, respectively.

In the experiments of Carnevale et al (1991b) the behavior of two essentially different types of vortices was studied: sink-produced vortices and stirring-produced vortices [in good approximation described by (15) and (16), respectively]. It was observed that both types of vortices show (when cyclonic) a definite drift in the northwest direction; i.e. they climb the slope under a certain angle, in agreement with the numerical results obtained by McWilliams & Flierl (1979), Mied & Lindemann (1979), and McWilliams et al (1986), although along somewhat different trajectories. A striking difference between the tracks of the isolated (Gaussian) stirring vortex and the sink vortex is that while the sink vortex moves along an almost straight path, that of the stirring-induced vortex is often highly curved, with a characteristic kink. This difference is due to the different vorticity distributions of these vortices: The sink vortex has a single-signed vorticity distribution which remains more or less compact during the course of the experiments, whereas the double-signed vorticity distribution of the stirring-induced vortex does not remain compact, but instead shows a considerable shedding of anticyclonic vorticity (see Figure 6). The behavior of these vortices over bottom topography agrees very well with numerical simulations carried out by Carnevale et al (1991b), both for the inviscid case and for the case in which viscous effects due to the bottom Ekman layer and lateral diffusion of vorticity are included. Although some minor differences are observed, the results of the inviscid and the viscous simulations agree very well, indicating that the basic mechanism behind the northwestward drift of cyclonic vortices is essentially *nonlinear* and *inviscid*.

The same authors also studied the behavior of vortices in a rotating tank with conical bottom topography. In their experiments with a *conical-hill* topography it was observed that cyclonic vortices generated at some distance away from the hilltop move toward the top along anticyclonic spiral-shaped trajectories. When the vortex was released too close to the hilltop, however, it showed a mere anticyclonic looping around the top, its radial motion being hampered by an induced anticyclonic vortex residing at the top of the conical topography. On the other hand, cyclonic vortices generated in a *conical valley* were found to climb out of the valley following cyclonic spiral paths. These observations clearly confirm the general tendency of cyclonic vortices to drift in the *local* northwest direction (defined by the local topography gradient), provided their spatial scale is small compared with that of the topography.

3. VORTEX INTERACTIONS

3.1 *Merging of Vortices*

Vortices, as occurring in two-dimensional turbulence, may show a very complicated interaction behavior. Although two like-signed potential vor-

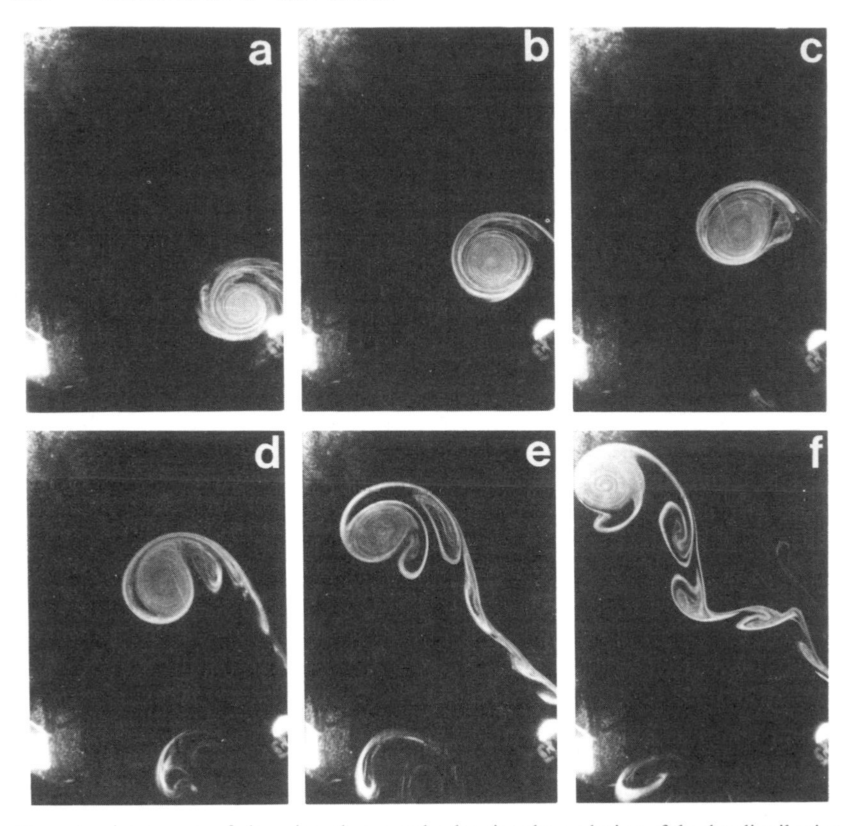

Figure 6 A sequence of plan-view photographs showing the evolution of the dye distribution in a stirring induced vortex moving uphill (in the northwest direction) over an inclined bottom in the rotating tank. The dye was initially confined in a bottomless inner cylinder used for the generation of the vortex; this cylinder was located in the deepest part of the tank (at the bottom-right in each photograph). In the plan-view pictures the topography gradient runs in "vertical" direction, the top of each photograph corresponding to shallow ("north") and the bottom to deep ("south"). (After Carnevale et al 1991b.)

tices will circle around a common center without ever getting closer, it is known from analytical and numerical studies based on the two-dimensional Euler equations that two like-signed vortices with a finite vorticity core will merge when their distance of separation is smaller than some critical value. This merger process is the predominant mechanism for the evolution of decaying two-dimensional turbulence, and has for this reason been studied extensively during the past decade. Theoretical and numerical investigations by e.g. Zabusky et al (1979), Overman & Zabusky (1982), and Melander et al (1988) have provided important information about the merging. In particular for the case of two equal circular patches (diameter $2R$) of equal uniform vorticity at initial separation distance d (measured between their centers) it was predicted that merging will occur when d is smaller than the critical distance d_c, given by

$$d_c/R = 3.2. \tag{22}$$

The validity of this merging criterion has been confirmed by a number of high-resolution numerical simulations of inviscid two-dimensional flows. Laboratory experiments on interacting barotropic vortices in a rotating fluid conducted by Griffiths & Hopfinger (1987) have demonstrated that the symmetric merger of anticyclonic vortices is in agreement with the theoretical prediction (22): They measured $d_c/R = 3.3 \pm 0.2$ for anticyclones. However, for cyclonic vortices their observations were at variance with the theoretical merging criterion, because these vortices eventually merged for all cases considered, even for distances as large as $d/R = 4.5$.

In the laboratory experiments of Griffiths & Hopfinger (1987), the vortices were generated by two identical sources (or sinks) at the free surface, positioned symmetrically about the center of the circular container, at a desired distance d apart. In order to visualize the flow evolution, dye of different colors was injected into the cores of the vortices, thus providing a useful tool to follow the sequence of events during the merging process in detail. Velocity measurements through streak photography of neutrally buoyant tracer particles revealed that the azimuthal velocity distribution of the stable (small Rossby number) initial vortices closely approaches that of a Rankine vortex. This implies that the vorticity is to good approximation uniform in a circular region with radius R, so that a good comparison with situations considered in the cited theoretical/numerical studies is possible. Griffiths & Hopfinger observed that, once two identical anticyclonic vortices were created, they began to orbit in an anticyclonic sense around the center of the tank. If their initial distance was small enough, each vortex became strongly asymmetric with characteristic cusps being formed prior to the merger. In that stage the cusps give the vortices the appearance of so-called corotating V-shapes (Overman & Zabusky 1982). Subsequently, the cusps are observed to grow until they meet each other, whereupon they get rapidly drawn around the other vortex. At the same time two spiral arms of dye are observed to be extended backwards, in the cyclonic direction. During the actual merging the combined core structure has an elliptical shape, which quickly relaxes to a circular structure containing two entwined spirals of the different dyes.

The merging process between two identical cyclonic vortices is identical, only with the azimuthal directions reversed.

The theoretical considerations and numerical simulations that led to the merging criterion (22) are based on the pure two-dimensional Euler equation, according to which cyclonic and anticyclonic vortices would behave identically. To explain the discrepancy between the theoretical prediction and the actually observed merging behavior of cyclonic vortices, Griffiths & Hopfinger (1987) suggested that three-dimensional effects due to the presence of the Ekman layer at the tank bottom are present. In a recent paper by Carnevale et al (1991a) it was shown that this effect is not substantial enough to explain the deviating behavior of cyclones entirely, and that the clue for the explanation lies in the parabolical shape of the free-surface of the fluid in the rotating container. As discussed in Section 2.4, bottom topography causes cyclonic vortices to drift to the local northwest. A curved free surface is dynamically equivalent to bottom topography, and will thus cause the vortices to drift. More specifically, a parabolic free surface will cause cyclones/anticyclones to translate along inward/outward spiral trajectories in anticyclonic/cyclonic directions. This induced drift will thus promote cyclone merger, whereas merger of anticyclones will be inhibited to some degree. Numerical simulations by Carnevale et al (1991a) in which these topographic free-surface effects were incorporated indeed show the cyclones to behave exactly as observed in the laboratory experiments by Griffiths & Hopfinger.

3.2 *Axisymmetrization*

It has been observed in the merger of two like-signed vortices that the core of the newly formed vortex is initially elliptical, and then shows a gradual relaxation towards an axisymmetric shape (Griffiths & Hopfinger 1987). The Kirchhoff vortex, consisting of an elliptical patch of uniform vorticity, is known to be an analytical solution of the two-dimensional Euler equations (see Lamb 1932). This vortex shows a steady rotation without shape changes, and is stable for eccentricity values less than 3. The presence of strain and shear in the ambient flow field in general leads to a complicated vortex behavior, although solutions are known that show under certain conditions a steady rotation or periodic shape changes (Kida 1981). The elongated Kirchhoff vortex with eccentricity larger than 3 is unstable to higher-order asymmetric modes, which lead to the expulsion of vorticity filaments in the form of spiral arms. Melander et al (1987) have numerically studied the axisymmetrization of elliptical vortices with a smoothed vorticity distribution, and their high-resolution simulation revealed that there is an essentially *inviscid* trend towards axisymmetry during the first half revolution of the vortex. In the next stage of the evolution thin vorticity filaments are stripped off in the form of two spiral arms, while the vorticity gradient of the core intensifies. The vorticity filaments have been found to have a dominant influence on the evolution of the vortex core, even when they contain little net vorticity. However, diffusive processes minimize their effect on the core vortex, and after typically a few core revolutions the inviscid axisymmetrization mechanism takes over, gradually bringing the core into a circular shape. On the basis of these observations, Melander et al (1987) conjectured that all isolated stable, spatially-smooth steady solutions of the Euler equations are (nearly) axisymmetric. Although detailed vorticity distributions during merging of laboratory vortices have as yet not been measured (due to serious resolution problems), the dye distributions in the merger experiments of Griffiths & Hopfinger (1987) show a remarkable correspondence with the vorticity distributions observed in the high-resolution numerical simulations of vortex merger (Melander et al 1988) and the subsequent axisymmetrization stage (Melander et al 1987), indicating that filamentation of vorticity plays a crucial role in the adjustment process.

4. TWO-LAYER STRATIFIED VORTICES

Stratification introduces an additional scale into the problem, which is known as the internal Rossby radius of deformation Λ. This radius represents the radial distance to which a perturbation of the isopycnal surfaces will spread before the spreading is balanced by the Coriolis force. We consider here the two-layer stratified quasi-geostrophic model which has all the essential ingredients of rotating stratified flows, but which is simpler than the constant-gradient flow for understanding physical processes such as vortex merging, vortex alignment, and baroclinic instability. The quasi-geostrophic assumption neglects local accelerations ($\partial \mathbf{v}/\partial t$) relative to Coriolis acceleration ($2\mathbf{\Omega} \times \mathbf{v}$), and takes the relative vorticity to be comparable to vorticity associated with stretching of fluid columns by variations in layer depth, but small compared with the background vorticity f. While departures from the geostrophic balance are allowed, the model does not include centrifugal forces ($\rho v^2/r$) that are associated with finite Rossby numbers ω/f and which can be significant in frontal ocean eddies, for example.

4.1 *Baroclinic Vortex Structure*

Solutions to the potential vorticity equation for discrete (potential) vortices in a two-layer stratified fluid were obtained by Gryanik (1983) and by Hogg & Stommel (1985a). These solutions nicely illustrate the stratification effect on the velocity field and on vortex interactions. Baroclinic, finite core vortices are characterized by two independent horizontal length scales: the internal Rossby radius Λ and the core radius R. Analytical solutions for a two-layer stratified geostrophic, finite-core, model vortex were obtained by Pedlosky (1985), Griffiths & Hopfinger (1987, from here on referred to as GH) and Helfrich & Send (1988).

The stream functions ψ_i, $(i = 1, 2)$ of the upper and lower layers of depths H_i satisfy the inviscid quasi-geostrophic potential vorticity conservation equations $DQ_i/D_it = 0$, where $D/D_it = \partial/\partial t + J(\psi_i, \ )$ and the potential vorticity is

$$Q_i = \nabla^2\psi_i + (-1)^i \frac{f^2}{g'H_i}(\psi_1 - \psi_2), \tag{23}$$

with the interface displacement given by

$$\eta_i - H = \pm\frac{f}{g'}(\psi_2 - \psi_1),$$

where $g' = g(\rho_2 - \rho_1)/\rho_2$ is the reduced gravity, with ρ_1 and ρ_2 the fluid densities of the upper and lower layers respectively. For simplicity, the layers are assumed to be of equal depth H. Further we consider the case in which the fluid having anomalous vorticity is confined to the upper layer 1 and forms a vortex core of radius R of uniform potential vorticity Q_0, different from the potential vorticity in the rest of the layer away from the vortex core, which is $Q = f/H$. In general, $Q = (\omega + f)/\eta$, where ω is the relative vorticity and η is the fluid layer depth, so that when $\eta = H$, the relative vorticity $\omega = 0$. With these assumptions Equations (23) take the form

$$\nabla^2\psi_1 + \tfrac{1}{2}\lambda^{-2}(\psi_2 - \psi_1) = Q_1,\ r < R,$$

$$\nabla^2\psi_1 + \tfrac{1}{2}\lambda^{-2}(\psi_2 - \psi_1) = 0,\ r > R,$$

$$\nabla^2\psi_2 + \tfrac{1}{2}\lambda^{-2}(\psi_1 - \psi_2) = 0, \tag{24}$$

where $\lambda = (1/\sqrt{2})(g'H)^{1/2}/f = \Lambda/\sqrt{2}$ and the right-hand side is simply the relative vorticity $Q_1 = \omega_0 = HQ_0 - f$, which vanishes everywhere outside the core $(r > R)$. In polar coordinates centered on $r = 0$, the solutions to Equations (24), subject to suitable boundary conditions, for the azimuthal velocities $v_i = \partial\psi_i/\partial r$ in each layer are (see GH)

$$\frac{v_i}{R\omega_0} = \frac{1}{4}\frac{r}{R} \pm \frac{1}{2}K_1\left(\frac{R}{\lambda}\right)I_1\left(\frac{r}{\lambda}\right), \quad r < R$$

$$\frac{v_i}{R\omega_0} = \frac{1}{4}\frac{R}{r} \pm \frac{1}{2}I_1\left(\frac{R}{\lambda}\right)K_1\left(\frac{r}{\lambda}\right), \quad r > R, \tag{25}$$

where K_1 and I_1 are the modified Bessel functions; the positive sign refers to the top layer velocity ($i = 1$), and the negative sign to the bottom layer velocity ($i = 2$). The baroclinic component of the vortex flow for $r > R$ is expressed by $I_1(R/\lambda)K_1(r/\lambda)$ and for $r < R$ by $K_1(R/\lambda)I_1(r/\lambda)$. At large r, such that $r/\lambda \gg 1$, the flow is barotropic. It is noteworthy that the shapes of the velocity profiles are not dependent on the sign of the vorticity in the core. The velocity profiles given by Equations (25) are plotted in Figure 7 for three values of the ratio λ/R.

In the limit $R/\lambda \to 0$, we have $I_1(R/\lambda) \to \frac{1}{2}R/\lambda$ and we recover, for $r > R$, the results of Hogg & Stommel (1985a) for a discrete vortex:

$$v_{ip} = \frac{1}{2}\frac{s}{r}\left[1 \pm \frac{r}{\lambda}K_1\left(\frac{r}{\lambda}\right)\right]. \tag{26}$$

The constant $s = (R^2\omega_0)/2$ is the vortex intensity and the vortex strength is $2\pi s$. The circulation Γ outside the vortex core is a function

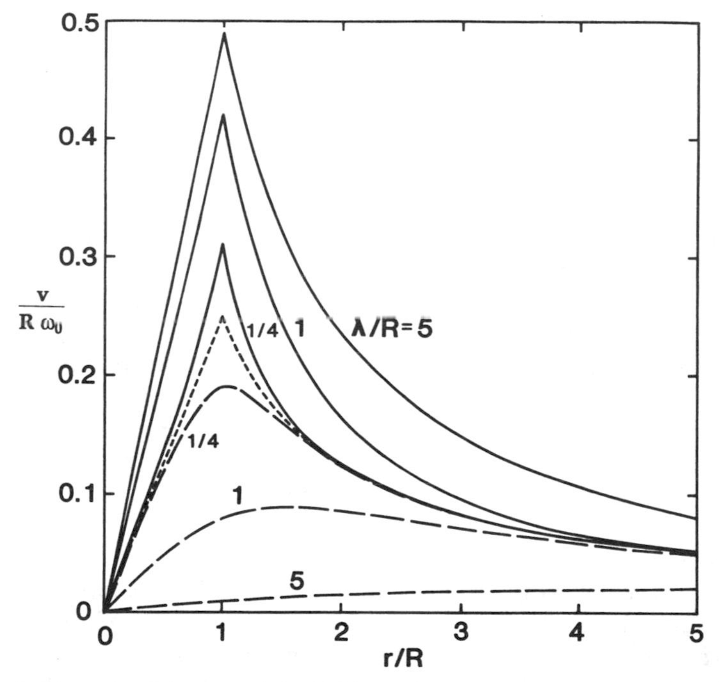

Figure 7 Examples of the azimuthal velocities given by Equation (25). Top layer velocities (*solid lines*), bottom layer velocities (*long dashed lines*), barotropic limit (*short dashed line*). (After Griffiths & Hopfinger 1987.)

of r except in the barotropic limit $\lambda = 0$, where $\Gamma = \pi s$. In the limit $\lambda/R \rightarrow 0$, where the influence of density gradients vanishes and the flow becomes independent of depth, the flow structure approaches the Rankine vortex:

$$v_1 = v_2 = \begin{cases} \dfrac{1}{2}\dfrac{s}{R^2}r, & r < R, \\[2ex] \dfrac{1}{2}\dfrac{s}{r}, & r > R. \end{cases} \tag{27}$$

At the opposite extreme, $\lambda/R \gg 1$, the maximum velocity at the edge ($r = R$) of the core approaches $v_1 = \frac{1}{2}R\omega_0 = s/R$, twice that in (27). The upper layer velocity again decreases with radius as $v_1 \sim r^{-1}$, but only over a distance comparable to the core radius. Over a much greater distance there is now an additional decrease such that at $r \gg \lambda$, the velocity asymptotically tends to the barotropic value (27). In this strongly stratified regime the velocities are independent of depth (barotropic) at $r \gg \lambda$ and strongly baroclinic at $r < \lambda$. At $r < \lambda$ the bottom layer velocity is much less than the velocity in the top layer and approaches a maximum value at $r \approx 1.1\ \lambda$. For intermediate values of λ/R, interfacial shear is always greatest at the outer edge of the core, but is smaller for smaller values of λ/R. At $\lambda/R \approx 1$, the top layer velocity immediately outside the core decreases with increasing radius more rapidly than in either the unstratified or strongly stratified cases.

GH have performed experiments in a circular tank 100 cm in diameter and 45 cm deep which rotates about a vertical axis through the center (see also Section 3.1). The rotation rate was $\Omega = 1.0$ rad s^{-1} in the (positive) anticlockwise direction. The fluid layers are of equal depth $H = 20$ cm. The density difference $\Delta\rho = \rho_2 - \rho_1$ between the layers was varied such that the Rossby radii $\Lambda = (g'H)^{1/2}/f$ were 0, 1.5, 5, 10, and 15 cm. Vortices were produced by sources (anticyclonic vortices) and sinks (cyclonic vortices) and visualized by dye injected through the sources or, in the case of cyclone generation, introduced before suction was started. Velocities were obtained from streak photographs.

In Figure 8 the velocity profiles in the top and bottom layers are shown for an anticyclonic vortex. The vortex was initiated in the top layer of the two-layer stratified fluid with $\Lambda = 5$ cm. Figure 8 shows a clear deviation from $v_1 \sim r^{-1}$ beyond the core radius. Furthermore, the motion in the bottom layer is much less than that in the top layer for $r < 2\Lambda$, while for $r > 3\Lambda$ the flow is practically independent of depth (nearly the same

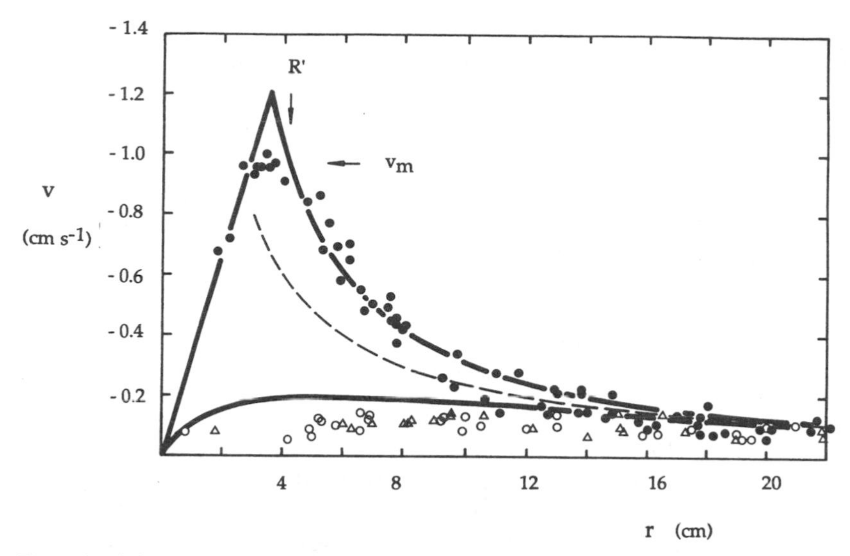

Figure 8 Azimuthal velocities measured in a two-layer anticyclone with $\Lambda = 5$ cm. Data in the top layer (●) were taken $9T_\Omega$ after vortex generation, and in the bottom layer after $6T_\Omega$(○) and $12T_\Omega$(△). Solid curves are the profiles of Equation (25) and the dashed line corresponds to the barotropic vortex $v = s/2r$. The core radius R' is defined by Equation (28). (After Griffiths & Hopfinger 1987.)

velocity in top and bottom layers). The comparison between the measured and model velocities has been made by fitting the predicted form (25) for v_1 $(r > R)$ to the data from the top layer by using an estimate for the ratio Λ/R, evaluating a multiplying constant s, and then using that constant to compute the bottom layer velocity v_2. The measured bottom layer velocities are satisfactorily modeled although they are somewhat smaller than those predicted. This deviation from the potential vorticity model (24) might be due to bottom friction and to finite interface thickness. Another possibility is that when injection is started the interface in the central part of the vortex core is displaced beyond the equilibrium state and then, when injection is stopped, relaxes back to reduce the anticyclonic velocity in the lower layer. For the cyclonic vortices the comparison between the model and observations is similar. The core radius R' was defined by the intersection of the predicted velocity v_1 with the maximum velocity observed v_m:

$$R' = \frac{1}{2}\frac{s}{v_m}\left[1+2I_1\left(\frac{R'}{\lambda}\right)K_1\left(\frac{R'}{\lambda}\right)\right]. \tag{28}$$

As seen in Figure 8, R' is very close to the radius at which the measured velocity profile meets that expected for a point vortex of the same strength. For unstratified vortices this gives $R' = \Gamma/2\pi v_m$.

4.2 *Baroclinic Vortex Interactions*

It is seen from Equation (25) or (26) that the two-layer flow is strongly baroclinic within a distance of the order of the Rossby radius. For point (potential) vortices the displacement velocity of the nth vortex due to N other vortices is in the x direction

$$x_{pn} = \sum_{n \neq k}^{k=N} \frac{s_k}{2}\left(\frac{y_n - y_k}{r_{nk}^2}\right)\left[1 + \Phi \frac{r_{nk}}{\lambda} K_1\left(\frac{r_{nk}}{\lambda}\right)\right] \tag{29}$$

and similar in the y direction, where $(r_{nk})^2 = (x_n - x_k)^2 + (y_n - y_k)^2$. The factor $\Phi = -1$ when the nth and kth vortices are in opposite layers and $\Phi = 1$ when in the same layer. The interaction of baroclinic pairs is determined by $K_1(r/\lambda)$ and is, therefore, of short range. According to Equation (29), two baroclinic vortices in the same layer propel each other more rapidly than barotropic vortices when the distance between them decreases. On the contrary, if in opposite layers their translational or orbital velocities go to zero as the separation goes to zero; there is vanishing coupling across the layers. Integral invariants for these vortices have been derived by Hogg & Stommel (1985a) and by Young (1985).

A property of baroclinic vortex pairs, first pointed out by Hogg & Stommel (1985a), is the ability of certain pairs to transport heat or other scalar quantities. This is because a vortex of negative sign in the top layer of a two-layer fluid will push the interface downward into the cooler (denser) layer and a vortex of positive sign in the lower layer will pull the interface downward; the hot fluid volumes corresponding to the interface displacements will, therefore, be carried by the vortex pair. For this reason this pair was named by Hogg & Stommel a "hot heton." When the signs are reversed the pair is a "cold heton." Displacement of the vortex pair occurs by self-propulsion when the vortices are not aligned. This is generally the case in a system containing more than one pair because splitting of a vortex pair by other pairs takes place. This then leads to heat transport and a decay of potential energy of the system. Hogg & Stommel calculated the amount of heat transported by discrete vortex hetons and found it to be inversely related with r/λ.

The interaction of heton pairs shows some interesting trajectories and exchanges of partners. Of particular interest is the interaction between two hot hetons. For a separation d of the heton pair larger than the Rossby radius the hetons repel each other, and the hetons will cohere when separated by a distance about equal to or less than the Rossby radius. This property has been used by Hogg & Stommel (1985b) to simulate the spreading of fronts by the disintegration of a cloud of hetons simulating a pool of warm water. The cloud of radius r_0 consisted of a lattice of 37 hot hetons separated by distance d with $r_0 = 3d$. For a heton separation larger or equal to Λ ($\Lambda/d \leq 1$) the cloud expands in time at a nearly constant rate. In this case the cloud breaks down into clumps of vortices with the number of clumps or modes proportional to d/Λ or r_0/Λ. At $r_0/\Lambda \approx 3$ the cloud explodes to form only two clumps. The marginal value for the $m = 2$ instability is $r_0/\Lambda = 1.28$, which is identical to the baroclinic instability criterion for a baroclinic vortex of radius R of Pedlosky (1985) and Helfrich & Send (1988), in good agreement with experiments by Saunders (1973) and Griffiths & Linden (1982). The linear rate of increase in mean radius also agrees with experiments on the spread of a patch of baroclinic turbulence by Griffiths & Hopfinger (1984). The predicted amplitude is, however, a factor of 5 too large. The discrepancy between prediction by the discrete vortex model and experiments is not surprising. For one thing, cold and hot vortex pairs coexist in the experiments and the finite core size in the experiments modifies the baroclinic contribution to the velocity field by $2I_1(R/\lambda)/(R/\lambda)$. This has an important effect on the selfpropulsion velocity when R/λ is not small (Polvani 1991). Furthermore, merging of like-sign vortices in the same layer, or alignment when in opposite layers, can occur and modify the dynamics. Nevertheless, it is clear that point vortex models do provide valuable insight.

The experiments by Griffiths & Hopfinger (1986) on heton pair interactions revealed some of the differences between finite core size hetons and discrete vortex hetons. Qualitatively, the interaction trajectories are similar at the start but effects due to finite core size rapidly cause deviations from the ideal trajectories. Dissymmetries in Ekman dissipation (bottom vortices spin down much more rapidly than the vortices in the top layer when there is a free surface) cause asymmetries in the configurations. The displacement velocities are generally smaller than expected from discrete vortex interactions. Vortex merging was also observed. Numerical calculations of the interaction between finite-core hetons and antihetons would help in understanding the different origins of the phenomena observed in the experiments.

4.3 *Merging and Alignment Conditions*

Density stratification can modify merging conditions of vortices through the effects of the baroclinic component of the velocity field and through the direct influence of buoyancy forces. Buoyancy forces oppose the increase in potential energy of the flow which must occur in the absence of sufficient dissipation, as vortices of the same sign and same density level approach or merge. A strong baroclinic component could favor merging.

In the experiments of GH, described in Section 4.1, it was observed that two vortices of like signs and intensities generated in the top layer merged whenever the distance between sources or sinks was less than a critical distance, but did not merge from larger distances. Figure 9 shows photographs of an experiment by GH with two anticyclones of $\Lambda/R' = 2.5$ and $d/R' = 4.5$, where d is the initial separation between vortices. The two vortices are observed to merge after $t \approx 5T_\Omega$ (second frame in Figure 9) or $ts/2\pi d^2 \approx 0.34$. Qualitatively, the appearance of the interacting baroclinic vortices and the merging process are similar to that observed for barotropic vortices. We note the formation of cusps and detrainment of vorticity via two spiral arms. The boundary of critical separation d_c/R' between stable and merging pairs as a function of Λ/R' obtained in the experiments by GH is shown as the thick solid line in Figure 10. Beginning at $\Lambda/R' = 0$, where the critical separation is $d_c/R' \approx 3.3$ (the numerical value depends on the definition of R), the stability boundary falls to smaller values of the dimensionless separation with increasing Rossby radius to reach a minimum of $d_c/R \approx 2.7$ at $\Lambda/R' \approx 1$ (typical of ocean eddies). For Rossby

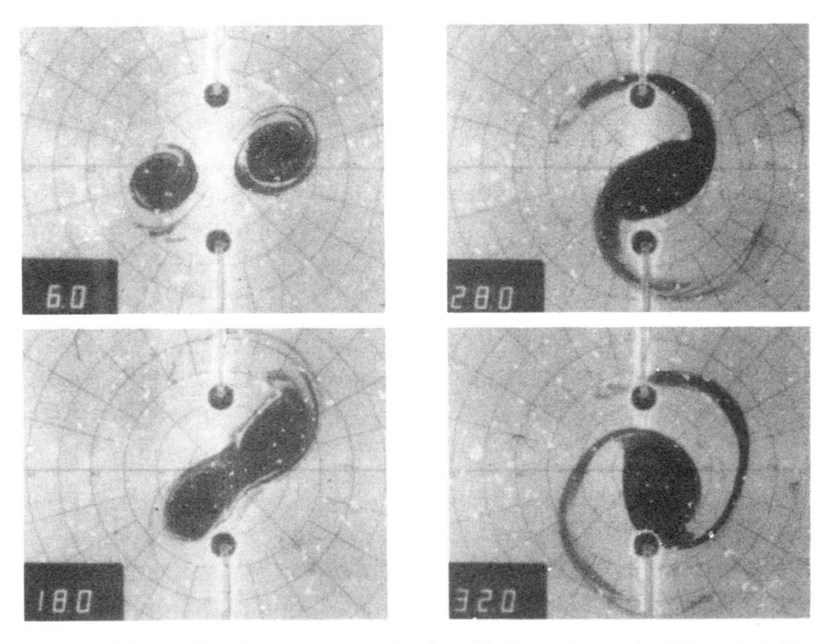

Figure 9 Photographs of two merging anticyclones in the top layer: $d = 18$ cm, $\Lambda = 10$ cm, $R' = 4.0$ cm, $s = 7.4$ cm^2 s^{-1}. The numbers indicate the elapsed orbital time in rotation periods. The orbital time is $44T_\Omega$.

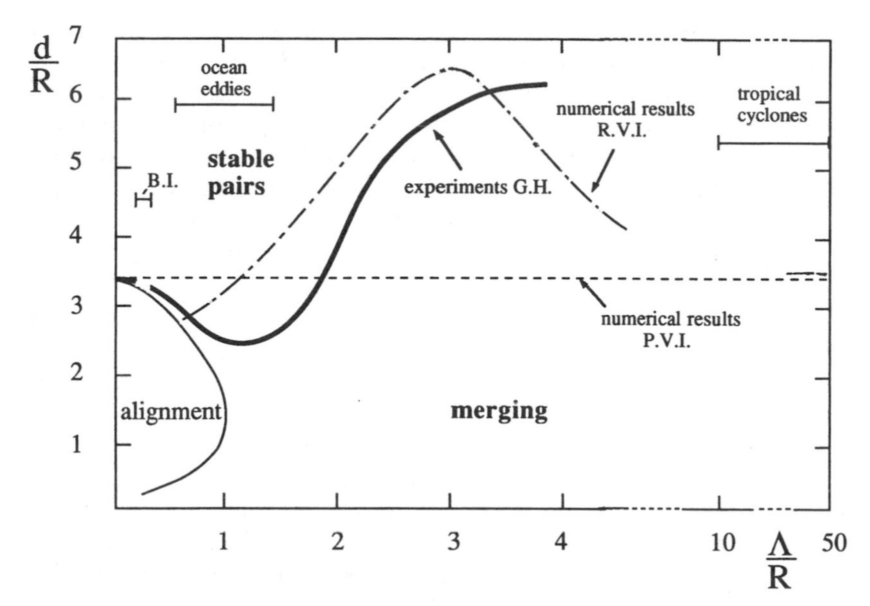

Figure 10 Merging and alignment boundaries as a function of Λ/R for a pair of identical baroclinic vortices. The thick solid line is the experimental curve of Griffiths & Hopfinger (1987) with $R = R'$. The horizontal dashed line is the numerical result for potential vorticity initialization PVI (Polvani et al 1989 and Verron et al 1990). The point-dash line represents the numerical results of Verron et al for RVI. The fine solid line is the alignment boundary (Polvani 1991). B.I. stands for baroclinic instability, which is likely to occur in the experiments in this range of Λ/R'.

radii greater than the core radius, the critical distance increases rapidly with Λ/R' and reaches $d_c/R' \approx 6.3 \pm 0.3$, corresponding to $d_c/\Lambda \approx 1.6$, at $\Lambda/R' \approx 4$. Thus, identical baroclinic vortex pairs can merge over much larger distances than barotropic vortex pairs. In the experiments it was not possible to reach values of $\Lambda/R \sim 10$ or more nor to determine the maximum value of d_c/R'. We only know that when $\Lambda/R' \to \infty$ and the two layers uncouple, the barotropic value must again be obtained. There are observations of tropical cyclones, where $\Lambda/R \approx 10$–50, which attract each other over distances of about 10 times their size (Brand 1970). At values of $\Lambda/R' < 0.4$ the curve in Figure 10 is discontinued because there are indications that baroclinic instability occurs on the eddy scale for these small values of Λ/R' (see Section 4.4). Merging occurs generally in a time t_c short compared with the orbital time $2\pi d^2/s$. Dimensionless merging times $t_c/(2\pi d^2/s)$ are generally 0.1 except when $\Lambda/R' < 1$ where $t_c/(2\pi d^2/s)$ can be as large as 2.3.

At present we have no good physical explanation for the dependence of merging conditions on Λ/R. The increased stability of vortex pairs when $\Lambda/R' \approx 1$ could be explained by less mutual straining because, when Λ/R' is small, the azimuthal velocity outside the core decreases more rapidly due to the baroclinic effect (see Figure 8). The origin of the long-range attraction for large Rossby radii could, at a first thought, also be attributed to the baroclinic effect on the velocity field of the vortex because for values of $\Lambda/R' \geq 3$, the velocity at $r/R' \approx 6$ differs from the barotropic value of a vortex of the same strength (Figure 8). Along the same line of thought heton repulsion and attraction could be invoked. This requires, however, the existence of an appreciable baroclinic vorticity component (vertical shear across the interface) in order to split the vortex pair (to offset the alignment of the top and bottom vortices). If the vortices are governed by the geostrophic model (24), splitting would not be possible.

Polvani et al (1989) have integrated numerically the quasi-geostrophic potential vorticity equations (23) using a contour dynamics algorithm. Two corotating V-state vortices of uniform potential vorticity patches were initiated in the upper layer in a way similar to the model with zero anomalous potential vorticity in the lower layer. Surprisingly, the merging conditions in the model, when the fluid layers are of equal depth, are insensitive to stratification if the layers are of identical depth with the critical separation $d_c/R \approx 3.3$. This merging boundary of Polvani et al is shown by the horizontal dashed line in Figure 10. In order to verify the validity of contour dynamics calculations in a stratified fluid, Verron et al (1990) have integrated the same equations by a finite difference code. The dissipation was modeled by a biharmonic term such that the energy decrease remained negligible during the time of integration ($Q_1 \cong Q_0$). The vortices initiated were of the form $Q_1 = \omega_0[1-0.5\ (\tanh\ X+1)]$ and $Q_2 = 0$, where $X = \alpha\ (r/R-1)$, with $\alpha = 4$. For this initialization, referred to as PVI (Potential Vorticity Initialization), the results are identical to those of the contour dynamics algorithm with no dependence of the critical separation on Λ when the layers are of equal depth. Verron et al (1990) have in addition considered an initial state of the form $Q_1 = \{\omega_0[1-0.5(\tanh X+1)]-\Lambda^{-2}\psi_{10}\}$ and $Q_2 = \Lambda^{-2}\psi_{10}$, referred to as "Relative Vorticity Initialization" (RVI). In this case the merging boundary, shown in Figure 10 by a point-dash line has a behavior closer to the experiments with a peak value of $d_c/R \approx 7$ and a minimum value lower than the barotropic critical state. The minimum and maximum occur, however, at lower values of Λ/R than in the experiment even when taking into account that R, in the simulation by Verron et al, does not correspond to the definition of R' ($R > R'$). Beyond the value $\Lambda/R = 3$ where d_c/R has a maximum, the critical separation falls back rapidly to the barotropic value. This sensitivity of the flow behavior to initial conditions has been further examined by Verron & Hopfinger (1991) who compared the differences in velocity profiles. The experimental velocity profiles shown in Figure 8 are closer to potential vorticity initialization than relative vorticity initialization, at least when $\Lambda/R \approx 1$ where velocity profiles were measured. As mentioned above, the velocity in the lower layer in the core region deviates, however, significantly from the geostrophic model (24). Further experiments on baroclinic vortex merging and in particular measurements of velocity profiles for larger values of Λ/R would be useful to clarify these discrepancies.

When the vortices of equal sign are not in the same layer we speak of vertical alignment. The question of alignment is as fundamental to stratified geostrophic turbulence as merging. Alignment is associated with the energy transfer from baroclinic to barotropic modes (Rhines 1979). In decaying stratified turbulence McWilliams (1989) found that vortices are mostly elongated in the vertical. This question has recently been considered in some detail by Polvani (1991) using a contour dynamics method and the same model as in Polvani et al (1989). Alignment is defined by the decrease in the intercentroid distance in a time of the order of the orbital period. Polvani considered circular initial states and V-states. The interesting point is that the V-states can be overlapping. Alignment takes place when stable V-states cease to be stable. The alignment boundary for circular vortices is close to but different from V-state vortices, the difference being related to an uncertainty about the stability of V-states in a certain parameter range of $d_c/R-\Lambda/R$ (see Polvani). The alignment boundary determined by Polvani for circular vortices is shown in Figure 10 by a fine solid line. This boundary does not intersect the d/R axis at zero, suggesting that baroclinic vortices of like sign are unstable to baroclinic, mode 1, disturbances of sufficiently large amplitude (Flierl 1988).

4.4 *Instability of Baroclinic Vortices*

We recall here only some aspects of the conditions under which stable vortices can exist. The linear stability problem for baroclinic vortices of equal depth has been investigated by Pedlosky (1985) and later generalized by Helfrich & Send (1988) to include layers of unequal depths and finite-amplitude evolution. The model considered by Pedlosky and by Helfrich & Send is based on the geostrophic potential vorticity equations (23), with Q_1 and Q_2 specified and for which normal mode solutions have been obtained. In this inviscid formulation, the dispersion relation relates the Froude number $\mathrm{F} = f^2R^2/g'H_1$, the layer depth ratio $\delta = H_1/H_2$, and the barotropic potential vorticities $Q_\mathrm{B} = Q_1+\delta^{-1}Q_2$. The stability analysis indicates that the growth rate of the perturbations depends on the velocity difference across the interface. The $m = 2$ mode is the first mode to become unstable as F increases. For the purely baroclinic case $Q_B = 0$, the critical Froude number is $F_c \approx 2$ for $\delta = 1$. Decreasing δ (increasing H_2) increases F_c. The increase in F_c closely follows $(H_2/H_1)^{1/2}$ known from Phillip's (1954) baroclinic instability model. The critical parameter for instability can be approximated by

$$\gamma_c = F_c\left(\frac{H_1}{H_2}\right)^{1/2} \approx 2. \tag{30}$$

Higher order modes appear for larger values of γ. The condition $Q_B = 0$ is approximately verified in the *constant-volume* vortex experiment by Griffiths & Linden (1981), where the vortices in the two layers are of opposite signs. Linear theory is in good agreement with these experiments (see Helfrich & Send). The experimental data have, however, a large scatter which is probably due to the presence of Ekman dissipation which affects the stability conditions (Pedlosky 1970).

The important point of the theory is that when $Q_B = Q_1$ ($Q_2 = 0$) the vortices are stable. This implies that in the baroclinic vortex merging experiment of GH the vortices should remain stable even for all values of Λ/R' if these vortices obey the potential vorticity model $Q_1 = \omega_0$ and $Q_2 = 0$. The way in which the vortices were created (that is by injection of fluid or by suction) has some resemblance with the *constant-flux* vortices of Griffiths & Linden (1981). There too, the vortex produced by the interface displacement is of the same sign as the primary vortex created by a source. Instability was nevertheless observed for values of $\gamma_c = F_c(H_1/H_2)^{1/2} \sim 50$, giving a corresponding critical value of $(\Lambda/R')_c \sim 0.15$. When this value is used to examine the experiments of GH it is necessary to take into account that merging of vortices roughly doubles their size and, therefore, stable merging events should only be observed for vortices with $\Lambda/R' \geq 0.3$. In the experiments of GH, the lowest value of Λ/R' for which merging was observed (other than for the one-layer limit $\Lambda = 0$) was 0.38. At this value the associated merging time was very large (more than twice the orbital period) which might be reminiscent of some competition between merging and instability. If baroclinic instability indeed exists in the experiments, it would imply that the experimental vortices can not be entirely represented by the potential vorticity model (24).

Baroclinic instability can also occur on the tank scale in experiments with a stratified rotating fluid. In a tank with a free surface, conditions of solid-body rotation (zero motion in the rotating frame) can never be obtained when the fluid is stratified (Greenspan 1968). This is because diffusion across the curved surfaces of constant density drives a meridional circulation and by conservation of angular momentum this circulation causes a persistent slow azimuthal drift. In a two-layer fluid experiment this drift creates a shear across the interface and hence an inclination of surfaces of constant density with respect to surfaces of constant pressure. Baroclinic instability of this slow mean motion relative to the tank rotation is possible (Griffiths & Linden 1985) and was also observed in the experiments by GH and Griffiths & Hopfinger (1986). For the larger density differences the differential azimuthal motion in the experiments of GH was about 3% of the tank rotation; this flow became intermittently unstable to a mode 2 wave instability and this mode lasted for a few hours. Experiments could only be conducted when the flow was axisymmetric or returned to an axisymmetric state so that the tank rotation could be adjusted before the experiment by slightly changing the tank rotation. The lower layer in contact with the bottom adjusted rapidly to the new rotation rate.

The vortex flow in a linear density gradient is a good model of the isolated anticyclonic eddies or lenses observed in the Atlantic Ocean. According to Griffiths & Linden (1981), the conditions of stability of eddies in a constant density gradient are similar to the two-layer vortices and in particular to the constant flux surface eddies discussed above. Here, the important parameter is the Prandtl ratio Nh/fR, where Nh/f is the Rossby radius of deformation, with h a scale height (half height of the vortex), R the radius, and $N = [-(g/\rho)\, d\rho/dz]^{1/2}$. Griffiths & Linden observed an asymmetric, $m = 2$, instability when $Nh_c/fR_c \sim 0.2$ or $(fR_c/Nh_c)^2 \sim 50$. Hedstrom & Armi (1988) have performed a more detailed experimental study of the same vortex structure as Griffiths & Linden in a constant density gradient fluid. In contrast to Griffiths & Linden's observation, the vortices, according to Hedstrom & Armi, seem to remain stable, which they attribute to a more careful, nonperturbing, injection technique in their experiments. It is of interest to mention that the aspect ratio and the velocity profiles measured by Hedstrom & Armi inside and outside a stable vortex are in good agreement with Gill's (1981) solution for an inviscid intrusion.

A vortex in a constant density gradient develops a series of density steps above and below, the number and horizontal extent of which grow in time. The origin of these layers is a finite-amplitude manifestation of the double-diffusive McIntyre (1970) instability. The two diffusing quantities are the scalar field (salt) and angular momentum.

5. VORTICES IN THE PRESENCE OF MEAN SHEAR

Often, vortices are imbedded in a mean horizontal shear and a brief mention should be made of the present state of knowledge. To illustrate the essential phenomena, we consider only the simplest case of the barotropic two-dimensional free shear flows and in particular the plane mixing layer. Without rotation it is well known that the primary instability is two-dimensional and the vorticity is concentrated in a row of two-dimensional vortices. These vortices are then destabilized by three-dimensional motions. When background rotation is added, with the rotation vector $\mathbf{k}f$ ($\mathbf{k}$ is unit vector) being either parallel or antiparallel to the horizontal mean shear vorticity vector $\mathbf{k}\partial U/\partial y$, these two-dimensional vortices are further destabilized or stabilized depending on the sign of the ratio

$$S = -f/(\partial U/\partial y). \tag{31}$$

If $S > 0$—that is if the shear vorticity vector is parallel to the background vorticity vector (cyclonic)—rotation is stabilizing; if $S < 0$ (anticyclonic) it is destabilizing. Note that this instability is a three-dimensional instability analogous to the centrifugal instability of the vortices discussed in Section 2.3. The important parameter which emerges from linear stability theory (Yanase et al 1990) is in fact a combination of S and $(\partial U/\partial y - f)$, expressed by an analogous Richardson number or Bradshaw number B (Bradshaw 1968)

$$\mathrm{B} = -\frac{f(\partial U/\partial y - f)}{(\partial U/\partial y)^2} = S(1+S). \tag{32}$$

Linear stability theory shows that modes with wave vector $\mathbf{k}_Z$ grow very rapidly when $\mathrm{B} < 0$, that is when $-1 < S < 0$. Maximum destabilization occurs for $S = 1/2$ or $\mathrm{B} = -1/4$. If $S \leq -1$ rotation is again stabilizing and the motions are two-dimensional.

These linear stability results give a good approximation for a turbulent mixing layer with rotation in a way analogous to what is observed for a stratified mixing layer. The rotating mixing layer experiments by Bidokhti & Tritton (1992) show destabilization for $-1 < S < 0$ and restabilization when $S < -1$. The crossover from destabilization to stabilization is indeed at $S \approx -1$. In these experiments S is expressed by $S_\omega = -f/(\partial U/\partial y)_{\max} = -f\delta_\omega/(U_2 - U_1)$, where U_2 and U_1 are the free stream velocities at y_2 and y_1 with $y_2 > y_1$. The Reynolds number is $\mathrm{Re} = \delta_\omega(U_2 - U_1)/\nu \approx 5 \times 10^3$. An open question is the growth of the mixing layer stabilized by rotation. Experiments seem to indicate that vortex merging and hence growth is stopped—a result which is hardly plausible since merging is a two-dimensional process and should not be affected by rotation.

Lesieur et al (1991) performed three-dimensional numerical simulations of a temporally developing mixing layer using a $48 \times 48 \times 24$ mesh grid

(Re $\sim 10^2$). The behavior of destabilization and restabilization is similar to what is observed in experiments. A measure of stabilization and destabilization is the spanwise velocity variance $\langle w'^2 \rangle$ compared with its value without rotation. The results of Lesieur et al are presented in terms of a Rossby number $\text{Ro} = 1/|S_\omega|$. Destabilization is observed for *anticyclonic* shear if $\infty > \text{Ro}_i > 1$, where Ro_i is the initial Rossby number. As, in time, the shear layer grows, the Rossby number decreases and rotation restabilizes the flow with respect to three-dimensional perturbations when $\text{Ro} < 1$.

The giant, long-lived eddies observed in the atmospheres of the major planets are also imbedded in a mean or zonal shear of the same sign as the vorticity in the eddy. In addition, there is a strong β-effect, possibly influenced by stratification (Dowling & Ingersol 1988). A large number of models have been proposed to explain the origin of these giant vortices, their long lifetime and their structure. A clear exposition of the different models which have been proposed, can be found in Read (1992). One of these models is the shallow water model and it is worth mentioning the clever experiments by Nezlin & Snezhkin (1991) and collaborators. These experiments demonstrate the asymmetry between large-scale anticyclonic and cyclonic eddies in a shallow water system where the dispersion by the β-effect for anticyclones is opposed by the nonlinear effect due to finite displacement of the interface. The shallow water model also permits one to simulate spiral vortices (Nezlin & Snezhkin 1991).

6. VORTICITY CONCENTRATION

6.1 *Observation of Vorticity Concentration*

Forcing of a fluid with background vorticity f causes vortex lines to contract or stretch and hence leads to the formation of vortices. The forcing can have very different origins: topographic effects, convection (local or distributed), or turbulent boundary layer forcing. Weak forcing gives rise to quasi-geostrophic vortices, whereas strong forcing can originate intense vortex formation. The formation of intense (tornado-like), cyclonic vortices, by distributed turbulent boundary forcing was demonstrated by Hopfinger et al (1982). An example of these vortices, visualized by small air bubbles which concentrate in the low pressure core, is shown in Figure 11. These vortices are only approximately aligned with the rotation axis because the Rossby number on the vortex scale is of order 10 and their lifetime is ten to twenty rotation periods. In the experiments of Hopfinger et al three-dimensional turbulence was produced in a rotating tank by an oscillating grid with oscillation frequency $n > f$. In oscillating-grid turbulence the rms turbulent velocity u and the turbulence scale l vary

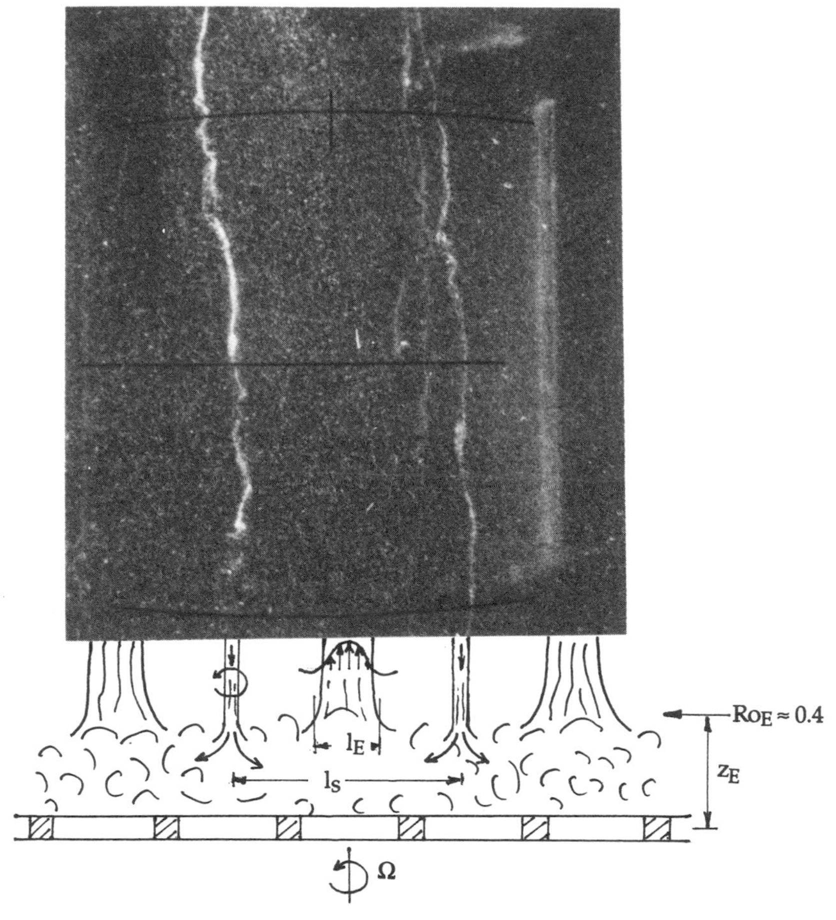

Figure 11 Image of concentrated vortices in the experiment of Hopfinger et al (1982) visualized by small air bubbles in the fluid which concentrate in the low pressure intense vortex cores. The sketch indicates the possible mechanism of vorticity concentration.

with distance z, measured from the grid, like $u \sim Kz^{-1}$ and $l \sim z$, where K is the grid action. A three-dimensional turbulence layer exists as long as the turbulence frequency $u/l > f$; otherwise turbulence gives way to inertial waves. The thickness z_E of the turbulent layer (an equivalent Ekman layer) determined by $u/fl \sim O(1)$ is then $z_E \sim (K/f)^{1/2}$ and the turbulent velocity at this level is $u_E \sim (Kf)^{1/2}$. Maxworthy (1992) gives an energy argument for $u/fl \sim O(1)$ where the energy of the swirl velocity acquired by a particle due to the Coriolis force is of the same order as the turbulent eddy kinetic energy. In the experiments of Hopfinger et al, the transition from three-dimensional turbulence to a rotation-dominated state occurs at a turbulent Rossby number $Ro_E = u_E/fl_E \approx 0.4$. The mean vortex spacing l_S scales on $l_E \sim z_E$ and, consequently, the number of vortices N per unit area scales as $N \sim f/K$. It is of interest to recall that these vortices support helical waves and are also subject to vortex breakdown.

In convection experiments by Boubnov & Golistyn (1986) and Fernando et al (1991) the same $N \sim f$ behavior was observed in the irregular vortex, high Rayleigh number, regime. In convection these irregular vortices are also concentrated vortices, though they rarely span the whole fluid layer depth (Fernando et al 1991). The Rayleigh number in these convection experiments is a flux Rayleigh number $Ra_q = BH^4/\kappa^2\nu$ and the Taylor number $Ta = f^2H^4/\nu^2 = 1/E^2$, where E is the Ekman number, κ the thermal diffusivity, ν the kinematic viscosity, and B the buoyancy flux per unit area $B = g\alpha q/\rho c_p$, with q the heat flux at the boundary per unit area, α the coefficient of thermal expansion, and c_p the specific heat. The critical Rayleigh number Ra_c for onset of convection is (Chandrasekhar 1961) a function of Ta and is approximated by $Ra_c \approx 8.7\ Ta^{2/3}$ for large Ta. A regular vortex pattern exists in the range $Ra_c < Ra_q < 20\ Ra_c$ with the vortex spacing $l_S \sim (H/f)^{1/3}$. For $Ra_q > (25–100)Ra_c$ the vortex pattern is irregular and $l_S \sim f^{-1/2}$ with a weak dependence on Ra_q. Linear theory does not, therefore, give the correct spacing and evolution of irregular vortices but it, nonetheless, provides some useful information.

6.2 *Models of Vortex Formation*

A physical mechanism by which vorticity can be concentrated was proposed by Maxworthy et al (1985). At the depth z_E, fingers form on the scale l_E and propagate away from the mixed region at the velocity u_E. By continuity, in between these fingers there is a down-flow with strong vorticity intensification as mass from the fingers converges in these down-flow regions. In heat convection these fingers or columns are driven by buoyancy as hot fluid rises away from the heated boundary and the down-flow in the concentrated vortices is cold fluid. In the oscillating-grid experiment where the fluid is of uniform density, it was conjectured that the columns are the envelope of inertial waves which feed energy from the mixed region throughout the fluid layer. It is, however, difficult to account for the mass flux observed in experiments (Mory 1988). Of interest is the stability model of Mory & Capéran (1987). By analogy with thermal convection in a rotating fluid, Mory & Capéran have proposed a marginal stability model of the turbulent rotating fluid layer where the gradient of the turbulent kinetic energy along the rotation axis $-\partial\langle w^2\rangle/\partial z$ is analogous to the mean temperature in thermal convection. The solution to the linear perturbation equations of this state, where the turbulent kinetic energy equation replaces the thermal diffusion equation, leads to a critical Reynolds number (the equivalent of the Rayleigh number) as a function of Ta. This model would indicate that a convective motion, as sketched in Figure 11, is driven by the gradient of the vertical component of the kinetic energy and that a preferred horizontal scale (scale l_S) of the motion is imposed. The existing experiments support the existence of a vertical gradient of $\langle w^2 \rangle$ but cannot be used to verify the theory. It would be necessary to construct experiments where the marginal conditions can actually be established.

7. CONCLUDING REMARKS

More than in any other area of research, the interplay of theory, numerical simulations, and experiments in studying vortices in rotating fluids has been very strong and fruitful. This is because in two-dimensional flows numerical simulations can use very fine mesh grids and can, therefore, capture the important processes. Contour dynamics/contour surgery methods are also very powerful tools in the study of the dynamics of two-dimensional vorticity patches.

Much is still incomplete or poorly understood. Two-dimensional barotropic vortex stability leading to dipole, tripole (and possibly quadrupole) formations, including topographic effects, needs refinements. In a stratified fluid, vortex stability and merger conditions depend crucially on the vortex structure and the vertical variability (changes across the interface for example) and models which aim at explaining the discrepancies between experiments and numerical simulations, concerning for instance merger conditions, will probably have to include the vortex structure. The dynamics of vortices in rotating shear flows or zonal flows raises many questions and should attract continued interest. Vorticity concentration in rotating turbulence and the formation of three-dimensional concentrated tornado-like vortices remain important problems to which we have only incomplete answers. The stability of these vortices brings in the concept of vortex breakdown which is one of the most difficult unsolved problems of vortex dynamics. The transport of a scalar quantity by vortices or eddies was mentioned only in the context of hetons. An important transport mechanism is provided by vortex interaction and merging and by Ekman pumping. There are statistical results on particle pair dispersion in two-dimensional or geostrophic turbulence but as yet little is known about the mechanisms and the role of coherent eddies.

Literature Cited

Adem, J. 1956. A series solution for the barotropic vorticity equation and its application in the study of atmospheric vortices. *Tellus* 8: 364–72

Bidokhti, A. A., Tritton, D. J. 1992. The structure of a turbulent free shear layer in a rotating fluid. *J. Fluid Mech.* 241: 469–502

Bjerknes, J., Holmboe, J. 1944. On the theory of cyclones. *J. Meteorol.* 1: 1–22

Boubnov, B. M., Golitsyn, G. S. 1986. Experimental study of convective structures in rotating fluids. *J. Fluid Mech.* 167: 503–31

Bradshaw, P. 1968. The analogy between streamline curvature and buoyancy in turbulent shear flows. *J. Fluid Mech.* 36: 177–91

Brand, S. 1970. Interaction of binary tropical cyclones of the western North Pacific Ocean. *J. Appl. Meteorol.* 9: 433–41

Carnevale, G. F., Cavazza, P., Orlandi, P., Purini, R. 1991a. An explanation for anomalous vortex merger in rotating-tank experiments. *Phys. Fluids* A3: 1411–15

Carnevale, G. F., Kloosterziel, R. C., van Heijst, G. J. F. 1991b. Propagation of barotropic vortices over topography in a rotating tank. *J. Fluid Mech.* 233: 119–39

Carton, X. J., Flierl, G. R., Polvani, L. M. 1989. The generation of tripoles from unstable axisymmetric vortex structures. *Europhys. Lett.* 9: 339–44

Carton, X. J., McWilliams, J. C. 1989. Barotropic and baroclinic instabilities of axisymmetric vortices in a quasi-geostrophic model. In *Mesoscale/Synoptic Coherent Structures in Geophysical Turbulence*, ed. J. C. J. Nihoul, B. M. Jamart, pp. 225–44. Amsterdam: Elsevier

Chandrasekhar, S. 1961. *Hydrodynamic and Hydromagnetic Stability*. Oxford: Clarendon

Cushman-Roisin, B., Heil, W. H., Nof, D. 1985. Oscillations and rotations of elliptical warm-core vortices. *J. Geophys. Res.* 90: 11,756–64

Dowling, T. E., Ingersol, A. P. 1988. Potential vorticity and layer thickness variations in a flow around Jupiter's Great Red Spot and White Oval BC. *J. Atmos. Sci.* 45: 1380–96

Dritschel, D. G. 1985. The stability and energetics of co-rotating uniform vortices. *J. Fluid Mech.* 157: 95–134

Dritschel, D. G., Legras, B. 1991. The elliptical model of two-dimensional vortex dynamics. Part II: Disturbance equations. *Phys. Fluids* A3: 855–69

Fernando, H. J. S., Chen, R. R., Boyer, D. L. 1991. Effect of rotation on convective turbulence. *J. Fluid Mech.* 228: 513–47

Firing, E., Beardsley, R. C. 1976. The behavior of a barotropic eddy on a β-plane. *J. Phys. Oceanogr.* 6: 57–65

Flierl, G. R. 1977. The application of linear quasi-geostrophic dynamics to Gulf Stream rings. *J. Phys. Oceanogr.* 7: 365–79

Flierl, G. R. 1988. On the instability of geostrophic vortices. *J. Fluid Mech.* 197: 349–88

Gent, P. R., McWilliams, J. C. 1986. The instability of circular vortices. *Geophys. Astrophys. Fluid Dyn.* 35: 209–33

Gill, A. E. 1981. Homogeneous intrusions in a rotating stratified fluid. *J. Fluid Mech.* 103: 275–95

Greenspan, H. P. 1968. *The Theory of Rotating Fluids*. Cambridge: Cambridge Univ. Press

Griffiths, R. W., Hopfinger, E. J. 1984. The structure of mesoscale turbulence and horizontal spreading at ocean fronts. *Deep Sea Res.* 31: 245–69

Griffiths, R. W., Hopfinger, E. J. 1986. Experiments with baroclinic vortex pairs in a rotating fluid. *J. Fluid Mech.* 173: 501–18

Griffiths, R. W., Hopfinger, E. J. 1987. Coalescing of geostrophic vortices. *J. Fluid Mech.* 178: 73–97

Griffiths, R. W., Linden, P. F. 1981. The stability of vortices in a rotating, stratified fluid. *J. Fluid Mech.* 105: 283–316

Griffiths, R. W., Linden, P. F. 1982. Laboratory experiments on fronts. Part I: Density-driven boundary currents. *Geophys. Astrophys. Fluid Dyn.* 19: 159–87

Griffiths, R. W., Linden, P. F. 1985. Intermittent baroclinic instability and fluctuations in geophysical circulation. *Nature* 316: 801–3

Gryanik, V. M. 1983. Dynamics of singular geostrophic vortices in a two-level model of the atmosphere or ocean. *Izv. Akad. Nauk USSR, Atmos. Oceanic Phys.* 19: 171–79

Hedstrom, K., Armi, L. 1988. An experimental study of homogeneous lenses in a stratified rotating fluid. *J. Fluid Mech.* 191: 535–56

Helfrich, K. R., Send, U. 1988. Finite-amplitude evolution of two-layer geostrophic vortices. *J. Fluid Mech.* 197: 331–48

Hogg, N. G., Stommel, H. M. 1985a. The heton, an elementary interaction between discrete baroclinic geostrophic vortices and its implications concerning eddy heat flow. *Proc. R. Soc. London Ser. A* 397: 1–20

Hogg, N. G., Stommel, H. M. 1985b. Hetonic explosions: the breakup and spread of warm pools as explained by baroclinic point vortices. *J. Atmos. Sci.* 42: 1465–76

Holton, J. R. 1979. *An Introduction to Dynamic Meterology*. New York: Academic

Hopfinger, E. J., Browand, F. K., Gagne, Y. 1982. Turbulence and waves in a rotating tank. *J. Fluid Mech.* 125: 505–34

Joyce, G., Montgomery, D. 1973. Negative temperature states for the two-dimensional guiding-centre plasma. *J. Plasma Phys.* 10: 107–21

Kida, S. 1981. Motion of an elliptic vortex in a uniform shear flow. *J. Phys. Soc. Jpn.* 50: 3517–20

Kloosterziel, R. C. 1990a. On the large-time asymptotics of the diffusion equation on infinite domains. *J. Eng. Math.* 24: 213–36

Kloosterziel, R. C. 1990b. *Barotropic vortices in a rotating fluid*. PhD thesis. Univ. Utrecht, The Netherlands

Kloosterziel, R. C., van Heijst, G. J. F. 1991. An experimental study of unstable barotropic vortices in a rotating fluid. *J. Fluid Mech.* 223: 1–24

Kloosterziel, R. C., van Heijst, G. J. F. 1992. The evolution of stable barotropic vortices in a rotating free-surface fluid. *J. Fluid Mech.* 239: 607–29

Lamb, H. 1932. *Hydrodynamics*. Cambridge: Cambridge Univ. Press. 6th ed.

Legras, B., Santangelo, P., Benzi, R. 1988. High-resolution numerical experiments for forced two-dimensional turbulence. *Europhys. Lett.* 5: 37–42

Leith, C. E. 1984. Minimum enstrophy vortices. *Phys. Fluids* 27: 1388–95

Lesieur, M., Yanase, S., Métais, O. 1991. Stabilization and destabilization effects of solid body rotation on quasi-two-dimensional shear layers. *Phys. Fluids* A3: 403–7

Maas, L. R. M. 1992. Nonlinear and free-surface effects on the spin-down of barotropic axisymmetric vortices. *J. Fluid Mech.* In press

Masuda, A., Marubayashi, K., Ishibashi, M. 1989. A laboratory experiment and numerical simulation of an isolated barotropic eddy in a basin with topographic β. *J. Fluid Mech.* 213: 641–55

Maxworthy, T. 1992. Convective and shear flow turbulence with rotation. In *Rotating Fluids in Geophysical and Industrial Applications, CISM Course Lect. Notes*, Vol. 329, ed. E. J. Hopfinger. Berlin: Springer-Verlag

Maxworthy, T., Hopfinger, E. J., Redekopp, L. 1985. Wave motions on vortex cores. *J. Fluid Mech.* 151: 141–65

McIntyre, M. I. 1970. Diffusive destabilization of the baroclinic circular vortex. *Geophys. Fluid Dyn.* 1: 19–57

McWilliams, J. C. 1984. The emergence of isolated coherent vortices in turbulent flow. *J. Fluid Mech.* 146: 21–43

McWilliams, J. C. 1989. Statistical properties of decaying geostrophic turbulence. *J. Fluid Mech.* 198: 199–230

McWilliams, J. C., Flierl, G. R. 1979. On the evolution of isolated, nonlinear vortices. *J. Phys. Oceanogr.* 9: 1155–82

McWilliams, J. C., Gent, P. R., Norton, N. J. 1986. The evolution of balanced, low-mode vortices on the β-plane. *J. Phys. Oceanogr.* 16: 838–55

Melander, M. V., McWilliams, J. C., Zabusky, N. J. 1987. Axisymmetrization and vorticity-gradient intensification of an isolated two-dimensional vortex through filamentation. *J. Fluid Mech.* 178: 137–59

Melander, M. V., Zabusky, N. J., McWilliams, J. C. 1988. Symmetric vortex merger in two dimensions: causes and conditions. *J. Fluid Mech.* 195: 303–40

Mied, R. P., Lindemann, G. R. 1979. The propagation and evolution of cyclonic Gulf Stream rings. *J. Phys. Oceanogr.* 9: 1183–1206

Montgomery, D., Matthaeus, W. H., Stribling, W. T., Martinez, D., Oughton, S. 1992. Relaxation in two dimensions and the "sinh-Poisson" equation. *Phys. Fluids* A4: 3–6

Mory, M. 1988. Coherent vortices in a turbulent and rotating fluid. *Fluid Dyn. Res.* 3: 299–304

Mory, M., Capéran, P. 1987. On the genesis of quasi-steady vortices in a rotating turbulent fluid. *J. Fluid Mech.* 185: 121–36

Nezlin, M. V., Snezhkin, E. N. 1991. *Rossby Vortices, Solitons and Spiral Structures*. New York: Springer-Verlag

O'Donnell, J., Linden, P. F. 1991. Free-surface effects on the spin-up of fluid in a rotating cylinder. *J. Fluid Mech.* 232: 439–53

Orlandi, P., van Heijst, G. J. F. 1992. Numerical simulation of tripolar vortices in 2D flow. *Fluid Dyn. Res.* 9: 179–206

Overman, E. A., Zabusky, N. J. 1982. Evolution and merger of isolated vortex structures. *Phys. Fluids* 25: 1297–1305

Pedlosky, J. 1970. Finite amplitude baroclinic waves. *J. Atmos. Sci.* 27: 15–30

Pedlosky, J. 1985. Instability of heton clouds. *J. Atmos. Sci.* 42: 1477–86

Pedlosky, J. 1987. *Geophysical Fluid Dynamics*. New York: Springer-Verlag. 2nd ed.

Phillips, N. A. 1954. Energy transformations and meridional circulations associated with simple baroclinic waves in a two level quasi-geostrophic model. *Tellus* 6: 273–86

Polvani, L. M. 1991. Two-layer geostrophic

vortex dynamics. Part II. Alignment and two-layer V-states. *J. Fluid Mech.* 225: 241–70

Polvani, L. M., Zabusky, N. J., Flierl, G. R. 1989. The two-layer geostrophic vortex dynamics. Part I: Upper layer V-states and merger. *J. Fluid Mech.* 205: 215–42

Read, P. L. 1992. Long-lived eddies in the atmospheres of the major planets. In *Rotating Fluids in Geophysical and Industrial Applications*, *CISM Course Lect. Notes*, Vol. 329, ed. E. J. Hopfinger. Springer-Verlag

Rhines, P. B. 1979. Geostrophic turbulence. *Annu. Rev. Fluid Mech.* 11: 401–41

Robert, R., Sommeria, J. 1991. Statistical equilibrium states for two-dimensional flows. *J. Fluid Mech.* 229: 291–310

Rossby, C. G. 1948. On displacements and intensity changes of atmospheric vortices. *J. Mar. Res.* 7: 175–87

Saffman, P. G., Baker, G. R. 1979. Vortex interactions. *Annu. Rev. Fluid Mech.* 11: 95–122

Saunders, P. M. 1973. The instability of a baroclinic vortex. *J. Phys. Oceanogr.* 3: 61–65

Simpson, J. J., Dickey, T. D., Koblinsky, C. J. 1984. An offshore eddy in the California Current System—I: Interior dynamics. *Prog. Oceanogr.* 13: 5–49

Stern, M. E. 1975. Minimal properties of planetary eddies. *J. Mar. Res.* 33: 1–13

Tojo, S. 1953. The dynamics of a vortex embedded in a constant zonal current. *J. Meteorol.* 10: 175–78

van Heijst, G. J. F., Kloosterziel, R. C. 1989. Tripolar vortices in a rotating fluid. *Nature* 338: 569–71

van Heijst, G. J. F., Kloosterziel, R. C., Williams, C. W. M. 1991. Laboratory experiments on the tripolar vortex in a rotating fluid. *J. Fluid Mech.* 225: 301–31

Verron, J., Hopfinger, E. J., McWilliams, J. C. 1990. Sensitivity to initial conditions in the merging of two-layer baroclinic vortices. *Phys. Fluids* A2: 886–89

Verron, J., Hopfinger, E. J. 1991. The enigmatic merger conditions of two-layer baroclinic vortices. *C. R. Acad. Sci. Paris* t. 313, Série II: 737–42

Wedemeyer, E. H. 1964. The unsteady flow within a spinning cylinder. *J. Fluid Mech.* 20: 383–99

Yanase, S., Riley, J. J., Lesieur, M., Métais, O. 1990. The effect of rotation on transition in a two-dimensional shear layer. *Bull. Am. Phys. Soc.* 35: 2277 (Abstr.)

Young, W. R. 1985. Some interactions between small numbers of baroclinic, geostrophic vortices. *Geophys. Astrophys. Fluid Dyn.* 33: 35–62

Zabusky, N. J., Hughes, M. H., Roberts, K. V. 1979. Contour dynamics for the Euler equations in two dimensions. *J. Comput. Phys.* 30: 96–106

Annu. Rev. Fluid Mech. 1993. 25: 291–323

BOUNDARY MIXING AND ARRESTED EKMAN LAYERS: ROTATING STRATIFIED FLOW NEAR A SLOPING BOUNDARY

Chris Garrett

Department of Physics and Astronomy, University of Victoria, Victoria, British Columbia V8W 3P6, Canada

Parker MacCready

Rosenstiel School of Marine and Atmospheric Science, Miami, Florida 33149-1098

Peter Rhines

School of Oceanography, University of Washington, Seattle, Washington 98195

KEY WORDS: ocean mixing, spin-up, spin-down

INTRODUCTION

Motivation

We are concerned here with the behavior of a rotating, stratified fluid near a sloping rigid boundary, with boundary conditions of zero normal buoyancy flux and no slip. Although this is an interesting fluid dynamical problem in its own right, we have been motivated by two major, and at first sight disparate, topics in physical oceanography. The first, known as "boundary mixing", is concerned with how turbulent mixing at the sloping sides of the density-stratified ocean affects the stratification in the interior. The second topic involves the way in which the combination of strati-

0066–4189/93/0115–0291$02.00

fication and bottom slope may affect the communication into the interior of the no-slip boundary condition at the seafloor.

BOUNDARY MIXING Spatially-smoothed temperature, salinity, and other oceanic properties can be mapped with respect to what are loosely called isopycnal surfaces, along which the potential density does not change locally. As no buoyancy forces arise in moving along them, they are preferred pathways for the flow, even if inclined to the horizontal, and so are important reference surfaces. For those isopycnals that outcrop at the sea surface in subtropical regions it is possible that the temperature and salinity, away from the surface mixed layer, are indeed largely determined by advection and stirring along isopycnal surfaces, without any need for diapycnal (cross-isopycnal) mixing. For the abyssal ocean, however, the constant supply of cold bottom water from polar regions must create a slow upwelling (of about 4 m y^{-1} on average) that is somehow balanced by downward mixing of heat (Munk 1966). This requirement is certainly well documented for some isolated abyssal basins (Whitehead & Worthington 1982, Hogg et al 1982, Saunders 1987, Whitehead 1989) where inflow of cold water has been directly measured and yet the temperature field is assumed not to be changing (Figure 1).

The mixing rate required is typically a few times 10^{-4} m^2 s^{-1}, much larger than the $\mathcal{O}(10^{-5})$ m^2 s^{-1} or less found from direct measurements of turbulence in the top one kilometer or so of the ocean (Gregg 1987, 1989). Theoretical arguments about the depth dependence of the interior mixing rate will eventually be settled by direct measurements in the abyss. In the meantime, the potential mismatch has revived interest in the suggestion by Munk (1966), pursued by Armi (1978, 1979a,b), that the required mixing occurs not in the ocean interior, but rather in turbulent boundary layers above the sloping bottom.

It was suggested by Armi (1978, 1979a) and Ivey (1987a,b) that the

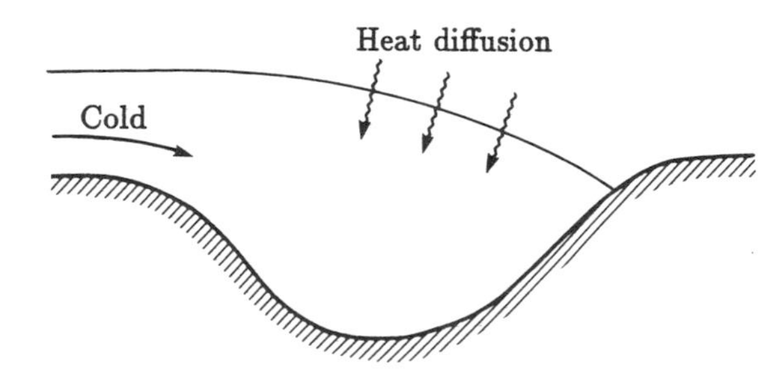

Figure 1 Cross-section of cold water entering a basin within which an isotherm maintains a fixed position due to a balance between advection and diffusion of heat.

effective vertical diffusivity, K_V, at a depth z in the ocean, is given by

$$K_V = K_{Vb} A_b(z, h)/A(z) \tag{1}$$

where K_{Vb} is the vertical diffusivity in the boundary layer of thickness h, $A_b(z, h)$ is the horizontal area occupied by the boundary layers at depth z, and $A(z)$ is the horizontal area of the ocean interior (Figure 2). Use of this formula with plausible values for h and K_{Vb} made an effective K_V of 10^{-4} $m^2 s^{-1}$ seem attainable, but Garrett (1979a,b) pointed out that (1) requires the boundary layer to be as stratified as the interior. This seemed unlikely in the presence of vigorous mixing unless there is rapid exchange between the boundary layer and the ocean interior, as in fact envisaged by Armi (1979a,b) on the basis of observations of detached mixed layers. Nonetheless, most density profiles to the seafloor (e.g. Armi & Millard 1976) showed a region of reduced stratification near the seafloor, so that a reduction factor to K_V from (1) did seem necessary.

Major new insight into the fluid dynamics of boundary mixing was provided by Phillips et al (1986). They showed that the distortion of isopycnals caused by mixing near a sloping boundary would give rise to buoyancy forces that would drive a secondary circulation (Figure 3), and that this would tend to drive dense water upslope and light water downslope, thus enhancing the upslope transport of density and perhaps making boundary mixing more effective than had been recognized.

Garrett (1990), however, pointed out that it is the vertical, rather than upslope, buoyancy flux that matters, and that the advective contribution to this is countergradient, with dense fluid slumping back down and light fluid rising. Thus boundary mixing appears to be less effective than implied even by reducing (1) to allow for reduced stratification in the boundary layer! On the other hand, the secondary circulation continually acts to restore the stratification on which turbulent mixing can act, without the

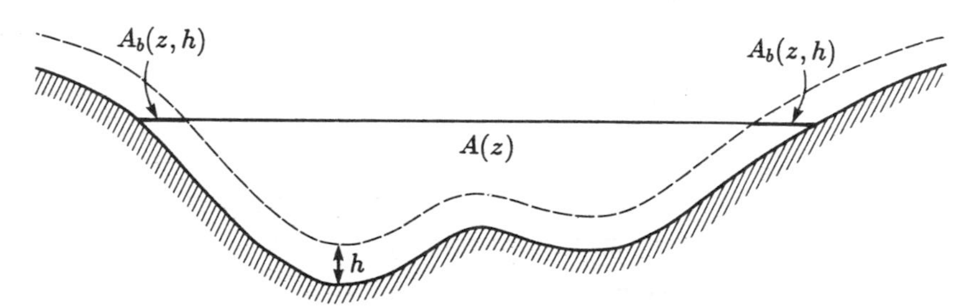

Figure 2 An ocean basin has a bottom boundary layer of thickness h. $A_b(z, h)$ is the horizontal area, at depth z, within the boundary layer; $A(z)$ is the interior area at depth z.

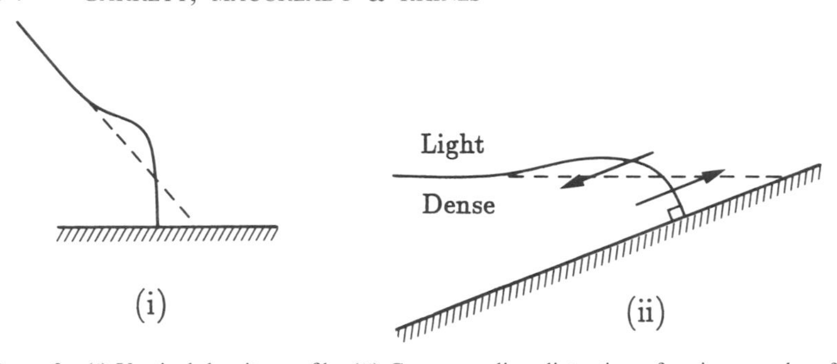

Figure 3 (*i*) Vertical density profile. (*ii*) Corresponding distortion of an isopycnal surface. The arrows indicate the secondary circulation driven by buoyancy forces.

need for mixed fluid to be ejected from the boundary layer and new, stratified, fluid entrained.

A quantitative, dynamical discussion of these two roles of the secondary circulation—opposing the vertical diffusive flux but restoring the stratification on which the mixing can act—will be presented later in this review. We turn next, though, to our other main motive for considering the behavior of a stratified, rotating fluid near a sloping boundary.

ARRESTED EKMAN LAYERS In a quasi-geostrophic flow in the ocean interior, the basic force balance is between a pressure gradient and the Coriolis force. Near the seafloor the current is reduced by friction so that the pressure gradient will be able to drive a cross-isobar flow. This is the basic physics of an Ekman layer. Lateral variations in the current cause variations of the flux in the Ekman layer and force flow into, or out of, the interior. Coriolis forces acting on this flow then reduce the initial currents in the interior, a process known as spin-down (Figure 4). The

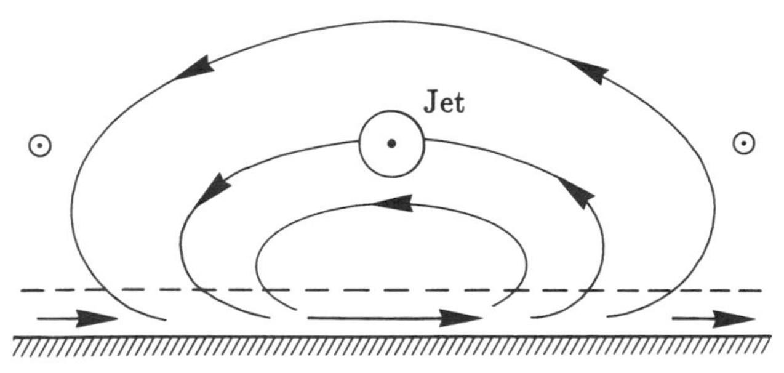

Figure 4 Schematic of the interior circulation driven by the convergence and divergence of the Ekman flux beneath a jet. The Coriolis force on the interior circulation decreases the strength of the jet.

vertical extent of the spun-down region is given by a factor f/N times the horizontal scale of the flow, where f is the Coriolis frequency and N is the buoyancy frequency (e.g. Pedlosky 1979).

When the interior flow is above a sloping bottom, however, this fast spin-down process may be disrupted by buoyancy forces (Figure 5). As the imbalance of pressure gradient and Coriolis force drives fluid cross-slope in the bottom Ekman layer, the advection of density gives rise to a buoyancy force which opposes further motion. If a final balance between pressure gradient, Coriolis force and cross-slope buoyancy force is achieved, there is no further Ekman flux, no fluid injected into, or sucked from, the interior, and no further spin-down of the interior flow. In such circumstances the interior flow sees a nearly free-slip boundary condition at the seafloor.

Rhines & MacCready (1989) drew attention to this important physics and estimated the "shut-down time" for a bottom slope $\tan\theta$ by assuming the cross-slope flow in the Ekman layer to be comparable with the interior along-slope geostrophic flow U. The cross-slope displacement after time t is then Ut, giving rise to a buoyancy perturbation of magnitude $N^2Ut\sin\theta$, where N is the interior buoyancy frequency. The cross-slope component $N^2Ut\sin^2\theta$ then balances the pressure gradient fU after a time

$$\tau_s = S^{-1}f^{-1}, \tag{2}$$

where $S = N^2\sin^2\theta/f^2$ is the Burger number based on the slope. As we shall review later, this is an underestimate of the true shut-down time as the average cross-slope flow in the Ekman layer is less than U and buoyancy forces act progressively as the Ekman layer moves cross-slope. Nonetheless, (2) gives the correct order of magnitude estimate of the time for a sloping seafloor to become "slippery". It may be much less than the

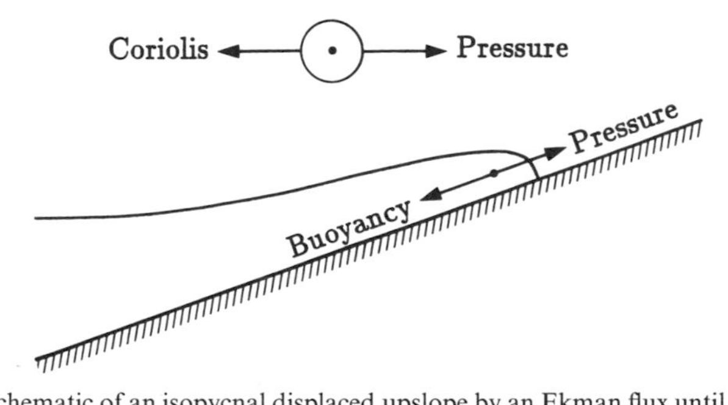

Figure 5 Schematic of an isopycnal displaced upslope by an Ekman flux until the buoyancy force balances the pressure gradient.

time required for Ekman divergence to spin-down the interior flow, with profound implications for ocean circulation. We return to this issue later but for the moment note that for a bottom slope of 0.01 and with $f = 10^{-4}\,s^{-1}$, then $S = 0.01$ if $N = 10^{-3}\,s^{-1}$ (representative of the abyssal ocean) or $S = 1$ for $N = 10^{-2}s^{-1}$ (on the continental shelf). Thus, allowing for smaller and larger slopes, we see that values of S greater than one as well as much less than one, and shut-down times from hours to years, are of oceanographic interest.

Difficulties

Boundary mixing is affected by rotation but does not depend on it. In a rotating reference frame, however, the Coriolis force on the cross-slope flow induced by buoyancy will tend to accelerate an along-slope flow and eventually be balanced by divergence of an along-slope stress. We will see this later in terms of the equations of motion, and show how the steady-state along-slope flow outside the boundary layer, in the ocean interior, becomes part of the solution rather than something which can be prescribed independently. We might therefore be concerned about the relevance of steady-state boundary mixing results in a situation where the along-slope interior flow is different from that required by the steady state solution. Conversely, the simple dynamical considerations involved in the arrest of an Ekman layer on a slope suggest that a steady state should be achievable for any along-slope interior flow. This is inconsistent with the steady state solution discussed above, possibly because the simple arrested Ekman layer arguments assume just advection of the density field and ignore its diffusion. In other words, the boundary mixing models ignore time-dependence and imply an along-slope flow that may not match the prescribed one, whereas the time-dependent Ekman layer models tend to ignore buoyancy diffusion and appear to predict a final steady state when it may not be possible.

Reconciliation

Both the boundary mixing and arrested Ekman layer theories deal with the behavior of a stratified, rotating flow near a sloping boundary. It should thus be possible to reconcile the two approaches within the context of a single conceptual model. Recent work has shown that this is largely the case. In brief, the answer is actually quite simple, at least for constant coefficients: The solution of the time-dependent problem adjusts to the steady-state "boundary mixing" solution near the slope. The implied along-slope flow in the ocean interior, if different from the imposed one, tends to diffuse into the interior according to a "slow diffusion" equation

(MacCready & Rhines 1991). For the turbulent case there may be more possibilities if the mixing coefficients adjust with time.

Outline

Although our oceanographic motivations have led us to consider the situation where mixing is largely confined near the boundary, we shall first review the steady-state and time-dependent solutions for constant coefficients of viscosity and diffusivity. Following that, we derive some general results for arbitrary profiles of the mixing coefficients, emphasizing the physics and introducing the key dimensionless parameters that govern the problem. Some results of numerical solutions will also be reviewed.

We then examine the relevance of the theories to the ocean, considering briefly the processes that might lead to enhanced mixing near slopes. We also summarize how the local solutions, which ignore cross-slope variations in properties, can be applied in an ocean with variable bottom slope or interior properties.

Many theoretical and observational problems remain; these will be discussed in the final section of the paper.

GOVERNING EQUATIONS

The basic problem to be considered has no variations in the cross-slope direction and so has governing equations

$$\frac{\partial U}{\partial t} - fV = \frac{\partial}{\partial z}\left(\nu \frac{\partial U}{\partial z}\right) \tag{3}$$

$$\frac{\partial V}{\partial t} + fU - \frac{1}{\rho_0}\frac{\partial \bar{p}}{\partial y} + B\sin\theta + \frac{\partial}{\partial z}\left(\nu \frac{\partial V}{\partial z}\right) \tag{4}$$

$$0 = -\frac{1}{\rho_0}\frac{\partial \bar{p}}{\partial z} + B\cos\theta - \frac{\partial}{\partial z}(\overline{w^2}) \tag{5}$$

$$\frac{\partial B}{\partial t} + VN^2\sin\theta = \frac{\partial}{\partial z}\left(\kappa \frac{\partial B}{\partial z}\right) \tag{6}$$

where the coordinates y, z are upslope and bottom-normal with respect to the plane bottom that is inclined at an angle θ to the horizontal (Figure 6). Here $U(z, t)$ and $V(z, t)$ are the mean flows alongslope and upslope respectively, $B(y, z, t) = -g(\bar{\rho} - \rho_0)/\rho_0$ is the mean buoyancy in terms of the mean density $\bar{\rho}(z, t)$ and a reference density ρ_0, and N^2 is the (constant) vertical buoyancy gradient in the interior of the fluid away from the slope.

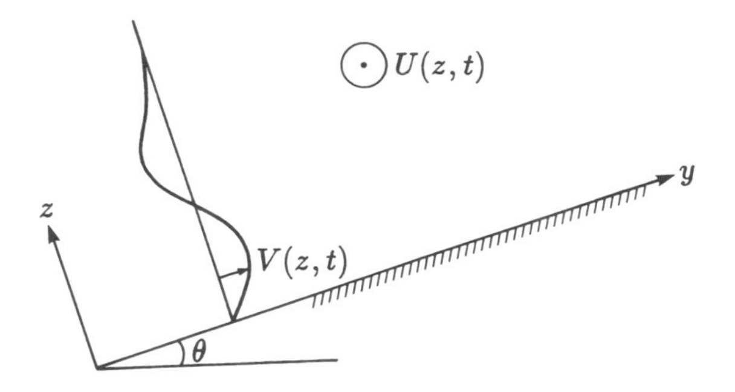

Figure 6 Definition sketch for the coordinate axes and current components above a sloping bottom at an angle θ to the horizontal.

(The mean is defined here as the average over the turbulent fluctuations, and so assumes a spectral gap between them and the slowly-varying mean flow.) For flat interior isopycnals we must have

$$\partial B/\partial z \to N^2 \cos\theta \qquad \text{as } z \to \infty \tag{7}$$

whereas

$$\partial B/\partial y = N^2 \sin\theta \qquad \text{for all } z. \tag{8}$$

The mean pressure $\bar{p}$ is taken with respect to the pressure in a fluid of density ρ_0 at rest. Eddy transports of x and y momentum normal to the slope are represented in terms of the same eddy viscosity $\nu(z, t)$; as this just represents the ratio of momentum flux to mean gradient we are not necessarily assuming a short mixing length, but we do simplify by taking the same coefficient in both x and y directions. The Reynolds stress component $\overline{w^2}$ has been retained but is independent of y. The eddy buoyancy transport normal to the slope is parameterized in terms of an eddy diffusivity κ, but again this is merely a ratio of eddy flux to mean gradient without requiring any assumption about the nature of the mixing. The Coriolis parameter f is twice the component of the Earth's rotation normal to the bottom.

This formulation omits any consideration at this stage of cross-slope variation or of the role of sloping interior isopycnals. It also ignores the response of the fluid to the Reynolds stresses and upslope buoyancy flux associated with even the inviscid reflection of internal waves from the slope (Wunsch 1971, Ou & Maas 1986), but the problem as formulated permits discussion of much of the key physics associated with mixing processes near the slope.

The governing equations above may be combined into a single equation

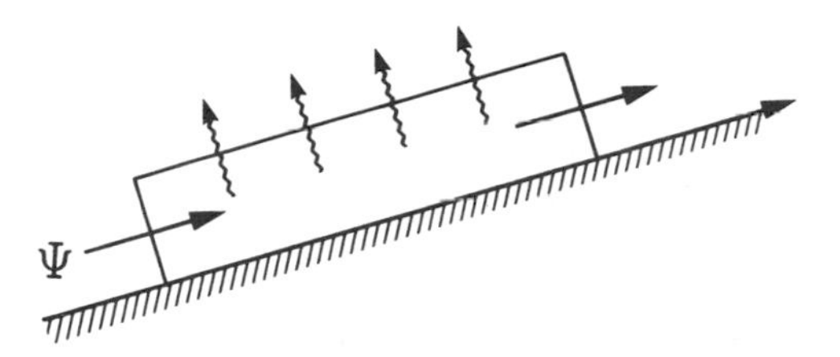

Figure 7 A steady state mass balance in the presence of upslope mass flux requires a diffusive flux through the top of a control volume.

for a single variable such as the along-slope flow (e.g. Garrett 1991). In the steady state case we note that it is first convenient to describe the upslope flow $V(z)$ in terms of a streamfunction $\Psi(z)$ by $V = d\Psi/dz$. Then (6) integrates to

$$\Psi N^2 \sin\theta = \kappa\, \partial B/\partial z \tag{9}$$

if we take $\Psi = 0$ on $z = 0$ where there is no bottom-normal buoyancy flux.

The physics of this is illustrated by considering the control volume in Figure 7: A volume flux Ψ leaves the upslope end of the box less dense than when it entered, and so must have lost mass by diffusion while in transit. Now $\partial B/\partial z \to N^2 \cos\theta$ (the interior stratification) as $z \to \infty$, so that $\Psi \to \kappa_\infty \cot\theta$, with κ_∞ the value of $\kappa(z)$ as $z \to \infty$ (Phillips et al 1986, Thorpe 1987).

In the steady state (3) may also be integrated once to give

$$\nu\, dU/dz = -f(\Psi - \kappa_\infty \cot\theta). \tag{10}$$

Physically, the Coriolis force on the upslope transport can only be balanced by an along-slope stress.

From (4) and (5), the steady state equation for the vorticity component parallel to the x-axis is

$$\frac{d^2}{dz^2}\left(\nu \frac{d^2\Psi}{dz^2}\right) + \left(\frac{f^2}{\nu} + \frac{N^2 \sin^2\theta}{\kappa}\right)\Psi = N^2 \sin\theta \cos\theta + f^2 \frac{\kappa_\infty \cot\theta}{\nu}. \tag{11}$$

Mixing tilts the isopycnals close to the boundary, causing a torque that drives upslope or downslope flow and is balanced by viscous forces.

CONSTANT MIXING COEFFICIENTS

Steady State

If ν, κ are independent of z the solution of (11) is (Weatherly & Martin 1978, Thorpe 1987)

$$\Psi = \kappa \cot \theta [1 - e^{-qz}(\cos qz + \sin qz)] \tag{12}$$

with

$$q^4 = \frac{1}{4}\left(\frac{f^2}{\nu^2} + \frac{N^2 \sin^2 \theta}{\nu\kappa}\right) = \frac{f^2}{4\nu^2}(1 + S\sigma), \tag{13}$$

where $S = N^2 \sin^2 \theta / f^2$ is the slope Burger number and $\sigma = \nu/\kappa$ is the Prandtl number. The effect of the boundary is thus confined to a distance, of order q^{-1}, that is less by a factor $(1 + S\sigma)^{-1/4}$ than the Ekman layer thickness above a flat bottom or for no interior stratification.

The net upslope flow $\kappa \cot \theta$ was originally discussed by Phillips (1970) and Wunsch (1970) for the nonrotating case; the requirement of no buoyancy flux through the slope causes the isopycnals to bend down to meet the slope at right angles (Figure 8), creating buoyancy forces that drive the upslope flow. The need for such a flow is clear from the control volume in Figure 7; the diffusive loss of mass out of the lid of the box can only be balanced by a divergence of the upslope mass transport. Phillips (1970) has shown that in a tilted tube the advective buoyancy flux in the boundary layers on the tilted top and tilted bottom can be much greater than the interior diffusive flux parallel to the axis of the tube.

The associated along-slope flow for the rotating case is given by

$$U = \frac{f\kappa}{q\nu} \cot \theta (1 - e^{-qz} \cos qz) \tag{14}$$

if we apply $U = 0$ at $z = 0$. We note that the along-slope flow outside the boundary layer is in a direction that might be described as "upwelling-favorable" in the Ekman sense, but its value $f\kappa(q\nu)^{-1} \cot \theta$ is part of the solution rather than being arbitrary.

Time Dependence

MacCready & Rhines (1991) have investigated the time-dependent response of the bottom boundary layer for constant ν, κ but with arbitrary

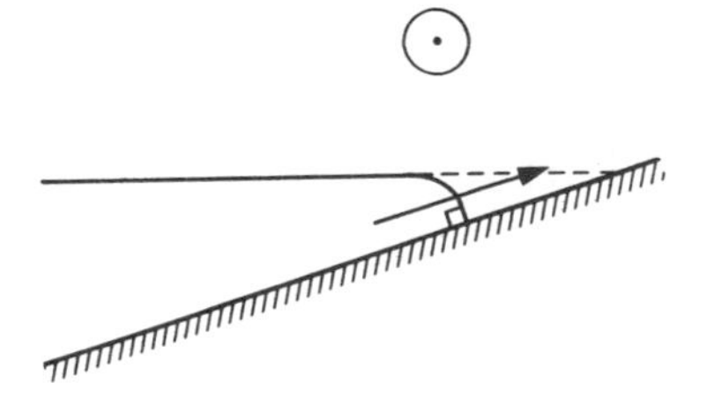

Figure 8 Schematic of the isopycnal distortion and resulting currents for a fluid with constant viscosity and diffusivity.

along-slope interior flow. They point out that this sets up an initial Ekman layer in a time of order f^{-1}, with a flux that is either upslope or downslope depending on the direction of the interior flow. They also argue that, outside this layer and at times significantly greater than f^{-1}, the acceleration and viscous terms in (4) may be neglected, giving a geostrophically balanced along-slope flow. Differentiation normal to the slope and the use of (5) gives

$$f\,dU/dz = (\partial B/\partial z - N^2 \cos\theta)\sin\theta. \tag{15}$$

This is the "thermal wind" equation connecting the vertical shear of the along-slope current to the horizontal density gradient, which involves $\partial B/\partial z$ due to our coordinate rotation. The basic physics is that the horizontal pressure gradient balancing the Coriolis force changes vertically if the hydrostatic vertical pressure gradient varies due to horizontal differences in density. We see from (15) that changes in U and B must be connected. Thus the viscous term in (3) may be partly balanced by a cross-slope flow V, rather than deceleration alone, in order that (6) should produce geostrophically compatible changes in B. Similarly, diffusion of B may also induce a cross-slope flow as well as changes in B. MacCready & Rhines (1991) show that combining (3), (6), and (15) leads to an equation

$$\frac{\partial U}{\partial t} = \nu\left(\frac{1/\sigma + S}{1+S}\right)\frac{\partial^2 U}{\partial z^2} \tag{16}$$

for the evolution of the along-slope flow. The same equation also applies to B, since U is assumed to be in thermal wind balance. It is remarkable that the complex processes of advection and diffusion in the y-z plane mimic simple one-dimensional diffusion of momentum or buoyancy normal to the boundary. Equation (16), previously applied by Gill (1981), Garrett (1982), and Flierl & Mied (1985) to the spin-down of geostrophic flows in the ocean interior, is termed the "slow diffusion equation" by MacCready & Rhines (1991) since for $\sigma > 1$ it implies slower adjustment of U than would be caused by viscosity alone. Foster (1989) finds an analogous equation for steady, nonlinear flow over topography.

The slow diffusion equation (16) may be used to help predict the behavior of the cross-slope transport (and its associated along-slope boundary stress) over time. Consider the cross-slope transport equation, formed by taking the z-integral of (6) and retaining the time dependence

$$\Psi|_\infty = \frac{-1}{N^2 \sin\theta}\int_0^\infty \frac{\partial B}{\partial t}\,dz + \kappa\cot\theta. \tag{17}$$

While the slow diffusion equation does not apply for $t < f^{-1}$ or for

$z < q^{-1}$, it is generally applicable at later times over most of the, now thickened, boundary layer. Hence (16) may be used to estimate the integral term in (17). Doing this, MacCready & Rhines (1991) find, for an initial value problem, that the transport decays as $t^{-1/2}$ from its starting value, given by standard Ekman theory, toward its steady value, $\kappa \cot \theta$. The decay time scale, or shut-down time, is given by

$$\tau_s' = S^{-2} f^{-1} (1/\sigma + S)(1+S)^{-1}. \tag{18}$$

In the limit $\sigma \to \infty$, $S \ll 1$, this is equal to the original estimate (2). For $\sigma = \mathcal{O}(1)$ and small S, the shut-down time is much increased over (2), owing to the diffusive thickening of the density boundary layer.

While the estimate (18) is supported by numerical experiments, its analytical derivation relies on the assumption that the along-slope flow is in thermal wind balance—a condition which may be invalid for various choices of the dimensionless parameters. In particular, with strong density diffusion (e.g. $\sigma = 1$) the flow near the boundary appears to be dominated by the steady balance given earlier. Numerical results from MacCready & Rhines (1991) in fact suggest that the steady solution (14) for the along-slope velocity sets the lower boundary condition for the slow diffusion equation, which then acts to bring the interior velocity to this value, as we know must eventually happen. Thus we are left with a fairly consistent picture of the chain of events, with the shut-down process bringing the cross-slope transport toward $\kappa \cot \theta$ over a time τ_s', and the interior velocity being altered to its final value $f\kappa(qv)^{-1}$ by slow diffusion.

VARIABLE MIXING COEFFICIENTS

If the mixing is much larger near the slope, as is likely in oceans and lakes, the behavior is different in many ways from that in the situation with constant coefficients.

General Steady State Results

The schematic of the distortion of an isopycnal surface by mixing confined near the bottom (Figure 3), compared with that for a constant mixing rate (Figure 8), suggests that the cross-slope secondary circulation will be bidirectional. In fact, as already remarked, the net upslope transport implied by (9) is $\kappa_\infty \cot \theta$, or zero if $\kappa \to 0$ as $z \to \infty$.

BUOYANCY FLUX The advective upslope buoyancy flux in the boundary layer is $\int_0^\infty BV\,dz$. With $V = d\Psi/dz$ this may be integrated by parts to obtain $-\int_0^\infty \kappa(\partial B/\partial z)^2 (N^2 \sin \theta)^{-1}\,dz$ for $\kappa_\infty = 0$ and using (9). Combining this with the upslope diffusive flux $\int_0^\infty -\kappa N^2 \sin\theta \, dz$, the total buoyancy flux F_B may be written

$$F_B = \int_0^\infty -\kappa\left[N^2 \sin^2\theta + \left(\frac{\partial B/\partial z}{N^2 \cos\theta}\right)^2 N^2 \cos^2\theta\right](\sin\theta)^{-1}\, dz. \quad (19)$$

The second, advective, contribution to F_B is essentially a "shear dispersion" addition to the first, diffusive, contribution and is dominant unless the average value of $\partial B/\partial z$ over the region of large κ is less than $N^2 \sin\theta$.

If we examine the buoyancy flux across a horizontal, instead of bottom-normal, line, we obtain the same total value F_B in a steady state (Figure 9), but the diffusive flux may be written

$$F_{\text{diff}} = \int_0^\infty -\kappa[N^2 \sin^2\theta + (\partial B/\partial z)\cos\theta]\, dz/\sin\theta \quad (20)$$

$$= \int_0^\infty -\kappa\left[N^2 \sin^2\theta + \left(\frac{\partial B/\partial z}{N^2 \cos\theta}\right) N^2 \cos^2\theta\right] dz/\sin\theta \quad (21)$$

and the advective flux is

$$F_{\text{adv}} = \int_0^\infty \kappa N^{-2}(\partial B/\partial z)(N^2 \cos\theta - \partial B/\partial z)\, dz/\sin\theta. \quad (22)$$

Thus, although $F_{\text{diff}} + F_{\text{adv}} = F_B$ as given by (19), F_{adv} is of the opposite sign to F_{diff} if the stratification in the boundary layer is reduced below its interior value so that $\partial B/\partial z < N^2 \cos\theta$. This curious result is easily understood by referring to Figure 3. With respect to a normal to the bottom, there is a tendency for the water moving upslope to be denser than the water moving downslope, thus augmenting the diffusive transport. On the other hand, with respect to a horizontal line the rising water tends to be less dense than

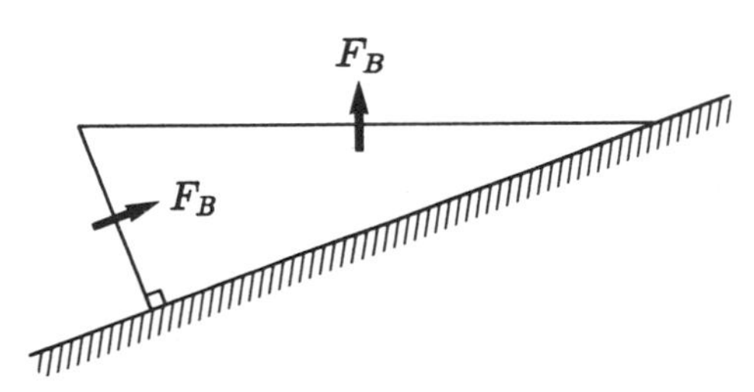

Figure 9 In a steady state the vertical buoyancy flux across a horizontal section through the boundary layer is the same as the upslope buoyancy flux across a normal to the slope.

the sinking water (as expected for a buoyancy-driven flow) thus opposing the diffusive flux.

This reduction of the total flux can also be recognized by comparing (19) for the total flux to (20) for the vertical diffusive flux; the reduction factor $(\partial B/\partial z)(N^2 \cos\theta)^{-1}$ occurs squared rather than linearly! It is convenient to define a *mixing effectiveness* I as the ratio of F_B to the buoyancy flux $\int_0^\infty -\kappa N^2\, dz/\sin\theta$ that would occur if the interior stratification $\partial B/\partial z = N^2 \cos\theta$ extended all the way to the boundary. This effectiveness is then the fraction by which a formula such as (1) must be modified to allow for reduced stratification near the boundary; it may be small.

ALONG-SLOPE FLOW A useful general result may also be derived for the along-slope current in the steady state case by combining (9) and (10) to obtain

$$dU/dz = \nu^{-1} f \cot\theta[\kappa_\infty - \kappa(\partial B/\partial z)/N^2 \cos\theta]. \tag{23}$$

The buoyancy gradient $\partial B/\partial z$ is only constrained to be greater than $-N^2 \sin^2\theta \sec\theta$ to maintain static stability in the boundary layer, but is more likely to be positive so that, with $\kappa(z)$ generally much greater than κ_∞, integrating (23) from $U = 0$ at $z = 0$ to the top of the boundary layer, where $\kappa = \kappa_\infty$ and $\partial B/\partial z = N^2 \cos\theta$, gives an along-slope flow U_∞ that is negative. This downwelling-favorable along-slope flow for the variable coefficient case is in contrast to the result (14) for the constant coefficient case. Here also, though, U_∞ seems to be a part of the solution, rather than arbitrary.

PARAMETER SPACE While (19) gives the total upslope or vertical buoyancy flux for any profile of $\kappa(z)$ that tends to zero as $z \to \infty$, the diffusivity and buoyancy gradient are not independent, but rather connected through a solution to (11). Garrett (1990) showed that the solution, including the mixing effectiveness I, depended on the Burger number S, the Prandtl number σ and also on the ratio of the distance h from the boundary over which significant mixing occurs to the boundary layer thickness, q^{-1} from (13), based on the values ν_0 and κ_0 of $\nu(z)$ and $\kappa(z)$ at $z = 0$.

In general we expect the region of greatly reduced stratification to be $\mathcal{O}(q^{-1})$, so that if qh is significantly less than one, the second, advective, contribution to (19) is small and the boundary mixing rather ineffective. On the other hand, for $qh \gg 1$ the mixing extends well into the region which can restratify under the influence of buoyancy forces and is likely to be effective. We can examine these expectations for specific models.

Extended mixing region If $qh \gg 1$, Garrett (1991) has argued that the first term in (11) is important only within a distance of $\mathcal{O}(q^{-1})$ from the boundary; outside this the along-slope flow is in thermal wind balance given by (15) as well as satisfying (23) required by the buoyancy equation. For $\kappa_\infty \to 0$ these imply that, through the mixing region of thickness h,

$$\partial B/\partial z = N^2 \cos\theta \, S\sigma(1+S\sigma)^{-1} \tag{24}$$

$$dU/dz = -f \cot\theta \, S(1+S\sigma)^{-1}. \tag{25}$$

These formulae apply even if the Prandtl number σ is a function of z. If σ is constant, however, integrating (25) over the mixing region leads to

$$U_\infty \approx -fh \cot\theta \, S(1+S\sigma)^{-1} = -Nh\cos\theta \, S^{1/2}(1+S\sigma)^{-1} \tag{26}$$

which could be $\mathcal{O}(1)$ m s^{-1} if $N = 10^{-2}$ s^{-1}, $f = 10^{-4}$ s^{-1}, and $\tan\theta = 10^{-2}$ so that $S = 1$, if h is $\mathcal{O}(10^2)$ m and σ is not large.

In this possible flow regime the Richardson number for a small slope is $(\partial B/\partial z)(\partial U/\partial z)^{-2} \approx \sigma(1+S\sigma)$ which is likely to be greater than 1, implying the need for some mechanism other than shear instability of the flow itself to maintain the mixing.

The mixing effectiveness I for this flow, assuming the first term in (19) to be small, is given approximately by $(S\sigma)^2(1+S\sigma)^{-2}$ if σ is independent of z. The cross-slope flow V is upslope close to the boundary, but weak and downslope in the extended mixing region if κ decreases with z. This solution is of particular interest if S is small, as it shows the possibility of an extended mixing region with weak stratification within which the along-slope flow is nonetheless in geostrophic balance. This contrasts with the situation for $qh \ll 1$ in which U, as well as B, is affected by the mixing.

Slab model A particular situation with $qh \ll 1$ is a slab model in which the mixing is vigorous within a height h above the bottom, and vanishingly small for $z \gg h$. Garrett (1991) showed that care is required in determining and applying the boundary conditions at $z = h$. In particular, he found that the along-slope flow in $z < h$ is very weak, but then jumps to a value

$$U_\infty = -\frac{3}{8} fh \cot\theta \, S(1+S\sigma_2)^{-1} \tag{27}$$

just outside the slab, in the limit of vanishing viscosity ν_2 and diffusivity κ_2 for $z \geq h$, where σ_2 is the Prandtl number ν_2/κ_2.

The buoyancy gradient is also very weak in the boundary layer, but has a jump across the top of the slab given by

$$\Delta B = \frac{3}{8} N^2 h \cos\theta \, S\sigma_2(1+S\sigma_2)^{-1}. \tag{28}$$

This is less than the value $\frac{1}{2}N^2 h\cos\theta$ that would arise from mixing alone, over a thickness h, of an original profile with buoyancy gradient N^2. It is thus qualitatively compatible with the downwelling-favorable, negative, value of U_∞ in (27). However, this slab solution for strong mixing not only fails to satisfy a thermal wind balance in the slab but also gives a jump in U across the top of the slab that is of opposite sign to the value $f^{-1}\Delta B \sin\theta$ that would be expected from (15).

For this model of a rather well-mixed slab ($qh \ll 1$), and with $f = 0$ and $\nu = \kappa$ for simplicity, the second, advective contribution to the upslope buoyancy transport F_B given by (19) is reduced by a factor $2.1 \times 10^{-4}(qh)^8$. This is very small due to the coefficient as well as the high power of qh. The calculation depends on assuming constant ν, κ in the slab, making it rather "stickier" than appropriate for a turbulent flow in which the effective mixing rates are reduced near the boundary. However, in the limit of free-slip boundary conditions the coefficient of $(qh)^8$ only increases by a factor of 4 or so. Thus the advective upslope contribution to F_B may be very small for a model with top-hat profiles of the mixing coefficients, and the purely diffusive transport, given by the first term of (19), may dominate. Salmun & Phillips (1992) have argued that this was the case in some laboratory experiments in which boundary mixing was induced by oscillations of a rough inclined plane.

On the other hand, the factor $\sin^2\theta$ in the first, diffusive, term in (19) suggests that at small slopes the second, advective, term is likely to dominate the integral if κ falls off gradually, permitting a significant $\partial B/\partial z$ to develop in the transition region. This was the case in calculations by Garrett (1990) for exponential profiles of ν, κ which showed dominance of the advective contribution to F_B even for fairly small values of qh.

More numerical integration of the governing equations for other profiles of ν, κ would be worthwhile. In particular, it would be interesting to see if the abrupt jump to U_∞ given by (27) still occurs if the top-hat mixing profile is smoothed. Such numerical integration, however, would probably best be carried out in the context of the full time-dependent problem.

Time Dependence

If the mixing is described by coefficients that vary with distance from the bottom but are independent of time, we might expect that the solution of the time-dependent equations would tend to the steady state solution close to the boundary, followed by slow diffusion into the interior of the downwelling-favorable along-slope flow given as U_∞ in the steady solution. On the other hand, if the fluid motion near a sloping boundary is partly driven by the along-slope flow in the interior, the mixing coefficients may themselves change with time in response to changing conditions of the mean stratification and shear and this could greatly affect the nature of the solution. In particular, we expect a marked asymmetry between upwelling and downwelling conditions, as seen in the numerical results in Figures 10 and 11.

If the initial Ekman flux is downslope, this is likely to lead to static instability and enhanced mixing. To avoid static instability of the mean density profile the boundary layer will need to continue to thicken as long

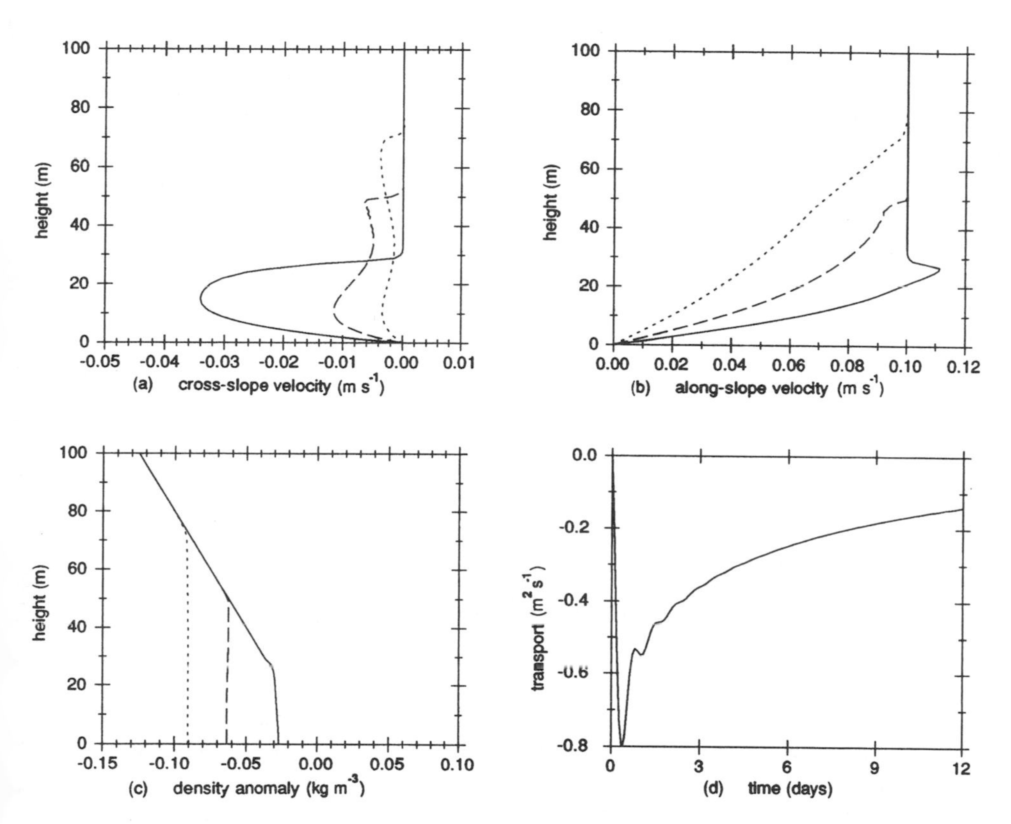

Figure 10 Results from MacCready & Rhines (1992) of a numerical integration of (3)–(6). The diffusivities ($\sigma = 1$) varied between 10^{-4} and 10^{-2} m^2 s^{-1} based on the gradient Richardson number. For this run $f = 10^{-4}$ s^{-1}, $N = 3.5 \times 10^{-3}$ s^{-1}, and $\sin\theta = 0.01$ so that $S = 0.12$. Initially the cross-slope velocity and B' were zero, and the along-slope velocity was downwelling-favorable with magnitude 0.1 m s^{-1}. (These runs used a different sign convention, with positive along-slope velocities being downwelling-favorable.) The velocities (*a*) and (*b*) and density anomaly $\rho - \rho_0$ (*c*), are plotted at 0.5 days (*solid*), 3 days (*long dashes*), and 9 days (*short dashes*). The integrated cross-slope transport is plotted in (*d*). Note the decreasing transport and the thickening of the mixed layer by static instability. The along-slope velocity approached a linear shear, nearly in thermal wind balance with the density field.

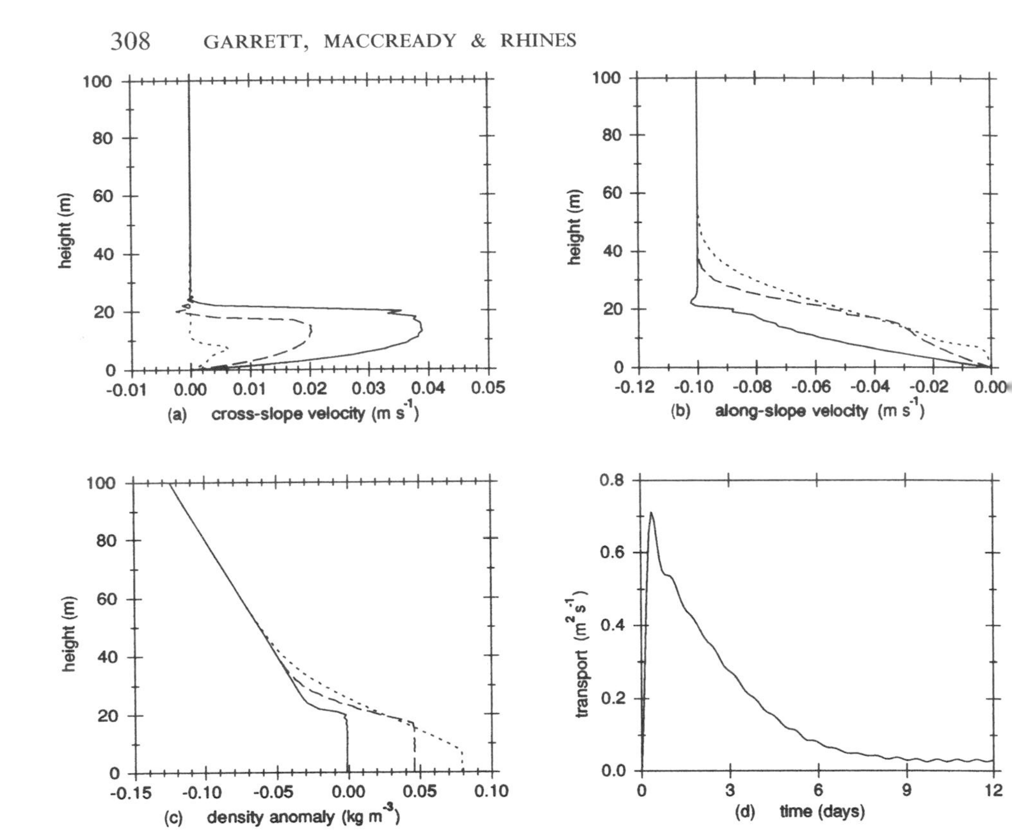

Figure 11 Same as Figure 10 except for an upwelling-favorable along-slope velocity. Note the rapid decrease in transport, the eventual thinning of the mixed layer, and the slow diffusive penetration of the along-slope velocity (and the buoyancy anomaly) into the interior.

as there is any downslope flux (although it is also possible that in reality intrusions will peel off into the interior). If the initial Ekman flux is upslope, however, there is no reason for the boundary layer to grow in thickness. It could, in fact, become thinner as the density difference across the top of a mixed layer near the boundary increases due to advection, suppressing the mixing and permitting buoyancy-driven restratification.

This asymmetry was clearly shown in the pioneering study by Weatherly & Martin (1978) which used the Level II turbulence closure model of Mellor & Yamada (1974). Very similar results to this were obtained by Jin (1990) and Trowbridge & Lentz (1991) using a slab model, and by MacCready & Rhines (1992). Field observations (Figure 12) also show a significantly thicker mixed layer when the along-slope flow is downwelling-favorable than when it is upwelling-favorable.

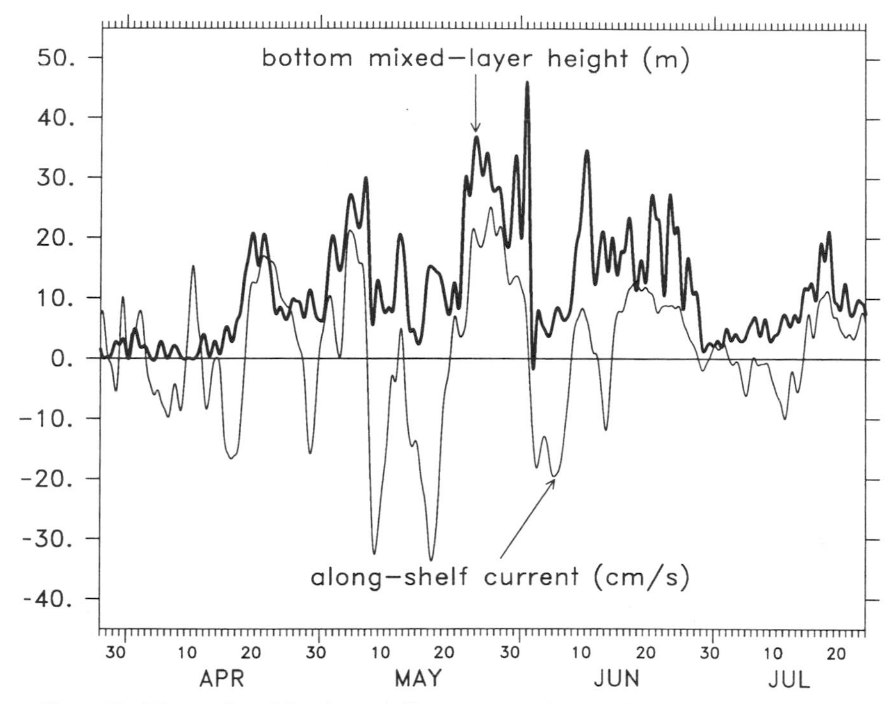

Figure 12 Time series of the along-shelf current 7 m above a sloping bottom off northern California compared with the thickness of a well-mixed bottom boundary layer (from Lentz & Trowbridge 1991). Positive current is poleward and downwelling-favorable.

Trowbridge & Lentz (1991) and MacCready & Rhines (1991) assumed flat interior isopycnals, a steady along-slope velocity U_e outside the boundary layer, and used a perturbation buoyancy

$$B' = B - N^2 y \sin\theta - N^2 z \cos\theta \tag{29}$$

to represent the change in buoyancy from the initial stratification. With the acceleration terms in the momentum equation also ignored (on the grounds that they contribute only to irrelevant initial inertial oscillations), the governing equations become

$$-fV = \frac{\partial}{\partial z}\left(\nu \frac{\partial U}{\partial z}\right) \tag{30}$$

$$f(U - U_e) = B' \sin\theta + \frac{\partial}{\partial z}\left(\nu \frac{\partial V}{\partial z}\right) \tag{31}$$

$$\frac{\partial B'}{\partial t} + VN^2 \sin\theta = \frac{\partial}{\partial z}\left(\kappa \frac{\partial B'}{\partial z}\right) + N^2 \cos\theta \frac{\partial \kappa}{\partial z}. \tag{32}$$

Assuming that the mixing is confined to a layer of thickness h, Trowbridge & Lentz (1991) integrate these equations over the boundary layer and assume that the bottom stress obeys a quadratic drag law based on the average velocity $(\bar{U}, \bar{V})$ over the layer. Hence

$$-fh\bar{V} = -C_d(\bar{U}^2 + \bar{V}^2)^{1/2}\bar{U} \tag{33}$$

$$fh(\bar{U} - U_e) = -C_d(\bar{U}^2 + \bar{V}^2)^{1/2}\bar{V} + \overline{B'} \sin\theta \tag{34}$$

$$d\overline{B'}/dt = -N^2 h \sin\theta \bar{V}, \tag{35}$$

where $\overline{B'}$ is the integrated extra buoyancy in the layer. They determine h from a closure condition based on the layer Richardson number given by

$$R_b = \frac{g(\Delta\rho/\rho_0)h}{(\Delta\mathbf{u})^2} = \frac{-\overline{B'} + \frac{1}{2}N^2h^2}{(\bar{U} - U_e)^2 + \bar{V}^2}, \tag{36}$$

where $\Delta\rho$ is the difference between the average density in the boundary layer and the density just above it; $\Delta\mathbf{u}$ is the difference between the average velocity in the layer and the imposed velocity outside the layer. If $R_b < 1$ they increase h to make it equal to 1, and otherwise leave h unchanged. Integration of this simple slab model, which is essentially that of Pollard et al (1973) and is also derived by Jin (1990), gives results, for both upwelling and downwelling situations with particular parameter choices, that are very close to those obtained by Weatherly & Martin (1978) from a more elaborate numerical model.

The time scale for significant reduction in the Ekman flux, based on the initial values of the cross-slope transport and mixed layer thickness, is $2^{-3/4}(C_dN/f)^{-1/2}S^{-1}f^{-1}$ for small values of C_dN/f (as is likely). Note that here again the fundamental time scale $S^{-1}f^{-1}$, derived in the introduction, appears. While the model solution evolves towards a steady state, with zero flow in the slab, Trowbridge & Lentz (1991) downplayed this on the grounds that the along-slope velocity component in the boundary layer would develop a thermal wind shear due to the horizontal buoyancy gradient associated with the tilted isopycnals of a mixed boundary layer—particularly for a downwelling-favorable initial flow. In this case, assuming slab-like behavior and expressing the bottom drag in terms of the average velocity over the layer become inappropriate. The thickening of the bottom mixed layer in the downwelling case may also substantially increase the time for arrest; we return to this issue later.

MacCready & Rhines (1992) integrated the full equations numerically, using a gradient Richardson number formulation for the turbulent mixing; examples of their results are shown in Figures 10 and 11. Guided by their results, they also developed an analytical model which assumes a well-mixed density profile throughout the boundary layer but also allows the velocity to develop significant shear, including the thermal wind shear. For a constant viscosity ν_0 in the mixing layer, they integrate (30) and (31) in the vertical to obtain

$$d\overline{B'}/dt = [2fh/(\delta \sin \theta)]\, dM/dt, \tag{37}$$

where $M = \int V\,dz = h\bar{V}$ is the total transport across the slope and $\delta = (2\nu_0/f)^{1/2}$ is the natural Ekman thickness. Combined with (35) which may be written $d\overline{B'}/dt = -N^2 \sin\theta\, M$, this gives

$$dM/dt = -\tfrac{1}{2}(\delta/h)M/\tau_s, \tag{38}$$

where $\tau_s = S^{-1}f^{-1}$ is the original estimate of the shut-down time in (2).

This analysis shows that the shut-down process is exponential (for constant ν_0 and h), but with a time scale that is increased to $(2h/\delta)\tau_s$ due to the dilution of the buoyancy force by mixing. It also suggests that downwelling boundary layers, which become thicker, may take longer than upwelling layers to come to a halt. The use of a constant eddy viscosity near the slope in this model makes possible comparisons between theoretical and numerical solutions, but leaves the boundary stress higher than would be produced by a log layer. Also, both Trowbridge & Lentz (1991) and MacCready & Rhines (1992) in their analytical model assume that it is possible at late time in the solution to have a well-mixed buoyancy layer within which the velocity field is in a mainly thermal wind balance; this is hard to reconcile with the steady state solutions described earlier which suggest that for $qh \ll 1$ the velocity as well as the buoyancy is well mixed, whereas for $qh \gg 1$ the along-slope velocity is in thermal wind balance but the buoyancy gradient normal to the slope is nonzero. Thus the evolution of the time-dependent solution may be hard to describe with simplified models, particularly if S is not small.

The most remarkable feature of the numerical solutions shown in Figures 10 and 11 is the tendency for thermal wind shear to replace viscous shear. Buoyancy advection tilts the isopycnals in this way, whether the flow is up- or downslope. Classic stratified spin-up carries out this same process, yet in that case the vertical scale of the thermal wind field is the Prandtl penetration scale fL/N, whereas here it is initially confined to the boundary layer, and then slowly diffuses upward. In the downwelling case (Figure 10) weak stratification persists in the boundary layer [though even weaker than given by (24) with $S = 0.12$]. If the boundary layer had no stratification, thermal wind would bring the along-slope velocity U_e to rest at the boundary over a vertical scale of $fU_e/N^2 \sin\theta$; this is 82 m for the parameters of Figure 10—not much more than the thickness of the boundary layer after 12 days and suggesting that an almost steady state is close to being achieved. In the upwelling case (Figure 11) the boundary layer becomes thinner as it progresses up the slope due to the suppression of mixing by the increasing density contrast, followed by buoyancy-driven restratification. This leaves behind some thermal wind and there is also some slow diffusion further into the interior.

Existing models have thus revealed the kinds of behavior to be expected in time-dependent problems. More exploration of parameter space would be worthwhile, however, to examine the evolution of flows forced directly by the mixing rather than as a response to external flows, and also to see how the residual stratification in the downwelled boundary layer changes with increasing values of the Burger number S.

Much of the work to date has been for the flat interior isopycnals, although Jin's (1990) slab model, Bird et al's (1982) extension of the work of Weatherly & Martin (1978) and Garrett's (1990, 1991) study of the steady state problem have allowed for sloping isopycnals in the ocean interior and a consequent thermal wind there. Future studies will take this further and include, as one limiting case, the situation when the bottom is flat and only the interior isopycnals slope. In this case the bottom boundary layer thickness is that of an Ekman layer, uninfluenced by stratification, and no buoyancy forces arise to arrest the Ekman flow. Nonetheless, steady state solutions may develop near the boundary and then slow-diffuse into the interior.

Arrest or Spin-Down?

We now return to the question raised in the introduction of whether an oceanic current above a sloping seafloor will adjust to the no-slip boundary condition there more quickly by Ekman layer arrest, involving just the current close to the bottom, or the usual spin-down process affecting more of the water column. Both time scales require a specification of the drag law at the bottom; if we take this to be quadratic with a drag coefficient C_d the arrest time is of order $(C_d N/f)^{-1/2} S^{-1} f^{-1}$ whereas the spin-down time is of order $C_d^{-1}(H/U)$. Here U is the along-slope current in the interior and H the vertical distance, of order f/N times its horizontal scale, over which it would be spun down.

The ratio of arrest time to spin-down time is $S^{-1} C_d^{1/2} (Nf)^{-1/2} U/H$ and so is critically dependent on the value of S. We will consider different situations later. We note that the ratio may be written as $S^{-1}\delta/H$, where $\delta = C_d^{1/2} U (Nf)^{-1/2}$ is the scale of the bottom boundary layer according to the model of Pollard et al (1973) for a flat boundary, and in fact the same value of the ratio applies in the laminar case with δ replaced by the Ekman layer thickness. For the downwelling case we have seen how the boundary layer may become much thicker due to static instability and that this may delay its arrest. A very rough estimate of the shut-down time comes from noting that the final thickness of the mixed layer required for thermal wind to bring the interior velocity U to rest is $Uf(N^2 \sin\theta)^{-1}$ if we assume that there is no density jump across the top of the mixed layer. The amount of water that has moved downslope is then $\frac{1}{2}U^2f^2(N^2\sin\theta)^{-2}(\tan\theta)^{-1}$. Taking an initial Ekman flux of $C_d U^2/f$ (though this will be reduced as the thermal wind develops) the shut-down time is then of order $\frac{1}{2}C_d^{-1}N^{-1}S^{-3/2}$ (taking $\cos\theta \simeq 1$ for small slopes), illustrating great sensitivity to S, and implicitly to N which appears as N^{-4}. We will derive some specific values for shut-down and spin-down times later; we note that an observational signal of arrest in the ocean would be either zero bottom stress, or, equivalently from integrating (3), no net cross-slope flow.

APPLICATIONS TO THE OCEAN

Regimes

In regions of the ocean where mean along-slope currents are weak, or have been spun down near the bottom by either Ekman layer arrest or by the interior circulation driven by Ekman layer divergence, we might expect the physics of boundary mixing to be dominant with weak net cross-slope flow. We discuss first the significance and influence of this regime in the ocean before considering the three-dimensional nature of ocean circulation which allows for local cross-slope Ekman flux driven by an adjusting along-slope current.

Mixing Processes Near Slopes

It is clear that the effectiveness of boundary mixing depends upon there being mixing processes that extend further from the slope than the distance over which restratification can occur. As found by Phillips et al (1986) and Salmun & Phillips (1992), this is unlikely for the turbulence generated by bottom friction right at the slope, which is likely to be suppressed in the restratified region and so operate only on water that is rather well-mixed. The result is a very small net buoyancy flux.

Of course intermittent mixing events at the seafloor, followed by the restratification that can occur on a slope (and distinguishes the boundary layer there from that over a flat bottom or at the sea surface), will lead to a net buoyancy flux. Garrett (1991) investigated the kinematic consequences of this, finding that, in the simplest case with no net motion up or down the slope during the event, half of the vertical buoyancy flux achieved in the event is lost in the restratification! Overall, though, he estimated that the process was unlikely to be important globally.

It has been increasingly recognized, however, that the reflection of internal gravity waves from the sloping seafloor can lead to enhanced mixing for a considerable distance from the slope. The dynamical reason for this is that internal wave rays are reflected at the same angle to the vertical, rather than to the normal, in order to conserve the wave frequency (e.g. Phillips 1977). Thus for incidence from deep water (Figure 13) the ray tube is narrowed leading to an increase in wave energy density to conserve the energy flux. This increase is enhanced by a reduction in the group velocity, and the wave shear increases still further due to the increased vertical wavenumber. Incident waves from some directions are correspondingly reduced, but, allowing for a full incident spectrum, Eriksen (1982, 1985) and Garrett & Gilbert (1988) show that there is a considerable net increase in shear and hence in the probability of shear instability and mixing in the stratified water near the boundary. The latter authors attempted to quantify the amount of energy available for mixing by determining the wavenumber such that the shear spectrum for lower wavenumbers gives a Richardson number of order one, and then arguing that waves with higher wavenumbers would break and give up their energy flux to mixing. Their results were sensitive to the ratio N/f and to the bottom slope but suggested that this form of near boundary mixing could well be significant in the deep ocean, particularly at low latitudes. Ivey & Nokes (1989) have shown in the laboratory that internal waves reflected from a sloping boundary can lead to vigorous mixing in a turbulent boundary layer. Evidence in the ocean for the enhancement of internal waves on reflection has been found by Eriksen (1982), though puzzles in interpreting available current meter data remain (Gilbert 1991). Moreover, following the theoretical analyses of Baines (1971), Gilbert & Garrett

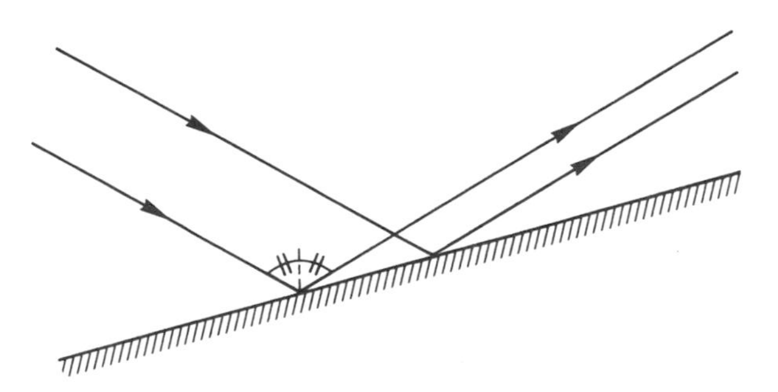

Figure 13 Internal wave rays incident on a sloping bottom are reflected at equal angles to the vertical.

(1989) argue that the mixing may be less above topography that is concave rather than convex.

More recently, fine-scale and microscale velocity data above the sloping sides of Fieberling Guyot, a seamount in the northeast Pacific Ocean, have shown evidence for significantly enhanced mixing up to 200 m or so above the bottom, well into the stratified water above a well-mixed boundary layer (J. Toole, R. Schmitt & K. Polzin, personal communication; Figure 14), and possibly associated with internal wave reflection. In another energetic flow regime, enhanced turbulence outside a bottom mixed layer has been found in the Florida Straits (D. Winkel & M. Gregg, personal communication; Figure 15).

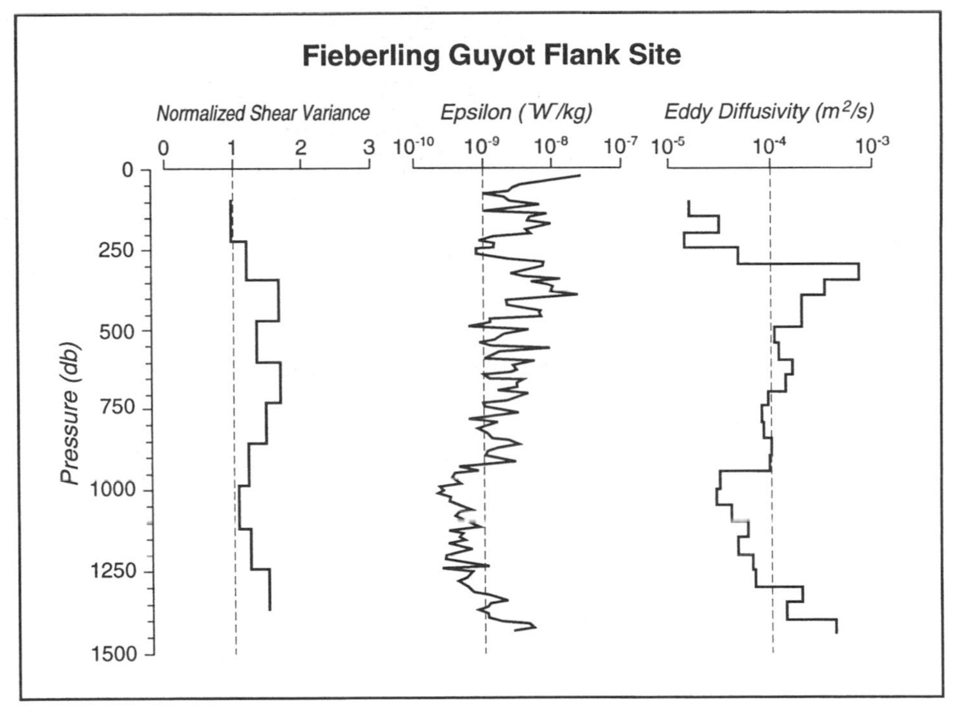

Figure 14 Vertical profiles, above the sloping sides of Fieberling Guyot, of (*left*) the variance of fine-scale shear (3 to 128 m wavelength), normalized by the local N^2, (*center*) the turbulent kinetic energy dissipation rate ε averaged over 10 m in the vertical, and (*right*) the corresponding vertical eddy diffusivity given by $\Gamma\varepsilon/N^2$ (Osborn 1980) using $\Gamma = 0.25$ (J. Toole, R. Schmitt & K. Polzin, personal communication). The turbulence is enhanced near the bottom, yet is surprisingly strong in the upper kilometer as well. (Typical mid-ocean values for ε correspond to the lowest values seen in this profile.) Internal wave enhancement by the seamount may be responsible for the large shear seen in the lefthand profile; a value of 2 corresponds to a Richardson number of $\frac{1}{4}$.

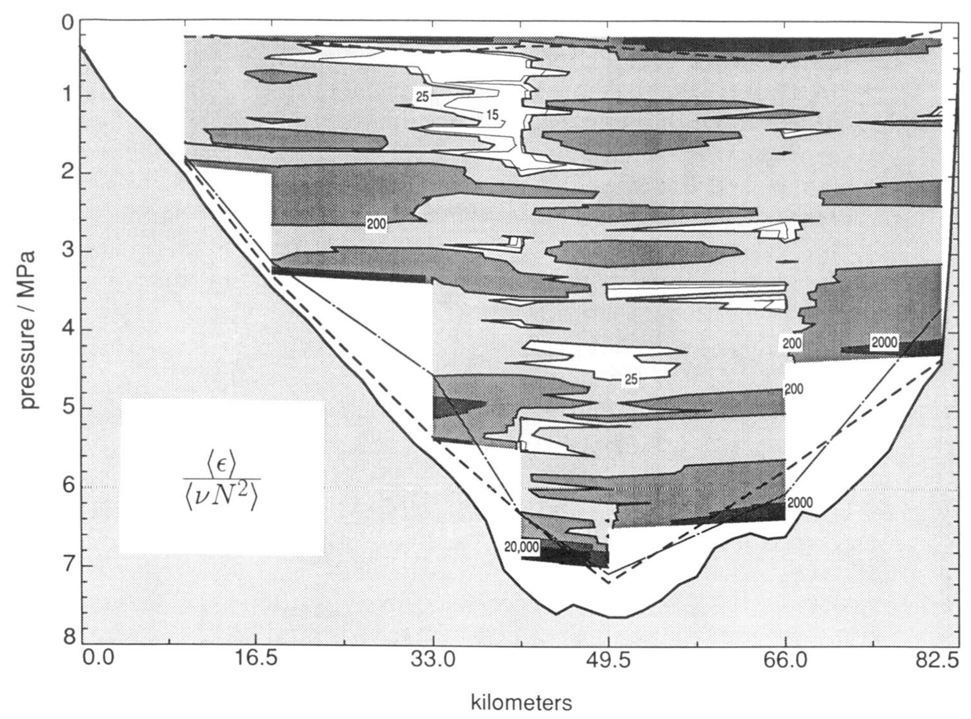

Figure 15 A cross-section of turbulent kinetic energy dissipation measured with a free-fall profiling instrument by D. Winkel & M. C. Gregg (personal communication). The site is the Florida Straits at 27°N, where the Florida Current, with speeds of typically 0.5 m s^{-1}, develops intense turbulence near the solid boundary as well as sporadically in the interior. The thickness of the actively mixing bottom layer (*solid-dashed curve*) often exceeds the thickness of the density-mixed layer (*dashed curve*). The profiles are nondimensionalized as $\varepsilon/\nu N^2$, where ε is the turbulent dissipation. Even without this nondimensionalization similar features appear.

While internal waves are an important mechanism for near boundary mixing, Thorpe's (1987) measurements on the abyssal continental slope southwest of Ireland showed larger signals at the period of the semidiurnal lunar tide; the generation of internal tides in the ocean may yet turn out to be the dominant mixing process near slopes.

Patching Local Solutions to the Interior

The discussion of boundary mixing earlier in this review assumed a constant interior stratification, a uniform bottom slope, and uniform mixing rates so that the whole problem becomes independent of the upslope coordinate. In practice all of these assumptions could be violated. Phillips et al (1986) considered the effect of boundary mixing on an interior pycnocline, or region of large vertical density gradient. They argued that the enhanced mixing at the slope leads to a spread of the isopycnals up and down the slope (Figure 16) and that the resulting buoyancy forces on this drive a tertiary flow (on top of the secondary circulation associated with the local boundary mixing). This tertiary flow converges at the pycnocline, causing an outflow into the ocean interior and spreading the isopycnals in the same way as if the mixing had occurred in the fluid interior in the first place.

Boundary mixing concepts and results for constant interior N can presumably still be applied locally provided that the tertiary flow is considerably weaker than the secondary flow. Garrett (1991) showed that this is the case if the interior pycnocline thickness is much greater than the thickness of the boundary layer, as is likely to be the case in most geophysical situations. While a formal expansion of the overall solution in terms of a small ratio of boundary layer thickness to vertical height of environmental variability can be pursued, or new boundary layer scalings considered if the interior pycnocline is very thin (Salmun et al 1991), the nature of the solution was described by McDougall (1989). He assumed that the tertiary circulation maintains flat isopycnals (after averaging across the distortion of the boundary layer) and that the interior stratification changes slowly, if at all. Conservation of density in the boundary layer at depth z then requires an average vertical velocity w_b in it given by

$$A_b w_b N^2 = d(A_b K_{eff} N^2)/dz, \tag{39}$$

where A_b is the horizontal area of the boundary layer at depth z and $K_{eff} N^2$ is the average vertical buoyancy flux per unit area, with K_{eff} reduced below the actual eddy diffusivity there by the effectiveness factor of boundary mixing discussed earlier. Vertical variations of $A_b w_b$ times the length C of

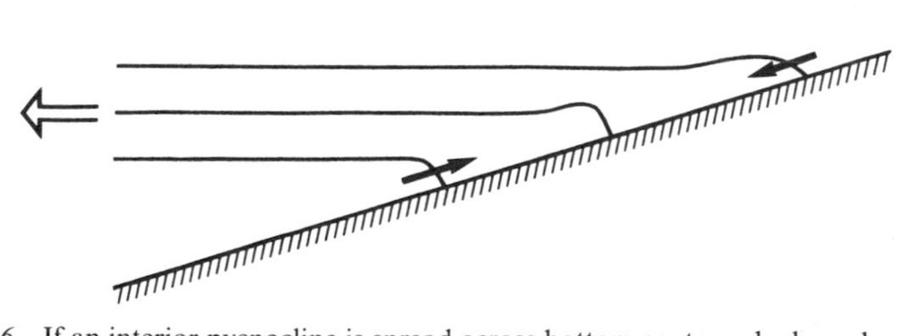

Figure 16 If an interior pycnocline is spread across bottom contours by boundary mixing, the convergent buoyancy-driven tertiary circulation in the boundary layer drives a flow into the interior and spreads the isopycnals there.

a depth contour then require net exchange between the boundary layer and the ocean interior given by

$$Cu = \frac{d(A_b w_b)}{dz} = \frac{d}{dz}\left[N^{-2}\frac{d}{dz}(A_b K_{eff} N^2)\right] \tag{40}$$

showing that exchange between the boundary layer and the interior may be driven by vertical changes in K_{eff} or A_b as well as N^2. As pointed out by McDougall (1989), this result implies inflow at the top of a seamount to supply a net flow down its sloping sides. The inflow rate, of course, depends on the strength and effectiveness of boundary mixing.

If the vertical diffusivity κ_∞ in the ocean interior is finite, consideration of the tertiary circulation may have to include the net upslope flow $\kappa_\infty \cot\theta$ (Woods 1991), but this is likely to be smaller than the tertiary flow induced by the turbulent boundary mixing. This discussion has also averaged over variations along a particular depth contour and so ignores the way that the tertiary flow might leave the boundary layer at some point to return further along on the same contour. Moreover, the above discussion of the tertiary flow ignores the way in which there may be vigorous lateral exchange between the boundary layer due to flow separation (Armi 1978) or just the turbulent eddies shed into the interior due to lateral instability of an along-slope current, much as in a river interacting with its bounding bed. This is certainly a likely process in energetic flows like the Florida Current (Figure 15), and in the deep ocean is suggested by measurements of chemical tracers and suspended sediments (Eittreim et al 1975). However, the possible overall constraint on the effectiveness of boundary mixing, due to reduced stratification in the mixing region close to the boundary, remains.

The above discussion has also assumed flat interior isopycnals. If this is not the case, as for an along-slope current that has been spun down, with tilted isopycnals and thermal wind shear near the boundary, further investigation of the effectiveness of boundary mixing may be necessary. The basic question, however, is whether a strong along-slope mean flow has been spun down.

Evolving Along-Slope Currents

The physics of the arrested Ekman layer might show up in the ocean in response to time varying interior flows (Figure 12). Perhaps more importantly, it might play a role in the dynamics of deep western boundary currents, such as that in the North Atlantic (Figure 17). Here the mean current averages about 0.2 m s^{-1} near the bottom, with pulses to twice that value, and is downwelling-favorable. It decreases upward with an *e*-

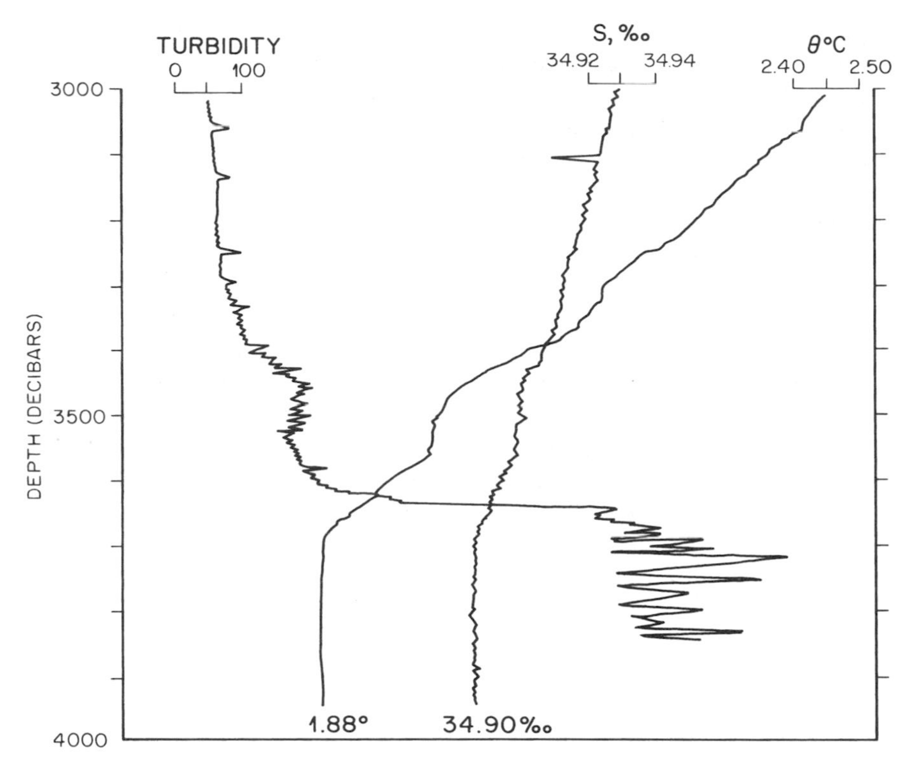

Figure 17 Profiles of salinity S, potential temperature θ, and turbidity at the Blake-Bahama Outer Ridge (30.3°N, 74.7°W) (Jenkins & Rhines 1980). Here the Deep Western Boundary Current, with average speed at two nearby moorings of 0.21 m s^{-1} and 0.13 m s^{-1} (200 m above bottom), winds southward from sources in the Greenland Sea, along steeply sloping topography (slope $\simeq 2\times10^{-2}$). The mixed layer of this downslope-favorable current has reached a thickness of 270 m, though more typically it ranges from 40 to 100 m in this region. There are signs of mixing, and intrusion from the boundary, above the bottom mixed layer. (The turbidity records light attenuation due to fine sediment carried from the boundary into the interior.)

folding scale of about 800 m; the shear is in thermal wind balance with a tilt of the density surfaces that is upward towards the slope, holding cold water at an unusual height. Small suspended particles taken up from the bottom appear in the turbidity profile, giving direct evidence of mixing through the density mixed layer and also suggesting a history of older mixing events intruding into the fluid above.

The parameters $N = 1.2\times10^{-3}$ s^{-1}, $f = 7.3\times10^{-5}$ s^{-1}, and $\sin\theta = 2\times10^{-2}$ here give a Burger number $S = 0.11$. With $C_d \simeq 2.5\times10^{-3}$

the arrest time $(C_d N/f)^{-1/2}S^{-1}f^{-1}$ for an upwelling situation would be 7 days, whereas the more appropriate downwelling formula $\frac{1}{2}C_d^{-1}N^{-1}S^{-3/2}$ gives 54 days. Spin-down via Ekman suction would involve most of the water column and have a time scale $C_d^{-1}(H/U) \simeq 70$ days if we take $H = 3000$ m (as for a current width of about 50 km) and $U = 0.2$ m s^{-1}. Thus, shut-down via arrest of the Ekman layer would seem to be at least as important as spin-down and it would seem surprising that in the presence of both the bottom velocity still persists. The resolution seems to be related to the upward tilt of the interior isopycnals with a slope almost equal to that of the bottom. In such circumstances little shut-down occurs and the current is regenerated against spin-down by the available potential energy of the interior density field (P. MacCready, submitted for publication).

On significantly smaller slopes in the deep ocean S would be much less, with spin-down becoming more important than shut-down even without the complication of sloping interior isopycnals. On the continental shelf, however, it is likely that shut-down occurs more rapidly than spin-down, but the ratio of time scales needs evaluation using the appropriate parameter values for each situation (see Trowbridge & Lentz 1991).

OUTSTANDING PROBLEMS

Theoretical Questions

Much has been learnt in recent years about the basic fluid dynamics of mixing near a sloping boundary. In particular, it is clear that the effectiveness of the process depends on whether the mixing mechanism can extend far enough from the boundary that buoyancy-driven flows can continually restore the stratification on which the mixing can act. It is this ability of the boundary layer to restratify that distinguishes it from boundary layers at the surface or above a flat bottom (unless there is an interior thermal wind in those cases). While flow separation is not necessary to bring fresh, stratified water into contact with the boundary, the role of exchange by lateral eddies requires further investigation.

For steady conditions with flat interior isopycnals the nature of the solutions is reasonably clear, and for time-dependent problems we do have a framework which recognizes the tendency for buoyancy forces to arrest an Ekman layer and for the along-slope flow at the edge of the layer to slowly diffuse into the ocean interior. We do not, however, have a complete quantitative understanding of the time for arrest to occur. Further work, clearly described in terms of the independent dimensionless parameters of the problem, is required. The situation becomes more complicated if, as is very likely, the mixing parameters evolve in response to changing conditions of flow and stratification. The model of MacCready & Rhines

(1992) points to the gradual replacement of viscous shear with thermal wind shear, but the stratification in the final state is rather uncertain.

Allowing for sloping interior isopycnals adds an extra richness to the problem. As discussed earlier, this has been included in some studies, but more investigation is required, with interesting possible applications to the particular example of an interior front near a flat surface.

We have argued that patching local solutions to the interior flow is straightforward if the vertical scale of parameter variation is greater than the boundary layer thickness. The effect of the Earth's rotation on the tertiary flow remains to be investigated.

Observational Needs

The theoretical discussion has largely depended on representing eddy fluxes of buoyancy and momentum in terms of mixing coefficients. We have pointed out that this need not imply that mixing length arguments are valid, but it is clear that future representations of the eddy fluxes will have to be guided by their measurement in the field. This presents considerable technical difficulties and also statistical concerns due to the small correlation coefficients expected in an environment dominated by internal waves; these are discussed by J. J. M. van Haren et al (submitted for publication) who present the results of a pilot experiment. This study, and the earlier work of Thorpe (1987) and Thorpe et al (1990), show that slopes can be regions of highly variable motion with overturns and other evidence for mixing occurring to a considerable height above the bottom.

A major concern, of course, is whether the mixing is associated with a negligible net upslope flow, as in steady-state boundary mixing theories, or whether there is still a net Ekman flux, upslope or downslope, due to the ocean interior currents. Instrument inaccuracies make it difficult in practice to establish the cross-slope component of the boundary layer flow even with good coverage (Trowbridge & Lentz 1991). There is also the need to discriminate clearly between Eulerian and Lagrangian mean flows (Ou & Maas 1986). Investigation of the spread of boundary-mixed fluid into the ocean interior, whether by the tertiary circulation discussed here or as part of flow separation and eddy activity, would also be useful.

This review has concentrated on the basic fluid dynamical processes that can be expected to occur at a sloping boundary of a rotating stratified fluid. The topic continues to present interesting problems of considerable importance for models of ocean circulation.

ACKNOWLEDGMENTS

We thank John Toole, Ray Schmitt, Kurt Polzin, Dave Winkel, and Mike Gregg for permission to show their unpublished data, Rosalie Rutka for assistance in the preparation of this review, and the Natural Sciences and Engineering Research Council (Canada) and the Office of Naval Research (U.S.A.) for supporting our research.

Literature Cited

Armi, L. 1978. Some evidence for boundary mixing in the deep ocean. *J. Geophys. Res.* 83: 1971–79

Armi, L. 1979a. Effects of variations in eddy diffusivity on property distributions in the oceans. *J. Mar. Res.* 37: 515–30

Armi, L. 1979b. Reply to comments by C. Garrett. *J. Geophys. Res.* 84: 5097–98

Armi, L., Millard, R. C. 1976. The bottom boundary layer of the deep ocean. *J. Geophys. Res.* 81: 4983–90

Baines, P. G. 1971. The reflexion of internal/inertial waves from bumpy surfaces. *J. Fluid Mech.* 46: 273–91

Bird, A. A., Weatherly, G. L., Wimbush, M. 1982. A study of the bottom boundary layer over the eastward scarp of the Bermuda Rise. *J. Geophys. Res.* 87: 7941–54

Eittreim, S., Biscaye, P. E., Amos, A. F. 1975. Benthic nepheloid layers and the Ekman thermal pump. *J. Geophys. Res.* 80: 5061–67

Eriksen, C. C. 1982. Observations of internal wave reflection off sloping bottoms. *J. Geophys. Res.* 87: 525–38

Eriksen, C. C. 1985. Implications of ocean bottom reflection for internal wave spectra and mixing. *J. Phys. Oceanogr.* 15: 1145–56

Flierl, G. L., Mied, R. P. 1985. Frictionally-induced circulations and spin down of a warm-core ring. *J. Geophys. Res.* 90: 8917–27

Foster, M. R. 1989. Rotating stratified flow past a steep-sided obstacle: incipient separation. *J. Fluid Mech.* 206: 47–73

Garrett, C. 1979a. Comments on "Some evidence for boundary mixing in the deep ocean" by Laurence Armi. *J. Geophys. Res.* 84: 5095

Garrett, C. 1979b. Mixing in the ocean interior. *Dyn. Atmos. Oceans* 3: 239–65

Garrett, C. 1982. On spin-down in the ocean interior. *J. Phys. Oceanogr.* 12: 1145–56

Garrett, C. 1990. The role of secondary circulation in boundary mixing. *J. Geophys. Res.* 95: 989–93

Garrett, C. 1991. Marginal mixing theories. *Atmos.-Ocean* 29: 313–39

Garrett, C., Gilbert, D. 1988. Estimates of vertical mixing by internal waves reflected off a sloping bottom. *Proc. Int. Liège Coll. Ocean Hydrodyn. 19th*, ed. J. C. J. Nihoul, B. M. Jamart, pp. 405–424. New York: Elsevier

Gilbert, D. 1991. Testing the critical reflection hypothesis. *Proc. 'Aha Huliko'a Hawaiian Winter Workshop, 6th*, pp. 53–70. Honolulu: Univ. Hawaii

Gilbert, D., Garrett, C. 1989. Implications for ocean mixing of internal wave scattering off irregular topography. *J. Geophys. Res.* 19: 1716–29

Gill, A. E. 1981. Homogeneous intrusions in a rotating stratified fluid. *J. Fluid Mech.* 103: 275–95

Gregg, M. C. 1987. Diapycnal mixing in the thermocline: a review. *J. Geophys. Res.* 92: 5249–86

Gregg, M. C. 1989. Scaling turbulent dissipation in the thermocline. *J. Geophys. Res.* 94: 9686–98

Hogg, N., Biscaye, P., Gardner, W., Schmitz, W. J. 1982. On the transport and modification of Antarctic Bottom Water in the Vema Channel. *J. Mar. Res.* 40: 231–63 (Suppl.)

Ivey, G. N. 1987a. The role of boundary mixing in the deep ocean. *J. Geophys. Res.* 92: 11,873–78

Ivey, G. N. 1987b. Boundary mixing in a rotating stratified fluid. *J. Fluid. Mech.* 121: 1–26

Ivey, G. N., Nokes, R. I. 1989. Vertical mixing due to the breaking of critical internal waves on sloping boundaries. *J. Fluid Mech.* 204: 479–500

Jenkins, W. J., Rhines, P. B. 1980. Tritium in the deep North Atlantic Ocean. *Nature* 286: 877–80

Jin, Y.-H. 1990. *Vertically integrated models of bottom mixed layer growth in the ocean.* PhD thesis. Memorial Univ. Newfoundland, St. John's. 144 pp.

Lentz, S. J., Trowbridge, J. H. 1991. The bottom boundary layer over the northern California shelf. *J. Phys. Oceanogr.* 21: 1186–201

MacCready, P., Rhines, P. B. 1991. Buoyant inhibition of Ekman transport on a slope and its effect on stratified spin-up. *J. Fluid Mech.* 223: 631–61

MacCready, P., Rhines, P. B. 1992. Slippery bottom boundary layers on a slope. *J. Phys. Oceanogr.* In press

McDougall, T. J. 1989. Dianeutral advection. *Proc. 'Aha Huliko'a Hawaiian Winter Workshop*, 5*th*, pp. 289–315. Honolulu: Hawaii Inst. Geophys.

Mellor, G. L., Yamada, T. 1974. A hierarchy of turbulence closure models for planetary boundary layers. *J. Atmos. Sci.* 31: 1791–806

Munk, W. H. 1966. Abyssal recipes. *Deep-Sea Res.* 13: 207–30

Osborn, T. 1980. Estimates of the local rate of vertical diffusion from dissipation measurements. *J. Phys. Oceanogr.* 10: 83–89

Ou, H. W., Maas, L. 1986. Tidal-induced buoyancy flux and mean transverse circulation. *Cont. Shelf Res.* 5: 611–28

Pedlosky, J. 1979. *Geophysical Fluid Dynamics*. New York: Springer-Verlag. 624 pp.

Phillips, O. M. 1970. On flows induced by diffusion in a stably stratified fluid. *Deep-Sea Res.* 17: 435–43

Phillips, O. M. 1977. *The Dynamics of the Upper Ocean*. Cambridge: Cambridge Univ. Press. 336 pp. 2nd ed.

Phillips, O. M., Shyu, J. -H., Salmun, H. 1986. An experiment on boundary mixing: mean circulation and transport rates. *J. Fluid Mech.* 173: 473–99

Pollard, R. T., Rhines, P. B., Thompson, R. O. R. Y. 1973. The deepening of the wind-mixed layer. *Geophys. Fluid Dyn.* 3: 381–404

Rhines, P., MacCready, P. 1989. Boundary control over the large scale circulation. *Proc. 'Aha Huliko'a Hawaiian Winter Workshop*, 5*th*, pp 75–97. Honolulu: Hawaii Inst. Geophys.

Salmun, H., Phillips, O. M. 1992. An experiment on boundary mixing, Part 2. The slope dependence at small angles. *J. Fluid Mech.* In press

Salmun, H., Killworth, P. D., Blundell, J. R. 1991. A two-dimensional model of boundary mixing. *J. Geophys. Res.* 96: 18,447–74

Saunders, P. M. 1987. Flow through Discovery Gap. *J. Phys. Oceanogr.* 17: 631–43

Thorpe, S. A. 1987. Current and temperature variability on the continental slope. *Phil. Trans. R. Soc. London Ser. A* 323: 471–517

Thorpe, S. A., Hall, P., White, M. 1990. The variability of mixing on the continental slope. *Phil. Trans. R. Soc. London Ser. A* 331: 183–94

Trowbridge, J. H., Lentz, S. J. 1991. Asymmetric behavior of an oceanic boundary layer above a sloping bottom. *J. Phys. Oceanogr.* 21: 1171–85

Weatherly, G. L., Martin, P. J. 1978. On the structure and dynamics of the oceanic bottom boundary layer. *J. Phys. Oceanogr.* 8: 557–70

Whitehead, J. A. 1989. Surges of Antarctic Bottom Water into the North Atlantic. *J. Phys. Oceanogr.* 19: 853–61

Whitehead, J. A., Worthington, L. V. 1982. The flux and mixing rates of Antarctic Bottom Water within the North Atlantic. *J. Geophys. Res.* 87: 7903–24

Woods, A. W. 1991. Boundary-driven mixing. *J. Fluid Mech.* 226: 625–54

Wunsch, C. 1970. On oceanic boundary mixing. *Deep-Sea Res.* 17: 293–301

Wunsch, C. 1971. Note on some Reynolds stress effects of internal waves on slopes. *Deep-Sea Res.* 18: 588–91

Annu. Rev. Fluid Mech. 1993. 25: 325–71

QUANTIZED VORTICES AND TURBULENCE IN HELIUM II

Russell J. Donnelly

Department of Physics, University of Oregon, Eugene, Oregon 97403

KEY WORDS: liquid helium, mutual friction, second sound

1. INTRODUCTION

The study of liquid helium has been underway since 1908, when helium was first liquefied. The superfluidity of helium II has been known since 1938, and has been intensively studied ever since. Surprisingly, there is still no fundamental microscopic theory widely accepted for superfluid ^{4}He, although there is for superfluid ^{3}He. Nevertheless, the phenomenological theory of superfluid helium is well understood and we are now finding liquid helium being used extensively in engineering applications. Indeed, the field of cryogenic engineering is becoming a considerably larger area of research than the basic physics effort.

In Volume 6 of this series, Paul Roberts and I wrote a review of the subject of superfluid mechanics, in an effort to make the subject accessible to the fluid mechanics community (Roberts & Donnelly 1974). At the time we did little with mutual friction and the structure of quantized vortices. Now the Editors have asked me to write an article on quantized vortices as such. Since I have just written a monograph on the subject (Donnelly 1991a) it seems that again the emphasis here should be on the information of interest to the fluid mechanics community, and that the subject of turbulence in helium II should be emphasized.

Topics of interest in cryogenic engineering include heat transfer (Pfotenhauer & Donnelly 1985), the transfer of helium in space (Snyder 1988; Kittel 1987, 1988), and the generation of high Reynolds number flows (Donnelly 1991b), and superfluid turbulence. These topics all require a deep understanding of the behavior of quantized vortices. I have tried to make the frontier subjects in superfluid turbulence clear by drawing ex-

0066–4189/93/0115–0325$02.00

amples from pure counterflow turbulence, where there is no net mass flow, and where the connection with classical turbulence is distant, to open flows, where the analogies with classical turbulence may be profound.

In order to make this article reasonably self-contained, I will repeat some material available in a number of other references (including Volume 6), then continue to describe the nature and behavior of quantized vortices. In general I will not be emphasizing microscopic issues, such as the structure and nucleation of vortices, which are covered extensively in the physics literature on liquid helium.

2. BACKGROUND ON LIQUID HELIUM II

The phase diagram of low-temperature helium is shown in Figure 1. The critical temperature is 5.2 K and the critical pressure is 2.26 bar. The normal boiling point of liquid helium is 4.2 K and, by means of pumping over the liquid, one can reduce the temperature to ~1 K. Other means can be found to further reduce the temperature, and in fact liquid helium will not solidify under its own vapor pressure at any temperature. It requires an imposed pressure of about 25 bars to solidify helium at low temperatures.

If we pump on a sample of liquid helium we reduce the pressure, lower

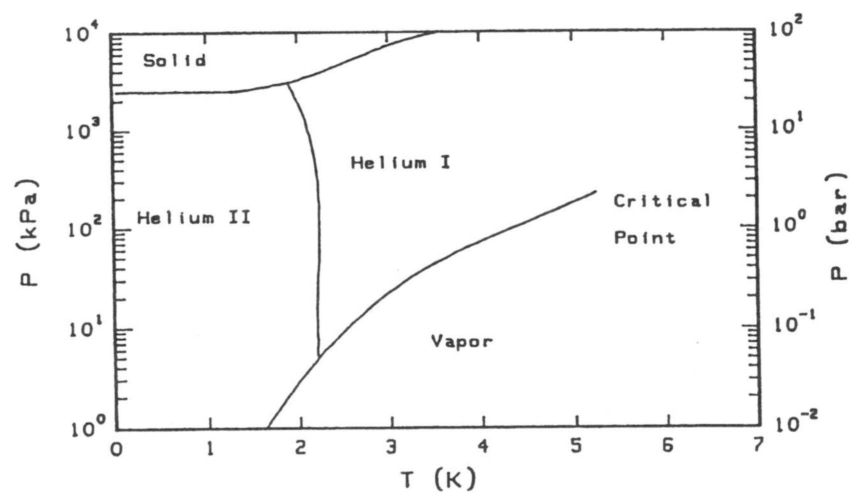

Figure 1 Phase diagram of helium showing the regions of He I, He II, and solid helium. The line separating He I and He II is called the lambda line; at saturated vapor pressure, the lambda point is at 2.172 K.

the temperature, and follow along the vapor pressure curve in Figure 1. From 5.2 K to 2.172 K we are in a liquid phase called helium I. Experimentally the liquid is seen to boil under reduced pressure just as any normal liquid would. Below $T_\lambda = 2.172$ K, the so-called lambda transition, the liquid enters a special state known as helium II, which exhibits superfluidity. Above the lambda transition, the liquid behaves as a classical fluid called helium I: "classical" in the sense that it obeys the Navier-Stokes equation and the expected boundary conditions.

When the viscosity of helium II is measured in a rotating-cylinder viscometer, the result is of order 20μp and is temperature dependent. On the other hand helium II will flow through fine capillary tubes with no pressure drop whatever. This apparent *viscosity paradox* is conventionally explained by imagining that helium II consists of some sort of mixture of two fluids: that is, a normal fluid of density ρ_n, velocity $\mathbf{v}_n$, and viscosity η, and a superfluid of density ρ_s, velocity $\mathbf{v}_s$, and zero viscosity. Thus the total density ρ is found by

$$\rho = \rho_n + \rho_s \tag{1}$$

and the mass flux $\mathbf{j}$ is

$$\mathbf{j} = \rho_n \mathbf{v}_n + \rho_s \mathbf{v}_s. \tag{2}$$

Physicists understand that the normal fluid, while used in a continuum sense, is really the sum of "elementary excitations" of the fluid, which are studied experimentally by inelastic neutron scattering. The predominant excitations are called *phonons* and *rotons*.

Evidence that the viscosity of the superfluid is truly zero is dramatically illustrated by the use of a superfluid gyroscope (Figure 2). Here a toroidal flow channel is packed with jeweler's rouge which increases superflow velocities, making detection easier. The entire apparatus is set in rotation above T_λ and gradually cooled while rotating. At some $T < T_\lambda$ the rotation is stopped but the superfluid continues to rotate. This is best seen by using the toroid as a fluid gyroscope—any attempt to tilt the channel is accompanied by a precessional motion. Such experiments show that superflows can persist indefinitely.

Thermodynamic evidence shows that the entropy, S, of helium II belongs entirely to the normal component; the entropy of the superfluid is zero.

There is in addition to the effects we have been discussing the *fountain pressure* (Figure 3*a*). Two volumes of liquid helium connected by a fine capillary exhibit the unusual behavior of raising the pressure when the temperature is increased on one side. The connection between the pressure and temperature gradients is

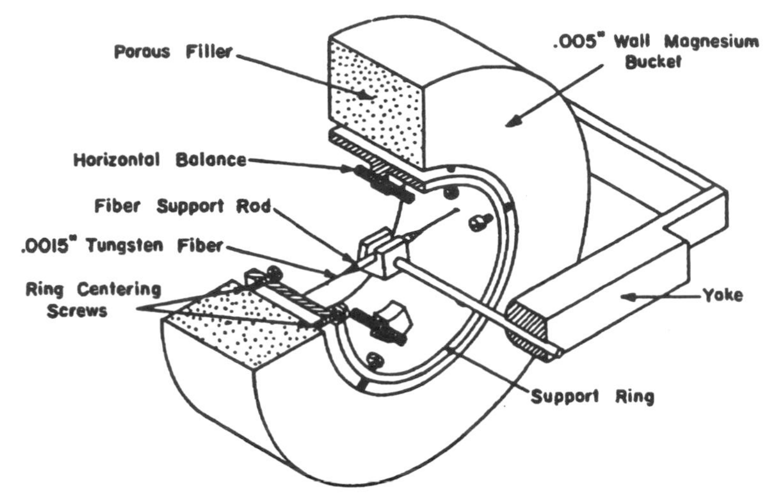

Figure 2 A superfluid gyroscope developed by Reppy's group at Cornell University. The porous filter is used to raise the characteristic velocity of superfluid circulation. The angular momentum of the circulating flow is measured by its gyroscopic effects. (After Kukich 1970.)

$$\Delta P/\Delta T = \rho S. \tag{3}$$

A modification of the fountain pressure apparatus is shown in Figure 3*b*, which depicts the celebrated helium fountain.

The classic device for measuring ρ_n is shown in Figure 4: the Andronikashvili pile-of-disks experiment. The torsion pendulum shown consists of a pile of disks with spacing small compared to the viscous penetration depth at the frequency of the pendulum. The moment of inertia of the pendulum consists of the disk assembly plus the entrained normal fluid. The period of oscillation is measured as a function of temperature and the results determine ρ_n and hence ρ_s from Equation (1). The results for ρ_n/ρ and ρ_s/ρ found by this and other methods are shown in Figure 5.

An unusual flow peculiar to helium II is illustrated in Figure 6. A channel, heated at one end and cooled at the other end draws superfluid to the heater, and since there is no net mass flux j, Equation (2) shows that normal fluid must counterflow to conserve mass. A vigorous submerged jet can be observed coming from the cool end of the channel, which, however, has no moving parts.

If the heater in Figure 6 is switched on and off periodically, the two fluids will set up a periodic reversing counterflow that amounts to a longitudinal standing wave of *second sound*. Second sound is a direct consequence of the two-fluid model and consists of temperature or entropy

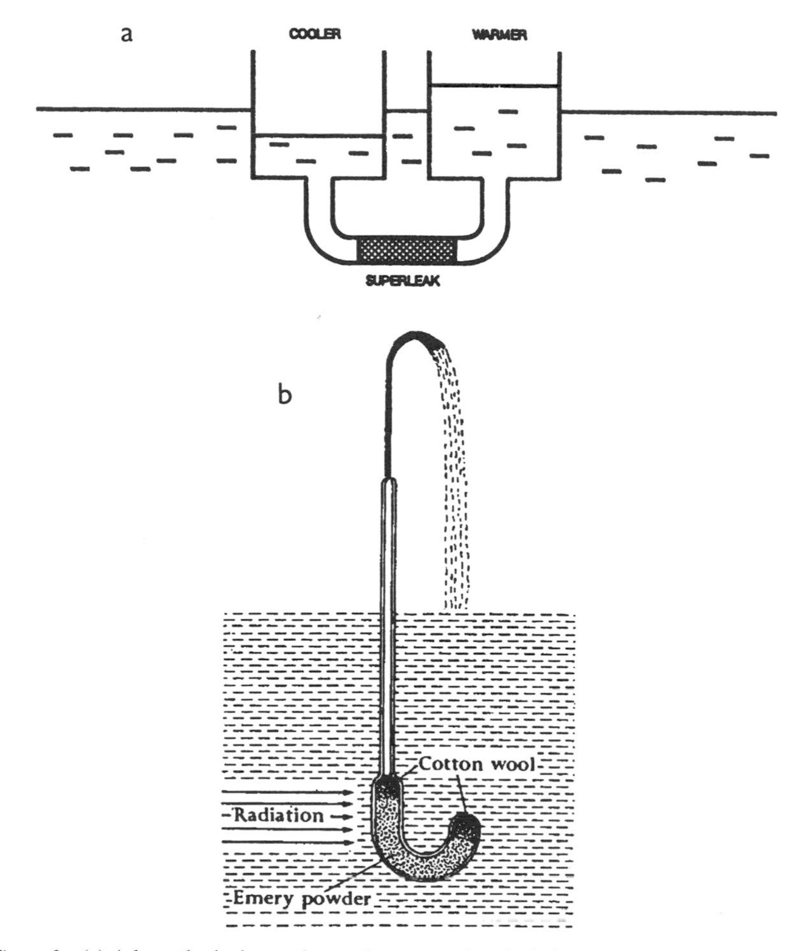

Figure 3 (*a*) A hypothetical experiment demonstrating the behavior of two vessels of He II connected by a superleak. (*b*) A helium fountain making use of the fact that the emery powder immobilizes the normal fluid, and the superfluid flowing toward a source of heat overshoots to produce a vigorous jet.

fluctuations (as distinct from density fluctuations for first sound). Second sound reaches velocities as high as ~ 20 m/sec at some temperatures. The velocity of second sound goes to zero at the lambda transition and, of course, does not exist in helium I. There can be second sound shock waves by analogy to ordinary shock waves in helium II.

The two-fluid model has equations of motion which can be written for small velocities:

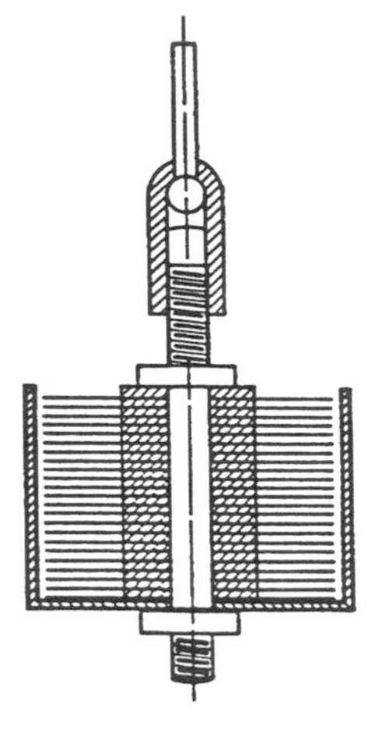

Figure 4 The pendulum used in the Andronikashvili pile-of-disks experiment.

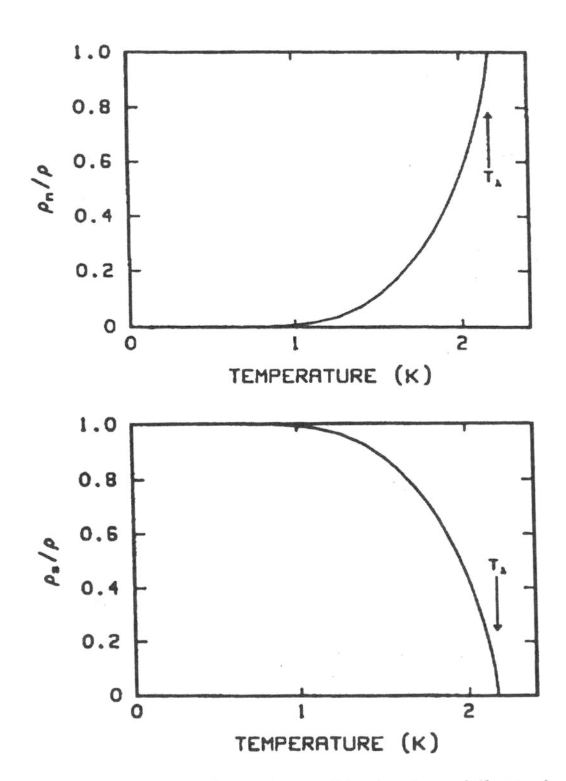

Figure 5 Temperature dependence of (*top*) ρ_n/ρ and (*bottom*) ρ_s/ρ.

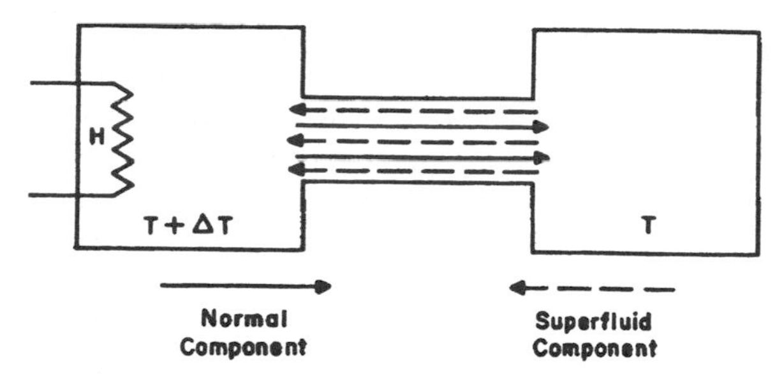

Figure 6 Closed tube containing helium heated at one end and in contact with a heat reservoir at the other.

$$\rho_s \frac{D\mathbf{v}_s}{Dt} = -\frac{\rho_s}{\rho}\nabla p + \rho_s S \nabla T \tag{4}$$

$$\rho_n \frac{D\mathbf{v}_n}{Dt} = -\frac{\rho_n}{\rho}\nabla p - \rho_s S \nabla T + \eta \nabla^2 \mathbf{v}_n \tag{5}$$

$$\text{curl}\,\mathbf{v}_s = \mathbf{0}. \tag{6}$$

Equation (4) is an Euler equation for the superfluid and (5) is a Navier-Stokes equation for the normal component. The terms $\rho_s S \nabla T$ represent the fountain pressure. When $\mathbf{v}_n = \mathbf{v}_s$, (4) and (5) add up to a Navier-Stokes equation for the total fluid. Ordinarily $\mathbf{v}_n \neq \mathbf{v}_s$ because of the irrotational restriction (6) which was put forward by Landau. It denies vorticity to the superfluid.

Experiments by various groups in the 1950s soon showed that (4–6) cannot be complete. The solution for steady rotation in a bucket, for example, (see Figure 8) shows that the depth of the meniscus must be ρ_n/ρ as deep as for a classical fluid. In practice, however, the meniscus is always its full classical depth. The cause of this discrepancy turned out to be the phenomenon of quantized vortices.

3. DISCOVERY AND PROPERTIES OF QUANTIZED VORTICES

3.1 *Onsager's Ideas*

The idea of quantized circulation in superfluid helium was first put forth to students and colleagues at Yale University by Lars Onsager beginning about 1946. Onsager enjoyed the drama of an important scientific announcement and made public his discovery in a remark following a paper by Gorter on the two-fluid model at the Conference on Statistical

Mechanics in Florence in 1949. He said, in part, "Thus the well-known invariant called the hydrodynamic circulation is quantized; the quantum of circulation is h/m. . . . In the case of cylindrical symmetry, the angular momentum per particle is $\hbar$" (Onsager 1949). Enormous ramifications of this single remark have come about and it has been observed more than once that the ratio of scientific insight to length of announcement must be a record in the history of science.

Onsager did not follow his announcement with detailed papers. His next useful remark on the subject appeared in London's book on superfluids in 1954. London quotes (London 1954, p. 151) an unpublished remark by Onsager at the 1948 Low Temperature Physics Conference at Shelter Island. There Onsager considered the situation in a rotating bucket and concluded that the liquid rotates in a series of concentric cylinders each with an integral number of quanta of circulation and with the central cylinder at rest. The multiple connectivity of the cylinders allowed him to retain the conditions of zero divergence and zero curl for the superfluid while supporting finite circulation. At reasonable rates of rotation, this array of concentric regions is indistinguishable from a classical rotation, and accounts for the observation of a full classical meniscus. This tendency for the superfluid to adopt an approximately classical flow distribution crops up in other experiments and proved to be a barrier in establishing the correctness of these ideas experimentally.

Onsager's calculation contains a number of elements common to many discussions on quantized circulation. First, the circulation $\kappa = h/m = 9.97 \times 10^{-4}\,\mathrm{cm}^2$ depends on Planck's constant h and the mass of the helium atom m. It is interesting that m is taken to be the exact mass of the bare helium atom. That is a surprising result in itself, considering how strongly coupled atoms really are in a liquid. This result is widely accepted as true, but has not been verified experimentally beyond the level of order 1% obtained in determining the quantum of circulation κ.

3.2 *Feynman Vortices*

As it happened, Feynman (1955) was working on the same problem and came to a somewhat different conclusion. He considered that the vortices in the superfluid might take the form of a vortex filament with a core of atomic dimensions—truly a vortex line. In this picture, the multiple connectivity of a vortex arises because the superfluid is somehow excluded from the core and circulates about the core in quantized fashion. Feynman considered what core radius would result if the core (of radius a) were hollow and determined by the surface tension of the fluid. This gave him a core radius of order an Ångstrom. He estimated the energy per unit length (tension) of such vortices as

$$\varepsilon = \int \tfrac{1}{2}\rho_s v_s^2 \, dr^2 = (\rho_s \kappa^2/4\pi) \ln b/a, \tag{7}$$

where ρ_s is the superfluid density, b is the radius of the bucket, or the mean distance between vortices, and v_s is the superfluid velocity. This is an enormous energy: assuming $b/a = 10^7$, it amounts to 1.85×10^{-7} erg/cm or 13.4 K/Å. The centrifugal force on each ring of fluid surrounding the core is balanced by a pressure gradient given by

$$\frac{dp}{dr} = \frac{\rho_s v_s^2}{r} = \frac{\rho_s \kappa^2}{4\pi^2 r^3}. \tag{8}$$

A sketch of the velocity and pressure distributions around a vortex line is shown in Figure 7.

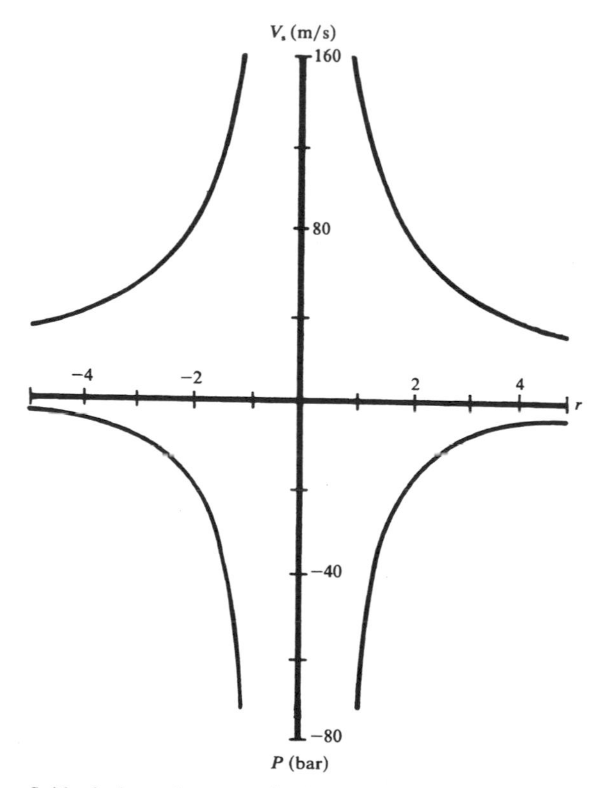

Figure 7 Superfluid velocity and pressure distributions about a rectilinear vortex line located at $r = 0$ with a core radius of 1 Å and circulation $\kappa = h/m$. In an actual quantized vortex, the velocity is unlikely to exceed the Landau critical velocity w_L of 60 m/s (after Glaberson & Donnelly 1986).

Feynman also conjectured how the vortex lines might be arranged. First, note that since the circulation κ enters squared in the energy, a doubly quantized-vortex line would have four times the energy of a singly quantized line, and would likely be unstable to breakup into four separate lines. No accepted experimental evidence for multiple or fractional quantization of vortex lines has ever appeared. Then Feynman noted that in uniform rotation the curl of the velocity is the circulation per unit area, and the curl is 2Ω for the normal fluid. He stated that there should be

$$n_0 = \frac{\text{curl } \mathbf{v}_s}{\kappa} = \frac{2\Omega}{\kappa} = 2 \times 10^3 \Omega \tag{9}$$

lines arranged parallel to the axis of rotation and with approximately uniform density. This equation applied to more general flows is called "Feynman's rule." It states that the number density of quantized-vortex lines is the ratio of the vorticity in the normal fluid to the quantum of circulation. We shall see in Section 4 below that there is increasing evidence in many flow experiments that vorticity in the superfluid tends to match vorticity in the normal fluid.

Figure 8 shows a rotating bucket with an array of quantized vortices, each having circulation $\kappa = h/m = 9.97 \times 10^{-4}$ cm^2/sec and arranged to give the superfluid the vorticity of classical rotation: $n_0\kappa = 2\Omega$, where n_0 is the area density of vortices. The density n_0 is about 2000 lines/cm^2/radian

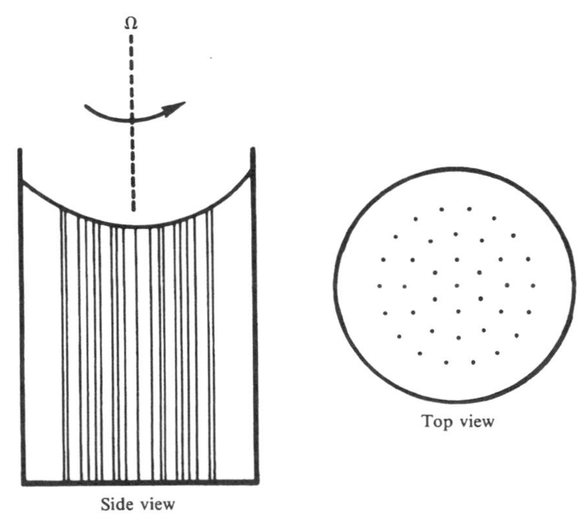

Figure 8 Array of quantized vortices in a rotating bucket. Some vortices are missing near the outer edge.

per second rotation rate. Thus the spacing between vortices is about 10^{-2} cm.

Feynman also thought about the nature of turbulence in the superfluid. His speculations were brief and to the point (Feynman 1955):

> In ordinary fluids flowing rapidly and with very low viscosity the phenomenon of turbulence sets in. A motion involving vorticity is unstable. The vortex lines twist about in an even more complex fashion, increasing their length at the expense of the kinetic energy of the main stream. That is, if a liquid is flowing at a uniform velocity and a vortex line is started somewhere upstream, this line is twisted into a long complex tangle further downstream. To the uniform velocity is added a complex irregular velocity field. The energy for this is supplied by pressure head.
>
> We may imagine that similar things happen in helium. Except for distances of a few Angstroms from the core of the vortex, the laws obeyed are those of classical hydrodynamics. A single line playing out from points in the wall upstream (both ends of the line terminate on the wall, of course) can soon fill the tube with a tangle of line. The energy needed to form the extra length of line is supplied by the pressure head. (The force that the pressure head exerts on the lines acts eventually on the walls through the interaction of the lines with the walls.) The resistance to flow somewhat above initial velocity must be the analogue in superfluid helium of turbulence, and a close analogue at that.

Although Feynman's (1955) conception of vortices uses Onsager's (1949) basic circulation hypothesis, the idea of discrete vortex lines leads to a different phenomenology. The logarithmic factor which we see in (7) will appear frequently in our discussions and arises because we are dealing with vortex lines and not simply quantized circulation.

Early experiments also revealed that the counterflow shown in Figure 6 soon fills with a homogeneous tangle of quantized vortices above some threshold heat flux.

4. VORTEX DYNAMICS AND MUTUAL FRICTION

In the case of rotation as in Figure 8, or counterflow, as in Figure 6, once vortices appear, they provide a mechanism to couple the two fluids together. This is called "mutual friction" and was first studied by Hall and Vinen at Cambridge in the 1950s (Hall & Vinen 1956). With mutual friction, the irrotational condition (6) is dropped and (4) and (5) become

$$\rho_s \frac{D\mathbf{v}_s}{Dt} = -\frac{\rho_s}{\rho}\nabla p + \rho_s S \nabla T - \mathbf{F}_{ns} \tag{10}$$

$$\rho_n \frac{D\mathbf{v}_n}{Dt} = -\frac{\rho_n}{\rho}\nabla p - \rho_s S \nabla T + \mathbf{F}_{ns} + \eta \nabla^2 \mathbf{v}_n. \tag{11}$$

Again, if $\mathbf{v}_n = \mathbf{v}_s$ these equations add up to a Navier-Stokes equation for the total fluid. A great deal is known about $\mathbf{F}_{ns}$, but the details need not concern us here (Donnelly 1991a).

4.1 *Measurement of Superfluid Vorticity: Second Sound and Ion Trapping in Helium II*

Second sound (a longitudinal wave) is greatly attenuated by the presence of quantized vortices. For the rotating bucket of Figure 8, propagation across the bucket is strongly attenuated while propagation parallel to the axis is scarcely affected. Hall & Vinen (1956) showed that the attenuation coefficient for second sound in rotating helium is

$$\alpha = B\Omega/2u_2, \tag{12}$$

where u_2 is the velocity of second sound, and B is a temperature-dependent coefficient of order unity which must be measured or obtained from theory. In such an experiment the vorticity is $\omega = 2\Omega$ and the vortex-line density is $n_o = 2\Omega/\kappa$ cm^{-2}. However n_o can also be thought of as the line density L in cm/cm^3 of fluid. The result (12) comes from assuming that the mutual-friction term $\mathbf{F}_{ns}$ in (10) and (11) can be written

$$\mathbf{F}_{ns} = -B(\rho_s\rho_n\omega/2\rho)(\mathbf{v}_n - \mathbf{v}_s) \tag{13}$$

since the vorticity $\omega = \kappa L = 2\Omega$. For a turbulent flow such as in a counterflow channel the equivalent assumption is that

$$\mathbf{F}_{ns} = -B\delta(\rho_n\rho_s\omega/2\rho)(\mathbf{v}_n - \mathbf{v}_s), \tag{14}$$

where again $\omega = \kappa L$ is the magnitude of the vorticity and δ is a constant of order unity which depends on the details of the flow and its measurement. In turbulence experiments, values of L of 0–1,500,000 cm^{-2} can be generated. The least line-density resolution one can currently achieve is about 20 cm^{-2} corresponding to detecting a change in vortex-core volume of about one part in 10^{14}.

We shall discuss second-sound attenuation for turbulent flows in Section 5.11.

Negative ions will stick to quantized vortices at temperatures below $\sim$1.7 K. In an array of vortices in rotating helium such as in Figure 8, a beam of ions produced by an α-emitting radioactive source will be attenuated passing across the bucket. The missing ions can be moved up the vortex lines by electric fields and collected at the top. Ions have also been used by Schwarz and his colleagues at IBM to measure the density L of vortex line in a turbulent counterflow (Awschalom et al 1984). Various strategies can be used to obtain spatial information on the line density.

We therefore have the remarkable situation that the magnitude of the vorticity can be measured directly in helium II, in contrast to the situation in classical fluids.

4.2 *Dynamics of Quantized Vortex Rings and the Localized Induction Approximation*

It is possible to gain some insight into vortex dynamics and the structure of the cores of vortices of constant circulation by considering the classical expressions for the energy and velocity of vortex rings in an inviscid fluid. We can show that such rings can be described by a total energy formally equivalent to a Hamiltonian H and that the velocity v and impulse P of the vortex rings are connected by Hamilton's equation (Roberts & Donnelly 1970)

$$u = \partial H/\partial P. \tag{15}$$

The simplest situation occurs when the core radius a is negligible in size compared to the radius of the vortex ring. Then the equations for a hollow-core model can be written

$$E = \tfrac{1}{2}\rho_s\kappa^2 R\,[\ln(8R/a) - \tfrac{3}{2}] \tag{16}$$

$$v = (\kappa/4\pi R)\,[\ln(8r/a) - \tfrac{1}{2}] \tag{17}$$

$$P = \rho_s\kappa\pi R^2. \tag{18}$$

Application of (15) to (16) using (18) gives (17). Quantized vortex lines obviously do not have hollow cores and in general vortex-ring formulae may be written

$$E = \tfrac{1}{2}\rho_s\kappa^2 R\,[\ln(8R/a) - \alpha]$$

$$v = (\kappa/4\pi R)\,[\ln(8r/a) - \beta], \tag{19}$$

where α and β depend on the specific core model used (see Donnelly 1991a, section 1.6).

4.2.1 VORTEX RING PROPAGATION BELOW 1 K One of the most important experiments ever carried out on helium II is the measurement of the properties of ions moving through helium II below 1 K performed by Rayfield & Reif (1964). They showed experimentally that when ions are drawn through an electric field at low temperatures the ions become attached to vorticity in the liquid and this vorticity takes the form of quantized vortex rings, with the ion located somewhere on the core of a vortex ring. Furthermore, the experiments showed that the circulation about the rings is exactly one quantum, as defined in Section 3 above. The experiments established that vortex rings obey equations of the form (16) to (18) and have drag on them at finite temperatures well accounted for by drag forces of the type discussed later in this review.

4.2.2 THE LOCALIZED INDUCTION APPROXIMATION A significant generalization of the vortex ring calculation was obtained by Arms & Hama (1965). Its importance lies in making possible approximate calculations of the motion of arbitrary configurations of very thin vortex lines. The fluid velocity at some point in space, induced by a vortex line, is given by an exact analogy to the theory of electromagnetism called the Biot-Savart law. In this analogy the fluid velocity corresponds to the magnetic field **H** and vorticity $\boldsymbol{\omega}$ corresponds to the current density **j**. The equation $\boldsymbol{\omega} = \text{curl}\,\mathbf{v}$ is analogous to Maxwell's equation $\text{curl}\,\mathbf{H} = (4\pi/c)\mathbf{j}$. The integration is over all of the vortex singularities in the fluid; boundary effects are included by extending the integral to the images of the singularities.

If an arbitrary segment of vortex line is parameterized as in Figure 9, then the Biot-Savart law can be written

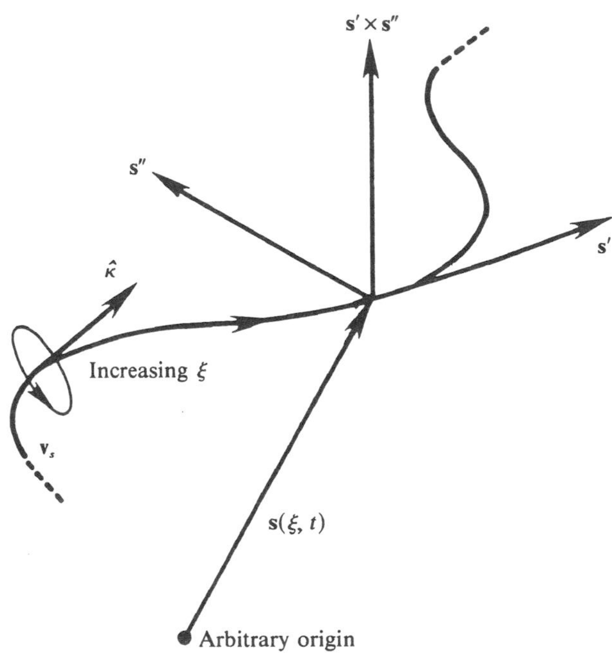

Figure 9 The space curve represents a vortex line with its position described as $\mathbf{s}(\xi, t)$. The local tangent $\mathbf{s}' = \hat{\boldsymbol{\kappa}}$, where $\hat{\boldsymbol{\kappa}}$ is a unit vector along the vortex line in the direction of $\boldsymbol{\kappa}$. Here $\mathbf{s}' = d\mathbf{s}/d\xi$ and $\mathbf{s}'' = d^2\mathbf{s}/d\xi^2$ is the local curvature vector (whose magnitude is $1/R$). The binormal $\mathbf{s}' \times \mathbf{s}''$ is approximately in the direction of the local induced velocity $\mathbf{v}$, and also has the magnitude $1/R$. The instantaneous velocity of the line is given by $\mathbf{v}_L = d\mathbf{s}(\xi, t)/dt = \dot{\mathbf{s}}$.

$$\mathbf{v} = (\kappa/4\pi) \int (\mathbf{s}_o - \mathbf{r}) \times d\mathbf{s}_o / |\mathbf{s} - \mathbf{r}|^3, \tag{20}$$

where $\mathbf{r}$ is any point in the fluid and the integral is over the relevant line segments. If $\mathbf{r} = \mathbf{s}_o$ is a point *on* the line, the integral in (20) diverges. If we expand $\mathbf{s}$ in a Taylor series about $\mathbf{s}_o$ then (20) becomes

$$\mathbf{v} \approx (\kappa/4\pi) \int (d\xi/2\xi) \mathbf{s}' \times \mathbf{s}'', \tag{21}$$

where the integral is over the whole vortex array except for a distance of the order of the core radius a on either side of $\mathbf{s}_o$. Ignoring "nonlocal" portions of the vortex (more than some distance L away from $\mathbf{s}_o$ measured along the arc) and approximating the cross-product by its value at $\mathbf{s}_o$, we get the local self-induced velocity

$$\mathbf{v}_i \approx (\kappa/4\pi) \ln (L/a) \mathbf{s}' \times \mathbf{s}'', \tag{22}$$

where $\mathbf{s}' \times \mathbf{s}''$ has the magnitude $1/R$, R being the radius of curvature at $\mathbf{s}_o$. In scalar terms

$$v_i \approx (\kappa/4\pi R) \ln (L/a). \tag{23}$$

The choice of L to best approximate (20) depends on the details of the vortex configuration.

If we wish to include the far field of the vortex we can make a rough approximation by integrating (20) over the line omitting a region along the vortex of length L on either side of $\mathbf{s}_o$. Then

$$\mathbf{v}_{\text{(nonlocal)}} = (\kappa/4\pi) \int' (\mathbf{s}_o - \mathbf{r}) \times d\mathbf{s}_o / |\mathbf{s} - \mathbf{r}|^3, \tag{24}$$

where the prime on the integral indicates that the local line element is omitted.

An immediate application of (23) is an intuitive understanding of how a vortex ring moves through a fluid: If L is replaced by R we have an approximation to (17). A vortex ring moves through the fluid principally because of its curvature. As an example a ring with $a = 10^{-8}$ cm and $R = 10^{-3}$ cm moves through the liquid at about 1 cm/s. A numerical simulation was carried out in our laboratory using a mesh of 100 points on the circumference. The localized induction, cut off at nearest neighbors from a selected origin, provided 73% of the total induced velocity.

The forces on vortex lines begin with a generalization of the lift force on a cylinder with circulation $\boldsymbol{\kappa}$ about it which is in a flow $\mathbf{v}_\infty$, $\mathbf{f}_M = \rho \mathbf{v}_\infty \times \boldsymbol{\kappa}$. This would apply to a vortex line at absolute zero with the modification that the line can move at a velocity $\mathbf{v}_L$ different from the local superfluid velocity. Thus $\mathbf{v}_\infty$ is replaced by $(\mathbf{v}_L - \mathbf{v}_{sl})$ where $\mathbf{v}_{sl}$ is the sum of the flow $\mathbf{v}_s$ plus any self-induced flow $\mathbf{v}_i$. Now suppose we are at finite temperatures. The only modification to the lift force from absolute zero is to use ρ_s in place of the total density ρ:

$$\mathbf{f}_M = \rho_s \kappa \mathbf{s}' \times (\mathbf{v}_L - \mathbf{v}_{sl}). \tag{25}$$

But now the vortex line will be moving through a gas of excitations, phonons, and rotons. Assuming that the friction owing to collisions with the line depends in a first approximation on the relative velocity between the line and the normal fluid, the drag force can be written

$$\mathbf{f}_D = -\gamma_o \mathbf{s}' \times [\mathbf{s}' \times (\mathbf{v}_n - \mathbf{v}_L)] + \gamma_o' \mathbf{s}' \times (\mathbf{v}_n - \mathbf{v}_L), \tag{26}$$

where the two components of the drag, and friction coefficients γ_o and γ_o' recognize that the circulation about the line is directional, and that collision of a roton with the line (in the sense of kinetic theory) will likely have cross sections for parallel and perpendicular exchange of momentum.

While (26) is expressed in terms of $(\mathbf{v}_s - \mathbf{v}_L)$ it is equally useful and valid to describe the drag in terms of $(\mathbf{v}_n - \mathbf{v}_{sl})$ where $\mathbf{v}_n$ and $\mathbf{v}_{sl}$ are macroscopically-averaged line, normal fluid, and superfluid velocities. Such definitions are interchangeable, but inevitably introduce extra friction coefficients α and α' as follows:

$$\mathbf{f}_D = -\alpha \rho_s \kappa \mathbf{s}' \times [\mathbf{s}' \times (\mathbf{v}_n - \mathbf{v}_{sl})] - \alpha' \rho_s \kappa [\mathbf{s}' \times (\mathbf{v}_n - \mathbf{v}_{sl})]. \tag{27}$$

If we neglect the inertia of the core, then the sum of the forces on each line element must vanish and we have from (25) and (27)

$$\mathbf{f}_M + \mathbf{f}_D = \mathbf{0} \tag{28}$$

which yields:

$$\begin{aligned} \mathbf{f}_M + \mathbf{f}_D &= \rho_s \kappa \mathbf{s}' \times (\mathbf{v}_L - \mathbf{v}_{sl}) - \alpha \rho_s \kappa \mathbf{s}' \times [\mathbf{s}' \times (\mathbf{v}_n - \mathbf{v}_{sl})] - \alpha' \rho_s \kappa \mathbf{s}' \times (\mathbf{v}_n - \mathbf{v}_{sl}) \\ &= \rho_s \kappa \mathbf{s}' \times [(\mathbf{v}_L - \mathbf{v}_{sl}) - \alpha \mathbf{s}' \times (\mathbf{v}_n - \mathbf{v}_{sl}) - \alpha' (\mathbf{v}_n - \mathbf{v}_{sl})] \\ &= \mathbf{0}. \end{aligned} \tag{29}$$

Thus the term in square brackets must be in the direction $\mathbf{s}'$ or equal to zero. The latter condition leads to

$$\mathbf{v}_L = \mathbf{v}_{sl} + \alpha \mathbf{s}' \times (\mathbf{v}_n - \mathbf{v}_{sl}) - \alpha' \mathbf{s}' \times [\mathbf{s}' \times (\mathbf{v}_n - \mathbf{v}_{sl})]. \tag{30}$$

Equation (30) is a fundamental result in vortex dynamics which can be used for a variety of purposes including simulations of the development of three-dimensional vortex motion in an arbitrary flow.

4.3 *Uniformly Rotating Helium II*

The various results and definitions of the previous section, by themselves, give little insight as to their origin and usefulness. The first application of mutual friction was to relate the attenuation of second sound in uniformly rotating helium II to a more microscopic view of frictional forces on an isolated vortex line. Consider rotation of a container which is sufficiently rapid to give us a uniform array of vortices such as is pictured in Figure 8. When equilibrium is reached the sum of the force per unit volume on the normal fluid plus the drag force per unit volume on the vortices must total zero:

$$\mathbf{F}_{ns} + n_o \mathbf{f}_D = \mathbf{0}. \tag{31}$$

From (27) we have

$$\mathbf{f}_D = -\alpha \rho_s \kappa \mathbf{s}' \times [\mathbf{s}' \times (\mathbf{v}_n - \mathbf{v}_s)] - \alpha' \rho_s \kappa \mathbf{s}' \times (\mathbf{v}_n - \mathbf{v}_s), \tag{32}$$

where $\mathbf{v}_s$ is used in the place of $\mathbf{v}_{sl}$ because the lines are assumed to be straight and there is no self-induced velocity. Equations (31) and (32) combine to give

$$\mathbf{F}_{ns} = \omega \rho_s \alpha \mathbf{s}' \times [\mathbf{s}' \times (\mathbf{v}_n - \mathbf{v}_s)] + \omega \rho_s \alpha' \mathbf{s}' \times (\mathbf{v}_n - \mathbf{v}_s), \tag{33}$$

where the vorticity $\omega = 2\Omega$. Second-sound resonance experiments give values of α and α'.

A more general case of mutual friction arises if the vortex lines are not straight, but curved. Under these conditions the self-induced velocity v_i will not be zero. If the curvature of the vortex lines is slow enough so that the local radius of curvature R can be used in the Arms-Hama approximation (22), then we can write $v_i = (\kappa/4\pi R)\ln(L/a)$ where L is some characteristic microscopic length: the radius of curvature, or the interline spacing, for example. Referring to Figure 9 the self-induced velocity will be in the direction of $\mathbf{s}' \times \mathbf{s}''$ which has the magnitude $1/R$:

$$\mathbf{s}' \times \mathbf{s}'' = (\mathbf{s}' \cdot \nabla)\mathbf{s}' \tag{34}$$

and we have approximately

$$\mathbf{v}_i \approx \nu (\mathbf{s}' \cdot \nabla)\mathbf{s}', \tag{35}$$

where by (7)

$$\nu = (\kappa/4\pi)\ln(L/a) = \varepsilon/\rho_s \kappa. \tag{36}$$

If the curved lines are dense enough to form a continuum, then (30) and (31) can be used in the equation of motion with

$$\mathbf{v}_{sl} = \mathbf{v}_s + \nu(\mathbf{s}\cdot\nabla)\mathbf{s}' \tag{37}$$

and (33) is generalized to read

$$\mathbf{F}_{ns} = \rho_s\omega\alpha\mathbf{s}'\times[\mathbf{s}'\times(\mathbf{v}_n-\mathbf{v}_s)] + \rho_s\omega\alpha'\mathbf{s}'\times(\mathbf{v}_n-\mathbf{v}_s) + \rho_s\omega\alpha\nu\,\mathrm{curl}\,\mathbf{s}' + \rho_s\omega\alpha'\nu(\mathbf{s}'\cdot\nabla)\mathbf{s}', \tag{38}$$

where the last two terms arise from v_i. The third term uses the vector relationship

$$(\mathbf{s}'\cdot\nabla)\mathbf{s}' = -\mathbf{s}'\times\mathrm{curl}\,\mathbf{s}'.$$

At $T = 0$, $\mathbf{v}_L = \mathbf{v}_{sl}$ which we now simply call $\mathbf{v}_s$ so that the superfluid equation of motion becomes

$$\partial\mathbf{v}_s/\partial t = \mathbf{v}_s\times\boldsymbol{\omega} + \nu\omega(\mathbf{s}'\cdot\nabla)\mathbf{s}' + \nabla\mu. \tag{39}$$

Absorbing $\mathbf{v}_s\times\boldsymbol{\omega}$ into $d\mathbf{v}_s/dt$, (10) is generalized to read

$$\rho_s\frac{d\mathbf{v}_s}{dt} = -\frac{\rho_s}{\rho}\nabla p + \rho_s S\nabla T + \rho_s\nu\omega(\mathbf{s}'\cdot\nabla)\mathbf{s}' - \mathbf{F}_{ns}. \tag{40}$$

The normal fluid equation remains unchanged in form. Equations (11), (38), and (40) are called the Hall-Vinen-Bekarevich-Khalatnikov equations (HVBK). They are the most general equations of motion in use today. They have been used successfully to study the Taylor-Couette stability problem in helium II (Barenghi & Jones 1988, Swanson & Donnelly 1991).

4.4 *Vortex Waves*

4.4.1 KELVIN WAVES Since vortex lines have an energy per unit length, they have tension, and therefore can sustain wave motions. In particular, we can consider a helical deformation of wavenumber k and amplitude d, where $d \ll k^{-1}$ (see Figure 10).

Ignoring the nonlocal contribution and effective mass of the core, the line moves with the Arms-Hama velocity (23), where L is reasonably taken as being of order k^{-1}. When $d \ll k^{-1}$, this velocity is perpendicular to the undisturbed line and to the displacement vector from the undisturbed line to the point considered. Each vortex-line element therefore executes motion about the undisturbed line in a circle of radius d with a frequency $\omega = v_i/d$. The radius R is given approximately by $R \sim 1/dk^2$ so that

$$\omega(k) \sim (\kappa k^2/4\pi)\ln(1/ka). \tag{41}$$

The correct expression was given by Lord Kelvin (Thomson 1880) and is

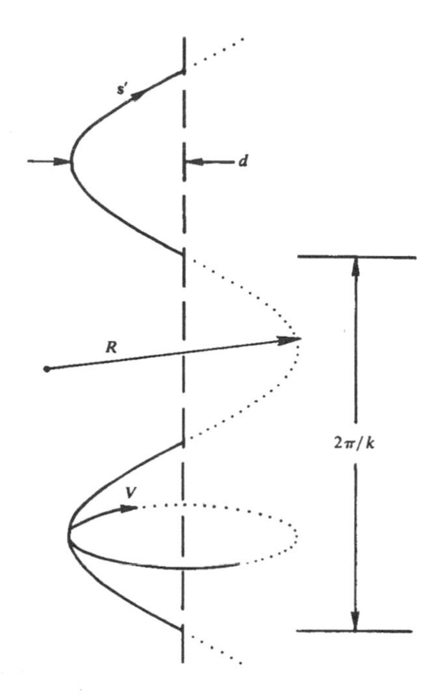

Figure 10 A vortex line deformed into a helix. The amplitude of the deformation is d and its wavenumber is k; R is the radius of curvature of the line at some point and $\mathbf{s}'$ is a unit vector along the vortex line indicating the direction of the circulation κ (after Glaberson & Donnelly 1986).

$$\omega^* = \frac{\kappa}{2\pi a^2}\left(1 \pm \left\{1 + ka\left[\frac{K_o(ka)}{K_1(ka)}\right]\right\}^{1/2}\right), \tag{42}$$

where K_n is a modified Bessel function of order n. The corresponding dispersion curve is shown in Figure 11, which is a purely classical result modified to match the quantized circulation of vortices in helium II.

4.4.2 TKACHENKO WAVES The behavior above is for an isolated vortex line. If we now consider an array of vortex lines, for example as generated in a rotating frame, then every vortex core will move in the field produced by all other vortex cores, and we can, in general, expect a very complicated interaction to occur between vortices, which is simple only in certain specialized cases.

One such case was first investigated by Tkachenko (1966a,b) who considered an infinite array of classical vortices; he found that triangular lattices were stable, and that the normal modes of such a lattice (plane waves) consisted of elliptical motions of the vortices about their equilibrium positions. The resulting wave motion is illustrated in Figure 12. Experimental evidence for the existence of these waves has been obtained by Andereck & Glaberson (1982).

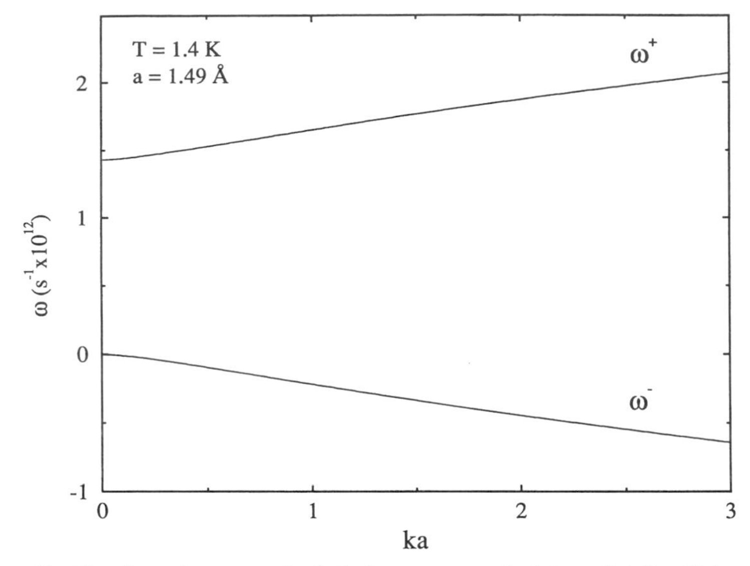

Figure 11 The dispersion curves for helical vortex waves in helium II (after Glaberson & Donnelly 1986).

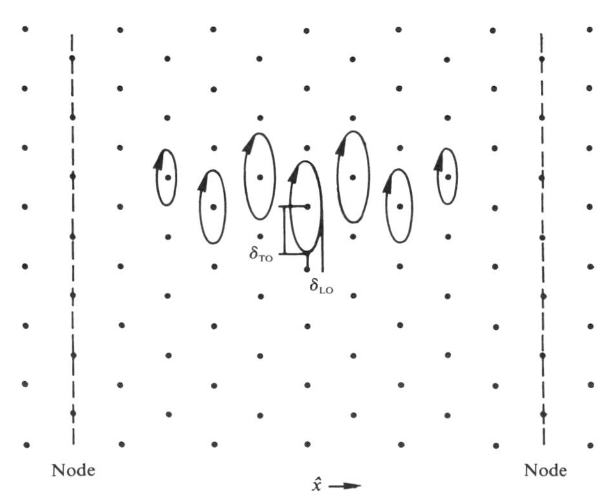

Figure 12 A Tkachenko standing wave mode. The dots are the undisturbed triangular vortex lattice points. δ_{T0} and δ_{L0} are the semimajor and semiminor axes of the elliptical paths executed by the vortices (after Glaberson & Donnelly 1986).

4.4.3 INSTABILITY OF KELVIN WAVES Samuels & Donnelly (1990) have carried out computer simulation studies of vortex waves. They were surprised to find that Kelvin waves at $T = 0$ K are unstable above a threshold amplitude. They used the Biot-Savart law and the Arms & Hama localized induction approximation as discussed above.

The system studied consisted of a vortex line extended between two parallel planes 10^{-3} cm apart. The boundary conditions chosen were that the vortex must meet the boundaries normally, and can slip along the boundary. These conditions were met by the method of images. The vortex line was modeled by $N = 128$ straight line vortices.

For initial conditions the simulations were begun with a superposition of a planar wave and two neighboring sidebands of small amplitude as a perturbation. As the equations of motion are integrated forward in time the amplitude of the sidebands starts to grow due to a Benjamin-Feir (1967) instability. After some time, however, the main harmonic begins to grow and the strength of the sidebands declines as shown in Figure 13. The recurrence continues in the sense of the Fermi-Pasta-Ulam phenomenon (Fermi et al 1965). If the initial amplitude is increased, the simple recurrence is not observed. Instead a state of "confined chaos" is observed as shown in Figure 13*b*.

4.4.4 MOTION OF A VORTEX RING WITH AN ASYMMETRIC FORCE The interpretation of vortex-ring experiments below 1 K (Section 4.2.1) is that the force on the ring due to the action of the electric field on the trapped ion is uniform. Actually, the ion remains confined to one location on the ring, so that the force on the ring owing to the ion is localized and hence asymmetric. Samuels & Donnelly (1991) investigated this problem numerically and analytically, and found that the localized force on the ion will combine with a specific series of vortex waves to give a uniform growth of the ring with a small component of drift velocity perpendicular to the applied electric field. An example of the lowest modes of waves on the vortex-ring core is shown in Figure 14. This type of energy transfer may also play an important role in superfluid turbulence. The same calculations should be valid for localized forces on thin-core classical vortices as well.

4.5 *Vortex Coupled Superfluidity*

The two-fluid behavior of helium II with normal and superfluid velocity fields moving completely separately is observed only at the lowest velocities. Almost all systems exhibit superfluidity at Reynolds numbers above something of order 10 to 100, but the two fluids move together, and the kinematic viscosity appears to be η/ρ rather than η/ρ_n. The most common example of this is the rotating bucket shown in Figure 8. Here the kinematic

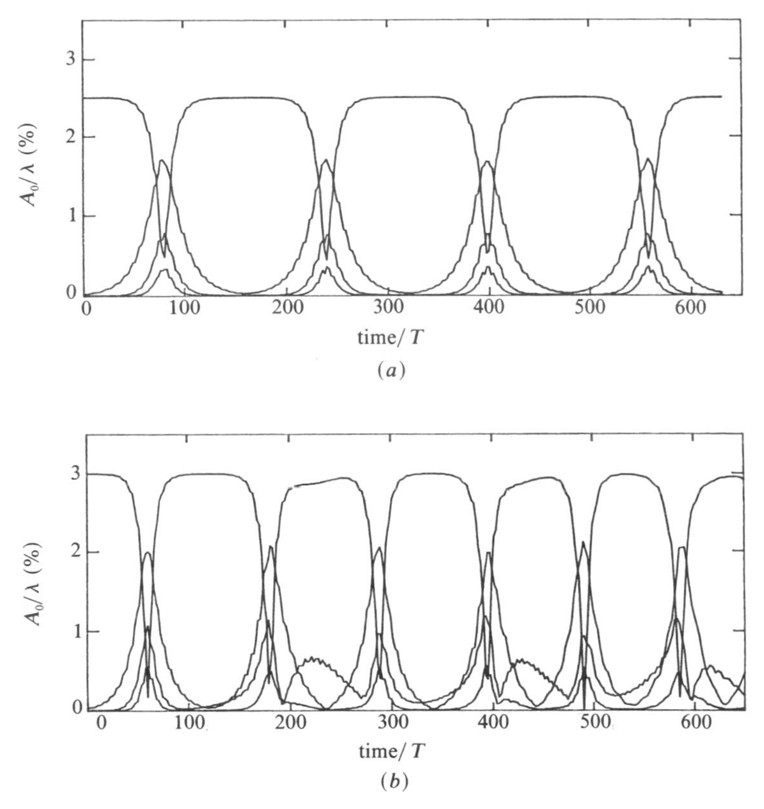

Figure 13 Sideband instability and recurrence phenomenon for a quantized vortex line. At $t = 0$ we impose a cosine wave of $n = 11$ half waves of amplitude $A_0 = 0.025\lambda$ with two sidebands (10 and 12 half waves with small amplitude). (*a*) Plotted are the amplitude of the main harmonic ($n = 11$) and the lower harmonics (10, 9, 8 in order of decreasing amplitude) of the three closest sidebands as a function of time. A plot of the upper harmonics ($n = 12$, 13, 14) looks very similar. (*b*) The same initial conditions as (*a*) except that the initial amplitude of the cosine wave is raised to produce what is called "confined chaos."

viscosity is known to be η/ρ from spinup experiments, but second sound can still be transmitted through the fluid. Other examples include the flow near an oscillating disk. At low amplitudes the superfluid ignores disk oscillations such as the pile shown in Figure 4. But at higher amplitudes, experiments show that the two fluids are coupled together, and the pile of disks measures the total density instead of the normal fluid density. Various experiments show that the flow of a submerged jet (such as the apparatus of Figure 3*b* with the exit submerged) has both fluids entrained. One of the most astonishing results is the measurement of the friction factor for flow through a pipe by Walstrom et al (1988), who were able to show that

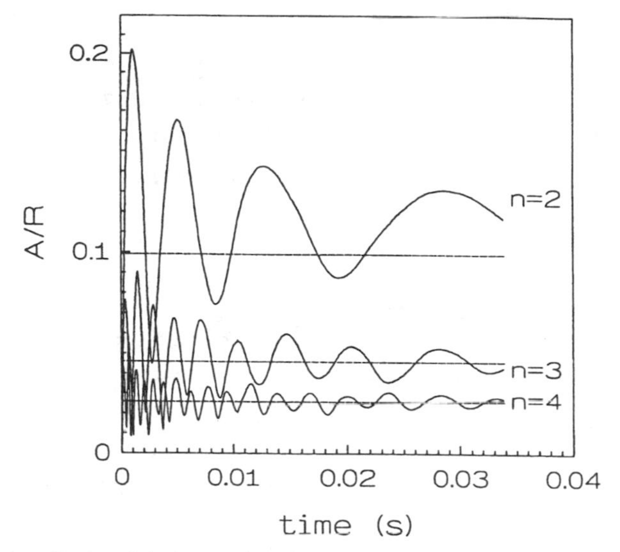

Figure 14 Amplitudes of the lowest three harmonics of waves on a vortex core in response to the force from an electric field on a trapped ion in the core. The straight lines are the amplitudes predicted analytically (Samuels & Donnelly 1991).

the friction factor obeys the classical correlation up to Reynolds numbers as high as 2,000,000. Their results are shown in Figure 15. At face value this result would suggest that the flow and boundary conditions for this geometry are completely classical. On second thought one realizes that superfluidity still exists in the sense of vortex-coupled superfluidity and therefore it is not intuitively obvious how this apparently classical result could occur. A thorough discussion of the situation is given by Donnelly (1991b) and the phenomenon is named "vortex-coupled superfluidity" because the two fluids are coupled together by vortices, yet superfluidity still exists.

The exact nature of vortex-coupled superfluidity is still a mystery, and very much a matter of current research in several laboratories.

4.6 *Vortex Ring Propagation Above 1 K*

One of the most famous experiments in liquid helium is the Rayfield-Reif study of quantized vortices in helium II (see Section 4.2.1). Below 1 K ions become attached to singly quantized vortex rings, whose propagation through helium II has revealed so much about the nature of superfluidity. Above 1 K, however, the situation is dramatically different. Figure 16 shows an apparatus constructed by Borner, Schmeling & Schmidt (1983)

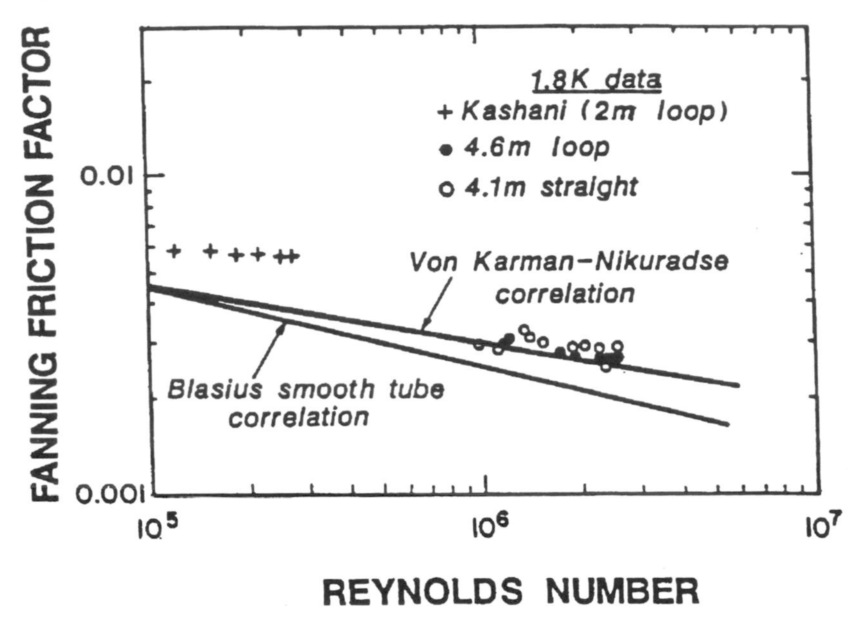

Figure 15 Friction factor for a smooth pipe in helium II. The solid line is the classical correlation; the crosses represent data from a rough pipe (Walstrom et al 1988).

to study the propagation of classical vortex rings created by a plunger. The rings are observed by means of beams of first and second sound and an array of sound receivers placed above. The method of circulation measurement uses the flow-induced variations (Doppler effect) of the traveling times of first-sound and second-sound shock waves. The evolution of large-scale vortex rings with nearly homogeneous vorticity distribution was observed and measured. Their translational velocity and normal and superfluid circulations were measured at temperatures $1.28 < T < 2.02$ K in the region 1–7 orifice diameters downstream from the orifice. The total normal circulation in the rings and superfluid circulation are identical at all temperatures studied and all piston velocities U_M studied as shown in Figure 17.

They found that the vortex ring circulations in both fluids are equal even at the closest measurable distance from the orifice of the vortex ring generator. The Reynolds numbers for the flow inside the ring generator were 20,000 to 40,000. Particle visualization experiments on vortex rings were carried out by Murakami and his students (see Ichikawa & Murakami 1991). They concluded that these vortex rings propagate classically.

This experiment suggests that in a flow beginning in a tube and ejected

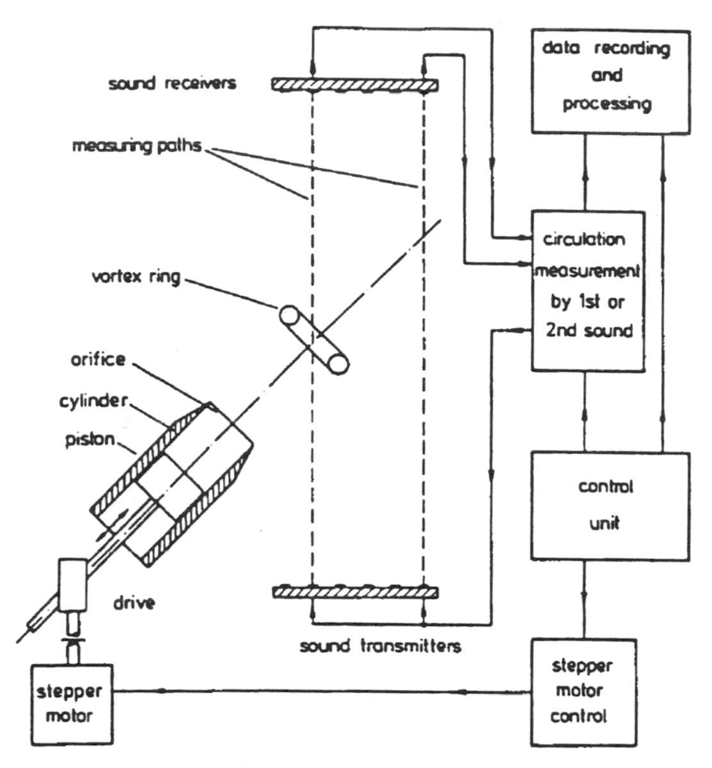

Figure 16 Schematic drawing of the test setup used by Borner et al (1983) and by Borner & Schmidt (1985) for the generation and measurement of large-scale vortex rings in helium II using first sound and second sound shock waves.

into quiescent fluid-free walls, the force of mutual friction can apparently equalize the normal and superfluid velocity fields, and, perhaps more fundamentally, the vorticity fields become matched.

4.7 *Studies of Vorticity Matching in Normal and Superfluid Flows*

If the flows are completely classical, then the vorticity in the normal and superfluid flows would have to "match" in some sense. How can such vorticity matching occur?

First, one should appreciate that much of the experimental data is limited to external flow variables such as pressure and temperature differences. It is probably correct to say that the mean flows are "classical" but one knows little about fluctuations in this exotic two-fluid environment.

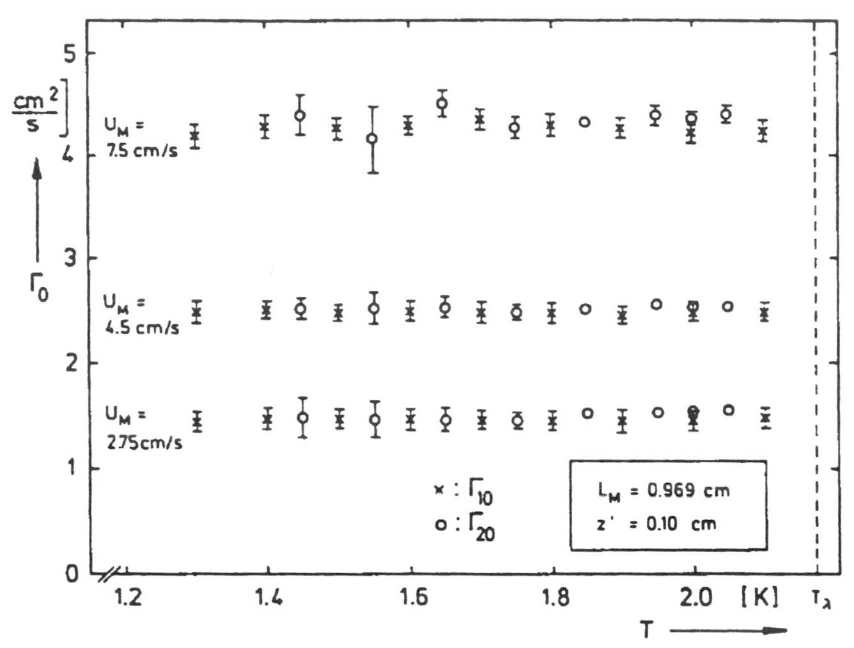

Figure 17 Normal Γ_{10} and superfluid circulations Γ_{20} in vortex rings as a function of temperature determined by the apparatus of Figure 16. Piston velocities are denoted U_M and correspond to a constant piston stroke L_M. The steady part of the vortex ring profile (that is, ignoring the behavior at the edges from acceleration and retardation of fluid) corresponds to $0.5L_MU_M$, the classical value.

David Samuels (private communication) has been examining this problem with vortex simulations. Samuels' study was to determine the physics responsible for vorticity matching. Since we know that not all types of helium II flow show vorticity matching (for example, at very low Reynolds numbers), examining the limits of this matching is also an objective of current research.

Samuels models vortex filaments by N nodes connecting straight vortex line segments. The velocity of each node is calculated using the Biot-Savart law and mutual friction. Boundary conditions are met by the method of images. The position of each node is updated using an adaptive time-step method (Runge-Kutta-Fehlberg) on a fast computer. When vortex filaments cross, provisions are made to reconnect the filament mesh. Since the Biot-Savart law is a nonlocal integral, a straightforward implementation is an order N^2 process. This limits one to a maximum N of a few thousand, running on a Cray Y-MP. The full Biot-Savart law is required for these simulations because a local approximation would ignore the nonlocal interactions that are the most important when the length of the vortex filament is large.

Samuels finds that there are two necessary conditions for vorticity matching:

(*a*) A one-dimensional (or higher) region where $\mathbf{v}_n = \mathbf{v}_s$ must initially exist in the fluid.

(*b*) A source of superfluid vorticity must be present. Residual vortex lines pinned to the walls is often assumed to fill this need.

The region with $\mathbf{v}_n = \mathbf{v}_s$ must be at least one dimensional so that a superfluid vortex filament can fit within this region. The normal fluid at the boundary of the matched velocity region will have some vorticity ω_n. Superfluid vortex filaments with a component of vorticity in the same direction as ω_n are transported by mutual friction from the superfluid vorticity source to the boundary of the region of matched velocities. As the filaments accumulate here, the superposition of their velocity fields extends the matched velocity region to a larger (three-dimensional) volume.

A useful example of a flow which satisfies condition (*a*) is the normal fluid vortex. Assuming for simplicity that $\mathbf{v}_n$ is a vortex flow with some core structure and $\mathbf{v}_s = \mathbf{0}$, then on the axis of the normal fluid vortex $\mathbf{v}_n - \mathbf{v}_s$ goes to zero and condition (*a*) is met. Superfluid vortex filaments with the same orientation of circulation as the normal fluid vortex will be attracted to the normal fluid vortex core by mutual friction and filaments of the opposite circulation will be repulsed. If some source of superfluid vorticity [condition (*b*)] exists then we would expect normal fluid vortices to show vorticity matching. This type of flow is important to our understanding of the vorticity matching of high Reynolds number flows in superfluid helium since it provides a simple model for the interaction of superfluid vortex filaments with concentrated vortex structures in the turbulent normal fluid flow.

A recent (unpublished) study by Samuels considers pipe flow of helium II where the normal fluid is considered to be in Poiseuille flow and the superfluid in plug flow (Figure 18). A single half ring shown at the bottom of Figure 19, presumably from some pinned residual vorticity in the pipe, grows in the flow as shown but can only grow in the region where the normal fluid velocity exceeds the superfluid velocity. The place where they are equal is called the nodal surface. The way in which the growth process works is complicated—especially if there are multiple locations that are growing—but simulations show that the result is an increasing accumulation of vortex rings in the vicinity of the nodal surface which spread

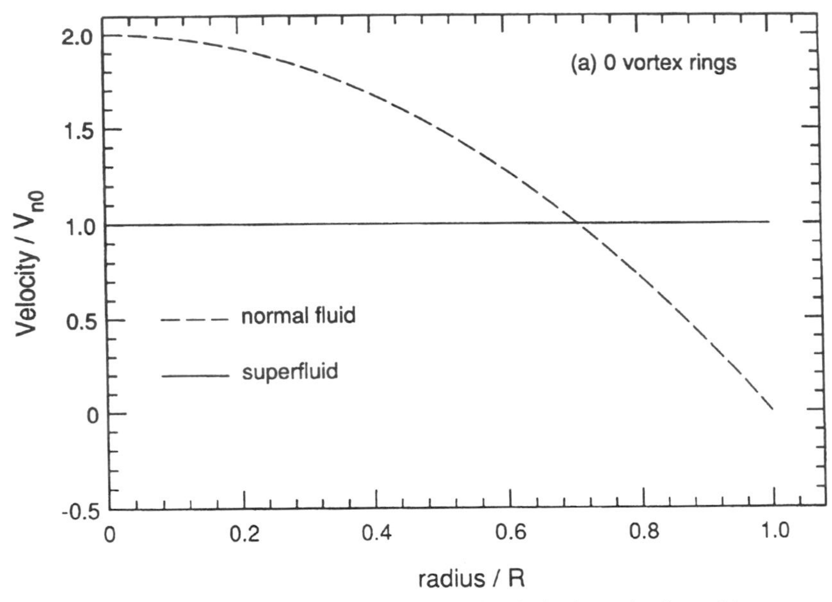

Figure 18 Superfluid and normal fluid velocity profiles in laminar pipe flow with no superfluid vortex filaments.

radially inward and outward. The ultimate effect of these rings is to make the superfluid profile start to match the normal profile as shown in Figure 20.

5. TURBULENCE IN SUPERFLUID HELIUM

5.1 *Counterflow Turbulence*

The simplest form of turbulence in helium II, and by far the most studied and understood, occurs with zero mass flow as suggested in Figure 6. The remarkable quote from Feynman given in Section 3.2 foresaw accurately what the nature of this form of turbulence might be. In the next few sections we discuss counterflow turbulence, then turn our attention to grid turbulence and turbulence in open flows, which have the greatest contact with turbulence in ordinary fluids.

A nontechnical review of superfluid turbulence produced by counterflow has appeared in Scientific American (Donnelly 1988) and a thorough review has been published by Donnelly & Swanson (1986).

Figure 19 Time lapse growth of a superfluid vortex filament in the spanwise plane of the pipe. The initial state is the half ring at the bottom of the figure. The filament does not grow through the nodal surface into the center of the pipe. (After D. Samuels, private communication.)

5.2 *Application of Vortex Dynamics to Superfluid Turbulence*

We have shown that if the inertia of the vortex cores is neglected, the motion of an element of line such as is shown in Figure 9 is given by Equation (30):

$$\mathbf{v}_{\mathrm{L}} = \mathbf{v}_{\mathrm{sl}} + \alpha \mathbf{s}' \times (\mathbf{v}_{\mathrm{n}} - \mathbf{v}_{\mathrm{sl}}) - \alpha' \mathbf{s}' \times [\mathbf{s}' \times (\mathbf{v}_{\mathrm{n}} - \mathbf{v}_{\mathrm{sl}})]. \tag{30}$$

Here $\mathbf{v}_{\mathrm{L}} = \dot{\mathbf{s}}, \mathbf{v}_{\mathrm{sl}}$ consists of the vector sum of the potential flow of the superfluid at large distances from any vortex line, and $\mathbf{v}_{\mathrm{i}}$ is the superflow induced by any curvature of the vortex line in the sense of our discussion of the localized induction approximation. In classical incompressible hydrodynamics two geometrically similar flows are dynamically similar at the same Reynolds number. This powerful concept underlies scaling of flows in all of classical incompressible fluid dynamics. The question then becomes, how can we accomplish scaling in quantum fluid dynamics? It is not difficult to make (30) dimensionless: We need only divide by v_{ns} which we define to be the absolute value of the spatial and temporal average of

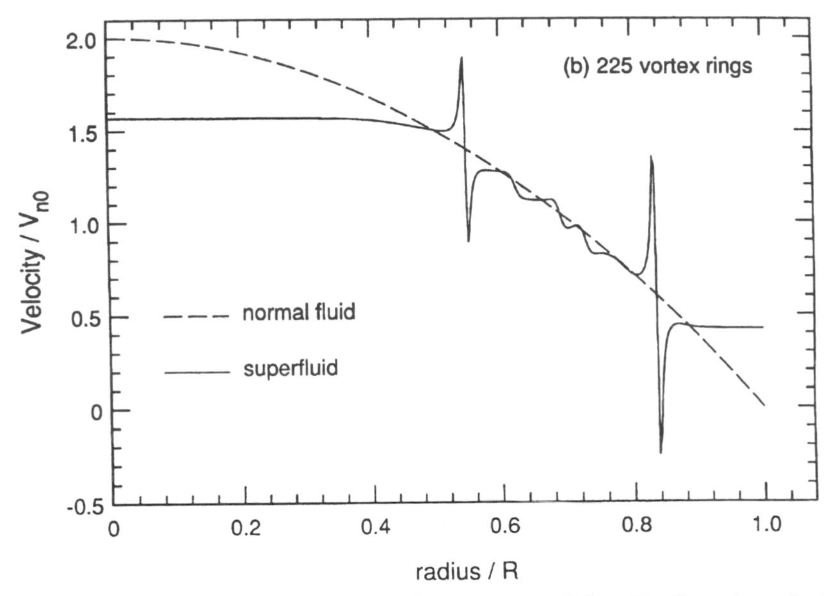

Figure 20 Velocity profiles with 225 rings present. (After D. Samuels, private communication.)

$\mathbf{v}_{ns} = \mathbf{v}_n - \mathbf{v}_s$ in the flow of interest. Such a procedure need not necessarily be restricted to flows with zero mass flow. We then have the dimensionless equation

$$\mathbf{u}_L = \mathbf{u}_i + \alpha \mathbf{s}' \times (\mathbf{u} - \mathbf{u}_i) - \alpha' \mathbf{s}' \times [\mathbf{s}' \times (\mathbf{u} - \mathbf{u}_i)], \tag{43}$$

where we have defined the dimensionless vortex-line velocity in the frame of the superfluid by

$$\mathbf{u}_L = (\mathbf{v}_L - \mathbf{v}_s)/v_{ns}, \tag{44}$$

a dimensionless tangle-induced velocity

$$\mathbf{u}_i = \mathbf{v}_i / v_{ns}, \tag{45}$$

and a dimensionless counterflow velocity

$$\mathbf{u} = \mathbf{v}_{ns} / v_{ns}. \tag{46}$$

5.3 *Dynamical Similarity*

Let us investigate dynamical similarity using Equation (43). Consider two geometrically similar flows whose characteristic dimension is d. Time will scale with d/v_{ns}, all tangle lengths will scale with d, and the vortex ratio $\mathbf{u}_i$

must be the same for similar points on the tangles in each flow. The friction parameters α and α' must be the same in each flow, requiring the temperature to be the same. Since, α and α' are slightly frequency- and velocity-dependent, dynamical similarity will be only approximate as we change velocities. This problem can be neglected in many practical situations. The core parameter a however, depends only on temperature and cannot scale with d. This fundamental difficulty is ameliorated somewhat by the fact that the ratio d/a appears in a logarithmic term, as we shall now see.

The induced velocity $\mathbf{v}_i$ is given by the Biot-Savart law (20) and can be approximated as in (21) when the integral is over the whole tangle except for some distance a_{eff} on either side of $\mathbf{s}_o$, where a_{eff} is of order a. Since the radius of curvature $R = |s''|^{-1}$ and $\Gamma = \kappa$ we have

$$v_i = \frac{\kappa}{4\pi R} \ln\left(\frac{R_n}{a_{\text{eff}}}\right) = \frac{\kappa}{4\pi R} l, \tag{47}$$

where we use R_n in place of L in (23) to avoid confusion with the line density, and $l = \ln(R_n/a_{\text{eff}})$.

Applying (47) to two geometrically similar flows will require that $\mathbf{s}' \times \mathbf{s}''$ and thus $\mathbf{v}_i$ and $\mathbf{u}_i$ will be parallel at similar points in the two flows. Thus we need only consider the magnitude of $\mathbf{u}_i$ which from (45) and (47) is

$$|\mathbf{u}_i| = \kappa l / 4\pi R v_{ns}. \tag{48}$$

Now let us consider a whole vortex tangle. Let $R_o = \langle R \rangle$ and $l_o = \langle l \rangle$ where the brackets denote an average over the whole tangle, and

$$u_i = \kappa l_o / 4\pi R_o v_{ns}. \tag{49}$$

If $l = l_o$ throughout the tangle, then the requirement that $\mathbf{u}_i$ be the same in two flows is replaced by a requirement that u_i be the same. If l varies, equality of u_i will give approximate scaling if the distribution of l is narrow enough. Swanson (1985) and Swanson & Donnelly (1985) show in Appendix B of their paper that the equation of vortex-line motion will not differ significantly in channels of different sizes even with a relatively wide distribution.

The ideas of scaling are not particularly useful unless the necessary quantities are experimentally observable. Unfortunately R_o, a natural length in the problem, is not observable. What can be measured is the average vortex-line length L per unit volume. Since L has dimensions cm^{-2}, a length scale $L^{-1/2}$ can be deduced, and we can define approximate time-averaged proportionality constants c_1 and c_2 by

$$\overline{|\mathbf{s}''|} = c_1 L^{1/2} = R_o^{-1} \qquad \overline{|\mathbf{s}''|^2} = c_2^2 L \tag{50}$$

which are independent of d for geometrically similar flows. We need in addition to find an expression for l_o, which is a function of the geometry of the tangle. Let us define the tangle-geometry parameter g by

$$l_o = \ln(R_o/ag). \tag{51}$$

When we use the Arms-Hama approximation and the assumption of a narrow curvature distribution, $g = R_o a_{\text{eff}}/aR_n$. The parameter g is configuration- and core-model-dependent. For a hollow vortex ring described by (15–18), $g = e^2/8$. We expect that both c_1 and g will be of order unity.

With the definitions of c_1 and g, we are in a position to relate our scaling parameters to measurable quantities. In the calculations described here most quantities are tangle averages. Let us define the dimensionless applied counterflow velocity

$$V = 4\pi v_{ns} d/\kappa l_o = L^{1/2} c_1 d/u_i \tag{52}$$

where the second equality is from the definition (48) and where now

$$l_o = \ln(R_o/ag) = \ln(1/agc_1^{1/2}) \approx \ln(1/L^{1/2}a). \tag{53}$$

For dynamical similarity $L^{1/2}d$, c_1, and u_i, are independent of d. Thus V is a dimensionless velocity describing tangle properties.

If we define the quantity

$$\beta = \kappa l_o/4\pi \tag{54}$$

then V has the form of a Reynolds number with β replacing the kinematic viscosity. Since c and g are of order unity, (52) will give useful scaling without a knowledge of c or g as long as $a \ll L^{-1/2}$, i.e. one can use $l \approx \ln(1/L^{1/2}a)$ with relatively good results.

5.4 *Numerical Simulation of a Vortex Tangle*

The dynamics of vortices in helium II forms a particularly attractive problem to simulate numerically. The equations of motion, which we have discussed in Section 4, are useful in the thin-filament approximation where the Arms-Hama localized induction hypothesis is often employed as a simplification. We have referred briefly to the results of simulation studies in Section 4.4 of solitary and Kelvin waves.

The idea of applying vortex dynamics to the full problem of superfluid turbulence is a formidable one. Schwarz (1988) has nevertheless developed this approach in considerable detail, and has presented results which compare favorably with experiments.

5.5 *Vortex Reconnections*

The equations of vortex dynamics developed in Section 4 neglect all dynamical effects arising from other vortices or boundaries. Many years ago Feynman (1955) speculated that vortices approaching each other closely would reconnect as shown in Figure 21*a*. On a small enough scale, this event is surely quantum mechanical. Schwarz has observed that vortex-vortex reconnection provides a mechanism by which vortex regularities can multiply (Figure 21*c*), and generalized this idea to include vortex-surface reconnections, as shown in Figure 21*b*. He estimated how closely a line of radius of curvature R must come to a boundary before its image-induced motion becomes comparable to its self-induced motion. When two vortices approach closer than this distance, a local instability occurs

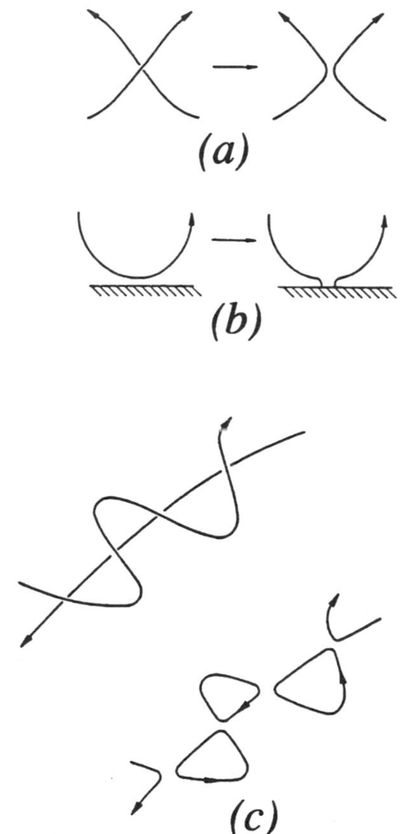

Figure 21 Illustrations of (*a*) a possible reconnection sequence between vortex filaments in a tangle, (*b*) a vortex filament reconnection at a surface, and (*c*) multiplication of singularities through the reconnection process. Here two vortex lines reconnect to form five. (After Schwarz 1988.)

in which the velocity field of each vortex acts to deform the other in such a way that the two vortices are driven together at a point where their vorticity vectors are oppositely directed. He assumed that the result of this is a sudden reconnection as shown in Figure 21 with immediate separation of the vortices to macroscopic distances.

Schwarz noted that his studies demonstrate that the details of when and how the vortices are reconnected or of how they behave immediately thereafter have no significant influence on the behavior of the vortex tangle. He referred to a simulation consisting of Equation (30) and the reconnection assumptions of Figure 21 as the *reconnecting vortex tangle model.*

From his experience Schwarz was able to give a qualitative picture of the self-sustaining vortex-tangle state (Schwarz 1988):

> The self-induced velocity causes a complicated three-dimensional internal motion of the vortex tangle, the whole thing being washed along by any applied superflow field v_s which may be present. Highly curved sections of line, and sections propagating opposite to v_{ns} decay. Simultaneously, other parts of the vortex tangle where the self-induced motion is being overtaken by the v_{ns} field grow by ballooning outwards. The cross-stream nature of the vortex growth implies that in the steady state at least a certain fraction of the singularities is constantly being driven toward the walls. The line-line reconnections which occur as the vortex tangle undergoes its complicated dance play several important roles. First, they provide a mechanism by which new vortex singularities can be created (Figure 21), allowing the vortex tangle to be established and sustained against the loss of singularities at the walls. Secondly, and more subtly, since the vortex amplification process is essentially a two-dimensional outward motion in the plane perpendicular to v_{ns}, the reconnections and the subsequent motions along v_{ns} which result are necessary to maintain the three-dimensional random nature of the vortex tangle. Finally, the reconnections occur more often as the tangle becomes denser. The increasing frictional line loss associated with the creation of a more and more highly curved vortex tangle is the factor which eventually limits the tangle density. All of these complicated dynamical features interact self-consistently to produce the turbulent steady state.

5.6 *Computational Considerations*

The evolution of a vortex tangle can be investigated beginning with an initial configuration such as is shown in Figure 22. The algorithms for stepping the vortex configuration forward in time are described by Schwarz (1985). Since the calculations are performed in a finite sample of fluid, it is necessary to specify how the boundaries are treated. The sample of fluid is always taken as a rectangular box with one set of faces perpendicular to v_{ns}. This set of faces is subject to periodic boundary conditions, i.e. one line leaving the box appears to reenter it from the opposite face. This makes the fluid appear infinite in the direction of the flow. The other boundaries are treated in one of three ways: (*a*) as periodic, if the intent

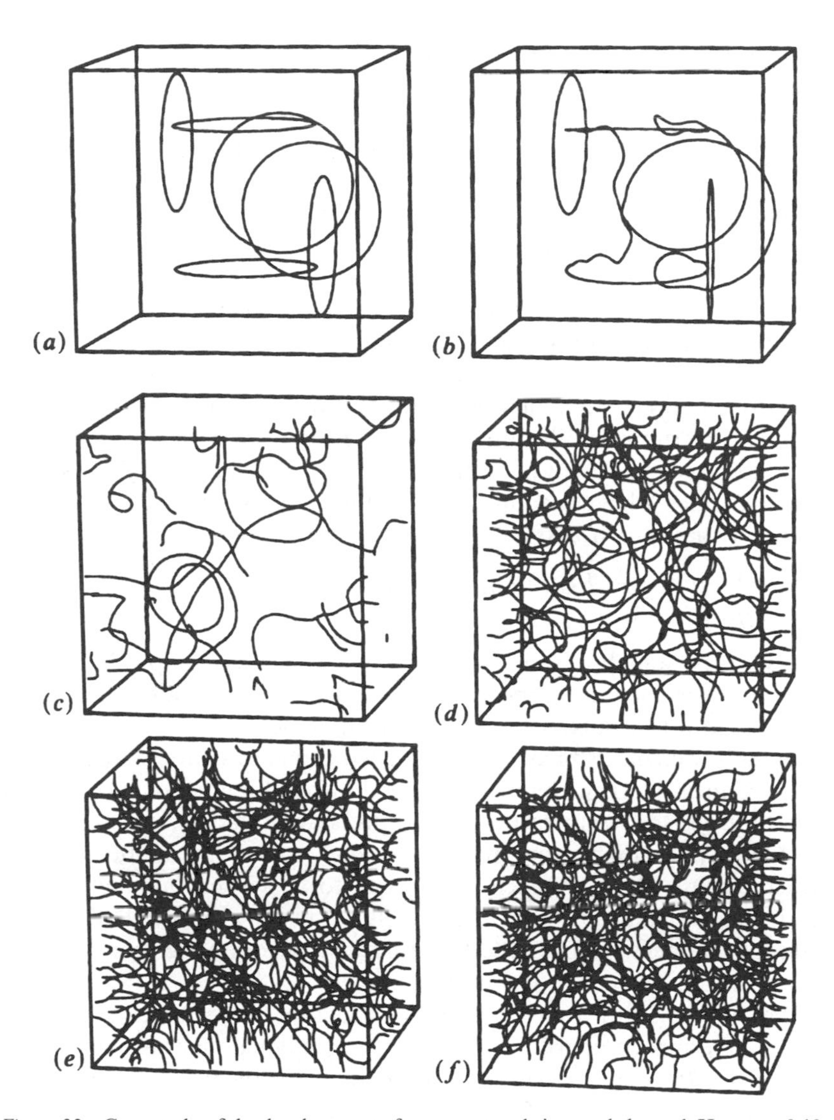

Figure 22 Case study of the development of a vortex tangle in a real channel. Here $\alpha = 0.10$, corresponding to a temperature of about 1.6 K, and $v_{s,0} = 75\ \text{cm}^{-1}$ into the front face of the channel section shown. The quantities t_o are dimensionless times. (*a*) $t_o = 0\ \text{cm}^2$, no reconnections; (*b*) $t_o = 0.0028\ \text{cm}^2$, three reconnections; (*c*) $t_o = 0.05\ \text{cm}^2$, 18 reconnections; (*d*) $t_o = 0.28\ \text{cm}^2$, 844 reconnections; (*e*) $t_o = 0.55\ \text{cm}^2$, 12,128 reconnections; (*f*) $t_o = 2.75\ \text{cm}^2$, 124,781 reconnections. (After Schwarz 1988.)

is to make the fluid unbounded in all directions, (*b*) as smooth, rigid boundaries, in which case a vortex line approaching the face will reconnect to the wall as shown in Figure 21*b*—the end then gliding freely along the wall, or (*c*) as rough rigid boundaries, in which case vortex lines terminating on the wall undergo a complicated pinning and depinning motion as they move along.

The evolution of a typical numerical experiment is shown in Figure 22. Here an initial configuration of six vortex rings is allowed to evolve in a rough channel under the influence of a pure superflow driving field. This situation is seen to evolve towards a self-sustaining chaotic steady state with well-defined average properties independent of the initial conditions.

The rough-wall calculations are straightforward, but computationally expensive. Periodic boundary conditions in all directions are much more efficient, but run into difficulties which are removed by a special mixing step (discussed in detail by Schwarz 1988). This procedure uses the assumption that lines which encounter each other closely will reconnect with essentially unit probability.

5.7 *Devices for Producing Turbulence in the Superfluid*

The simplest and by far the most common method of production of superfluid turbulence occurs when a channel, heated at a closed end, is operated at a heat flux exceeding a critical value, as shown in Figure 6. Counterflow channels break down into two broad categories: narrow and wide channels. Narrow channels such as shown in Figure 23 can be studied only by means of temperature and/or pressure gradients; wide channels have the space to be equipped for second sound attenuation or ion measurements. Wide channels, then, tend to be several millimeters to a centimeter in diameter. Narrow channels are usually capillary tubes. The range of channel sizes in counterflow turbulence is dramatic: from microns to centimeters, or four orders of magnitude!

The principles of wide-channel construction are simple. Figure 24 shows a thermal counterflow apparatus for studying quantum turbulence equipped with second sound transducers to probe the vorticity transverse to the axis of the channel. Second sound attenuation is observed to begin only above a critical heat flux q_c (where q is the ratio of power W supplied to the heater to the cross-sectional area of the channel A). The channel shown can be rotated to produce a regular array of vortices for calibration purposes or to conduct experiments with combined heat and rotation. The fully instrumented wide channel shown in Figure 24 can be used for both transverse and second-sound attenuation measurements as well as axial temperature difference measurements.

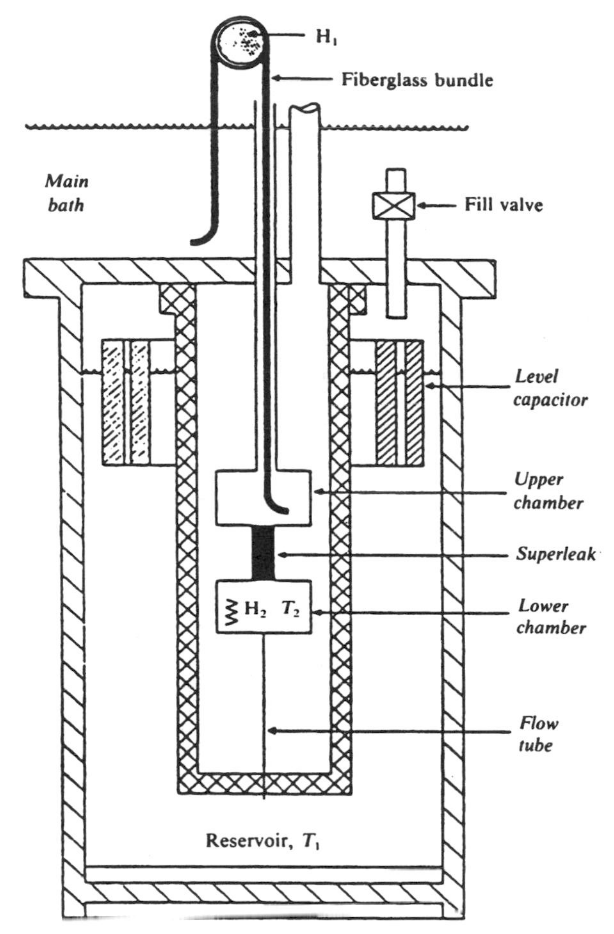

Figure 23 A modern apparatus designed by Courts & Tough (1988) to produce flows with v_s and v_n in the same or opposite directions (after Courts & Tough 1988).

Further experimental details and results of many investigations are contained in Donnelly (1991a).

5.8 *Grid Turbulence*

We have recently been conducting some experiments with a towed grid apparatus in superfluid helium which shed light on the use of helium II for high Reynolds number turbulence research (Smith & Donnelly 1991).

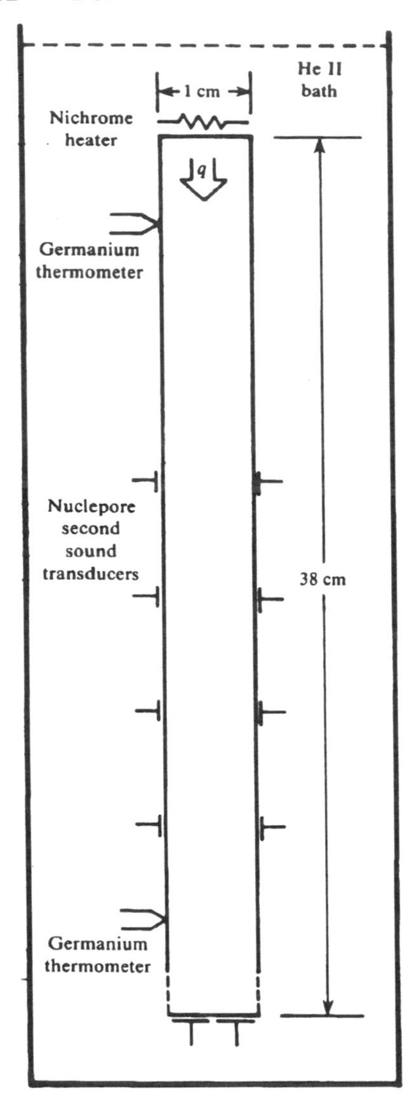

Figure 24 Schematic of one possible arrangement of wide channel interchangeable sections used at the University of Oregon. Both transverse and axial second sound transducers are installed; thermometers allow the axial temperature gradient to be determined.

5.8.1 APPARATUS The apparatus shown in Figure 25*a* consists of a brass channel whose interior cross section is 1 cm square. It is approximately 60 cm in length, and instrumented along its length with pairs of second sound transducers approximately 5 cm apart.

The grid was constructed by machining a uniform series of slots, in a 1.5 mm thick square brass wafer, approximately 1 cm on a side. The

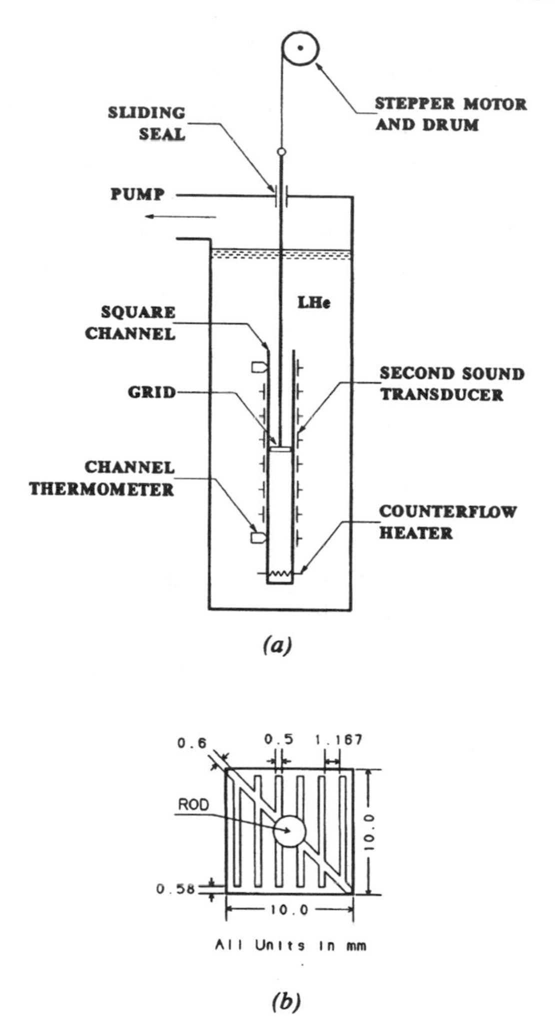

Figure 25 (*a*) Apparatus used to study the behavior of quantized vorticity behind a towed grid. (*b*) Detail of the grid construction.

remaining solid portions are approximately 0.5 mm wide and we take the mesh size as being $l_M = 0.167$ cm as shown in Figure 25*b*. We selected the slot configuration so as to minimize the contact area with the channel around the perimeter of the grid. This in turn minimizes the contribution to the turbulent field due to the shear stress created at the walls. The grid is suspended at the end of a 3/32″ rod, and the entire assembly's cross-sectional area is over 65% open. The rod extends outside the bath through a vacuum seal, and is connected to a stepper motor via a cable and drum assembly. The stepper motor allows us to locate the grid to well within 0.5 mm with our protocol. At 1.65 K, this arrangement is capable of achieving grid Reynolds numbers in the range $1.3 \times 10^3 < \mathrm{Re}_{\mathrm{grid}} < 5.2 \times 10^5$. We thus have a dynamical range extending over nearly three orders of magnitude.

The protocol was to tow the grid at various velocities from the bottom to the top of the apparatus and observe the decay at any transducer pair as shown in Figure 25*a*. The time $t = 0$ was taken to be the time the grid just passed the detector. This was not a simple matter because the hole in the channel wall exposing a nuclepore transducer is 9 mm in diameter. When the grid is occulting the transducer, the signal is degraded. We made a profile of this signal loss by drawing the grid upwards quasi-statically and taking the reference point for $t = 0$ as the location where the signal went back to 90% of its full value. Experience has shown this to be a reliable and reproducible procedure.

5.8.2 ATTENUATION OF SECOND SOUND Second sound is strongly attenuated by the presence of quantized vortices. The technique is to establish a resonance in the region of the flow of interest using porous nuclepore membranes working as capacitive loudspeakers and microphones. The pores are of order 0.1 micron in diameter and easily pass superfluid, but block the normal fluid. When set into vibration the transducers generate second sound, which is a fluctuation in the relative densities of the normal and superfluid. The resonant response is shown in the example in Figure 26, where the full width at half maximum Δ_o and the amplitude A are taken from a best fit.

The attenuation of second sound is given by

$$\alpha = \frac{\pi \Delta_o}{u_2}\left(\frac{A_o}{A} - 1\right), \tag{55}$$

where u_2 is the velocity of second sound, and Δ_o and A are taken from the second sound response curve as explained in Figure 26. The quantized vortex line density is given by the theory of mutual friction as

$$L = \frac{4u_2\alpha}{\kappa B} = \frac{4\pi\Delta_o}{\kappa B}\left(\frac{A_o}{A} - 1\right), \tag{56}$$

where B is a dimensionless coefficient of mutual friction. For a careful discussion of these equations see Swanson (1985) and for the theory of mutual friction see Donnelly (1991a).

The root mean square vorticity in the superfluid is given by

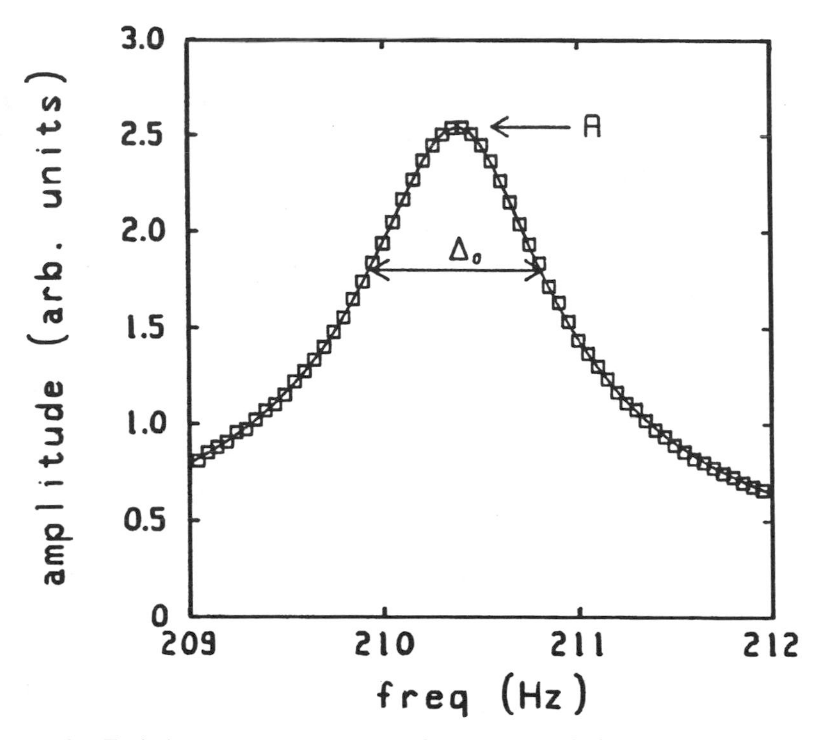

Figure 26 Typical resonance response curve for second sound of the type used to deduce line density. The quantity Δ_o is the full width at half power, and the amplitude is deduced from the peak power. The squares are data points and the curve is a best fit of a Lorentzian function.

$$\omega = \langle(\nabla \times \mathbf{v}_s)^2\rangle^{1/2} = \frac{\kappa}{2} L \tag{57}$$

and since the dimensions of L are cm^{-2}, we can deduce a characteristic length

$$l \cong \frac{1}{\sqrt{L}}. \tag{58}$$

These relationships have been used successfully for many years in the study of counterflow-induced turbulence.

5.8.3 DECAY OF GRID TURBULENCE Experimentally, homogeneous turbulence is generally created behind a grid (at least 65% open) placed within a wind tunnel test section. The turbulence thus generated has a mean velocity, and the flow is examined at various distances behind the grid. In this configuration, it is then straightforward to measure the overall time decay of the mean square velocity fluctuations, $\langle u^2 \rangle$, at a particular scale. In classical wind tunnel studies (Gence 1983, Comte-Bellot & Corrsin 1966, Bennett & Corrsin 1978) it has been found that $\langle u^2 \rangle \propto t^{-m}$, where u is a small-scale velocity fluctuation and m is a decay exponent dependent upon Reynolds number (here denoted $\mathrm{Re}_{\mathrm{grid}}$, and based upon mean oncoming velocity and grid mesh length). Experimentally it is found that $m \sim 2.51$ for $1300 < \mathrm{Re}_{\mathrm{grid}} < 1800$ and $m \sim 1.2$ for $17{,}000 < \mathrm{Re}_{\mathrm{grid}} < 135{,}000$. In fully developed turbulence, one can show that (Landau & Lifshitz 1987)

$$\langle \omega^2 \rangle \sim -\frac{1}{\nu}\frac{\partial \langle u^2 \rangle}{\partial t} \tag{59}$$

(where $\langle \omega^2 \rangle$ is the mean square vorticity, a characteristic of the turbulent field, and ν is the kinematic viscosity) so that the mean square vorticity is observed to decay approximately as $t^{-3.51}$ and $t^{-2.2}$ for low and high Reynolds numbers respectively. These decays may be compared with the inviscid (Proudman & Reid 1954) and zero Reynolds number (von Karman & Howarth 1938) solutions in which the mean square vorticity decays as t^{-2} and $t^{-3.5}$ respectively. Vinen's theory (Vinen 1957) suggests that the free decay of line density in turbulent flows of helium II goes as

$$\frac{dL}{dt} = -\chi_2 \frac{\hbar}{m_4} L^2, \tag{60}$$

where χ_2 is a constant of order unity, and $\hbar = h/2\pi$. Integration yields

$$L = \frac{L_0}{1 + \dfrac{L_0 \chi_2 \hbar}{m_4} t}, \tag{61}$$

where L_0 is the initial line density. Note that Equation (56) provides a means of relating the line density (and hence ω_{rms}) to the measured resonant amplitude at a given instant, while Equation (61) provides a possible model for its decay. Theoretically then, one expects this superfluid turbulence to decay as t^{-1} to first order. Milliken, Schwarz & Smith (1982) arrived at a similar decay law by somewhat different physical reasoning. They generated turbulence between ultrasonic transducers, and their results are in good agreement with the predicted t^{-1} time dependence. However, since there is no clear characteristic length scale or velocity, there does not seem to be any way of associating a Reynolds number with their data for the sake of comparison with classical experiments.

Experimental results expressed as a function of mesh Reynolds number seem to level off between 10,000 and 100,000 at about 1.35. The data in Figure 27 also show quite a departure from classical results at low Reynolds number. This may be due to unlocking of the two fluids, and needs more investigation.

5.8.4 HOMOGENEOUS TURBULENCE The turbulence produced by our towed-grid apparatus can reach very high line densities compared to most counterflow experiments. It is tempting to try to compare our data to classical experiments on homogeneous turbulence performed in wind tunnels over the years.

In order to progress further we must make two key assumptions which are still not fully supported by experimental and theoretical evidence. Much of the evidence for them has been discussed above. However, we shall use them since they lead to suggestive connections with the theory of homogeneous turbulence, and we can only judge their usefulness by the results. We assume first that the normal fluid acquires vorticity in the usual way for viscous fluids, but that the vorticity in the superfluid somehow matches it. This has been known for many years, but has not been exploited in turbulence studies. Thus

$$\langle(\nabla \times \mathbf{v}_s)^2\rangle = \langle(\nabla \times \mathbf{v}_n)^2\rangle = \langle(\nabla \times \mathbf{v})^2\rangle, \ \mathrm{Re} \to \infty. \tag{62}$$

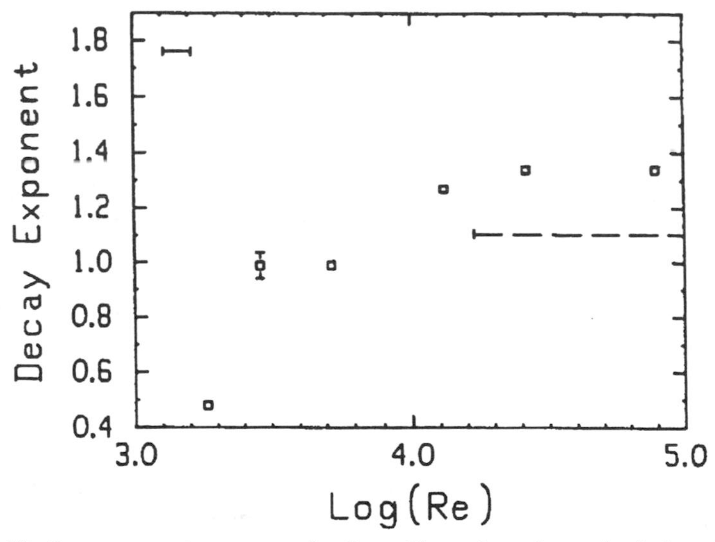

Figure 27 Decay exponent m vs generating Reynolds number, where a simple decay law of the form t^{-m} has been assumed. Horizontal dashed lines indicate classical results.

The experimental evidence at this time suggests that Re of 100 to 1000 is sufficiently high (Donnelly 1991b). In the present paper we implement this assumption by dropping the two-fluid velocity fields and using only a velocity $\mathbf{v}$. The second assumption is that the total density and not the normal-fluid density enters into the definition of the kinematic viscosity:

$$\nu = \frac{\eta}{\rho}. \tag{63}$$

It has been known for many years that a rotating bucket of liquid helium rotates classically both in spinup and in steady state (Donnelly 1956). The experimental evidence for this is summarized by Donnelly (1991b).

Once one has the vorticity, the energy dissipation per unit mass can be calculated directly from

$$\omega^2 = \langle(\nabla \times \mathbf{v})\rangle^2 = \varepsilon/\nu \tag{64}$$

(Landau & Lifshitz 1987, Section 34).

Kolmogorov scaling connects the dissipation per unit mass to the velocity fluctuations (Δu) and the mean eddy size l by

$$\varepsilon = \varepsilon_0 \frac{(\Delta u)^3}{l}, \tag{65}$$

where ε_0 is a proportionality constant to be determined by a separate experiment or by theory.

If the turbulence is homogeneous and isotropic, then

$$q = \tfrac{1}{2}\langle(u^2 + v^2 + w^2)\rangle = \tfrac{3}{2}(\Delta u)^2. \tag{66}$$

Thus for homogeneous, isotropic turbulence of sufficiently high Reynolds number, we find the measurement of second sound attenuation yields the mean eddy size (58), l, the dissipation per gram, and the kinetic energy per gram, provided ε_0 can be determined. Note that even if vorticity could be determined classically, the mean eddy size would not automatically be available. It is the quantization of circulation which makes the determination possible in helium II.

5.9 *Propagation of a Turbulent Burst*

Referring to Figure 25*a*, it is possible to position the grid above one of the transducers and draw the grid suddenly up to the top of the apparatus. One then sees attenuation in a few seconds, corresponding to the arrival of a turbulent burst propagating down into quiescent fluid. An example of the burst, plotted in terms of line density is shown in Figure 28. At first

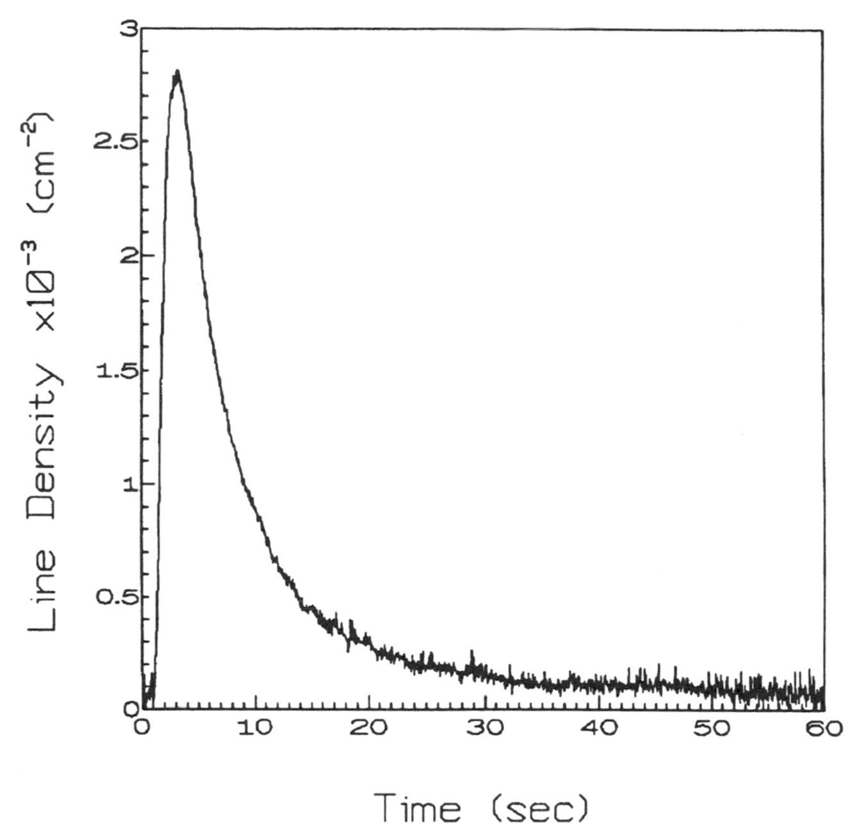

Figure 28 Typical profile of a turbulent front obtained by averaging a number of trials. The line density is deduced from the attenuation of second sound.

we had anticipated that the turbulence would spread by diffusion, but the data, which we shall describe below, are *not* consistent with diffusion spreading according to the law

$$\langle h(t) \rangle = \sqrt{2Dt}, \tag{67}$$

where $h(t)$ is the distance of the turbulent front from its initial position. Existence of such a front was anticipated in a paper by Barenblatt (1983) and extended in a recent paper by Chen & Goldenfeld (1992). Our experiments soon showed that, as anticipated theoretically, fronts produced by the towed grid appear more rapidly as the towing speed is increased. It would be premature to interpret these current results further, but it is interesting and exciting that helium II can be used to generate flows which can be compared to high Reynolds number predictions for classical fluids.

6. CONCLUSIONS

The flow of helium II is now understood over a wide range of conditions. At the lowest velocities the pure two-fluid theory of Landau (which we outlined in Volume 6) applies. At higher velocities quantized vortex lines appear in the flow and profoundly modify it. The equations of motion are now the Hall-Vinen-Bekarevich-Khalatnikov equations and problems such as the stability of Taylor-Couette flow can be tackled. The larger subject is, of course, turbulent flow. The peculiar form of turbulence that appears in counterflow experiments is now well understood. However more general flows are just beginning to be seriously studied. When the two fluids flow in the same direction, there appears to be a coupling between the fluids and a state called vortex-coupled superfluidity exists. This state appears to be closely related to classical Navier-Stokes fluids at high Reynolds numbers. Just how this coupling comes about is only beginning to be studied.

We can be confident that there are exciting times ahead using liquid helium for high Reynolds number research.

ACKNOWLEDGMENTS

This research is sponsored by the National Science Foundation under grant DMR-91-18924. The towed-grid experiments were also supported through a grant from the Office of Naval Research ONR N00014-89-1274.

Literature Cited

Andereck, C. D., Glaberson, W. I. 1982. Tkachenko waves. *J. Low. Temp. Phys.* 48: 257–96

Arms, R. J., Hama, F. R. 1965. Localized-induction concept on a curved vortex and motion of an elliptic vortex ring. *Phys. Fluids* 8: 553–59

Awschalom, D. D., Milliken, F. P., Schwarz, V. W. 1984. Mutual friction in He II near the lambda transition. *Phys. Rev. Lett.* 22: 54–56

Barenblatt, G. I. 1983. Selfsimilar turbulence propagation from an instantaneous plane source. In *Nonlinear Dynamics and Turbulance*, ed. G. I. Barenblatt, G. Iooss, D. D. Joseph. Boston: Pitman

Barenghi, C. F., Jones, C. A. 1988. The stability of the couette flow of helium II. *J. Fluid Mech.* 197: 551–69

Benjamin, T. B., Feir, J. E. 1967. The disintegration of wave trains on deep water. *J. Fluid Mech.* 27: 417–30

Bennett, J. C., Corrsin, S. 1978. Small Reynolds number nearly isotropic turbulence in a straight duct and a contraction. *Phys. Fluids* 21: 2129–40

Borner, H., Schmeling, T., Schmidt, D. W. 1983. Experiments on the circulation and propagation of large scale vortex rings in He II. *Phys. Fluids* 26: 1410–16

Borner, H., Schmidt, D. W. 1985. Investigation of large-scale vortex rings in He II by acoustic measurements of circulation. *Lect. Notes Phys.* 235: 135–46

Chen, L. Y., Goldenfeld, N. 1992. Renormalization group theory for the propagation of a turbulent burst. *Phys. Rev.* A45: 5572–77

Comte-Bellot, G., Corrsin, S. The use of a contraction to improve the isotropy of grid generated turbulance. *J. Fluid Mech.* 25: 657–82

Courts, S. C., Tough, J. T. 1988. Transition to superfluid turbulence in two-fluid flow of He II. *Phys. Rev.* B38: 74–80

Donnelly, R. J. 1956. *On the hydrodynamics of superfluid helium*. PhD thesis. Yale Univ.

Donnelly, R. J. 1988. Superfluid turbulence. *Sci. Am.* 256: 100–8

Donnelly, R. J. 1991a. *Quantized Vortices in Helium II*. Cambridge: Cambridge Univ. Press

Donnelly, R. J., ed. 1991b. *High Reynolds Number Flows Using Liquid and Gaseous Helium*. New York: Springer-Verlag

Donnelly, R. J., Swanson, C. E. 1986. Quantum turbulence. *J. Fluid Mech.* 173: 387–429

Fermi, E., Pasta, J., Ulam, S. 1965. In *Collected Papers of Enrico Fermi*, ed. E. Segre, pp. 978–88. Chicago: Univ. Chicago Press

Feynman, R. P. 1955. Application of quantum mechanics to liquid helium. In *Progress in Low Temperature Physics*, ed. C. J. Gorter, vol. I. Amsterdam: North-Holland

Gence, J. N. 1983. Homogeneous turbulence. *Annu. Rev. Fluid Mech.* 15: 201–22

Glaberson, W. I., Donnelly, R. J. 1986. Structure, distributions and dynamics of vortices in helium II. In *Progress in Low Temperature Physics*, ed. C. J. Gorter, chap. 1.II. Amsterdam: North-Holland

Hall, H. E., Vinen, W. F. 1956. The Rotation of liquid helium II. I. Experiments on the propagation of second sound in uniformly rotating helium II. *Proc. R. Soc. London Ser. A* 238: 204–14

Ichikawa, N., Murakami, M. 1991. Application of flow visualization technique to superflow experiment. See Donnelly 1991b, pp. 209–14

Kittel, P. 1987. Liquid helium pumps for in-orbit transfer. *Cryogenics* 27: 81–87

Kittel, P. 1988. Operating characteristics of isocaloric fountain-effect pumps. *Adv. Cryog. Eng.* 33: 465–70

Kukich, G. 1970. *The decay of persistent currents of superfluid helium*. PhD thesis. Cornell Univ., Ithaca, NY

Landau D., E. M. Lifshitz. 1987. *Fluid Mechanics*. London: Pergamon. 2nd ed.

London, F. 1954. *Superfluids*, Vol. II. New York: Wiley

Milliken, F. P., Schwarz, K. W., Smith, C. W. 1982. Free decay of superfluid turbulence. *Phys. Rev. Lett.* 48: 1204–7

Onsager, L. 1949. (Discussion on a paper by C. J. Gorter). *Nuovo Cimento Suppl.* 6: 249–50

Pfotenhauer, J. M., Donnelly, R. J. 1985. Heat transfer in liquid helium. In *Advances in Heat Transfer*, ed. J. P. Hartnett, T. F. Irvine, Jr., Vol. 17, pp. 66–157. New York: Academic

Proudman, L., Reid, W. 1954. On the decay of a normally distributed and homogeneous turbulent velocity field. *Phil. Trans. R. Soc. London Ser. A* 247: 163–89

Rayfield, G. W., Reif, F. 1964. Quantized vortex rings in superfluid helium. *Phys. Rev.* A136: 1194–1208

Roberts, P. H., Donnelly, R. J. 1970. Dynamics of vortex rings. *Phys. Rev. Lett.* A31: 137–38

Roberts, P. H., Donnelly, R. J. 1974. Superfluid mechanics. *Annu. Rev. Fluid Mech.* 6: 179–225

Samuels, D. C., Donnelly, R. J. 1990. Sideband instability and recurrence of kelvin waves on vortex cores. *Phys. Rev. Lett.* 64: 1385–88

Samuels, D. C., Donnelly, R. J. 1991. Motion of charged vortex rings in helium II. *Phys. Rev. Lett.* 67: 2505–8

Schwarz, K. W. 1985. Three dimensional vortex dynamics in superfluid ^{4}He: line-line and line-boundary interaction. *Phys. Rev.* B31: 5782–804

Schwarz, K. W. 1988. Three-dimensional vortex dynamics in superfluid ^{4}He: homogeneous superfluid turbulence. *Phys. Rev.* B38: 2398–417

Smith, M. R., Donnelly, R. J. 1991. A study of homogeneous turbulence in superfluid helium. See Donnelly 1991b, pp. 231–42

Snyder, H. A. 1988. Dewar to dewar model for superfluid helium transfer. *Cryogenics* 28: 86–89

Swanson, C. E. 1985. *A study of vortex dynamics in counterflowing helium II*. PhD thesis. Univ. Oreg., Eugene

Swanson, C. E., Donnelly, R. J. 1985. Vortex dynamics and scaling in turbulent counterflowing helium II. *J. Low Temp. Phys.* 61: 363–99

Swanson, C. J., Donnelly, R. J. 1991. Instability of Taylor-Couette flow of helium II. *Phys. Rev. Lett.* 67: 1578–81

Tkachenko, V. K. 1966a. On vortex lattices. *Sov. Phys. JETP* 27: 1282–86

Tkachenko, V. K. 1966b. Stability of vortex lattices. *Sov. Phys. JETP* 23: 1049–51

Vinen, W. F. 1957. Mutual friction in a heat current in liquid helium II, III. Theory of mutual friction. *Proc. R. Soc. London Ser. A* 242: 493–515

von Karman, T., Howarth, L. 1938. On the statistical theory of isotropic turbulence. *Proc. R. Soc. London Ser. A.* 164: 192–215

Walstrom, P. L., Weisend, J. G. II, Maddocks, J. R., Van Sciver, S. W. 1988. Turbulent flow pressure drop in various He II transfer system components. *Cryogenics* 28: 101

Annu. Rev. Fluid Mech. 1993. 25 : 373–97

WAVE BREAKING IN DEEP WATER

M. L. Banner

School of Mathematics, University of New South Wales, P.O. Box 1, Kensington, N.S.W. 2033, Australia

D. H. Peregrine

School of Mathematics, University of Bristol, University Walk, Bristol BS8 1TW, England

KEY WORDS: water waves, white caps, wind waves

INTRODUCTION

Every mariner is aware that dangerous large breaking water waves occur on the world's oceans. The scope of this review is somewhat greater. Wave breaking occurs at a large range of scales and we do not restrict ourselves to the deep ocean. "Deep water" in the context of water wave studies implies water sufficiently deep that the surface waves are unaffected by the direct effects of variations in bed topography. Thus even a small pond can support breaking deep-water waves. Shallow water breaking is reviewed in Peregrine (1983).

Some comments on the visual aspect of breakers are in order, since direct observation still has a role to play in the study of this complex phenomenon. The most dramatic breakers are plunging breakers where the breaking commences by the wave overturning and forming a forward moving sheet of water which plunges down into the water in front causing splashes, air entrainment, and eddies. Although plunging breakers are common on beaches they are less common on deep water, so much so that some people have argued that they do not occur naturally. However, read Coles (1991) for a distillation of an experienced yachtsman's account of waves at sea.

Most other breakers are described as spilling breakers. From their

0066–4189/93/0115–0373$02.00

initiation, "white water" falls down the front face of the wave. The falling water appears white because of entrained air bubbles and drops created at the surface. On a small scale, say wave crests less than 4 cm high, surface tension is sufficiently strong to prevent air entrainment. Such small breakers are as a result often overlooked. However, recently their significance has been realized and the term "microscale breakers" is being used (see Figure 1).

The water waves are usually generated by wind with breaking playing a significant part in the growth of waves. In addition, the turbulence directly associated with breaking is dominant in mixing processes beneath the free surface and thus is crucial to the transfer of heat and mass. This transfer is vital—on small scales for aquatic life and water quality, and on global scales it is an important factor in the Earth's weather and climate. Transfer of CO_2 to the oceans is one influential factor in the debate on global warming.

Of more obvious concern to the layperson is the safety of vessels and structures at sea. Modern ships are not immune to severe damage, or total loss due to breaking waves. Smaller ships such as trawlers capsize; larger ships suffer structural damage which may be life threatening. For routing commercial vessels away from severe wave conditions remote sensing of the sea surface gives valuable data. Radar signals backscattered from the

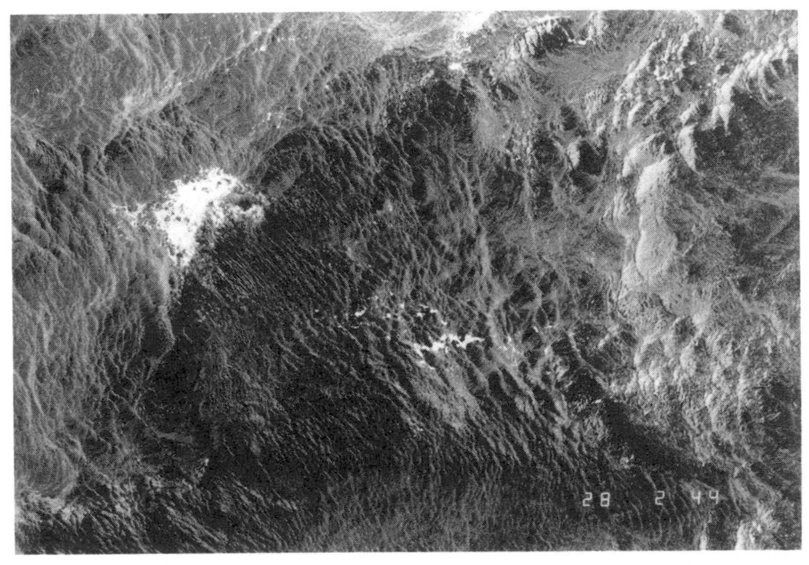

Figure 1 An area of wind blown water surface 4 m × 3 m, showing a small spilling breaker and a lot of microscale breaking. (M. L. Banner)

sea surface to satellite-borne instruments can be used to provide directional data on wave heights and wind speeds. The role of enhanced local radar backscatter from breaking waves is yet to be resolved.

Wind-generated waves are not the only ones that break. Ships generate waves when they move, and it is usual for these waves to be breaking in the vicinity of the ship. In some ways these are simpler than wind waves since in an otherwise calm sea the forcing motion is well-defined, and except for water involved in the breaking process the main part of the water has no vorticity.

The bulk of water-wave theory is for linear waves. This only provides a good approximation for waves with gentle slopes and hence gives little or no insight into breaking. However, the linear wave properties of superposition, energy density, and group velocity are at the core of one range of theoretical approaches. These describe the sea surface with a Fourier spectrum and use theoretical and empirical terms to describe the evolution of the sea state. These often include nonlinear interactions, wind input, and breaking. This last term is, perhaps, the least well known and may simply include all dissipative effects necessary to fit the data used to refine the model.

Direct study of the hydrodynamics of breaking is difficult since it is an unsteady phenomenon. However, weakly nonlinear theory can give an indication of when breaking is likely to occur. Numerical approximations can do better. "Numerically exact" solutions for two-dimensional steadily traveling waves provide the basic flow for stability calculations. Direct numerical modeling of the wave overturning is also possible. In all cases this work is limited to flows with zero or constant vorticity.

Once a wave breaks, splashing, turbulence, and air entrainment make theoretical modeling difficult. Even so, some simple models of spilling breakers do seem to give reasonable results.

MEASUREMENT IN THE FIELD

Detection Methods and Problems of Quantification

It is generally recognized that an individual wave breaking event usually starts when water particles near a wave crest develop a velocity in the wave propagation direction sufficiently large for them to fall down the front of the wave. However, the surface fluid speed is difficult to measure in the field and consequently there have been a number of indirect approaches used to detect and quantify wave breaking. These depend on a surface geometry signature such as a jump in the slope of the water surface at the toe of the breaker, or on locally enhanced properties associated with the breaking such as optical contrast, high frequency energy, radar reflectivity, or acoustic output. While whitecapping provides a familiar visual signature, widespread breaking of very short gravity wind waves occurs without air entrainment in the form of microscale breakers, and their lower visual contrast makes them far more difficult to detect unambiguously using optical methods as is evident in Figure 1. Progress with detection techniques is reviewed below, together with the closely related question of quantification, for which the introduction of more detailed classification criteria such as length scale and directionality further complicates the detection problem.

Phillips (1985, section 6) provides a useful discussion on the statistical quantification of breaking in the wind-wave spectrum, based on the underlying probability distribution $L(\mathbf{c})$, such that $L(\mathbf{c})\mathbf{dc}$ represents the average total length per unit surface area of breaking fronts that have intrinsic velocities in the range $\mathbf{c}$ to $\mathbf{c}+d\mathbf{c}$. Various statistical measures of interest may be derived from this basic distribution, such as the total number of breakers past a fixed point, the fraction of sea surface turned over per unit time, the whitecap cover, and the momentum and energy fluxes associated with breaking events. However, techniques for measuring $L(\mathbf{c})$ are not well-established, and the available data on wave breaking relies on the less direct methods mentioned earlier which are discussed more fully below.

Optical Detection Methods

Of the above derived statistics, the whitecap cover has received the most attention through systematic measurements of its variation with wind speed and atmospheric stability. Whitecap cover is an important parameter that influences the sea surface microwave brightness temperature and shortwave albedo which are both important in passive remote sensing of the sea surface. It is also a useful measure of the rate at which bubbles are injected into the oceanic mixed layer. Modern video recording and image processing techniques have been particularly useful for whitecap cover measurements. However, interpretation of whitecap cover statistics in terms of active breaking is not straightforward, owing to the persistence of "fossil" foam. Efforts have been made in this direction through visible albedo classification as type A (young) whitecaps and type B (mature) whitecaps (Monahan & Woolf 1989). Interestingly, in fresh water, the onset of whitecapping requires a higher wind speed and the whitecap cover at a given wind speed is lower by a significant margin when compared with ocean data (Monahan 1969). These effects appear to be due to the surface chemistry influence of salt on the formation and coalescence characteristics of air bubbles (Scott 1986), which results in the greater persistence of the smaller bubbles that occur in salt water (Monahan 1969, Scott 1975). In the open ocean, type A whitecap cover is observed to be a little stronger than cubic in the wind speed, with a weak dependence on the atmospheric stability (air-sea temperature difference) and on the sea surface temperature. Type B whitecap coverage is similar, but with a somewhat weaker wind speed and atmospheric stability dependence. Monahan & O'Muircheartaigh (1986) give a useful overview of whitecaps in the context of passive remote sensing of the sea surface and recent research aspects are described in the research symposium monograph edited by Monahan & MacNiocaill (1986).

Wave Gauge Detection Method

Although whitecap cover is accessible with relative convenience, more fundamental parameters associated with breaking are often of interest. Other studies of breaking statistics have involved visual detection of whitecaps passing a fixed location at which wind and wave parameters are monitored. This technique was used by Holthuijsen & Herbers (1986), whose study provided interesting data on the joint breaking probability distribution with respect to wind speed, wave period, and wave height. They demonstrated the inadequacy of using a simple local wave slope criterion to detect breaking and the rather marginal significance of the joint wave height and period distribution as a basis for extracting breaking statistics. An innovative point detection method used by Longuet-Higgins & Smith (1983) and Thorpe & Humphries (1980) relies on the rapid jump in surface elevation at the leading edge of the spilling region of a breaker. This technique used a floating spar with a fine wire wave gauge (called a "jump-meter') to make the detections, but the choice of jump threshold is not known a priori and may be responsible for the relatively low breaking probabilities reported by this method.

Weissman et al (1984) investigated the use of a high frequency analysis of wave height data from a fixed wire gauge at short wind fetches and low wind speeds to detect the passage of breaking events. According to their findings, increased wave energy levels in the high frequency range are associated with breaking events, and a threshold level can be used to detect the passage of a breaking crest past the sensor. Recently, this technique has been refined by Katsaros & Atakturk (1991). However, further field studies are needed to examine the validity of this technique under open ocean conditions, where high frequency wave measurements are more difficult and Doppler distortion of the high frequency elevation spectrum is likely. Figure 7 in Holthuijsen & Herbers (1986), reproduced here as Figure 2, summarizes the observed fraction of breaking waves as a function of wind speed reported by these various detection methods; it is evident that there is considerable scatter among the statistics based on these methods. Thus appropriate care needs to be exercised when interpreting such data and future work will need to reconcile these differences.

Radar Methods

Narrow-beam Doppler radars with footprint dimensions much smaller than the dominant waves have also been used to detect large-scale breaking events. The technique is based on measuring the significant increase in scatterer speed (from about the orbital speed to the phase speed) within breaking events (Keller et al 1986). A significant goal of such research is to estimate the contribution of reflections from breaking waves to the radar backscatter cross section in remote sensing applications. The underlying physical mechanisms and choice of the most suitable threshold to identify breaking events have been examined in a series of recent papers by Jessup and co-workers (1990, 1991a, 1991b). Depending on the spike classification criterion adopted, the contribution from spike events due to large-scale

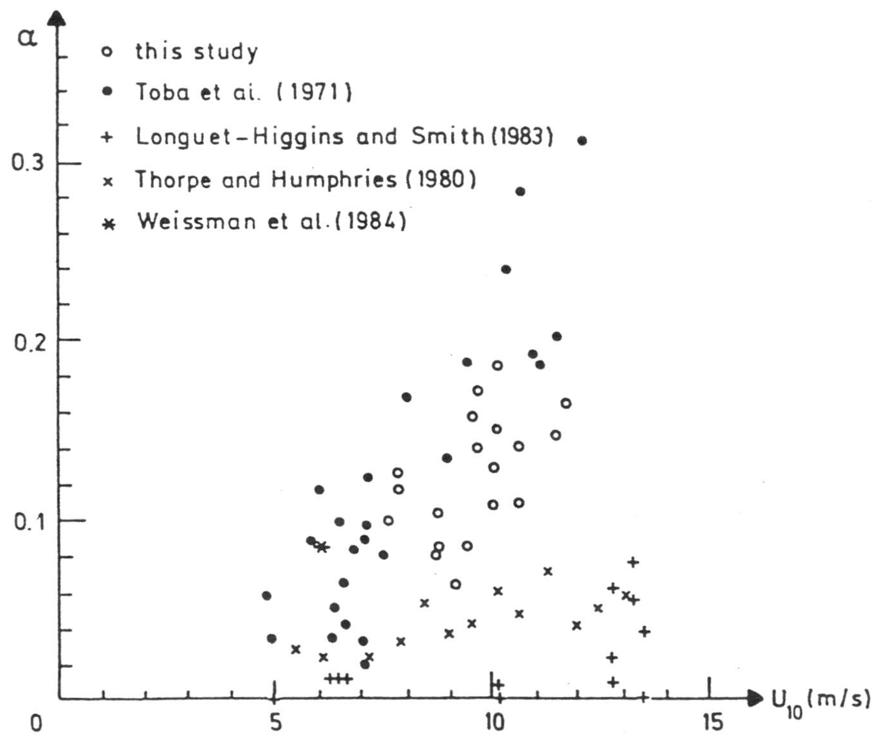

Figure 2 The observed fraction of breaking waves (α) as a function of wind speed (U_{10}) from Holthuijsen & Herbers (1986), with permission of American Meteorological Society.

whitecap events was assessed as 10–20% of the radar cross section. Based on their detailed analysis of radar returns and collocated video images of the sea surface, Jessup et al (1991a) concluded that the increased Doppler bandwidth provided the most consistent signature of sea spikes associated with large-scale wave breaking, but the spatial resolution limitation inherent in the $O(1\ \text{m})$ scale of the radar footprint precludes a complete understanding of the physical mechanisms involved. Thus while considerable progress has been made, further effort is still needed to refine our understanding of microwave reflectivity associated with breaking waves at sea. In particular, the possible contribution to the backscattered cross section from microscale breaking waves remains to be addressed.

Acoustic Methods

It is a familiar experience that whitecaps produce noise associated with the dynamics of the entrained air bubbles. A detailed field study from a tower in the Bight of Abaco was reported by Snyder et al (1983) in which the acoustic output from large-scale whitecaps was used to trigger a rapid sequence of photographs, covering an area of 10 m x 10 m. A wave gauge array in the field of view monitored the directional wave spectrum. This study allowed a detailed examination of the correlation between the onset of whitecapping with a threshold based on the local vertical acceleration threshold, providing qualified support for this concept. It also provided very useful statistics of low order temporal and geometrical whitecap statistics. Breaking also contributes significantly to the ambient underwater noise spectrum which can be measured with a hydrophone. This provides a potentially useful method for remotely sensing the wind speed since whitecapping is strongly dependent on wind speed (e.g. Lemon et al 1984). The ambient noise produced by whitecaps has been exploited to extend our present knowledge of their basic properties. Farmer & Vagle (1988) deployed a vertical hydrophone array to investigate the properties and distribution of whitecap events and the influence of their bubble clouds on the ambient noise field generated by the whitecaps. More recently, within the SWAPP (Surface Waves Processes Program, Weller et al 1991), Farmer has deployed an acoustic drifter instrument package with the capability of monitoring bubble cloud distributions and directional ambient sound characteristics. Such studies are providing substantial new information on the fundamental properties of ocean whitecaps and their associated acoustic properties.

Present Status

Field results on breaking statistics for whitecaps from the various observational techniques show well-defined trends with wind speed, but are characterized by very considerable scatter. Some of this is undoubtedly due to the breaking detection criterion adopted, while some is likely to be due to environmental influences. Insufficient field data are available on the distribution of whitecapping with wavelength and none appear to exist on the directionality properties of whitecaps or on any aspect of microscale breaking. However, more data on such breaking wave statistics will follow with the presently increasing interest and attention of air-sea interaction investigators.

LABORATORY EXPERIMENTS

To circumvent the difficulties of field measurement, several investigators have reported laboratory studies of fundamental properties of breaking waves. Some studies have been concerned primarily with breaking detection and statistics, while other studies have focused on investigating basic properties of breaking waves.

Detection

While some authors have exploited the "jump-meter" approach (e.g. Xu et al 1986, Banner 1990), other authors have investigated the use of local wavetrain properties derived from the Hilbert transform of the wave elevation signal (Melville 1982, 1983; Hwang et al 1989). Visual detection techniques have been reported as well, using wave-by-wave flow visualization, (e.g. Koga 1984) and optical backscatter (Ebuchi et al 1987) to detect and describe characteristics of breaking wind waves. More directly, Melville & Rapp (1988) used horizontal surface particle velocity—detected by a novel use of laser anemometry—to register breaking occurrences in modulating wave groups. They investigated a range of fundamental aspects, including surface current enhancement and the validity of linking breaking events with necessarily large associated wave slopes. While highly desirable, extension of this direct detection scheme to the field is not yet feasible. Microwave backscatter signatures of breaking waves have also been investigated in laboratory studies (e.g. Kwoh & Lake 1984, Banner & Fooks 1985, Melville et al 1988, Loewen & Melville 1991) to assist with the interpretation of microwave backscatter from wind waves at sea. Such studies have motivated and guided investigation of sea spike returns in the ocean remote sensing context described earlier in this article.

A complementary goal of the laboratory approach has been to isolate and study the influence of wave breaking on a range of fundamental air-sea interfacial properties. Two modes—propagating and quasi-stationary—have been used. The use of the latter allows for considerable simplification in the instrumentation, avoiding the need for a surface-following servo-mechanism in certain classes of investigations.

Steady Flows

Steady breakers usually occur in the form of spilling breakers. In the laboratory, Banner & Melville (1976), Banner & Fooks (1985), Banner & Cato (1988), and Banner (1990) used a subsurface hydrofoil in an otherwise steady, uniform current to create such breakers as shown in Figure 3. These authors investigated, respectively, the existence of separation of the overlying air flow, the properties of the fluctuating flow in the spilling crest region and consequent increased radar reflectivity, mechanisms of underwater noise generation from breaking waves, and the augmented form drag associated with breaking waves when compared with unbroken waves. Battjes & Sakai (1981) used a similar configuration to investigate the mean velocity field and turbulence structure of the trailing wake from a spilling breaker, noting its similarity to a self-preserving turbulent wake.

In two comprehensive studies, Duncan (1981, 1983) towed a submerged hydrofoil at constant speed, examining in detail the relative contributions to the wave resistance from the breaking and nonbreaking trailing surface waves generated by the obstacle. Duncan found that the breaking wave contribution could exceed the latter by a significant margin. He also investigated detailed geometric and hydrodynamic properties of the break-

Figure 3 Quasi-steady spilling breaker generated in a laboratory flume by the submerged hydrofoil seen at the lower right side. (M. L. Banner)

ing region, observing that the drag associated with the breaking region was proportional to the downslope component of its weight. He also noted a persistent low frequency oscillation in the length of the breaking region, at about four times the intrinsic wave period. Banner (1987) also noted this surging characteristic in a laboratory investigation of the perturbation response of a quasi-steady breaking wave. Models that describe the steady state and such transient response properties of quasi-steady breaking waves are described later in this review. Based on the detailed measurements of Duncan (1981, 1983), Cointe (1987) published a more detailed and comprehensive theory of the dynamics of steady breaking waves, together with an analysis of their stability properties, and also described a numerical model for calculating unsteady breaker motion.

Unsteady Flows

Experimental studies of the details of unsteady breaking waves fall into various categories. The first are those where kinematic information that might be useful in predicting or estimating wave breaking occurrences is sought. The work of Kjeldsen & Myrhaug (1978), Bonmarin & Ramamonjiarsoa (1985), and Bonmarin (1989) fall into this category. Perhaps Bonmarin (1989) gives the most detail of how wave steepness and speed change as a deep-water wave approaches breaking. These studies define steepness in terms such as (wave height)/(crest to trough distance) and thus do not necessarily relate directly to wave overturning, or to portions of wave surface exceeding the maximum theoretical slope for steady waves of just over 30 degrees to the horizontal.

Descriptive studies of three-dimensional breaking waves have been made by Kjeldsen (1984) for a spectral distribution and by She et al (1992) where breaking of waves was caused by wave focusing in a wide tank.

Experimental studies that include comparisons between experiments and equivalent potential flow computations have been made by Kjeldsen & Myrhaug (1980), Dommermuth et al (1987), Skyner et al (1990), and Skyner & Greated (1992). In each case wave breaking was induced by focusing two-dimensional waves in space-time. That is, the frequency of the wave generated was smoothly reduced in such a manner that, if linear theory for slowly-varying waves were to hold, all wave energy would arrive at the same place at the same time. Agreement with computation is fairly good, though Dommermuth et al only computed one example and Kjeldsen & Myrhaug (1980) made only a general comparison. Skyner and co-workers measure velocity fields with Particle Image Velocimetry, and compute many examples, though with a linear time-to-space transformation of the initial waves. They find that the details of wave breaking are very sensitive to initial conditions, but in Skyner & Greated (1992) they report a remarkably good agreement between theory and measurement, even in the jet.

The most fundamental studies have been made by Rapp & Melville (1990). They have used the two-dimensional focused waves approach to examine in detail a range of isolated breaking wave events. From many viewpoints the most significant dynamical aspect of wave breaking is its efficacy in transferring momentum and energy from the surface wave motion into the underlying water motion. For example in wind-driven currents, the major transfer of stress is from wind to waves and waves to current; wave breaking is significant and dominant in the two transfers respectively. Rapp & Melville give measurements of the surface motion, momentum flux, energy changes, breaking induced currents, turbulent fluctuations, and surface mixing. For example, Figure 4 shows the ensemble mean velocity field generated by 10 repeats of a plunging breaker at 1, 4, 6, 10, 20, and 50 periods after the breaking event. Although it is too early to assess the full impact of this work, it should prove valuable for many purposes.

THEORY

Almost all theoretical studies of wave hydrodynamics relevant to breaking are for irrotational flow. For a long time there was little more than Stokes' (1880; see Lamb 1932, section 250) hypothesis that the steepest steadily traveling wave train would have a 120 degree angle at its crests. We now know that regular two-dimensional wave trains in deep water are liable to a number of hydrodynamic instabilities. All of these instabilities can lead to wave breaking if the initial wave train is steep enough.

Instabilities of Uniform Wave Trains

Tanaka's (1983, 1985) instability is more closely related to wave breaking than any other. Uniform wave trains have an energy density that does not increase monotonically with wave steepness (Longuet-Higgins 1975). At a steepness of $ak = 0.43$ ($H/L = 0.137$), the energy density has a maximum, and Tanaka showed steeper waves are unstable. Jillians (1989) showed how the unstable eigenfunctions are concentrated near the wave crest and used numerical methods to show how the instability eventually leads to the wave breaking.

Another instability which occurs for a wide range of wave steepnesses was found by Benjamin & Feir (1967) and Benjamin (1967): Infinitesimal long modulations of a wave train grow in amplitude until strongly modulated wave groups occur. Lake et al (1977) found further that, in a narrow wave flume the wave groups "demodulate" and a uniform wave train

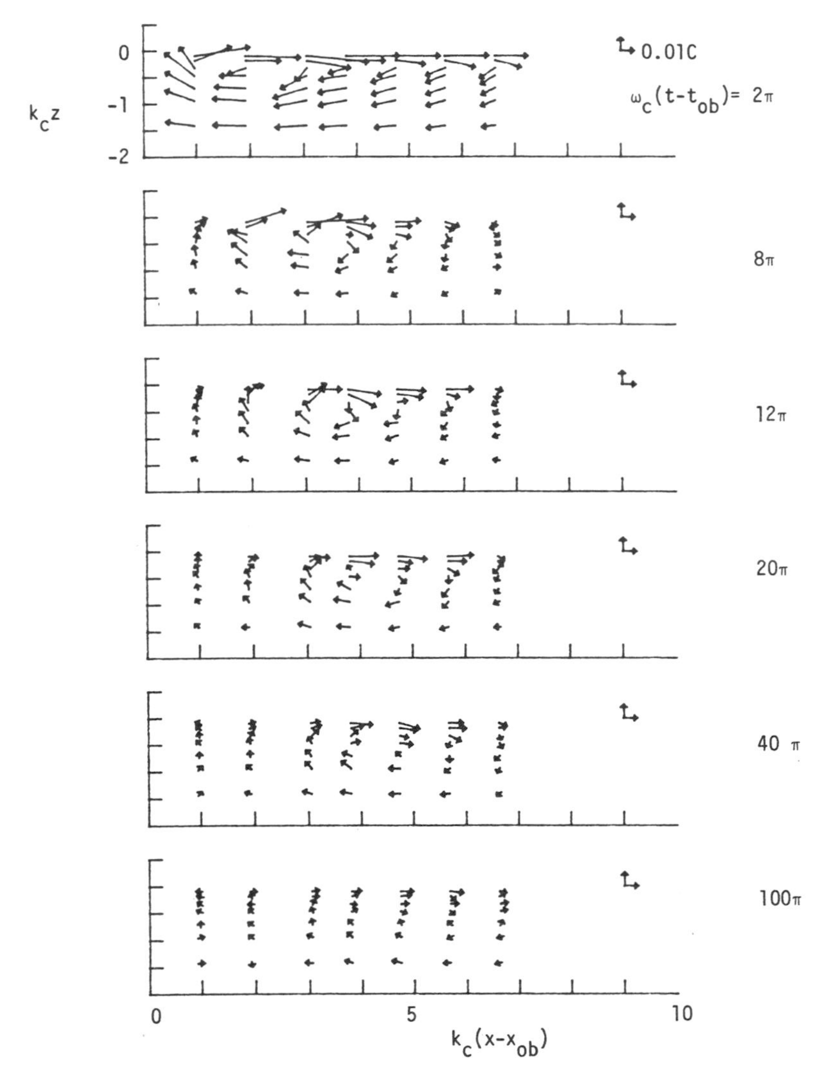

Figure 4 The mean velocity field at the following times after a plunging breaker occurs (from top to bottom at 1, 4, 6, 10, 20, and 50 periods); from Rapp & Melville (1990) with permission of The Royal Society, London.

recurs. For sufficiently steep waves the new wave train has a longer period than the original one, i.e. the waves experience a "frequency downshifting." This appeared to occur only when the waves at maximum modulation were breaking or close to breaking. Further experiments have been conducted by Su (1982), Su et al (1982), Melville (1982, 1983), and Chereskin & Mollo-Christensen (1985).

Computations for fully nonlinear irrotational examples of gently modulated waves show wave breaking occurring for wave trains of moderately gentle steepness—for example at a steepness which is one-quarter the maximum, which corresponds to an energy density of only one sixteenth of the steepest wave train of that period. A preliminary account of this work is given in Dold & Peregrine (1986). Interestingly, except for the increasing magnitude of the modulation most of the wave development is as one would expect from linear theory. The wave groups travel at half the phase velocity. Individual wave crests pass through the modulation and breaking first occurs when the modulation has grown "too big" and a crest passes through the maximum of the modulation. The details of wave breaking depend on the growth of the modulation and the precise timing of crests passing through the maximum so that details vary from example to example. In one case Dold & Peregrine obtained a wave with a crest very close indeed to the 120 degree angle which Stokes (1880) found for the limiting traveling wave.

Steeper deep-water wave trains also suffer an instability in which alternate crests grow at the expense of those in between. This instability was found by Longuet-Higgins (1978) and later shown to be a special case of a three-dimensional instability investigated by McLean (1982) and called Class II. Benjamin-Feir instability is a special case of McLean's Class I instability. The further evolution of the alternate crest instability, beyond the exponential growth of the linear hydrodynamic instability analysis, was computed by Longuet-Higgins & Cokelet (1978) who also found that the waves broke. Su (1982, figure 10) shows experimentally generated waves which are breaking after developing a three-dimensional instability which appears to be an example of Class II instability.

Rayleigh-Taylor instability of an interface has often been cited as a possible cause of wave breaking. This instability occurs if pressure in the denser fluid is less than in the lighter fluid, as when water is at rest over air. Numerous computational studies (P. McIver & D. H. Peregrine, unpublished) revealed no examples for irrotational waves, other than very weak, poorly defined, small regions at the tip of jets. On the other hand, study of steady waves on flows with vorticity shows that with sufficiently strong vertical shear, waves are unstable. See Teles da Silva & Peregrine (1988).

The instabilities which clearly contribute to the variable amplitude and short crests of many deep-water waves make it difficult to advance a "breaking criterion" such as is often used for the rather different circumstance of waves on beaches.

Unsteady Flow

Despite the fact that overturning waves are unsteady two-dimensional flows with a free surface, some progress has been made towards an analytical description. Longuet-Higgins (1980) suggests that a solution for a rotating hyperbola falling under gravity could represent the motion at the tip of the jet. New (1983) found that the curve of the face of waves underneath a jet is often well described by an ellipse, and found unsteady solutions for flow around an elliptical free surface. More comprehensive, but approximate solutions have been described by Greenhow (1983) and Jillians (1988). One feature of all these analytical solutions is that they have several free parameters. In addition, details of computed solutions show that the similarity of breaking waves apparent to the eye does not survive close inspection. The initial size, direction, and velocity of the jet can vary substantially as well as its size relative to the rest of the wave.

Full details of the wave profile, velocity field, and pressure fields in waves as they overturn and form jets are obtainable from detailed unsteady numerical computations. These all follow Longuet-Higgins & Cokelet (1976) in using boundary integral methods, but more robust, more accurate, and more efficient integration schemes have been developed. The most striking feature discovered from those computations (Peregrine et al 1980) is that the water rising up the front of the wave into the jet is subject to large accelerations. Typical computed maxima are around $5g$, where g is the gravitational acceleration. New et al (1985) give details of profiles, velocities, and accelerations of a few waves.

The traditional criterion for wave breaking is that horizontal water velocities in the crest must exceed the speed of the crest. This appears self-evident, but from detailed flow fields it is found that since the crest shape is changing there is often no precisely relevant crest velocity. Rather, there is a range of velocities which roughly correspond to crest speed, and water velocities usually exceed these by appreciable margins.

Much work has been done with deterministic two-dimensional computation of waves, especially for the case where waves are caused to break by focusing of components with differing frequencies (Dommermuth et al 1987, Skyner et al 1990, Skyner & Greated 1992). Despite this type of activity, surprisingly little progress has been made in developing quantitative descriptions of wave breaking. The simple descriptions: "plunging breaker" and "spilling breaker" cannot yet be quantified. In part this is due to difficulty of modeling the splashing phase of a breaker after an overturning jet plunges into the water.

Spilling Breakers

Spilling breakers are more amenable to modeling since steady examples can be generated, and other spilling breakers are sometimes quasi-steady. Models are mostly at a very elementary level simply representing the spilling breaker as a "roller" riding passively on the water. (See Figure 5*a* which shows mean streamlines.) However, after study of a range of experiments, Peregrine & Svendsen (1978) suggest that these flows may be best modeled by considering the whole region of turbulence, as in Figure 5*b*. In a reference frame moving with the wave, the turbulent velocities are of the same order of magnitude as the wave velocity. The fluid content of the "roller" is continually mixing with the rest of the turbulent fluid in the wave.

In considering the source of the turbulence, we note that the water falling/spilling down the front of the wave is clearly contributing to the turbulence by losing its potential energy. On the other hand Peregrine & Svendsen come to the conclusion that this is relatively unimportant in quasi-steady waves and the falling water is more important when it con-

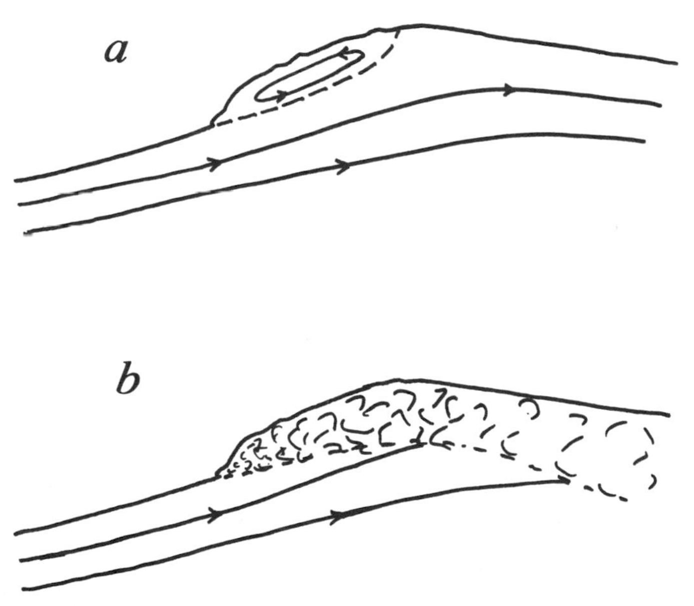

Figure 5 Spilling breakers: (*a*) The traditional view with a surface "roller," (*b*) viewed as the source of a turbulent region.

tacts previously undisturbed water. Between the water masses there is a large velocity difference tangential to the smooth surface in front of the breaker, which suggests an analogy to the well-studied shear layer between two streams of different velocities, i. e. the turbulent mixing layer. Such measurements as are available from hydraulic jumps supported this view, e.g. Hoyt & Sellin (1989). The structure of quasi-steady waves is thus an initial mixing-layer region, followed by the region beneath the crest of the wave where gravity influences and restrains the turbulent motions near the surface. Finally trailing behind the breaker is a turbulent wake which has a momentum deficit relative to the breaking wave. When viewed in a frame of reference where the wave is propagating this wake contains the momentum lost from the wave during the breaking process. This same momentum is of importance in setting up currents, especially wind-driven currents.

The main application of these ideas to practical modeling has been for shallow-water waves (Madsen & Svendsen 1982, Svendsen & Madsen 1984). For deep-water waves the "roller" concept has been used by Banner (1987) in a "lumped mass" approximation to obtain ordinary differential equations which give a reasonable simulation of his measurements of the unsteady response of a steady breaker to a disturbance. The work that has been carried furthest at present (Tulin & Cointe 1986 and Cointe 1987) uses a modeling scheme intermediate between the above mentioned. Here, linear wave theory is combined with a passive hydrostatic model of the roller, and results compare favorably with experiments. All these models require refinement, but constitute a promising start to understanding a complex flow.

The process of initiation of spilling breakers on deep water is not entirely clear. A small plunging event at the wave crest does sometimes occur but other effects may be more important. This is particularly so in the presence of wind or previous breaking waves, where the flow is rotational and may well have current which is greatest at the surface in the direction of wave propagation, e.g. if the wind is generating the waves and a wind drift layer has formed, then, as indicated by Phillips & Banner (1974), the surface shear leads to a substantial reduction in the maximum height that a steady wave can have. However, the strongest shear under wind is at the surface and has a small length scale, which makes it seem unlikely that such a thin layer will strongly influence the dynamics of a large wave. However spilling might start as a small breaking event with capillary action inhibiting white water formation (i. e. no bubbles are created), but the breaker could then grow rapidly in intensity so that the more readily visible "whitecap" breaker appears to be spilling ab initio. This is not the only case where spilling may start without plunging. If a wind is blowing small capillary ripples are always present. Large wave crests are continually catching them up. The effect of the flow field in the large wave is to shorten and steepen those ripples it overtakes, and analysis shows (Popat 1989) that for a wide range of large gravity waves, the small ripples steepen up to breaking. This breaking may then trigger larger scale spilling on suitably steep waves as described above. Limiting capillary waves are thought to break in an entirely different manner from gravity waves. The trough steepens until it overhangs and a bubble may pinch off, with strong shears and circulation as in the gravity wave case (Crapper 1957, Longuet-Higgins 1992).

For wind blowing against the direction of wave propagation, the effects of wind-drift surface shear are different. Breaking waves become higher and more likely to plunge, as any surfer knows. Kjeldsen & Myrhaug (1980) show a laboratory example of this effect. Teles da Silva & Peregrine (1988) give theoretical examples of steep steady waves with constant vorticity.

Occurrence of Deep Water Breaking

A popular approach to predicting deep-water breaking for a directional wave spectrum has evolved from studies of modulating wave trains and narrow-band random waves, as described in a previous section. The actual threshold for breaking in a random wave spectrum is not well understood. In the spectral context, Longuet-Higgins (1969) presented a simple statistical model for the loss of energy by wave breaking in a random sea, based on a crest downward acceleration threshold of $0.5\,g$ for the sharp-crested limiting Stokes wave. More recently, Longuet-Higgins (1985, 1986) pointed out that careful distinction between Lagrangian and Eulerian acceleration is necessary, and Lagrangian accelerations calculated for steep, irrotational wave trains may be used as a basis for a breaking threshold. Investigation of this class of threshold forms the basis of theoretical papers by Snyder & Kennedy (1983) and Kennedy & Snyder (1983) for ocean waves near the spectral peak; they describe a theoretical framework and numerical simulations for various moments of the whitecap geometry, including the whitecap cover. Qualified support is given for the use of an acceleration threshold for the statistics of whitecapping of waves near the spectral peak. Ochi & Tsai (1983) also proposed a model for breaking statistics based on the joint wave amplitude-frequency distribution, examining one-dimensional, non-narrow band deep-water waves with various frequency spectra. This class of model was developed further in studies by Srokosz (1986), Yuan et al (1986), and Huang (1986). Papadimitrakis et al (1988) extended these previous analyses to embrace broader spectral bandwidths. By relating their findings to previous models and observations, they provide insight on the implications for wave energy dissipation. These studies point out the underlying importance of the fourth moment of the spectrum, which is strongly dependent on the high wavenumber tail of the spectrum. Although the form of the latter is not well established, it is presently an area of active concern and ongoing investigation. Attention to this issue is also drawn by the work of Glazman (1986), who examined the relation between the geometry of the sea surface, higher order moments of the wave spectrum, and the theory of random fields. This study examines two-dimensional wave groups, considering the wave envelope and wave slope statistics in modeling breaking wave occurrence.

Future theoretical progress will need to address mechanisms related to breaking that have been documented in laboratory and field observations. In addition to saturation from direct wind input, modulation of very short wind waves by longer waves is a mechanism contributing to the breaking of short wind waves. Several laboratory investigations have reported the marked attenuation of the short wave spectrum as the modulating wave steepens (e.g. Phillips & Banner 1974, Donelan 1987). Phillips & Banner (1974) modeled this effect as enhanced breaking due to wind drift layer influence, but this mechanism was questioned by Wright (1976). Donelan (1987) suggested that modification of the nonlinear wave-wave interactions was responsible for the observed behavior. Longuet-Higgins (1987) proposed a two-scale model which considered randomness in both long and short waves and examined the effects of breaking of the short waves under conditions where the short waves were regenerated by the wind, and reported predictions in qualitative agreement with observations. When the large-scale wind wave is itself involved in breaking, it produces a marked local attenuation of the entire short wind-wave spectrum in its wake (Banner et al 1989, figure 5).

In summary, there has been considerable progress with regular wave trains and narrow-band random waves. However, theoretical modeling of breaking statistics in broad-band directional wind seas embracing major whitecaps down to the ubiquitous microscale breakers is not well-established and remains a challenging and elusive goal. In turn, this compromises our ability to provide a reliable model for the spectral dissipation through wave breaking in sea state prediction, as discussed below.

Wind-Wave Modeling

The capability of making reliable sea state predictions for a prescribed wind field has been a long standing oceanographic goal with significant scientific, engineering, and economic benefits. While earlier wind-wave generation models focused on the behavior of the significant wave height, more recent models have pursued the prediction of the full directional spectrum of the wave height, based on the numerical solution of the radiative transfer equation (e.g. the WAMDI Group 1988). According to this formulation, the rate of evolution of the spectrum at a given wavenumber results from the net influence of source terms due to wind input $S_{in}(\mathbf{k})$, nonlinear wave-wave interactions $S_{nl}(\mathbf{k})$, and wave dissipation processes $S_{diss}(\mathbf{k})$. While the wind input source term is reasonably well modeled from measurements and the nonlinear spectral transfer term is known theoretically for homogeneous seas (Hasselmann 1962, 1963a, 1963b), the form of $S_{diss}(\mathbf{k})$ which includes wave breaking, is not well-understood, either observationally or theoretically.

The paucity of knowledge of dissipative processes occurring within the wave spectrum, particularly the inherent complexity of representing wave breaking, has resulted in very few models for $S_{diss}(\mathbf{k})$. The form proposed by Hasselmann (1974) treats the breaking events as an ensemble of pressure impulses which are weak-in-the-mean. The resulting form for $S_{diss}(\mathbf{k})$ is quasi-linear in the wave spectral density, with the coefficient a functional of the whole wave spectrum, weighted towards higher wave numbers. Other approaches to represent $S_{diss}(\mathbf{k})$ have also been proposed (e.g. Duffy 1991), but the Hasselmann form appears to have been most widely adopted in operational wave models.

These terms have been incorporated in the radiation transfer equation and solved numerically for simplified situations such as homogeneous wind and wave fields with fetch-limited growth, for which a reasonable observational base exists. Such cases have served to tune the level of the dissipation source term (e.g. see Komen et al 1985). However, with the advent of more detailed information on the shape of the directional wave number spectrum both near the spectral peak (e.g. Donelan et al 1985) and in the high wavenumber tail region (e.g. Banner et al 1989), it is becoming possible to subject the wave model predictions to closer scrutiny and possible refinement of the form of $S_{diss}(\mathbf{k})$.

Wind-wave models have been extended to handle more complex situations such as turning winds and refraction by horizontally sheared currents, although observational support for such calculations is not widely available. Large-scale ocean experiments are required with well-defined wind fields. While this is difficult to realize, the recent Surface Waves Dynamics Experiment (SWADE) conducted off the U.S. East Coast during 1990–1991 (Weller et al 1991) will provide such data and serve as a very valuable basis for testing wave models in complex wind fields and currents, particularly the validity of the adopted forms for $S_{diss}(\mathbf{k})$.

Most present wind-sea models use a prescribed wind field and a standard drag coefficient relationship which depends only on the wind speed to infer the wind stress. Recent observational investigations (e.g. Donelan 1982,

Smith et al 1992) reveal a sea state dependence in the wind stress, in addition to the dependence on wind speed. These studies found that young wind seas are associated with significantly higher drag coefficients than old wind seas. Interest in this problem has heightened in recent years, motivated by the need to provide the best estimate for the wind stress in models for sea state and wind-driven circulation. In this context, with the observed large augmentation of the local wind stress and wave form drag over breaking waves (Banner 1990), it is of interest to estimate the incremental impact of wave breaking in the spectrum on the wind stress. This is difficult to answer at present because of a lack of knowledge of the spectral distribution of breaking probability, but some initial efforts have been made in this direction by Phillips (1985, 1988) for the equilibrium range of wave numbers. When combined with detailed knowledge of the local energy dissipation and momentum flux associated with individual breakers, which is becoming available from laboratory studies (e.g. Melville & Rapp 1985, Rapp & Melville 1990), knowledge of this distribution will also provide a refinement of $S_{\mathrm{diss}}(\mathbf{k})$ as well as the momentum flux from breaking waves to the ocean currents.

SECONDARY ASPECTS

Air Entrainment, Bubble Clouds

Whitecapping produces clouds of air bubbles which are advected downwards by the surface layer turbulence. The formation and interaction of the air bubbles in the upper ocean mixed layer provides vertical and horizontal distributions of bubbles. The bubble cloud shapes are detectable using various sonar techniques (e.g. Thorpe 1986) and serve to label the surface layer turbulence, play a role in the exchange of gases between the atmosphere and the ocean, and influence the ambient noise spectrum. Progress in these areas is reflected in recent research symposium proceedings by Monahan & MacNiocaill (1986) and Kerman (1988).

More locally, a better understanding of the physical role played by the air bubbles entrained by breaking waves in basic processes such as wave energy dissipation and ambient underwater noise generation is becoming available through detailed laboratory investigations (e.g. Melville et al 1988, Loewen & Melville 1991a, Lamarre & Melville 1991) and modeling (e.g. see Loewen & Melville 1991b).

Spray

Associated with high winds, breaking waves cast off clouds of spray into the atmosphere, and it has been proposed in model studies (e.g. Ling et al 1980, Bortkovskii 1987) that this mechanism greatly enhances the net water vapor flux into the atmosphere. However, using data from the recent HEXOS experiment, DeCosmo (1991) found no such increase in the water vapor transfer coefficient with increasing wind speeds up to 18 m/s, despite the attendant increase in whitecapping, and suggested that while sea spray production and evaporation might increase at low levels, the enhanced moistening and cooling of the air near the interface would act to reduce the interfacial moisture flux and possibly the net flux, due to the reduction in the near-surface saturation vapor pressure. So even if the vertical turbulent transport is enhanced by the increased breaking activity, the net effect might be insignificant at the measurement elevation of several meters.

Influence on Remote Sensing of the Ocean

Satellite-borne active and passive microwave instruments presently in use (or scheduled for imminent deployment) have been shown to have the potential to provide routine, cost-effective monitoring of basic air-sea interfacial variables such as the global distribution of ocean wind stress, dominant wave height and direction, and sea surface temperature. As described above, breaking waves may well have an impact on the interpretation of ocean data from these remote sensing instruments, and their influence needs to be understood and quantified in order to improve the reliability of the algorithms used to interpret the microwave returns. Research in this direction is continuing.

ACKNOWLEDGMENTS

M. L. B. and D. H. P. gratefully acknowledge support for their research in this area from the Australian Research Council and from the U.K. Science and Engineering Research Council respectively.

Literature Cited

Banner, M. L. 1987. Surging characteristics of spilling zones of quasi-steady breaking waves. *Proc. IUTAM Symp. Nonlinear Water Waves*, ed. K. Horikawa, H. Maruo, pp. 151–58. Berlin: Springer-Verlag

Banner, M. L. 1990. The influence of wave breaking on the surface pressure distribution in wind wave interactions. *J. Fluid Mech.* 211: 463–95

Banner, M. L., Cato, D. H. 1988. Physical mechanisms of noise generation by breaking waves—a laboratory study. In *Sea Surface Sound—Natural Mechanisms of Surface Generated Noise in the Ocean*, ed. B. R. Kerman, pp. 429–36. Dordrecht: Kluwer. 639 pp.

Banner, M. L., Fooks, E. H. 1985. On the microwave reflectivity of small-scale breaking water waves. *Proc. R. Soc. London Ser. A* 399: 93–109

Banner, M. L., Melville, W. K. 1976. On the separation of air flow over water waves. *J. Fluid Mech.* 77: 825–91

Banner, M. L., Jones, I. S. F., Trinder, J. C. 1989. Wavenumber spectra of short gravity waves. *J. Fluid Mech.* 198: 321–44

Battjes, J. A., Sakai, T. 1981. Velocity field in a steady breaker. *J. Fluid Mech.* 111: 421–37

Benjamin, T. B. 1967. Instability of periodic wavetrains in nonlinear dispersive media. *Proc. R. Soc. London Ser. A* 299: 59–75

Benjamin, T. B., Feir, J. E. 1967. The disintegration of wavetrains in deep water, Pt. I, Theory. *J. Fluid Mech.* 27: 417–30

Bonmarin, P. 1989. Geometric properties of deep-water breaking waves. *J. Fluid Mech.* 209: 405–33

Bonmarin, P., Ramamonjiarsoa, A. 1985. Deformation to breaking of deep water gravity waves. *Exp. Fluids* 3: 11–16

Bortkovskii, R. S. 1987. *Air-Sea Exchange of Heat and Moisture during Storms*. Dordrecht: Reidel. 194 pp.

Chereskin, T. K., Mollo-Christensen, E. 1985. Modulational development of nonlinear gravity-wave groups. *J. Fluid Mech.* 151: 337–65

Cointe, R. 1987. *A theory of breakers and breaking waves*. PhD dissertation. Univ. Calif., Santa Barbara

Coles, K. Adlard 1991. *Heavy Weather Sailing*. London: Adlard Coles. 4th ed.

Crapper, G. D. 1957. An exact solution for progressive capillary waves of arbitrary amplitude. *J. Fluid Mech.* 2: 532–40

DeCosmo, J. 1991. *Air-sea exchange of momentum, heat and water vapor over whitecap sea states*. PhD thesis. Univ. Wash., Seattle. 212 pp.

Dold, J. W., Peregrine, D. H. 1986. Water-wave modulation. *Proc. 20th Int. Conf. Coastal Eng. Taipei, ASCE* 1: 163–75

Dommermuth, D. G., Yue, D. K. P., Rapp, R. J., Chan, F. S., Melville, W. K. 1987. Deep water breaking waves; a comparison between potential theory and experiments. *J. Fluid Mech.* 89: 432–42

Donelan, M. A. 1982. The dependence of the aerodynamic drag coefficient on wave parameters. *First Int. Conf. on Meteorol. and Air-Sea Interaction of the Coastal Zone*, pp. 381–87. Boston: Am. Meteorol. Soc.

Donelan, M. A. 1987. The effect of swell on the growth of wind waves. *Johns Hopkins APL Tech. Dig.* 8: 18–23

Donelan, M. A., Hamilton, J., Hui, W. H. 1985. Directional spectra of wind-generated waves. *Phil. Trans. R. Soc. London Ser. A* 315: 509–62

Duffy, D. G. 1991. The application of NASA's third generation wave model to LEWEX. In *Directional Ocean Wave Spectra*, ed. R. C. Beal, pp. 177–81. Baltimore: Johns Hopkins Univ. Press. 218 pp.

Duncan, J. H. 1981. An experimental investigation of breaking waves produced by a towed hydrofoil. *Proc. R. Soc. London Ser. A* 377: 331–48

Duncan, J. H. 1983. The breaking and non-breaking wave resistance of a two-dimensional hydrofoil. *J. Fluid. Mech.* 126: 507–20

Ebuchi, N., Kawamura, H., Toba, Y. 1987. Fine structure of laboratory wind-wave surfaces studied using an optical method. *Boundary-Layer Meteorol.* 39: 133–51

Farmer, D. M., Vagle, S. 1988. On the distribution of breaking surface wave distributions using ambient sound. *J. Geophys. Res.* 93: 3591–600

Glazman, R. E. 1986. Statistical characterization of sea surface geometry for a wave slope field discontinuous in the mean square. *J. Geophys. Res.* 91: 6629–41

Greenhow, M. 1983. Free-surface flows related to breaking waves. *J. Fluid Mech.* 134: 259–75

Hasselmann, K. 1962. On the nonlinear energy transfer in a gravity wave spectrum I. *J. Fluid Mech.* 12: 481–500

Hasselmann, K. 1963a. On the nonlinear energy transfer in a gravity wave spectrum II. *J. Fluid Mech.* 15: 273–81

Hasselmann, K. 1963b. On the nonlinear energy transfer in a gravity wave spectrum III. *J. Fluid Mech.* 15: 385–98

Hasselmann, K. 1974. On the spectral dissipation of ocean waves due to white capping. *Boundary-Layer Meteorol.* 6: 107–27

Holthuijsen, L. H., Herbers, T. H. C. 1986. Statistics of wave breaking observed as whitecaps in the open sea. *J. Phys. Oceanogr.* 16: 290–97

Hoyt, J. W., Sellin, R. H. T. 1989. The hydraulic jump as a mixing layer. *J. Hydraul. Div. ASCE* 115: 1607–14

Huang, N. E. 1986. An estimate of the influence of breaking waves on the dynamics of the upper ocean. In *Wave Dynamics and Radio Probing of the Sea Surface*, ed. O. M. Phillips, K. Hasselmann, pp. 295–313. New York: Plenum

Hwang, P. A., Xu, D., Wu, J. 1989. Breaking of wind generated waves: measurements and characteristics. *J. Fluid Mech.* 202: 177–200

Jessup, A. T., Keller, W. C., Melville, W. K. 1990. Measurements of sea spikes in microwave backscatter at moderate incidence. *J. Geophys. Res.* 95: 9679–88

Jessup, A. T., Melville, W. K., Keller, W. C. 1991a. Breaking waves affecting microwave backscatter, 1. Detection and verification. *J. Geophys. Res.* 96: 20,547–59

Jessup, A. T., Melville, W. K., Keller, W. C. 1991b. Breaking waves affecting microwave backscatter, 2. Dependence on wind and wave conditions. *J. Geophys. Res.* 96: 20,561–69

Jillians, W. J. 1988. *The overturning of steep gravity waves*. PhD dissertation. Cambridge Univ.

Jillians, W. J. 1989. The superharmonic instability of Stokes waves in deep water. *J. Fluid Mech.* 284: 563–79

Katsaros, K. B., Atakturk, S. S. 1991.

Dependence of wave breaking statistics on wind stress and wave development. Presented at IUTAM Breaking Waves Symp., Sydney

Keller, W. C., Plant, W. J., Valenzuela, G. R. 1986. Observations of breaking ocean waves with coherent microwave radar. In *Wave Dynamics and Radio Probing of the Sea Surface*, ed. O. M. Phillips, K. Hasselmann, pp. 295–313. New York: Plenum

Kennedy, R. M., Snyder, R. L. 1983. On the formation of whitecaps by a threshold mechanism. Part II. *J. Phys. Oceanogr*. 13: 1483–92

Kerman, B. R., ed. 1988. *Sea Surface Sound—Natural Mechanisms of Surface Generated Noise in the Ocean*. Dordrecht: Kluwer. 639 pp.

Kjeldsen, S. P. 1984. Whitecapping and wave crest lengths in directional seas. *Symp. on Description and Modelling of Directional Seas, Danish Hydraul. Inst*., Paper B-6, 16 pp.

Kjeldsen, S. P., Myrhaug, D. 1978. Kinematics and dynamics of breaking waves. *Rep. STF60 A78100. Ships in Rough Seas, Pt*. 4. Trondheim: Norwegian Hydrodyn. Labs.

Kjeldsen, S. P., Myrhaug, D. 1980. Wave-wave interactions, current-wave interactions and resulting extreme waves and breaking waves. *Proc. 17th Conf. Coastal Eng. ASCE, Sydney* 3: 2277–303

Koga, M. 1984. Characteristics of a breaking wind-wave field in the light of the individual wind-wave concept. *J. Oceanogr. Soc. Jpn*. 40: 105–14

Komen, G., Hasselmann, S., Hasselmann, K. 1984. On the existence of a fully developed wind-sea spectrum. *J. Phys. Oceanogr*. 14: 1271–85

Kwoh, D. S., Lake, B. M. 1984. A deterministic, coherent, and dual polarized laboratory study of microwave backscattering from water waves, Part 1: short gravity waves without wind. *IEEE J. Ocean. Eng*. 5: 291–308

Lake, B. M., Yuen, H. C., Rungaldier, H., Ferguson, W. E. 1977. Nonlinear deep-water waves: theory and experiment. Pt. 2, Evolution of a continuous wave train. *J. Fluid Mech*. 83: 49–74

Lamarre, E., Melville, W. K. 1991. Air entrainment and dissipation in breaking waves. *Nature* 351: 469–72

Lamb, H. 1932. *Hydrodynamics*. Cambridge: Cambridge Univ. Press, 6th ed. 738 pp.

Lemon, D. D., Farmer, D. M., Watts, D. R. 1984. Acoustic measurements of wind speed and precipitation over a continental shelf. *J. Geophys. Res*. 89: 3462–72

Ling, S. C., Kao, T. W., Saad, A. I. 1980. Micro-droplets and transportation of moisture from ocean. *J. Eng. Mech. Div., ASCE* 106: 1327–39

Loewen, M. R., Melville, W. K. 1991a. Microwave backscatter and acoustic radiation from breaking waves. *J. Fluid Mech*. 224: 601–23

Loewen, M. R., Melville, W. K. 1991b. A model of the sound generated by breaking waves. *J. Acoust. Soc. Am*. 90: 2075–80

Longuet-Higgins, M. S. 1969. On wave breaking and the equilibrium spectrum of wind-generated waves. *Proc. R. Soc. London Ser. A* 310: 151–59

Longuet-Higgins, M. S. 1975. Integral properties of periodic gravity waves of finite amplitude. *Proc. R. Soc. London Ser. A* 342: 157–74

Longuet-Higgins, M. S. 1978. The instabilities of gravity waves of finite amplitude in deep water. II. Subharmonics. *Proc. R. Soc. London Ser. A* 360: 489–505

Longuet-Higgins, M. S. 1980. On the forming of sharp corners at a free surface. *Proc. R. Soc. London Ser. A* 371: 453–78

Longuet-Higgins, M. S. 1985. Accelerations in steep gravity waves. *J. Phys. Oceanogr*. 15: 1570–79

Longuet-Higgins, M. S. 1986. Eulerian and Lagrangian aspects of surface waves. *J. Fluid Mech*. 173: 683–707

Longuet-Higgins, M. S. 1987. A stochastic model of sea surface roughness. I: wave crests. *Proc. R. Soc. London Ser. A* 410: 19–34

Longuet-Higgins, M. S. 1992. Capillary rollers and bores. *J. Fluid Mech*. 240: 659–79

Longuet-Higgins, M. S., Cokelet, E. D. 1976. The deformation of steep surface waves on water. I. A numerical method of computation. *Proc. R. Soc. London Ser. A* 358: 1–26

Longuet-Higgins, M. S., Cokelet, E. D. 1978. The deformation of steep surface waves on water. II. Growth of normal-mode instabilities. *Proc. Roy. Soc. London Ser. A* 364: 1–28

Longuet-Higgins, M. S., Smith, N. D. 1983. Measurements of breaking by a surface jump meter. *J. Geophys. Res*. 88: 9823–31

McLean, J. W. 1982. Instabilities of finite-amplitude water waves. *J. Fluid Mech*. 114: 315–30

Madsen, P. A., Svendsen, I. A. 1982. Turbulent bores and hydraulic jumps. *J. Fluid Mech*. 129: 1–25

Melville, W. K. 1982. The instability and breaking of deep-water waves. *J. Fluid Mech*. 115: 163–85

Melville, W. K. 1983. Wave modulation and breakdown. *J. Fluid Mech*. 128: 489–506

Melville, W. K., Rapp, R. J. 1988. The surface velocity field in steep and breaking waves. *J. Fluid Mech.* 189: 1–22

Melville, W. K., Loewen, M. R., Felizardo, F. C., Jessup, A. T., Buckingham, M. J. 1988. Acoustic and microwave signatures of breaking waves. *Nature* 336: 54–59

Monahan, E. C. 1969. Fresh water whitecaps. *J. Atmos. Sci.* 26: 1026–29

Monahan, E. C., MacNiocaill, G., eds. 1986. *Ocean Whitecaps and Their Role in Air-Sea Exchange Processes.* Dordrecht: Reidel. 294 pp.

Monahan, E. C., O'Muircheartaigh, I. G. 1986. Whitecaps and the passive remote sensing of the ocean surface. *Int. J. Remote Sensing* 7: 627–42

Monahan, E. C., Woolf, D. K. 1989. Comments on "Variations of Whitecap Coverage with Wind Stress and Water Temperature." *J. Phys. Oceanogr.* 19: 706–11

New, A. 1983. A class of elliptical free surface flows. *J. Fluid Mech.* 130: 219–39

New, A., McIver, P., Peregrine, D. H. 1985. Computations of breaking waves. *J. Fluid Mech.* 150: 233–51

Ochi, M. K., Tsai, C. H. 1983. Prediction of occurrence of breaking waves in deep water. *J. Phys. Oceanogr.* 13: 2008–19

Papadimitrakis, Y. A., Huang, N. E., Bliven, L. F., Long, S. R. 1988. An estimate of wave breaking probability for deep water waves. In *Sea Surface Sound—Natural Mechanisms of Surface Generated Noise in the Ocean*, ed. B. R. Kerman, pp. 71–83. Dordrecht: Kluwer. 639 pp.

Peregrine, D. H. 1983. Breaking waves on beaches. *Annu. Rev. Fluid Mech.* 15: 149–78

Peregrine, D. H., Cokelet, E. D., McIver, P. 1980. The fluid mechanics of waves approaching breaking. *Proc. 17th Coastal Eng. Conf. ASCE, Sydney* 1: 512–28

Peregrine, D. H., Svenden, I. A. 1978. Spilling breakers, bores and hydraulic jumps. *Proc. 16th Coastal Eng. Conf. ASCE, Hamburg* 1: 540–50

Phillips, O. M. 1985. Spectral and statistical properties of the equilibrium range in wind-generated gravity waves. *J. Fluid Mech.* 156: 505–31

Phillips, O. M. 1988. Radar returns from the sea surface-Bragg scattering and breaking waves. *J. Phys. Oceanogr.* 18: 1065–74

Phillips, O. M., Banner, M. L. 1974. Wave breaking in the presence of wind drift and swell. *J. Fluid Mech.* 66: 625–40

Popat, N. 1989. *Steep capillary waves on gravity waves.* PhD dissertation. Bristol Univ.

Rapp, R. J., Melville, W. K. 1990. Laboratory measurements of deep water breaking waves. *Phil. Trans. R. Soc. London Ser. A* 331: 735–80

Scott, J. C. 1975. The role of salt in whitecap persistence. *Deep Sea Res.* 22: 653–57

Scott, J. C. 1986. The effect of organic films on water surface motions. In *Ocean Whitecaps and Their Role in Air-Sea Exchange Processes*, ed. E. C. Monahan, G. MacNiocaill, pp. 159–165. Dordrecht: Reidel. 294 pp.

She, K., Greated, C. A., Easson, W. J. 1992. Experimental study of three-dimensional wave. Preprint, Edinburgh Univ. Phys. Dept.

Skyner, D. J., Greated, C. A. 1992. The evolution of a long-crested deep-water breaking wave. *Second Int. Offshore & Polar Eng. Conf., San Francisco*

Skyner, D. J., Gray, C., Greated, C. A. 1990. A comparison of time-stepping numerical predictions with whole-field flow measurement in breaking waves. In *Water Wave Kinematics*, ed. A. Tørum, O. T. Gudmestad, pp. 491–508. Dordrecht: Kluwer

Smith, S. D., Anderson, R. J., Oost, W. A., Kraan, C., Maat, N., et al 1992. Sea surface wind stress and drag coefficients: the HEXOS results. *Boundary-Layer Meteorol.* 60: 109–42

Snyder, R. L., Kennedy, R. M. 1983. On the formation of whitecaps by a threshold mechanism. Part I: basic formalism. *J. Phys. Oceanogr.* 13: 1482–92

Snyder, R. L., Smith, L., Kennedy, R. M. 1983. On the formation of whitecaps by a threshold mechanism. Part III: field experiment and comparison with theory. *J. Phys. Oceanogr.* 13: 1505–18

Srokosz, M. A. 1986. On the probability of wave breaking in deep water. *J. Phys. Oceanogr.* 16: 382–85

Stokes, G. G. 1880. Considerations relative to the greatest height of oscillatory irrotational waves which can be propagated without change of form. *Math. & Phys. Papers* 1: 225–28

Su, M.-Y. 1982. Evolution of groups of gravity waves with moderate to high steepness. *Phys. Fluids.* 25: 2167–74

Su, M.-Y., Bergin, M., Marlev, P., Myrick, R. 1982. Experiments on nonlinear instabilities and evolution of steep gravity-wave trains. *J. Fluid Mech.* 124: 45–72

Svendsen, I. A., Madsen, P. A. 1984. A turbulent bore on a beach. *J. Fluid Mech.* 148: 73–96

Tanaka, M. 1983. The stability of steep gravity waves. *J. Phys. Soc. Jpn.* 53: 3047–55

Tanaka, M. 1985. The stability of steep gravity waves II. *J. Fluid Mech.* 156: 281–89

Teles da Silva, A. F., Peregrine, D. H. 1988. Steep, steady, surface waves on water of finite depth with constant vorticity. *J. Fluid Mech.* 195: 281–302

Thorpe, S. A. 1986. Bubble clouds: a review of their detection by sonar, of related models and of how Kv may be determined. In *Ocean Whitecaps*, ed. E. C. Monahan, G. MacNiocaill, pp. 57–68. Dordrecht: Reidel. 294 pp.

Thorpe, S. A., Humphries, P. N. 1980. Bubbles and breaking waves. *Nature* 283: 463–65

Tulin, M. P., Cointe, R. 1986. A theory of spilling breakers. *Proc. 16th Symp. Naval Hydrodyn.* Washington DC: Natl. Acad.

WAMDI Group (Wave Model Development and Implementation Group). 1988. The WAM model—a third generation ocean wave prediction model. *J. Phys. Oceanogr.* 18: 1775–810

Weissman, M. A., Katsaros, K. B., Atakturk, S. S. 1984. Detection of breaking events in a wind-generated field. *J. Phys. Oceanogr.* 14: 1608–19

Weller, R. A., Donelan, M. A., Briscoe, M. G., Huang, N. E. 1991. Riding the crest: a tale of two wave experiments. *Bull. Am. Meteorol. Soc.* 72: 163–83

Wright, J. W. 1976. The wind drift and wave breaking. *J. Phys. Oceanogr.* 6: 402–5

Yuan, Y., Tung, C. C., Huang, N.E. 1986. Statistical characteristics of breaking waves. In *Wave Dynamics and Radio Probing of the Sea Surface*, ed. O.M. Phillips, K. Hasselmann, pp. 265–72. New York: Plenum

Xu, D., Hwang, P. A., Wu, J. 1986. Breaking of wind-generated waves. *J. Phys. Oceanogr.* 16: 2172–78

Annu. Rev. Fluid Mech. 1993. 25 : 399–453

ORDER PARAMETER EQUATIONS FOR PATTERNS

Alan C. Newell, Thierry Passot,[1] *and Joceline Lega*[2]

Arizona Center for Mathematical Sciences, University of Arizona, Tucson, Arizona 85721

KEY WORDS: amplitude equations, convection, defects, phase transitions, stability

1. INTRODUCTION

Patterns of an almost periodic nature appear all over the place. One sees them in cloud streets, in sand ripples on flat beaches and desert dunes, in the morphology of plants and animals, in chemically reacting media, in boundary layers, on weather maps, in geological formations, in interacting laser beams in wide gainband lasers, on the surface of thin buckling shells, and in the grid scale instabilities of numerical algorithms. This review deals with the class of problems into which these examples fall, namely with pattern formation in spatially extended, continuous, dissipative systems which are driven far from equilibrium by an external stress. Under the influence of this stress, the system can undergo a series of symmetry breaking bifurcations or phase transitions and the resulting patterns become more and more complicated, both temporally and spatially, as the stress is increased. Figures 1 through 3 show examples of patterns in lasers, binary and ordinary fluids, and liquid crystals. The goal of theory is to provide a means of understanding and explaining these patterns from a macroscopic viewpoint that both simplifies and unifies classes of problems which are seemingly unrelated at the microscopic level.

Convection in a large aspect ratio horizontal layer of fluid heated from below is the granddaddy of canonical examples used to study pattern formation and behavior in spatially extended systems. For low values of the vertical temperature difference, which is the external stress parameter in this case and whose non-dimensional measure is called the Rayleigh

[1] Observatoire de Nice, BP 229, 06304 Nice Cedex 4, France.
[2] Institut Non Linéaire de Nice, BP 71, 06108 Nice Cedex 02, France.

0066–4189/93/0115–0399$02.00

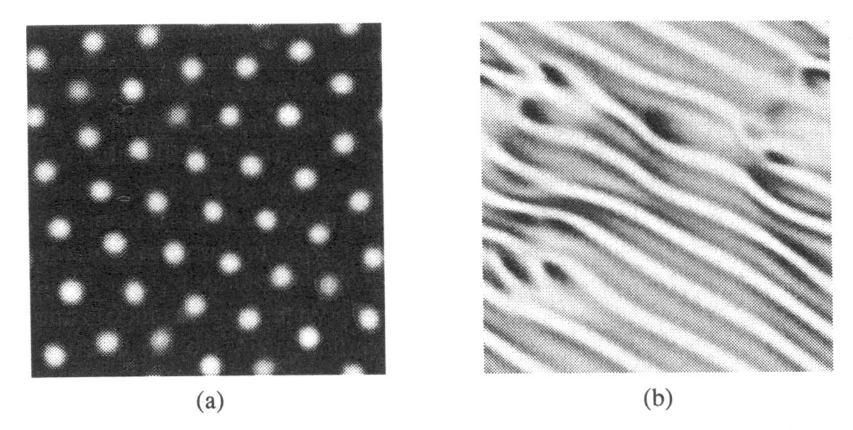

Figure 1 (*a*) Near field, stationary light intensity pattern of two counterpropagating beams at subcritical values of the beam intensity in a weakly nonlinear, Kerr focusing medium. The basic pattern is hexagonal. Also observe the hepta-penta defect (bright spots surrounded by seven and five neighbors). Along two directions 120° apart, one observes dislocations as shown in Figure 8*d*. (*b*) Snapshots of a supercritical traveling wave state of a "Raman" 3-level laser with a modulational instability developing along the phase contours.

Figure 2 Two localized wavepackets. The outer ring is an annular region of a binary fluid mixture showing a stationary wavepacket. The arrow denotes the direction of phase propagation. The center shows a localized two-dimensional wavepacket in a liquid crystal. Each localized pulse probably owes its presence to the same ingredients.

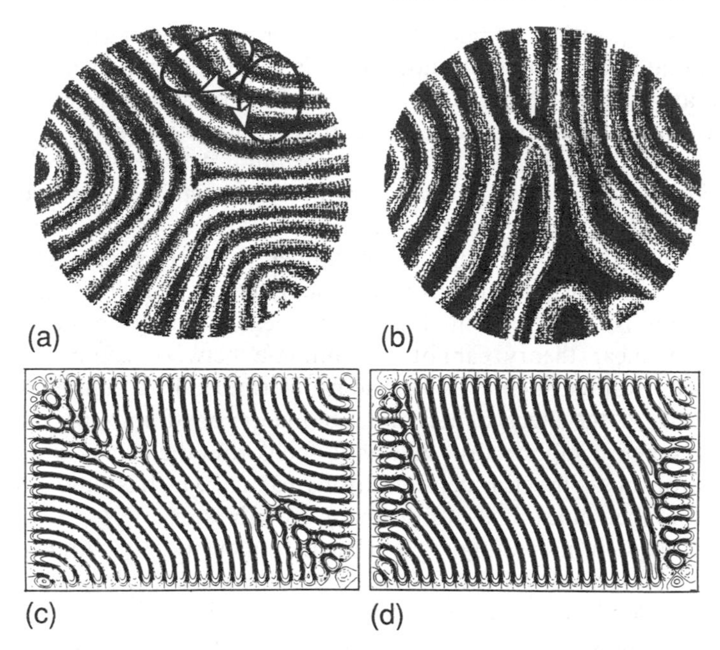

Figure 3 (*a*) A stationary pattern at $R = 2.61R_c$ containing one concave and three convex (foci) disclinations with two cells of the dipolar mean drift circulation drawn in. (*b*) A snapshot taken shortly after a dislocation pair has been nucleated; $R = 4.84R_c$. [(*a*) and (*b*) from Heutmaker & Gollub (1987).] (*c*), (*d*) from Greenside & Coughran (1984), show the results of a numerical simulation of the Swift-Hohenberg equation after τ_H and $\Gamma\tau_H$ respectively.

number R, heat is transported across the layer by conduction; the fluid remains at rest and possesses symmetries of time and horizontal space translation (in the limit of an infinite horizontal geometry), rotation, and reflection (at least about vertical axes). If the conditions on the top and bottom boundaries are that the shear stresses are zero, then the system is also Galilean invariant with respect to horizontal velocities. If the boundaries are rigid, this symmetry is no longer present. At a certain value R_c of the external stress, the stability of the conduction state is lost and various new shapes and structures that involve nonzero velocity fields which advect the heat are preferentially amplified. In the case of an infinite horizontal geometry, these structures are often cellular rolls whose horizontal projections are stationary or traveling wave periodic patterns which break one or more but rarely all of the symmetries enjoyed by the unstable conduction state. For example, a stationary pattern of periodic convecting rolls breaks (retains) the continuous translation symmetry in the direction perpendicular (parallel) to the roll axes.

What is important to understand is that, because of degeneracies, the pattern that finally emerges is not uniquely identified by a linear stability investigation of the simple state. The set of states which are equally (or almost equally) linearly amplified (e.g. the convection rolls) or which remain neutrally stable (e.g. the zero horizontal velocity state when the boundary conditions are stress-free rather than rigid) as the stress parameter exceeds critical can contain many modes. In the convection example, while there is a single wavenumber k_c which is most amplified, any linear superposition of rolls of wavelengths $\lambda = 2\pi k_c^{-1}$ with arbitrary orientations is equally admissible from linear stability considerations. Therefore linear theory cannot discriminate between competing configurations. Rather, the final outcome is determined by a combination of *external biases*, the effects of horizontal boundary conditions in convecting fluids, geometric imperfections in elastic shells, and the *nonlinear coupling* between various competing configurations, each of which is almost equally likely from the point of view of its linear growth rate.

As a first step in creating a theoretical framework, we should like to reduce the dimension of the system by choosing a coordinate system (a basis) that clearly separates the modes which are dynamically active from those which play a passive role. Specifically, we want to divide the space S of all possible fields into two subsets, which we will call A and P. A, connoting *active*, will contain all these modes that tap directly into the external stress source and are bona fide competitors in the battle for survival. P, connoting *passive*, will contain all the nonstarters. They will be present in the final state, but only to the extent that they are regenerated by nonlinear interactions of the active modes. They are *slaved* in the sense that their amplitudes are determined algebraically in terms of the amplitudes of the members of A. They have no direct access to the external stress source. Sometimes, the division of S into A and P is unambiguous and the identification of which belongs where is straightforward. Other times, the dividing line is less sharp and it is these situations that are most challenging.

Near onset (that is when the value R of the stress parameter is close to the critical value R_c at which the transition takes place), the choice of A and P can be made by examining the linear stability of the state about to destabilize. If the spectrum of the stability operator (which corresponds to the growth rates of various configurations) is discrete, then the division into A and P is clear. A contains those modes whose growth rates are positive, zero, or weakly negative for R close to R_c and P contains all modes that are strongly damped. The coordinates of A are simply the amplitudes of the discrete modes or configurations belonging to A. They are called *amplitude order parameters* because their relative values tell us about the degree of order and structure in the system. For R close to its critical value R_c, the amplitudes of the passive modes in the set P will very quickly relax to a manifold, called the *center manifold*

$$P = P(A), \tag{1.1}$$

determined algebraically as a balance between each linearly decaying passive mode and its regeneration by nonlinear interactions involving members of A. On this manifold, the amplitudes will evolve on a time scale proportional to $|R - R_c|^{-1}$ according to a set (usually finite) of coupled ordinary differential equations, called *amplitude equations*

$$\frac{dA}{dt} = G(A) \tag{1.2}$$

often associated with the names of Landau (1944), Stuart (1960), and Schlüter et al (1965). Their form is universal in the sense that $G(A)$ consists of terms which are simply products of active amplitudes. All relevant information about the underlying microscopic system is contained in coefficients, which involve integrals and averages of products of the various mode shapes. Therefore, detailed structure is far less important than overall symmetry properties. The original microscopic fields can be reconstructed to a good approximation by combining the solution of (1.2) with the configurations which are the basis vectors for A, together with the graph (1.1). Since solving (1.2) is much easier than solving the original microscopic equations, considerable simplification in both qualitative and quantitative understanding has been achieved.

If the spectrum of growth rates is continuous (or almost so, in the sense that nearest neighbors are closer than $|R - R_c|$), then things are more subtle. In some cases (in others, the problem is still open), we overcome the ambiguity of continuous bands of competing modes by introducing as order parameters the *envelopes* of the various short-scale mode shapes. The envelopes depend slowly on space and time, in contrast to the amplitudes used in the discrete spectrum case which depended only on time. For the continuous spectrum situation, A is infinite dimensional and the envelopes obey a class of coupled nonlinear *partial* differential equations often called the Newell-Whitehead-Segel (NWS) (Newell & Whitehead 1969, 1971; Segel 1969) or Ginzburg-Landau equations.[3] The advantage

[3] Historical note: Landau wrote down amplitude equations (o.d.e.'s) as in (1.2) to describe the post-bifurcation behavior of unstable modes. In their theory of superconductivity, Ginzburg and Landau introduced the notion of a space dependent order parameter which again satisfies an o.d.e. Although, to our knowledge, they never combined these ideas to write down a p.d.e. for the evolution of an envelope order parameter, the use of canonical equations with universal application closely parallels the spirit of their work.

of this simplification is that the envelope order parameters depend only on slow scales and satisfy equations which, like the amplitude equations, have a universal structure. Their form depends principally on symmetry considerations and not on microscopic details. Again, in principle, the asymptotic state of the system can be reconstructed from the solutions of these equations and the corresponding analog of the graph (1.1), but, as we shall see, the notion of a center manifold is on a much shakier footing. There are many open questions.

Far from onset, it is not at all clear how to decompose the space of all configurations into active and passive modes although, in recent years, there have been major efforts to generalize the notion of a center manifold to what is called a global center manifold or inertial manifold. An inertial manifold is a finite dimensional Euclidean space in the phase space of the system which contains the attractor (over a reasonably large parameter range!) and to which orbits are rapidly attracted. Unfortunately, in most cases to which this notion has been applied, the lack of appropriate coordinates has precluded its utility as an analytical tool since the use of inefficient coordinates means that the embedding space is far too big an object to deal with easily. To date, therefore, most pattern dynamicists have resorted to simpler and more manageable means of description which go some distance towards identifying what the correct coordinatization of patterns should be.

The first approach is linked permanently with the name of Busse (Busse 1967a,b, 1978, 1981; Clever & Busse 1979). The idea is to follow a branch of solutions which begins at R_c to finite values of $R-R_c$ and examine the linear stability of the finite amplitude states. In particular, Busse and his coworkers have exhaustively investigated the stability of a family of straight parallel convective rolls parametrized by the Rayleigh number R, wavenumber k and Prandtl number ν/κ—the ratio of molecular and thermal diffusivities. For a fixed Prandtl number, the domain of stable solutions in R, k space is called the *Busse balloon*. The nature of the instabilities at its borders has been documented and the predictions of the theory have proven to be very valuable in understanding much observed behavior (Figure 4).

But natural convective patterns, i.e. those which arise from the amplification of initial inhomogeneities, are rarely a small variation of a straight roll pattern on a global scale. Instead, a combination of the property of rotational invariance, which makes rolls of all orientations equally likely, and the boundary conditions, which choose the roll direction locally, conspires to give fairly complicated pattern textures, consisting of patches of curved rolls almost circular in shape, orthogonal patches of almost straight rolls, and defects such as foci, grain boundaries, disclinations, and

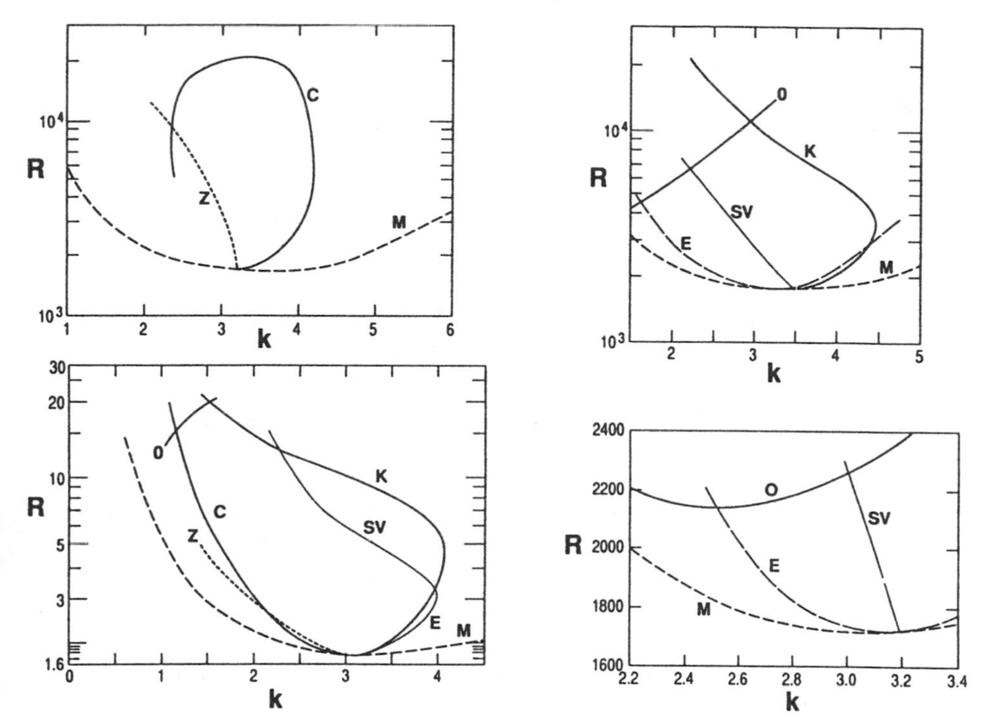

Figure 4 Schematic drawings of the Busse balloon at Prandtl numbers (*upper left*) Pr = 70, (*lower left*) Pr = 2.5, (*upper right*) Pr = 0.71, (*lower right*) Pr = 0.1. The borders are marked: C (cross-roll), Z (zigzag), SV (skew-varicose), E (Eckhaus), O (oscillations), K (knot), and denote the various instabilities. M denotes the neutral stability curve of the conduction solution. Only Z, SV, and E correspond to long wave instabilities of the underlying roll pattern. C, K, and O require the addition of new phases.

dislocations. The challenge is to develop a macroscopic field-particle theory that takes advantage of some underlying periodic structure of the microscopic field and that, at the same time, can handle the pattern defects which are singularities of (and the price paid for) the macroscopic description.

The starting point is the finite amplitude roll solution $w = f(\theta)$, $\mathbf{k} = \nabla\theta$, of the Busse analysis, except that instead of taking the wavevector $\mathbf{k}$ to be constant, it is allowed to vary slowly over the container so that $\nabla \cdot \mathbf{k} = O(\Gamma^{-1})$, where $\Gamma^{-1} = d/\mathrm{L}$ is the inverse aspect ratio of roll wavelength to container size. If the local planform contains more than one wavevector, additional phases $\theta_1, \theta_2, \ldots, \nabla\theta_1 = \mathbf{k}_1, \nabla\theta_2 = \mathbf{k}_2, \ldots$, must be introduced. If traveling waves are involved, the corresponding phase will have a slowly varying frequency $\omega = -\theta_t$ as well as wavevector $\mathbf{k}$. Because $f(\theta)$ is no longer an exact solution of the field equations, corrections of orders Γ^{-n}, $n = 1, 2, \ldots$ are necessary. The solvability condition for these corrections leads to the *phase diffusion equation*—the finite amplitude analog of the phase component of the NWS equation. Sometimes the phase diffusion equation is nonlocal and involves a mean drift field $\mathbf{U}$. The imposition of a fixed periodicity means that, almost everywhere, the amplitude (the norm of w) is slaved to $|\mathbf{k}|$; near defects, however, it becomes an independent order parameter.

We summarize the four categories [(*i*) near onset, discrete spectrum, (*ii*) near onset, continuous spectrum, (*iii*) far from onset, finite amplitude rolls and their stability, and (*iv*) far from onset, modulated roll patterns] with Table 1, which lists the names of people who have made contributions to each of these approaches. The idea of using an envelope order parameter for systems with continuous spectra came directly from the use of envelopes to describe the general propagation properties of nonlinear wave envelopes (Benney & Newell 1967). Indeed the conservative part of the complex Ginzburg-Landau equation is nothing other than the nonlinear Schrödinger equation and one of the most interesting aspects of pattern dynamics is how the dispersive wave and nonlinear focusing character of the system gives rise to a much richer set of behaviors than one would expect from the gradient flow component associated with diffusion and saturable nonlinearities. The method for treating slowly varying finite amplitude patterns followed from the seminal work of Whitham (1974) on nonlinear wavetrains.

It should also be stressed that experiments, involving greatly improved

Table 1 The four categories of order parameter equations

	$R-R_c$ small	$R-R_c$ of order one
$\Gamma = O(1)$ or periodic quasiperiodic	Amplitude equations Landau 1944 Koiter 1963 Gor'kov 1957 Stuart 1960 Schlüter et al 1965	Finite Amplitude rolls and their stability Busse (1967a,b, 1978, 1981) Busse & Whitehead 1974 Clever & Busse 1979
$\Gamma \gg 1$	Envelope equations Newell & Whitehead (1969, 1971) Segel 1969 Pomeau & Manneville 1979	Slowly Modulated Patterns Whitham 1974 Howard & Kopell 1977 Cross & Newell 1984 Newell et al 1990

techniques of parameter control and data acquisition and analysis, have played a vital role in increasing our understanding. In particular, we mention the work of Busse & Whitehead (1974) on convection patterns and defects, Fenstermacher et al (1979) on the Taylor-Couette problem, Libchaber & Maurer (1978) on convection in small aspect ratio boxes, Ahlers & Behringer (1978, 1982), Ahlers et al (1985), Dubois & Bergé (1978), Croquette (1989), Pocheau et al (1985), Pocheau (1988), and Heutmaker & Gollub (1987) on natural convection patterns, and Kolodner et al (1986, 1990), Joets & Ribotta (1989), Rehberg et al (1989), Steinberg et al (1989) and Sano et al (1992) on traveling wave convection and localized structures in convecting binary fluid mixtures and nematic liquid crystals. Indeed one of the gratifying features of the field of pattern dynamics is that progress has been made by the collective and friendly competitive effects of computational, experimental, and theoretical scientists.

The outline of the review is as follows. In Section 2 we discuss the ideas behind the derivation of amplitude, envelope, and phase-diffusion equations and make clear that while much has been achieved, even more remains to be done. Section 3 covers pattern singularities. In Section 4 we describe some of the most important properties of order parameter equations and in particular point out interesting differences between gradient and nongradient flows. Section 5 discusses two applications. Useful review articles are listed in the following references: Getling (1991), Haken (1979), Swinney & Gollub (1981), Wesfreid et al (1988), Coullet & Huerre (1990), Busse & Kramer (1990), Joseph (1976), Cross & Hohenberg (1992), Rabinovich (1992), and Murray (1989). One of the aims of this article is to focus attention on open problems—on challenges rather than successes. Because patterns are manifested in all the sciences—physical, life, and behavioral—these questions and their answers have ramifications way beyond the context in which they were initially posed. The subject is alive!

2. ORDER PARAMETER EQUATIONS

Here we concentrate on a few of the main ideas. We will use the model

$$\frac{\partial w}{\partial t}+\left(\frac{\partial^2}{\partial x^2}+\frac{\partial^2}{\partial y^2}+1\right)^2 w+2\beta\frac{\partial^2 w}{\partial x^2}-\mathrm{R}w+\alpha w^2+w^3=0 \tag{2.1}$$

for illustrative purposes. The "conduction" solution $w = 0$ becomes linearly unstable when $\mathrm{R} = \mathrm{R_c}$ where $\mathrm{R_c}$ depends on the parameter β and the boundary conditions. Our goals are to understand what happens both *near onset*, that is when $\mathrm{R} = \mathrm{R_c}+\mu$, $|\mu| \ll 1$, and *far from onset*, that is when μ

is of order one. We study two cases (i) $\partial/\partial y = \alpha = \beta = 0$ with $w = \partial^2 w/\partial x^2 = 0$ at $x = 0, L$ and (ii) w bounded as $x^2 + y^2 \to \infty$.

Near Onset: Identification of A and P

The decomposition of the phase space into active and passive modes is achieved by studying the linear stability of the simple state. In general, solutions will be sought in the form $\exp(\sigma t + i\mathbf{k}\cdot\mathbf{x})\phi(\mathbf{z}, \mathbf{n})$ where $\phi(\mathbf{z}, \mathbf{n})$ is the set of shapes compatible with boundary conditions and satisfying the linear stability problem in the directions $\mathbf{z}$ of finite extent, and $\exp i\mathbf{k}\cdot\mathbf{x}$, for continuous $\mathbf{k}$, is the Fourier basis element compatible with the property of translation symmetry associated with the directions of infinite extent. From the linear stability problem we find the complex dispersion relation $\mathsf{L}(\sigma = \nu - i\omega, i\mathbf{k}, \mathbf{n}, \mathbf{R})$ from which we obtain

$$\nu = \nu(\mathbf{k}, \mathbf{n}, \mathbf{R}), \quad \omega = \omega(\mathbf{k}, \mathbf{n}, \mathbf{R}) \tag{2.2}$$

giving the growth rate ν and frequency ω (possibly multivalued with multiplicity m; e.g. left and right traveling waves) as functions of the continuous wavevector $\mathbf{k}$, discrete wavevector $\mathbf{n}$, and external parameters $\mathbf{R}(\mathrm{R}, L, \alpha, \beta, \ldots)$. The neutral stability manifold is the set of surfaces $\nu(\mathbf{k}, \mathbf{n}, \mathbf{R}) = 0$. If we choose the first member R of $\mathbf{R}$ to be the stress parameter, then this manifold can be written as a set of surfaces $\mathrm{R} = \mathrm{R}(\mathbf{k}, \mathbf{n}, L, \alpha, \beta, \ldots)$. The simple solution loses stability as R increases through R_c. We call the point or set of points $\mathbf{k}, \mathbf{n}$ on these surfaces for which R attains its minimum value R_c the locus of critical modes K. If there is more than one point $\mathbf{k}, \mathbf{n}$ in K, we say we have a degeneracy. Degeneracies, which can be discrete or continuous, are a direct reflection of symmetries. Sometimes, as in both examples (i) and (ii), the growth rate σ of an unstable configuration remains real as it increases through zero for increasing values of the stress parameter R. The transition from a damped state to an amplified state without oscillations is called an exchange of stabilities. Other times, as in lasers and, for certain parameter ranges, in convection in binary fluids and liquid crystals, the real part ν of the growth rate σ first becomes positive at finite values of ω—a situation called overstable or a Hopf bifurcation.

For example in (i), $\mathbf{k} = \mathbf{0}$, $\nu = \mathrm{R} - [(n^2\pi^2/L^2) - 1]^2$, x is a direction of finite extent, $\phi(x, n) = \sin n\pi x/L$, and the set of neutral stability curves $\nu = 0$ have the shapes of scalloped shells with minima $\mathrm{R} = 0$ at $L = n\pi$. (See Figure 5.) Except at locations where the neutral stability curves cross, K consists of one point. For (ii) with $\beta = 1$, the set K consists of the points $k_x = \pm\sqrt{2}$, $k_y = 0$, and $\mathrm{R}_c = -3$. For (ii) with $\beta = 0$, the set K is a circle, $k^2 = k_x^2 + k_y^2 = k_c^2$, and $\mathrm{R}_c = 0$ (see Figure 6). The circular shape of K is a direct reflection of the rotational symmetry. If the spectrum of growth

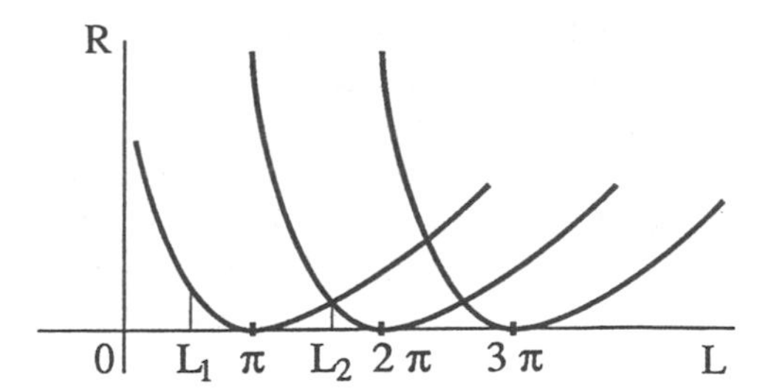

Figure 5 Neutral stability curves $\sigma^2 = \mathrm{R} - (n^2\pi^2 L^{-2} - 1)^2$, $n = 1, 2, \ldots$

rates is discrete as it is in (*i*), we see that, for a given L, as R increases, ν first becomes positive for a single mode (when $L = L_1$) or at most two (when $L = L_2$). For finite L, the spectrum of growth rates is well separated so that if $\mathrm{R} = \mathrm{R_c}(L_1) + \mu$ where $0 < \mu \ll 1$, then the mode $\sin \pi x/L_1$ is weakly amplified and all others $\sin n\pi x/L_1$, $n > 1$ are strongly damped. In

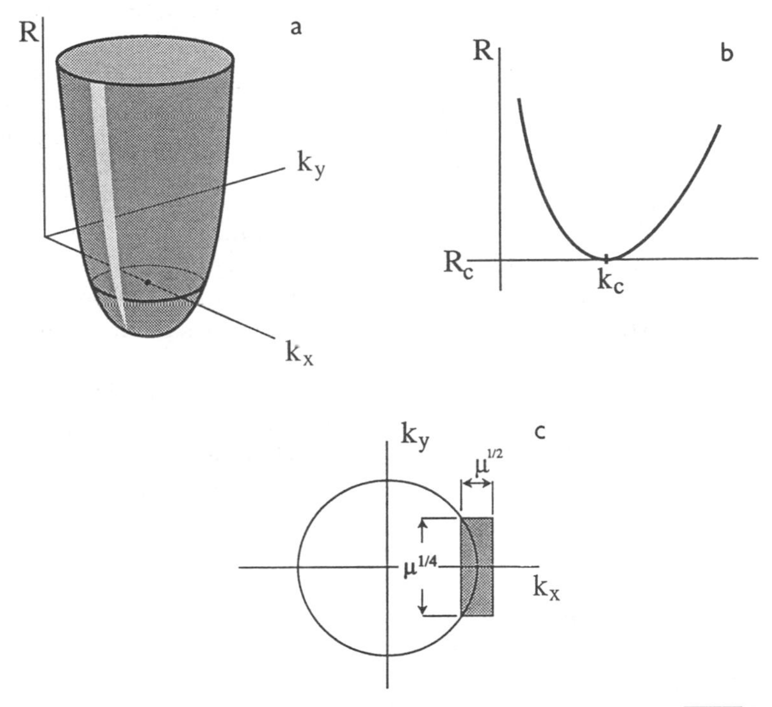

Figure 6 Neutral stability surfaces (*a*) R vs k_x, k_y when $\beta = 1$, (*b*) R vs $k = \sqrt{k_x^2 + k_y^2}$ when $\beta = 0$. (*c*) Projection of (*b*) in k_x, k_y plane with rectangle R of active modes in the neighborhood of $(k_c, 0)$.

this case, for a given small μ, it is easy to define A and P. For $L = L_1$, A is spanned by $\sin \pi x/L_1$ and P is spanned by $\sin n\pi x/L_1$, $n > 1$. For L near L_2, say $L = L_2 + \mu\beta$, A is spanned by $\sin \pi x/L$, $\sin 2\pi x/L$ and P is spanned by $\sin n\pi x/L$, $n > 2$. The order parameter in the first case is the time dependent amplitude A_1 of $\sin \pi x/L_1$. The amplitudes A_n, $n > 1$ very quickly relax to values $A_n = A_n(A_1)$, the center manifold. The order parameters for the second case are $A_1(t)$, $A_2(t)$—the coefficients of $\sin j\pi x/L$, $j = 1, 2$, and the center manifold is $A_n = A_n(A_1, A_2)$, $n > 2$. For L large, however, the spectrum of growth rates becomes more dense. For example, if $L = n\pi$ and $\mathrm{R} = \mu$, the growth rate of $\sin n\pi x/L$ is μ and the decay rates of its neighbors $\sin[(n \pm r)\pi x/L]$ are $\mu - (4r^2\pi^2/L^2)$. In order that neighboring modes are strongly damped, we require $\mu L^2 \ll 1$. Only in that case can we consider the spectrum truly discrete. For μL^2 greater than or equal to unity, the subdivision of S into A and P is less clear. In this case, we follow a strategy of assuming the worst: take L to be infinite and think of the direction x as one of infinite extent. Similar considerations apply when treating the problem of convection in a horizontal layer of fluid. When $\Gamma\sqrt{\mathrm{R} - \mathrm{R}_c}$ is of order one or greater, we must abandon the discrete mode analysis, and seek another way of defining A and P.

Consider example (*ii*), the anisotropic case. For $\beta = 1$, the neutral stability surface is a parabolic cylinder and it is clear that when $\mathrm{R} = \mathrm{R}_c + \mu$, there will be an order $\sqrt{\mu}$ bandwidth of modes in the k_x and k_y directions around the preferred wavevector $\mathbf{k}_c = (\sqrt{2}, 0)$. To include all potentially amplified, neutral, and weakly damped modes, we choose as order parameter the *slowly varying envelope* $W(x, y, t)$ of the real field w written as $W \exp ik_c x + (*)$ where, for $\beta = 1$, $W^{-1}(\partial W/\partial x, \partial W/\partial y, \partial W/\partial t)$ are of orders $\mu^{1/2}$, $\mu^{1/2}$, and μ respectively. $W^{-1}(\partial W/\partial t)$ will be of order $\mu^{1/2}$ in the case of a Hopf bifurcation to a traveling wave to account for envelope advection by the group velocity. This description means that A includes modes both within, and a finite distance outside, the amplified band. While it may be reasonable to conjecture that all modes in the remaining region are slaved, this conclusion is by no means obvious or justified a priori. For example, in the one-dimensional case, we will find that finite amplitude solutions of the form $W(x, t) \propto \sqrt{\mu - 8(k - k_c)^2}\; e^{i(k - k_c)x}$ are unstable if $24(k - k_c)^2 > \mu$ [the Eckhaus instability (Eckhaus 1965)]; the finite amplitude response of the system is to eliminate or inject a roll pair by locally forcing the amplitude to zero (Kramer & Zimmermann 1985, Kramer et al 1988). Near this point, the local wavenumber $k = \partial\phi/\partial x$, $(W = A \exp i\phi)$ becomes very large and clearly runs through the region we have designated as P. Indeed at the point in space and time where $A = 0$, k becomes singular and $\int k\,dx$ undergoes a jump of $\mp 2\pi$. In the two-dimensional case, the finite amplitude stage of the Eckhaus instability usually leads to the nucleation of a dislocation pair so that the amplitude becomes zero for a finite time at two particular (x, y) locations rather than along a whole line, $x =$ constant, at one particular time t. The two dislocations then move apart so as to eliminate a roll pair. The thorny question is, then: In what sense does the solution of the envelope equation for $W(x, t)$ $[W(x, y, t)]$ correctly describe the behavior of the original field $w(x, t)$ $[w(x, y, t)]$? Some rigorous work on this question has been done by Collet & Eckmann (1991) when k remains in the Eckhaus stable band. Even during the wavenumber adjustment process described above, it is still likely that the envelope equation gives a reliable (in a well-defined sense) description of the real dynamics because in 1-D the finite amplitude stage of the roll pair is very fast and follows a low dimensional "singular" solution for k and, in 2-D, both dislocation formation and travel are correctly captured. However, nothing has yet been proved.

The division of S into A and P is much more difficult in the rotationally degenerate isotropic case because then there is an annulus [of width $O(\sqrt{\mu})$ about $k = k_c$] of wavevectors whose corresponding modes are all linearly amplified. In principle, we can take all the modes lying in an annular neighborhood K of width $O(\sqrt{\mu})$ containing both amplified and weakly damped modes to be the active set A and the generalized Fourier amplitudes $\hat{w}(\mathbf{k}, t)$ of $w(x, y, t)$ where $\mathbf{k} \in \mathsf{K}$ to be the system's order parameters. It is straightforward to write down in a formal way the integro-differential amplitude equations for $\hat{w}(\mathbf{k}, t)$ (Besterhorn & Haken 1990). However, since $w(x, y, t)$ is only bounded as $x^2 + y^2 \to \infty$, the Fourier amplitudes $\hat{w}(\mathbf{k}, t)$ are generalized functions of $\mathbf{k}$ and we have to make clear in what way the representation $w(x, y, t) = \int_{\mathbf{k} \in \mathsf{K}} \hat{w}(\mathbf{k}, t) e^{i\mathbf{k}\cdot\mathbf{x} - i\omega(\mathbf{k}, \mathrm{R}_c)t} d\mathbf{k}$ makes sense. First, we might assume that A consists of an arbitrarily chosen set of modes $\exp(i\mathbf{k}_j \cdot \mathbf{x})$, $|\mathbf{k}_j| = k_c$, lying on the critical circle so that $\hat{w}(\mathbf{k}, t) = \sum_j \delta(\mathbf{k} - \mathbf{k}_j) W_j(t)$, $j = 1, \ldots, N$, and obtain amplitude equations for $W_j(t)$. If the attractor for these equations is a fixed point corresponding to a single roll (e.g. W_1 finite, $W_j = 0$, $j = 2, \ldots, N$) or a single hexagon ($|W_1| = |W_2| = |W_3|$, $\mathbf{k}_1 \cdot \mathbf{k}_2 = \mathbf{k}_2 \cdot \mathbf{k}_3 = \mathbf{k}_3 \cdot \mathbf{k}_1 = k_c^2 \cos(2\pi/3)$, $W_j = 0$, $j \neq 1, 2, 3$), then this is a good indication (although not a rigorous proof) that the pattern *locally* will look like a field of straight parallel rolls or a field of hexagons with a fixed orientation. The particular orientation in each case would be chosen by some external bias. We can go further if, after establishing that a single roll pattern dominates locally, we also assume it dominates globally so that everywhere in space the local wavevector of the pattern lies close to a given wavevector $\mathbf{k}_c = (k_c, 0)$, namely in the (*shaded*) rectangle shown in Figure 6*c*. In that case, we can take A to be the set of modes in rectangle R and use as order parameter the envelope $W(x, y, t)$ of the field $We^{i(k_c x - \omega_c t)} + (*)$ where $W^{-1}(\partial W/\partial x) = O(\mu^{1/2})$,

$W^{-1}(\partial W/\partial y) = O(\mu^{1/4})$, and $W^{-1}(\partial W/\partial t)$ will be of order $\mu^{1/2}$ if $\omega \neq 0$ and of order μ if $\omega = 0$. Patterns which are a slow modulation of square [e.g. $\mathbf{k}_1 = (k_c, 0)$, $\mathbf{k}_2 = (0, k_c)$] or hexagonal [e.g. $\mathbf{k}_1 = k_c(1,0)$, $\mathbf{k}_2 = k_c(-\frac{1}{2}, \sqrt{3}/2)$, $\mathbf{k}_3 = k_c(-\frac{1}{2}, -\sqrt{3}/2)$] planforms or involve two or more roll patterns separated by grain boundaries can be similarly represented. However, we must be careful with such a representation. While it is true that the space independent [i.e. $W(x,y,t) = W(t)$] solution is stable, it is not true that a solution which represents a field of straight parallel rolls with a slightly bigger or smaller wavenumber will necessarily be stable to disturbances which have wavevectors which lie outside the rectangular neighborhood. For example, a set of rolls with wavevector $(k_c \pm K, 0)$ can become unstable to either (*a*) a long wavelength instability (whose wavevector belongs to A) or (*b*) to a cross roll disturbance $(0, k_c)$ (which does not belong to A). Therefore it is implicitly assumed when we use an envelope description that if the pattern wavevector initially is in A, it will stay there. Otherwise, we must introduce a second envelope order parameter. No situation that involves a revolt of the designated "slaves" can be tolerated! The presence of other potentially active rolls (for example, the almost perpendicular rolls joined by a grain boundary to the primary set of rolls in Figure 11*a*) must be anticipated.

As the reader can see, we are a long way from establishing a rigorous theory which ensures the uniform validity of the approximation (in any suitable norm) of the field by a linear combination of configurations, chosen from linear stability considerations, with complex envelopes. Nevertheless, despite the lack of rigor, the envelope equations we will shortly meet have had considerable success in helping us understand pattern behavior in a wide variety of situations, including the behavior of defects in both roll and hexagonal planforms. The most serious criticism of the envelope approach is that a pattern that naturally arises in isotropic systems can rarely be described globally as a small deviation from a fixed set of parallel rolls. Rather, as we have already mentioned in the introduction, while the pattern may locally appear to consist of straight parallel rolls, the wavevector changes significantly, albeit slowly, in both direction and amplitude as we traverse the convection layer. For such situations, an alternative strategy, described in the *Far from Onset* subsection, must be followed.

Near Onset: Weakly Nonlinear Analysis

Having established the structure of the active set A, in some cases rigorously, in others at best plausibly, we now address the question of deriving equations which describe the motion of the system on its center manifold.

The reader should consult references (Schlüter et al 1965; Newell & Whitehead 1969, 1971; Busse 1967b; Newell 1989; Joseph 1976) for details. A brief outline is sketched here. We write the field $w(\mathbf{x}, \mathbf{z}, t)$ as an asymptotic expansion $\sum \varepsilon_n(\mu) w_n(\mathbf{x}, \mathbf{z}, t)$ where $\{\varepsilon_n(\mu)\}$ is an asymptotic sequence as $\mu \to 0$ and $\varepsilon_0 w_0$ is a linear combination of members of A whose amplitudes or envelopes are the order parameters. $\varepsilon_0(\mu)$ is chosen to balance the linear growth rate μw with the most relevant nonlinear correction. If cubic, then ε_0 is $\mu^{1/2}$. The solvability conditions (obtained either (*a*) by insisting that the asymptotic series for w is uniformly valid in time and space or (*b*) by applying the Fredholm alternative theorem) for the iterates $w_1, w_2, \ldots,$ which satisfy $\mathsf{L}(\partial/\partial t, \nabla_{\mathbf{x}}, \nabla_{\mathbf{z}}, \mathrm{R})w = g(w_r)$, $r < n$, give the equations for the order parameters. The solutions w_n will belong to P and determine the graph (1.1) perturbatively. Each equation has a universal character that depends on the underlying symmetries.

REMARK 1 The linear terms which arise in the envelope order parameter equations are simply a reflection of the fact that the linear stability operator $\mathsf{L}(\partial/\partial t, \nabla, \partial/\partial z, \mathrm{R})$, (which gives rise to the complex dispersion relation), applied to $W(x, y, t) \exp(ik_c x - i\omega_c t)\phi(z)$ is $\exp(ik_c x - i\omega_c t)\phi(z)\mathsf{L}(-i\omega_c + \partial/\partial t, ik_c + \partial/\partial x, \partial/\partial y, \mathrm{R}_c + \mu) W(x, y, t)$. The last factor is

$$\left(\frac{\partial \mathsf{L}}{\partial \sigma}\right)_c \left\{\frac{\partial W}{\partial t} - \nu\left(k_c - i\frac{\partial}{\partial x}, -i\frac{\partial}{\partial y}, \mathrm{R}_c + \mu\right) W \right.$$

$$\left. + i\omega\left(k_c - i\frac{\partial}{\partial x}, -i\frac{\partial}{\partial y}, \mathrm{R}_c + \mu\right) W - i\omega(k_c, 0, \mathrm{R}_c) W \right\} \quad (2.3)$$

where ν and ω are expanded in a Taylor series in derivatives $\partial/\partial x$ (order $\mu^{1/2}$), $\partial/\partial y$ [order $\mu^{1/2}(\mu^{1/4})$ in the anisotropic (isotropic) case], and $\partial/\partial t$ (order $\mu^{1/2}$ if $\omega \neq 0$, order μ if $\omega = 0$). In the isotropic case ν and ω are functions of k_x, k_y though $k^2 = k_c^2 - 2ik_c \partial/\partial x - \partial^2/\partial y^2 - \partial^2/\partial x^2$. Note that to leading approximation, $k^2 - k_c^2$ is represented by the operator $-2ik_c \partial/\partial x - \partial^2/\partial y^2$.

REMARK 2 The strongest nonlinear interaction is quadratic (Palm 1960). In the case that a quadratic product of members of A has no projection in A, the quadratic term in the amplitude equation is absent. In the rotationally invariant case of convection, the members of A are $\exp(i\mathbf{k}_j \cdot \mathbf{x})\phi(z)$, $|\mathbf{k}_j| = k_c$. Hexagons arise because the product of two members of A whose wavevectors have an angular separation of $2\pi/3$ again gives another member of A providing the appropriate quadratic functional of the vertical structure [e.g. $\phi(d\phi/dz)$] has a nonzero projection in $\phi(z)$. [If the members of A are traveling waves, then nontrivial quadratic interaction also requires the satisfaction of the additional resonance condition $\pm\omega(\mathbf{k}_1)\pm\omega(\mathbf{k}_2)\pm\omega(\mathbf{k}_3) = 0$, $|\mathbf{k}_j| = k_c$.] The coefficient of the quadratic term directly measures the size of this projection. For example, it is zero if there is symmetry about the midplane of the convection layer as is the case in the convection experiments we discuss in Section 5. Any asymmetry about the midplane means that up and down are distinguished, and this up-down asymmetry is a property of hexagonal planforms (the center can be upflow or downflow) not shared by rolls. In writing down the amplitude equations we will assume that the quadratic coefficient is small, of order $\sqrt{\mu}$, so that each of the linear, quadratic, and cubic terms is of order $\mu^{3/2}$. Of course, that is not always the case. When a thin spherical shell collapses through a hexagonal planform, the quadratic terms are of order one, the subcritical solution which balances linear and quadratic terms is unstable, and the collapse is not arrested until the deformation has reached finite amplitudes at which stage its spatial structure is very different. Similar remarks obtain for parallel shear flows (Orszag & Patera 1983).

REMARK 3 Symmetry considerations imply that only certain nonlinear terms can be present (Coullet et al 1985). For example, translation symmetry means that if W satisfies a single order parameter equation, so must $W \exp i\phi_0$. Therefore the nonlinear cubic terms can have the form $W_1^2 W_1^*$ or $W_2 W_2^* W_1$ but not W_1^3, $W_2^2 W_1$ or W_1^{*3}; quadratic terms, proportional to $W_2^* W_3^*$, can only arise from hexagonal planforms where $\mathbf{k}_1+\mathbf{k}_2+\mathbf{k}_3 = \mathbf{0}$.

REMARK 4 In situations where a constant field is solution of $\mathsf{L}(\partial/\partial t, \nabla, \partial/\partial z, \mathrm{R})w = 0$, slowly varying mean fields can be driven by curvature effects and by slow gradients of the pattern intensity. In thermal convection, these mean fields are manifested as a mean drift velocity field that can produce an order one effect on the pattern by advecting its phase contours. For example, in the case of stress-free rather than rigid boundary conditions on the upper and lower plates, the Oberbeck-Boussinesq equations are Galilean invariant with respect to the horizontal direction. A constant horizontal velocity $\mathbf{U}$ is also marginal in the sense that it is a zero growth rate solution of the linear stability problem. Although it is not possible to add a net global momentum to the fluid, the slowly varying property of the envelope W means that it is possible for slow gradients of the pattern intensity WW^* and other pattern densities like the current $i(W\nabla W^* - W^*\nabla W)$ to drive slowly varying mean drift fields which still respect conservation of total momentum. The mean drift field is most easily calculated by considering the equation for the vertical vorticity field

$\zeta = (\nabla \times \mathbf{u})\cdot\mathbf{z} = -\nabla_1^2\psi(\mathbf{u} = \nabla \times \psi\mathbf{z})$ which typically reads (Siggia & Zippelius 1981a) for the rotationally invariant case,

$$\left(-\frac{1}{\Pr}\frac{\partial}{\partial t}+\nabla^2\right)\zeta = -\frac{\gamma}{\Pr}\frac{\partial}{\partial y}\left[\frac{\partial}{\partial x}WW^* - \frac{i}{2k_c}\frac{\partial}{\partial y}\left(W^*\frac{\partial W}{\partial y} - W\frac{\partial W^*}{\partial y}\right)\right]. \tag{2.4}$$

In (2.4), Pr is the Prandtl number and γ is a positive constant produced from quadratic products of the vertical structure and its derivatives whose vertical average $\bar{\gamma}$ is nonzero. ∇_1^2 is the horizontal Laplacian. In turn, the divergence-free mean drift $\mathbf{U}$ affects the advection of the phase contours of W through the addition of the term in the envelope equation, that is by adding $i\mathbf{k}_c\cdot\mathbf{U}W$ to the bracketed term in (2.3). Note that the effect of the mean drift is to produce a space dependent Doppler correction $\mathbf{k}_c\cdot\mathbf{U}(x, y, t)$ to the frequency.

In binary fluid convection, the relatively weak boundary conditions on the alcohol concentration field (the diffusion of the concentration field is much smaller than the diffusion of heat) means that, even near onset, a mean concentration field driven by gradients in the pattern intensity can have far reaching consequences, a situation we discuss in Section 5.

Near Onset: Canonical Amplitude-Envelope Equations

In each case, the equations are written in terms of the original independent variables x, y, t. Amplitudes are of order $\sqrt{\mu}$.

(*a*) For a single mode, real order parameter $W(t)$ such as one would find in example (*i*), one finds, writing $w(z, t) = W(t)\phi(z) + \ldots$ that

$$\frac{dW}{dt} = I + \mu W + \alpha W^2 - \beta W^3 = -\frac{\partial F}{\partial W}. \tag{2.5}$$

The term I models bias or geometric imperfections (Koiter 1963). The flow is gradient. The graph of stationary solutions W versus μ is given in Figure 7*a*.

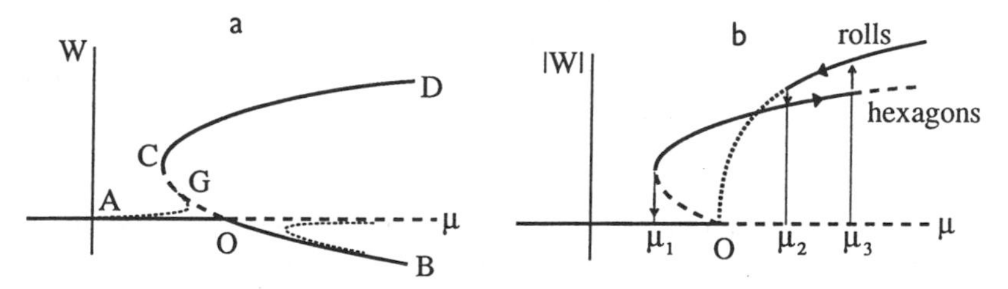

Figure 7 (*a*) Amplitude-stress parameter diagram for (2.5) showing the stable (*undashed*) and unstable (*dashed*) solutions for $I = 0$. When $I > 0$, the unfolding of the diagram is shown with a dotted line. Imperfections cause pitchfork bifurcations to be continuous; for transcritical and subcritical bifurcations they trigger early onset. (*b*) Bifurcation diagram showing regions of existence and stability (*undashed*) of roll and hexagon planforms.

(*b*) For multimode, complex amplitude order parameters, $w(x, y, z, t) = \Sigma_j [W_j(t)e^{i\mathbf{k}_j \cdot \mathbf{x}} + (*)]\phi(z) + \ldots$, $|\mathbf{k}_j| = k_c$, the equations are:

$$\frac{dW_j}{dt} = \mu W_j + \alpha_{jlm} W_l^* W_m^* - \sum_l \beta_{jl} W_l W_l^* W_j, \quad j = 1, \ldots, N. \tag{2.6}$$

The coefficient α_{jlm} is $O(\sqrt{\mu})$ when $\mathbf{k}_j + \mathbf{k}_l + \mathbf{k}_m = \mathbf{0}$ and $\pm\omega(\mathbf{k}_j, \mathrm{R_c}) \pm \omega(\mathbf{k}_l, \mathrm{R_c}) \pm \omega(\mathbf{k}_m, \mathrm{R_c}) = 0$, and zero otherwise. In general, but not always, the presence of the quadratic term will lead to subcritical solutions of transcritical type with a hexagonal planform. If β_{jl} is real and symmetric, and $\alpha_{123} = \alpha_{231} = \alpha_{312} = \alpha$, Busse (1967b) has shown that the RHS of (2.6) is gradient and equal to $-\partial F/\partial W^*$. If further $\beta_{jl} > \beta_{jj} = \beta > 0$, then the stable planforms are either rolls $W_1 = \sqrt{\mu\beta^{-1}}$, $W_j = 0, j = 2, \ldots, N$ or hexagons $W_j = A \exp i\phi_j, j = 1, 2, 3$ corresponding to $\mathbf{k}_j$'s 120 degrees apart whose amplitudes are given by the larger of the two roots of $\beta A^2 - \alpha s A - \mu = 0$ with $s = \exp - i(\phi_1 + \phi_2 + \phi_3) = \pm 1$ being a reflection of translational invariance. For subcritical and a small range of supercritical values, $\mu_1 < \mu < \mu_2$, $\mu_1 < 0$, hexagons are the stable attractors; for $\mu_2 < \mu < \mu_3$, the two states can coexist although the fact that the flow is gradient means that the state with the lowest free energy F will invade the other; for $\mu > \mu_3$, a field of rolls is the stable attractor. For β_{jl} not symmetric, both roll and hexagon states and all other stationary states can be unstable and a time-dependent state in which the system repeatedly but unsuccessfully attempts to access each of the three roll components of a hexagonal structure can be observed (Busse 1981). Mixed roll-hexagon states involving hexagons whose roll components have different amplitudes are also possible (Ciliberto et al 1990). Rhombus states (two non-perpendicular rolls of equal amplitudes) have been observed in chemical patterns (Ouyang & Swinney 1991). Even for symmetric matrix coefficients β_{jl} and α_{jlm}, i.e. gradient flows, stable quasiperiodic (quasicrystals) solutions can be realized. Furthermore, spatially turbulent (but time independent) states for which the energy is distributed uniformly to all modes with wavevectors on $\mathbf{k} = |k_c|$ are possible (Newell & Pomeau 1992). At any point, the field would have a Gaussian distribution. The correlation length would be the roll wavelength. Very little attention has been paid to these "spin glass" ground states of the free energy F.

(*c*) For a single mode, the envelope order parameter W_+ and W_- [(*) denotes complex conjugate] are related to the flow field W by

$$w = [W_\mathrm{R} \exp i\theta_+ + W_\mathrm{L} \exp i\theta_- + (*)]\phi(z), \quad \theta_\pm = \pm k_c x - \omega(k_c, \mathrm{R})t. \tag{2.7}$$

(i) In the anisotropic case, the single traveling wave envelope equation ($W_{\mathrm{L}} = 0$) without mean drift (Newell 1979) is

$$\frac{\partial W}{\partial t}+\omega_{\mathrm{c}}'\frac{\partial W}{\partial x}=\frac{1}{2}\left(\frac{\partial}{\partial x}\;\frac{\partial}{\partial y}\right)$$

$$\times\begin{bmatrix}\dfrac{\partial\nu}{\partial \mathrm{R}}\dfrac{\partial^2 \mathrm{R}}{\partial k_x^2}+i\dfrac{\partial^2\omega}{\partial k_x^2} & \dfrac{\partial\nu}{\partial \mathrm{R}}\dfrac{\partial^2 \mathrm{R}}{\partial k_x\partial k_y}+i\dfrac{\partial^2\omega}{\partial k_x\partial k_y}\\[2ex] \dfrac{\partial\nu}{\partial \mathrm{R}}\dfrac{\partial^2 \mathrm{R}}{\partial k_x\partial k_y}+i\dfrac{\partial^2\omega}{\partial k_x\partial k_y} & \dfrac{\partial\nu}{\partial \mathrm{R}}\dfrac{\partial^2 \mathrm{R}}{\partial k_y^2}+i\dfrac{\partial^2\omega}{\partial k_y^2}\end{bmatrix}$$

$$\times\begin{bmatrix}\dfrac{\partial}{\partial x}\\[2ex]\dfrac{\partial}{\partial y}\end{bmatrix}W+\mu W-(\beta_{\mathrm{r}}+i\beta_i)W^2W^*. \tag{2.8}$$

(ii) In the one-dimensional case, the envelope equations for counter-propagating waves (γ, β, δ complex) are

$$\frac{\partial W_{\mathrm{R}}}{\partial t}+\omega_{\mathrm{c}}'\frac{\partial W_{\mathrm{R}}}{\partial x}=\gamma\frac{\partial^2 W_{\mathrm{R}}}{\partial x^2}+\mu W_{\mathrm{R}}-\beta|W_{\mathrm{R}}|^2W_{\mathrm{R}}-\delta|W_{\mathrm{L}}|^2W_{\mathrm{R}}, \tag{2.9}$$

$$\frac{\partial W_{\mathrm{L}}}{\partial t}-\omega_{\mathrm{c}}'\frac{\partial W_{\mathrm{L}}}{\partial x}=\gamma\frac{\partial^2 W_{\mathrm{L}}}{\partial x^2}+\mu W_{\mathrm{L}}-\beta|W_{\mathrm{L}}|^2W_{\mathrm{L}}-\delta|W_{\mathrm{R}}|^2W_{\mathrm{L}}. \tag{2.10}$$

Equation (2.9) with $W_{\mathrm{L}} = 0$ is usually referred to as the complex Ginzburg-Landau (CGL) equation (Newell & Whitehead 1971, Stewartson & Stuart 1971):

$$\frac{\partial W}{\partial t}+\omega_{\mathrm{c}}'\frac{\partial W}{\partial x}=(\gamma_{\mathrm{r}}+i\gamma_i)\frac{\partial^2 W}{\partial x^2}+\mu W-(\beta_{\mathrm{r}}+i\beta_i)W^2W^*. \tag{2.11}$$

If the real and imaginary parts of the matrix in (2.8) are positive and can be diagonalized simultaneously, then the two-dimensional CGL equation is (2.11) with $\partial^2/\partial x^2$ replaced by ∇_1^2. Note that in the anisotropic case (or if the unstable state is spatially uniform, i.e. $\mathbf{k}_{\mathrm{c}} = \mathbf{0}$), the x and y directions in the envelope scale the same and the envelope equation looks isotropic. In contrast, for the isotropic case, the parallel and perpendicular directions scale differently and the envelope equation looks anisotropic.

(iii) In the isotropic case, the envelope equation for a traveling wave with mean drift effect included (Brand et al 1986a,b) is

$$\frac{\partial W}{\partial t}+\omega_c'\left(\frac{\partial}{\partial x}-\frac{i}{2k_c}\frac{\partial^2}{\partial y^2}\right)W=-ik_cUW+\mu W+\frac{i\omega_c'}{2k_c}\frac{\partial^2 W}{\partial x^2}$$

$$+\frac{1}{2}\left(\frac{\partial v}{\partial \mathrm{R}}\mathrm{R}''+i\left(\omega_c''-\frac{\omega_c'}{k_c}\right)\right)\left(\frac{\partial}{\partial x}-\frac{i}{2k_c}\frac{\partial^2}{\partial y^2}\right)^2 W-(\beta_r+i\beta_i)W^2W^*. \quad (2.12)$$

The NWS equation for stationary patterns is (2.12) with $\omega\equiv 0$, $\beta_i=0$, $\gamma=\frac{1}{2}v_{\mathrm{R}}\mathrm{R}''$, i.e.

$$\frac{\partial W}{\partial t}-\gamma\left(\frac{\partial}{\partial x}-\frac{i}{2k_c}\frac{\partial^2}{\partial y^2}\right)^2 W=-ik_cUW+\mu W-\beta W^2W^*. \quad (2.13)$$

Many of the predictions of this equation have been successfully tested in experiments carried out by Wesfreid et al (1978). The mean drift U is found from (2.4). For hydrodynamic convection with stress free boundaries, the vertical average of (2.4) eliminates $\partial^2\xi/\partial z^2$, and using $\xi\simeq(\partial U/\partial y)$, we obtain

$$-\frac{1}{\mathrm{Pr}}\frac{\partial U}{\partial t}+\frac{\partial^2 U}{\partial y^2}=-\frac{\bar{\gamma}}{\mathrm{Pr}}\left[\frac{\partial}{\partial x}WW^*-\frac{i}{2k_c}\frac{\partial}{\partial y}\left(W^*\frac{\partial W}{\partial y}-W\frac{\partial W^*}{\partial y}\right)\right]. \quad (2.14)$$

From (2.14), we see U is of order μ and the mean drift advection term in (2.13) is $O(\mu^{3/2})$ and comparable with the other terms in the envelope equation. When the principle of exchange of stabilities obtains, $\partial/\partial t$ is of order μ and U is no longer an independent order parameter but is slaved to W through the nonlocal relation (2.14). For rigid boundary conditions, the vertical average of the left-hand side of (2.4) is dominated by $(\partial^2\xi/\partial z^2)=(\partial/\partial y)(\partial^2 U/\partial z^2)$ which is nonzero. As a result, U is of order $\mu^{3/2}$ and the mean drift correction in (2.13) is of higher order (μ^2) than the order terms and comparable to the next correction [e.g. $W(\partial/\partial x)|W|^2$] to the envelope equation. Both the mean drift and higher order derivative corrections generally make (2.13) a nongradient flow (Pomeau et al 1983).

Equations (2.8–2.13) are all asymptotic expansions for the evolution of the envelope W over long times. For example, for times $t\sim\mu^{-1/2}$, the envelope W in the CGL equation (2.11) moves without change of shape with the group velocity; for times μ^{-1}, linear and nonlinear dispersion, amplification and nonlinear saturation (or growth) modify its shape; for longer times, $t\sim\mu^{-3/2}$ we must take account of such terms as higher order dispersion $\partial^3 W/\partial x^3$, and nonlinear gradients $W(\partial/\partial x)|W|^2$. What is important to remember is that each of the successive terms in the asymptotic expansion for $\partial W/\partial t$ is smaller than the one before. If we take them equal in amplitude, then there is in general no reason to exclude the next correction. The only circumstances in which one can take $\omega_c'(\partial W/\partial t) = O(\mu)$ and $\gamma_i(\partial^2 W/\partial x^2) = O(\mu^{3/2})$ to be of the same order is when ω_c' itself is small (of order $\mu^{1/2}$). Sometimes, however, the small parameter μ can be removed altogether. For example, in (2.11), the change of variables $W(x, t) = \mu^{1/2} q[X = \mu^{1/2}(x - \omega_c' t), T = \mu t]$ removes the group velocity and gives the μ-independent [the corrections include μ e.g. $i\mu^{1/2} q(\partial/\partial x)|q|^2$] equation

$$\frac{\partial q}{\partial T} = (\gamma_r + i\gamma_i)\frac{\partial^2 q}{\partial X^2} + q - (\beta_r + i\beta_i)q^2 q^* + O(\mu^{1/2}).$$

Likewise in (2.9) and (2.10), if the group velocity is small (of order $\mu^{1/2}$), the small parameter μ can be removed. However if ω_c' is of order one, then the right traveling envelope W_R will see the average effect of the intensity $|W_L|^2$ of the left going envelope as it sweeps to the left. Moreover, there is no way to change variables to remove the strong linear y dispersion in (2.12) without hopelessly scrambling the next order terms, nor of removing the μ in the NWS equations for a slowly varying hexagonal structure ($\partial/\partial x \to \mathbf{k} \cdot \nabla$, $\partial/\partial y \to \mathbf{k} \times \nabla$). Not to worry! While there are both aesthetic (the equation is canonical) and practical (the simulation can use larger grid spacing) advantages of removing the small parameter, it is not always possible to do so. Its presence is an inconvenience but not an obstacle to understanding.

Far from Onset

We start by assuming the existence of a family (parametrized by the wavenumber k) of fully nonlinear 2π-periodic solutions of the governing equations each member of which represents either a linearly stable stationary periodic pattern or a traveling wave, an assumption usually justified by the observation that such configurations locally dominate the pattern. A modulated wave can be considered to be a vector-valued function V that locally resembles a plane traveling wave whose local wavevector and local frequency vary on scales $\mathbf{X} = \varepsilon \mathbf{x}$, $T = \varepsilon t$, much larger than those of the underlying wavelength or period. It will have a representation of the form,

$$V(x, t, \varepsilon) = U\left[\theta = \frac{\Theta(X, T, \varepsilon)}{\varepsilon}\right] + \sum \varepsilon^j U_j. \tag{2.15}$$

The fast variation of U occurs through the phase variable θ, i.e. $\nabla_{\mathbf{x}}\theta = \nabla_{\mathbf{X}}\Theta = \mathbf{k}(X, T)$, $-\theta_t = -\Theta_T = \omega(X, T)$, and the local wavevector and frequency vary slowly in space and time.

What are the differences with the near onset case? First of all, the basic state that we are going to start from with the perturbation analysis is not the zero state but rather a fully nonlinear wave, exactly periodic. As a consequence, for a fixed wavenumber, the null eigenspace of the linearized operator L_s about this basic state is now of dimension one, corresponding to the free translation mode, and not dimension two as it is the case near onset where the amplitude mode is also marginal. Remember, close to onset, $\partial U/\partial W$ (where W is the complex amplitude) is in the kernel of the linearized operator: Far from onset and at finite amplitudes, only $\partial U/\partial \theta$ is in the kernel. As an immediate result we can see that the order parameter of the system is not a complex quantity but rather a scalar—the phase mode θ. For all wavenumbers in a certain range, the amplitude of the rolls A (e.g. norm of the vertical velocity) is determined algebraically from the wavenumber k: It is slaved. Another important difference from the case close to onset is the definition of the small parameter. Obviously ε is not the deviation from the critical value of the stress parameter, but rather it is linked to the aspect ratio Γ of the container. Its inverse gives the number of rolls that have to be crossed before the local wavevector varies by order one in norm or direction. Such a theory must be able to describe the dynamics of rolls of any direction and thus must be rotationally invariant. The directions X and Y are identically related to their corresponding "small scales" by $X = \varepsilon x$, $Y = \varepsilon y$. While the method we describe is primarily aimed at analyzing situations far from onset, it is also illuminating to examine the phase dynamics of each of the complex envelope equations (2.8–2.13) treating μ as an order one parameter and introducing ε as the inverse aspect ratio. For example, the phase equation for the CGL equation will, in a certain limit, give rise to the Kuramoto-Sivashinsky equation (Kuramoto 1984).

In what follows, we briefly describe the derivation of a rotationally invariant equation for the total phase Θ (and not its perturbation $\phi = \theta - \mathbf{k}\cdot\mathbf{x} + \omega t$ from a globally defined field of straight parallel waves) and show how to couple it to the dynamics of other marginal modes (such as mean flows). We then discuss the reintroduction of the amplitude as a free parameter when the stress parameter is close to critical or for values of the wavenumber k close to the marginal stability curve at which values the amplitude is close to zero and the slaving relation $A = A(k)$ fails. Given a nonlinear partial differential equation of the form: $V_t = F(V)$ where $F(V)$ is a nonlinear function of V and its derivatives with respect to x, we first have to look for a traveling wave, 2π-periodic solution of the form $V(x, t) = V_0(\theta; A)$ indexed by $k = |k|$. The frequency $\omega(k)$ and amplitude $A(k)$ are determined by the wavenumber k. For example, for the CGL equation (2.11), the solutions are $W = A\exp(ikx - i\omega t)$, where

$\mu - \gamma_r k^2 - \beta_r A^2 = 0$ and $\omega = \omega_c' k + \gamma_i k^2 + \beta_i A^2$. In contrast, for conservative systems (take $\gamma_r = \beta_r = \mu = 0$), both the amplitude and wavenumber are free parameters. Allowing for large-scale modulations of the phase leads us to search for solutions in the form (2.15). The local wavevector and frequency satisfy a consistency relation, $k_t + \nabla\omega = 0$, corresponding to the conservation of phase. Each evolves on slow scales X, T and for convenience it will be useful to replace an expansion for the phase $\Theta = \Theta_0 + \varepsilon\Theta_1 + \ldots$ by an expansion for ω: $\omega = \omega_0 + \varepsilon\omega_1 + \ldots$, the wavenumber k remaining unchanged. The period of the wave cannot be allowed to depend on X and T. The choice of fixed period (usually taken as 2π) leads to the determination of the dependence of ω and A on k. In the case of stationary patterns, this choice determines A as a function of k.

The derivatives are transformed according to $\nabla_x \to k\partial_\theta + \varepsilon\nabla_x$, $\partial_t \to -\omega(\partial/\partial\theta) + \varepsilon(\partial/\partial T)$ or as $\partial_t \to \varepsilon\Theta_T\partial_\theta + \varepsilon^2\partial_T$ if the system is purely diffusive. Substitution of these representations into the governing equation yields equations for the iterates,

$$\omega_n \partial_\theta U_0 + \mathsf{L}_s U_n + N_n(U_0, U_1, \ldots, U_{n-1}, \omega_0, \ldots, \omega_{n-1}) = 0. \qquad (2.16)$$

The linear operator L_s depends only on U_0, ω_0, and $\partial/\partial\theta$; N_n is a nonlinear function of its arguments and their derivatives. Equation (2.16) contains two unknowns, ω_n and U_n. It is a Floquet problem as L_s acts only in θ and has 2π-periodic coefficients in θ. Since the right-hand side contains only coefficients which are also 2π-periodic in θ, one can look for solutions which have the same periodicity.[4] However one knows that, due to the translational invariance of the system, L_s has a null eigenmode, namely $\partial_\theta U_0$. Therefore a solvability condition must be imposed on the r.h.s. so that the inhomogeneous terms lie in the range of L_s. Defining the inner product as $\langle U | U' \rangle = \int^{2\pi} UU' \, d\theta$, the Fredholm alternative condition $\omega_n \langle U^\dagger | \partial_\theta U_0 \rangle + \langle U^\dagger | N_n \rangle = 0$, where $U^\dagger$ is the left eigenmode of L_s, gives ω_n. It is then possible to solve (2.16) for U_n. At first order, the phase equation obtained is $k_T + \nabla_X \omega_0 = -\varepsilon\nabla_X\omega_1$ and corrections of any order can be likewise calculated. In general ω_n will depend on gradients of k to order n. In the case of a traveling wave, the phase equation can develop shocks because $\omega_0(k)$ is nonlinear in k, and this can give rise to the development of multivaluedness in wavenumber which diffusive effects will regularize into wavenumber shocks. This fact makes the traveling wave problem very difficult but some useful results have been obtained by Howard & Kopell (1977) and Bernoff (1988). Here we will restrict ourselves to the results of

[4] One must watch for subharmonics. Far from onset, the damping rates of the subharmonics $k/2$, $k/3$ of the basic pattern wavenumber k can become small and finite amplitude perturbations [e.g. dislocations (Newell & Passot 1992)] can trigger their growth.

the purely diffusive case. The reader should consult references (Cross & Newell 1984, Newell et al 1990) for details. The complex Swift-Hohenberg (CSH) equation $w_t+(\nabla^2+1)^2w-\mathrm{R}w+w^2w^*=0$ was used by Cross & Newell (1984) as an illustrative example because the exact solution is $w=A\exp i\theta$ and all calculations can be carried out explicitly. The program has been carried out for the full Oberbeck-Boussinesq equations by Newell et al (1990). The basic method is the same but its implementation is much more complicated. At leading order we find the amplitude relation

$$A^2-A_0^2(k,\mathrm{R})=0. \tag{2.17}$$

At order ε, we obtain the phase diffusion equation which, because of rotational invariance, can always be written in the form,

$$\tau(k)\Theta_{\mathrm{T}}+\nabla_{\mathbf{X}}\cdot\mathbf{k}B(k)=\tau(k)\theta_t+\varepsilon\nabla_{\mathbf{x}}\cdot\mathbf{k}B=0, \tag{2.18}$$

where, for the CSH equation, $A_0^2=\mathrm{R}-(k^2-1)^2$, $\tau=A_0^2$, and $B(k)=A_0^2(dA_0^2/dk^2)$. So far, so good. Things have been relatively straightforward. Now we arrive at two difficult parts of the theory: the role of mean flows and the reintroduction of the amplitude as an active order parameter near pattern singularities.

In the context of Rayleigh-Bérnard convection with stress-free top and bottom boundaries close to onset, the generation of mean flows is relatively easy to understand. In addition to the complex amplitude of the rolls, there is another marginal mode corresponding to a horizontal velocity with no vertical dependence. This mode, which is driven by roll curvature, is important and responsible for both the oscillatory and skew varicose instabilities (see Figure 4). In the case of rigid boundary conditions however, the situation is less clear. Any horizontal flow satisfying the boundary conditions is damped. Far from onset, there is only one (physical) marginal mode, namely the phase mode, but that does not necessarily imply that no mean flow is generated at leading order. The linearized operator L_s for fluctuations about the finite amplitude roll solution of the Oberbeck-Boussinesq equations contains two null eigenmodes, the first one corresponding to the translational invariance and leading to the phase diffusion equation and the second one corresponding to the conservation of mass or more simply to the indeterminacy of the pressure up to a constant. To this latter mode corresponds an adjoint eigenvector (velocity, temperature, pressure) $=(\mathbf{0},0,1)$. This eigenmode is not physical in the sense that the pressure can always be eliminated; for an incompressible flow it is a Lagrange multiplier expressing the constraint of incompressibility and this is why we do not obtain two time dependent coupled order parameter equations.[5] We shall see that the mean drift mode is slaved to the phase mode. The solvability condition associated with this marginal mode simply demands that all the successive iterates of the horizontal velocity must be divergence free on the large scales. This constraint is trivially satisfied at order ε. But at order ε^2, the slow gradients of the Reynolds stresses give rise to a nontrivial constraint on the first iterate $\mathbf{u}_1$ of the velocity field which is calculated at order ε. This constraint can be satisfied because $\mathbf{u}_1$ depends on the arbitrary constant P_s added to the pressure at leading order. The large-scale gradient of this arbitrary slowly varying pressure P_s enters the phase equation in such a way that the phase contours are advected by the resulting solenoidal horizontal velocity. The net result is that a term $\rho \mathbf{V} \cdot \mathbf{k}$ is added to (2.18) and $\mathbf{V} = \nabla \times \psi \hat{\mathbf{z}}$ is recovered from an equation for ψ. For illustrative purposes, we will use the Cross-Newell model for calculating ψ, namely

$$\nabla^2 \psi = \hat{\gamma} \mathbf{z} \nabla \times (\mathbf{k} \nabla \cdot \mathbf{k} A^2). \tag{2.19}$$

The expression for the Oberbeck-Boussinesq equations is given in Newell et al (1990).[6] We stress again that the difficulty with the mean drift in no-slip boundary conditions is that the two slow modes are not both order parameters, since one (the mean drift) is slaved to the other (the phase mode). There is a large-scale coupling which gives nonlocal equations and renders difficult (if not impossible) the derivation of these equations with the usual ideas of center manifold theory.

We now turn to the question of the regularization of (2.18) and the reintroduction of the amplitude as an active order parameter. The ideas are new (Passot & Newell 1992), not yet fully tested, but appear to be promising. The difficulty is that (2.18) is ill-posed in the sense that it is only stable when both B and $(d/dk)kB$ are negative. This occurs for the wavenumber band $k_Z(\mathrm{R}) < k < k_E(\mathrm{R})$ known as the Busse balloon. When mean drift effects are included, the Eckhaus boundary $k_E(\mathrm{R})$ becomes the skew-varicose boundary $k_{SV}(\mathrm{R})$ and the zigzag instability boundary $k_Z(\mathrm{R})$ can move to lower values. The natures of the instabilities when k is less than k_Z or when k is greater than k_E are different. At the zigzag border k_Z, the instability is a supercritical bifurcation and it saturates when the rolls develop a zigzag pattern. The regularization of (2.18) can simply be

[5] If one uses potential variables $\mathbf{u} = \nabla \times \nabla \times \phi \hat{\mathbf{z}} + \nabla \times \psi \hat{\mathbf{z}}$, then pressure is eliminated from the Oberbeck-Boussinesq equations. However the boundary condition is $\psi \equiv$ const. and, in this formulation, this free constant plays the role of P_s.

[6] Because the first nontrivial manifestation of mean drift appears at order ε^2 one must solve exactly the order ε problem after applying the solvability condition which leads to the phase diffusion equation. This is best done (see Newell et al 1990) by using a SVD decomposition of the matrix obtained from the projection of L_s on the appropriate Galerkin basis.

achieved by going to the next order of the expansion by adding to the phase diffusion equation a term proportional to $\varepsilon^2\nabla^2\nabla\cdot\mathbf{k}$, the Laplacian of roll curvature. There are other nonlinear terms but this one is the most important, capable of achieving a balance with $\nabla\cdot\mathbf{k}B$ after introducing the scaling $\tilde{Y}=\varepsilon^{1/2}y$ in the "along the roll" direction. In the appropriate limit, it reduces to the phase component of the NWS equation. The saturated stationary states often contain disclinations which are regularized shock solutions, involving discontinuities in wavevector, of the hyperbolic system $\nabla\cdot\mathbf{k}B(k)=0$ when $k<k_Z$ and $B>0$ (Passot & Newell 1992). On the other hand, at the Eckhaus-skew varicose-border, the instability leads to either a readjustment of the roll wavelength or to the formation of dislocations. During either process the amplitude is driven to zero and locally the wavenumber crosses the marginal stability boundary $k=k_r$ at which the slaving relation $A=A_0(k,\mathrm{R})$ ceases to hold. Indeed, at this point, a second null vector of L_s (the first was $V_1=\partial U/\partial\theta$), $V_2=(dU/dA)=(\partial U/\partial A)+(\theta/2k^2)(\partial k^2/\partial A)(\partial U/\partial\theta)$, becomes periodic (because $\partial k^2/\partial A$, proportional to A, is zero there) and a second solvability condition (which is not necessary when V_2 is nonperiodic) is now required. This introduces terms depending on slow time derivatives and gradients of A into the "eikonal" equation (2.17) which then becomes a partial differential rather than algebraic equation. Ignoring mean drift, the regularized phase-amplitude equations are (2.18) with $\varepsilon^2\eta\nabla^4\Theta$ added and (2.17) becomes $\varepsilon^2A_T+g(k)A[A^2-A_0^2(k,\mathrm{R})]-\xi\nabla^2A=0$, where $\xi,\eta,g(k)$ are calculable. These equations are valid for wavenumbers $k<k_r$, close to and inside the marginal stability curve. In order to obtain a description valid for a general wavenumber (not k_c) on the marginal stability curve we introduce a complex order parameter $W=Ae^{i\theta}$ satisfying $W_t+\Lambda(-\nabla^2)W+WW^*+WW^*N(-\nabla^2)W=0$ and choose $\Lambda(k^2)$, $N(k^2)$ and a time-stretching variable in terms of $\tau(k)$, $B(k)$, and $\lambda_s(k^2,\mathrm{R})$ in order that the Cross-Newell reduction of the equation for W matches the phase-amplitude equations above. It is important to note that $\Lambda(-\nabla^2)$ is the physical space representation of $\lambda_s(k^2,\mathrm{R})$, the spectrum associated with the operator L_s obtained by linearizing the original microscopic equations about the finite amplitude rather than the zero state. Near the marginal stability border $k=k_r$ however, $\lambda_s(k^2,\mathrm{R})$ and $\lambda(k^2,\mathrm{R})$, the spectrum of the operator obtained by linearizing about the zero state, are directly related by $-2\lambda(k^2,\mathrm{R})=\lambda_s(k^2,\mathrm{R})$. Because of this, the complex order parameter equation reproduces the NWS equation near onset. It is also important to emphasize that the complex order parameter equation is only valid when amplitudes are small. For wavenumbers k well inside the marginal stability curve, we would like to have a description that involves only one order parameter. The natural candidate is the real part u of W which satisfies

$$u_t+\Lambda(-\nabla^2)u+(u^2+u^{*2})N(-\nabla^2)u = 0. \tag{2.20}$$

In (2.20), u^* must be chosen so that u^2+u^{*2} represents the amplitude A^2—a slowly varying quantity. In general, the conjugate variable cannot be chosen as the imaginary part u^* of a complex variable $W = u+iu^*$ which also satisfies (2.20) because, far from onset, the two solutions u and u^* will not in general evolve so as to keep the intensity u^2+u^{*2} slowly varying. Therefore, we need another algorithm to define u^*. In one-spatial dimension, an analytical signal can be constructed from an almost monochromatic real-valued signal by use of the Hilbert transform. In two dimensions, we use wavelet analysis to define the decomposition of a real signal u into its slowly varying amplitude and phase components. Wavelet analysis also captures local pattern behavior (such as is found near dislocation instabilities; see next section) when more than one wavenumber is present.

We claim that (2.20), where $A^2 = u^2+u^{*2}$ is determined as a functional of u by a wavelet algorithm, is the canonical equation for patterns. It can support all the observed singularities. Its Cross-Newell reduction leads to precisely the same Cross-Newell equation and therefore the same Busse balloon as the original equations. Moreover, it is much simpler than the original equations even though it does contain small scales. It has a rational derivation, whereas other models, such as the Swift-Hohenberg equation, are phenomenological. Preliminary simulations (Passot & Newell 1992) are very promising. A technical difficulty remains, however. Because the algorithm for solving (2.20) is best carried out in Fourier space, it is presently most suitable for periodic boundary conditions. Most experimental situations require a combination of Dirichlet and Neumann conditions on the boundary and a modification of the algorithm to reflect these constraints has yet to be developed.

3. SINGULARITIES OF MACROSCOPIC PATTERN FIELDS

In most natural patterns, it is impossible to define an almost everywhere continuous vector field to serve as a measure of the phase gradient which has the properties that its singularities correspond directly to the simplest point defects called disclinations (e.g. see Figure 3*a*). On the other hand, a director field (Kléman 1983) (vectors without arrows) arising from real fields w or real order parameters $w = A\cos\theta$, whose phase gradient $\nabla\theta = \pm[1-(w^2/A^2)]^{-1/2}\nabla w$ can be positive or negative, has this property.[7]

[7] The use of a periodic function of θ rather than θ itself helps remove the multivaluedness associated with the phase variable.

Disclinations are characterized by the behavior of the local director field in their vicinity and the relevant measure is the angle through which the director twists as its midpoint circumscribes the disclination in a counter-clockwise direction on a contour C. In Figures 8*a*,*b* we show a convex (concave) dislocation which has twist $+\pi(-\pi)$. The foci on the boundaries of the convection cell in Figure 3*a* are convex disclinations, whereas the interior disclination is concave. Twists are independent of C, additive in the sensę that the twist in the director field about several disclinations is the sum of the individual twists. Since $\nabla\theta$ is only a director field in the neighborhood of disclinations, the circulation $\int_C \nabla\theta \cdot \mathbf{ds}$, makes no sense without introducing an artificial boundary. Moreover, they are not local perturbations of a field of straight parallel rolls. Observe that the center (separating) lines of convex (concave) disclinations are always maximum upflows or downflows. There are, however, special configurations of disclinations for which a vector field $\mathbf{k} = \nabla\theta$ can be defined almost everywhere. One simple example is a target pattern shown in Figure 8*c* which can be considered as a superposition of two convex disclinations and has a twist of 2π. Another such arrangement, shown in Figure 8*d*, involves a

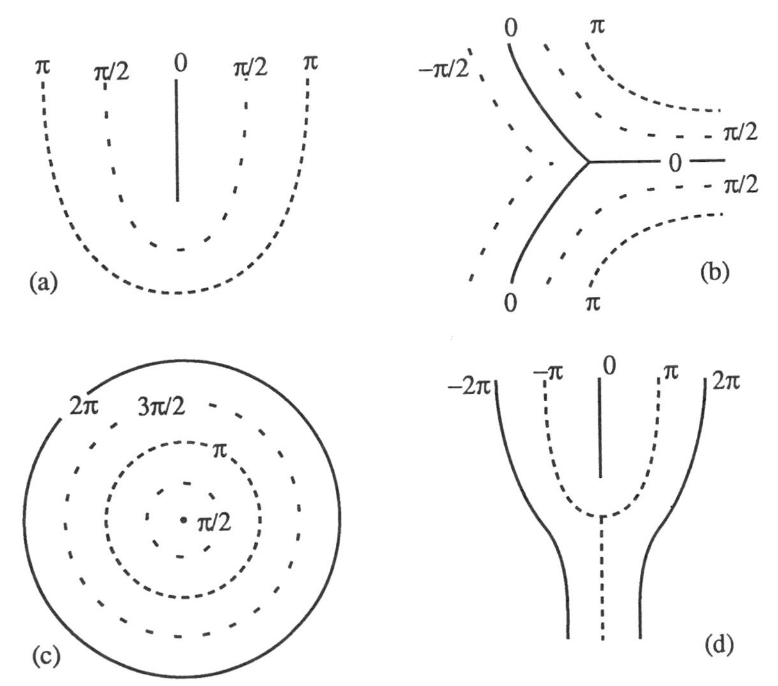

Figure 8 (*a*) Convex disclination, (*b*) concave disclination, (*c*) target pattern, (*d*) dislocation.

special superposition of a convex and concave disclination and is called a dislocation. It is local in the sense that in the far field it is equivalent to a vortex superposed on a field of straight parallel rolls. Its far field phase is $k \cdot x + \text{Im}[\ln(x+iy)]$. Using the labeling shown in Figure 8*d*, the phase undergoes a 2π jump along the axis below the dislocation. The director, now a vector $\mathbf{k}$, has zero twist. Moreover, a circulation can be defined and is equal to -2π. It gives a measure to the extra roll pair above the dislocation. Circulation is also independent of C and is additive. Whereas the dislocation is topologically equivalent to two disclinations, there is no reason for it to be energetically equivalent and, in general, the free energy (if the flow is gradient) of two free disclinations will be different from that of a dislocation.

The roll curvature that makes necessary the presence of concave disclinations is induced by boundary constraints. Consider Figure 3*a*. We suggest that the structure in the vicinity of the concave disclination at the center of the cylinder is initiated by a zigzag instability which saturates in the far field because $k = k_Z$ and in the near field because the unstable growth is balanced (regularized) by the next correction $\varepsilon^2\eta\nabla^2\nabla\cdot\mathbf{k}$, the Laplacian of the curvature, in the phase diffusion equation. For small ε, we suggest (and have reasonably concrete evidence to prove) that the stationary state is simply a regularization of the shock solution of the Cross-Newell equation in its hyperbolic region. In the absence of mean drift, the Cross-Newell equation is $\nabla\cdot\mathbf{k}B(k) = 0$ and it is easy to see that for $k < k_Z$, so that $B > 0$, $(kB)' < 0$, characteristics of the same family generally will intersect and form discontinuities in k_x, k_y, the components of the wavevector $\mathbf{k}$. The shock jump conditions are $s[k_x]+[k_y] = 0$ (from $\nabla\times\mathbf{k} = 0$) and $-[k_xB]+s[k_yB] = 0$ (from $\nabla\cdot\mathbf{k}B = 0$) where $s(y) = dx/dy$ is the curve across which the discontinuities occur. For solutions where the wavenumber is continuous, as would be the case when its far field value is selected by other influences (e.g. roll curvature), $B([k_x]^2+[k_y]^2) = 0$ and this means $k = k_Z$ where $B(k_Z) = 0$. One example of this solution is: $\theta = k_Z y$, $0 < \phi < \pi/3$; $\theta = k_Z[(\sqrt{3}/2)x+\frac{1}{2}y)]$, $\pi/3 < \phi < \pi$; $\theta = k_Z[(\sqrt{3}/2)x-\frac{1}{2}y)]$, $\pi < \phi < 5\pi/3$; $\theta = -k_Z y$, $5\pi/3 < \phi < 2\pi$, where ϕ is measured from the horizontal. The angular slope of the shock line is given by $s = -[k_y]/[k_x]$ which is $1/\sqrt{3}$ on $\phi = \pi/3$, ∞ on $\phi = \pi$ and 2π, and $-1/\sqrt{3}$ on $\phi = 5\pi/3$. Note that the smallest wavenumbers occur along the shock directions and are equal to $\frac{1}{2}k_Z$. The zigzag instability of such rolls gives rise to two sets of rolls of wavenumber k_Z with an angular separation of 120°. The regularized version of this solution is very close to what is seen in Figure 3*a*. Another solution is: $\theta = k_Z y$, $-5\pi/4 < \phi < \pi/4$; $\theta = k_Z x$, $\pi/4 < \phi < \pi/2$; $\theta = -k_Z x$, $\pi/2 < \phi < 3\pi/4$ with shocks along $\phi = \pi/4$, $\pi/2$, $3\pi/4$. This solution is likely to be seen when the cross-roll instability is more important than the zigzag. When mean drift effects are included, these solutions are modified a little. The choice of solution depends on whether the local wavenumber first crosses the zigzag or cross-roll instability boundary. Similar remarks obtain for convex disclinations. Of particular interest are solutions joining constant states with different wavenumbers, each chosen, say, by a different wavenumber selection mechanism. In these cases, we conjecture that the shock is time-dependent. More details are given in Passot & Newell (1992). Although it is not very useful for shock solutions that join constant wavevector states, it should be pointed out that the Cross-Newell equation can be simplified by a hodograph transformation to the separable linear equation

$$\frac{\partial}{\partial k}\left(kB\frac{\partial\Phi}{\partial k}\right)+\frac{1}{k}\left[\frac{\partial}{\partial k}(kB)\right]\frac{\partial^2\Phi}{\partial\psi^2}=0,\quad x=\frac{\partial}{\partial k_x}\Phi,\quad y=\frac{\partial}{\partial k_y}\Phi$$

and $\mathbf{k} = (k\cos\psi, k\sin\psi)$ (Pomeau & Rica 1992).

As the stress parameter (e.g. Rayleigh number) is lowered towards the onset value, concave disclinations separating patches of curved rolls will destabilize into patches of straight rolls mediated by grain boundaries such as is observed in Figure 11*d*. We observe this behavior in numerical simulations of the Swift-Hohenberg equation where the concave disclination at the center of the cylinder destabilizes for values of the stress parameter μ about 0.1. Thus, far from onset, the extra contribution to the free energy necessitated by the roll curvature and disclination is smaller than the penalty required by not satisfying the boundary conditions with orthogonal patches of straight parallel rolls and grain boundaries. Near onset, the latter is smaller than the former.

Whereas disclinations are singularities associated with real order parameters, dislocations and grain boundaries can be associated with the vector field $\mathbf{k} = \nabla\theta = \mathrm{Im}\,\nabla W/W$ of a complex order parameter $W = Ae^{i\theta}$. A dislocation is simply a zero of the complex field W and a grain boundary is a curve across which the complex envelopes W_1, W_2 of two almost parallel roll fields with distinct average wavevectors $\mathbf{k}_1$ and $\mathbf{k}_2$ decay exponentially. The complex order parameter equations near onset support dislocations and grain boundaries but not disclinations, because, except at the point when $W = 0$, $\mathbf{k}$ is uniquely defined. Moreover, being zeros of a complex field in two spatial dimensions, near onset dislocations are topologically stable (Coullet et al 1989a). However, they are also solutions of the real order parameter equation (2.20) valid far from onset, but in this context are no longer constrained to be stable. Far from onset, dislocations represent large deformations of the underlying periodic pattern of wavenumber k and, because of this, the subharmonics $k/2$ and

$k/3$, although linearly damped, can be excited through subcritical finite amplitude instabilities (Newell & Passot 1992). The consequence is that, far from onset, dislocations can temporarily destabilize and give rise to new structures and new behavior dominated locally by $k/3$ and $k/2$ modes. The new structures include bridges, each supported by two convex and two concave disclinations dominated by $k/3$. The new behavior includes sudden roll disappearance (the amplitude analog to the Eckhaus phase instability) and the gliding of dislocations. The latter process is greatly facilitated by the formation of disclinations. An important point to stress is that this behavior is a property of real rather than complex order parameter equations.

Near onset, dislocations are vortex solutions of the NWS or CGL equations (2.8)–(2.13). In Figure 9, we draw the contours of (*a*) $\mathrm{Re}\, W = \mathrm{Im}\, W = 0$ and (*b*) $w = We^{ik_c x} + (*) = 0$, where W satisfies the isotropic CGL valid for anisotropic systems,

$$W_t - \gamma \nabla^2 W = \mu W - \beta W^2 W^*, \quad (\gamma = \gamma_r + i\gamma_i, \beta = \beta_r + i\beta_i). \tag{3.1}$$

When γ_i, β_i are nonzero, a finite nonzero value is chosen for the far field wavenumber (Hagan 1982). If $\gamma_i = \beta_i = 0$, the far field is the constant state $\sqrt{\mu\beta_r^{-1}} \exp i\phi_0$. Dislocations move as a result of (*a*) the effects of the elasticity of the pattern (Peach-Köhler force) which depends on the difference $K = k - k_c$ of the background wavenumber k from bandcenter k_c and (*b*) mutual interaction. In the case of gradient flow ($\beta_i = \gamma_i = 0$), they move in the direction so as to bring the pattern wavevector back to bandcenter

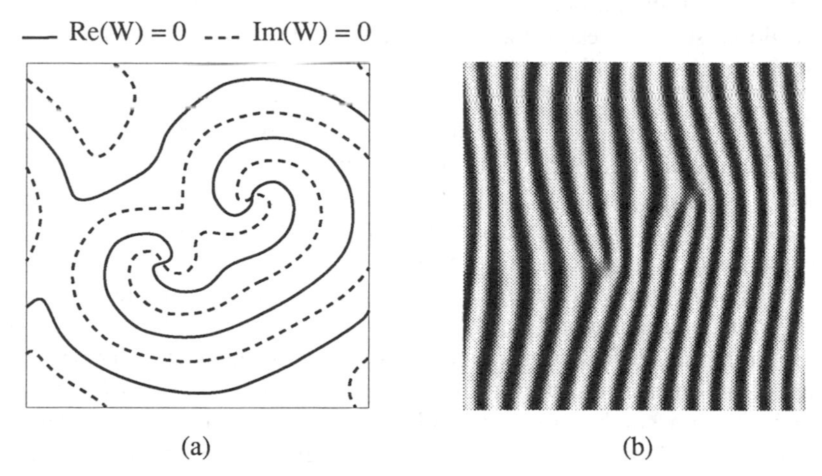

Figure 9 (*a*) Contours of $\mathrm{Re}\, W = 0$ and $\mathrm{Im}\, W = 0$. (*b*) Snapshot of contours of the zeros of $\mathrm{Re}\, W \exp(ik_c x - i\omega_c t)$. (From Lega 1989.)

$K = 0$ and their velocity is proportional to K (actually $v \ln|v| \propto K$) (Pomeau et al 1983; Bodenschatz et al 1988, 1991; Neu 1990). For isotropic systems, the defect speed v for NWS is proportional to $K^{3/2}$ (Siggia & Zippelius 1981b). Much work needs to be done on: (*a*) source-sink defects in patterns of oppositely traveling one-dimensional waves (Coullet et al 1989a), (*b*) mean drift effects on defect velocity, (*c*) mutual interaction of defects (Kawasaki 1984, Neu 1990, Pismen & Rodriguez 1990, Rica & Tirapegui 1991, Elphic & Meron 1991), and (*d*) defects in more complicated planforms (Ciliberto et al 1990). Near onset, the hepta-penta defect seen in Figure 1*a* occurs at a point where the complex amplitudes of two of the three rolls making up the hexagon are zero. Question: If rolls are supercritical, and at a hepta-penta defect, $W_2 = W_3 = 0$, can such defects remain stationary at subcritical values of the stress parameter? Presumably yes, but only as long as the W_1 roll is sustained against linear decay by a bias I from the far field.

4. PROPERTIES OF ENVELOPE EQUATIONS

We divide the discussion of order parameter equation properties into three categories. First, we review the nature of instabilities of simple patterns like straight parallel rolls and target patterns. For straight rolls, we will map out all the long wave instability boundaries of the Busse balloon, namely those boundaries which involve perturbations with wavevectors close to that of the original pattern, zigzag, Eckhaus, or skew-varicose. Second, we discuss *gradient* flows $W_t = -\delta F/\delta W^*$ for which $dF/dt \leq 0$. The global stability result that the free energy decreases until it reaches a minimum does not allow different solutions to coexist. This means that different wavenumber selection mechanisms choose the same result and mitigates against a time-dependent turbulent flow on a complicated attractor. There is no reason, however, for the ground state of the free energy itself to be spatially periodic, although in many cases it is. Third, we discuss *nongradient* flows which are richer and much less predictable. In particular, we shall be interested in two origins of nongradient behavior: the presence of mean drift flows which are usually a manifestation of the nonlocal nature of the envelope or phase diffusion equations and the presence of dispersion and intensity dependent phase modulation. The order parameter equation for envelopes of traveling waves, $W_t = -\delta F/\delta W^* + i\delta H/\delta W^*$, is a combination of gradient and Hamiltonian components. Moreover, the latter is sometimes exactly integrable (e.g. the nongradient part of CGL is the NLS equation), and more importantly, the behavior of the total system is often dominated by the behavior of its conservative component. As a result, fully turbulent (temporal and spatial) states can be found near the onset of the pattern (Steinberg et al 1989, Newell 1979, Brand et al 1986a,b, Bretherton & Spiegel 1983).

Instabilities of Simple Patterns

The straight parallel roll solution $W = \sqrt{(\mu-\gamma K^2)/\beta}\exp iKx$ of the NWS equation (2.13) (assuming rigid boundary conditions so that mean drift corrections are of order $\mu^{1/2}$) is linearly unstable either if (*i*) $K^2 > K_E^2 = \mu/3\gamma$ (Eckhaus instability) or (*ii*) $K < 0$ (zigzag instability) with fastest growing perturbations (*i*) $\exp i(K \pm L)x$, $4\gamma^2K^2L^2 = (3\gamma K^2-\mu)\times(\gamma K^2+\mu)$ and (*ii*) $\exp(iKx \pm iMy)$, $2k_cK+M^2 = 0$, respectively. The phase diffusion equation for (2.13) has the form (2.20) with $\tau = A^2 = \mu-\gamma[K_x+(1/2k_c)K_y^2]^2$, $B = A^2$. When linearized about the state $\Theta = K_0x$, one obtains the Pomeau-Manneville equation (Pomeau & Manneville 1979), (set $\mu = K_0x+\psi$)

$$\psi_t - \frac{\gamma}{A_0^2}\left[\frac{d(KA^2)}{dK}\right]_0 \psi_{xx} - \frac{\gamma}{k_c}K_0\psi_{yy} = 0. \tag{4.1}$$

Clearly the roll solution $\sqrt{(\mu-\gamma K_0^2)/\beta}\exp iK_0x$ is unstable whenever either the parallel or the perpendicular diffusion coefficient,

$$D_\parallel(K_0) = \frac{\gamma}{A_0^2}\left(\frac{d}{dK}KA^2\right)_0 = \gamma\frac{\mu-3\gamma K_0^2}{\mu-\gamma K_0^2}, \quad D_\perp(K_0) = \frac{\gamma K_0}{k_c} \tag{4.2}$$

is negative, corresponding to the Eckhaus and zigzag instabilities respectively. More generally, for systems far from onset, one can also linearize the phase-diffusion equation (2.18) about a straight parallel roll state $\Theta = k_0x$ and then the parallel and perpendicular diffusion coefficients $D_\parallel(k_0)$ and $D_\perp(k_0)$ are $-\tau^{-1}(k_0)[d/dk(kB)]_0$ and $-\tau^{-1}(k_0)B(k_0)$ respectively. When mean drift effects are included by adding $\tau\mathbf{U}\cdot\mathbf{k}$ to (2.18) with $\mathbf{U} = \nabla\times\psi\hat{\mathbf{z}}$ found from (2.19), we find that the zigzag instability boundary $k = k_Z$ is moved to the left where $D_\perp(k_Z) = -\hat{\gamma}k^2A^2(k)$ and $\hat{\gamma}$ is the coefficient in (2.19). The Eckhaus instability becomes the skew-varicose instability for which the fastest growing mode has the form exp $[i(K_0+K\cos\rho)x+iK(\sin\rho)y]$ for $\rho \neq 0$ (cf Figure 4). Perturbations along neighboring phase contours are out of phase and lead to a more rapid necking of the former and the nucleation of dislocation pairs. Instabilities of stationary circular target patterns which choose wavenumber k_B, $B(k_B) = 0$, have been investigated experimentally by Steinberg et al (1985), Pocheau et al (1985), Pocheau (1988) and theoretically by Newell et al (1990). It is important to note that these instabilities can occur for

(R, k) values inside the Busse balloon. The stability characteristics of straight and circular rolls are different.

For traveling waves whose envelopes W satisfy the CGL equation (2.11), the analogue of the Eckhaus instability is the Eckhaus-Benjamin-Feir instability (Newell 1974, Lange & Newell 1974, Stuart & DiPrima 1978) of a monochromatic wavetrain $W = A_0 \exp(iK_0 x - i\Omega_0 t)$ found by writing $W = A_0 \exp i\theta$, $\theta = K_0 x - \Omega_0 t + \psi$, $\Omega_0 = \gamma_i K_0^2 + \beta_i A_0^2$, $\beta_r A_0^2 = \mu - \gamma_r K_0^2$, and linearizing to find

$$\psi_t + 2\left(\gamma_i - \frac{\beta_i \gamma_r}{\beta_r}\right) K_0 \psi_x = \left[\gamma_r + \frac{\beta_i \gamma_i}{\beta_r} - \frac{2\left[1 + \left(\frac{\beta_i}{\beta_r}\right)^2\right] \gamma_r^2 K_0^2}{\mu - \gamma_r K_0^2}\right] \psi_{xx} + \text{h.o.t.}, \tag{4.3}$$

where h.o.t. refers to higher order spatial derivatives of ψ. Note that when $\beta_i = \gamma_i = 0$, we recover the y-independent part of (4.1) and the Eckhaus instability. The coefficient of ψ_x is the group velocity Ω_0'. The monochromatic wavetrain is unstable whenever

$$D_{\parallel}(K_0) = \gamma_r + \frac{\beta_i \gamma_i}{\beta_r} - 2[1 + (\beta_i \beta_r^{-1})^2] \gamma_r^2 K_0^2 (\mu - \gamma_r K_0^2)^{-1}$$

is negative. The set of wavenumbers K_0 for which $D_{\parallel}(K_0) > 0$ is called the EBF band. However the wavetrain with $K_0 = 0$ can also be unstable whenever the Newell (1974) criterion,

$$\beta_r \gamma_r + \beta_i \gamma_i < 0, \tag{4.4}$$

holds. This last instability is a generalization of the Benjamin-Feir instability for deep water waves (Benjamin & Feir 1967)—the modulational and focusing instabilities describing the onset of the collapse of Langmuir waves in plasmas and filaments in optical beams respectively (Kosmatov et al 1991). Each is a manifestation of the fact that wavetrain solutions $W = W_0 \exp -i\beta_i |W_0|^2 t$ of the NLS equation $W_t - i\gamma_i \nabla^2 W - i\beta_i W^2 W^* = 0$ are unstable when $\beta_i \gamma_i < 0$ (Benney & Newell 1967). The reason for the instability is easiest to understand in the optics context where NLS can be viewed as the paraxial approximation for the evolution of the wavepacket $W(x, y, t)$ of an electromagnetic carrier wave $\exp i\omega[(n_0 z/c) - t]$ with γ_i (positive) proportional to inverse Fresnel number and $-\beta_i |W|^2$ proportional to δn, the deviation of the refractive index from its background value. Clearly if $\beta_i < 0$, any local increase in light intensity is accompanied by an elevation in refractive index which helps gather even more light in that region and reinforce the increase in refractive index. In one transverse dimension, the growth of the instability is arrested by diffraction and the resulting balance between nonlinear focusing and diffraction gives rise to the bright soliton solution

$$W(x,t) = 2\eta \operatorname{sech} 2\eta \sqrt{\frac{-\beta_i}{2\gamma_i}}[x-\bar{x}(t)] \exp(iKx - i\sigma), \tag{4.5}$$

where $\bar{x}_t = 2K\gamma_i$ and $\sigma_t = \gamma_i K^2 + 2\beta_i \eta^2$. In two transverse directions [cf (3.1)], diffraction is unable to arrest nonlinear focusing and singular filaments are formed in finite time (Kosmatov et al 1991). When the effects of group velocity dispersion γ_i and nonlinear dispersion β_i dominate those of diffusion γ_r and saturation β_r then even solutions of the CGL equations can develop *large-amplitude* fluctuations.

These considerations are very important for understanding the behavior of systems of traveling waves near onset. The first point to stress is that if (4.4) holds, there are no constant amplitude traveling wavetrains. Wavetrains break up into localized pulses [cf (4.5)]; we will meet examples of these in Section 5. Moreover, in places where the wave amplitude of a right traveling wave is small, a left traveling wave can appear and source-sink defects dividing regions of left and right traveling waves can nucleate (Coullet et al 1991). In two dimensions, the consequence of (4.4) is the nucleation of vortices and the onset of a *defect mediated turbulence* (Coullet et al 1989b). The instability of traveling waves to long wavelength perturbations along their crests is even more pronounced in the rotationally invariant case (2.12) when $\beta_i < 0$. An example of this type of instability is seen vividly in Figure 1*b* which shows traveling wave patterns in a "Raman" laser. Of further interest is a result of Bartuccelli et al (1990) who found for (3.1) a second boundary in the $\beta = \beta_i/\beta_r$, $\gamma = \gamma_i/\gamma_r$ plane beyond the Newell hyperbola $1+\beta\gamma < 0$, beyond which they were unable to bound a combination of $H^{(1)}$ and $L^{(4)}$ norms. To date, no such boundary has been detected either numerically or experimentally. Nevertheless, it is worth pursuing because the mathematics suggests that a qualitative change of behavior (from large-phase to large-amplitude fluctuations) may occur in this region. At the very least, the occurrence of nonlinear focusing-driven large-amplitude fluctuations may trigger the early onset of subcritical instabilities. This effect is very evident in the destabilization of certain numerical algorithms (Stuart 1989, Briggs et al 1983).

It is important to distinguish between *convectively* and *absolutely* unstable flows. In some parameter regimes, even the Newell and Eckhaus-Benjamin-Feir instabilities belong to the former category (Weber et al 1992) and, as a result, have markedly different consequences. A solution is said to be linearly unstable if its small perturbations grow exponentially under the dynamics. However, when an external flow is present and the solution is advected (as the envelope of a traveling wave is by the group velocity), it may happen that the small disturbances applied at a given point are swept away by the flow before they become noticeable. In this case, one speaks of a convective instability (Landau & Lifshitz 1959); otherwise, the instability is said to be absolute. As a consequence, convectively unstable solutions may be of little experimental relevance, since the instability will only show up downstream (if ever). The nature of an instability is clearly related to the relative sizes of two time scales in the problem—namely the inverse of the growth rate of the small perturbations, and the time required for them to travel across the support of the original perturbation. According to a general criterion (Bers 1975), an unstable flow is convectively unstable if the perturbations with zero group velocity (taken in a general sense with complex wavenumber and frequency) are all damped. While the distinction between absolute and convective instabilities is argued to be important for open flows (Huerre & Monkewitz 1985), it may also turn out to be crucial for the stability of localized objects, like defects. For instance, the far fields of many defects are traveling waves. The latter are also solutions to the corresponding envelope equation, and may be phase unstable. Nevertheless, if the wavenumbers of the finite amplitude waves in the far field are such that the phase instability is of a convective nature, the defect may still be stable (Weber et al 1992) because the small perturbations would be advected away and absorbed either by other localized stable structures or boundaries. A second example is the behavior of the wavetrain [with selected wavenumber; see (4.7)] which is left behind an advancing front that invades the unstable zero state of the CGL equation where (4.4) holds. As the length L of the finite amplitude wake increases so that more and more allowable wavenumbers $n\pi L^{-1}$ fall under the gain band of the Newell instability [whose width is found by setting $K_0 = 0$ in (4.3) and balancing the destabilizing ψ_{xx} term with the stabilizing higher order derivative ψ_{xxxx}], the finite amplitude wavetrain destabilizes. In the laboratory frame, the wake is unstable but in the frame of reference moving with the front, all turbulent disturbances decay and so the front is left unaffected (Nozaki & Bekki 1983).

Gradient Flows, Fronts, and Wavenumber Selection

The existence of a free energy F gives a "moral" direction to the flow. It moves in phase space so as to achieve some optimal state defined by those configurations which minimize F. For example, it is clear that the free energy of NWS (2.13) (where A is area)

$$F = \lim_{|A|\to\infty} |A|^{-1} \int_A \left(\gamma \left| \left(\frac{\partial}{\partial x} - \frac{i}{2k_c} \frac{\partial^2}{\partial y^2} \right) W \right|^2 - \mu |W|^2 + \frac{\beta}{2} |W|^4 \right) dx\, dy \tag{4.6}$$

is minimized for $W = \sqrt{\mu\beta^{-1}} \exp i\phi_0$ with ϕ_0 constant, corresponding to the bandcenter $K = 0$. Given the choice of all paths, the system will move so as to realize one of these states, usually by attaining the amplitude $\sqrt{\mu\beta^{-1}}$ locally and by ironing out phase differences by the slow motion of Bloch wall defects which separate states with different phases. In the one-dimensional case where W is real, the preferred states $W = \pm\sqrt{\mu\beta^{-1}}$ are connected by Ising walls $\sqrt{\mu\beta^{-1}} \tanh \sqrt{\mu}(x-x_0)/\sqrt{2\gamma}$ (unstable when W is complex) which also annihilate and eventually eliminate the positive contribution of the first term in the integrand of F. Domain walls which separate metastable states with different free energy values F, $F+\Delta F$ move at speeds that depend on ΔF.

Boundary conditions can constrain the states available to the system. For example, the trivial looking conservation of twist law $(\phi_x)_t = (\phi_t)_x$ has the nontrivial consequence that, providing the phase at the endpoints of an interval $x = \pm L$ is stationary (or periodic if $\phi_t \neq 0$), the total phase jump $[\phi] = \int_{-L}^{L} \phi_x \, dx$ is conserved. Suppose one has Eckhaus-stable roll states $W_j = \sqrt{(\mu - \gamma K_j^2)\beta^{-1}} \exp iK_j x$, $3\gamma K_j^2 < \mu$, $j = 1, 2$ with W_1 in $-L < x < 0$ and W_2 in $0 < x < L$. The total twist is $(K_1 + K_2)L$. For a roll state of wavenumber K, $F = -(2\beta)^{-1}(\mu - \gamma K^2)^2$. The graph of F vs K has an inflection point at $K = \pm K_E = \pm\sqrt{\mu(3\gamma)^{-1}}$. For $K^2 < K_E^2$, F is convex so that $F(K_1) + F(K_2) > F(K_1 + K_2/2)$. Therefore the free energy can be lowered by the phase diffusing so as to achieve the average wavenumber $K_1 + K_2/2$ over $-L < x < L$. The twist is conserved. By contrast, for nongradient flows, wavenumber jumps can be sustained and two Eckhaus-Benjamin-Feir stable states can be connected by the analogue of a dark soliton of NLS (Bekki & Nozaki 1985). In one space dimension, an Eckhaus unstable state with wavenumber K will realize a new state with wavenumber

$$K - \frac{\sqrt{(3\gamma K^2 - \mu)(\gamma K^2 + \mu)}}{2\gamma K}$$

(Kramer et al 1988, Kramer & Zimmermann 1985). In two dimensions, there is an alternative path in the phase space which is initially less steep but which eventually leads to an average wavenumber closer to bandcenter and to a lower free energy. In contrast to the one-dimensional case where the amplitude is driven to zero over a whole line $x = \text{const.}$, the relaxation in two dimensions involves the nucleation of isolated dislocation pairs which move so as to remove extra rolls from the pattern (Bodenschatz et al 1991).

Fronts between metastable states move at speeds depending on the difference of free energies. Fronts between stable and unstable states fall in two categories, the Kolmogorov front and the nonlinear front. The characteristics of the former are found from purely linear stability considerations; the latter only occurs near subcritical bifurcations and is the "continuation" to $\mu > 0$ of the front between the stable uppermost branch and metastable lowest branch when $\mu < 0$. Kolmogorov fronts select a wavenumber in their wake. For example, for CGL in its group velocity frame, the front between the unstable state $W = 0$ and its stable wake moves with velocity $V = 2\sqrt{\mu\gamma_r(1+\gamma_i^2/\gamma_r^2)}$ and chooses from the one parameter family of finite amplitude states,

$$W = \sqrt{(\mu-\gamma_r K^2)\beta_r^{-1}}\exp(iKx-i\Omega t), \quad \Omega = \gamma_i K^2 + \frac{\beta_i}{\beta_r}(\mu-\gamma_r K^2), \qquad (4.7)$$

the wavetrain whose wavenumber K has the same frequency Ω as that of the low amplitude wave at the leading edge, $\Omega = V^2\gamma_i(\gamma_r^2+\gamma_i^2)^{-1}$. In the gradient case, $\beta_i = \gamma_i = 0$, the bandcenter $K = 0$ is chosen. For a small range of μ, $0 < \mu < \mu_c$ near a subcritical bifurcation, the nonlinear front, which in this range has a faster decaying leading edge, is realized (Dee & Langer 1983, van Saarloos 1989, van Saarloos & Hohenberg 1992). Its velocity is not found from linear considerations. However, it turns out that its shape and velocity can be found by a method developed for integrable systems (Powell et al 1991). Although plausible arguments for this coincidence can be put forward, the result is not properly understood. Furthermore, the role of the free energy in choosing between front solutions is also not yet understood (because $F \to -\infty$ as $t \to \infty$). In Powell et al (1991), some conjectures were offered. It would be interesting, therefore, both for physical and theoretical reasons, to examine the fronts which join the Eckhaus-unstable rolls [the instability is subcritical (Tuckerman & Barkley 1990)] and study their final attracting states to try to understand the physical manifestation of the differences in the nature of the fronts (grain boundaries of dislocation pairs) as μ (here $3K^2-1$) crosses μ_E.

Fronts are not the only solutions that select wavenumbers. Grain boundaries, such as we see in Figure 3 (Manneville & Pomeau 1983, Tesauro & Cross 1987), roll curvature (Pomeau & Manneville 1981, Cross & Newell 1984), dislocations (Pomeau et al 1983), tapered geometries (Kramer et al 1982), and boundary conditions (Cross et al 1983, Pomeau & Zaleski 1981), are also responsible for wavenumber selection. For gradient flows, all these selection mechanisms choose the wavenumber that minimizes F.

Does this mean that there can be no nontrivial states of gradient systems? No! Certainly the final states are likely to be time independent (up to a group action, e.g. rotation, compatible with symmetries) but when the set of degenerate states K is open, then, depending on the coefficients β_{jl} in (2.6), the ground states of F can be quasicrystals or even spatially turbulent.

Nongradient Flows

The differences between gradient and nongradient flows are as great as the differences between behaviors of solutions of linear and nonlinear systems of equations. Whereas for gradient flows we have global stability results based on free energy minimization, for nongradient flows we have in general only local stability results. To the settled mind, nongradient and nonlinear flows are unsettling. To the pioneer, they promise the adventure of undiscovered beauty, a universe of the unknown to be explored and understood.

In nontraveling wave patterns, the presence of mean drift and dislocations far from onset are principally responsible for the nongradient character of the system. No longer is there a free energy to insist that all selection mechanisms (grain boundaries, roll curvature, stationary defects, zigzag instability, tapering) choose the same wavenumber. This immediately introduces the possibility of frustration and persistent time dependence. For example, if boundary or other external conditions were to conspire so that a given pattern had both a grain boundary and roll curvature, then the system would respond by attempting to satisfy simultaneously both wavenumber selection demands. Time dependence will persist and the attractors will be more complicated.

However, just as nonlinear is not a homogeneous domain in the space of all equations, as nonelephant does not categorize the animal species on the North American continent, nongradient flows are open to categorization. Moreover, it is possible that there will be similar success in analyzing some of these categories as there has been in understanding the behavior of soliton and purely hyperbolic systems of nonlinear partial differential equations. Indeed the CGL model, which is one of the most ubiquitous of envelope equations, has the interesting property that the flow is the sum of a purely gradient component (the NWS equation) and, in 1-D, an exactly integrable Hamiltonian flow (the NLS equation). The Newell criterion (4.4) shows that in some cases the behavior of the solutions of CGL can be dominated by the Hamiltonian and exactly integrable part; indeed it is this part which is responsible for the localized pulses found recently by Thual & Fauve 1988. One might ask: For systems such as (4.2) where the exactly integrable Hamiltonian component dominates the gradient component, can one obtain stronger results than local stability using a coordinatization of phase space suggested by the exactly integrable part?

For example, if we consider the gradient terms in the CGL equation as a perturbation of the NLS equation, then the soliton solution (4.6) of the latter deforms according to (Newell 1978):

$$\eta_t = \eta\left[2\mu - \left(\frac{16}{3}\beta_r + \frac{4\gamma_r\beta_i}{(-3\gamma_i)}\right)\eta^2 - 2\gamma_r K^2\right], \quad K_t = \frac{-16\gamma_r\beta_i}{(-3\gamma_i)}K\eta^2. \tag{4.8}$$

Observe: 1. There is a global attractor for this system, 2. the envelope wavenumber K and the pulse velocity $2\gamma_i K$ tend to zero; the pulse will move with the linear group velocity ω_c' of the most unstable wave k_c, and 3. the nonlinear focusing ($\beta_i\gamma_1 < 0$) character of the soliton pulse and the fact that its width is inversely proportional to its amplitude η mean that diffusion can overcome even a negative β_r as long as $4\beta_r + [\gamma_r\beta_i/(-\gamma_i)] > 0$. Of course, one can argue (correctly) that away from the support of the pulse, the solution W will grow exponentially in time. However, a periodic array of soliton pulses (even when $\beta_r < 0$) should be possible and are indeed observed in simulations (Schöpf & Kramer 1990). One might conjecture that such solutions can be realized when $4\beta_r + [\gamma_r\beta_i/(-\gamma_i)] > 0$ and the time for the Benjamin-Feir instability to form pulses from finite time, singular spatially uniform states is much shorter than the blow-up time. To date, however, no results of a general nature have been found.

5. APPLICATIONS: TWO PROBLEMS

Convection in Binary Fluid Mixtures

Convection in binary fluid mixtures (either He^3/He^4 or alcohol/water) is interesting and nontrivial because of the Soret effect and the different diffusion rates of heat and concentration. Temperature gradients induce concentration gradients, sometimes so as to enhance convection by inducing the heavier fluid to separate from the lighter in the direction of the temperature gradient (destabilizing) and sometimes so as to inhibit convection by having the opposite effect. Here we are most interested in the latter case. When convection finally gets going, the different diffusion rates mean that rather than being stationary in space, the most unstable state can consist of traveling waves. There are two principal parameters, the *Rayleigh number* $\mathrm{R} = (\alpha g d^3/\nu\kappa)\Delta T$, where $\alpha = -1/\rho(\partial\rho/\partial T)_{p,c}$, is the coefficient of cubic expansion and the *separation ratio* $\psi = -(\kappa/T)(\beta/\alpha)$, where $\beta = -1/\rho(\partial\rho/\partial c)_{p,T}$. The ratios of diffusion rates are the Prandtl $\mathrm{Pr} = \nu/\kappa$ and Lewis $\mathrm{Le} = D/\kappa$ numbers where D is the concentration diffusion. In the R, ψ plane, the stability diagram is shown in Figure 10*a*.

The separation ratio parameter ψ measures the influence of Soret-induced concentration diffusion on density stratification. As ψ becomes increasingly negative,[8] the onset of convection is delayed to higher and

[8] In alcohol-water mixtures, the interesting alcohol level falls in the range between beer and whiskey and so the disappointed or exhuberant experimentalist always has his source of consolation or celebration right at hand.

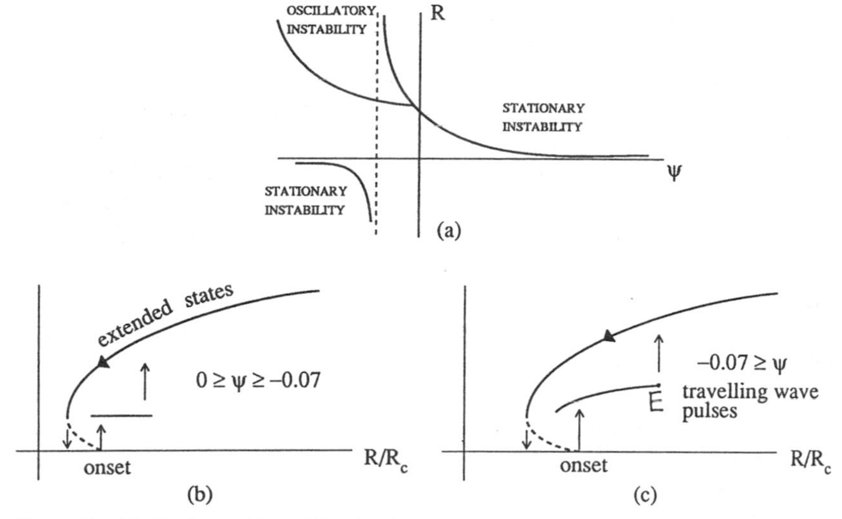

Figure 10 (*a*) Regions of instability in the (R, ψ) plane. (*b*), (*c*) Bifurcation diagrams for $|\psi| \leq 0.07$, $|\psi| \geq 0.07$.

higher Rayleigh numbers and when convection is finally triggered the system acts like a pure fluid at high Rayleigh numbers. At these values, the finite amplitude response of the system does not saturate the resulting convection at small amplitudes and the system is highly nonlinear. The magnitude of the nonlinearity in the system is therefore controlled by varying R and $|\psi|$. At low R and small $|\psi|$, near onset the nonlinearities are relatively weak and a description involving the near onset envelope order parameters is valid. What makes the system so rich, however, is that the linear and nonlinear dispersive characters of the finite amplitude states seem to be at least as important as the diffusion and nonlinear saturation effects. Therefore a wide variety of outcomes is possible and fully developed turbulent states can, in principle, exist near threshold (Newell 1979, Bretherton & Spiegel 1983, Riecke 1991).

Here we focus on two regimes of behavior. Figures 10*b* and 10*c* are schematic graphs of the experimentally measured convection amplitudes versus Rayleigh number for the weakly nonlinear regimes (*i*) $0 \geq \psi \geq -0.07$ and (*ii*) $-0.07 \geq \psi \geq -0.12 \sim -0.24$. The higher curve is the same for both figures and represents an extended convective state in which rolls fill the entire cell. Depending on Rayleigh number, they may be traveling or steady. Even in the former case, their frequency is greatly reduced from that of the low amplitude traveling wave states. In regime (*i*), amplitude levels are clearly compatible with weakly nonlinear theories.

Several theorists, notably Mike Cross (1986), have used coupled traveling wave envelope models to describe the behavior in finite domains with endwalls, and many of the predictions have been confirmed, at least qualitatively, by Fineberg et al (1988) and Steinberg et al (1989). Two features of the Cross analysis are important to stress. First, the presence of a finite group velocity means that these states have a convective rather than absolute instability character. The energy of a right traveling disturbance is swept to the right with speed ω_c' and attains maximum amplitude near the end of the convection cell. Second, boundary reflection is important and provides a seed for a left traveling wave which reaches its finite amplitude state at the left end of the cell. In the finite amplitude region, left and right growing waves cannot coexist because the real part of the cross-coupling coefficient δ_r in (2.9) is greater than the self-coupling coefficient β_r. The competition between left and right growing states and their mutual seeding through boundary reflections give rise to both steady and blinking states (Cross 1986). At larger values of μ, the front speed $2\sqrt{\mu\gamma_r[1+(\gamma_i^2/\gamma_r^2)]}$, connecting the finite amplitude regime to the zero regime, becomes comparable to the group velocity. The instability becomes absolute and then the finite amplitude regime of one or other of the traveling states fills the entire cell. We should point out that Cross's theory, which assumes that all coefficients in the CGL equations are real and that the bifurcation is supercritical, does not quite correspond with observation. Nevertheless it includes two of the most important features—group velocity advection and boundary reflections—and it is reasonable to suppose that improved models would compare even more favorably with experiment.

In an annular geometry, endwall influences are removed and the pattern can consist of predominantly right or left traveling wave disturbances. In such circumstances, two kinds of behavior are seen. In regime (*i*), a series of right traveling disturbances with envelopes that form local pulses are observed. Sometimes they relax to a periodic array of fixed pulses. More often, the pulses interact with each other chaotically, disappearing locally, shifting, and reappearing. There is also a more violent response in which pulses will grow locally in a singular fashion and then collapse. In all cases, the pulses are observed to travel more slowly than the linear group velocity. This behavior has been termed *dispersive chaos* (Kolodner et al 1990) connoting that linear dispersion and nonlinear phase modulation are important ingredients in controlling behavior. In regime (*ii*), stable localized traveling waves have been observed by several investigators (Niemela et al 1990, Anderson & Behringer 1990, Kolodner 1991, Glazier & Kolodner 1991). They exist for a range of subcritical and supercritical values of Rayleigh number. These waves are almost stationary in the laboratory frame. Certainly their speeds are very small compared to the linear group velocity of the underlying traveling wave. As ψ is decreased even further, large amplitude, stationary, confined states of arbitrary lengths exist at subcritical values of the Rayleigh number. The challenge to theory is to explain localization, slow speed, and, for larger $|\psi|$, arbitrary length.

Theory has not yet succeeded in describing all this behavior but there are promising indications that a coupled system of envelope equations for $A(x, t)$—the envelope of the fluctuating wave field, and $\bar{C}(x, t)$—the mean concentration averaged over wave period and depth, will greatly help understanding. We briefly outline the ideas. Consider first the power balance of the CGL equation (2.11). Observe that even if β_r is negative (as it is in these Soret number regimes), the power balance is

$$\frac{\partial}{\partial t}\int |W|^2\,dx = 2\mu\int |W|^2\,dx - 2\gamma_r\int\left|\frac{\partial W}{\partial x}\right|^2 dx - 2\beta_r\int |W|^4\,dx, \tag{5.1}$$

and localized pulses whose widths are proportional to inverse amplitude need not blow up in finite time because diffusion can overcome nonlinear growth. For small β_r/β_i, γ_r/γ_i ratios, that condition requires $4\beta_r + [\beta_i\gamma_r/(-\gamma_i)] > 0$ [see (4.8)]. This fact was first pointed out by Newell (1978, 1989) and it has been confirmed by theoretical work and numerical simulations of Schöpf & Kramer (1990) and by a most interesting result of Powell & Jakobsen (1992). The latter authors find classes of bounded and unbounded solutions which divide the γ_r/γ_i, β_r/β_i plane into two regions. The bounded solutions and their behavior mimic rather well some of the numerical simulations of Bretherton & Spiegel (1983) and Schöpf & Kramer (1990). Moreover, although the unbounded solutions have infinite power, they would appear to describe rather well the rapid growth and collapse behavior observed in experiments in the dispersive chaos regime. We point out that in this regime, the ratio γ_i/γ_r is about -0.03 (from the calculations of Schöpf & Zimmermann 1989) while $\beta_i/|\beta_r|$ is about 8. Unbounded solutions would be regularized by quintic saturation terms [add $-\delta A^3 A^{*2}$, where $\delta = \delta_r + i\delta_i$; $\delta_r > 0$, to the RHS of (2.11)] and Thual & Fauve (1988) have indeed found generalizations of the soliton solution (4.6) in both one and two [replace $\partial^2/\partial x^2$ by $\partial^2/\partial x^2 + \partial^2/\partial y^2$ in (2.11)] spatial dimensions. One might then conjecture that each lower branch of localized solutions shown in Figures 10*b* and 10*c* is simply a continuous deformation of the upper branch obtained schematically by replacing the constant intensity wavetrain solutions $2|W|^2(\mu - \beta_r|W|^2 - \delta_r|W|^4) = 0$, $\beta_r < 0$ by the solutions for the amplitude η of the localized pulse

$$\eta\left[2\mu-\frac{1}{3}\left(16\beta_r+\frac{4\gamma_r\beta_i}{(-\gamma_i)}\right)\eta^2-\frac{32}{3}\delta_r\eta^4\right]=0.$$

As the coefficient $4\beta_r+\gamma_r\beta_i/(-\gamma_i)$ increases, the stable upper branch moves toward the localized branch. But the localized branch stops at E. We suggest this may happen because, as the amplitude increases and dispersive quintic terms such as $\delta_i A^3 A^{*2}$ are included, the unstable band of the Benjamin-Feir instability disappears just as it does for water waves of higher amplitude. Once the focusing mechanism for localization is no longer present, the localized pulse will destabilize and become a traveling wavetrain on the upper branch. However, all these ideas require a large negative focusing factor $\beta_i\gamma_i$. Unfortunately while β_i is large for most of the Soret number range, linear dispersion γ_i is fairly small and, according to the Zimmermann & Schöpf (1989) calculation, even changes sign about $\psi \sim -0.05$.

In addition, one still has to explain the slow velocity of the pulses. In order to obtain a picture consistent with observation, one should include the mean concentration field which is driven by gradients in the pattern intensity. The importance of this effect has been illustrated by the work of Barten et al (1991), who simulated the full Navier-Stokes equations and found that wave action moves concentration laterally so as to inhibit convection at the leading edge of the pulse. Moreover, because Le is very small, a constant concentration field is in the kernel of the linear stability operator of the conduction solution. Therefore a slowly varying concentration should be included as an active order parameter. Riecke (1991) has included this effect and noted that averaging the equation gives rise to a term $\partial/\partial x|A|^2$ involving the gradient of the wave intensity. Averaging over z, and assuming that the boundary contributions give rise to damping, one obtains

$$\frac{\partial \bar{C}}{\partial t}=b\frac{\partial}{\partial x}|A|^2+L\frac{\partial^2 \bar{C}}{\partial x^2}-v\bar{C}. \tag{5.2}$$

The effect of a nonzero $\bar{C}$ is to add a term $\alpha\bar{C}W$ to the RHS of (2.11). Because dispersion becomes small as $|\psi|$ increases, and amplitudes become larger, it is important also to include quintic nonlinearity $-\delta W^3 W^{*2}$ and nonlinear gradient terms $-pW(\partial/\partial x)|W|^2-q(\partial/\partial x)W^2W^*$, although the caveat given in the last paragraph of the *Near Onset* part of Section 2 should be borne in mind. Numerical simulations [Riecke (1991) ignores higher order terms in the envelope equation; Levine & Rappel (1992) ignore mean concentration] appear to give qualitatively promising results. We suggest that velocity slowdown is analogous to the "Raman" wavenumber and frequency downshift observed in soliton pulses in optical fibers. In that context, the delay effect of the nonlinear refractive index adds a term $-i\beta_2 W(\partial/\partial x)|W|^2$ to the governing NLS equation whose effect is to decrease the wavenumber and the velocity of the pulse from ω_c' (roughly c/n_0) to $\omega_c'-\beta_2 I$, where I is the pulse intensity. Likewise, in this context, it is the term $-i(p_i+q_i)W(\partial|W|^2/\partial x)$ that achieves the same result; or if one ignores these terms, the term $\alpha_i \bar{C} W$ introduces a similar effect. Inclusion of this term and a treatment of the resulting system by soliton perturbation methods gives (4.8) with an additional term $64(p_i+q_i)/15|\beta_i/\gamma_i|\eta^4$ added to the second equation. The solution approaches a new fixed point in which the velocity $(2\gamma_i K)$ in the group velocity reference frame is proportional to η^2. Thus the new pulse velocity is $\omega_c'+[\gamma_i(p_i+q_i)/\gamma_r]\eta^2$ which, if signs are right, decreases as the pulse amplitude increases. In regime (*ii*), where the group velocity is larger, the absence of growing perturbations away from the pulse when $\mu>0$ is probably due to the fact that the zero state is convectively unstable and fluctuations are swept into and damped by the stable pulse itself. In regime (*i*), the linear group velocity is smaller, giving fluctuations in regimes of small amplitude more time to grow and cause the localized pulses to exhibit the kind of time dependence referred to as dispersive chaos. For large values of $|\psi|$, the localized pulse regime is entirely subcritical and close to the extended state boundary in the bifurcation diagram. The localized pulse of arbitrary length would appear to correspond to the extended state joined to the metastable zero state through stationary leading and trailing edge fronts—a situation only possible for nongradient flows. While these ideas may go some way towards developing a reliable model, there is much that still needs to be nailed down, including employing a more realistic approximation to the vertical structure of the fields. The ball is in the theoretician's court.

For wider channels, the strong dispersion in the direction along the wavecrests [see (2.12)] should lead to even richer dynamics. Because $\beta_i/|\beta_r|$ is relatively large, one can, to a good approximation, consider the NLS portion (the LHS and $-i\beta_i W^2 W^*$) of (2.12) to be dominant and use perturbation theory to calculate the distortion of solutions [and in particular dark soliton (Bekki & Nozaki 1985) solutions] of the defocusing NLS equation.

Convection in Cylindrical Containers

Over the past two decades, a series of sophisticated and careful experiments have given us reliable details about pattern evolution in both rectangular and cylindrical boxes with large aspect ratios ($\Gamma = L/d$ varying from 14 to 50), covering all ranges of Prandtl number—high (Pr >7), moderate

(1 < Pr < 7), and low (Pr < 1). The control of the parameters has been such that a given experiment can be continued for up to 50 and in some cases 100 horizontal diffusion times, often of the order of a week or more. The experiments in the high Prandtl number range have reproduced the Busse balloon in circumstances where straight rolls were originally forced. They also display no time dependence until very large Rayleigh numbers are reached—typically greater than ten times critical. Furthermore, the time-dependent behavior does not occur till after the bimodal state, which consists of the original rolls superimposed with rolls at 90°, occurs. For uncontrolled initial conditions, the evolution of a pattern occurs in two stages. On the horizontal diffusion scale, the pattern consists of approximately circular patches surrounding foci singularities which in rectangular boxes often sit in the corners (see Figures 3*c*,*d*). Wavenumber selection is important and it would appear that the wavenumber in the circular patches is indeed k_B where $B(k_B) = 0$. At this value, however, for infinite Prandtl number fluids, $D_\perp(k_B) = -(1/\tau)B(k_B) = 0$ and the rolls lose their resistance to lateral bending. Because of this, the rolls can develop undulations along their axes which scale as $\Gamma^{1/2}d$ (rather than $\Gamma d = L$) and this means higher order correction terms must be added to (2.18). Therefore, theory predicts and experiments confirm that the earliest time scale on which the pattern can settle down is $\Gamma\tau_H$ [where $\tau_H = L^2/\kappa = \Gamma^2(d^2/\kappa)$ is the horizontal diffusion time scale]. In rectangular geometries the pattern can become stationary whereas in circular geometries, as first noted by Ahlers & Behringer (1978), it can stay permanently time dependent. Two questions emerge: Why does the pattern evolve from 3*c* to 3*d* by replacing the circular patch by two sets of rolls that are almost orthogonal rolls and separated by a grain boundary? What causes persistent time dependence in circular geometries? We will attempt some explanation below but stress that these questions are largely open.

At moderate to low aspect ratios, mean drift effects become important. Here we give a brief account of some of the experimental observations, relying principally on the articles of Ahlers et al (1985), Heutmaker & Gollub (1987), and Croquette (1989). Heutmaker & Gollub (1987) investigated convection in water in a cylindrical dish of depth $d = 3$ mm, diameter $L = 41.7$ mm, Prandtl number 2.5, and aspect ratio $\Gamma = L/d$ of about 14. They found three regimes of behavior. In the first $R < 1.2R_c$, they observed, consistent with Ahlers et al (1985), that the pattern remains time dependent and aperiodic. After several horizontal diffusion time scales, the patterns simplify and appear to consist of orthogonal patches of straight parallel rolls connected through grain boundaries (see Figure 11). The width of the interior patch seems to be just compatible with the known fact (proved by stability analysis) that in rectangular boxes, rolls

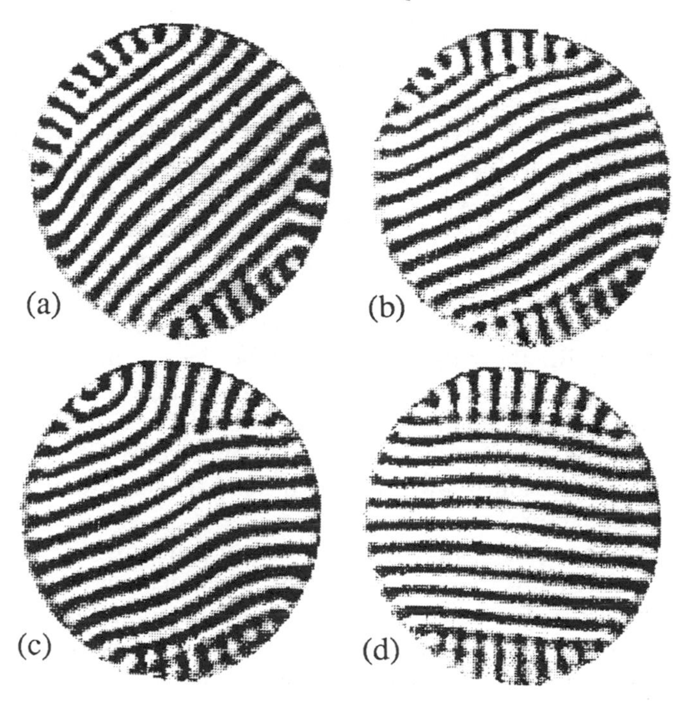

Figure 11 Snapshot, from Ahlers et al (1985), illustrating persistent time dependence near $R = R_c$. The pattern evolves from (*a*) two sets of orthogonal rolls separated by a grain boundary to (*b*), (*c*) the formation of a circular patch and interior concave disclination to (*d*) the original pattern rotated.

line up parallel to the shorter side. The pattern is not stationary and the time dependence seems to be associated with the weakly circular patch surrounding a boundary focus at the end of each grain boundary. Notice that for these low values of $(R-R_c)/R_c$, the roll axes and the boundary normal are not parallel and a significant angle can occur between these two directions. The shape of the box would seem to have some influence. In a square cell of the same aspect ratio, some of the runs appear to stabilize after 100 horizontal diffusion times. This regime is not observed at lower Prandtl numbers ($Pr = 0.71$) in the experiments of Pocheau et al (1985).

In the second regime, $1.2R_c < R < 4.5R_c$, the pattern will always stabilize but the final structure may not be unique, although nonuniqueness may only occur at values of the Rayleigh number where the zigzag and left cross-roll instability boundaries intersect (see Figure 4*b*). Again, in both the rectangular and circular geometries, the textures are dominated by circular patches which (usually) surround sidewall foci and are separated by concave disclinations (e.g. Figure 3*a*). The rolls are more bent and the roll axes are almost everywhere perpendicular to the boundary. The pattern takes several horizontal diffusion times to become time independent. The band of wavenumbers is almost wholly contained in the stable portion of the Busse balloon between zigzag and skew-varicose instability boundaries. Figure 3*a* is a stationary pattern. After transients, it would appear that grain boundaries disappear and the pattern singularities which mediate circular patches are concave disclinations. Why? How do orthogonal straight roll patches mediated by grain boundaries evolve to circular patches with disclinations as R increases? Finally, in the third regime, $R > 4.5R_c$, the pattern remains time dependent via repetitive nucleation of dislocation pairs (see Figure 3*b*) owing to what appears to be extra phase production at sidewall foci. Although the dynamics is not periodic, it has a quasi-period of about 20–40 vertical diffusion times. Similar behavior at much lower Rayleigh numbers ($R \sim 1.15R_c$) is seen by Pocheau et al (1985).

How might theory help interpret and understand this observed behavior? Since the pattern textures consist mostly of circular and straight roll patches, the first suggestions as to what might be going on come from examining the linear stability character of these planforms. An important observation of Heutmaker & Gollub (1987) is that for $R_c < R < 1.2R_c$, the distribution of wavenumbers in the pattern lies outside the zigzag instability border; for $1.2R_c < R < 4.5R_c$, it lies entirely within the Busse balloon whereas when $R > 4.5R_c$, the short wavelength tail of the distributions lies beyond the skew-varicose stability boundary (Figure 12). Therefore we might conjecture that the pattern shown in Figure 11*a* consisting of two sets of mutually orthogonal rolls is unstable because the grain boundaries select a wavenumber for the interior rolls to the left of the zigzag stability boundary. The resulting bending of interior rolls allows a circular patch to form, which connects with the residue of the grain boundary and the interior rolls through a concave disclination (Figures 11*b*,*c*). The circular patch selects a wavenumber well inside the Busse balloon and the roll bending in the interior meets increased resistance. Because roll bending cannot be sustained by the boundaries so as to eliminate the rest of the grain boundary, there seems to be a reversal of the process in which the crests of successive rolls in the circular patch disconnect, glide, and reconnect so as to reform the original grain boundary, rotated through an angle (Figure 11*d*). The dynamics would appear to be low dimensional and, in the right coordinates, should be captured by understanding the way in which the unstable and stable manifolds of solutions corresponding to grain boundaries and circular patches intersect. But there is no solid quantitative evidence that this thinking is correct, nor

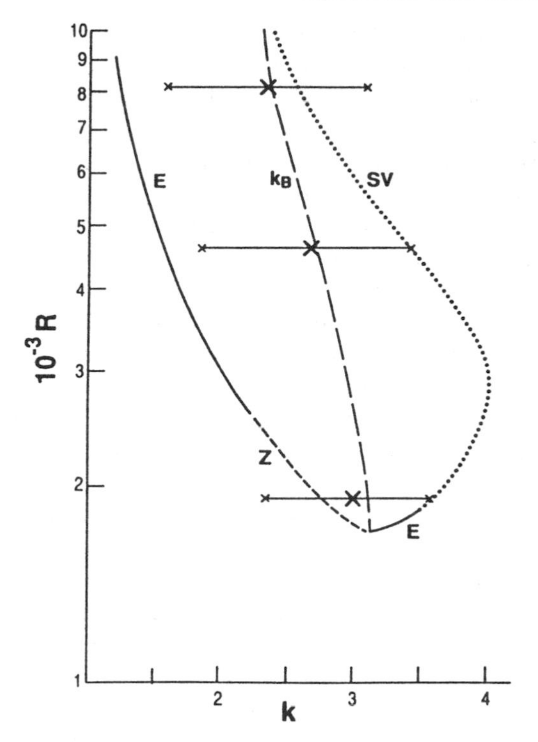

Figure 12 R, k plane showing Eckhaus (E), skew-varicose (SV) and zigzag (Z) boundaries and the curve k_B(R)—the wavenumber selected by circular roll patches. The bands at $R = 1.1R_c$, $2.61R_c$, $4.64R_c$ are the ranges of wavenumbers (approximately distributed normally about the points marked × which, for $R = 2.61R_c$ and $4.64R_c$, is k_B) measured experimentally by Heutmaker & Gollub (1987).

has there been any attempt to coordinate these configurations which makes obvious the low dimensional character of the dynamics.

As the Rayleigh number increases to range from $1.2R_c$–$4.5R_c$, the bending influence of the boundary increases and circular roll patches surrounding sidewall foci and mediated by concave disclinations dominate. Their presence is consistent with the phase dynamics picture and as discussed in Section 3, the wavevector field appears to be given by shock solutions of the Cross-Newell equation. We note from Figure 12 that the center of the wavenumber distribution is precisely k_B, the wavenumber selected by circular rolls. [For circular rolls, the induced mean drift is zero even at low Prandtl numbers, and so the selected wavenumber is the zero of $B(k)$. For infinite Prandtl numbers, rolls with any finite curvature select

k_B.] Target patterns are also stable in this regime. But there are many unanswered questions. How and why does the transition between the weakly time-dependent state for $R < 1.2R_c$, consisting of orthogonal roll patches and grain boundaries, to the time independent state, consisting of circular patches and disclinations, occur? How does the transition depend on the Prandtl number? What is the role of mean drift in helping or hindering the transition? What is the cause of nonuniqueness in these stationary patterns? Further, for $R < 1.18R_c$, Ahlers et al (1985) find in a cylindrical geometry that while patterns relax to a fixed shape, that shape continues to rotate, slowly and almost rigidly. (We might note that in these cases, remnants of grain boundaries remain between the circular patches.) What causes this? Why do Heutmaker & Gollub (1987) not observe this? Is it a function of Prandtl number? Again, what is the role of mean drift, which becomes stronger at lower Prandtl numbers, in maintaining these stationary patterns?

Beyond $4.5R_c$, Heutmaker & Gollub report that the pattern again develops time dependent behavior on a much shorter time scale. The most obvious cause is that, through some mechanism, the rolls at the center of the cylinder between circular roll patches become compressed, nucleate dislocations which (quickly on the vertical diffusion time scale) climb to the sidewalls, and (slowly on the horizontal diffusion time scale) glide back into the foci. A roll pair is lost. The pattern stress is relieved. But the process repeats. Somehow, the foci become active and generate new rolls which migrate and again compress the rolls in the center of the cell. The process is repetitive but not periodic: a wimpy turbulence involving a broadband temporal power spectrum but spatial order. As already noted, exactly the same behavior is observed at low Prandtl numbers by Pocheau et al (1985) and Croquette (1989). What pushes the wavenumber across the skew-varicose boundary? Again, the center of the wavenumber distribution lies exactly at k_B. Two mechanisms have been suggested. One (Newell et al 1990) is that the production of new rolls by the focus is a manifestation of the instability of target patterns which occurs for values of (R, k) inside the Busse balloon and at a value of R close to the observed onset of time dependence. But how relevant is this result for the circular patches attached to the sidewall? A complementary suggestion is that the boundaries reinforce the dipolar nature of the mean flow. In a very clever experiment, Daviaud & Pocheau (1989) have shown that the onset of instability is delayed if the mean flow enhancement by the boundary is reduced by introducing a permeable boundary which greatly diminishes the mean drift velocity in the region from the focus to the center of the cell. Again, from the phase diffusion equation $\theta_t = -\rho \mathbf{V} \cdot \mathbf{k} - (1/\tau)\nabla \cdot \mathbf{k} B$ one can see that weakening $\mathbf{V}$ diminishes the rate at which new phase is produced. Daviaud

& Pocheau (1989) introduce approximate solutions of the Cross-Newell model equations which support this picture. Whichever mechanism, or combination of the two mechanisms, is relevant, there remain many open challenges and questions. For example, by appropriately coordinating the configuration, can one reproduce the observed low dimensional dynamics? What happens when the wavenumber k_B selected by circular roll patches itself crosses the skew-varicose stability boundary? Do we get a macho turbulence (broadband spatial and temporal power spectra) because dislocation pairs will be nucleated almost everywhere? What is the role of boundaries? What are the correct boundary conditions for the phase diffusion and mean drift equations and for the amplitude driven regularization of the former? Will the regularized theory correctly capture the behavior of all the defects?

The list of open questions goes on and on. Patterns with features similar to those observed in hydrodynamics are now being observed in chemistry, materials science, and optics. The ideas are universal; the challenges are endless. The subject is alive!

ACKNOWLEDGMENTS

We want to thank Guenther Ahlers, Mike Cross, Paul Kolodner, Lorenz Kramer, Alain Pocheau, and Victor Steinberg for valuable input and Yves Pomeau for extensive and detailed comments on the manuscript. This work has been supported by AFOSR Contract FQ8671-900589 and NSF grants DMS 8922179 and DMS 9021253.

Literature Cited

Ahlers, G., Behringer, R. P. 1978. Evolution of turbulence from the Rayleigh-Bénard instability. *Phys. Rev. Lett.* 40: 712–16

Ahlers, G., Behringer, R. P. 1982. Heat transport and temporal evolution of fluid near the Rayleigh-Bénard instability in cylindrical container. *J. Fluid Mech.* 125: 219–58

Ahlers, G., Cannell, D. S., Steinberg V. 1985. Time dependence of flow patterns near the convective threshold in a cylindrical container. *Phys. Rev. Lett.* 54: 1373–76

Anderson, K. E., Behringer, R. P. 1990. Long time scales in traveling wave convection patterns. *Phys. Lett. A* 145: 323–28

Barten, W., Lücke, M., Kamps, M. 1991. Localized travelling wave convection in binary fluid mixtures. *Phys. Rev. Lett.* 66: 2621–24

Bartuccelli, M., Constantine, P., Doering, C. R., Gibbon, J. D., Giselfält. 1990. On the possibility of soft and hard turbulence in the complex Ginzburg-Landau equation. *Physica D* 44: 421–44

Bekki, N., Nozaki, K. 1985. Formation of spatial patterns and holes in the generalized Ginzburg-Landau equation. *Phys. Rev. Lett.* 51: 133–35

Benjamin, T. B., Feir, J. E. 1967. The disintegration of wave trains on deep water. Pt. 1. Theory. *J. Fluid Mech.* 27: 417–30

Benney, D. J., Newell, A. C. 1967. Propagation of nonlinear wave envelopes. *J. Math. Phys.* 46: 113–39 (Now *Stud. Appl. Math.*)

Bernoff, A. 1988. Slowly varying fully nonlinear wavetrains in the Ginzburg-Landau equation. *Physica D* 30: 363–91

Bers, A. 1975. Linear waves and instabilities. In *Physique des Plasmas*, ed. C. DeWitt, J. Peyraud, pp. 117–213. New York: Gordon and Breach

Besterhorn, M., Haken, H. 1990. Traveling

waves and pulses in a two-dimensional large-aspect-ratio system. *Phys. Rev. A* 42: 7195–203

Bodenschatz, E., Pesch, W., Kramer, L. 1988. Structure and dynamics of dislocations in anisotropic pattern-forming systems. *Physica D* 32: 135–45

Bodenschatz, E., Weber, A., Kramer, L. 1991. Interaction and dynamics of defects in convective roll patterns of anisotropic fluids. *J. Stat. Phys.* 64: 1007–15

Brand, H. R., Lomdahl, P. S., Newell, A. C. 1986a. Benjamin-Feir turbulence in convective binary mixtures. *Physica D* 23: 345–61

Brand, H. R., Lomdahl, P. S., Newell, A. C. 1986b. Evolution of the order parameter in situations with broken symmetry. *Phys. Lett. A* 118: 67–73

Bretherton, C. S., Spiegel, E. A. 1983. Intermittency through modulational instability. *Phys. Lett. A* 96: 152–56

Briggs, W., Newell, A. C., Sarie, T. 1983. A mechanism for instability of nonlinear finite difference equations. *J. Comput. Phys.* 51: 83–106

Busse, F. H. 1967a. On the stability of two-dimensional convection in a layer heated from below. *J. Math. Phys.* 46: 149–50 (Now *Stud. Appl. Math.*)

Busse, F. H. 1967b. The stability of finite amplitude cellular convection and its relation to an extremum principle. *J. Fluid Mech.* 30: 625–49

Busse, F. H. 1978. Non-linear properties of thermal convection. *Rep. Prog. Phys.* 41: 1929–67

Busse, F. H. 1981. Transition to turbulence in Rayleigh-Bénard convection. In *Topics in Applied Physics*, ed. H. L. Swinney, J. P. Gollub, 45: 97–137. Berlin/Heidelberg/New York: Springer-Verlag

Busse, F. H., Kramer, L., eds. 1990. *Nonlinear Evolution of Spatio-Temporal Structures in Dissipative Continuous Systems, NATO Series B* 225. New York: Plenum

Busse, F. H., Whitehead, J. A. 1974. Oscillatory and collective instabilities in large Prandtl number convection. *J. Fluid Mech.* 66: 67–79

Ciliberto, S., Coullet, P., Lega, J., Pampaloni, E., Peréz-Garcia, C. 1990. Defects in roll-hexagon competition. *Phys. Rev. Lett.* 65: 2370–73

Clever, R. M., Busse, F. H. 1979. Instabilities of convection rolls in a fluid of moderate Prandtl number. *J. Fluid Mech.* 91: 319–714

Collet, P., Eckmann, J. P. 1991. Diffusive regime for the Ginzburg-Landau equation. Preprint

Coullet, P., Huerre, P., eds. 1990. *New Trends in Nonlinear Dynamics and Pattern Forming Phenomena: The Geometry of Nonequilibrium, NATO Series* 237, New York: Plenum

Coullet, P., Fauve, S., Tirapegui, E. 1985. Large scale instability of nonlinear standing waves. *J. Phys. (Paris) Lett.* 46: 787–91

Coullet, P., Elphick, C., Gil, L., Lega, J. 1989a. Topological defects of wave patterns. *Phys. Rev. Lett.* 59: 884–87

Coullet, P., Gil, L., Lega, J. 1989b. Defect-mediated turbulence. *Phys. Rev. Lett.* 62: 1619–22

Coullet, P., Frisch T., Plazza F. 1991. Preprint

Croquette, V. 1989. Convective pattern dynamics at low Prandtl number, Part I. *Contemp. Phys.* 30: 113–33; Convective pattern dynamics at low Prandtl number, Part II. *Contemp. Phys.* 30: 153–71

Cross, M. C. 1986. Traveling and standing waves in binary-fluid convection in finite geometries. *Phys. Rev. Lett.* 57: 2935–38

Cross, M. C., Daniels, P. G., Hohenberg, P. C., Siggia, E. D. 1983. Phase winding solutions in a finite container above the convective threshold. *J. Fluid Mech.* 127: 155–83

Cross, M. C., Newell, A. C. 1984. Convection patterns in large aspect ratio systems. *Physica D* 10: 299–328

Daviaud, F., Pocheau, A. 1989. Inhibition of phase turbulence close to onset of convection by permeable lateral boundary condition for the mean flow. *Europhys. Lett.* 9: 675–80

Dee, G., Langer, J. 1983. Propagating pattern selection. *Phys. Rev. Lett.* 50: 383–86

Dubois, M., Bergé, P. 1978. Experimental study of the velocity field in Rayleigh-Bénard convection. *J. Fluid Mech.* 85: 641–53

Eckhaus, W. 1965. *Studies in Nonlinear Stability*. New York: Springer-Verlag

Elphick, C., Meron, E. 1991. Dynamics of phase singularities in two-dimensional oscillating systems. *Physica D* 53: 385–99

Fenstermacher, P. R., Swinney, H. L., Gollub, J. P. 1979. Dynamical instabilities and the transition to chaotic Taylor vortex flow. *J. Fluid Mech.* 94: 103–29

Fineberg, J., Moses, E., Steinberg, V. 1988. Spatially and temporally modulated traveling-wave pattern in convecting binary mixtures. *Phys. Rev. Lett.* 61: 838–41

Getling, A. V. 1991. Formation of spatial structures in Rayleigh-Bénard convection. *Sov. Phys. Usp.* 34: 737–76

Glazier, J. A., Kolodner, P. 1991. Interactions of nonlinear pulses in convection in binary fluids. *Phys. Rev. A* 43: 4269–80

Gor'kov, L. P. 1957. Stationary convection in a plane liquid layer near the critical heat transfer point. *Sov. Phys. JETP* 33: 311–15

Greenside, H. S., Coughran, W. M. 1984. Nonlinear pattern formation near the onset of Rayleigh-Bénard convection. *Phys. Rev. A* 30: 398–428

Hagan, P. S. 1982. Spiral waves in reaction-diffusion equations. *SIAM J. Appl. Math.* 42: 762–86

Haken, H., ed. 1979. *Pattern Formation and Pattern Recognition*. New York: Springer-Verlag

Heutmaker, M. S., Gollub, J. P. 1987. Wave-vector field of convective flow patterns, *Phys. Rev. A* 35: 242–60

Howard, L. N., Kopell, N. 1977. Slowly varying waves and shocks structures in reaction-diffusion equations. *Stud. Appl. Math.* 56: 95–145

Huerre, P., Monkewitz, P. A. 1985. Absolute and convective instabilities in free shear layers. *J. Fluid Mech.* 159: 151–68

Joets, A., Ribotta, R. 1989. Propagative patterns in the convection of a nematic liquid crystal. *Liquid Crystals* 5: 717–24

Joseph, D. 1976. Stability of fluid motions I & II. In *Springer Tracts in Natural Philosophy*, vols. 27, 28. Berlin/Heidelberg/New York: Springer-Verlag

Kawasaki, K. 1984. Topological defects and non-equilibrium. *Prog. Theor. Phys. Suppl.* 79: 161–90

Kléman, M. 1983. *Points, Lines and Walls*. Chichester, New York: Wiley

Koiter, W. T. 1963. Elastic Stability and Postbuckling Behavior In *Proceedings, Symposium on Nonlinear Problems*, ed. R. T. Langer. Madison: Univ. Wisconsin Press

Kolodner, P., Passner, A., Surko, C. M., Walden, R. W. 1986. Onset of oscillatory convection in a binary fluid mixture. *Phys. Rev. Lett.* 56: 2621–24

Kolodner, P. 1991. Stable and unstable pulses of travelling wave convection. *Phys. Rev. A* 43: 2827–32

Kolodner, P., Glazier, J. A., Williams, H. 1990. Dispersive chaos in one-dimensional travelling-wave convection. *Phys. Rev. Lett.* 65: 1579–82

Kosmatov, N. E., Shvets, V. F., Zakharov, V. E. 1991. Simulation of wave collapse in the nonlinear Schrödinger equation. *Physica D* 52: 16–35

Kramer, L., Ben Jacob, E., Brand, H., Cross, M. C. 1982. Wavelength selection in systems far from equilibrium. *Phys. Rev. Lett.* 49: 1891–94

Kramer, L., Zimmermann, W. 1985. On the Eckhaus instability for spatially periodic patterns *Physica D* 16: 221–32

Kramer, L., Schober, H. R., Zimmermann, W. 1988. Pattern competition and the decay of unstable patterns in quasi-one-dimensional systems *Physica D* 31: 212–26

Kuramoto, Y. 1984. *Chemical Oscillations, Waves and Turbulence*, Springer Ser. Synergetics, vol. 19. Berlin: Springer-Verlag

Landau, L. D. 1944. On the problem of turbulence. *C. R. Acad. Sci. URSS* 44: 311

Landau, L. D., Lifshitz, E. M. 1959. *Fluid Mechanics*. London: Pergamon

Lange, C. G., Newell, A. C. 1974. A stability criterion for envelope equations. *SIAM J. Appl. Math.* 27: 441–56

Lega, J. 1989. *Topological defects associated with the breaking of time-translation invariance*. PhD thesis. Univ. Nice

Levine, H., Rappel, W. J. 1992. Pulses of arbitrary length in the complex Ginzburg-Landau equation. Preprint

Libchaber, A., Maurer, J. 1978. Local probe in a Rayleigh-Bénard experiment in liquid helium. *J. Phys. (Paris) Lett.* 39: 369–72

Manneville, P., Pomeau, Y. 1983. A grain boundary in cellular structures near the onset of convection. *Phil. Mag. A* 48: 607–21

Murray, J. D. 1989. *Mathematical Biology*. Berlin: Springer-Verlag

Nasuno, S., Sano, M., Sawada, Y. 1989. Phase wave propagation in the rectangular convective structure of nematic liquid crystal. *J. Phys. Soc. Jpn.* 58: 1875–78

Neu, J. C. 1990. Vortices in complex scalar fields. *Physica D* 43: 385–406

Newell, A. C., 1974. Envelope equations. In *Lectures in Applied Mathematics, Nonlinear Wave Motion* 15: 157–63. Providence: Am. Math. Soc.

Newell, A. C. 1978. The inverse scattering transform, nonlinear waves, singular perturbations and synchronized solitons. *Rocky Mtn. J. Math.* 8: 25–52

Newell, A. C. 1979. Bifurcation and nonlinear focusing. In *Pattern Formation and Pattern Recognition*, ed. H. Haken. New York: Springer-Verlag

Newell, A. C. 1989. The dynamics and analysis of patterns. In *Complex Systems*, Santa Fe Inst. Studies in the Sciences of Complexity Vol. VII, ed. D. Stein. Menlo Park, CA: Addison-Wesley

Newell, A. C., Passot, T., Souli, M. 1990. The phase diffusion and mean drift equations for convection at finite Rayleigh numbers in large containers. *J. Fluid Mech.* 220: 187–252

Newell, A. C., Passot, T. 1992. Instabilities of dislocations in fluid patterns. *Phys. Rev. Lett.* 68: 1846–49

Newell, A. C., Pomeau, Y. 1992. Turbulent crystals. Preprint

Newell, A. C., Whitehead, J. A. 1969. Finite

bandwidth, finite amplitude convection. *J. Fluid Mech.* 38: 279–303

Newell, A. C., Whitehead, J. A. 1971. Review of the finite bandwidth concept. *Proc. IUTAM* 1969 *Symp. Instability of Continuous Systems*, Harrenalb, pp. 284–289, ed. H. Leipholz. Berlin: Springer-Verlag

Niemala, J. J., Ahlers, G., Cannell, D. S. 1990. Localized traveling-wave states in binary-fluid convection. *Phys. Rev. Lett.* 64: 1365–68

Nozaki, K., Bekki, N. 1983. Pattern selection and spatio-temporal transition to chaos in the Ginzburg-Landau equation. *Phys. Rev. Lett.* 51: 2171–73

Orszag, S. A., Patera, A. T. 1983. Secondary instability of wall-bounded shear flows. *J. Fluid Mech.* 128: 347–85

Ouyang, O., Swinney, H. L. 1991. Transition to chemical turbulence. *Chaos* 1: 411–19

Palm, E. 1960. On the tendency towards hexagonal cells in steady convection. *J. Fluid Mech.* 8: 183–92

Passot, T., Newell, A. C. 1992. Regularization of the phase diffusion equation for natural convection patterns. Preprint

Pismen, L., Rodriguez, J. D. 1990. Mobility of singularities in the dissipative Ginzburg-Landau equation. *Phys. Rev. A* 42: 2471–74

Pocheau, A. 1988. Transition to turbulence of convective flows in a cylindrical container. *J. Phys. (Paris)* 49: 1127–45

Pocheau, A., Croquette, V., Le Gal, P. 1985. Turbulence in a cylindrical container of argon near threshold. *Phys. Rev. Lett.* 55: 1094–97

Pomeau, Y., Manneville, P. 1979. Stability and fluctuations of a spatially periodic convective flow, *J. Phys. (Paris) Lett.* 40: 609–12

Pomeau, Y., Manneville, P. 1981. Wavelength selection in axisymmetric cellular structures. *J. Phys. (Paris)* 42: 1067–74

Pomeau, Y., Rica S. 1992. Preprint

Pomeau, Y., Zaleski, S. 1981. Wavelength selection in one-dimensional cellular structures. *J. Phys. (Paris)* 42: 515–28

Pomeau, Y., Zaleski, S., Manneville, P. 1983. Dislocation motion in cellular structures. *Phys. Rev. A* 27: 2710–26

Powell, J. A., Jakobsen, P. K. 1992. Localized states in fluid convection and multiphoton lasers. *Physica D*. Submitted

Powell, J. A., Newell, A. C., Jones, K. R. T. 1991. Competition between generic and nongeneric fronts in envelope equations. *Phys. Rev. A* 44: 3636–52

Rehberg, I. Rasenat, S., Steinberg, V. 1989. Traveling waves and defect-initiated turbulence in electroconvecting nematics. *Phys. Rev. Lett.* 62: 756–59

Rica, S., Tirapegui, E. 1991. Analytical description of a state dominated by spiral defects in two-dimensional systems. *Physica D* 48: 396–424

Riecke, H. 1991. Self-trapping of travelling wave pulses in binary mixture convection. Preprint

Sano, M., Sato., K., Janiaud, B. 1991. Oscillatory instabilities of a 2-d periodic pattern in EHD convection. Preprint

Schöpf, W., Kramer, L. 1990. Small amplitude solutions of the Ginzburg-Landau equation. *Phys. Rev. Lett.* 66: 2316–19

Schöpf, W., Zimmermann, W. 1989. Multicritical behavior in binary fluid convection. *Europhys. Lett.* 8: 41–46

Schlüter, A., Lortz, D., Busse, F. H. 1965. On the stability of steady finite amplitude convection. *J. Fluid Mech.* 23: 129–44

Segel, L. A. 1969. Distant sidewalls cause slow amplitude modulation of cellular convection. *J. Fluid Mech.* 38: 203–24

Siggia, E. D., Zippelius, A. 1981a. Pattern selection in Rayleigh-Bénard convection near threshold. *Phys. Rev. Lett.* 47: 835–38

Siggia, E. D., Zippelius, A. 1981b. Dynamics of defects in Rayleigh-Bénard convection. *Phys. Rev. A* 24: 1036–49

Steinberg, V., Ahlers, G., Cannell, D. S. 1985. Pattern formation and wavenumber selection by Rayleigh-Bénard convection in a cylindrical container. *Phys. Scr.* 32: 534–47

Steinberg, V., Fineberg, J., Moses, E., Rehberg, I. 1989. Pattern selection and transition to turbulence in propagative waves. *Physica D* 37: 359–83

Stewartson, K., Stuart, J. T. 1971. A nonlinear instability theory for a wave system in plane Poiseuille flow. *J. Fluid Mech.* 48: 529–45

Stuart, A. 1989. Nonlinear instability in dissipative finite difference schemes. *SIAM Rev.* 31: 191–220

Stuart, J. T. 1960. On the non-linear mechanics of wave disturbances in stable and unstable parallel flows, Pt. 1. The basic behaviour in plane Poiseuille flow. *J. Fluid Mech.* 9: 353–70

Stuart, J. T., DiPrima, R. C. 1978. The Eckhaus and Benjamin-Feir resonance mechanisms. *Proc. R. Soc. London Ser. A* 362: 27–41

Swinney, H. L., Gollub, J. P., eds. 1981. *Hydrodynamic Instabilities and the Transition to Turbulence. Topics in Applied Physics* 45. Berlin/Heidelberg/New York: Springer-Verlag

Tesauro, G., Cross, M. C. 1987. Grain boundaries in models of convective patterns. *Phil. Mag. A* 56: 703–24

Thual, O., Fauve, S. 1988. Localized struc-

tures generated by subcritical instabilities. *J. Phys. (Paris)* 49: 1829–33
Tuckerman, L., Barkley, D. 1990. Bifurcation analysis of the Eckhaus instability. *Physica D* 46: 57–86
van Saarloos, W. 1989. Front propagation into unstable states. II. Linear versus nonlinear marginal stability. *Phys. Rev. A* 39: 6367–90
van Saarloos, W., Hohenberg, P. C. 1992. Fronts, pulses, sources and sinks in generalized complex Ginzburg-Landau equations. *Physica D* 56: 303–67
Weber, A., Kramer, L., Aranson, I. S., Aranson, L. B. 1992. Stability limits of traveling waves and the transition to spatio-temporal chaos in the complex Ginzburg-Landau equation. Preprint
Wesfreid, J. E., Brand, H. R., Manneville, P., Albinet, G., Boccara, M., eds. 1988. *Propagation in Systems Far from Equilibrium.* Berlin/Heidelberg: Springer-Verlag
Wesfreid, J. E., Pomeau, Y., Dubois, M., Normand, C., Bergé, P. 1978. Critical effects in Rayleigh-Bénard convection. *J. Phys. (Paris)* 39: 725–31
Whitham, G. B. 1974. *Linear and Nonlinear Waves.* New York: Wiley-Interscience

References added in proof
Cross, M. C., Hohenberg, P. C. 1992. Pattern formation outside of equilibrium. Preprint
Rabinovich, M. I., Fabrikaut, A. L., Tsimning, L. Sh. 1992. Finite dimensional spatial disorder. *Usp. Fiz. Nauk.* To appear

Annu. Rev. Fluid Mech. 1993. 25: 455–84

PERSPECTIVES ON HYPERSONIC VISCOUS FLOW RESEARCH

H. K. Cheng

Department of Aerospace Engineering, University of Southern California, Los Angeles, California 90089-1191

KEY WORDS: hypersonic boundary layer instability, hypersonic flow computation, hypersonic flight, theoretical gas dynamics, viscous interaction

1. INTRODUCTION

The National Aerospace Plane (NASP) and several other space programs initiated during the past decade in the U.S. and abroad (see Williams 1986, Burns 1989, Parkinson & Conchie 1990, Koelle 1990, Ito et al 1990, Lozino-Lozinsky & Neiland 1989) have rekindled considerable interest in *hypersonics*. Almost one quarter of a century separates the present from the dynamical era of hypersonic flow research in the mid-1950s and early 1960s, during which critical flow physics problems posed by atmospheric reentry were identified and solved while many aspects of aerodynamic and aerothermodynamic theories were established. What, then, are the issues and advances in this field as perceived in the modern setting? The immense impact of the computer revolution on the design concept and analysis strategy, the experience with the Space-Transportation-System (Space Shuttle) program, as well as advances in material and propulsion technologies since the 1970s should all have made the modern research environment and progress vastly different from those of the Sputnik-Apollo era. This article examines issues and advances in current hypersonic flow research perceived to be of interest in theoretical fluid/gas dynamics. The scope and depth of the review are necessarily limited, and so is the list of cited references, although the latter turns out to be quite extensive owing to the diverse nature of the field. Helpful are two recent texts by Anderson (1989) and Park (1990) which provide useful background material for the

0066–4189/93/0115–0455$02.00

discussion of current issues. [See the reviews by Cheng (1990) and Treanor (1991).]

The nature of this diverse field may perhaps be appreciated by considering simplistically the flight Mach number M_∞ and the Reynolds number Re_∞ (or the Knudsen number $Kn = M_\infty/Re_\infty$) as two driving parameters which control the high-temperature real-gas properties and the molecular-transport processes. A lowering of Re_∞ (increasing Kn) as the vehicle ascends to the more rarefied atmosphere must bring about non-equilibrium in the internal molecular excitations and flow chemistry, and in the translational motion of the particles as well, since these are controlled mainly by particle-collision processes, such as viscous and diffusive processes. The speed and altitude ranges of the Space Shuttle and the NASP ascent/descent corridors encompass most such domains. Thus, apart from the fluid dynamic aspect of hypersonic viscous flows, one must address issues of nonequilibrium gas dynamics affecting the flows of interest, hence the use of "perspectives" in the article's title.

In the present framework, the study of viscous hypersonic flow will face transition problems of two kinds which represent, in fact, the two major areas of current research: the turbulence transition at the high Re range and, at the other end, the transition to the free-molecule limit. Work on fully-developed turbulent boundary/shear layers lies outside the scope of this review; some recent work applicable to turbulence transition in hypersonic boundary layers will nevertheless be noted. Readers may find helpful insight on turbulence modeling and CFD for aerodynamic flows offered in recent articles by Chapman (1992), Moin (1992), Cheng (1989), and Mehta (1990). Towards the rarefied-gas regime, there are quite a few Direct Simulation Monte Carlo (DSMC) calculations of varying themes to be studied, and several issues on continuum extension are in need of clarification.

The present article is part of a review on recent work in viscous hypersonic flows and nonequilibrium high-temperature gas dynamics. For material addressing the current development and issues in modeling nonequilibrium aerothermodynamics and rarefied gas dynamics, the reader is referred to Cheng (1992).

2. HYPERSONIC AIRCRAFT AS WAVERIDER

On hypersonic vehicle design and research, Townend (1991) lists three recurrent themes: 1. replacement of expendable ballistic space launchers with reusable aerospace planes, 2. hypersonic airliners, and 3. trans-atmospheric orbital transfer vehicles. Central to all three is research which aims at integrating air-breathing propulsion into an aerodynamic design called the "waverider." This term refers to a concept evolved from Nonweiler's (1963, 1990) study which utilized the streamlines behind a known shock wave for generating examples of three-dimensional (3-D) lifting bodies in a supersonic flow—a particular example of which is the caret wing generated from streamlines behind a plane shock (see Küchemann 1978, pp. 74–79, 450–514; Stollery 1990).

2.1 *Waverider as a Generic Design: the Breguet Range*

A great number of recent studies and overviews on waveriders were presented in the proceedings of an international waverider symposium (Anderson et al 1990), where substantial improvement in lift-to-drag ratio (L/D) and other aerodynamic features over standard configurations is reported. The article by Eggers et al (1990) therein reviews hypersonic aerodynamics and the waverider concept, covering a wide scope on their classical development. At this juncture, it will be refreshing to recall a discussion by Küchemann on aircraft cruising range and his vision of hypersonic flight.

Küchemann (1978, pp. 7–9) anticipated a trend of increasing propulsive efficiency η_p with flight speed, and a corresponding decreasing trend in L/D of waverider aircraft, so that the product $\eta_p L/D$ remains roughly constant—close to the value π. With this, and the provision that the fuel carried is not too small a fraction of the total weight, Küchemann concluded from the Breguet range formula that a nonstop flight to the farthest point on the globe is feasible even if hydrocarbon fuel is used, *irrespective* of flight speed. For a Mach-8 Orient Express, or the projected NASP vehicle X-30 at the same speed, the cruise would take about two hours. This conclusion is made explicit in Küchemann's (1978, p. 551) "spectrum of aircraft" reproduced here in Figure 1. It shows the maximum ranges of four types of aircraft designed for cruising at four very different Mach numbers, each having a two-hour flight time.

2.2 *Waverider Wing Studies*

THE VISCOUS CORRECTIONS Skin friction must be included in the performance analysis of a waverider wing. Results of optimization which take into account skin friction have been referred to as "viscous optimized" (Bowcutt et al 1987, Corda & Anderson 1988). The viscous-optimized waveriders obtained are seen to differ considerably in shape depending on (turbulence) transition locations assumed in the calculations, signifying the critical need of a reliable transition prediction method (see Section 5).

BOUNDARY-LAYER DISPLACEMENT, FLOW CHEMISTRY, AND HIGH-ALTITUDE EFFECTS Owing to the boundary-layer displacement effect at sufficiently

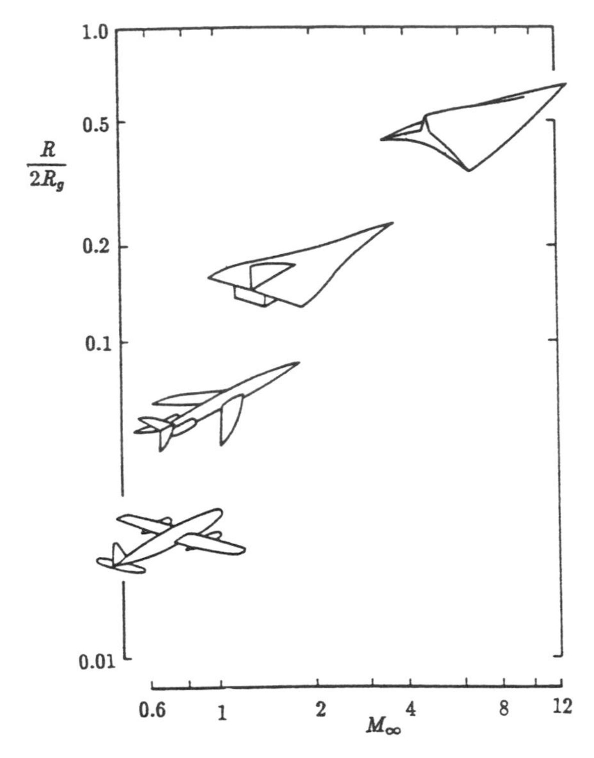

Figure 1 The "Spectrum of Aircraft" (from Küchemann 1978): Range covered by a two-hour flight as a function of cruise Mach number for four different aircraft designs. The value $R/2R_g = 0.5$ represents the distance to the farthest end of a great circle route.

low Reynolds numbers, wing loading, skin friction, and surface heating rate may increase significantly for a thin wing at low incidence. Anderson et al (1992) studied examples of waveriders with a 60-meter chord optimized for this viscous-interaction (displacement) effect. The displacement effects on lift and drag are affected little by the optimization performed in the study, which nevertheless, alters the waverider planform and its thickness distribution drastically. This observation signifies a configuration insensitivity at a given Re and M_∞, which could be translated to a greater degree of freedom for the designers. The effect of air chemistry on waverider aerodynamics has also been studied (Anderson et al 1992) but found to be small for the examples considered.

For higher altitude applications, Anderson et al (1991) consider waverider wings 5 m in length at altitudes of 80–120 km, corresponding to a Knudsen-number range from 10^{-3} to a unit order, even though the wing was generated by an inviscid procedure. "Bridging functions" (Warr 1970, Wilhite et al 1985, Potter 1988) were introduced to empirically correlate rarefied hypersonic flow data. The usefulness of this approach and its alternatives are better examined in the context of low-density hypersonic flows (Cheng 1992).

ISSUES WITH A SHARP LEADING EDGE A question on the practicality of the waverider design arises, which concerns the *sharp* leading edge inherent to the inviscid solution procedure used. An important issue was raised by Nonweiler: namely, whether a genuinely sharp leading edge made of available materials can survive the heat flux from hypersonic flight without the aid of active cooling. An answer was offered by the "conducting plate" theory and experiment (Nonweiler et al 1971, Nonweiler 1990) which demonstrate that solid-body conductivity and radiative cooling can together be effective in limiting the temperature on a sharp-edged wing. For a 14° wedge-shaped leading edge built from material with a conductivity comparable to graphite, the maximum temperature on a 75° swept wing at a speed of 6.5 km/sec is not expected to exceed 2000 K, according to the study. Recent progress in material research (e.g. Sanzero 1990) could make this passive-cooling approach more attractive. A thin/slender configuration with or without a sharp leading edge is apparently preferred over a nonslender/blunt shape in the quest for a high L/D. This may be essential for economic operations at cruising speeds as well as "cross range" consideration in transatmospheric operation (Walberg 1985).

2.3 *Integrated Aerodynamic Design*

The merit of a waverider or any aerodynamic design cannot be assessed without considering the constraints placed by the power plant installation, propulsion concept, and other details in an integrated design (Küchemann 1978, Townend 1991). Figure 2 indicates the various parts of the external and internal flows of a scramjet engine of generic design and shows the need for an integrated analysis (Billig 1992). The surface pressure on the ramp would add substantially to the total lift L and drag D; the rear portion where the burned gas exits takes the form of a "half nozzle" where the thrust T is principally derived; and the pressure also contributes to the lift and pitching moment. As an integrated system, one may speak of the net thrust $(T-D)$ available for acceleration. As the scramjet vehicle ascends to higher altitudes, the ability to accelerate further depends on the precarious balance between the diminishing T and D.

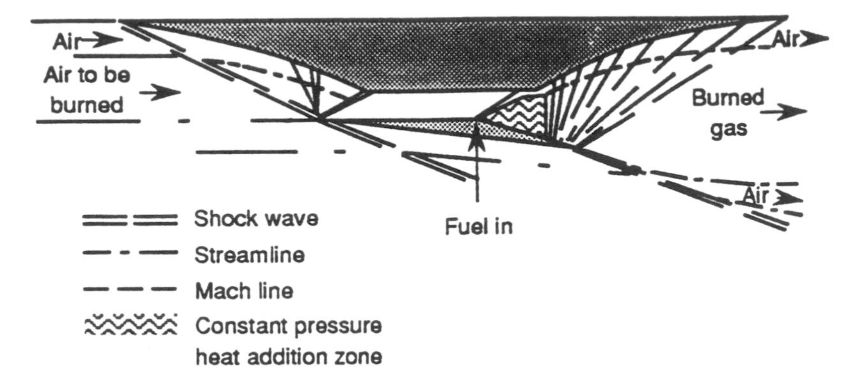

Figure 2 Schematic of a hypersonic ramjet (from Billig 1992).

3. VISCOUS INTERACTION: COMPUTATIONAL METHODS

The fluid dynamics of hypersonic flows is complicated by the interaction of the boundary layer and shear layer with shock waves, leading to flow separation and instability not amenable to straightforward analyses. The need for numerical solutions to the Navier-Stokes (NS) or other full equation systems has been made apparent in Figure 2, where significant interaction of boundary layers with shock/expansion waves occurs in most regions. Note that in the straight precombustion passage (called "isolater") in Figure 2, a shock train (not shown) must form through wave reflection and viscous interaction. Complicated shock-shock interaction patterns can create a supersonic jet impinging on the cowl lip, causing an unexpectedly high local heating rate as was first investigated by Edney (1968), and a stagnation heating rate thirty times the normal value was reported (Holden et al 1988, Glass et al 1989) but has yet to be adequately explained by a viscous interaction analysis (cf Figure 3, reproduced from Weiting 1990). As a prelude to the discussions of the following sections, several major approaches to viscous-flow calculations underlying much of the current hypersonic flow studies will be noted; their application and extension to high-temperature aerothermodynamics are best examined in the context of nonequilibrium and rarefied hypersonic flows but will not be discussed here (Cheng 1992). Some of the basic computational procedures for compressible viscous flow calculations have been elucidated in texts and monographs (e.g. Anderson et al 1984, Hoffman 1989). The CFD approaches of interest here will be discussed in three categories according to the level of approximation for the governing equations.

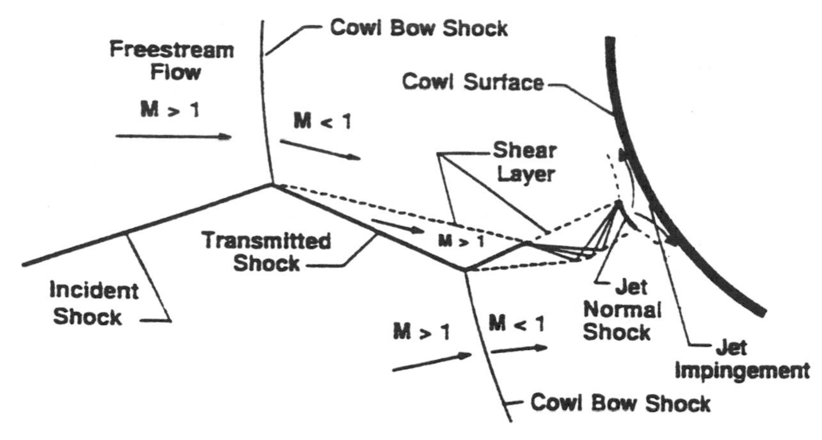

Figure 3 Interaction of incident shock and bow shock near engine cowl leading edge: Edney's Type IV supersonic jet interference pattern (sketch reproduced from Weiting 1990).

3.1 *Interacting Boundary Layer Equations*

Unlike in standard boundary layer theory, the streamwise pressure gradient in the boundary layer equations is not given a priori and is determined in a system coupled to the Euler equations governing the outer flows. The steady-state version of the equations allows upstream influence and flow separation and is comparable to a composite form of the PDEs in triple-deck theory (e.g. Werle & Vasta 1974, Burggraf et al 1979; also Anderson et al 1984).

3.2 *Parabolized Navier-Stokes (PNS)*

Even if the viscous interaction is strong, the thin viscous boundary layer permits the deletion of most streamwise partial derivative terms in the viscous, heat-conducting, and diffusive parts of the NS equations. For steady-state applications, the reduced nonlinear equations give the appearance of PDEs of parabolic type, hence the name PNS. With a hypersonic free stream, the upstream influence can be omitted in the less viscous part of the external flow, and the PNS system may then be integrated simply as an initial boundary-value problem by marching in the downstream direction. Documented results show excellent capability of the methods in shock-capturing, and in describing the interactions on the global scale as well as capturing cross-flow separations. As methods for analyzing viscous interaction, however, a shortcoming of these marching procedures is their preclusion of the upstream influence of the downstream condition and reverse flows owing to the special treatments of the streamwise pressure gradient term needed to suppress the *departure solution*, which would otherwise be a solution to the ill-posed initial boundary-value problem (Vigneron et al 1978; Anderson et al 1984, pp. 433–40; for recent PNS implementation in hypersonic flow, see Butta et al 1990, Tannehill et al 1990, Krawczyk et al 1989).

In passing, one notes that the equations in viscous shock-layer theories (Davis 1970, Moss 1976, Cheng 1963, 1966; also Gupta et al 1992) may also be regarded as simpler versions of the PNS using shock fitting. They may best be discussed in the context of the continuum extension to the rarefied gas dynamic regime (Cheng 1992), inasmuch as a considerable number of their applications have been made during the past decade in studies comparing the continuum and particle-simulation models (Moss et al 1987; Moss & Bird 1985; Cheng et al 1989, 1991).

3.3 *Iterative/Time-Accurate Navier-Stokes*

THIN-LAYER NAVIER-STOKES To retain the upstream-influence capability, one may restore the time-dependent terms to the PNS equations, solving them as a time marching, initial boundary-value problem, or apply an iterative procedure to the (steady) PNS equations with suitable outflow boundary condition. This version is often referred to as the *thin-layer* NS and is believed to be basic to several codes in current use: ARC3D (Pulliam & Steger 1980), F3D (Ying et al 1986), NS3D (Blottner 1990), and CFL3D (Vatsa et al 1989). A space-marching iterative procedure called the "supra-characteristic method" (Stookesberry & Tannehill 1986) also belongs to this class. In the first two codes, flux-vector splitting with upwind differencing (Steger & Warming 1981, van Leer 1982) or similar techniques are used; these prove to be robust in many shock-capturing calculations (Roe 1986, van Leer et al 1987). The F3D code has been further developed and used successfully in 3-D hypersonic flow analyses (Ryan et al 1990), nonequilibrium hydrogen-air reactions (Lee & Deiwert 1990), and hypersonic flow through an expansion slot in a 3-D ramp (Hung & Barth 1990). The TVD (Total Variation Diminishing) and similar schemes were used in many of these works to enhance the shock-capturing capability (cf Yee 1987). Compared to the full NS calculations, the thin-layer version may represent considerable savings in computer resources and programming effort. The global convergence to the steady state can be accelerated with some zonal strategy, as seen from the cited examples.

FULL NAVIER-STOKES CALCULATIONS Similar remarks apply to the full NS equations and calculations. Among the full 3-D NS codes used in current hypersonic flow analyses is the LAURA (Langley Aerothermodynamic

Upwind Relaxation Algorithm). Designed for finite-volume formulation (Gnoffo 1990), LAURA adapts Roe's averaging to flux components across cell boundaries for the convective terms (see Roe 1986) and Harten's (1983) symmetric TVD scheme (see also Yee 1987). Unique in this relaxation procedure is the *point-implicit* strategy, which is believed to render the procedure stable for an arbitrary Courant number without the need of solving large, block-tridiagonal matrix equations. The LAURA code has been applied to a variety of nonequilibrium aerothermodynamic problems as well as rarefied, hypersonic flow studies (cf Gnoffo 1990, Greendyke et al 1992).

The procedure of MacCormack's explicit, time-split, predictor-corrector method (MacCormack & Baldwin 1975, Hung & MacCormack 1975) and the implicit version (MacCormack 1982) solve equations in finite-volume conservation-law form and are supposedly second-order accurate in space and time, as elucidated in the text of Anderson et al (1984). Application of the explicit 2-D version by Hung & MacCormack (1975) to flow past a flat plate with a compression corner at $M_\infty = 14.1$, $Re_\infty = 1.04 \times 10^5$, agrees quite well with experimentally measured surface pressure, heating rate, and skin friction (Holden & Moselle 1969) for deflection angles $\alpha = 0$–$18°$. The same set of experimental data was also compared with solutions by the supra-characteristics method in Stookesberry & Tannehill (1986). However, the comparison for $\alpha = 24°$ was not satisfactory in either study for the reason to be noted shortly. A 3-D version of a similar procedure was successfully applied to a complete reentry configuration at Mach 6 by Shang & Scherr (1986), assuming $\gamma = 1.40$ and a Baldwin-Lomax turbulence model (cf Anderson 1989, pp. 353–59). In the implicit version, a stage is added to each of the predictor and corrector steps, where an approximately factorized time-dependent operator is applied to implicitly update the unknowns by simply inverting bidiagonal matrices as was done in the explicit version. Whereas the procedure is unconditionally stable for unbounded time steps Δt according to linear model analyses, the quotients $\mu\Delta t/\rho(\Delta x)^2$ and $\mu\Delta t/\rho(\Delta y)^2$ are required to be bounded to maintain accuracy. One observes, however, that the latter requirements are similar to, and almost as restrictive as, the stability condition for an explicit method applied to a diffusion/heat equation, and that the density ρ in these products can cause severe problems in rarefied hypersonic flow applications.

The second-order accurate, compressible NS solver recently proposed by MacCormack & Candler (1989) is virtually a *relaxation* procedure and appears to be extremely promising for 3-D applications, according to MacCormack (1990). Type-dependent procedures have been effectively implemented according to the flux-vector splitting algorithm in solving equations in conservation-law form (Steger & Warming 1980). The splitting, applied mainly to the streamwise flux, is believed to help relaxation convergence by virtue of the increased weight of the diagonal elements in the block tridiagonal matrix of the difference equations. In MacCormack & Candler's (1989) procedure, Gauss-Seidel line relaxation is adopted to solve the unfactored matrix equations, thereby avoiding unwarranted errors from the approximate factorization, which slows down convergence. The new procedure allows large time steps, and the calculation can be performed on a common workstation. It is unclear if convergence acceleration routines (e.g. Cheung et al 1991) would be helpful in further enhancing the method's performance. Recently, this procedure has been adapted to solve the (full) Burnett (1936) equations for rarefied hypersonic flows (Zhong et al 1991a,b), for which issues with boundary conditions and the solution's uniqueness remain unresolved (Cheng 1992). More recently, a similar procedure has been implemented with an adaptive grid which successfully captures the deflagation front behind a bow shock in a premixed hydrogen-air supersonic flow (Wilson & MacCormack 1992).

IMPORTANCE OF 3-D INFLUENCE We return now to the comparison of Hung & MacCormack's (1975) calculation with Holden & Mosselle's (1969) measurement for the case with the ramp angle $\alpha = 24^\circ$ mentioned earlier. In this case, Rudy et al (1991) used the thin-layer CFL3D (Vatsa et al 1989) code to demonstrate that spanwise (global) 3-D effects can resolve all the noticeable discrepancies. The computed surface oil flow and pressure contours in the symmetry plane and a downstream plane are reproduced in Figure 4. This provides perhaps an excellent example in which 3-D computation has proven crucial in settling a fluid dynamics issue which would have been perceived as being 2-D in origin. Among other computer programs currently being used in viscous hypersonic flow study are those of Edwards & Flores (1990), Thomas & Neier (1990), and Liu & Jameson (1992).

4. VISCOUS INTERACTION: THEORETICAL DEVELOPMENT

We turn next to the development of viscous interaction theory in this section and later to the investigation of the related instability problem of hypersonic boundary layers (Section 5).

4.1 *Viscous Interaction on Triple-Deck Scales*

Significant *global* interaction of a laminar boundary layer with an external hypersonic flow ($M_1 \gg 1$) has been the subject of extensive investigation in the past (e.g. Hayes & Probstein 1959, Moore 1964, Cox & Crabtree

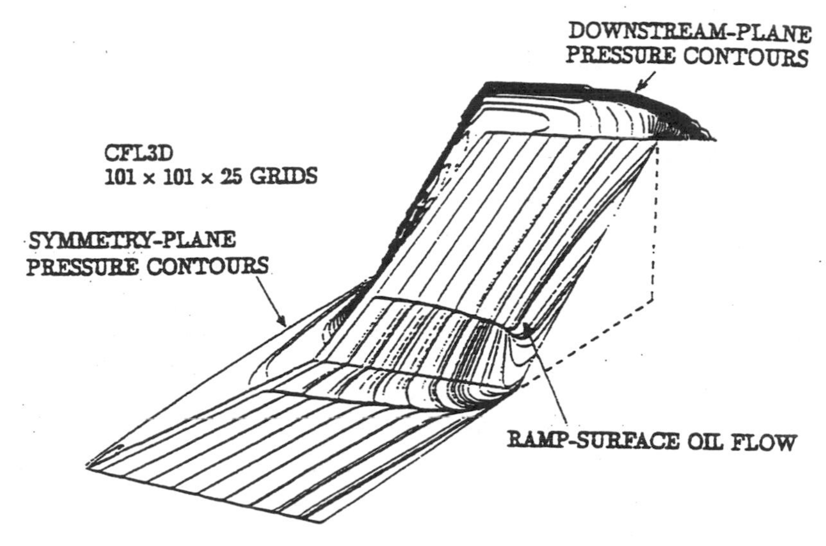

Figure 4 Surface oil flow pattern and pressure contours in the symmetry plane and downstream plane computed by a thin-layer NS code (CFL3D) for a ramp on a flat plate at $M_1 \sim 14$, $Re_1 \sim 10^5$ (Rudy et al 1991). The analysis resolved discrepancies between experiment and earlier 2-D calculations.

1965, and the review by Mikhailov et al 1971). There is yet another more universal and important interactive feature of a boundary layer occurring on a much shorter scale noted earlier by Lighthill (1953), which permits upstream influence and separation and became the focus of a vast number of theoretical studies two decades later (see reviews by Stewartson 1974, 1981; Smith 1982, 1986; Sychev 1987). Central to all the recent work is the triple-deck theory which stipulates a three-tier structure made up of lower, main, and upper decks, with the streamwise scale short enough that a small self-induced pressure rise is sufficient to provoke flow reversal and separation.

4.2 *Triple-Deck Theory Applied to Hypersonic Flow*

The basic parameter controlling the triple-deck structure for a locally supersonic external flow can be written for the present purpose as (Stewartson 1974)

$$\varepsilon \equiv \Gamma\left(\frac{\gamma-1}{2}\right)\frac{M_1}{\sqrt{\gamma(M_1^2-1)}}\left[M_1^3\sqrt{\frac{C}{Re_1}}\right]^{1/4}, \tag{4.1}$$

where Γ is a function of wall temperature and wall shear immediately upstream of the interaction zone, and the product inside the square brackets is simply the Lees-Stewartson global-interaction parameter

$$\chi \equiv M_1^3 \sqrt{\frac{C}{\text{Re}_1}} \tag{4.2}$$

familiar from the classical theory. In the preceding, the subscript "1" refers to conditions immediately upstream of the triple deck and the constant C is the Chapman-Rubesin coefficient $\mu^* T_1/\mu_1 T^*$, where the asterisks refer to the reference temperature of the hypersonic boundary layer. It is apparent from (4.1) that χ remains the important parameter controlling the viscous interaction on *both* global and triple-deck scales. Characterizing the theory for this flow structure are the orders of magnitudes of the thickness ratios of the lower, the main, and the upper decks, and also the normalized pressure and streamwise-velocity perturbations, which are representable, respectively, as

$$\varepsilon^5, \varepsilon^4, \varepsilon^3, \varepsilon^2, \varepsilon. \tag{4.3}$$

(The streamwise length scale Δ for the triple deck is the same as that of the upper deck.) This version of the theory is to be referred to as the *standard* version and requires the ε in (4.1) to be asymptotically small, and is clearly inapplicable to a regime where χ is not small. There is however a Newtonian version of this approach which considers $(\gamma-1)/2$ to be asymptotically small in addition to M_1 being large, while allowing an unbounded χ (Section 4.4). The upstream influence through the lower deck may be best seen from the formulation of Rizzetta et al (1978) using the shear $\tau \equiv \partial u/\partial y$ as a dependent variable, in which a Neumann boundary condition for τ at the wall ($y = 0$), after eliminating the pressure gradient, is

$$\frac{\partial \tau}{\partial y} = \frac{d^2}{dx^2} \int_0^\infty (\tau - 1)\, dy, \tag{4.4}$$

where the second x-derivative makes the elliptic nature of the problem apparent.

Among the examples (see Stewartson 1974; Smith 1982, 1986) is the *free-interaction* solution which is an eigen/departure solution that is admissible if provoked. The latter leads to separation and flow reversal in the lower deck, and reaches a pressure plateau downstream; it represents physically the precursor at the head of a large recirculation region. For a ramp angle in a suitably small range, solutions with recirculation and reattachment on the ramp downstream were obtained by Rizzeta et al (1978).

4.3 *Is a Departure Solution Admissible at Large χ?*

The foregoing discussion would suggest that departure solutions of the triple-deck theory are unlikely at large χ (strong global interaction). A classical example of global interaction at an unbounded χ is that of an aligned flat plate, for which the self-similar solution at a uniform wall temperature yields a self-induced pressure p/p_∞ proportional to χ (Stewartson 1955, Hayes & Probstein 1959). Nieland (1970) found, however, that an indeterminancy exists for an expansion of this solution in descending powers of χ, i.e.

$$\frac{p}{p_\infty} = D_\infty \chi [1 + \cdots + a_1 \chi^{-2n} + \cdots], \tag{4.5}$$

where for a certain exponent n, the constant a_1 cannot be determined. The finding suggests an upstream influence excluded by the solution procedure. Using a tangent-wedge pressure formula, and assuming a unit Prandtl number and an insulated wall, Nieland found $n = 50.6$. This value was confirmed subsequently in the analysis of Werle et al (1973) which considers a wide range of wall temperatures, and in Brown & Stewartson's (1975) investigation where the eigen solution was found to be insensitive to the approximation made on the outer flow. There, the exponent n was shown to be a function of the specific-heat ratio γ and of the wall-to-stagnation temperature ratio. These features may nevertheless be reconciled with the triple-deck formalism discussed below.

4.4 *The Theory for $\gamma \to 1$*

The impasse in the triple-deck theory posed by large χ is overcome by the theory of Brown et al (1975) based on small $(\gamma - 1)/2$ and high M_1^2, which could be called a Newtonian theory (Hayes & Probstein 1959) but which is less restrictive than the latter since the assumption of a strong shock is not strictly required. Let ε, ε_p, and Δ gauge the orders of magnitude of the velocity and pressure perturbations, and the streamwise length scale of the triple deck, respectively. These can be expressed in this case for a nonvanishing χ explicitly as

$$\varepsilon = (\gamma - 1)(T_w/T_o)^2 C \chi^{-1/2}, \quad \varepsilon_p = (\gamma - 1)(T_w/T_o)^6,$$

$$\Delta = (\gamma - 1)^3 (T_w/T_o)^6, \tag{4.6}$$

showing that a triple-deck structure is possible for $\gamma \to 1$. They also suggest that wall cooling should make the theory much more accurate. Using the tangent-wedge approximation, the crucial pressure-displacement relation in Brown et al (1975) needed for closure of the interaction problem can be written as

$$\mu p = -\frac{d}{d\xi'}(A+p), \tag{4.7}$$

where A is a displacement due to the lower deck, and μ is a constant of the order $(\gamma-1)(T_w/T_o)^6\chi^2$ in the case of small χ. This may be compared with $p = -dA/dx$ in the (standard) supersonic triple-deck theory. Computational studies with this version of the theory have been made for a compressive ramp (Rizzetta et al 1978) and for free interaction (Gajjar & Smith 1983).

On the other hand, for finite χ and small $(\gamma-1)/2$, (4.7) leads to

$$\frac{d}{d\xi'}(A+p) \approx 0, \tag{4.8}$$

which implies that the boundary-layer outer edge, hence the flow in the upper deck, is little affected by the interaction. In this connection, one may examine whether the triple-deck result can be reconciled with Nieland's algebraic eigen solution (4.5). The latter may now be interpreted as

$$a_1\chi^{-2n} = b_1\left(\frac{x}{x_1}\right)^n \sim b_1 \exp\left(n\frac{x-x_1}{x_1}\right). \tag{4.9}$$

Note that $\chi \propto x^{-1/2}$ and that the triple deck is centered at x_1. Now the free-interaction solution in the theory of Brown et al (1975) gives a pressure precursor of the same form as (4.9) with the exponent n being identified as $n = (0.8273)/\Delta = O[(\gamma-1)^{-3}(T_w/T_o)^{-6}]$, which is indeed a large number for the γ and T_w/T_o of interest, as was anticipated. The Newtonian version of the analysis (Brown et al 1975) remains to be completed with the inclusion of the centrifugal correction in the Busemann pressure formula; this is expected to alter substantially the pressure-displacement relation (4.7).

4.5 *Critical Influence of Wall Cooling*

For hypersonic flight applications, theory and analysis must take into consideration the effect of a low wall-to-stagnation temperature ($T_w/T_o \ll 1$). It may be noted that the assumption $T_w/T_o = O(1)$ is implicit in the standard theory, and the wall temperature need not fall too far below the stagnation/recovery level before a significant departure from the standard theory can occur, as the following will confirm.

The analysis of Brown et al (1990) of the triple deck for small χ identifies a critical wall temperature level T_w^*:

$$s_w^* \equiv \frac{T_w^*}{T_o} \sim \left[\lambda^5 \gamma^{-1/2} \left(\frac{2}{\gamma - 1}\right)^2 \chi_1\right]^{1/(4\omega+2)}, \tag{4.10}$$

where λ is a normalized undisturbed wall shear (equal to 0.332 for an aligned flat plate), and ω is the exponent in the viscosity-temperature relation $\mu \propto T^\omega$; the Newtonian factor $2/(\gamma - 1)$ is included to indicate its influence but the limit $\gamma \to 1$ was *not* taken. Depending on the ratio T_w/T_w^*, three distinct wall-temperature ranges exist:

(*i*) supercritical—$T_w \gg T_w^*$,

(*ii*) transcritical—$T_w = O(T_w^*)$, and

(*iii*) subcritical—$T_w \ll T_w^*$. (4.11)

For the supercritical and transcritical ranges, the set of scale factors ε, ε_p, and Δ is not basically different from that of the standard theory

$$\varepsilon = \lambda^{-2}\frac{\gamma - 1}{2} s_w^{2\omega} v_1, \quad \varepsilon_p = \lambda^{-2}\frac{\gamma - 1}{2} s_w^{2\omega+1} v_1^2, \quad \Delta = \lambda^{-5}\gamma\left(\frac{\gamma - 1}{2}\right)^2 s_w^{4\omega+2} v_1^3, \tag{4.12}$$

where

$$s_w = \frac{T_w}{T_o}, \quad v_1 \equiv \left(\frac{s_w^*}{s_w}\right)^{\omega + 1/2}. \tag{4.13}$$

The relation between the pressure rise and the lower-deck displacement for the ranges (*i*) and (*ii*) can be reduced to

$$P_1 = -\frac{d}{d\xi'}(A + \nu P_1), \tag{4.14}$$

where $\nu = kv_1$, and k is a constant of order unity determined by the boundary-layer profiles just upstream of the triple deck, independent of χ_1. The term νP_1 in (4.14) is absent from the standard theory, and represents a *transcritical* (cold-wall) effect. For the subcritical range ($s_w \leq s_w^*$), the gauging parameters of (4.12) must change in order to remain small, to keep the reduced PDE in canonical form, and to avoid degeneracy in the $P - A$ relation. This is accomplished simply by replacing v_1 therein by $1/k$, and (4.14) changes over the subcritical case to

$$\frac{d}{d\xi'}(P_1 + A) = -v_1^{-4} P_1. \tag{4.15}$$

Interestingly, the relative scales of the triple deck, i.e. ε and Δ, no longer depend on the Reynolds and Mach numbers in this case and vanish with s_w.

Figure 5 reproduces the results of Brown et al (1990) for the overpressure in a free interaction for $\sigma \equiv v_1^4$ in the range of $1 < \sigma < \infty$, with the origin of ξ' located at the separation point. The existence of the transcritical and subcritical s_w-ranges was anticipated in Nieland (1990, private communication)—the length scales therein differ however from those in Brown et al (1990). The analysis of Brown et al shows clearly the drastic reduction in the triple-deck length scales, hence in the extent of the upstream influence as s_w vanishes. This means that laminar separation can occur but becomes more abrupt under a strong cooling.

5. BOUNDARY-LAYER INSTABILITY AND TRANSITION STUDIES

Many investigations of flow instability and turbulence transition in hypersonic boundary layers have been undertaken recently. The development is helped substantially, perhaps, by the sequence of analyses on compressible boundary-layer instability made decades earlier by Mack and others (see reviews by Mack 1984, 1987a,b; Reshotko 1976).

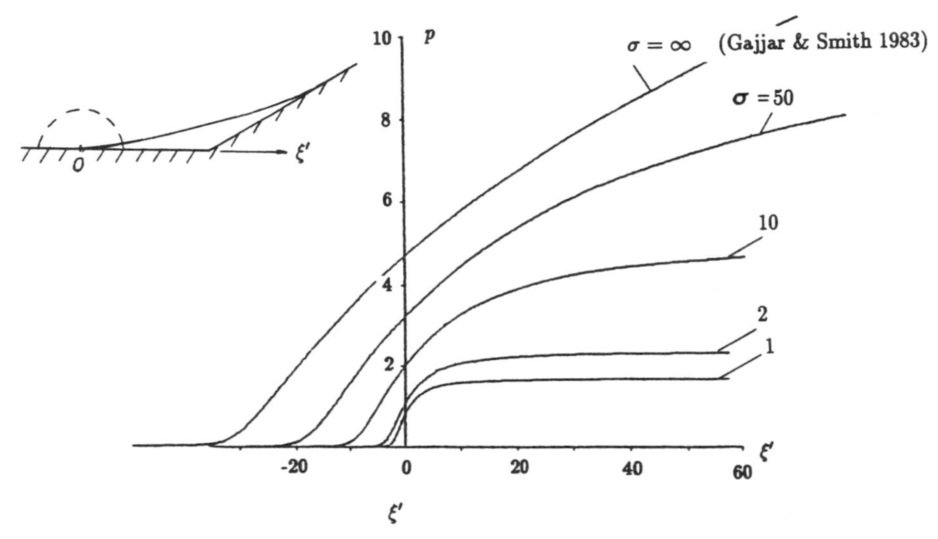

Figure 5 Normalized over pressure in a compressive free-interaction for $\sigma \equiv (T_*/T_w)^{4\omega+2} = 1, 2, 10, 50, \infty$ (Brown et al 1990). Note the plateau pressure depends on surface temperature, Mach and Reynolds numbers through σ and T_* defined in the text.

Some caution should be exercised at this juncture on the use of the viscosity-temperature ($\mu - T$) relation in extending the stability analysis to high-temperature real-gas flow, apart from other more obvious considerations. According to a recent study (Kang & Kunc 1991), for example, the viscosity of dissociating iodine at $T = 1000$–2000 K will have a *negative* slope in the $\mu - T$ relation, i.e. $d\mu/dT < 0$; similar properties may occur in other dissociating/ionizing gases and their impact on the stability analysis needs to be ascertained. The other aspect in need of caution is the assumption of translational equilibrium in certain stability and transition calculations, where the combination of low Re and high M_1 makes the gas-rarefaction effect important. Take for example, a hypersonic boundary layer on a slender/thin body, which may have a boundary-layer thickness δ of 2% the global scale L, or larger; in this case, it can be shown that the local mean free path is of the order of δ.

5.1 *Parallel-Flow Instability Applied to Compressible Boundary Layers*

Lees & Lin (1946) extended the viscous (Tollmien-Schlichting waves) and inviscid (Rayleigh theorem) results of parallel-flow instability to the compressible case. They noted that the condition $D(\rho DU) = 0$ (with $D \equiv d/dy$), signifies a maximum angular momentum and plays the same role in compressible theory as does $D^2U = 0$ (an inflection point) in incompressible theory. Unlike in the incompressible case, this generalized inflection point can be found at some $U = U_s$ in the compressible boundary layer on a flat plate, and therefore neutrally stable waves with phase velocity $c = U_s$ can exist. Lees & Lin limited their consideration to 2-D subsonic relative waves, i.e. $|U(y) - c| < a(y)$. This rules out the supersonic relative waves with $|U(y) - c| > a(y)$ and the possibility that, in a supersonic boundary layer, the TS type waves are most amplified at some oblique (wave) angles. Allowance of supersonic relative waves would render possible acoustic wave propagation and reflection within the boundary layer, admitting a sequence of higher modes for each phase velocity, as Mack (1984, 1987a,b) subsequently found. [The modes are designated/ordered by a number "n" according to the sign changes (zero crossings) occurring in the pressure profile.] The second mode turns out to be the most unstable for flat plates and slender cones at high Re (inviscid) and also for all Re (viscous) at $M_1 > 4$, as was confirmed by subsequent experiments (discussed below). The neutral stability waves, both inflectional and noninflectional, are significant (as they are in classical theory) in that they identify with parts of the boundaries delimiting the instability/stability domains of interest. Figures 6 and 7 (reproduced from Mack 1984) present these curves of neutral instability in the domain of wave number α and Mach number M_1

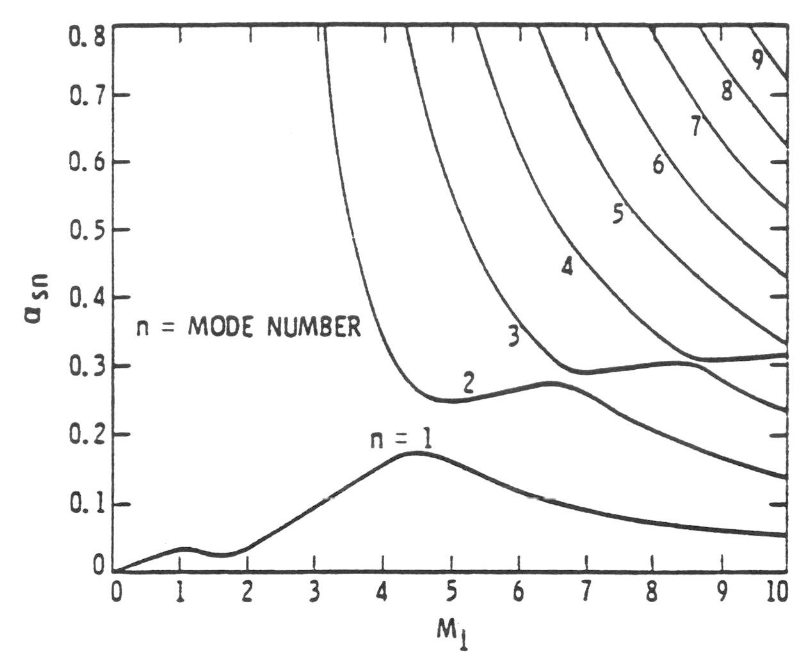

Figure 6 Wave number as a function of outer-edge Mach number of the neutral inflectional instability modes admissible to the inviscid stability equations of a compressible boundary layer on an aligned flat plate (from Mack 1984).

for 2-D inflectional and noninflectional waves, respectively. The calculations were made for an insulated flat plate. Note that a sequence of noninflectional waves of neutral stability can exist for each c in the entire range $U_1 \leq c \leq U_1 + a_1$, but the results for $c = U_1$ shown in Figure 7 are more important since each curve therein forms a part of the boundary for some genuinely unstable domain.

Among several peculiar features of Figures 6 and 7 are the similarity of the two graphs in trends at high and low α, and the drastic slope change together with what appears to be a mode-switching behavior in Figure 6, to be delineated in Section 5.3 below. One unique feature of a boundary layer with high M_1 is the progressive movement of the generalized inflection point towards the boundary-layer outer edge as M_1 increases. Thus at high M_1, this location falls inside the "edge layer" (Bush 1966, Bush & Cross 1967, Lee & Cheng 1969), and the stability analysis must take into consideration the appropriate $\mu - T$ law. An adverse effect of wall cooling must be noted. At $M_1 = 10$, Mack's (1987b) calculations revealed that the

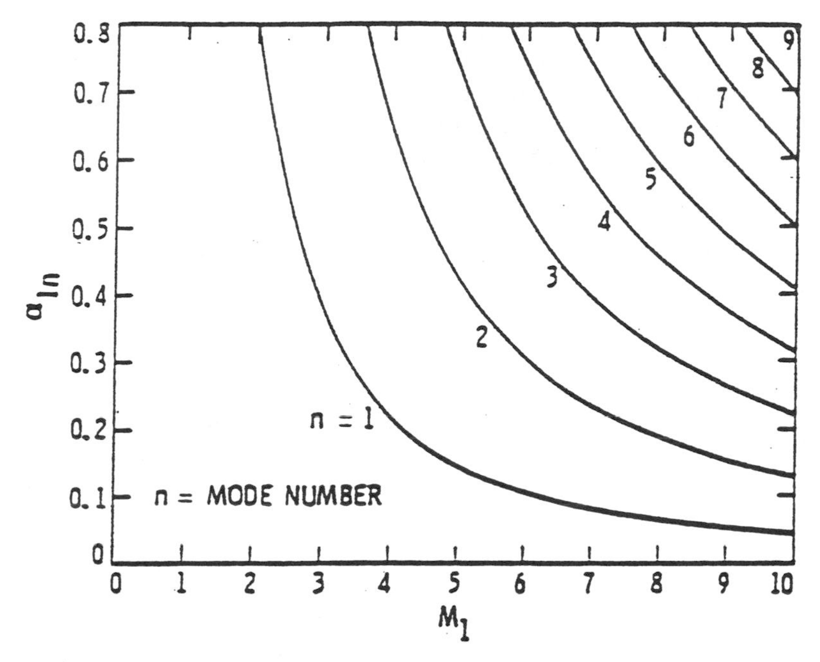

Figure 7 Wave number as a function of outer-edge Mach number of the neutral non-inflectional instability modes admissible to the inviscid stability equations of a compressible boundary layer on an aligned flat plate (from Mack 1984).

temporal amplification rates of the second, third, and fourth modes at $T_w/T_o \sim 0.05$ are almost twice the corresponding rates for an insulated wall. This was confirmed experimentally, at least for the second mode.

5.2 *Experimental Studies of Hypersonic Boundary-Layer Transition*

There have been primarily three sets of experimental studies on hypersonic boundary-layer instabilities at $M_1 = 4.5$–8.5—reported in Kendall (1975), Demetriades (1978), and Stetson (1988). These focused on flat plates and cones in wind tunnels and employed hot-wire anemometer techniques. Kendall's experiments confirmed the existence of the second mode and its dominance in a hypersonic boundary layer, Demetriades verified Mack's findings on the adverse wall cooling effect on the second mode, and Stetson et al investigated tip-bluntness, wall cooling, and other effects on slender cones. The latter studies and related works are comprehensively reviewed in Stetson & Kimmel (1992) who also note the existence of a harmonic of the second mode unaccounted for by the theory. Figure 8, reproduced in part from Malik et al (1990), shows good agreement of Stetson's cone data at $M_\infty = 8$ with Mack's calculation for the corresponding outer-edge Mach number $M_1 = 6.8$ in the second-mode frequencies near the maximum growth rate. The noticeable difference in the magnitude of the peak growth rate was believed to be caused by the inadequate accuracy of the mean flow represented by the boundary-layer solution, but the results based on the PNS-generated mean flow in the study of Malik et al (1990) were still

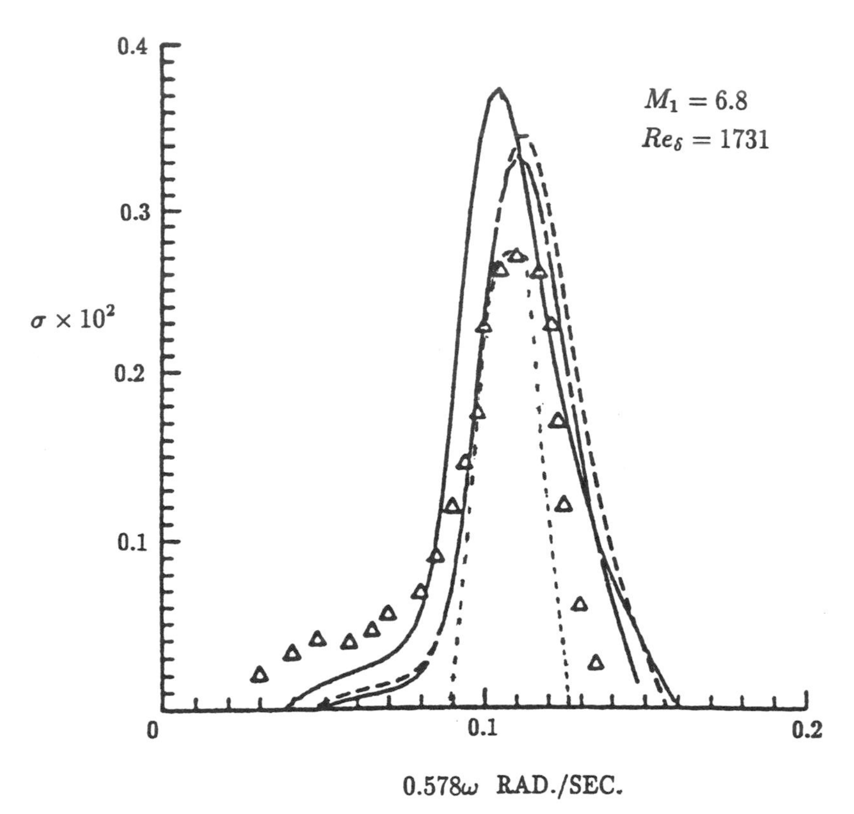

Figure 8 Comparison of amplification rates of the second mode as a function of frequency using four different basic-state equations with experimental data for the boundary layer on a cone in a Mach 8 wind tunnel, for which $M_1 = 6.8$, $Re_\delta = 1731$. (Results and data from Malik et al 1990, Simen & Dallman 1992, Stetson 1988.) Shown are the boundary layer (*solid line*), PNS blunt cone (*dashed line*), PNS sharp cone (*long and short dashed line*) from Malik et al (1990); the thin-layer NS calculation (*dotted line*) of Simen & Dallmann (1992); and the experimental data (Δ) of Stetson (1988).

far from the mark (cf *dashed line* and *long and short dashed line* in Figure 8). However, a more recent analysis by Simen & Dallmann (1992) produces growth rates (*dotted line*) rather close to the measurements near the peak, attributed principally to the merit of a version of thin-layer NS used in the mean-flow analysis.

5.3 *Asymptotic Properties at High* M_1

NEAR-MODE CROSSING Mack's result shown in Figure 6 indicates the existence of a segment on the neutral curve of each (inflectional) mode where the slope $d\alpha/dM_1$ becomes positive, while the slopes are negative elsewhere and on the noninflectional neutral curves in Figure 7. This segment has been alluded to as the "vorticity mode" (cf Mack 1984), which will be named hereafter the *vorticity submode* (of each inflectional neutral mode). The parts on the neutral inflectional curves with negative $d\alpha/dM_1$ will be called *acoustic submodes* and those of the noninflectional neutral waves are called *acoustic modes*. Another feature noticeable from comparing Figure 6 to Figure 7 is the close proximity of the acoustic submode of an inflectional nth mode (in Figure 6) to the acoustic noninflectional $(n-1)$th mode or nth mode of Figure 7, depending on whether the acoustic submode is to the left or to the right of the vorticity submode. Furthermore, the segments of vorticity submodes tend to form a continuous curve at high M_1, as suggested by Figure 6.

This and the feature of *near-mode crossing* of the acoustic submodes noted above is best explained by Smith & Brown's (1990) asymptotic result which illustrates the switching from a vorticity to an acoustic submode along the nth inflectional neutral curve:

$$\left(\alpha-\frac{1}{4}\Gamma_1\right)\left[\alpha-\frac{\pi(4n+1)}{1.788M_1^2}\right]=E(M_1^2,\gamma), \tag{5.1}$$

where Γ_1 stands for $(\ln M_1^2)^{1/2}$, the value of 1.788 was arrived at from $\gamma = 1.40$, and E has a magnitude comparable to $M_1^{-2}\exp(-2\alpha M_1^2)$. Thus the two distinct submodes are separated by an exponentially small amount, and the vanishing of the two factors on the left leads, respectively, to the vorticity and the acoustic submodes. While the vorticity and acoustic submodes differ little in their phase velocities at high M_1, i.e. $c = U_s+O(U_sM_1^{-2})$, they differ substantially in the wave numbers as the foregoing discussion and (5.1) have indicated. Supported by their analysis, Smith & Brown (1990) propose to classify instability modes at high M_1, including the neutral waves, into two main kinds: One is the acoustic mode with wave number and growth rate given, respectively, by

$$\alpha = O(n\mathrm{M}_1^{-2}), \quad \alpha c_i/U_1 = O(\mathrm{M}_1^{-6}\Gamma_1^{-1}), \tag{5.2}$$

and the other is the single vorticity mode with

$$\alpha = O(\Gamma_1), \quad \alpha c_i/U = O(\mathrm{M}_1^{-2}\Gamma_1). \tag{5.3}$$

The latter is far more unstable and significant than the acoustic mode at high Mach number (see also Brown et al 1991). The result (5.2) was also noted by Cowley & Hall (1990). A vorticity mode similar to (5.3) was found in the mixing layer considered by Balsa & Goldstein (1990).

VISCOSITY-TEMPERATURE LAW DEPENDENCE The foregoing asymptotic study was based on a linear viscosity-temperature law. Assuming $\mu \propto \sqrt{T}$, Blackaby et al (1992) found, instead of (5.2), for the acoustic mode:

$$\alpha = O(n\mathrm{M}_1^{-3/2}), \quad \alpha c_i/U_1 = O(\mathrm{M}_1^{-7/2}), \tag{5.4}$$

and for the vorticity mode, instead of (5.3):

$$\alpha = O(1), \quad \alpha c_I/U_1 = O(\mathrm{M}_1^{-2}). \tag{5.5}$$

The reasons for the change may be traced partly to the difference in the edge-layer behavior of the mean-flow structure for $\mu \propto T$ (Lee & Cheng 1969) and for $\mu \propto T^{\omega}$, $\omega < 1$ (Bush & Cross 1967).

INSTABILITY AT LARGE χ Blackaby et al also made an instability analysis for the flat-plate problem in the strong-interaction regime ($\chi \to \infty$), assuming $\mu \propto \sqrt{T}$, and found

$$\alpha = O(\mathrm{M}_1^{7-3\lambda}), \quad \alpha c_i/U_1 = O(\mathrm{M}_1^{\lambda-1}), \tag{5.6}$$

where $\lambda = 9\gamma/(6\gamma - 1)$, with which both wave number and amplification rates are seen to increase with M_1 for all $\gamma > 1$, quite unlike (5.2)–(5.5) which correspond to $\chi \to 0$. This is rather surprising since little supporting experimental evidence of instability can be found in the literature in this case.

THE FIRST-MODE/TS WAVES In a triple-deck formalism, Smith (1989) found that the parallel-flow assumption at high M_1 *cannot* hold for TS waves even at a rather low $\mathrm{M}_1 = O(\mathrm{Re}^{1/16})$, and that, in order to keep the TS waves effectively subsonic for their validity, they must be directed outside the wave-Mach-cone ψ, i.e. $\tan \psi > (\mathrm{M}_1^2 - 1)^{1/2}$. Cowley & Hall (1990) studied the influence of a shock on the TS waves in the hypersonic boundary layer over a wedge, using a triple-deck approach. To simplify the analysis they introduce a special kind of hypersonic Newtonian approximation, in which the entire shock-layer thickness becomes narrow enough to be comparable to the upper deck. Seddougui et al (1991) found an adverse wall-cooling influence on the *spatial* growth rate of the (viscous) TS mode, and showed destabilization of the otherwise stable modes.

NONLINEAR EVOLUTION OF THE ACOUSTIC MODE Nonlinear spatial evolution of the unstable waves in the acoustic mode was studied by Goldstein & Wundrow (1990). As seen from Equation (5.2), the amplification rate of this mode/submode at high M_1 is so weak that even a pressure fluctuation of the order $M_1^{-4}(\ln M_1)^{-1}$ suffices to initiate the nonlinear evolution.

5.4 *Hypersonic Boundary-Layer Transition*

Transition prediction requires the identification of the free-stream disturbance field, and determination of the boundary/shear-layer response, as well as of the (linear and nonlinear) amplifications of these internalized disturbances prior to the breakdown to turbulence. Following Malik et al (1990), the events up to the nonlinear breakdown will be called the stage of "transition onset," to be distinguished from the downstream "transitional zone" that follows.

TRANSITION-ONSET STAGE The e^N method and variants for locating the onset of transition may still work at high M_1 (though with N being substantially reduced from the magic "9"—see Malik et al 1990); their application requires knowledge of the more dominant first and second (if not all) instability modes, which provide the (spatial) amplification rate σ in

$$N = \ln(A_t/A_o) = \int_{x_o}^{x_t} \sigma(x)\,dx, \tag{5.7}$$

where A_o is the amplitude of the internalized disturbance at the onset of instability and the subscript "t" signifies the end of the transition-onset stage. An alternative to the e^N method is to assign an amplitude level to A_t at the end of the onset stage, instead of assigning a level for N which controls the amplitude ratio A_t/A_o. This calls for the computation of A_o from the free-stream disturbance amplitude via (linear) receptivity theory. A more viable method appears to be combining the receptivity, linear stability, and secondary instability theories [in the sense suggested by Herbert (1988)]. Görtler vortices in the boundary layer of a Mach 5 nozzle have been observed by Beckwith & Holly (1981); with Görtler vortices as primary disturbances, for example, the secondary instability waves can develop amplitudes comparable to the primary level at high M_1 (Spall & Malik 1989, Malik & Hussaini 1990) as was found at low M_1.

TRANSITIONAL ZONE MODELING In modeling the transition-zone flow, the Reynolds averaged Navier-Stokes (RANS) approach deals with an equation system developed from higher-order moments of the *ensemble*-averaged NS equations (e.g. Cebeci & Bradshaw 1988) whereas the Large-Eddy simulation (LES) models the turbulence in the subgrid scale and deals with numerical solutions to the *spatially*-averaged NS equations (e.g. Reynolds 1976, Lesieur 1990). Although the LES is still in a developing stage, its prospect as a reliable flow-data source seems high, even for NASP applications (Zang et al 1989). Using meshes and time steps small enough to resolve the Kolmogoroff scale, direct numerical simulation (DNS) may need no subgrid modeling and yield data with adequate details for some highly idealized, otherwise costly, transitional-flow computations. The latter are essential for the LES and RANS calibration. There is also an ONERA/CERT version of the RANS which receives encouraging support from comparison with DNS results (cf Malik et al 1990). Among the RANS arsenal in current development is the k-ω model (Wilcox 1988, 1991) which captures certain transitional-zone properties well.

Distinct from thc RANS and LES approaches, and perhaps more appealing, is the application of a *nonlinear* transition theory by Ng et al (1990) to the analysis of a secondary instability of the (primary) second mode on a cylinder at high M_1 in the early stages of the transition (cf Malik et al 1990, Figure 17). Their results on the Reynolds-stress profile agree quite well with the DNS data and reveal the predominance of a secondary instability in the vicinity of the critical layer attributed to a 3-D nonlinear effect. The analysis provides a possible explanation for the "rope-like structure" observed in the vicinity of the critical layer in hypersonic transition experiments (e.g. Potter & Whitfield 1965).

6. CONCLUDING REMARKS

In this review, an attempt was made to reflect on the current focus in hypersonic flow research by examining the recent works and their issues. The field is a diverse one even with the exclusion of nonequilibrium aerothermodynamics and rarefied hypersonic flows that underlie much of the modern design analyses for hypersonic flight; many works, particularly the more recent contributions abroad, have either escaped the writer's notice or were omitted for the sake of space. As noted earlier, a student of hypersonic viscous flow must face the turbulent transition and the transition to the collision-free limit, which represent two domains where unresolved issues in modeling and computation abound.

The present article does not cover the modern development in rarefied hypersonic flows corresponding to the latter transition. This and the recent work in nonequilibrium aerothermodynamics omitted here represent advances that have vastly expanded the scope of fluid/gas dynamics research and its applications. Along with these advances are critical issues on (*a*) how best to model the intermolecular and interatomic potentials in order to establish a concrete theory for determining the individual rates of state-to-state (rotational, vibrational, and electronic) transitions, and (*b*) the need for a thorough validation of the inelastic collision models in the particle-simulation techniques. These unresolved issues remain the major hurdle on the way to *quantitative* predictions. Outstanding among the key fluid dynamics problems in scramjet propulsion is perhaps turbulent mixing, which is compounded further with multi-scaled, complex reactions and their control.

Apart from many omissions, the view and interest expressed in this article have been limited to the theoretical side. However, the lack of pertinent experimental data in the energy and density ranges of interest is believed to be among the major obstacle to progress in hypersonic aerothermodynamics. Helpful in this respect are the flight experiments performed during the Space-Shuttle missions and the proposed Aero-assisted Flight Experiment Program (Hamilton et al 1991). Facilities for ground experiments in the form of free-piston shock tunnels—simulating nonequilibrium effects with enthalpy and Reynolds-number ranges corresponding to transatmospheric flight (Hornung 1988)—are being completed and preliminary testing has begun (Hornung et al 1992). Another study worthy of note, as well as theoretical support, is a nonequilibrium gas dynamics experiment with a hypersonic flow of iodine vapor which has low activation energies for vibrational-mode excitation and dissociation, and can be studied in the laboratory with relatively modest resources (Pham-Van-Diep et al 1992).

ACKNOWLEDGMENTS

This study has been supported by NASA/DOD Grant NAGW-1061 and by the AFOSR Math Information Science Program. Many individuals have helped the author in one way or another during the course of this review study; among them are S. N. Brown, D. Bushnell, J. A. Domaradzki, D. A. Erwin, M. M. Hafez, J. A. Kunc, C. J. Lee, R. E. Melnik, J. N. Moss, E. P. Muntz, C. Park, C. E. Treanor, P. L. Varghese, and H. T. Yang; also to be thanked are T. Austin, Y. Bao, C. Holguin, D. Wadsworth, D. Weaver, and E. Y. Wong for their invaluable assistance.

Literature Cited

Anderson, D. A., Tannehill, J. C., Pletcher, R. H. 1984. *Computational Fluid Mechanics and Heat Transfer*. Hemisphere

Anderson, J. D. Jr. 1989. *Hypersonic and High Temperature Gas Dynamics*. New York: McGraw-Hill

Anderson, J. D., Chang, J., McLaughlin, T. A. 1992. Hypersonic waveriders: effects of chemically reacting flow and viscous interaction. *AIAA Pap. 92-0302*

Anderson, J. D., Ferguson, F., Lewis, M. J. 1991. Hypersonic waveriders for high altitude applications. *AIAA Pap. 91-0530*

Anderson, J. D., Lewis, M. J., Corda, S., Blankson, I. M., eds. 1990. *Proc. 1st Int. Hypersonic Waverider Symp.*, Univ. Maryland, Oct. 17–19, 1990; to appear as *AIAA Prog. Astronaut. Aeronaut. Ser.*

Balsa, T. F., Goldstein, M. E. 1990. On the instabilities of supersonic mixing layers: a high Mach number asymptotic theory. *J. Fluid Mech.* 216: 585–611

Beckwith, I. E., Holley, B. B. 1981. Görtler vortices and transition in wall boundary layers of two Mach 5 nozzles. *NASA TP-1869*

Billig, F. S. 1992. Research on supersonic combustion. *AIAA Pap. 92-0001*

Blackaby, N. D., Cowley, S. J., Hall, P. 1992. On the instability of hypersonic flow past a flat plate. *J. Fluid Mech.* In press

Blottner, F. G. 1990. Accurate Navier-Stokes calculation for hypersonic flow over a spherical nosetip. *J. Spacecr. Rockets* 27: 113–22

Bowcutt, K. G., Anderson, J. D., Capriotti, D. 1987. Viscous optimized hypersonic waveriders. *AIAA Pap. 87-0272*

Brown, S. N., Cheng, H. K., Lee, C. J. 1990. Inviscid-viscous interaction on triple-deck scales in a hypersonic flow with strong wall cooling. *J. Fluid Mech.* 220: 309–37

Brown, S. N., Khorrami, A. F., Neish, A., Smith, F. T. 1991. On hypersonic boundary layer interaction and transition. *Philos. Trans. R. Soc. London Ser. A* 335: 139–52

Brown, S. N., Stewartson, K. S. 1975. A nonuniqueness of the hypersonic boundary layer. *Quart. J. Mech. Appl. Math.* 28: 75–90

Brown, S. N., Stewartson, K. S., Williams, P. G. 1975. Hypersonic self-induced separation. *Phys. Fluids* 18: 633–39

Burggraf, O. R., Rizzetta, D., Werle, M. J., Vasta, V. N. 1979. Effect of Reynolds number on laminar separation on a supersonic stream. *AIAA J.* 17: 336–43

Burnett, D. 1936. The distribution of molecular velocities and the mean motion in a nonuniform gas. *Proc. London Math. Soc. Ser. 2* 40(3): 382–430

Burns, B. R. A. 1989. Aerodynamic design challenge of a single stage to orbit, reusable launch vehicle. *Proc. Int. Conf. Hypersonic Aerodynamics*. Univ. Manchester, UK, Pap. No. 1

Bush, W. B. 1966. Hypersonic strong-interaction similarity solutions for flow past a flat plate. *J. Fluid Mech.* 25: 51–64

Bush, W. B., Cross, A. K. 1967. Hypersonic weak-interaction similarity solutions for flow past a flat plate. *J. Fluid Mech.* 29: 349–59

Butta, B. A., Song, D. J., Lewis, C. H. 1990. Nonequilibrium viscous hypersonic flows over ablating Teflon surfaces. *J. Spacecr. Rockets* 27: 194–204

Cebeci, T., Bradshaw, P. 1988. *Physical and Computational Aspects of Convective Heat Transfer*. Berlin: Springer-Verlag

Chapman, D. R. 1992. A perspective on aerospace CFD. *Aerosp. Am.* Jan. 16–59

Cheng, H. K. 1963. The blunt-body problem in hypersonic flow at low Reynolds number. *Cornell Aero. Lab. Rep. AF-1285-A-10*

Cheng, H. K. 1966. Viscous hypersonic blunt-body problem and the Newtonian theory. In *Fundamental Phenomena in Hypersonic Flow*, ed. G. J. Hall, pp. 91–132. Ithaca: Cornell Univ. Press

Cheng, H. K. 1990. Book review: Hypersonic and High-Temperature Gas Dynamics by J. D. Anderson, Jr. *AIAA J.* 28: 766–68

Cheng, H. K. 1992. Perspectives on hypersonic viscous and nonequilibrium hypersonic flow research. *Univ. So. Calif. Dept. Aerospace Eng. Rep. USCAE 151*

Cheng, H. K., Lee, C. J., Wong, E. Y., Yang, H. T. 1989. Hypersonic slip flows and issues on extending continuum model beyond the Navier-Stokes level. *AIAA Pap. 89-1663*

Cheng, H. K., Wong, E. Y., Dogra, V. K. 1991. A shock-layer theory based on thirteen-moment equations and DSMC calculations of rarefied hypersonic flows. *AIAA Pap. 91-0783*

Cheng, S. I. 1989. Hypersonic combustion. *Proc. Energy Combust. Sci.* 15: 183–202

Cheung, S., Cheer, A., Hafez, M., Flores, J. 1991. Convergence acceleration of viscous and inviscid hypersonic flow calculations. *AIAA J.* 29: 1214–23

Corda, S., Anderson, J. D. Jr. 1988. Viscous optimized hypersonic waveriders designed from axisymmetric flow fields. *AIAA Pap. 88-0369*

Cowley, S. J., Hall, P. 1990. The instability of hypersonic flow past a wedge. *J. Fluid Mech.* 214: 17–42

Cox, R. N., Crabtree, L. F. 1965. *Elements of Hypersonic Aerodynamics*. Cambridge: Cambridge Univ. Press

Davis, R. T. 1970. Numerical solution of the hypersonic viscous shock-layer equations. *AIAA J.* 8: 843–51

Demetriades, A. 1978. New experiments on hypersonic boundary layer stability including wall temperature effects. *AIAA Pap. 74-535* (1974)

Edney, B. 1968. Anomalous heat transfer and pressure distributions on blunt bodies

at hypersonic speeds in the presence of an impinging shock. *Aeronaut. Res. Inst. Sweden, Stockholm, Rep. 115*; also *AIAA J.* 6: 15–21
Edwards, T. A., Flores, J. 1990. Computational fluid dynamics nose-to-tail capability: hypersonic unsteady Navier-Stokes code validation. *J. Spacecr. Rockets* 27: 123–30
Eggers, A. J., Ashley, H., Springer, G. 1990. Waverider configuration from the 1950's to the 1990's. *Proc. 1st Int. Hypersonic Waverider Symp.*, Univ. Maryland, Oct. 17–19, 1990
Gajjar, J., Smith, F. T. 1983. On hypersonic self-induced separation: hydraulic jumps and boundary layers with algebraic growth. *Mathematika* 30: 77–93
Glass, C. E., Weiting, A. R., Holden, M. S. 1989. Swept shock-on-lip—A comparison of analytical and experimental results. *NASA TN 1085*
Gnoffo, P. A. 1990. Code calibration program in support of the aeroassisted flight experiment. *J. Spacecr. Rockets* 27: 131–42
Goldstein, M. E., Wundrow, D. W. 1990. Spatial evolution of nonlinear acoustic mode instabilities on hypersonic boundary layers. *J. Fluid Mech.* 219: 585–607
Greendyke, R. B., Gnoffo, P. A., Lawrence, R. W. 1992. Electron number density profile for the aeroassist flight experiment. *AIAA Pap. 92-0804*
Gupta, R. N., Lee, K. P., Zobby, E. V. 1992. Enhancements to viscous-shock-layer technique. *AIAA Pap. 92-2897*
Hamilton, H. H., Gupta, R. N., Jones, J. J. 1991. Flight stagnation-point heating calculations of AFE vehicle. *J. Spacecr. Rocket* 28: 125–28
Harten, A. 1983. High resolution schemes for hypersonic conservation laws. *J. Comput. Phys.* 49: 357–93
Hayes, W. D., Probstein, R. F. 1959. *Hypersonic Flow Theory*. New York: Academic
Herbert, Th. 1988. Secondary instability of boundary layers. *Annu. Rev. Fluid Mech.* 20: 487–526
Hoffman, K. A. 1989. *Computational Fluid Dynamics for Engineers*. Austin, Tex.: Eng. Ed. Sys.
Holden, M., Moselle, J. R. 1969. Theoretical and experiment studies of the shock wave-boundary layer interaction on compression surfaces in hypersonic flow. *CALSPAN Rep. AF-2410-A-1*
Holden, M., Wieting, A. R., Moselle, J., Glass, C. 1988. Studies of aerothermal loads generated in regions of shock/shock interaction in hypersonic flow. *AIAA Pap. 88-0477*; also *NASP TN 1085* (1989)
Hornung, H. 1988. Lanchester memorial lecture—experimental real-gas hypersonics. *Aeronaut. J.* 92: 379–89; also *Z. Flugwiss. Weltraumf* 12: 293
Hornung, H., Sturtevant, B., Bélanger, J., Sanderson, S., Brouillette, M., Jenkine, M. 1992. Performance data of the new free-piston shock tunnel T5 at GALCIT. Cal. Inst. Tech. Grad. Aero. Lab. Memo.
Hung, C. M., Barth, T. J. 1990. Computation of hypersonic flow through a narrow expansion slot. *AIAA J.* 28: 229–35
Hung, C. M., MacCormack, R. W. 1975. Numerical solutions of supersonic and hypersonic laminar flows over a two-dimensional compression corner. *AIAA Pap. 75-2*
Ito, T., Akimoto, H., Miyaba, H., Kano, Y., Suzuki, N., Sasaki, H. 1990. Concept and technology development for HOPE spaceplane. *AIAA Pap. 90-5223*
Kang, S. H., Kunc, J. A. 1991. Viscosity of high-temperature iodine. *Phys. Rev. A* 44: 3596–3604
Kendall, J. M. 1975. Wind tunnel experiments relating to supersonic and hypersonic boundary layer transition. *AIAA J.* 13: 240–99; see also *Aerosp. Corp. Rep. BSD-TR-67-213,2* (1967)
Koelle, D. 1990. Sänger advanced space transportation system—progress report 1990. *AIAA Pap. 90-5200*
Krawczyk, W., Harris, T., Rajendran, N., Carlson, D. 1989. Progress in the development of PNS technology for external and internal flows. *AIAA Pap. 89-1828*
Küchemann, D. 1978. *The Aerodynamic Design of Aircraft*. Oxford: Pergamon
Lee, R. S., Cheng, H. K. 1969. On the outer-edge problem of a hypersonic boundary layer. *J. Fluid Mech.* 38: 161–79
Lee, S. H., Deiwert, G. S. 1990. Flux-vector splitting calculation of nonequilibrium hydrogen-air reactions. *J. Spacecr. Rockets* 27: 167–74
Lees, L., Lin, C. C. 1946. Investigation of the stability of the laminar boundary layers in a compressible fluid. *NACA TN 1115*
Lesieur, M. 1990. *Turbulence in Fluids*. Dordrecht: Kluwer
Lighthill, M. J. 1953. On boundary layer and upstream influence II. Supersonic flow without separation. *Proc. R. Soc. London Ser. A* 217: 478–507
Liu, F., Jameson, A. 1992. Multi-grid Navier-Stokes calculations for 3-D cascades. *AIAA Pap. 92-0190*
Lozino-Lozinsky, Ye. G., Neiland, V. Ya. 1989. The convergence of the Buran orbiter flight test and preflight study results and the choice of a strategy to develop a second-generation orbiter. *AIAA Pap. 89-5019*
MacCormack, R. W. 1982. A numerical method for solving the equations of com-

pressible viscous flow. *AIAA J.* 20: 1275–81
MacCormack, R. W. 1990. Solution of the Navier-Stokes equations in three dimensions. *AIAA Pap. 90-1520*
MacCormack, R. W., Baldwin, B. S. 1975. A numerical method for solving the Navier-Stokes equations with application to shock-boundary layer interaction. *AIAA Pap. 75-1*
MacCormack, R. W., Candler, G. V. 1989. The solution of the Navier-Stokes equations using Gauss-Seidel line relaxation. *Comput. Fluids* 17: 135–50
Mack, L. M. 1984. Boundary layer linear stability theory. *AGARD Rep. 709*; also see *AIAA J.* 13: 278–89 (1975)
Mack, L. M. 1987a. Review of linear compressible stability theory. In *Stability of Time Dependent and Spatially Varying Flow*, ed. D. L. Dwoyer, M. Y. Hussaini, pp. 164–87. New York: Springer-Verlag
Mack, L. M. 1987b. Stability of axisymmetric boundary layer on sharp cones at hypersonic Mach number. *AIAA Pap. 87-1413*
Malik, M. R., Hussaini, M. Y. 1990. Numerical simulation of interactions between Görtler vortices and Tollmien-Schlichting waves. *J. Fluid Mech.* 210: 183–99
Malik, M. R., Zang, T., Bushnell, D. 1990. Boundary layer transition in hypersonic flows. *AIAA Pap. 90-5232*
Mehta, U. B. 1990. Computational requirements for hypersonic performance estimates. *AIAA J.* 27: 103–12
Mikhailov, V. V., Neiland, V. Ya., Sychev, V. V. 1971. The theory of viscous hypersonic flow. *Annu. Rev. Fluid Mech.* 3: 371–96
Moin, P. 1992. The computation of turbulence. *Aerosp. Am.* Jan: 42–46
Moore, F. K. 1964. Hypersonic boundary layer theory. In *Theory of Laminar Flow*, ed. F. K. Moore, Sec. E: 439–527. Princeton: Princeton Univ. Press
Moss, J. N. 1976. Radiative shock layer solutions with coupled ablation injection. *AIAA J.* 14: 1311–17
Moss, J. N., Bird, G. A. 1985. Direct simulation of transitional flow for hypersonic re-entry conditions. *Progr. Astro. Aeronaut.* 96: 113–39
Moss, J. N., Cuda, V., Simmonds, A. L. 1987. Nonequilibrium effects for hypersonic transitional flows. *AIAA Pap. 87-0404*
Ng, L., Erlebacher, G., Zang, T. A., Pruett, D. 1990. Compressible secondary instability theory—parametric studies and prospects for predictive tool. *8th NASP Symp., Pap. no. 23*
Neiland, V. Ya. 1970. Propagation of perturbation upstream with interaction between a hypersonic flow and a boundary layer. *Akad. Nauk. SSSR* 3: 19
Nonweiler, T. R. F. 1963. Delta wings of shapes amenable to exact shock-wave theory. *J. R. Aeronaut. Soc.* 67: 39
Nonweiler, T. R. F. 1990. The waverider wings in retrospect and prospect. *Proc. 1st Int. Waverider Symp.*, Univ. Maryland, Oct. 17–19, 1990
Nonweiler, T. R. F., Wang, H. Y., Aggarwall, S. R. 1971. The role of heat conduction in leading edge heating. *Ing. Arch.* 40: 107; also see *Aero. Res. Council C. D. 1126* (1970)
Park, C. 1990. *Nonequilibrium Hypersonic Aerothermodynamics*. New York: Wiley
Parkinson, R., Conchie, P. 1990. HOTOL. *AIAA Pap. 90-5201*
Pham-Van-Diep, G. C., Muntz, E. P., Weaver, D., Dewitt, T. G., Bradley, M. K., et al. 1992. An iodine hypersonic wind tunnel for study of nonequilibrium reacting flows. *AIAA Pap. 92-0566*
Potter, L. J. 1988. Procedure for estimating aerodynamics for 3-D bodies in transitional flow. *Prog. Aeronaut. Astronaut.* 118
Potter, L. J., Whitfield, J. D. 1965. Boundary layer transition under hypersonic conditions. *AGARDograph 97*, pb. 3: 1–61
Pulliam, T. H., Steger, J. L. 1980. Implicit finite-difference simulations of three-dimensional compressible flow. *AIAA J.* 18: 159–67
Reshotko, E. 1976. Boundary-layer stability and transition. *Annu. Rev. Fluid Mech.* 8: 311–50
Reynolds, W. C. 1976. Computation of turbulent flows. *Annu. Rev. Fluid Mech.* 8: 183–208
Rizzetta, D. P., Burggaf, O. R., Jenson, R. 1978. Triple-deck solution for viscous supersonic and hypersonic flow past corners. *J. Fluid Mech.* 89: 535
Roe, P. L. 1986. Characteristic-based schemes for the Euler equations. *Annu. Rev. Fluid Mech.* 18: 337–65
Rudy, D. H., Thomas, J. L., Kumar, A., Gnoffo, P. A., Chakravathy, S. R. 1991. Computation of laminar hypersonic compression-corner flows. *AIAA J.* 29: 1108–13
Ryan, J. S., Flores, J., Chow, C. Y. 1990. Development and validation of a Navier-Stokes code for hypersonic external flow. *J. Spacecr. Rockets* 27: 160–66
Sanzero, G. 1990. *AIAA Pap. 90-5264*
Seddougui, T. R. F., Wang, H. Y., Aggarwal, S. R. 1991. Surface-cooling effects on compressible boundary-layer instability, and on upstream influence. *Eur. J. Mech. B/Fluids* 10: 117–45

Shang, J. S., Scherr, S. J. 1986. Navier-Stokes solution for a complete re-entry configuration. *J. Aircr.* 23: 881–88

Simen, M., Dallmann, U. 1992. On instability of hypersonic flow past a pointed cone—comparison of theoretical and experimental results at Mach 8. *AGARD Symp. Theor. Exp. Methods Hypersonic Flows*, Torino, Italy, May 4–7, 1992

Smith, F. T. 1982. On the high Reynolds number theory of laminar flow. *IMA J. Appl. Math.* 82: 207–81

Smith, F. T. 1986. Steady and unsteady boundary-layer separation. *Annu. Rev. Fluid Mech.* 18: 197–220

Smith, F. T. 1989. On first-mode instability in subsonic, supersonic and hypersonic boundary layers. *J. Fluid Mech.* 198: 127–53

Smith, F. T., Brown, S. N. 1990. The inviscid instability of a Blasius boundary layer at large values of the Mach number. *J. Fluid Mech.* 219: 499–518

Spall, R. E., Malik, M. R. 1989. Görtler vortices in supersonic and hypersonic boundary layers. *Phys. Fluid A* 1: 1822–35

Steger, J. L., Warming, R. F. 1981. Flux vector splitting of the inviscid gasdynamic equations with application to finite-difference methods. *J. Comput. Phys.* 40: 263–93; see also *NASA TM D-78605*

Stetson, K. F. 1988. On nonlinear aspects of hypersonic boundary-layer stability. *AIAA J.* 26: 883–85

Stetson, K. F., Kimmel, R. L. 1992. On hypersonic boundary-layer stability. *AIAA Pap. 92-0737*

Stewartson, K. S. 1955. On motion of a flat plate at high speed in a viscous compressible fluid: part II, steady motion. *J. Aeronaut. Sci.* 22: 303–9

Stewartson, K. S. 1974. Multistructure boundary layers on flat plate and related bodies. *Adv. Appl. Mech.* 14: 146–239

Stewartson, K. S. 1981. D'Alembert's paradox. *SIAM Rev.* 23: 308–43

Stollery, J. L. 1990. A review of force measurement on delta and caret wings made at Imperial College. *Proc. 1st Int. Hypersonic Waverider Symp.*, Univ. Maryland, College Park Md., Oct. 17–19, 1990

Stookesberry, D. C., Tannehill, J. C. 1986. Computation of separated flow on a ramp using the space marching supra-characteristics method. *AIAA Pap. 86-0564*

Sychev, V. V. 1987. *Asymptotic Theory of Separated Flows* (in Russian). Moscow: Moscow Sci. Pub., Physico-Math. Lit. (Distrib. USSR Nat. Comm. Theor. Appl. Mech.)

Tannehill, J. C., Buelow, P. E., Levalts, J. O., Lawrence, S. L. 1990. Three-dimensional upwind parabolized Navier-Stokes codes for real-gas flows. *J. Spacecr. Rockets* 27: 150–59

Thomas, P. D., Neier, K. L. 1990. Navier-Stokes simulation of three-dimensional hypersonic equilibrium flows with ablation. *J. Spacecr. Rockets* 27: 143–49

Townend, L. H. 1991. Research and design of hypersonic aircraft. *Philos. Trans. R. Soc. London Ser. A* 335: 201–24

Treanor, C. E. 1991. Book review: Nonequilibrium Hypersonic Aerothermodynamics, by C. Park. *AIAA J.* 29: 857–58

van Leer, B. 1982. Flux-vector splitting for the Euler equations. *Lec. Notes Phys.* 170: 507–12

van Leer, B., Thomas, J. L., Roe, P. L., Newsome, R. W. 1987. A comparison of numerical flux formulas for the Euler and Navier-Stokes equations. *AIAA Pap. 87-1104-CP*

Vatsa, V. N., Thomas, J. L., Wedan, B. W. 1989. Navier-Stokes computations of a prolate spheroid at angle of attack. *J. Aircr.* 26: 986–93

Vigneron, Y. C., Rakich, J. V., Tannehill, J. C. 1978. Calculation of supersonic viscous flow over delta wings with sharp subsonic leading edges. *AIAA Pap. 78-1137*

Walberg, G. D. 1985. A survey of aeroassisted orbital transfer. *J. Spacecr.* 22(1): 3–18; also see *AIAA Pap. 82-1370*

Warr, J. 1970. Orbital aerodynamic computer program to calculate force and moment coefficients on complex vehicle configurations. *LMSC/HREC D 1624 98 TM 54-20-275*, Lockheed, Houston, TX.

Weiting, A. 1990. Shock interference heating in scramjet engines. *AIAA Pap. 90-5238*

Werle, M. J., Dwoyer, D. L., Hankey, W. L. 1973. Initial conditions for the hypersonic/boundary-layer problems. *AIAA J.* 11: 525–30

Werle, M. J., Vasta, V. N. 1974. A new method for supersonic boundary layer separation. *AIAA J.* 12: 1491–97

Wilcox, D. C. 1988. Reassessment of the scale determining equations for advanced turbulence models. *AIAA J.* 26: 1299–1310

Wilcox, D. C. 1991. Progress in hypersonic turbulence modelling. *AIAA Pap. 91-1785*

Wilhite, A. W., Arrington, J. P., McCandless, R. S. 1985. Performance aerodynamics of AOTV. In *Thermal Design of Aeroassisted Orbital Transfer Vehicle*, ed. H. F. Nelson, *Progr. Astronaut. Aeronaut.* 96: 165–97

Williams, R. M. 1986. National Aerospace Plane technology for America's future. *Aerosp. Am.* 24: 18–22

Wilson, G. J., MacCormack, R. W. 1992. Modeling supersonic combustion using a fully implicit numerical method. *AIAA J.* 30: 1008–15

Yee, H. C. 1987. Upwind and symmetric shock-capturing schemes. *NASA Tech. Memo. 89464*; also see *NASA Tech. Memo. 100097*

Ying, S. X., Steger, J. L., Shift, L. B., Baganoff, D. 1986. Numerical simulation of unsteady, viscous, high-angle-of-attack flows using a partially flux split algorithm. *AIAA Pap. 86-2179*; also see Ying, S. X. 1986. PhD thesis. Stanford Univ.

Zang, T. A., Dinavahi, S., Piomelli, U. 1989. Reynolds-averaged and subgrid-scale models of transitional flows. *7th NASP Symp.*, Pap. no. 26

Zhong, X., MacCormack, R. W., Chapman, D. R. 1991a. Stabilization of the Burnett equations and application to high-altitude hypersonic flows. *AIAA Pap. 91-0770*

Zhong, X., MacCormack, R. W., Chapman, D. R. 1991b. Evaluation of slip boundary conditions for the Burnett equations with application to hypersonic leading edge flow. *Proc. 4th Int. Symp. Comput. Fluid Dynamics*, Davis, Calif.

Annu. Rev. Fluid Mech. 1993. 25: 485–537

AERODYNAMICS OF ROAD VEHICLES

Wolf-Heinrich Hucho

Ostring 48, D-6231, Schwalbach (Ts), Germany

Gino Sovran

General Motors Research and Environmental Staff, Warren, Michigan 48090-9055

KEY WORDS: aerodynamic design, aerodynamic testing, aerodynamic forces, flow fields

1. INTRODUCTION

In fluid mechanical terms, road vehicles are bluff bodies in very close proximity to the ground. Their detailed geometry is extremely complex. Internal and recessed cavities which communicate freely with the external flow (i.e. engine compartment and wheel wells, respectively) and rotating wheels add to their geometrical and fluid mechanical complexity. The flow over a vehicle is fully three-dimensional. Boundary layers are turbulent. Flow separation is common and may be followed by reattachment. Large turbulent wakes are formed at the rear and in many cases contain longitudinal trailing vortices.

As is typical for bluff bodies, drag (which is a key issue for most road vehicles—but far from the only one) is mainly pressure drag. This is in contrast to aircraft and ships, which suffer primarily from friction drag. The avoidance of separation or, if this is not possible, its control are among the main objectives of vehicle aerodynamics.

With regard to their geometry, road vehicles comprise a large variety of configurations (Figure 1). Passenger cars, vans, and buses are closed, single bodies. Trucks and race cars can be of more than one body. Motorcycles and some race cars have open driver compartments. With the race car being the only exception, the shape of a road vehicle is not primarily

0066–4189/93/0115–0485$02.00

Figure 1 With respect to geometry, road vehicles comprise a wide variety of shapes. Race cars and, even more so, motorcycles have to be studied with the driver in place.

determined by the need to generate specific aerodynamic effects—as, for instance, an airplane is designed to produce lift.

To the contrary, a road vehicle's shape is primarily determined by functional, economic and, last but not least, aesthetic arguments. The aerodynamic characteristics are not usually generated intentionally; they are the consequences of, but not the reason for, the shape. These "other than aerodynamic" considerations place severe constraints on vehicle aerodynamicists. For example, there are good reasons for the length of a vehicle being a given. Length for a passenger car is a measure of its size, and thus its class. To place a car in a specific market niche means recognizing length as an invariant in design. Furthermore, mass and cost are proportional to length. In the same sense all the other main dimensions of a vehicle, such as width and height (which define frontal area), are frozen very early in the design process. Even the details of a car's proportions are prescribed to close limits for reasons of packaging and aesthetics (Figure 2). Of course, some maneuvering room must be left to the aerodynamicists (the hatched regions). Otherwise, they would do no more than just measure the aerodynamic characteristics of configurations designed by others.

Depending on the specific purpose of each type of vehicle, the objectives of aerodynamics differ widely. While low drag is desirable for all road vehicles, other aerodynamic properties are also significant. Negative lift is decisive for the cornering capability of race cars, but is of no importance for trucks. Cars and, even more so, vans are sensitive to cross wind, but heavy trucks are not. Wind noise should be low for cars and buses, but is of no significance for race cars.

While the process of weighing the relative importance of a set of needs from various disciplines is generally comparable to that in other branches of applied fluid mechanics, the situation in vehicle aerodynamics is unique in that an additional category of arguments has to be taken into account: art, fashion, and taste. In contrast to technical and economic factors, these additional arguments are subjective in nature and cannot be quantified.

Exterior design (the term "styling" that was formerly used is today usually avoided) has to be recognized as extremely important. "Design is what sells" rules the car market worldwide. While design gives technical requirements a form that is in accord with fashion, the fundamental nature

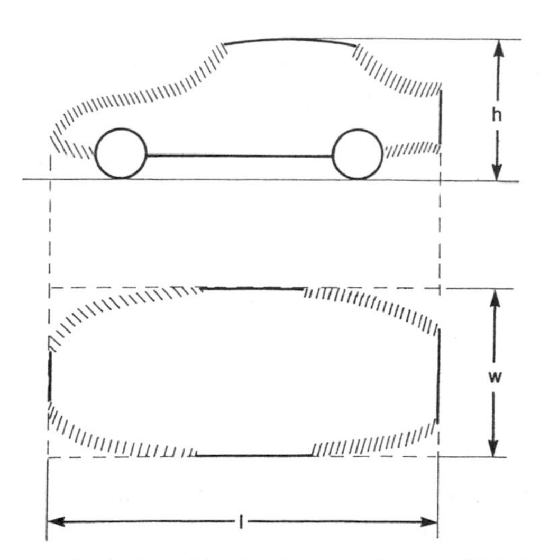

Figure 2 Right from the beginning of the development of a new vehicle, its main dimensions and detailed proportions are frozen. The limited maneuvering room for aerodynamics is identified by the hatched lines.

of fashion is change. Consequently, although vehicle aerodynamics is getting better and better, it is not progressing toward a single ultimate shape as in the case, for instance, of subsonic transport aircraft. To the contrary, it must come to terms with new shapes again and again.

There is no question, however, that aerodynamics does influence design. The high trunk typical of notchback cars with low drag is the most striking example. Despite the fact that it tends to look "bulky," it had to be accepted by designers because of its favorable effect on drag—and the extra luggage space it provides. Today's cars are streamlined more than ever, and an "aero-look" has become a styling feature of its own.

2. HISTORY

When the carriage horse was replaced by a thermal engine more than 100 years ago, nobody thought about aerodynamics. The objective of the body shell of the now horseless carriage was, as before, to shelter the driver and passengers from wind, rain, and mud. The idea of applying aerodynamics to road vehicles came up much later, after flight technology had made considerable progress. For both airships and aircraft, streamlined shapes were developed which lowered drag significantly, thus permitting higher cruising speeds with any given (limited) engine power.

The early attempts (Figure 3) to streamline cars were made according to aeronautical practice and by adapting shapes from naval architecture. These failed for two reasons. First, the benefits of aerodynamics were simply not needed. Bad roads and low engine power only permitted moderate driving speeds. Second, the approach of directly transplanting (with almost no change) shapes which had been developed for aeronautical and marine purposes was not appropriate. These streamlined shapes could be accommodated only if some important details of car design were subordinated, e.g. engine location, or the layout of the passenger compartment.

The long road from those days to today's acceptance of aerodynamics in the automobile industry has been described in great detail (Kieselbach 1982a,b, 1983; Hucho 1987b). From this history, only those events which were decisive will be highlighted here. Acknowledging the danger of being superficial, only five will be identified.

1. The recognition that the pattern of flow around half a body of revolution is changed significantly when that half body is brought close to the ground (Klemperer 1922, Figure 4).
2. The truncation of a body's rear end (Koenig-Fachsenfeld et al 1936, Kamm et al 1934, Figure 4).

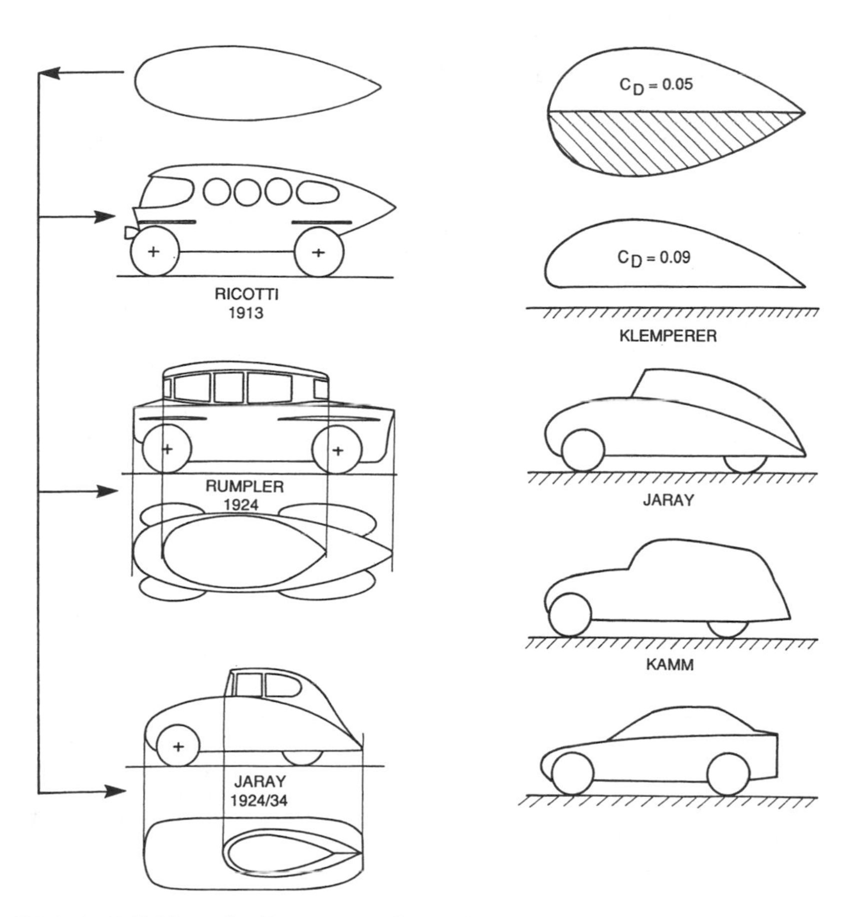

Figure 3 (*left*) The early attempts to apply aerodynamics to road vehicles consisted of the direct transfer of shapes originating from aeronautical and marine practice. The resulting shapes differed widely from those of contemporary cars and were rejected by the buying public. This unsuitable transfer procedure was very embarrassing for later attempts to introduce aerodynamics into vehicles.

Figure 4 (*right*) Klemperer (1922) recognized that the flow over a body of revolution, which is axisymmetric in free flight, changed drastically and lost symmetry when the body came close to the ground. By modifying its shape, however, he was able to reduce the related drag increase. Despite their extreme length, flow separates from the rear of streamlined cars. By truncating the rear shortly upstream of the location where separation would take place, shapes of acceptable length were generated with no drag penalty. This idea was first proposed by Koenig-Fachsenfeld for buses, and was transferred to cars by Kamm.

3 The introduction of "detail-optimization" into vehicle development (Figure 5, Hucho et al 1976).
4. The deciphering of the detailed flow patterns at car rear ends (Section 4.1).
5. The application of "add-ons" like underbody air dams, fairings, and wings to passenger cars, trucks, and race cars.

With these five steps, aerodynamics has been adapted to road vehicles, rather than road-vehicle configurations being determined by the demands of aerodynamics. The shape of cars changed in an evolutionary rather than a revolutionary manner over the years (Figure 6), and at first for reasons other than aerodynamic ones. Taste, perhaps influenced by the fascinating shapes of aircraft, called for smooth bodies with integrated headlamps and fenders (the "pontoon body"), and production technology made them possible. Better flow over the car and thus lower drag was only a spinoff. But, finally, the two oil crises of the 1970s generated great pressure for improving fuel economy drastically, and provided a break-

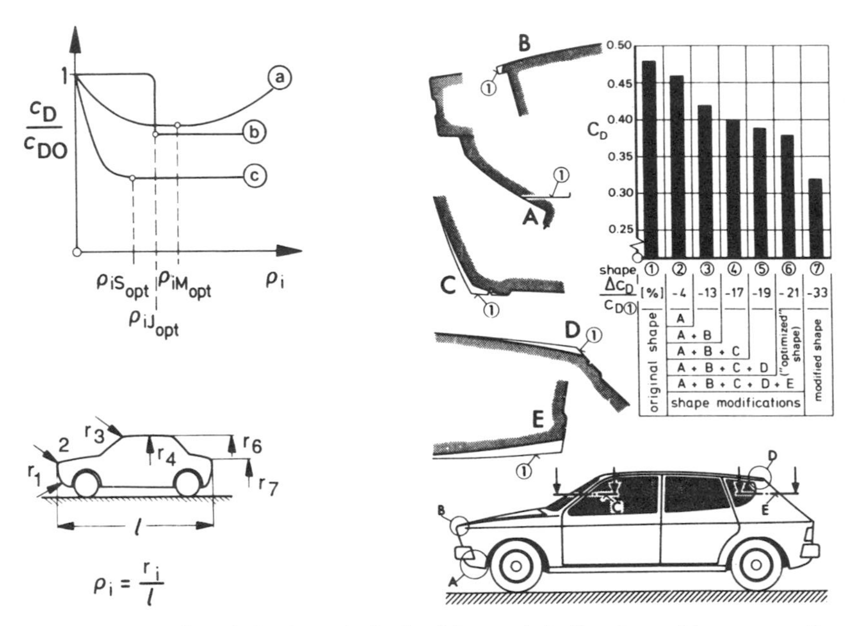

Figure 5 In detail optimization, a body detail is rounded off or tapered by no more than what is necessary to produce a drag minimum. In general, there are different types of drag variation that lead to such an optimum: (*a*) minimum; (*b*) jump; (*c*) saturation. Following this philosophy, it has been possible to significantly reduce the drag of hard-edged cars without altering their style.

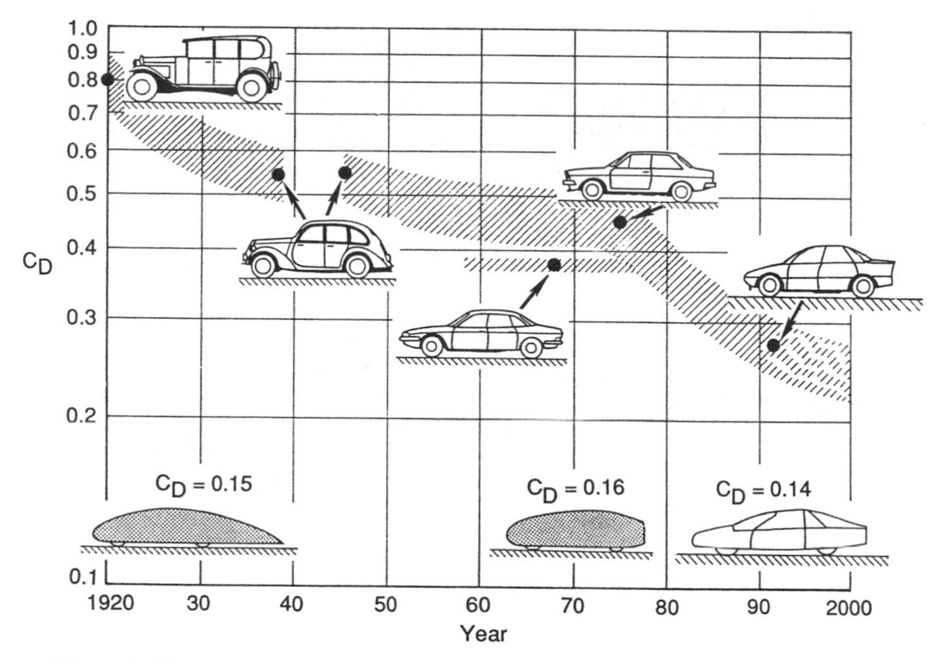

Figure 6 The drag history of cars. Using a logarithmic scale for drag emphasizes how difficult it is to achieve very low drag values. Research has been far ahead of what has been realized in production.

through for vehicle aerodynamics. Since then, drag coefficients have come down dramatically. This has been a major contributor to the large improvements in fuel economy that have been realized.

Research in road-vehicle aerodynamics has always been far ahead of practical application. Drag values demonstrate this. A drag coefficient as low as $C_D = 0.15$ was demonstrated for a body with wheels as early as 1922 (Klemperer, Figure 6), but it took more than 40 years to reproduce this value with an actual car—and then only with a research vehicle. Nevertheless, blaming the automobile industry for not taking advantage of the full potential of this technology is not justified, since the "concept" of any car is influenced by a variety of factors (Figure 7) which, collectively, are summarized by the term "market." However, the long-known potential for reducing drag (which relates to one of these factors) is now being exploited more and more. How far this trend will go depends on the future course of fuel prices and, perhaps, of emissions regulation (e.g. the possible regulation of CO_2 to control global warming).

In the following, the subject of road-vehicle aerodynamics will be treated

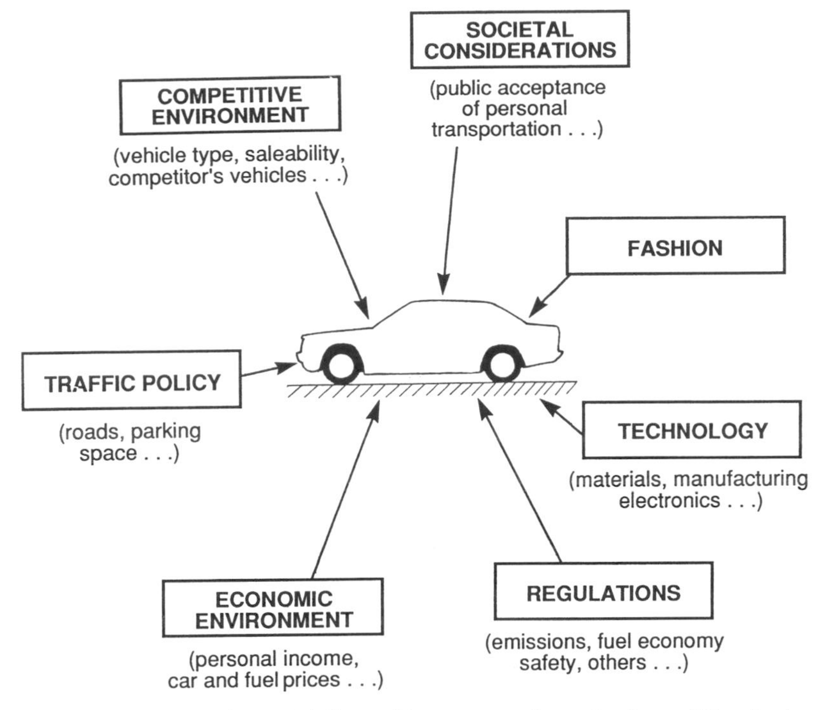

Figure 7 The concept of a car is influenced by many requirements of very different nature. A careful balance between them is required by the market.

in four sections. In the first (Section 3), the way that aerodynamics influences the operation of vehicles will be described, and without delving into the related fluid-mechanic mechanisms. These mechanisms will be discussed in a second section (Section 4). Then (Section 5), the aerodynamic-development process and the tools that are used in it will be described. In the final section (Section 6), open issues that remain and future trends will be discussed.

3. VEHICLE ATTRIBUTES AFFECTED BY AERODYNAMICS

3.1 *Performance and Fuel Economy*

The motivation for allowing aerodynamics to influence the shape of vehicles, if not their style, is the market situation, and this changes with time. Fuel economy and, increasingly, global warming are the current key arguments for low drag worldwide. In Europe, particularly Germany, top speed is still considered an important sales feature despite the rapidly increasing traffic density which largely prohibits fast driving even in the absence of speed limits.

Vehicle fuel consumption is a matter of demand and supply. On the demand side are the mechanical energies required for propulsion and by accessories. On the supply side is the efficiency with which this energy can be generated by the powerplant and delivered to the points of application. The influence of aerodynamics on this demand-supply relationship is through the drag force, which affects the propulsive part of the demand side.

Commercial aircraft, trains, ships, and highway trucks typically operate at a relatively constant cruising speed. In typical automobile driving, however, vehicle speed varies with time or distance. An analysis of the factors affecting automobile fuel economy can best be made if the driving pattern is prescribed. In the U.S., two speed variations of particular relevance are the Environmental Protection Agency (EPA) Urban and Highway schedules that are the basis for the fuel-economy and exhaust-emissions regulations. They represent the two major types of driving, and a combination of their fuel consumptions is used for regulatory purposes. In Europe, regulation is based on the Euromix cycle, a combination of a simple urban schedule and two constant-speed cruising conditions.

At any instant, the tractive force required at the tire/road interface of a car's driving wheels is (Sovran & Bohn 1981)

$$F_{TR} = \underbrace{R+D}_{\text{Road Load}} + \underbrace{M\frac{dV}{dt}}_{\text{Inertia}} + \underbrace{Mg\sin\theta}_{\text{Grade}},$$

where F_{TR} is the tractive force, R the tire rolling resistance, D the aerodynamic drag, M the vehicle mass, g the acceleration of gravity, and θ the inclination angle of the road.

The corresponding tractive power is

$$P_{TR} = F_{TR}V,$$

and the tractive energy required for propulsion during any given driving period is

$$E_{TR} = \int_0^T P_{TR}\,dt$$

for positive values of the integrand. The main reason that fuel is consumed in an automobile is to provide this tractive energy.

Writing an equation for instantaneous fuel consumption, integrating it over a total driving duration, and using the mean-value theorem to introduce appropriate averages for some of the integrands, the following fundamental equation for the average fuel-consumed-per-unit-distance-traveled, $\tilde{g}$, can be obtained (Sovran 1983):

$$\tilde{g} = \frac{k}{\eta_b \eta_d S} \underbrace{[E_{TR} + E_{ACC}]}_{\substack{\text{propulsion} \\ (P_{TR} > 0)}} + \underbrace{\quad g_u \quad}_{\substack{\text{braking and} \\ \text{idle}\,(P_{TR} \leq 0)}},$$

where k is a fuel-dependent constant, η_b is the average engine efficiency during propulsion, η_d is the average drivetrain efficiency, S is the total distance traveled, E_{ACC} is the energy required by vehicle accessories, and g_u is the fuel consumption during idling and braking. Typical units are gallons per mile in the U.S. and liters per 100 kilometers in Europe.

Aerodynamic drag is responsible for part of E_{TR}. However, E_{TR} is only part of the propulsive fuel consumption which is only part of the total fuel consumption. The impact of drag on total vehicle fuel consumption therefore depends on the relative magnitudes of these contributions.

For the U.S. driving schedules, E_{TR} can be described accurately (Sovran & Bohn 1981) by linear equations of the form

$$\frac{E_{TR}}{S} = [\underbrace{\;\alpha r_0\;}_{\text{Tire}} + \underbrace{\;\gamma\;}_{\text{Inertia}}]M + \underbrace{\beta(C_D A)}_{\text{Aero}},$$

where the vehicle descriptors M, C_D, A, and r_0 are the mass, drag coefficient, frontal area, and tire-rolling-resistance coefficient, respectively, and α, β, and γ are known constants which are different for each schedule. For a typical mid-size American car, drag is responsible for 18% of E_{TR} on the Urban schedule and 51% on the Highway.

The energy part (square brackets) of the propulsion term in the fuel-consumption equation is dominated by E_{TR}, which contributes $\simeq$94% for both Urban and Highway for the midsize car. The propulsion term itself accounts for $\simeq$81% of the total fuel consumption on the Urban schedule and $\simeq$96% on the Highway.

These quantifications permit the influence coefficient relating a percentage change in $C_D A$ to a percentage change in $\tilde{g}$ to be established. In general, this coefficient is vehicle as well as driving-schedule dependent (Sovran 1983), but for the midsize car being considered they are $\simeq$0.14 and $\simeq$0.46 for Urban and Highway, respectively. For the Euromix cycle, a typical influence coefficient for cars powered by spark-ignition engines is $\simeq$0.3, while for diesel engines it is $\simeq$0.4 (Emmelmann 1987b). In all cases, these values presume that the drivetrain gearing is rematched so that the road load power-requirement curve runs through the engine's brake-specific-fuel-consumption map in the same manner at the lower drag as at the higher drag.

If no other changes are made in a vehicle, the benefits of reduced drag are actually threefold: reduced fuel consumption, increased acceleration capability, and increased top speed. When maximum fuel-economy benefit is the objective the increased acceleration and top-speed capabilities can be converted to additional reductions in fuel consumption. Conversion of the increased acceleration capability is accomplished by regearing the drivetrain, as discussed above. Conversion of the increased top speed requires a reduction in installed engine power, and a corresponding percentage reduction in vehicle mass so that the acceleration capability of the vehicle is not diminished.

The preceding discussions have presumed the absence of ambient wind while driving. In the presence of wind a vehicle's wind speed is generally different than its ground speed, and its yaw angle is generally not zero. This affects the operating drag force, and therefore vehicle fuel economy (Sovran 1984). On the average, the result is a reduction in fuel economy.

3.2 *Handling*

While traveling along a road, a vehicle experiences more than just drag. The resultant aerodynamic force has components in all six degrees of freedom (Figure 8). In principle, they all influence a vehicle's dynamics. Traditionally, test engineers distinguish between a vehicle's behavior in still

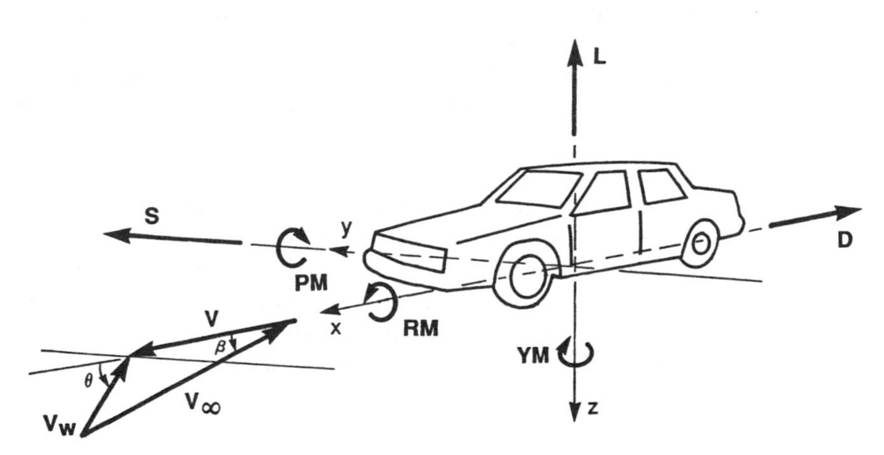

Figure 8 Aerodynamic forces and moments acting on a vehicle. Definitions of yaw angle β and the coordinate system; the latter is different from that used for flight dynamics.

air (handling), and its behavior in the presence of a crosswind (crosswind sensitivity).

The flow over a vehicle moving through still air is nominally symmetric about the vehicle's plane of symmetry. Lift, pitching moment and, of course, drag are therefore the only aerodynamic components. Unless special measures are taken, the vertical force on a bluff body close to the ground is positive, i.e. it tends to lift the vehicle. The accompanying reduction in load on the tires is, in principle, disadvantageous to handling. This is because the maximum side force that a tire can generate decreases when wheel load is reduced. However, the effect is negligible for most vehicles except race cars—at least at reasonable driving speeds. For a typical European car with a typical lift coefficient of 0.3, lift amounts to less than 3% of the vehicle's weight at a speed of 60 mph, and only 10% at 120 mph.

It is the pitching moment rather than total lift that counts in vehicle dynamics because it changes the load distribution between the front and rear axles, which alters the steering properties of a vehicle. With increasing speed, a negative (nose down) pitching moment promotes a tendency to oversteer, which is undesirable. But, this effect is hardly noticeable to the average driver. Nevertheless some European manufacturers design their cars for as low a rear-axle lift as possible, even if it involves a drag penalty.

In contrast, race cars live on negative lift. Their cornering capability has been improved dramatically by aerodynamic downforce. This downforce has to be balanced against the accompanying increase in drag, with the nature of the balance depending on the racetrack. High-speed courses with only a few bends, like LeMans (France), call for cars with low drag because top speed and fuel economy are decisive (the latter determines the number of fuel stops); speed while cornering is less important. On the other hand, courses with numerous bends, like Brands Hatch (Great Britain), require high downforce (Flegl & Rauser 1992) for short lap times.

3.3 *Crosswind Sensitivity*

In a crosswind, and while passing another vehicle in still air, the flow around a vehicle becomes asymmetric and so a side force, a yawing moment, and a rolling moment are produced. Also, the components of drag, lift, and pitching moment are altered, and normally they are increased. From both experience and numerical simulations of vehicle dynamics, it is known that only two of these components are significant to a vehicle's behavior in crosswind; these are yawing moment and side force.

The yawing moment referred to a vehicle's center of gravity gives a first indication of sensitivity to crosswind. For almost any vehicle, the yawing moment is unstable, i.e. it tends to twist the vehicle further away from the wind. As a result, the angle of yaw, the yawing moment, and the side force are increased even further.

In recent years, the center of gravity of passenger cars has, on average, moved steadily forward. This is mainly for two reasons. First, rear-engine cars have become rare; second, the move to front-wheel drive has resulted in vertical load being shifted from the rear to the front axle. Consequently, while the yawing moment referred to the center of the wheelbase has remained relatively constant, that referred to the center of gravity has decreased. As a result, the matter of crosswind sensitivity seems to have ceased to be of concern to the driving public.

Nevertheless, crosswind sensitivity remains a subject for research. Concern is focused on the influence of the various aerodynamic and vehicle parameters on a driver's reaction in sudden crosswind gusts (Sorgatz & Buchheim 1982). Driving simulators allow for changing all these parameters independently of each other (Willumeit et al 1991). On-road driving, computations, and tests on the simulator have all led to a result which is important for aerodynamic development: Yawing moment is more disturbing than side force.

There are two reasons why crosswind sensitivity might require future attention in production vehicles: decreasing vehicle weight and increasing yaw moment. Currently, the weight of passenger cars is increasing rather than decreasing. But new government regulations on fuel economy and emissions will soon force vehicle engineers to build lighter cars, because this makes significant improvements in fuel economy possible (Piech 1992). With regard to yawing moment, it is increased by the rounded rear-end shape which is in fashion at the present time, and it will go up even further unless countermeasures are taken.

In principle, with active kinematics for the rear axle (Donges et al 1990) the twist induced by crosswind can be compensated by an appropriate steering angle of the rear wheels; this would produce dynamic stability.

3.4 *Functionals*

The flow over a vehicle not only produces aerodynamic forces and moments, but also many other effects that can be summarized under the term *functionals*. During the development of a new vehicle they require at least the same attention as the forces and moments. These functional effects are:

FORCES ON BODY PARTS Vehicle bodies are made up of large and comparatively flat panels, and these have to withstand considerable aerodynamic loading. Hoods, doors, and frameless windows have to be tight under all conditions. Modern lightweight structures are prone to flutter. Special attention has to be paid to add-on parts like air shields and the various types of spoilers.

WIND NOISE The more that the formerly dominant noise sources (i.e. engine and tires) have been attenuated, the more that wind noise has become objectionable inside vehicles. Passenger cars and buses are of particular concern. In 1983, the vehicle speed at which wind noise was of the same annoyance as the noise from those other sources was 100 mph (Buchheim et al 1983); in 1992 it is only 60 mph. Open sun roofs and side windows can also cause low-frequency noise (booming) which is extremely annoying.

BODY-SURFACE WATER FLOW AND SOILING Water flow on a body's surfaces can impede visibility. Water streaks and droplets accumulating on the forward side windows prevent a clear view into the outside rearview mirrors. Soil diminishes the function of headlights and taillights. Thc sides of vans and buses are often used for advertising purposes and therefore need to be kept clean.

INTERIOR FLOW SYSTEMS Several interior flow systems pass through a vehicle. For passenger and sports cars the design of engine-cooling ducting has become extremely difficult because of increased engine power and less underhood space. Race cars are the most demanding. According to Flegl & Rauser (1992) up to 12 separate flow ducts have to be provided; these are for the "air box" which conducts the combustion air to the engine, several coolers for water, oil, and the turbocharged air, brake cooling, and cockpit ventilation. Passenger vehicles require ducting for proper ventilation and heating of the passenger compartment. Buses require high rates of air exchange free of drafts.

4. AERODYNAMIC CHARACTERISTICS

4.1 *Drag*

The physics of drag can be addressed from two different perspectives: 1. from that of the vehicle; 2. from that of the fluid through which it moves. From Newton's third law, the forces on the body and on the stream are equal and opposite to each other.

VEHICLE PERSPECTIVE The drag that a fluid stream exerts on a vehicle is the integral, over all surfaces exposed to the stream, of the local streamwise component of the normal (pressure) and tangential (skin friction) surface forces, i.e. $D = D_p + D_f$. The physics of drag is in the pressure (D_p) and friction (D_f) components. Measurement of only total drag (D) per se does not provide information on their relative magnitudes, nor on their distributions over the body surfaces. However, a degree of understanding of the origin and nature of drag contributions can be and has been gained by making systematic, parametric changes in the body surfaces. This has even been accomplished in the course of normal vehicle development (e.g. detail optimization, Section 5.1).

Direct evaluation of D_p and D_f requires knowledge of the detailed stress distributions over all vehicle surfaces. To obtain this experimentally is very difficult, except for very simple shapes, even in a reasearch sense. However, direct evaluations of components of D_p have been made in selected regions of a body. This is feasible in regions where the surface pressure is reasonably uniform, e.g. in base regions where the flow is fully separated. In general, however, the direct evaluation of D_p and D_f for bodies with the complex geometries of typical road vehicles is not practical. On the other hand, detailed surface-stress distributions are the specific output of computational fluid dynamics (CFD). Adequately validated computational codes can therefore contribute greatly to better understanding of road-vehicle aerodynamics.

STREAM PERSPECTIVE The wind-axis drag acting on a vehicle can be determined by applying the streamwise momentum equation to a large control volume containing the vehicle. One of the most significant results of the past 20 years of R&D in vehicle aerodynamics has been the identification of streamwise trailing vortices as a dominant feature of wakes. These originate in a number of different regions on a car body, and from the fluid-stream perspective are the producers (or the consequences?) of drag. In recognition of this, the momentum equation can be written in a form that expresses drag in terms of three integrals across the fluid stream downstream of the body: the downstream defects in stagnation pressure and streamwise dynamic head, and the dynamic head of the crossflow velocities in the wake which can be interpreted as a vortex drag. With uniform swirl-free inlet flow (characterized by V_∞ and $p_{t,\infty}$),

$$D = \iint_A (p_{t,\infty} - p_t)\,da + \frac{\rho}{2}\iint_A (U^2 - u^2)\,da + \frac{\rho}{2}\iint_A (v^2 + w^2)\,da,$$

where u, v, and w are the local x, y, and z velocity components, respectively, in the downstream cross-section, and p_t the corresponding stagnation (total) pressure.

This method of drag evaluation and breakdown requires data from extensive and detailed traverses behind vehicles. By reducing the crossflow velocities to a vorticity field, vortex structures are better discriminated, facilitating the process of tracking them back to their origin. The method is used primarily in research (e.g. Ahmed 1981, Ahmed et al 1984, Bearman 1984, Onorato et al 1984, Hackett & Sugavanam 1984). However, it has also been reduced to a fairly routine test procedure in one particular automotive tunnel (Cogotti 1987, 1989), where it is used as an aid in car-shape development. With it, the vortical structures in the wake are identified and their drag contributions evaluated. By tracing them back to their origin, local body-shape modifications can be explored for reducing drag. An inherent premise of this approach is that drag is minimized by minimizing vortex drag.

VORTEX DRAG In the early years of vehicle aerodynamics the direct application of concepts and knowledge from the aircraft field seemed reasonable, and this tendency has persisted even to this day (although not usually with practicing vehicle aerodynamicists). The wing that makes flight possible is a lifting body, streamlined (i.e. without extensive flow separation), and of large aspect ratio (i.e. with nominally 2-D flow, but with 3-D end effects). Conceptual models such as circulation and induced drag have proved to be very useful for it. On the other hand, the typical road vehicle is not designed to produce lift, is bluff (i.e. has regions of locally separated flow and a large wake), and is of very small aspect ratio in planview (width/length $\simeq 1/3$ for cars). Consequently, its flow field is highly 3-D (nearly all end effect), not a perturbation of a nominally 2-D one. The concepts of circulation and induced drag (i.e. a drag component related to lift) are therefore in question for such flows (e.g. Jones 1978, Hucho 1978). Nevertheless, airfoil thinking is still used (although not usually by workers in the automobile industry) to suggest that the inadvertent lift of passenger cars produces an induced-drag component.

Trailing streamwise vortices are generated at the tips of wings by the pressure difference between pressure and suction surfaces that produces lift (Figure 9a, where a wing is represented by a flat plate with an automobile-like aspect ratio of 1/3 at an angle of attack). The strength of these vortices is lift dependent. If two identical such wings at equal and opposite angles of attack are connected together (Figure 9b) each generates tip vortices, but the net lift of the system is zero. The drag represented by the vortices is clearly not related to net lift. However, identifiable lifting-body components exist, so the vortex-related drag can be attributed to the lift of the individual wings.

Trailing vortices can also be generated by nonlifting solid bodies at zero angle of attack (Figure 9c and 9d). They are generated by local pressure gradients at edges that are oblique to the local flow. The net lift is zero and there are no identifiable lifting-body components, so the drag associated with these vortices cannot be related to a lift. This demonstrates that

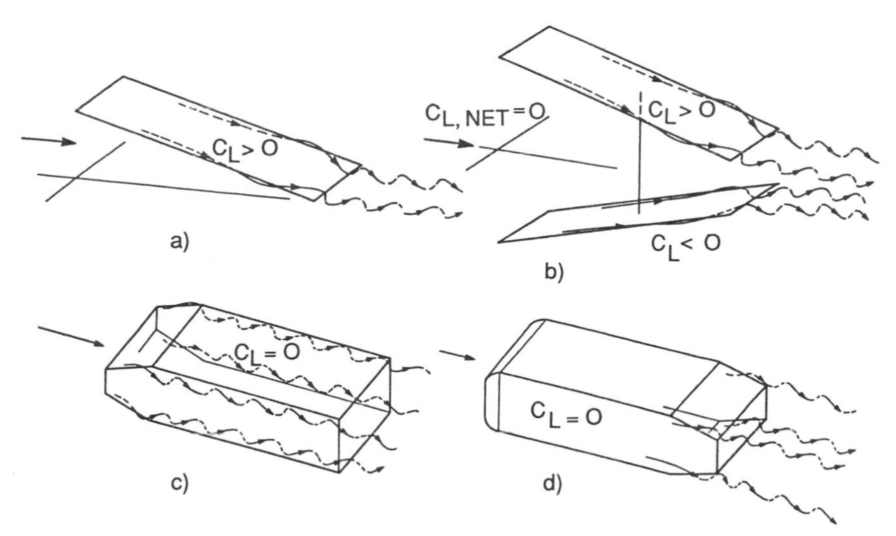

Figure 9 Longitudinal streamwise vortices (schematic) generated by lifting and nonlifting bodies.

the induced-drag concept for the contribution of trailing vortices to drag is not universal. The geometries and vortices of Figures 9c and 9d are typical of road vehicles.

CRITICAL AFTERBODY GEOMETRY The flow over the front part of any body moving subsonically through the air is easier to manage than that over the rear. This is especially true for automobiles now that free-standing headlights, radiators, and fenders are no longer in fashion. For any current car that has received aerodynamic attention, the contribution of the fore-body to drag is usually small. This does not even require streamline shaping, just careful attention to detail. The major aerodynamic problem is at the rear.

The flow over and from the upper-rear surface of vehicles is particularly interesting. The influence of the slant angle of this surface on drag has been extensively investigated over the past 20 years (e.g. Janssen & Hucho 1975; Morel 1978a,b; Ahmed 1984). The nature of the flow behavior is illustrated in Figure 10, where the slant angle is that measured from the horizontal. As the rear roof of the simple body is slanted from the square-back baseline (point A, Figure 10a), trailing vortices are formed at its lateral edges which are drag producing. However, the downwash generated between them promotes attached flow on the central portion of the slanted surface, generating a pressure recovery that is drag reducing. The net effect is a drag reduction.

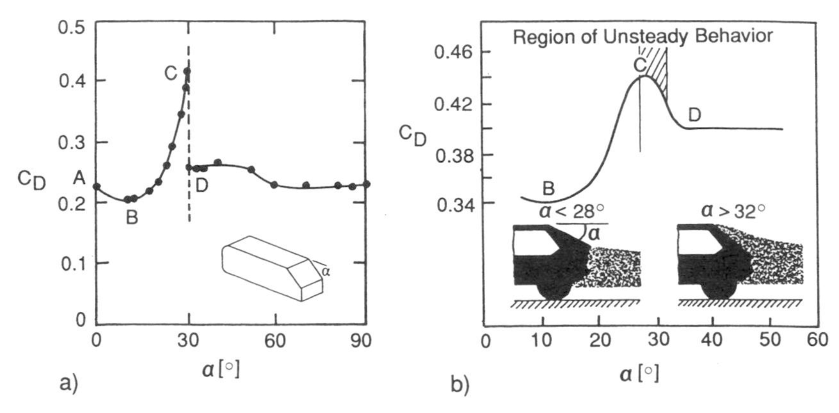

Figure 10 Dependence of drag, C_D, on the slant angle α of the upper-rear surface of fastbacks, and the existence of a critical phenomenon; (*a*) simple body, (*b*) automobile.

As the angle continues to increase the competing mechanisms grow individually, and the net drag reaches a minimum at $\simeq 15°$ (point B). It is to be noted that vortex drag is not a minimum at this point. At a sufficiently larger angle the drag reduction has decreased to zero, and further increases in angle result in a net increase in drag even though nominally attached flow is maintained on the centerline of the slanted surface.

At point C the drag is a maximum, and the trailing vortices are of maximum strength. At a slightly greater angle (the critical angle) the vortices lift off and/or burst, and the induced downwash is no longer able to produce attached flow on the centerline. The flow detaches abruptly, and a fully-separated flow develops on the slanted surface (point D). Its drag level is comparable to that of the initial square-back configuration. For this simple body with sharp rear edges, the transition is discontinuous and unidirectional. The flow patterns corresponding to C and D have been measured, and that of C is shown in Figure 11 (Ahmed 1984).

The same general behavior is observed for an actual automobile, where it was originally discovered (Figure 10b, Janssen & Hucho 1975). This has rounded edges in the slanted-roof area, and the flow pattern at transition can be bistable, switching slowly (with a period of many seconds) and randomly between that of C and D. As the slant angle increases the bistability eventually disappears, and only the D-pattern persists. For angles greater than this, the drag of the D-pattern is relatively insensitive to angle.

A drag maximum as a function of rear-upper-body geometry has also been observed for notchback cars (e.g. Figure 12, Nouzawa et al 1990). The relevant characteristic angle (β) is that between the horizontal and the

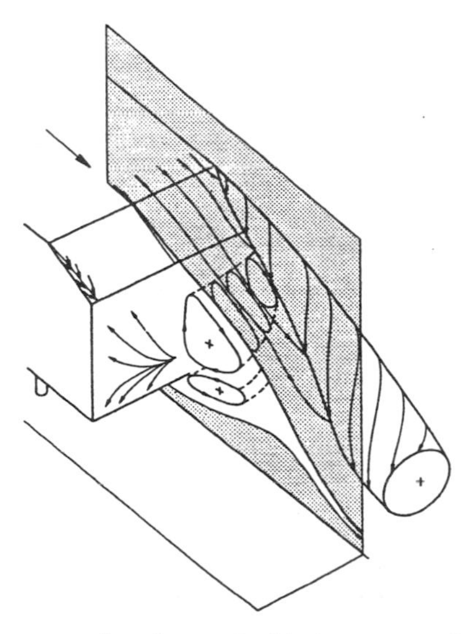

Figure 11 Wake flow pattern for a large subcritical upper-rear slant angle of fastbacks.

line connecting the end of the roof to the end of the trunk deck. For rear-window angles (α) that are $\geq 25°$, a drag maximum occurs at $\beta \simeq 25°$. At this maximum the wake pattern consists of a separation bubble behind the rear window, and trailing vortices. Measurements of instantaneous

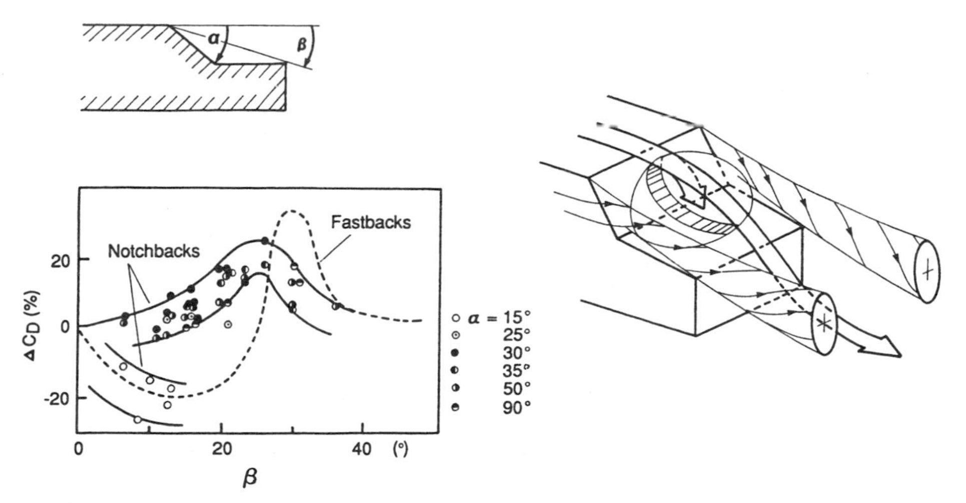

Figure 12 Dependence of drag on the afterbody geometry of notchbacks, and the wake flow pattern for high subcritical drag.

velocity in the wake and pressure on the trunk deck indicate significant bubble and wake unsteadiness (Nouzawa et al 1992). Corresponding measurements of the instantaneous drag force show that its fluctuations correlate with the unsteadiness of the separation bubble.

In vehicle design, the generic type of rear geometry (i.e. squareback, fastback, or notchback) is selected by the stylist, not the aerodynamicist. The choice is based on function and design theme, as well as aesthetics. The role of the aerodynamicist is to achieve low drag for the configuration that has been selected.

EFFECT OF AMBIENT WIND On the average, there is always an ambient wind and the wind is not aligned with the road along which a vehicle is traveling. Therefore, road vehicles operate at a nonzero angle of yaw, on the average (Figure 13). It is the aerodynamic-force component along the direction of travel that resists vehicle motion, so this is the relevant drag. This is in the body-axis frame of reference (when a vehicle is at zero pitch angle). Some typical variations of C_D with yaw angle are shown in Figure

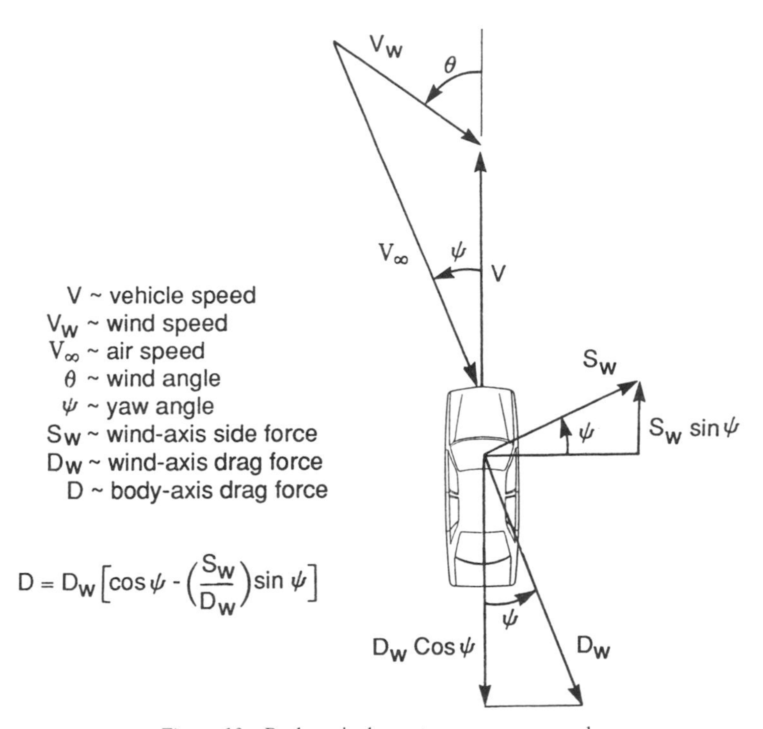

Figure 13 Body-axis drag at nonzero yaw angle.

14. Driving conditions with ψ large enough to reach maximum C_D are rare, except in gusts, and average values of ψ are said to be less than 10° for typical driving.

As is shown in Figure 13, the aerodynamics is in the wind-axis reference frame. The variation of body-axis drag (D) with yaw angle depends on the rate of increase of wind-axis drag (D_w) and the rate of decrease of the bracketed term containing the wind-axis ratio of side force to drag (S_w/D_w). Interpreting the plan view of the vehicle as the cross-section of a bluff airfoil, this ratio is analogous to the lift/drag ratio of an airfoil. The larger S_w/D_w, the greater the rate of decrease of the bracketed term and hence the smaller the rate of increase of C_D. In an extreme case, C_D can actually decrease with ψ, i.e. the wind can contribute an aerodynamic push. This has been accomplished with an experimental, small, commuter-type, high-fuel-economy vehicle having a truncated-airfoil planform (Retzlaff & Hertz 1990).

OTHER DRAG CONTRIBUTIONS The bodies of road vehicles are characterized by two types of open cavity that communicate freely with the external flow field. One is the engine compartment, with throughflow against internal resistances. Its drag characteristics are well understood by vehicle aerodynamicists (Wiedemann 1986). The other type consists of the wheel wells; these also contain large rotating tires that pump air. The wells contribute to drag, and the exposed wheels themselves generate bluff-body drag.

A measure of the state of the art in vehicle aerodynamics is the typical C_D of production cars. In the U.S. in 1975, this was about 0.55. Today, it is about 0.35. This large improvement reflects not only the skill of the aerodynamicists, but also the greatly increased concern about vehicle fuel

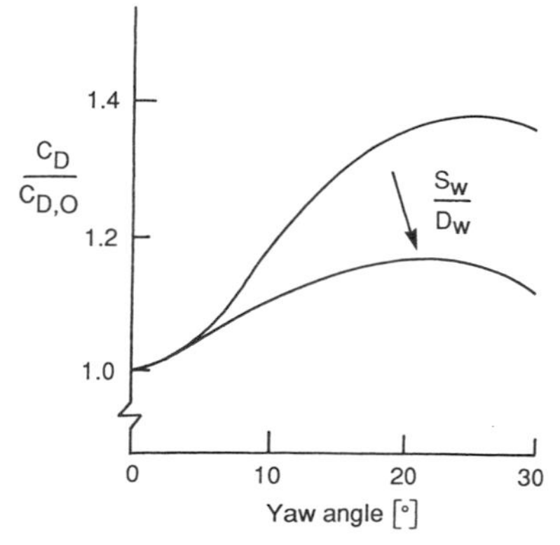

Figure 14 Typical variations of body-axis drag coefficient (normalized) with yaw angle.

economy. In Europe where fuel economy has always been important, a significant reduction in C_D has also taken place. In 1975 a typical value for production cars was 0.46. Today, several cars are below 0.30, and one car (Emmelmann et al 1990) is listed at 0.26.

4.2 *Lift and Pitching Moment*

The lift on a body close to the ground is determined by two counteracting effects (e.g. Stollery & Burns 1969). Although a symmetrical body has no lift when in free air, when close to the ground the body and the ground-plane—or in the terms of potential flow, the body and its mirror image—form a narrowing channel (Venturi nozzle) between them in which the local flow is accelerated and then decelerated. This results in an average negative gauge pressure on the body's underside, tending to generate a downforce. At the same time, flow resistance in the channel displaces more flow to the upper side of the body creating an effective positive camber of the body. This makes a negative contribution to the upper-surface pressure, tending to produce an upward force.

Lift effects are pronounced on race cars, because they are generated intentionally. Two means are used to produce a large downforce.

1. A Venturi channel in the body's underside. The effectiveness of this measure is enhanced by preventing lateral flow with side skirts. A large downforce can be generated with only a small penalty in drag.
2. Negatively cambered wings at the car's front and rear, even equipped with double flaps. In this case, there is a large increase in drag due to the induced drag of the wings, and mainly from the wing at the rear (Figure 15, Rauser & Eberius 1987).

Both means for producing large downforce have been developed to such an extent that lift coefficients on the order of $C_L = -3$ have been achieved with Formula One cars (Wright 1983). This has permitted transverse accelerations beyond a driver's endurability. Therefore, in 1982 regulations prescribed only flat underbodies, and side skirts were ruled out. Now, lift coefficients on the order of $C_L = -2$ are typical, producing a downforce at 300 km/h that is approximately twice a car's weight.

To be most effective, downforce-generating elements should produce only a minimum increase in drag. One means of accomplishing this would be to make the flaps of the wings movable. In a curve, they could be deflected to a high angle to generate maximum downforce, while in the straights they could be swiveled to zero deflection for minimum drag. However, for safety reasons, movable body parts have been prohibited on race cars. A means of circumventing this regulation would be to use a jet-flap for downforce/drag control, but this has not yet been tried.

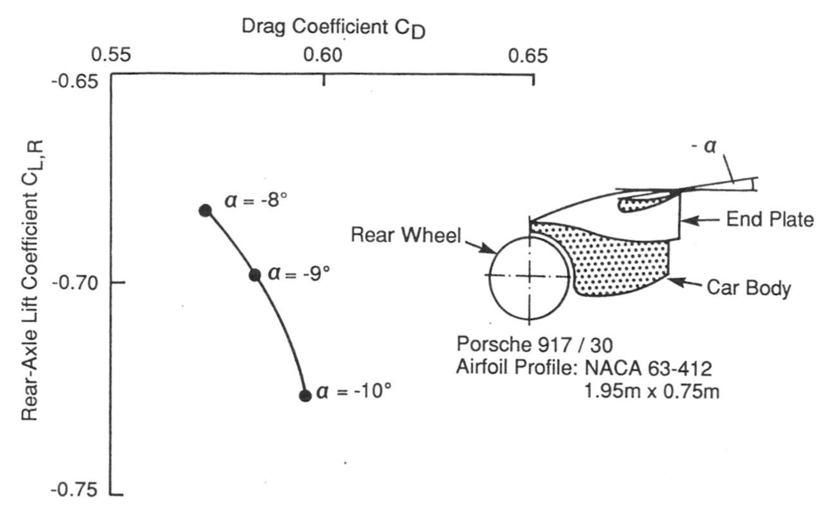

Figure 15 Rear-axle lift versus drag for three different angles of attack of the rear wing of a race car.

Passenger and sports cars use knowledge gained with race cars. Front lift is reduced by an underbody air dam; rear lift is reduced by either a rear spoiler (sometimes shaped as a wing), an underbody diffuser at the rear, or by a combination of the two. A few sports cars are equipped with a movable rear wing which is withdrawn at low speed to improve visibility to the rear.

4.3 *Yawing Moment and Side Force*

The generation of side force and yawing moment on cars is very much the same as the generation of lift and pitching moment on a wing at angle of attack (Figure 16, Squire & Pankhurst 1952, Barth 1960). When viewed from above, a car body can be regarded as a blunt airfoil that can generate lift (side force for the car) when its angle of attack (yaw angle for the car) is greater than zero. However, to generate a pressure diagram comparable to that for a wing the car needs a yaw angle much greater than the angle of attack required by the wing. In addition to its poor cross-sectional shape for a lifting body, this is because the aspect ratio of the car when viewed from the side (height over body length) is <1, while that of the wing in Figure 16 is $\gg 1$. The center of pressure of the area enclosed by the pressure distributions on the lee- and windward sides is located forward of the midpoint of the car's length, and therefore a yawing moment is generated.

Similar to airfoils, cars are aerodynamically unstable, i.e. the yawing

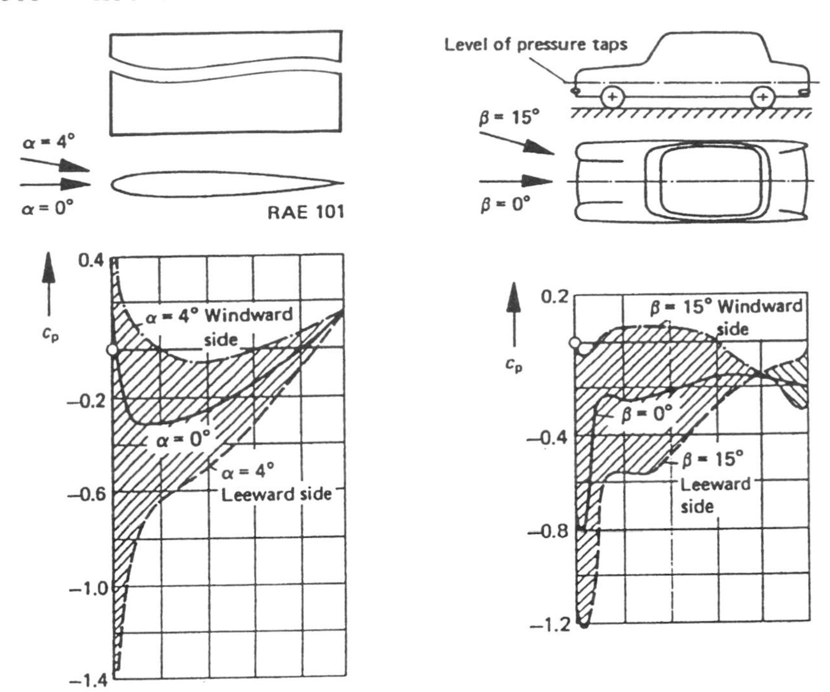

Figure 16 Comparison of the pressure distribution around an airfoil and a horizontal section of a car.

moment tends to turn the car away from the wind. Both their yawing moment and their side force increase almost linearly with yaw angle up to $\beta = 20^\circ$, and side force is linear to even larger angles (Figure 17, Emmelmann 1987a). The rate of increase is mainly determined by the side-view cross-sectional shape of a vehicle. Generally, a higher yawing moment goes with a lower side force, and vice versa.

Under yawed conditions, a rolling moment is also generated, lift is increased, and pitching moment is altered. In the speed range of passenger cars, the influence of these components on vehicle dynamics is minor. However, the subjective perception of a driver has to be considered, and this is currently being investigated in a vehicle simulator (H. Goetz, private communication).

Measures for reducing aerodynamic instability, or even reversing it to stability, have been known for a long time (Koenig-Fachsenfeld 1951). The only really effective method is the tail fin. Race cars make use of it,

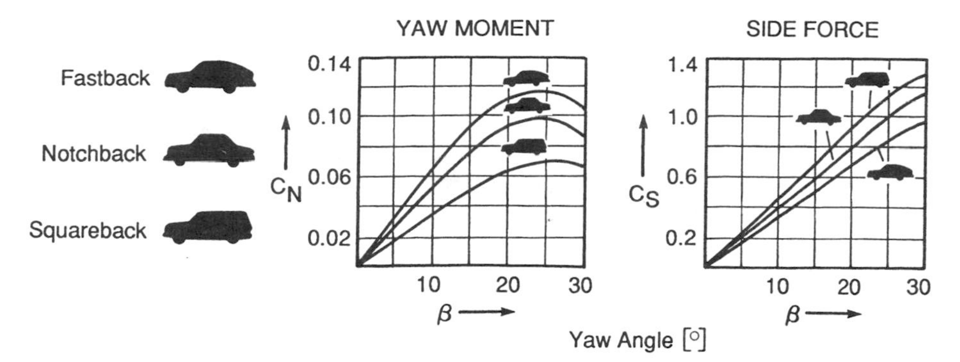

Figure 17 Yaw-moment and side-force coefficients for a family of cars with different rear-end shapes.

since the endplates of their rear wing act as such. But for passenger cars, airplane-like tail fins cannot be applied for practical reasons.

With the comparatively small shape modifications usually feasible in car development, only a limited reduction of yawing moment is possible (Sorgatz & Buchheim 1982, Gilhaus & Renn 1986). Rounded rear-end contours are unfavorable for crosswind sensitivity. Under yawed conditions the flow around the windward side's rear corner remains attached longer for a rounded corner than for a sharp edge, producing high negative pressure at this location and increased yawing moment. Some European cars have a sharp trailing edge on the rear pillars of the passenger-compartment "greenhouse" (the C pillars) to reduce yawing moment (Gilhaus & Hoffmann 1992). The accompanying increase in drag at yaw is tolerable because the large angles at which it occurs are infrequent.

Implicitly, all the preceding considerations are based on the assumption of an idealized wind, i.e. steady, with a spatially uniform velocity. In reality, this is never the case. Natural wind has a boundary layer character. When its velocity profile is combined vectorially with a vehicle's forward speed, a skewed oncoming relative-wind profile is generated (Figure 18, Hucho 1974) which cannot be reproduced either in a wind tunnel or by a side-wind facility at a test track.

Because of varying landscape along a road, the strength and local direction of wind can vary with distance along it. Furthermore, natural wind is gusty. Therefore, steady-state data obtained in test facilities cannot be exactly applicable to actual windy-day driving. The transient side force and yawing moment computed with an analytical model based on slender-body theory (Hucho & Emmelmann 1973) are significantly

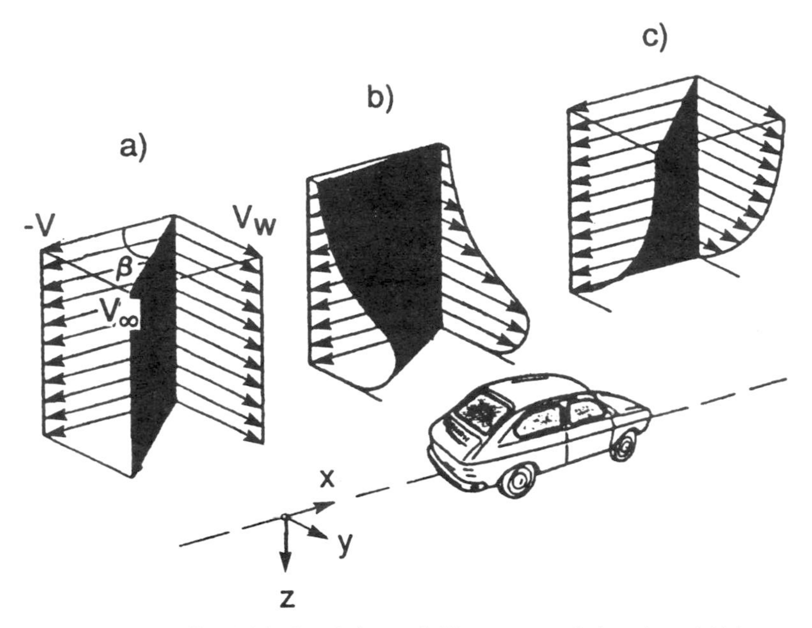

Figure 18 Wind profiles: (*a*) in the wind tunnel, (*b*) on a crosswind track, and (*c*) in natural wind.

larger than their steady state values, the magnitude of the overshoots depending on the gradient of the crosswind velocity along the direction of travel, i.e. on the gradient at the edge of the simulated wind gust.

4.4 *Detailed Surface Flows*

Detailed surface flows are investigated for a number of reasons: to properly locate openings for air inlets and outlets; to ascertain the forces on particular body parts; to determine the sources of wind noise; and to find means for controlling water flow and soil deposition on surfaces.

The openings for ducted air flows are placed in regions of high pressure. Therefore, the inlet for engine-cooling air has moved downward over the years along with the lower hoodlines, and is now close to the stagnation point. Consequently, large inlet grilles have become unnecessary and, to the complaint of many buyers, cars have lost their distinctive faces.

The inlet for passenger-compartment ventilation air is normally placed in the cowl area at the base of the windshield, and most commonly in the central part. The disadvantage of this location is that the ram-air volume flow rate depends on driving speed and so, therefore, does the flow velocity inside the passenger compartment. Placing the inlet away from the centerline in an area of zero pressure coefficient avoids this speed dependency, but then a continuously running fan is needed to achieve desired interior airflow levels.

Outlet openings were formerly placed at locations of high negative surface pressure. But, due to the high flow-field velocities at such locations the local wind noise was high and was communicated to the inside of a car. Ventilation outlets are now placed in regions of low to moderate negative pressure to avoid this problem. However, the larger outlet areas required must be hidden from view; for example, behind the rear bumper.

Knowledge of the aerodynamic forces acting on body parts such as doors and the hood is needed to properly select the position of hinges and locks. Flexible structures tend to flutter, and so locations for reinforcement also have to be determined.

Wind noise, as perceived inside the passenger compartment, is receiving increasing attention. The basic phenomena are now well understood (Stapleford & Carr 1971, George 1989). However, based on buyer complaints, the application of this knowledge to specific cars still seems to be a major difficulty.

Investigations of wind noise have to deal with three factors: sources, paths, and receivers. Sources may be such things as a leaking window seal, a large-scale separation bubble on the hood, high local flow velocities, or a small-scale separation in a gap between two doors (Figure 19, Hucho 1987b). By comparing the actual flow with one where separation has been avoided by artificial means, such as by sealing door gaps, the contribution of individual noise sources can be identified (Buchheim et al 1983, Ogata et al 1987). Aerodynamic noise can also be generated by the separated flow over protuberances on a car body, such as side rearview mirrors and radio antennae (Figure 20, Ogata et al 1987). Paths have to be treated according to their function. Panels can be dampened with absorbing materials, but windows can't. Openings for ventilation must remain open. Highly turbulent separated flow from the front part of a vehicle reattaches further downstream and may excite vibrations of the structure in the reattachment area; the wake of the side-mounted rearview mirror is a typical example. The receivers are human beings. Their individual sensitivity to noise (level, spectrum, discrete tones) must be accommodated.

It might be expected that low-drag cars would produce low wind noise, but no such correlation exists for today's cars. Even aerodynamically-well-designed cars have many flow separations (of large and small scale), areas of high local velocity, and leaking seals. Further advances are needed in the ability to identify noise sources. Recent research on high-speed trains

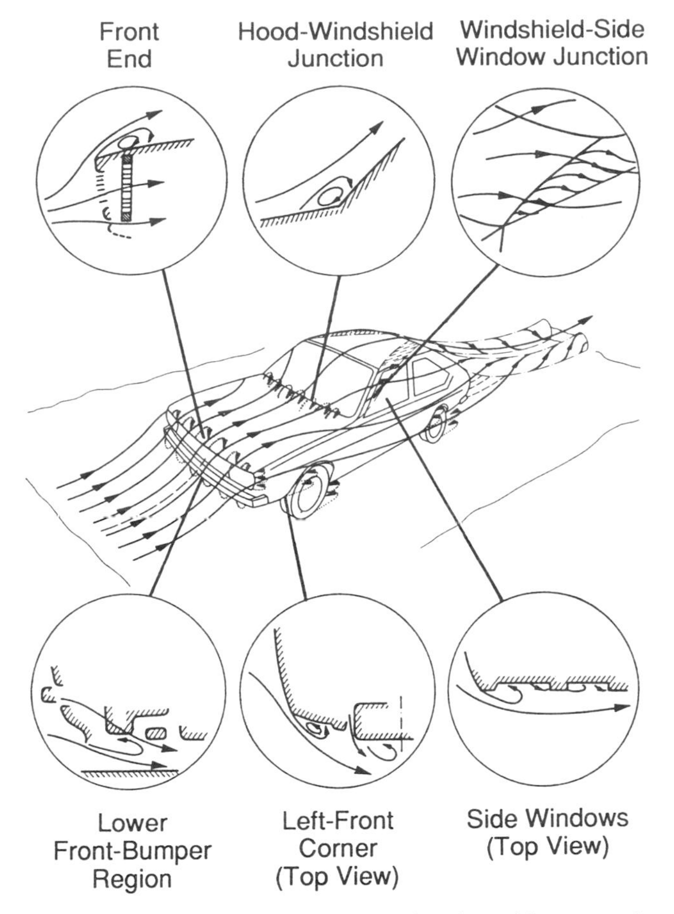

Figure 19 Flow around a car, and major locations of flow separation.

(Mackrodt & Pfitzenmaier 1987) should be applicable to cars. Progress in analytical modeling of sources and paths is also a necessity (Haruna et al 1992).

The air through which a vehicle moves is not always dry and clean. Water drops (either directly from rain or from splash and spray whirled up from the road) and dirt particles (either wet or dry) are heavier than air and so do not follow streamlines. The resulting two-phase flow has to be managed with consideration for both safety and aesthetics. Two points of view deserve consideration. Whirled-up water and dirt are annoying and dangerous not only for the vehicle generating them, but also for other vehicles in its vicinity. The latter problem is still very much neglected.

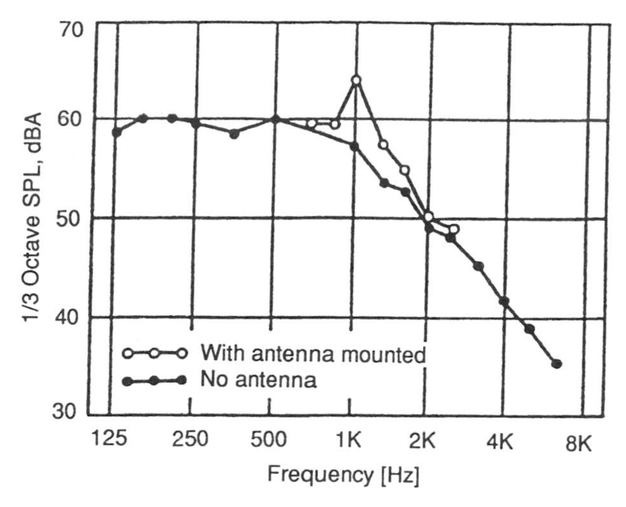

Figure 20 Aerodynamic noise generated by a radio antenna.

Water streaks that spill over from the windshield to the A-pillar (the post between windshield and front door) are broken up into droplets by an A-pillar vortex. These droplets tend to collect on the forward part of the side windows, obscuring a driver's view in the side-mounted rearview mirrors. By integrating a rain gutter into the A-pillar, the water from the windshield can be trapped and led down and away. Similarly, water from the roof can be trapped and kept away from the rear window, either by a trapping strip in the rubber seal (Piatek 1987) or, in the case of a hatchback door, by controlling the gap between roof and door (Figure 21). If these traps are carefully faired into a body's contour, they will not cause an increase in drag.

The deposition of dirt is more difficult to prevent. Those parts of the body close to the stagnation point are the most susceptible to dirt deposition, e.g. the headlamps. Up to now, no aerodynamic alternative to the mechanical cleaning of headlamps, either by high-pressure water spraying or by brushing, has been found. Dirt deposition on the rear window of a squareback car can be reduced by a guide vane at the end of the roof. However, the air-curtain effect has to be balanced against the increase in drag that accompanies it. The sides of delivery vans can be kept free of soiling, at least partly, by impeding the dirty water that spills out of the front wheel houses. This can be achieved by generating a low pressure underneath the vehicle that reverses the direction of the wheel-house outflow from outboard to inboard.

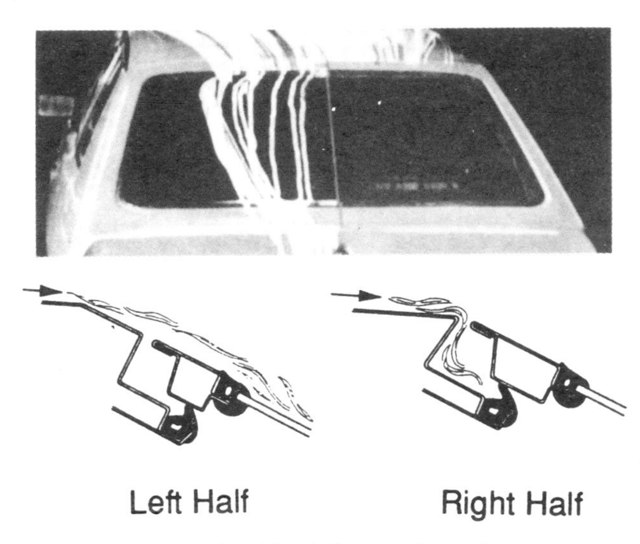

Figure 21 The gap between roof and hatch door can be used as a water trap. On the left side the large trapping gap has been sealed for the photograph, on the right side it is open and functional.

Splash and spray whirl-up is also a safety hazard for other vehicles. This is especially true for trucks and should no longer be ignored, particularly since solutions are already on hand. On wet roads, trucks with their almost uncovered wheels can raise opaque curtains of water higher than the eye-point of a neighboring car's driver (Figure 22, Goehring & Kraemer 1987). Up to now, the only way to impede this has been to completely cover all nonsteering wheels. This measure is even rewarded with a drag reduction, mainly under crosswind conditions, and is also useful in preventing cars from going under the truck in a crash.

Figure 22 Splash and spray can be kept under a vehicle by shielding the non-steered wheels. (*Left*) Wheels uncovered; the passing car can hardly be seen. (*Right*) Wheels and sides shielded, as well as the rear of the truck. (Photos courtesy of Mercedes-Benz.)

4.5 *Internal Flow Systems*

The flows in internal and recessed cavities that communicate freely with the external flow contribute to drag and have been discussed in Section 4.1. The flow in the passenger compartment does not make a significant drag contribution, but it has to be carefully designed for passenger comfort. High air-exchange rates have to be realized, but with a flow free from drafts. Consequently, flow velocities are kept low. The influence of buoyancy has to be recognized because of significant differences in temperature between heating and cooling. During heating, a temperature gradient is preferred (warm feet, cool head). During cooling, the conditioned air is directed to the face and chest. By making use of the physiological effect that a temperature lower than ambient can be simulated by blowing ambient air at moderate velocity over passengers, the use of air conditioning per se can be avoided in moderate climates (Gengenbach 1987).

5. DEVELOPMENT PROCESS AND TOOLS

5.1 *Aerodynamic Development of a Vehicle*

The aerodynamic development of a typical road vehicle follows a procedure different from that used for other types of fluid-dynamic machines, e.g. aircraft, compressors, or turbines. While the latter are designed to meet specific aerodynamic performance criteria—otherwise they wouldn't work—the aerodynamic properties of a road vehicle are largely by-products of the design process.

The typical design of a fluid-dynamic machine is accomplished in three major steps:

1. The main overall dimensions of the machine, be it the planform and span of an aircraft wing or the diameter and number of stages of a turbine or compressor, are determined by using nondimensional coefficients (derived from similarity laws) for which optimum values are known from accumulated experience. Typical coefficients are ones such as the lift coefficient (C_L) of a wing during aircraft cruise, or the specific speed (N_S) of a fan.
2. The detailed design of the machine's components follows an iterative process between analytical design and experimental verification. Computational fluid dynamics (CFD) has accelerated the convergence of this process significantly; furthermore, it permits interference effects among the components to be taken into account.
3. The separately optimized components are put together in a prototype system which is then tested. If the final device is very large, this is

preceded by testing with small-scale models. When design performance targets are not achieved, modifications are made.

The development of a road vehicle is accomplished differently. Aerodynamic development is performed in a closed loop containing aesthetic, packaging, and aerodynamic considerations. Both the number of iterations necessary and the quality of the final result depend on the ability of the aerodynamicist to recognize the intentions of the exterior designer (stylist), and to find solutions within the designer's limits of acceptability—which are extremely difficult to define because of their subjective nature.

In general, the aerodynamic development of a road vehicle can proceed from two different and opposite starting points. Typically, the designer makes the first proposal for a vehicle configuration. Then the aerodynamicist tries to improve the shape, according to his goals as described in Sections 3 and 4. One strategy which has proved to be very effective is the so-called detail-optimization (Hucho et al 1976). In this approach, details like corner and edge radii, and taper or boat-tailing are modified in small increments in order to find optimum values (Figure 5). As long as the drag coefficient of vehicles was still fairly high, say $C_D = 0.40$, the modifications necessary were frequently so small that aesthetics was hardly affected (Janssen & Hucho 1975).

The opposite approach starts with a simple body of very low drag which has the desired overall dimensions of the subsequent car. This so-called basic body is modified step by step, in cooperation with the exterior designer, until a final shape materializes which has both low drag and the desired appearance. Drag-coefficients below $C_D = 0.30$ have been achieved this way (Buchheim et al 1981). This is called *shape-optimization.*

Today the actual aerodynamic development process is usually a mixture of these two extremes. The original design of the stylist is influenced by his general knowledge of vehicle aerodynamics, more or less, and the shape modifications which are eventually necessary are constrained within rather narrow limits.

5.2 *Test Facilities*

Aerodynamic development per se is done almost entirely experimentally, guided by empiricism based on the knowledge described in Section 4. The primary test facility is the wind tunnel. Water tunnels and towing tanks are sometimes also used, but only as supplements. They lend themselves to flow visualization (Williams et al 1991), and also have the capability for reproducing the relative motion between vehicle and road, including wheel rotation (Larsson et al 1989, Aoki et al 1992). However, only those towing tanks can be used which have an extremely level and smooth bottom on which the submerged test vehicle can roll.

All vehicle characteristics influenced by aerodynamics are finally evaluated on the road. This also permits tests that cannot be made in a wind tunnel, e.g. measuring a vehicle's behavior in a cross wind.

Historically, the wind tunnel testing of automobiles started with small-scale models. Scales like 1:4 or 1:5 were preferred in Europe, the somewhat larger 3/8-scale in the U.S. The advantages of small-scale testing are that the models are cheaper than full-scale ones, are easy to handle, and can be quickly modified. Furthermore, only small wind tunnels are needed, and these are more generally available and can be rented at moderate cost.

For two reasons, small-scale testing was eventually only rarely used. At first, this was because test results from partial-scale models very often did not reproduce full-scale values with the accuracy needed. This deficiency was partly due to a lack of geometric similarity in the models, and partly to the unpredictable effects of Reynolds number. However, geometric similarity is not a fundamental problem but rather a matter of the skill and care of the model maker. Also, a Reynolds-number gap on the order of two can be bridged by artificially increasing the turbulence level of the wind tunnel (Wiedemann & Ewald 1989). Consequently, small-scale testing has again come into favor and is used by some car manufacturers with great success.

The second and even stronger objection to small-scale testing is non-aerodynamic in nature. Vehicle exterior design is done in full scale, because shapes in small scale cannot be adequately assessed aesthetically. Therefore, a full-scale model always exists, and if it is built on a realistic chassis, such as the one from the preceding model year, it can also be used as the wind-tunnel model.

5.3 *Limitations of Wind-Tunnel Testing*

Two major limitations exist when cars are tested in tunnels. First, due to their pronounced bluntness, cars disturb the flow in the test section of a wind tunnel significantly, the situation being similar to that with aircraft models under extreme stall (e.g. fighter aircraft during combat maneuvers). Second, the relative motion between vehicle and road and the rotation of the wheels are very difficult to reproduce, and are therefore usually neglected. However, great progress has recently been made in overcoming both of these limitations.

BLOCKAGE As a carryover from aircraft terminology, the ratio of a car's frontal area to the cross-sectional area at the tunnel-nozzle exit is called the blockage ratio. For road vehicles, this ratio can be extremely large compared to aircraft. Blockage ratios up to 20% are sometimes used. Thermal tests (e.g. engine cooling) are done with blockage ratios even higher than this. For a long time, a blockage ratio on the order of 5% was said to be appropriate for the aerodynamic testing of cars, this figure being borrowed from aeronautical practice. For a typical car, this would lead to a test-section cross-sectional area of 40 m^2. Very few automotive wind tunnels are this large; 25 m^2 is a typical value for the wind tunnels built recently in Europe.

With regard to the boundaries of the test section airstream, two different types of configuration are used. The closed-throat test section with solid walls is preferred in the U.S. The open-throat test section with freestream boundaries is common in Europe. The latter is characterized more precisely as only three-quarters open, the remaining quarter of the boundary being the ground floor which represents the road. A third type of boundary configuration which lies between the open- and closed-throat versions is one with slotted walls. Some automotive wind tunnels in Europe use this type (Eckert et al 1990).

Testing in an open test section is affected by a body as bluff as a car in three different ways:

1. The determination of the wind speed may be in error. If the model is too close to the nozzle's exit its flow field extends upstream far enough to modify the pressure at the location of the downstream pressure taps inside the nozzle (Figure 23, Kuenstner et al 1992).
2. The flow into the collector may be disturbed if the length (L) of the test section is insufficient and the cross-sectional area (A_C) of the collector is too small compared to the nozzle's cross-sectional area (A_N) (Figure 24, v. Schultz-Hausmann & Vagt 1988).
3. The streamlines in the vicinity of a vehicle are diverted more than in an air stream of infinite cross-section. Consequently, the drag is lower. Only this effect is subject to a blockage correction, not the preceding two.

While the first two effects are now well understood, blockage corrections for the third which are generally valid for open test sections still do not exist (SAE 1990). They are often said to be small and are therefore, for simplicity, neglected. It is doubtful whether this holds for blockage ratios of 10% or more.

The effect of body bluffness on closed-throat tunnel testing is primarily one of model blockage, although a possible influence on the wind-speed measurement cannot be ignored. In contrast to open throats, the streamline divergence around a model is less than that in an infinite stream. This causes the measured values of C_D to be larger than those in an infinite stream, rather than smaller, and the magnitude of the discrepancy is significantly greater than that for open test sections.

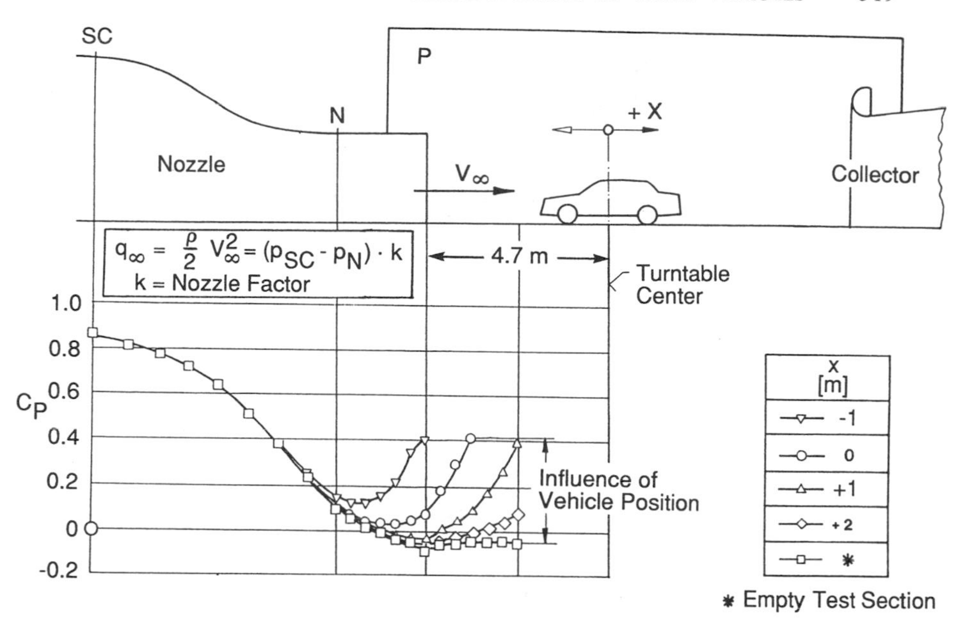

Figure 23 Pressure distribution on the centerline of the floor inside the nozzle of the BMW 10 m^2 acoustic wind tunnel. If the model is positioned too close to the nozzle's exit, the static pressure at location N is increased.

All classical blockage corrections are all-inclusive in nature, i.e. the velocity head of the oncoming wind is first corrected, and then the coefficients corresponding to all the measured forces, moments, and pressures are adjusted using this corrected head. The assumption underlying any concept of data correction (it is, more properly, data adjustment) is that the detailed nature of the flow field (in a nondimensional sense) is unchanged by blockage. This does not necessarily hold if the blockage ratio is large, or if flow separations and reattachments are affected.

Blockage correction for road vehicles in closed test sections has been the subject of an ongoing cooperative study by aerodynamicists active in vehicle testing (SAE 1991). From the many procedures considered, three have been established as acceptable for zero-yaw-angle testing in this application. These are: a wall-pressure signature method (Hackett & Wilsden 1975), a less comprehensive wall-pressure method (Hensel 1951), and a semi-empirical method (Mercker 1986). All three give approximately the same blockage correction, even at nonzero yaw angle. The pressure-signature method is the most comprehensive, requires the least subjective user input, and is readily applicable to vehicles with unknown flow patterns. It has the disadvantage of being demanding in terms of instru-

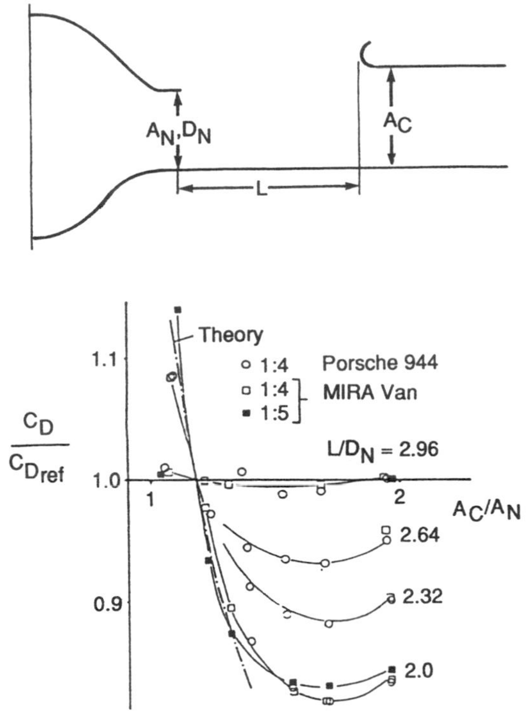

Figure 24 Influence of test-section length L (relative to the hydraulic diameter D_N of the nozzle) and the ratio of collector area (A_C) to nozzle area (A_N) on measured drag; Porsche 22 m^2 wind tunnel, open test section.

mentation and computation. The method has been modified and extended for tractor-trailer trucks by Schaub et al (1990), and the improvement is applicable to automobiles.

ROAD REPRESENTATION AND WHEEL ROTATION From the numerous suggestions for simulating the road in a wind tunnel (Figure 25) the following four have been reduced to practice (Mercker & Wiedemann 1990): a solid floor without any boundary layer control (*a*), with tangential blowing (*h*), or with boundary layer suction, (*d* and *g*); a moving belt in combination with an upstream scoop or a suction slot to remove the oncoming boundary layer (*c*).

The simplest and most common way to represent the road is with a solid and fixed ground floor. This permits a vehicle to be easily connected to a balance under the floor through four pads, one under each wheel, which can be adjusted for wheelbase and track. The wheels of a vehicle are not rotated during conventional testing. As long as the displacement thickness

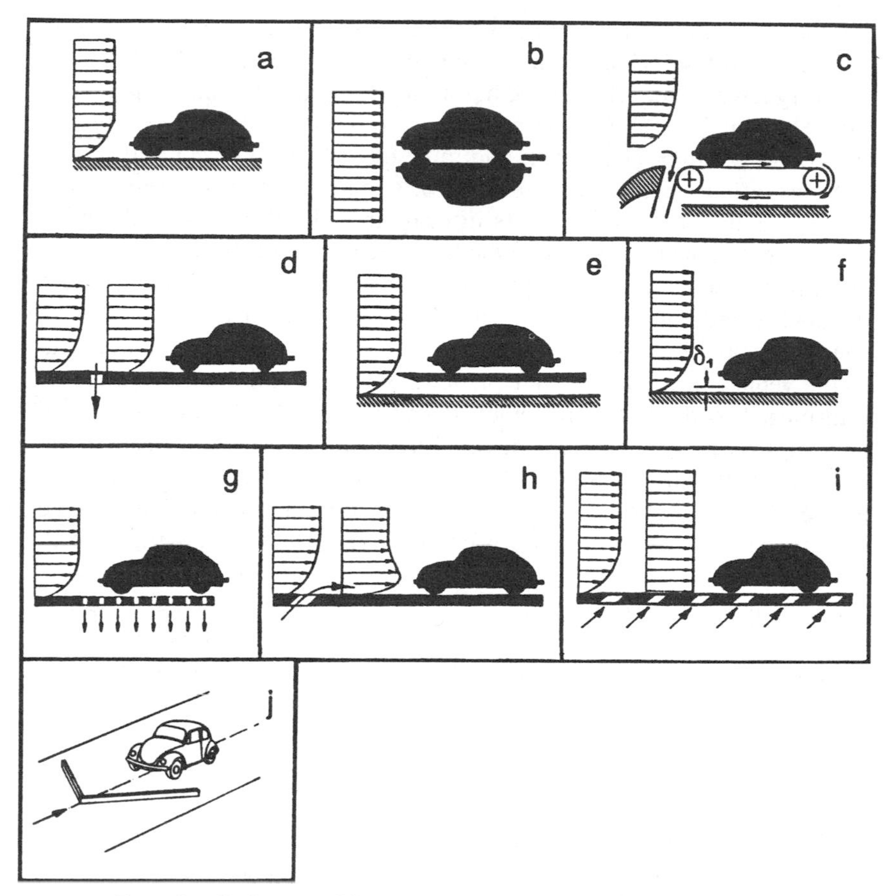

Figure 25 Various possibilities for representing the road in a wind tunnel.

of the floor boundary layer, as measured in an empty test section, is less than 10% of the vehicle's ground clearance, this kind of ground simulation is adequate for passenger-car development at drag-coefficient levels of $\simeq 0.40$ and above (Hucho et al 1975).

The moving belt represents (in principle at least) an almost perfect way to simulate the road. There are a surprising number of moving-belt wind tunnels in the world used for automotive testing. Nearly all of them (20, by one count) are for partial-scale models, and many are used for testing Formula One race cars. Quite a number are at educational institutions. There are only two full-scale facilities.

In moving-belt tunnels the belt's ability to support the weight of the test vehicle is limited. Therefore, in nearly all of them the vehicle is suspended from a force balance located above the test section, usually by a vertical strut. This strut is shielded from aerodynamic loads by a streamlined fairing. In a very few facilities the vehicle has an internal balance and is supported from the rear by a sting. The resultant aerodynamic interference of the strut or sting can be significant, and must be quantified by calibration.

Nonrotating wheels over a moving belt require a gap between wheels and belt, and this leads to errors in drag and lift (Beauvais et al 1968). Consequently, in all moving-belt facilities the wheels of the test vehicle are in contact with the belt, and driven by it. While the two-fold effect increases the fidelity with which the wind tunnel simulates on-road driving, it also confounds the aerodynamic measurements. The wheels are either attached to the vehicle in a manner that prevents their vertical loads from being transmitted to the body, or detached and independently supported in a manner that permits the streamwise force on them to be measured. In either case, the lift of the wheels is not measured. Furthermore vehicle drag can be inferred only if appropriate correction is made for the rolling resistance of the tires. The latter is determined by a tare test with the belt moving but the wind off.

A means for measuring the lift of a complete vehicle (i.e. including the contributions of the wheels) as well as that of only its body has been developed for full-scale cars. The vehicle is fitted with an internal balance and supported from the rear on a sting (Figure 26, Mercker et al 1991).

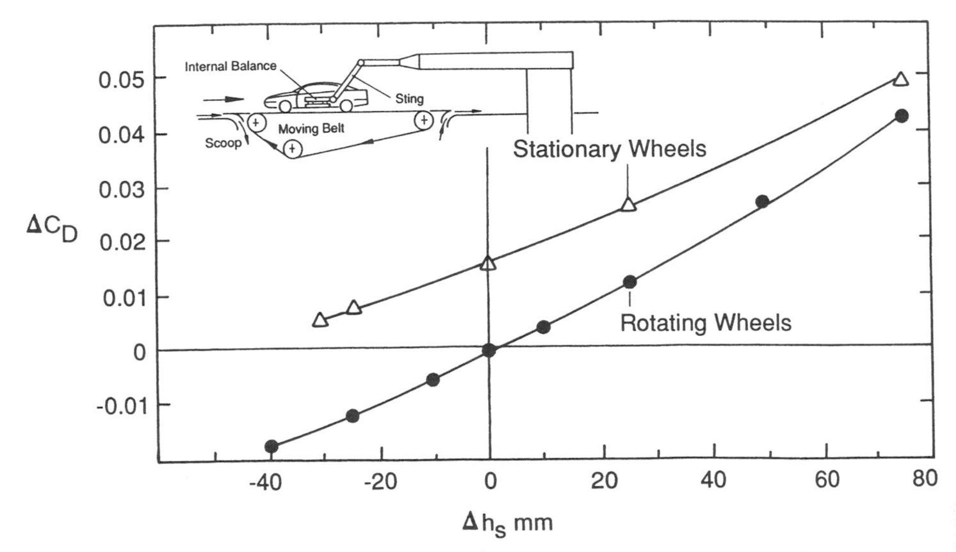

Figure 26 Influence of wheel rotation on drag, as measured for an Opel Calibra over a moving belt in the German-Dutch wind tunnel (DNW): Δh_s is the difference in underbody clearance height from the design level.

The suspension of all four wheels is replaced by a pneumatic system which permits accurate control of the vertical load at each wheel. This is necessary to permit measurement of the total vehicle lift, and to properly simulate the tire rolling resistance which depends on vertical load. The technique requires considerable model preparation, and setting up the test and performing the necessary calibrations requires much valuable time in the test section. Therefore, it is not well suited for routine testing, and is primarily restricted to research.

The influence of wheel rotation per se on the drag of passenger cars has been evaluated using this technique (Figure 26, Mercker et al 1991). Rotating wheels make a smaller contribution to vehicle drag than stationary ones. In addition, wheel rotation increases vehicle lift significantly.

It is argued that tangential blowing over a fixed floor comes close to a moving-floor simulation (Figure 27, Mercker & Knape 1989). In order to cope with wheel rotation, special wheel pads are under consideration. Either a miniature moving belt or a pair of little rollers integrated into the pads under each wheel may provide a practical solution.

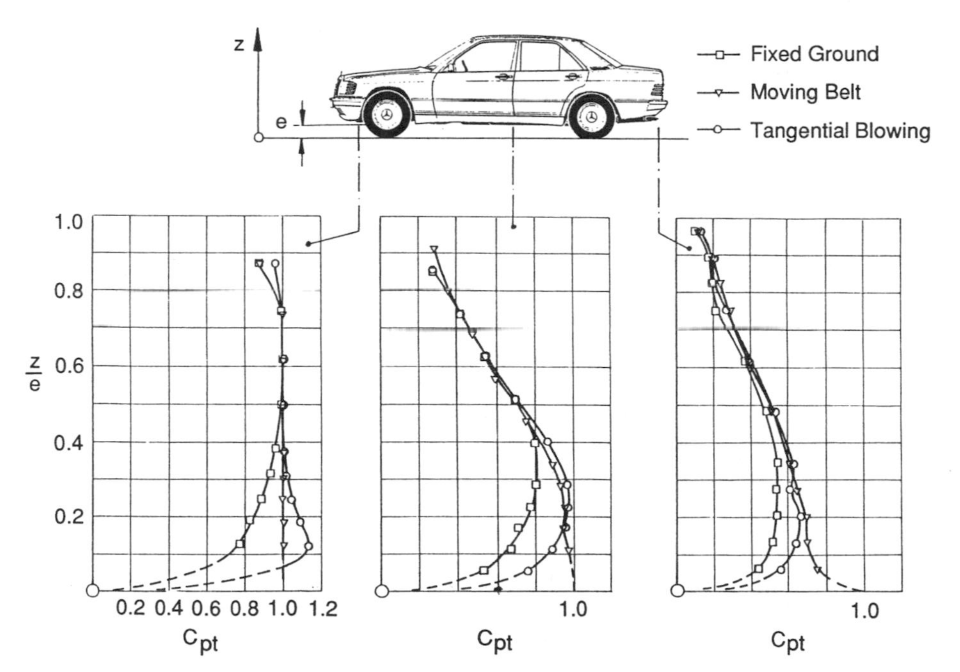

Figure 27 Dynamic-head distribution [$C_{pt} = (P_t - p_\infty)/q_\infty$] in the centerplane under a vehicle in a wind tunnel with three different types of ground plane simulation: fixed ground, moving belt, and tangential blowing.

In some fixed-floor automotive wind tunnels, boundary layer suction is applied, most frequently as concentrated suction. Air is sucked off through a strip of perforated sheet metal at the nozzle exit. The alternative, distributed suction, is used in only one tunnel (Vagt & Wolff 1987, Eckert et al 1992). In this case air is sucked through a large porous area of the ground floor in front of and under the car. The use of several suction chambers permits spatial variation of the suction rate over the surface of the floor.

In comparison to tangential blowing, suction has several disadvantages. If the boundary layer is to be reduced by the same amount, a much larger volume of air must be removed than is added when blowing is employed. Consequently, the negative angle of attack of the oncoming flow induced by the suction may exceed a tolerable value for car tests, say 0.2°. Forces and moments on a car are very sensitive to angle of attack.

With distributed suction, it is also not clear how much air should be sucked off, nor where. In comparison to the natural floor boundary layer in an empty test section, the thickness of the boundary layer on the ground underneath a moving vehicle is small (Hucho et al 1975), but not known a priori. Therefore, it is difficult to define the proper suction rate and distribution. In the case of tangential blowing, it has been empirically determined that the blowing rate required for zero displacement thickness in an empty tunnel at the location of the front wheels of a car will result in aerodynamic measurements in good agreement with those obtained with a moving belt (Mercker & Knape 1989).

5.4 *New Test Facilities*

Aerodynamic noise can be investigated in a wind tunnel only if there is a sufficient difference in sound-pressure level between the noise produced by the flow around (and through) a vehicle and the background noise of the tunnel. The formerly required 10 dB(A) difference (Buchheim et al 1983) may be sufficient for objective measurement of the external sound pressure field, but it is inadequate for subjective assessment of wind noise inside a car. Aero-acoustic wind tunnels have now been built which provide a 30 dB(A) sound-pressure-level difference (Ogata et al 1987, Hucho 1989a). Existing tunnels are now being equipped with additional sound dampers, even at the expense of a lower maximum wind speed, to make them better suited for aero-acoustic investigations.

Thermal tests make up the other half of vehicle aerodynamics. Adequate comfort for passengers and proper functioning of the engine and accessories have to be guaranteed under all operating conditions. One approach to thermal testing is to perform all aerodynamic test work in the same facility (Moerchen 1968). The other is to use separate specialized facilities for each kind of test. The latter has turned out to be the more efficient. Facilities for thermal tests can tolerate a less accurate simulation of the flow field around a car, so they can be built much smaller. Furthermore, the number of aerodynamic and thermal tests has grown so large that more than one facility is needed in any case. For air conditioning work in passenger cars, climatic tunnels with a test-section area between 6 and 10 m^2 are suitable (Buchheim et al 1986). An airstream of even smaller cross-section can be used for engine-cooling tests, and 2 to 4 m^2 are sufficient (Basshuysen et al 1989).

5.5 *Measurement Techniques*

The motivations for improving measurement techniques have been, and still are, threefold: time saving, increased accuracy, and more-detailed information for reaching the ever-higher aerodynamic targets.

Routinely, the frontal area of every model has to be determined. Formerly, this was done by projecting the shadow of a vehicle on a screen using a spotlight a long distance in front of it, and integrating the area of the shadow by hand. This procedure has now been replaced by several fully automated techniques using laser light (Buchheim et al 1987) which are far faster and more accurate.

For a long time, only three kinds of measurement were performed during routine vehicle-shape development: forces and moments; pressure distribution on specific sectors of the surface; and flow visualization by various techniques (Hucho & Janssen 1977). Information on the flow field was rather limited, leaving shape modification to the imagination and experience of the test engineer.

Deeper insight into flow mechanisms should provide a more rational basis for development work, but this requires additional types of data. Boundary layer profiles on the external surfaces and underhood velocity distributions are now being measured by laser-Doppler anemometry (LDA) (Buchheim et al 1987, Cogotti & Berneburg 1991). Wake surveys are being made with multi-hole probes (Cogotti 1987, 1989) and evaluated on the basis of momentum considerations (see Section 4.1).

5.6 *Computational Fluid Dynamics*

The primary reason for the automobile industry's interest in numerical methods is to save time during product development. The ability to quickly react to the ever-changing needs of the market has even higher priority than cost saving, which is very important in its own right. Consequently, all numerical methods need to satisfy two conditions—the first one is necessary, but only the second is sufficient:

- They have to reproduce the related physics with adequate accuracy;
- They have to be faster than experiment.

Only in some narrow niches does computational fluid dynamics as applied to vehicles fulfill both these conditions today. CFD is still more a subject of research than a development tool.

What accuracy means in this context becomes evident from the following requirement: CFD must be able to discriminate a change in drag as small as $\Delta C_D = 0.002$ (Ahmed 1992). Also, computing time has to be competitive with the wind tunnel (Hucho 1989b). The time needed to carry out one specific modification to a model and make the related measurement in a wind tunnel is only from five to ten minutes in partial scale, and from ten to fifteen minutes in full scale.

However, when it finally becomes practical, CFD will have much to offer. First of all, it can generate information before a testable model even exists. Secondly, numerical methods are not necessarily burdened with the limitations of the wind tunnel. For example, computational space can be made large enough to eliminate blockage effects, although not without cost. Also, simulation of the relative motion between vehicle and road and rotation of the wheels are comparatively easy to accommodate. Finally, once the equations have been solved, there is much more information available than from a routine experiment.

Exterior designers would be happy if reliable predictions of aerodynamic characteristics could be made right now. More and more, their creative design process is taking place on the computer. The selection of candidate designs is being made from the computer screen, and aerodynamic data (e.g. C_D) could contribute to these pre-hardware decisions, but only if CFD is able to generate results on time. Ultimately, it should only be necessary to build one or two clay models to finalize a configuration.

Consequently, a large effort is under way to improve CFD, and this is documented in numerous recent publications (e.g. Ahmed 1992, Kobayashi & Kitoh 1992). Four different routes are being followed in the application to road vehicles. They are based on:

- Laplace's equation,
- Reynolds-averaged Navier-Stokes equations (RANS),
- instantaneous Navier-Stokes equations, called direct numerical simulation (DNS),
- zonal models (hybrids).

The Laplace equation is being solved by the well-known panel method; only the surface of the model (and the road) have to be discretized. The vortex-lattice method has only been applied infrequently to cars (Stafford

1973), but was recently used to simulate the trailing vortices at their slanted base (Hummel & Ramm 1992). By definition, only nonviscous flow can be treated with the panel method, and hence no information on drag can be expected. Results are improved if the wake is modeled (Ahmed & Hucho 1977). Its geometry has to be taken from experiment, or from intuition. Pressure distribution can be predicted fairly well on all surfaces where the flow is attached. Forces and moments on body panels like doors, hood, and windows can be calculated. The panel method has been successfully applied to the development of the German high-speed train ICE (Inter City Express, Mackrodt et al 1980). The shape of the train's front end had to be designed to create the smallest possible pressure wave, in order to avoid problems when entering tunnels, passing through stations, or when two trains pass going in opposite directions.

The Reynolds-averaged Navier-Stokes equations (RANS) need a turbulence model for closure, and there is much controversy in the fluid mechanics community as to which one is the most appropriate. In engineering applications the k-ε model is the one most widely used. Close to a wall the empirical logarithmic law of the wall is usually applied, but has only been validated for plane two-dimensional boundary layers. As previously described, the flow over a road vehicle is anything but 2-D.

A very good result has been reported with the RANS concept (Hutchings & Pien 1988). Drag was predicted correctly for a generic car model; however, there were significant differences between computed and measured pressure distributions on the upper and lower rear surfaces. This highlights the need to validate more than the prediction of overall aerodynamic forces, since these might fortuitously camouflage the presence of large, but compensating, errors in local flows.

To date, none of the RANS codes has reproduced the large changes in flow pattern and drag that take place at a critical slant angle of the upper-rear surface of vehicles (see Section 4.1). A computation of such critical behavior has, however, been made using the full instantaneous Navier-Stokes equations (Figure 28, Tsuboi et al 1988). The longitudinal slender cylinder with slanted base of Morel (1978a) was used, and it was located out of ground effect. This configuration was used by Morel to explore the drastic change in drag at a slant angle of $\simeq 30^\circ$ discovered during the development of the Volkswagen Rabbit (Figure 10b, Janssen & Hucho 1975). Numerical results (drag coefficient and flow pattern) agree well with the measurements. Whether or not all the turbulence scales at the test Reynolds number were adequately resolved with the computational grid employed is an open question.

This result has stimulated numerous applications of DNS to actual vehicle development problems in Japan. With almost 10^6 grid points for

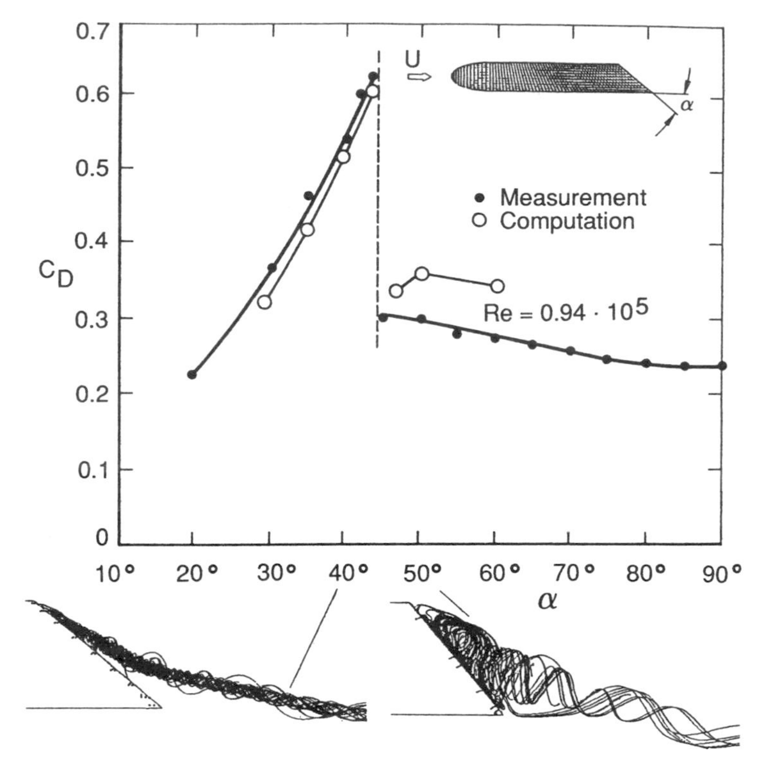

Figure 28 Drag versus rear slant angle α computed with a DNS code and compared to measurements at the same Reynolds number (measurement: Morel 1978a; computation: Tsuboi et al 1988). Bottom: computed flowpaths of the vortices emanating from the sharp edge of the slant.

a half model, DNS has been able to discriminate the effect of several aerodynamic devices (spoilers, flaps) on the drag and lift of a sports car (Figure 29, Kataoka et al 1991). Grid generation is said to require only three days, and CPU-time for a single configuration between 10 and 20 hours on a supercomputer. Calculated drag compares to measurement within 5%.

Hybrid computational schemes need some a priori knowledge of the flow field. The field around a model is partitioned (Larsson et al 1991) into zones of attached flow and zones where separation is expected to occur. The flow in the first type of zone is computed iteratively with a panel method and a boundary layer code. The second zone is treated with either RANS or DNS. An alternative approach is to compute the attached flow with an Euler code and a boundary layer code, and to model the wake.

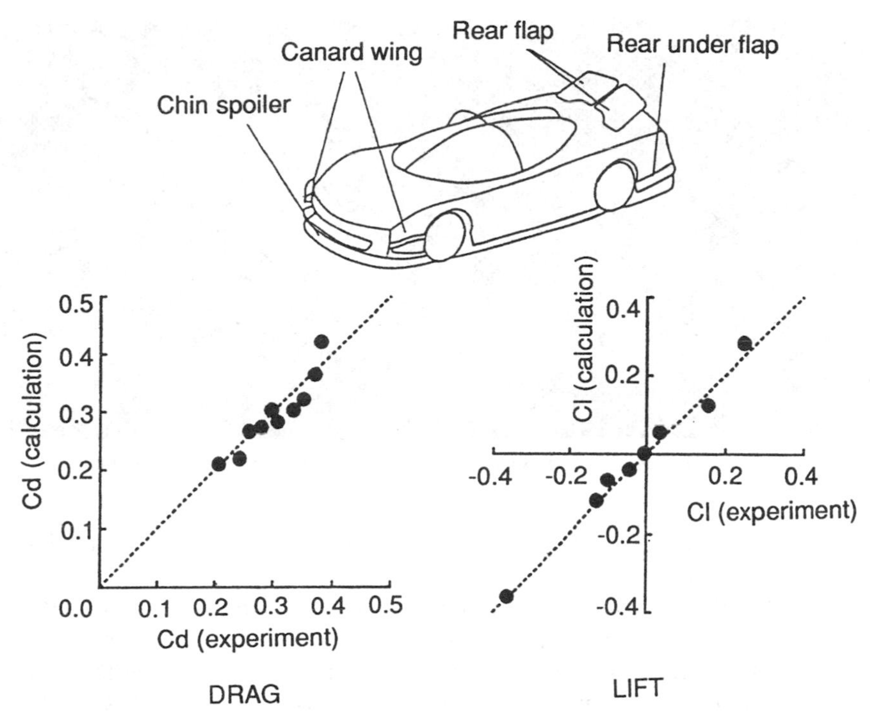

Figure 29 Comparison of computed (DNS) and measured drag and lift coefficients for a sports car with various aerodynamic devices.

Vorticity is generated within the boundary layer, and its transport can be computed with the Euler equations. A promising result has been achieved for two-dimensional flow around a circular cylinder (Krukow & Stricker 1990). No 3-D results and no details of the wake model have been published.

CFD has also been applied to internal flow systems, whose common feature is that they are nonisothermal. Underhood flow has been computed with RANS using the k-ε turbulence model (see e.g. Kuriyama 1988). Interaction between the interior and exterior flows has been treated by simultaneous computation of both flow fields (Ono et al 1992). The flow pattern inside the passenger compartment of cars and buses attracts increasing attention. For moderate air fluxes the velocities are sufficiently small that a laminar treatment is appropriate (Figure 30, Glober & Rumez 1990); for greater fluxes, turbulent flows have been computed (e.g. Ishihara et al 1991). Temperature and velocity distributions inside the compartment have been used to compute the thermal comfort of passengers, with

Figure 30 Streamlines inside a passenger compartment without passengers; laminar flow.

humans being described by a discretized thermo-physiological model (Dick & Stricker 1988).

CFD lends itself to the development of corrections for wind-tunnel data, but has been used for that purpose in only a few cases. However, the contours of interference-free streamline-shaped tunnel walls, either fixed (L. G. Stafford 1979, personal communication) or adjustable (Goenka 1990), have been computed. The geometries of the vehicles in the test section were comparatively simple because only their far-field flow was needed. A blockage ratio of up to 20% can be accommodated with properly shaped walls.

5.7 *Empirical Methods for Drag Estimation*

Early in the recent history of automative aerodynamics it was perceived as a shortcoming that the aerodynamic characteristics of a vehicle could only be established by measurement. The Motor Industry Research Association (MIRA) tried to overcome this with a rating method for drag (White 1967). It made use of nine descriptors of a vehicle's shape. Each geometric feature, such as the front end or the A-pillar, had to be subjectively rated with regard to its drag contribution. The drag of cars was high in those days, the coefficient ranging from 0.50 to 0.60. Freestanding headlamps and fenders were still in vogue. The method had the virtue of identifying those areas that were most important for drag, but it was unsuited for shapes with lower drag. Later, when CFD was still not recognized as being applicable to vehicle design, MIRA made a second attempt at a rating method (Carr 1987). Similar to the first one, drag is broken down according to the location of generation: forebody, afterbody, underbody, wheel and wheel wells, protuberances (e.g. mirrors), and engine cooling system. The contribution of each of these is determined using empirical functions which correlate geometric parameters (similar to those shown in Figure 5) that can be read from a vehicle drawing with local drag. Some 50 generic details are taken into account, 27 empirical (confidential) constants are applied, and no subjective rating is required. Validation of the method with 26 cars (Rose 1984) resulted in a fairly high accuracy of $\pm 5\%$. The method could be integrated into a CAD system, but its limits of applicability are difficult to define. Also, a major drawback is its linearity: No interaction between the various regions of a car body is allowed for. Work is currently under way at MIRA to extend this rating method to lift and lateral coefficients.

An attempt to develop an expert system for drag prediction failed despite promising first results (Buchheim et al 1989), and has been abandoned.

6. OPEN ISSUES AND FUTURE TRENDS

Using drag coefficient as the measure of the state of the art in vehicle aerodynamics, further progress appears to be possible. Today, well-tuned cars have drag coefficients of about 0.30. However lower values are possible (Figure 6), and have already been realized with production cars (Emmelman et al 1990). Whether or not they are feasible for any particular car is more a question of consistency with the vehicle's design concept than it is of aerodynamic capability.

Drag will be of even greater importance for some of the specialty vehicles likely to be on the market in the future (e.g. electric, hybrid). The greatest problem of electric vehicles—limited driving range—is directly impacted by drag. Drag coefficients less than 0.20 are feasible, and electrics can provide the incentive for achieving them. However, the development process will be of increased difficulty because there appears to be significant interaction between the local flows in different areas of a car body at these low drag levels.

If future fuel economy regulation forces cars to become much lighter, the other components of the resultant aerodynamic force will become more important. Directional stability can be improved by reducing lift, mainly at the rear axle. Sensitivity to crosswinds must be kept under control by reducing yawing moment and side force.

The management of underhood airflow will become more and more difficult. Fuel economy regulation is forcing more compact body shapes, and a current styling feature is lower hood lines. Together, they are dictating smaller, and hence more tightly packed, engine compartments. At the same time, the airflow requirement is increasing. Firstly, ozone-layer depletion has mandated a worldwide elimination of the chlorofluorocarbon refrigerant (freon) currently used in automotive air conditioning systems. (in 1991, ≃95% of the new cars sold by U.S. manufacturers had air conditioning). Its environmentally friendly replacement is a less effective refrigerant and requires more airflow for condenser cooling. Secondly, higher underhood temperatures are having a deleterious effect on the durability of engine and vehicle accessories in the engine compartment, and so these temperatures must be controlled. The combination of increased airflow requirement, higher-resistance air path, and under-bumper air inlets will make thermal management of the underhood compartment even more difficult than it is today. Consequently, it will require an even larger fraction of the aerodynamic effort expended in new-vehicle development.

Other parts of the flow field will also call for closer attention. Splash and spray have to be kept under control—not only for trucks where technically feasible solutions are waiting for application. Wind noise has to be reduced as the engine and tires become quieter.

The designers of vehicle shape need more room in which to maneuver. Sometimes they create new aesthetic shapes that are unfavorable for low drag. The required aerodynamic compensation has to come from places that do not affect appearance. Such regions are the vehicle's underside, the wheels, and the wheel wells.

From experiments in both research (Cogotti 1983) and development (Mercker et al 1991) it can be concluded that wheels and wheel wells contribute approximately half the drag of a low-drag car. Major contributors are the front wheels. Due to an outwardly spreading flow under the front part of a vehicle, the local flow approaches the front wheels at a large yaw angle, resulting in high wheel drag. Wake measurements (Cogotti 1987) also confirm this. Because cars must have wheels, and unconventional wheel arrangements are unacceptable, the only way to improve the situation is to suitably control the lateral spreading of the flow under a vehicle. However, only very modest attempts at this have been made so far (Emmelmann et al 1990).

With respect to wind-tunnel test techniques, ground simulation has to be improved for routine testing. Miniature moving belts or mini-dynos, one under each wheel, together with tangential blowing seems to be a feasible approach. Better boundary corrections (adjustments) for open-jet test sections are to be expected by using CFD.

Computer technology per se continues to improve at an almost astonishing rate. More and more aerodynamic computations (CFD) will become possible. They will definitely have a role in research. The question is the degree to which they will be feasible in vehicle development (accuracy, cost, time). How will CFD stack up against intelligent experimentation, e.g. as exemplified by the detail-optimization technique? Will it cope better with the increasingly interactive nature of the flow fields associated with very-low-drag shapes which is making experimental implementation of that technique more difficult? In principle, DNS can do it all. But are computational grids fine enough to discriminate all the scales of turbulent flow at realistic operating Reynolds numbers possible and/or feasible? Even if they are, is the huge amount of detailed data generated actually needed for good aerodynamic design?

Despite some fascinating published results from CFD, the technology is still applicable in only limited areas during the daily course of aerodynamic engineering. From the authors' point of view, the major reservations on existing CFD codes are that, with the exception of the panel method, their accuracy and limits are not well known. Many published comparisons with experiment have been made for body geometries having flows that are too complex. Consequently, it is not possible to determine whether the differences between experiment and CFD are due to the physics of the equations (e.g. turbulence model), numerical characteristics (e.g. mesh geometry, numerical algorithm, computation scheme), or a combination of the two.

As has been common practice in other areas of fluid dynamics, computational codes should be tested—and developed—by comparing their results with either exact (analytical) solutions (which don't exist for complex turbulent flows) or with experiments, but one should start with simple flows. Only when these are fully understood is the step to more complex flows justified. Configurations like cars and trucks have extremely complex flow fields. They should be tackled at the end of a code development process, not at the beginning.

Literature Cited

Ahmed, S. R. 1981. Wake structure of typical automobile shapes. *Trans. ASME, J. Fluids Engrg*. 103: 162–69

Ahmed, S. R. 1984. Influence of base slant on the wake structure and drag of road vehicles. *Trans. ASME, J. Fluids Engrg*. 105: 429–34

Ahmed, S. R. 1992. Numerische Verfahren. See Hucho 1992, Chap. 14

Ahmed, S. R., Hucho, W.-H. 1977. The calculation of the flow field past a van with the aid of a panel method. *SAE Pap. 770 390*

Ahmed, S. R., Ramm, R., Faltin, G. 1984. Some salient features of the time averaged ground vehicle wake. *SAE Pap. 840 300*

Aoki, K., Miyata, H., Kanai, M., Hanaoka, Y., Zhu, M. 1992. A water-basin test technique for the aerodynamic design of road vehicles. *SAE Pap. 920 348*

Barth, R. 1960. Der Einfluss unsymmetrischer Stroemung auf die Luftkraefte an Fahrzeugmodellen und aehnlichen Koerpern. *Automobiltech. Z.* 62: 80–95

Basshuysen, R. v., Chemnitzer, E., Stock, D. 1989. Der neue 3-m2- Klimawindkanal mit Allrad-Rollenpruefstand bei Audi. *Automobiltech. Z.* 91: 646–52

Bearman, P. W. 1984. Some observations on road vehicle wakes. *SAE Pap. 840 301*

Beauvais, F. N., Tignor, S. C., Turner, T. R. 1968. Problems of ground simulation in automotive aerodynamics. *SAE Pap. 680 121*

Buchheim, R., Deutenbach, K.-R., Lueckhoff, H.-J. 1981. Necessity and premises for reducing the aerodynamic drag of future passenger cars. *SAE Pap. 810 185*

Buchheim, R., Dobrzynski, W., Mankau, H., Schwabe, D. 1983. Vehicle interior noise related to external aerodynamics. *Int. J. Veh. Des.* SP 3

Buchheim, R. Schwabe, D., Roehe, H. 1986. Der neue 6 m2- Klimawindkanal von Volkswagen. *Automobiltech. Z.* 88: 211–18, 389–92

Buchheim, R., Durst, F., Beeck, M. A., Hentschel, W., Piatek, R., Schwabe, D. 1987. Advanced experimental techniques and their application to automotive aerodynamics. *SAE Pap. 870 244*

Buchheim, R., Knorr, G., Mankau, H., Tang, T. 1989. Expertensystem zur Unterstuetzung der Aerodynamikoptimierung von Fahrzeugen. *GMD-Stud.* 160: 60–83

Carr, G. W. 1987. New MIRA drag reduction prediction method for cars. *Auto. Eng.* June/July 1987: 34–38

Cogotti, A. 1983. Acrodynamic characteristics of car wheels. *Int. J. Veh. Des.* SP 3: 173–96

Cogotti, A. 1987. Flow-field surveys behind three squareback car models using a new fourteen-hole probe. *SAE Pap. 870 243*

Cogotti, A. 1989. A strategy for optimum surveys of passenger-car flow fields. *SAE Pap. 890 374*

Cogotti, A., Berneburg, H. 1991. Engine compartment airflow investigations using a laser-doppler velocimeter. *SAE Pap. 910 308*

Dick, A., Stricker, R. 1988. Zur Bewertung inhomogenen Klimas im Pkw durch ein thermophysiologisches Insassenmodell. *VDI-Bere.* 699: 247–64, Duesseldorf

Donges, E., Mueller, B., Seifdenfuss, T. 1990. Design of the BMW-safety concept for active rear-axle kinematics. *23rd FISITA Congr., Torino,*Pap. 905 059

Eckert, W., Vagt, J.-D., Wolff, B. 1990. Die Messtrecke mit geschlitzten Waenden im Porsche Windkanal. *Automobiltech. Z.* 92: 286–97

Eckert, W., Singer, N., Vagt, J.-D. 1992. The Porsche wind tunnel floor-boundary-layer control—a comparison with road data and results from moving belt. *SAE Pap. 920 346*

Emmelmann, H.-J. 1987a. Driving stabilty in side winds. See Hucho 1987, Chap. 5

Emmelmann, H.-J. 1987b. Performance of cars and light vans. See Hucho 1987, Chap. 3

Emmelmann, H.-J., Berneburg, H., Schulze, J. 1990. Aerodynamic development of the Opel Calibra. *SAE Pap. 900 327*

Flegl. H., Rauser, M. 1992. Hochleistungsfahrzeuge. See Hucho 1992, Chap. 7

Gengenbach, W. 1987. Heating, ventilation and air conditioning of motor vehicles. See Hucho 1987, Chap. 10

George, A. R. 1989. Automobile aeroacoustics. *AIAA Pap.* 89–1067, San Antonio, TX

Gilhaus, A., Renn, V. 1986. Drag and driving-stability related aerodynamic forces and their interdependence—results of measurements on 3/8 scale basic car shapes. *SAE Pap. 860 211*

Gilhaus, A., Hoffmann, R. 1992. Richtungsstabilitaet. See Hucho 1992, Chap. 5

Glober, S., Rumez, W. 1990. Berechnung der Stroemung und Temperaturverteilung in Fahrzeugkabinen mit einem Navier-Stokes-Verfahren. *VDI-Ber.* 816: 335–44

Goehring, E., Kraemer, W. 1987. Seitliche Fahrgestellverkleidungen fuer Nutzfahrzeuge. *Automobiltech. Z.* 89: 481–88; 659–66

Goenka, L. N. 1990. A numerical model to determine vehicle pressure-simulation accuracy in a slotted-wall automotive wind tunnel. *SAE Pap. 900 188*

Hackett, J. E., Wilsden, D. J. 1975. Determination of low speed wake blockage correction via tunnel wall static pressure measurements. *AGARD Fluid Dynamics Panel, Symp. on Wind Tunnel Des. Testing Techniques, London*

Hackett, J. E., Sugavanam, A. 1984. Evaluation of a complete wake integral for the drag of a car-like shape. *SAE Pap. 840 577*

Haruna, S., Kamimoto, I., Okamoto, S. 1992. Estimation method for automobile aerodynamic noise. *SAE Pap. 920 205*

Hensel, R. W. 1951. Rectangular wind tunnel corrections using the velocity ratio method. *NACA TN* 2372

Hucho, W.-H. 1974. *Versuchstechnik in der Fahrzeugaerodynamik. Koll. Industrieaerodynamik, Teil 3: Aerodynamik von Strassenfahrzeugen.* Aachen

Hucho, W.-H. 1978. The aerodynamic drag of cars—current understanding, unresolved problems and future prospects. See Sovran et al 1978, pp. 7–44

Hucho, W.-H., ed. 1987a. *The Aerodynamics of Road Vehicles.* London: Butterworth

Hucho, W.-H. 1987b. Aerodynamic drag of passenger cars. See Hucho 1987a, Chap. 4

Hucho, W.-H. 1989a. Neuartiger Akustik-Windkanal bei BMW. *Auto. Rev.* 41/5.10.1989: 43–45

Hucho, W.-H. 1989b. Numerischer Windkanal—Stromlinienautos aus dem Supercomputer? *Comput. Technol.* 1989/10: 44–62

Hucho, W.-H., ed. 1992. *Aerodynamik des Automobils.* Duesseldorf: VDI. In press

Hucho, W.-H., Emmelmann, H.-J. 1973.

Theoretical prediction of the aerodynamic derivatives of a vehicle in cross wind gusts. *SAE Pap. 730 232*

Hucho, W.-H., Janssen, L. J., Schwarz, G. 1975. The wind tunnel's ground floor boundary layer—its interference with the flow underneath cars. *SAE Pap. 750 066*

Hucho, W.-H., Janssen, L. J., Emmelmann, H.-J. 1976. The optimization of body details—a method for reducing the aerodynamic drag of vehicles. *SAE Pap. 760 185*

Hucho, W.-H., Janssen, L. J. 1977. Flow Visualization techniques in vehicle aerodynamics. *Int. Symp. on Flow Visualization, Tokyo*

Hummel, D., Ramm, G. 1992. A panel method for the computation of the flow around vehicles including side-edge vortices and wakes. *Proc. ATA-Conf. Innovation and Reliability in Auto. Des. and Testing, Florence*

Hutchings, B. J., Pien, W. 1988. Computation of three-dimensional vehicle aerodynamics using FLUENT/BFC. See Marino 1988, pp. 233–55

Ishihara, Y., Hara, J., Sakamoto, H., Kamemoto, K., Okamoto, H. 1991. Determination of flow velocity distribution in a vehicle interior using visualization and computation techniques. *SAE Pap. 910 310*

Janssen, L. J., Hucho, W.-H. 1975. Aerodynamische Entwicklung von VW-Golf und VW-Scirocco. *Automobiltech. Z.* 77: 1–5

Jones, R. T. 1978. Discussion. See Sovran et al 1978, pp. 40–44

Kamm, W., Schmid, C., Riekert, P., Huber, L. 1934. Einfluss der Reichsautobahn auf die Gestaltung der Kraftfahrzeuge. *Automobiltech. Z.* 37: 341–54

Kataoka, T., China, H., Nakagawa, K., Yanagimoto, K., Yoshida, M. 1991. Numerical simulation of road vehicle aerodynamics and effect of aerodynamic devices. *SAE Pap. 910 597*

Kieselbach, R. J. F. 1982a. *Streamline Cars in Germany. Aerodynamics in the Construction of Passenger Vehicles* 1900–1945. Stuttgart: Kohlhammer Ed. Auto und Verk.

Kieselbach, R. J. F. 1982b. *Streamline Cars in Europe and USA. Aerodynamics in the Construction of Passenger Vehicles* 1900–1945. Stuttgart: Kohlhammer Ed. Auto und Verk.

Kieselbach, R. J. F. 1983. *Aerodynamically Designed Commercial Vehicles* 1931–1961 *Built on the Chassis of: Daimler Benz, Krupp, Opel, Ford.* Stuttgart: Kohlhammer Ed. Auto und Verk.

Klemperer, W. 1922. Luftwiderstandsuntersuchungen an Automodellen. *Z. Flugtech. Motorluftschiffahrt.* 13: 201–6

Kobayashi, T., Kitoh, K. 1992. A review of CFD methods and their application to automobile aerodynamics. *SAE Pap. 920 338*

Koenig-Fachsenfeld, R. v. 1951. *Aerodynamik des Kraftfahrzeugs.* Frankfurt: Umschau

Koenig-Fachsenfeld, R. v., Ruehle, D., Eckert, A., Zeuner, M. 1936. Windkanalmessungen an Omnibusmodellen. *Automobiltech. Z.* 39: 143–49

Krukow, G., Stricker, R. 1990. Einsatzpotential gekoppelter Verfahren zur Simulation von 3D-Fahrzeugstroemungen. *VDI Ber.* 816: 367–77, Duesseldorf

Kuenstner, R., Deutenbach, K.-R., Vagt, J.-D. 1992. Measurement of reference dynamic pressure in open-jet automotive wind tunnels. *SAE Pap. 920 344*

Kuriyama, T. 1988. Numerical simulation on three dimensional flow and heat transfer in the engine compartment using "STREAM." See Marino 1988, pp. 283–92

Larsson, L., Nilsson, L. U., Berndtsson, A., Hammar, L., Knutson, K., Danielson, H. 1989. A study of ground simulation-correlation between wind-tunnel and water-basin tests of a full-scale car. *SAE Pap. 890 368*

Larsson, L., Broberg, L., Janson, C.-E. 1991. A zonal method for predicting external automoblie aerodynamics. *SAE Pap. 910 595*

Mackrodt, P.-A., Steinheuer, J., Stoffers, G. 1980. Entwicklung aerodynamisch optimaler Formen fuer das Rad/Schiene-Versuchsfahrzeug II. *Arch. Eisenbahntech.* 35: 67–77

Mackrodt, P.-A., Pfitzenmaier, E. 1987. Aerodynamik und Aeroakustik fuer Hochgeschwindigkeitszuege. *Phys. Unserer Zeit.* 18: 65–76

Marino, C., ed. 1988. Supercomputer applications in automotive research and development. *Proc. 2nd Int. Conf. on Supercomputing Appl. in the Auto. Ind., Seville.* Minneapolis: Cray Res.

Mercker, E. 1986. A blockage correction for automotive testing in a wind tunnel with closed test section. *J. Wind Eng. Ind. Aerodyn.* 22: 149–67

Mercker, E., Knape, H. W. 1989. Ground simulation with moving belt and tangential blowing for full scale automotive testing in a wind tunnel. *SAE Pap. 890 248*

Mercker, E., Wiedemann, J. 1990. Comparison of different ground simulation techniques for use in automotive wind tunnels. *SAE Pap. 900 321*

Mercker, E., Breuer, N., Berneburg, H., Emmelmann, H.-J. 1991. On the aerodynamic interference due to the rolling wheels of passenger cars. *SAE Pap. 910 311*

Moerchen, W. 1968. The climatic wind tunnel of Volkswagenwerk AG. *SAE Pap. 680 120*

Morel, T. 1978a. The effect of base slant on the flow pattern and drag of three-dimensional bodies with blunt ends. See Sovran et al. 1978, pp. 191–226

Morel, T. 1978b. Aerodynamic drag of bluff body shapes characteristic of hatch-back cars. *SAE Pap. 780 267*

Nouzawa, T., Haruna, S., Hiasa, K., Nakamura, T., Sato, H. 1990. Analysis of wake pattern for reducing aerodynamic drag of notchback model. *SAE Pap. 900 318*

Nouzawa, T., Hiasa, K., Nakamura, T., Kawamoto, A. 1992. Unsteady-wake analysis of the aerodynamic drag of a notchback model with critical afterbody geometry. *SAE Pap. 920 202*

Ogata, N., Iida, N., Fuji, Y. 1987. Nissan's low-noise full scale wind tunnel. *SAE Pap. 870 250*

Ono, K., Himeno, R., Fujitani, K., Uematsu, Y. 1992. Simultaneous computation of the external flow around a car body and the internal flow through its engine compartment. *SAE Pap. 920 342*

Onorato, M., Costelli, A. F., Garrone, A. 1984. Drag measurement through wake analysis. *SAE Pap. 840 302*

Piatek, R. 1987. Operation, safety and comfort. See Hucho 1987a, Chap. 6

Piech, F. 1992. 3 Liter/100 km im Jahr 2000. *Automobiltech. Z.* 94: 20–23

Rauser, M., Eberius, J. 1987. Verbesserung der Fahrzeugaerodynamik durch Unterbodengestaltung. *Automobiltech. Z.* 89: 535–42

Retzlaff, R. N., Hertz, P. B. 1990. Airfoil plan-view body shapes to reduce drag at yaw. *SAE Pap. 900 314*

Rose, M. J. 1984. Appraisal and modification of an empirical method for predicting the aerodynamic drag of cars. *MIRA Released Rep.* 1984/1

SAE 1990. Aerodynamic testing of road vehicles: open-jet wind tunnel boundary interference. *SAE Inform. Rep. J* 2071

SAE 1991. Aerodynamic testing of road vehicles: closed-test-section wind tunnel boundary interference. *SAE Inform. Rep. J* 2085

Schaub, U. W., Olson, M. E., Raimondo, S. 1990. Correction of wind tunnel force data for yawed full and half-scale truck models using a modified pressure-signature method. *SAE Pap. 900 187*

Sorgatz, U., Buchheim, R. 1982. Untersuchungen zum Seitenwindverhalten zukuenftiger Fahrzeuge. *Automobiltech. Z.* 84: 11–18

Sovran, G. 1983. Tractive-energy-based formulae for the impact of aerodynamics on fuel economy over the EPA driving schedules. *SAE Pap. 830 304*

Sovran, G. 1984. The effect of ambient wind on a road vehicle's aerodynamic work requirement and fuel consumption. *SAE Pap. 840 298*

Sovran, G., Morel, T., Mason, W. T., eds. 1978. *Aerodynamic Drag Mechanisms of Bluff Bodies and Road Vehicles.* New York: Plenum

Sovran, G., Bohn, M. S. 1981. Formulae for the tractive-energy requirements of vehicles driving the EPA schedules. *SAE Pap. 810 184*

Squire, H. B., Pankhurst, R. C. 1952. *ARC CP* 80

Stafford, L. G. 1973. A numerical method for the calculation of the flow around a motor vehicle. *Proc. Adv. Road Veh. Aerodyn. BHRA Fluid Eng., Cranfield*, pp. 167–83

Stapleford, W. R., Carr, G. W. 1971. Aerodynamic noise in road vehicles. *MIRA Rep.* 1971/2

Stollery, J. L., Burns, W. K. 1969. Forces on bodies in the presence of the ground. *Proc. 1st Symp. on Road Veh. Aerodynam.*, London, Nov. 6–7

Tsuboi, K., Shirayama, S., Oana, M., Kuwahara, K. 1988. Computational study of the effect of base slant. See Marino 1988, pp. 257–72

Vagt, J.-D., Wolff, B. 1987. Das neue Messzentrum fuer Aerodynamik—Zwei neue Windkanaele bei Porsche. Teil 1: *Automobiltech. Z.* 89: 121–29; Teil 2: *Automobiltech. Z.* 89: 183–89

von Schulz-Hausmann, K., Vagt, J.-D. 1988. Influence of test section length and collector area on measurements in 3/4-open-jet automotive wind tunnels. *SAE Pap. 880 251*

White, R. G. S. 1967. A rating method for assessing vehicle aerodynamic drag coefficients. *MIRA Rep.* 167/9

Wiedemann, J. 1986. Optimierung der Kraftfahrzeugdurchstroemung zur Steigerung des aerodynamischen Abtriebes. *Automobiltech. Z.* 88: 429–31

Wiedemann, J., Ewald, B. 1989. Turbulence

manipulation to increase effective Reynolds numbers in vehicle aerodynamics. *AIAA J.* 27: 763–69

Williams. J. E., Hackett, J. E. Oler, J. W. Hamar, L. 1991. Water flow simulation of automotive underhood airflow phenomena. *SAE Pap. 910 307*

Willumeit, H. P., Matheis, A., Mueller, K. 1991. Korrelation von Untersuchungsergebnissen zur Seitenwindempfindlichkeit eines Pkw im Fahrsimulator und Prueffeld. *Automobiltech. Z.* 93: 28–35

Wright, P. G. 1983. The influence of aerodynamics on the design of Formula-One racing cars. *Int. J. Veh. Des.*, SP 3: 158–72

Annu. Rev. Fluid Mech. 1993. 25: 539–75

THE PROPER ORTHOGONAL DECOMPOSITION IN THE ANALYSIS OF TURBULENT FLOWS

Gal Berkooz, Philip Holmes, and John L. Lumley

Cornell University, Ithaca, New York 14853

KEY WORDS: coherent structures, empirical eigenfunctions, modeling, turbulence

1. INTRODUCTION

1.1 *The Problems of Turbulence*

It has often been remarked that turbulence is a subject of great scientific and technological importance, and yet one of the least understood (e.g. McComb 1990). To an outsider this may seem strange, since the basic physical laws of fluid mechanics are well established, an excellent mathematical model is available in the Navier-Stokes equations, and the results of well over a century of increasingly sophisticated experiments are at our disposal. One major difficulty, of course, is that the governing equations are nonlinear and little is known about their solutions at high Reynolds number, even in simple geometries. Even mathematical questions as basic as existence and uniqueness are unsettled in three spatial dimensions (cf Temam 1988). A second problem, more important from the physical viewpoint, is that experiments and the available mathematical evidence all indicate that turbulence involves the interaction of many degrees of freedom over broad ranges of spatial and temporal scales.

One of the problems of turbulence is to derive this complex picture from the simple laws of mass and momentum balance enshrined in the Navier-Stokes equations. It was to this that Ruelle & Takens (1971) contributed with their suggestion that turbulence might be a manifestation in physical

0066–4189/93/0115–0539$02.00

space of a strange attractor in phase space. Since 1971 we have witnessed great advances in dynamical-systems theory and many applications of it to fluid mechanics, with, alas, mixed results in turbulence—despite the attractive notion of using deterministic chaos in resolving the apparent paradox of a deterministic model (Navier-Stokes) that exhibits apparently random solutions. This is due not solely to the technical difficulties involved: Proof of global existence and a finite-dimensional strange attractor for the 3-D equations in a general setting would be a great mathematical achievement, but would probably be of little help to specific problems in, say, turbomachinery. For a start, rigorous estimates of attractor dimension (Téman 1988) indicate that any dynamical system which captures all of the relevant spatial scales will be of enormous dimension. Advances in such areas will most probably necessitate a dramatic reduction in complexity by the removal of inessential degrees of freedom.

The first real evidence that this reduction in complexity might be possible for fully developed turbulent flows came with the experimental discovery of coherent structures around the outbreak of the second world war, documented by J. T. C. Liu (1988). The existence of these structures was probably first articulated by Liepmann (1952), and was thoroughly exploited by Townsend (1956). Extensive experimental investigation did not take place until after 1970, however (see Lumley 1989). Coherent structures are organized spatial features which repeatedly appear (often in flows dominated by local shear) and undergo a characteristic temporal life cycle. The proper orthogonal decomposition, which forms the subject of this review, offers a rational method for the extraction of such features. Before we begin our discussion of it, a few more general observations on turbulence studies are appropriate.

1.2 *Experiments, Simulations, Analysis, and Understanding*

In analytical studies of turbulence, two grand currents are clear: statistical and deterministic. The former originates in the work of Reynolds (1894). The latter is harder to pin down; linear stability theory is felt to have little to do with turbulence. Nonlinear stability, however, and such things as amplitude equations, definitely are relevant, so perhaps L. D. Landau and J. T. Stuart should be credited with the beginnings of an analytical nonstatistical approach. Lorenz' work was certainly seminal. Over the past twenty years a third stream has emerged and grown to a torrent which threatens to carry everything in its path: computational fluid dynamics.

Both analytical approaches have drawbacks. Statistical methods, involving averaged quantities, immediately encounter closure problems (Monin & Yaglom 1987), the resolution of which, even in sophisticated renormalization group theories (cf McComb 1990) usually requires use of empirical data (Tennekes & Lumley 1972). Nonetheless, they are intended for and are used for fully developed turbulence. Analytical methods have so far been unable to deal with the interaction of more than a few unstable modes, usually in a weakly nonlinear context, and thus have been restricted to studies of transition or pre-turbulence. Most of the dynamical systems studies have been limited to this area. Computational fluid dynamics bypasses the shortcomings of these methods by offering direct simulation of the Navier-Stokes equations. However, unlike analysis, in which logical deductions lead stepwise to an answer, simulation provides little understanding of the solutions it produces. It is more akin to an experimental method, and no less valuable (or less confusing) for the immense quantity of data it produces, especially at high spatial resolution.

Proper orthogonal decomposition (POD), while lacking the broad sweep of the approaches mentioned above, nonetheless has something to offer all three of these. 1. It is statistically based—extracting data from experiments and simulations. 2. Its analytical foundations supply a clear understanding of its capabilities and limitations. 3. It permits the extraction, from a turbulent field, of spatial and temporal structures judged essential according to predetermined criteria and it provides a rigorous mathematical framework for their description. As such, it offers not only a tool for the analysis and synthesis of data from experiment or simulation, but also for the construction, from the Navier-Stokes equations, of low-dimensional dynamical models for the interaction of these essential structures. Thus, coming full circle, we have a statistical technique that contributes to deterministic dynamical analysis.

In Sections 3 and 4 we review applications of the proper orthogonal decomposition, after developing its key features in Section 2. The latter is necessarily mathematical in style and while space limitations preclude a complete treatment, we include some of the new and lesser known results. Proofs are omitted; see Berkooz (1991b, Chapter 2) for details. Section 5 explores relations to some other techniques used in turbulence studies and Section 6 contains a concluding discussion. The remainder of this introductory section contains an historical survey.

1.3 *The Proper Orthogonal Decomposition*

The proper orthogonal decomposition is a procedure for extracting a basis for a modal decomposition from an ensemble of signals. Its power lies in the mathematical properties that suggest that it is the preferred basis to use in many circumstances. The POD was introduced in the context of turbulence by Lumley (1967, cf 1981). In other disciplines the same procedure goes by the names Karhunen-Loève decomposition or principal components analysis and it seems to have been independently rediscovered several times, cf Sirovich (1987). According to Lumley, quoting A. M. Yaglom (personal communication), the POD was suggested independently by several scientists, e.g. Kosambi (1943), Loève (1945), Karhunen (1946), Pougachev (1953), and Obukhov (1954). For use of the POD in other disciplines see: Papoulis (1965)—random variables; Rosenfeld & Kak (1982)—image processing; Algazi & Sakrison (1969)—signal analysis; Andrews et al (1967)—data compression; Preisendorfer (1988)—oceanography; and Gay & Ray (1986, 1988)—process identification and control in chemical engineering. Introductory discussions of the method in the context of fluid mechanical problems can also be found in Sirovich (1987, 1989, 1990) and Holmes (1990).

The attractiveness of the POD lies in the fact that it is a linear procedure. The mathematical theory behind it is the spectral theory of compact, self-adjoint operators. This robustness makes it a safe haven in the intimidating world of nonlinearity; although this may not do the physical violence of linearization methods, the linear nature of the POD is the source of its limitations, as will emerge from what follows. However, it should be made clear that the POD makes no assumptions about the linearity of the problem to which it is applied. In this respect it is as blind as Fourier analysis, and as general.

2. FUNDAMENTALS OF THE PROPER ORTHOGONAL DECOMPOSITION

2.1 *The Eigenvalue Problem*

For simplicity we introduce the proper orthogonal decomposition in the context of scalar fields: (complex-valued) functions defined on a interval Ω of the real line. The interval might be the width of the flow, or the computational domain. We restrict ourselves to the space of functions which are square integrable (or, in physical terms, fields with finite kinetic energy) on this interval. We need an inner product $(f, g) = \int_\Omega f(x)g^*(x)dx$, and a norm $\|f\| = (f, f)^{1/2}$. We start with an ensemble of realizations of the function $u(x)$, and ask which single (deterministic) function is most similar to the members of $u(x)$ on average? We need an averaging operation $\langle\ \rangle$, which may be a time, space, ensemble, or phase average. We suppose that the probabilistic structure of the ensemble is such that the average and limiting operations can be interchanged (cf Lumley 1971). Mathematically, the notion of "most similar" corresponds to seeking a function ϕ such that

$$\max_\psi \langle|(u, \psi)|^2\rangle/(\psi, \psi) = \langle|(u, \phi)|^2\rangle/(\phi, \phi). \tag{2.1}$$

That is, we find the member of the $\psi(=\phi)$ which maximizes the (normalized) inner product with the field u, which is most nearly parallel in function space. This is a classical problem in the calculus of variations. A necessary condition for (2.1) to hold is that ϕ is an eigenfunction of the two-point correlation tensor

$$\int_{\Omega} \langle u(x)u^*(x')\rangle \phi(x')dx' = \lambda\phi(x). \tag{2.2}$$

We define the average $R(x, x') = \langle u(x)u^*(x')\rangle$. That the maximum in (2.1) is achieved, and corresponds to the largest eigenvalue λ_1 of (2.2) is a consequence of spectral theory (Reisz & Sz. Nagy 1955). Moreover Hilbert-Schmidt theory assures us that there is not one, but a denumerable infinity of solutions of (2.2), as long as Ω is bounded. We will call these the empirical eigenfunctions, and denote them by $\{\phi_k\}$ and normalize them so that $\|\phi_k\| = 1$. Note that the subscript k does not denote a vector, but a member of the sequence. We order the eigenvalues by $\lambda_i \geq \lambda_{i+1}$, observing that the non-negative definiteness of $R(x, x')$ assures that $\lambda_i \geq 0$. We also have a diagonal decomposition:

$$R(x, x') = \Sigma_k \lambda_k \phi_k(x)\phi_k^*(x'). \tag{2.3}$$

In (2.3) and hereafter the sum is from 1 to infinity unless explicity indicated otherwise. As we will see in Section 2.2, almost every member (in a measure sense) of the ensemble may be reproduced by a modal decomposition in the eigenfunctions:

$$u(x) = \Sigma_k a_k \phi_k(x). \tag{2.4}$$

The diagonal representation of the two-point correlation tensor R ensures that the modal amplitudes are uncorrelated:

$$\langle a_k a_{k'}^* \rangle = \delta_{kk'}\lambda_k; \tag{2.5}$$

see Section 2.3 below. Here (2.4) is the proper orthogonal decomposition, and the set $\{\phi_k\}$ is called an empirical basis.

2.2 *The Span of the Empirical Basis*

The first step in understanding what can be done with the sequence $\{\phi_k\}$ is to characterize the set $S = \{\Sigma a_i \phi_i | \, \Sigma |a_i|^2 < \infty\}$, which is the span of the set $\{\phi_i\}$. That is, what functions can be represented by convergent sequences of empirical eigenfunctions? Note that we retain only eigenfunctions with nonzero eigenvalues, so that the $\{\phi_k\}$ need not form a complete basis. If one adds all the eigenfunctions with zero eigenvalue, one obtains a complete basis, but loses some of the advantages of the POD, as we shall see.

The first proposition describes the ability of the empirical basis to reconstruct the ensemble from which it was generated. The propositions are stated a little loosely, and we do not provide proofs here; tighter statements and proofs may be found in Berkooz (1991b).

Proposition 2.1 *If $R(x, x')$ is continuous, then almost every member u of the ensemble belongs to S.*

We denote by X the subset of u's of the ensemble that belongs to S, i.e. where Proposition 2.1 holds.

Corollary 2.1 *Let $\{b_i\}$ be an infinite sequence of real numbers, and u_i an infinite sequence of ensemble members in X. If $v = \Sigma_i b_i u_i$ is square integrable on Ω, then v lies in S.*

The following proposition together with the corollary above will characterize S.

Proposition 2.2 *If θ is in S, then there exist infinite sequences $\{u_i\}$ in X and scalars $\{b_i\}$ such that $\theta(x) = \Sigma_i b_i u_i$.*

We thus have a complete characterization of the span of the eigenfunctions: It is exactly the span of all the realizations of $u(x)$, with the exception of a set of measure zero. In particular, with the exception of a set of measure zero, every member of the ensemble that generated the eigenfunctions can be represented in terms of the eigenfunctions. A special case of this result (when the u_i take on discrete values, as would be the case in a computer experiment) was observed independently in Aubry et al (1991a).

From this we see that the sequence $\{\phi_i\}$ need not be complete. It is complete only if one includes the kernel of the operator R, that is, all the (generalized) eigenfunctions with zero eigenvalues. Of course, if R is positive definite, there are no zero eigenvalues, and one does get a complete basis. However, in many applications one can argue on physical grounds that the realizations $u(x, t)$ do not span the space of square integrable functions on Ω; for example, see Section 2.4 below. This result highlights a strong property of the POD. It a priori limits the space studied to the smallest linear subspace that is sufficient to describe the observed phenomena. This can be stated as a corollary:

Corollary 2.2 *If all the square integrable functions u on Ω having a certain property form a closed linear subspace, then the empirical eigenfunctions have the same property, and the converse is also true.*

The classical example is incompressibility. If $R(x, x')$ is formed from realizations of divergence-free vector fields u, then the eigenfunctions $\{\phi_i\}$ are also divergence-free.

2.3 *Optimality*

Suppose we have a signal $u(x, t)$ and a decomposition with respect to an (arbitrary) orthonormal basis $\{\psi_i\}$:

$$u(x, t) = \Sigma_i b_i(t)\psi_i(x). \tag{2.6}$$

If the $\{\psi_i\}$ have been nondimensionalized and normalized to give $(\psi_i, \psi_j) = \delta_{ij}$, then the coefficients b_i carry the dimension of the quantity u. If $u(x, t)$ is a velocity, the average kinetic energy per unit mass over the experiment is given by

$$\int_\Omega \langle uu^* \rangle dx = \Sigma_i \langle b_i b_i^* \rangle. \tag{2.7}$$

Hence, we may say that $\langle b_i b_i^* \rangle$ represents the average kinetic energy in the i-th mode. The following proposition establishes what is called the optimality of the POD or Karhunen-Loève decomposition.

Proposition 2.3 *Let $u(x, t)$ be an ensemble member square integrable on Ω for almost every t and $\{\phi_i, \lambda_i\}$ be the POD orthonormal basis set with associated eigenvalues. Let*

$$u(x, t) = \Sigma_i a_i(t)\phi_i(x) \tag{2.8}$$

be the decomposition with respect to this basis, where equality is almost everywhere. Let $\{\psi_i\}$ be an arbitrary orthonormal set such that

$$u(x, t) = \Sigma_i b_i(t)\psi_i(x). \tag{2.9}$$

Then the following hold:

1. *$\langle a_i(t)a_j^*(t) \rangle = \delta_{ij}\lambda_i$, i.e. the POD coefficients are uncorrelated.*
2. *For every n we have $\Sigma_i^n \langle a_i(t)a_i^*(t) \rangle = \Sigma_i^n \lambda_i \geq \Sigma_i^n \langle b_i(t)b_i^*(t) \rangle$.*

This proposition is the basis for the claim that the POD or Karhunen-Loève decomposition is optimal for modeling or reconstructing a signal $u(x, t)$. It implies that, among all linear decompositions, this is the most efficient, in the sense that, for a given number of modes the projection on the subspace used for modeling will contain the most kinetic energy possible in an average sense. In addition, the time series of the coefficients $a_i(t)$ are uncorrelated.

2.4 *Symmetries and Homogeneity*

We start by describing a particular kind of symmetry. We say that the two-point correlation $R(x, x')$ is *homogeneous* or *translation invariant* if $R(x, x') = r(x-x')$, i.e. R depends only on the difference of the two coordinates. In general, homogeneity of a system is defined through multi-point moments, but we only need second-order moments here. Assuming Ω is bounded and u is periodic, we may develop r in a Fourier series,

$$r(x-x') = \Sigma C_k e^{2\pi ik(x-x')}. \tag{2.10}$$

One can then solve the eigenvalue problem via the unique representation

$$R(x, x') = \Sigma C_n e^{2\pi inx} e^{-2\pi inx'}, \tag{2.11}$$

which implies that the $e^{2\pi inx}$ are exactly the eigenfunctions with eigenvalues C_n. Conversely, if the eigenfunctions are Fourier modes we can write (2.11) which implies (2.10). In summary, we can state

Proposition 2.4 $R(x, x') = r(x-x')$ *if and only if the eigenfunctions of R are Fourier modes.*

This observation is especially useful in systems where the domain Ω is of higher dimension. For example, if Ω is 2-D, then we have

$$R(\mathbf{x}, \mathbf{x}') = R(x, y, x', y'), \tag{2.12}$$

and if the x-direction is homogeneous, the problem of finding eigenfunctions in a 2-D domain is decoupled into a set of 1-D problems by writing

$$R = R(x-x', y, y') \tag{2.13}$$

and performing the same procedure as above, yielding a 1-D eigenvalue problem for every Fourier wavenumber. Examples of such applications can be found in Herzog (1986), Moin & Moser (1989); also see Lumley (1971). The observations above can be generalized to other cases where part of the domain has a more general symmetry group structure, see Berkooz & Titi (1992).

Unfortunately, in the context of coherent structures in the turbulent boundary layer, the above observation leads to Fourier structures which are, of course, not localized in all space directions, unlike the events observed. In an attempt to avoid this, and reintroduce locality, it is necessary to introduce phase relationships among the Fourier modes. In the following treatment, we adapt Lumley's application of the shot-noise decomposition (Rice 1944; Lumley 1971, 1981). Imagine a building block, which is the basic coherent structure, and a process that sprinkles the units randomly on the real line. (See Figure 1.) If $f(x)$ is the building block, with 0 as a reference point, in order to move the structure so that its reference point is at y we perform the convolution

$$u(x) = \int \delta(\xi - y) f(x - \xi) d\xi, \tag{2.14}$$

where $\delta(\xi)$ is the Dirac delta function (working in the space of generalized functions or distributions). This prompts us to the following:

Definition 2.1 *A convolution of the type*

$$u(x, t) = \int g_t(\xi) f(x - \xi) d\xi, \tag{2.15}$$

where $g_t(\xi)$ is a random process in the space of generalized functions, will be called a shot-noise decomposition of $u(x, t)$.

The goal is to reconstruct f from statistics of the system. To develop intuition for Definition 2.1, assume for simplicity that both f and u have an upper bound on frequencies in their Fourier decompositions, and that they are periodic. Then g need not be a generalized function, and the Fourier transform of g will be well-defined. If u, g, and f are the Fourier transforms of u, g, and f respectively, then clearly $\mathsf{u} = \mathsf{g}\mathsf{f}$. We see that, in general, a shot-noise decomposition is always possible, and that moreover

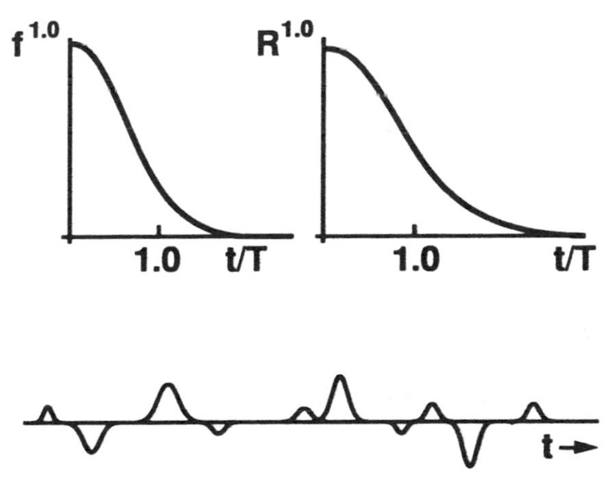

Figure 1 (*Top*) The basic building block and its autocorrelation. (*Bottom*) The result of random sprinkling with Gaussian amplitudes.

it is far from unique; one also has freedom in choosing f and g. To remove the ambiguity in the decomposition, and to formalize the notion that g "randomly" sprinkles f's, we make the following assumption on the process g.

Assumption 2.1 *Leg g be uncorrelated in nonoverlapping intervals, i.e.* $\langle g(\xi)g(\xi')\rangle = \delta(\xi-\xi')$.

This assumption removes part of the ambiguity of the decomposition, as follows:

Proposition 2.5 *If* R *is the Fourier transform of the two-point velocity correlation for a homogeneous process, then* $\mathsf{R} = |\mathsf{f}|^2$.

We see that Assumption 2.1 prescribes the power spectrum of the building block; the phase angles are yet to be determined. Before we discuss the retrieval of the phase information, we will examine some physical aspects of the shot-noise decomposition with the above assumption. It seems that this approach formalizes rather well the stochastic sprinkling of structures in physical space. An extension of this formalism to include stochastic sprinkling in time is also possible, the assumption being extended to non-correlatedness in time as well as in physical space. The structure of the building block in time corresponds to the life cycle, or evolution, of the coherent structure. The assumptions of noncorrelatedness in physical space and in time are, of course, simplistic approximations; coherent structures cannot be too close to each other, in either space or time, and hence there must be a short-range correlation of g. In addition, adjacent coherent structures affect each other dynamically, and hence g must be statistically dependent at different places and times, even if uncorrelated. In addition, the assumption of a single building block may also be restrictive, since we may expect to meet more than one form of coherent structure. In multiple dimensions, we should include the possibility that the building blocks will occur with different orientations. Some of these deficiencies will come back to haunt us when we try to retrieve the phase information for f.

A rational procedure was suggested by Lumley (1971, 1981) to obtain the phase information from the bi-spectrum (Brillinger & Rosenblatt 1967, Lii et al 1976). Our goal is to find the phase angle of the Fourier coefficients of f. We already have the moduli. We want to find $\theta(\kappa)$ such that

$$\mathsf{f} = \mathsf{R}^{1/2}e^{2\pi i\theta(\kappa)}. \tag{2.16}$$

Consider the triple correlation:

$$\langle u(x)u(x+r_1)u(x+r_2)\rangle = \iiint f(x-\xi)f(x+r_1-\xi')$$

$$\times f(x+r_2-\xi'')\langle g(\xi)g(\xi')g(\xi'')\rangle d\xi d\xi' d\xi''. \qquad (2.17)$$

We now extend the assumption on g and require that it be triply uncorrelated on nonoverlapping intervals, thus obtaining for the right hand side

$$\int f(x)f(x+r_1)f(x+r_2)dx. \qquad (2.18)$$

If we designate the triple correlation in (2.17) by $B(r_1, r_2)$ and its Fourier transform by $\mathsf{B}(\kappa_1, \kappa_2)$, we obtain

$$\mathsf{B}(\kappa_1, \kappa_2) = \mathsf{R}^{1/2}(\kappa_1)\mathsf{R}^{1/2}(\kappa_2)\mathsf{R}^{1/2}(\kappa_1+\kappa_2)$$

$$\times \exp\{2\pi i[\theta(\kappa_1)+\theta(\kappa_2)-\theta(\kappa_1+\kappa_2)]\}. \qquad (2.19)$$

The known quantities are $\mathsf{B}(\kappa_1, \kappa_2)$, and $\mathsf{R}(\kappa)$. As Lumley (1971) observed (see also Moin & Moser 1989), this problem is, in general, not solvable exactly, since $\mathsf{B}(\kappa_1, \kappa_2)$ may not be factorable as the right hand side prescribes. In fact, if our assumption (that g is triply uncorrelated) is correct, B will be so factorable, and not otherwise. There is very little information on B in turbulence, and none on its factorability. Moin & Moser observe that this problem is encountered in other disciplines as well; see Bartlet et al (1984) and Matsuoka & Ulrych (1984). Lumley suggested the following simple solution: reduce B to a one-dimensional domain by looking at $\mathsf{B}(c, c\kappa)$, where c is an arbitrary number, and solve the above equation on a finite number of points. This is not very convincing, however, and it is exactly at this point that our assumptions on g come back to haunt us. The lack of an exact general solution to the bi-spectrum equation suggests that our assumptions may have been too simplistic, either regarding the existence of a single building block, or regarding the statistical behavior of g.

An alternative strategy for representing localized structures in homogeneous directions involves the use of wavelet decompositions (Meyer 1987). While not optimal (see Proposition 2.4), it has recently been shown that periodic spline wavelets (Perrier & Basdevant 1989, cf Farge 1992) are not much less efficient than Fourier modes in capturing kinetic energy on the average. More specifically, Berkooz et al (1991a) show that, if the autocorrelation $R(x-x')$ is reconstructed with average error ε using N Fourier modes, then a wavelet reconstruction with a similar number of modes will incur error $3\varepsilon + c/2^{\delta n}$, where n is the order of the splines used, and the constants δ and c depend on how the wavelet octaves relate to N.

There are some restrictions on this result, and it relies on the rapid decay in Fourier space of spline wavelets, but it suggests a promising new direction to explore.

We can use the POD to examine the ergodicity of a dynamical system. Physically, if the phase space of the dynamical system is partitioned into disjoint closed regions, so that a system trajectory starting in one of these never enters the others, the system is not ergodic. Now, if a system has certain symmetries, these symmetries should appear as symmetries of the invariant measure. However, starting from a given initial condition, it is possible that the solution will not explore all the states associated with the symmetry group. If this occurs, we can say that the system is not ergodic. For example, suppose a 2-D map is invariant under the symmetry $(x, y) \rightarrow (y, x)$, and has two disjoint attractors lying to either side of the $x = y$ symmetry axis (and hence individually not invariant under the symmetry). A typical realization—an orbit of this map—will explore just one attractor, and will not have the symmetry of the full system. The invariant measure concentrated on the two attractors is not ergodic because of the disjointedness, and we can see this because of the lack of symmetry. As a result, if the empirical eigenfunctions obtained from a single run of the experiment (in time) have less symmetry than the problem as a whole, we conclude that the system is not ergodic. Note that it is also possible for there to be two disjoint attractors, each of which displays the full symmetry, and hence there may be no telltale lack of symmetry to point to the lack of ergodicity. More formally, we have:

Proposition 2.6 *Let S be a stationary ensemble of realizations, and m be the invariant measure associated with it. Let G be a linear symmetry group for S. Then a necessary condition for m to be ergodic is that, for almost every realization, each of the finite-dimensional eigenspaces corresponding to a given eigenvalue* (*which results from the time averages of that experiment*) *is invariant under G.*

The way one would go about checking this condition in an experiment would be:

1. Perform the experiment and measure $R(x, x')$.
2. Obtain the $\{\phi_i\}$ from $R(x, x')$.
3. Check that every ϕ_i satisfies the symmetry condition.

(Here, for simplicity, we consider a system with distinct eigenvalues.)

Recently Aubry et al (1991b) performed numerical integrations of the Kuramoto-Sivashinsky equation and computed the POD basis. By using the results above and the calculations cited, they conclude that for certain values of the bifurcation parameter, the system is not ergodic. On the other hand, if one wants to assume that the system is ergodic, one may use the symmetries of the system to increase the size of the ensemble. This approach has been advocated by Sirovich (1987) and applied in many studies (e.g. Sirovich & Park 1990). However, one should be cautious, as there are examples (Berkooz 1990b) where the partition into ergodic components is finer than the partition into symmetric components; in this case the image of the basis obtained by one experiment under a symmetry group will not produce the basis obtained by the ensemble average measure. See Berkooz & Titi (1992) for further discussion of this point. Caution is particularly warranted in cases of small systems or special geometries. For example, in a square Rayleigh-Bénard cell there is a possibility that a preferred rotation direction of the single roll may be chosen at random at the ime of onset and may never change throughout the life of the system. This indicates that there are at least two distinct and disjoint parts for the support of the invariant measure, each associated with a rotation direction, much as in the simple map example. Similar phenomena evidently can occur in the minimal flow unit of Jimenez & Moin (1991). The additional symmetry imposed on the ensemble of flows by artificial addition of images of flows under symmetry group elements, as advocated by Sirovich (1987), may therefore obscure the true nature of a particular system.

2.5 *The Nature of Attractors*

We first describe a geometrical consequence of the phase space description of asymptotic behavior afforded by the POD. In particular, we can give a probabilistic-geometric interpretation of the location of dynamics in phase space using Chebyshev's inequality. This generalizes the result in Foias et al (1990), who essentially reproduced the proof of Chebyshev's inequality in a specific case. Aubry et al (1991b) independently observed that the type of picture described in Foias et al (1990) is due to Chebyshev's inequality.

We will sketch the proof here, because it is geometrically instructive. First, recall Chebyshev's inequality.

Theorem 2.1 *Let* $\mathbf{X}$ *be a vector-valued random variable with zero mean and variance* $\sigma^2 = \langle |\mathbf{X}|^2 \rangle$. *Then for any* $\varepsilon > 0$

$$P\{|\mathbf{X}| \geq \varepsilon\} \leq \sigma^2/\varepsilon^2 \tag{2.20}$$

where $P\{\bullet\}$ *is the probability of that event.*

Now, denote by $\mathbf{X}_n$ *the vector-valued random variable*

$$\mathbf{X}_n = \{a_{n+1}, a_{n+2}, \ldots, a_\infty\} \tag{2.21}$$

so that the $\mathbf{X}_n$ *have zero mean and variance* $\sigma^2 = \lambda_{n+1} + \lambda_{n+2} + \cdots = \Sigma_{m=n+1}^{\infty} \lambda_m$.

Then Chebyshev's inequality gives

$$P\{|\mathbf{X}_n| \geq \varepsilon\} \leq \Sigma_{m=n+1}^{\infty} \lambda_m / \varepsilon^2. \tag{2.22}$$

In the space of functions u, square integrable on Ω, the coefficients a_n may be regarded as coordinates. The space spanned by $\{\phi_1, \ldots, \phi_n\}$ may be thought of as a surface in the function space, with associated coordinates $\{a_1, \ldots, a_n\}$. Containing this surface, and extending ε on each side of it, is a slab of thickness 2ε, defined by $|\mathbf{X}_n| < \varepsilon$. The inequality (2.22) tells us how likely it is to be outside that slab. Inequality (2.22) is useless for fixed m and $\varepsilon \to 0$. The way to extract something useful is to take a sequence $\varepsilon_n \to 0$ such that

$$\Sigma_{n+1}^{\infty} \lambda_m / \varepsilon_n^2 \to 0. \tag{2.23}$$

In other words, the ε_n are chosen so that their squares go to zero slower than the decay of the norms of the residual modes. This will give a series of slabs with thickness going to zero, while the probability of the solutions being in those slabs goes to one.

The problem is now shifted to computing the rate of decay of the residual energy $\Sigma_{m=n+1}^{\infty} \lambda_m$. There is analytical evidence that suggests that when the POD basis is used for turbulent flows, this residual decays at least exponentially fast asymptotically, as we argue later in this section. This enables us to take a series $\varepsilon_n^2 \to 0$ with a slightly smaller exponent. The result will be a series of slabs with thickness going exponentially to zero, and the probability of being in that slab going exponentially to one. This gives rise to a picture in which the attractor is very thin, albeit high or even infinite dimensional. Thus, the essentials of the dynamics may be controlled by a finite number of modes, as the dynamical models discussed in Section 4 suggest.

We turn now to a related matter. We show that if the POD spectrum decays fast enough (which is the case for systems that interest us), practically all the support of the invariant measure is contained in a compact set; that is, roughly speaking, all the likely realizations in the ensemble can be found in a relatively small set of bounded extent.

Proposition 2.7 *Consider a dynamical system whose solutions are continuous and square integrable on* Ω. *If* $\lambda_n = O[\exp(-cn)]$, *then for any* $\varepsilon > 0$ *there exists a compact set* B_ε *such that* $P\{B_\varepsilon\} > 1 - \varepsilon$.

This is quite interesting. If one performs a POD decomposition on a system about which little is known a priori, and gets a discrete spectrum that decays rapidly enough (see also the next section), Proposition 2.7 allows us to conclude that most of the likely realizations can be found in a compact domain. It is surprising that such fundamental information can be obtained from such a simple procedure.

We next describe a very reasonable situation in which the POD spectrum is likely to fall off exponentially and thus in which Proposition 2.7 will hold. "Regularity of solutions" is a mathematical property describing, essentially, the rate of decay of the tail of the wavenumber spectrum of instantaneous solutions of a partial differential equation (PDE). It appears that solutions of many chaotic systems have very high regularity, meaning that the instantaneous wavenumber spectrum decays rapidly (exponentially in most cases)—see Promislow (1991). We note that wavenumber spectra in fluid turbulence are generally believed to fall off exponentially (see Tennekes & Lumley 1972). We will establish here the relation between regularity results and the POD. We start by explaining what is meant by regularity, quote regularity results for the Navier-Stokes equations, and relate regularity to the eigenvalue spectrum of the POD. These results appeared in Foias et al (1990), but our treatment differs from theirs in that we use the optimality property of the POD to establish the connection, whereas they use the uncorrelatedness of the random coefficients. A closer examination of their proof shows that the estimates obtained from regularity results hold not only for the empirical bases but also for a basis of eigenfunctions of the Stokes operator or any other basis with similar orthogonality properties.

We start with regularity in a simple setting. Let the domain of our flow be a rectangular box, in which the real velocity field $\mathbf{u}$ is periodic and incompressible. We have a representation in terms of Fourier modes:

$$\mathbf{u}(\mathbf{x}, t) = \Sigma \mathbf{u}_{\mathbf{j}}(t) e^{i\mathbf{j}\cdot\mathbf{x}}. \tag{2.24}$$

We assume the kinetic energy is finite:

$$\Sigma |\mathbf{u}_{\mathbf{j}}|^2 = |\mathbf{u}|^2/(2\pi)^3 < \infty. \tag{2.25}$$

The Stokes operator is simply $A = -\nabla^2$. We can define fractional powers of A via

$$A^{\alpha}\mathbf{u} = \Sigma \mathbf{u}_{\mathbf{j}} |\mathbf{j}|^{2\alpha} e^{i\mathbf{j}\cdot\mathbf{x}}. \tag{2.26}$$

Similarly, we can define

$$\exp[\tau A^{s}]\mathbf{u} = \Sigma \mathbf{u}_{\mathbf{j}} \exp[\tau |\mathbf{j}|^{2s}] e^{i\mathbf{j}\cdot\mathbf{x}}. \tag{2.27}$$

If this sum converges, we say that $\mathbf{u}$ is Gevrey class regular. The relation between Gevrey regularity and the instantaneous turbulent spectrum is as follows. We can define a 3-D spectrum (see Tennekes & Lumley 1972) as

$$E(m) = \Sigma_{m \le |\mathbf{j}| < m+1} |\mathbf{u}_{\mathbf{j}}|^2. \tag{2.28}$$

This is the energy in a spherical shell. For Gevrey class regular fields, we have

$$\infty > \Sigma|\mathbf{u_j}|^2 \exp[2\tau|\mathbf{j}|^{2s}] \geq \Sigma E(m) \exp[2\tau m^{2s}]. \tag{2.29}$$

Therefore, for Gevrey class regular velocity fields the instantaneous spectrum decays at least exponentially fast. As we will see later, the last expression can be manipulated to obtain the asymptotic rate of decay of the random coefficients, which are nothing but $|\mathbf{u_j}|^2$ in a Fourier representation.

To convince the reader that regularity results are useful, we must do the following: Define regularity for an arbitrary problem (not necessarily with periodic boundary conditions); present regularity results for the Navier-Stokes equations; and show how regularity results, which are instantaneous information, transform to average information such as a rate of decay for the POD eigenvalues.

The definition of regularity for an arbitrary domain is based on the Stokes operator. The only difference is that, instead of developing the velocity field in terms of Fourier modes, one develops it in terms of the eigenfunctions of the Stokes operator. These eigenfunctions form a complete basis. The asymptotics of the eigenvalues of the Stokes operator are the same for all reasonable domains (Constantin & Foias 1989). Rigorous regularity results for the Navier-Stokes equations in two dimensions are given in Foias & Témam (1989). For the 3-D Navier-Stokes equations, to simplify the discussion, we assume that the vorticity is bounded above uniformly throughout the flow and in time. This may be a very large bound, but it nevertheless needs to be assumed. This enables us to bypass the blowup problem for the Navier-Stokes equations (see Anderson et al 1984). In this case, Foias & Témam (1989) showed that, after a transient period, there exists a σ such that the solutions u satisfy

$$|A^{1/2} \exp(\sigma A^{1/2})u| < C < \infty. \tag{2.30}$$

Four remarks are in order. First, some uniform bound for the averages is essential; as shown in Berkooz (1990b), one can have an ensemble with all members exhibiting exponentially decaying tails, yet with an average spectrum not exponentially decaying. See also Novikov (1963) and Monin & Yaglom (1987). Second, to get a comparison to the POD, one has to define some order on wavenumber space. The order we choose is through wavenumber shells, that is, through the 3-D spectrum. Third, as discussed in Berkooz (1990b) and Foias et al (1990), regularity results are relevant only to the far dissipative range of turbulence, or to very high order empirical eigenfunctions. (See Figure 2.) Fourth, the decay of the tail of the empirical eigen-spectrum will always be as fast, or faster, than that of the tail of the spectrum with respect to any other basis, in particular the

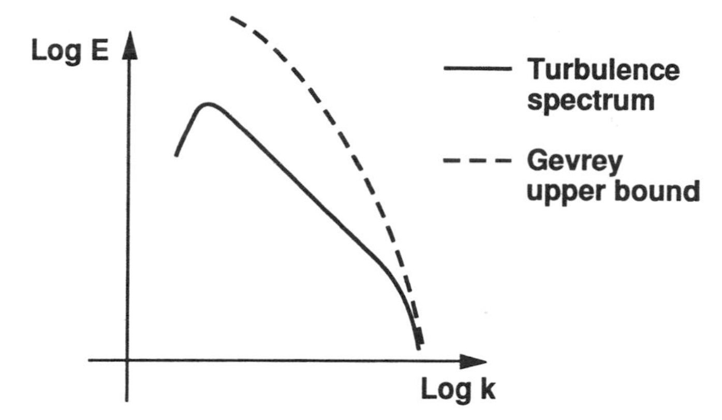

Figure 2 Spectrum of turbulence compared with upper bound for the spectrum from Gevrey regularity results.

Fourier spectrum. This is a straightforward consequence of the maximality of energy principle for the POD. (See Figure 3.)

We can now state our main results for this section.

Proposition 2.8 *If a solution* $\mathbf{u}(\mathbf{x}, t)$ *in a domain of dimension n is uniformly bounded in time in the norm* $|A^{\alpha} \exp[\beta A^{\gamma}]\mathbf{u}|$ *then the eigenvalues in the tail of the POD spectrum will satisfy*

$$\mu_k = o[k^{-2\alpha/n} \exp(-2\beta k^{2\gamma/n})]. \tag{2.31}$$

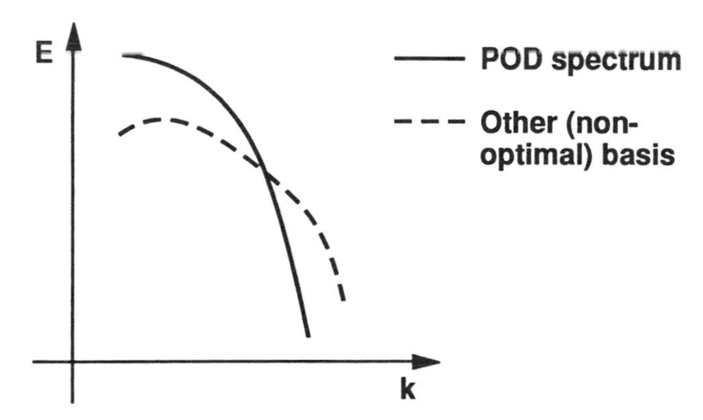

Figure 3 Typical spectrum of a dissipative system in POD and another basis. Note the "cross over" which will occur due to the equality in total energy (area under the curves).

Proposition 2.9 *Under the assumptions of the previous proposition, the empirical eigenfunctions have the same regularity as the solutions.*

2.6 *Computational Schemes and Further Results*

In this section we discuss some additional points of interest regarding the POD.

THE METHOD OF SNAPSHOTS This method was proposed by Sirovich (1987). It is a numerical procedure which can save time in computation of empirical eigenfunctions. Suppose one is performing a numerical simulation on a large number of grid points N, the number of ensemble members deemed adequate for a description of the process is M, and $N \gg M$ (the fundamental question of determining M is not part of Sirovich's treatment). In general the eigenfunction computation will become an $N \times N$ problem. However, this may be reduced to an $M \times M$ problem. Berkooz (1991b) gives an argument for the equivalence of the method of snapshots to the original formulation of the eigenvalue problem, as well as a linear independence condition omitted by Sirovich (1987).

DIMENSION AND THE CONDITIONAL POD An appealing concept is to define a dimension through the POD. The obvious thing to do is to define a dimension as the number of nonzero eigenvalues in the POD decomposition, as done in Aubry et al (1991a). However, this is merely the dimension of the smallest linear subspace containing the dynamics, and has, a priori, nothing to do with the Hausdorff dimension of the attractor. Even if the latter is finite, the former may not be so. This is underlined by the example in Berkooz (1990b) of a system having a limit cycle, which has Hausdorff dimension 1, with an infinite number of nonzero POD eigenvalues. Indeed, as Sirovich (1989) realized, this definition is practically useless, and he suggested the following working definition: "... the number of actual eigenfunctions required so that the captured energy is at least 90% of the total ..., and that no neglected mode, on the average, contains more than 1% of the energy contained in the principle eigenfunction mode." The concept of entropy introduced by Aubry et al (1991a) based on interpreting the POD eigenvalues as probabilities is plagued by the same problem.

In Berkooz (1991b) a connection between the various dimensions is found by using the concept of a conditional POD. Suppose (for simplicity of explanation) that all the trajectories of a system are confined to a bounded region of phase space. Pick a location in that region, u, and consider a ball of radius ε around it. Consider the POD conditioned on being in the ball. Now define a local dimension as the smallest number of conditional eigenfunctions that will return more than 1-δ of the energy.

Now take the upper limit as ε goes to zero, and then as δ goes to zero, and finally take the upper bound as u ranges over the bounded region of phase space. This dimension can be shown to lie between the ordinary dimension and the Hausdorff dimension. Note that it is not obvious that this dimension should lie below the ordinary dimension; as noted in Berkooz (1990b), finite dimension of an attractor does not guarantee that there are only a finite number of energy-containing POD modes.

ASYMPTOTICS OF THE POD EIGENFUNCTIONS For the sake of completeness we mention the interesting result of Sirovich & Knight (1985) on the asymptotic form of the POD eigenfunctions (under certain assumptions). Foias et al (1990) conclude from the results of Sirovich & Knight (1985) that the asymptotic form of the POD eigenfunctions is that of Fourier modes. However, the results of Section 2.2 on the span of the eigenfunctions show that this cannot be completely general. Take, for example, an ensemble of realizations which are all constant on part of the domain, in which case, by Corollary 2.2, the eigenfunctions will also have this property and so cannot be asymptotically close to Fourier modes in that region.

THE POD AND THE PDF IN FUNCTION SPACE We would like to mention the connection between the POD and the PDF in functional space. The invariant measure in functional space is an object of great interest; if one could obtain it explicitly one would have "a solution to turbulence," since all multi-point (single time) statistics would be available. From this point of view the POD is seen as the linear change of basis which turns the coordinates into uncorrelated (although probably dependent) random variables. As shown by Hopf's theory of turbulence, the characteristic functional of the PDF in functional space may be obtained by multi-point correlations (Hopf 1957). This leads us to propose a very simple model for the PDF in functional space: Using the representation

$$u(x, t) = \sum_k a_k(t)\phi_k(x), \tag{2.32}$$

we assume that the a_k's are independent and normally distributed with variance λ_k, $a_k \sim N(0, \lambda_k)$. While this is consistent with the picture the POD gives of the flow (the coefficients are uncorrelated and the spectrum is correct), it clearly implies a strong assumption on the modal dynamics. Nonetheless, in Section 5 we will see that this model is closely related to other statistical approaches that describe coherent motions in turbulence.

3. POD IN DATA DESCRIPTION AND ANALYSIS

In this and the following section we survey applications of the POD. As pointed out in the introduction, we make a distinction between applications of the POD to analysis of experimental results and applications of the POD to data analysis and we have grouped the studies according to the classes of problems treated: wall-bounded flows, free shear flows, convection problems, and mathematical model equations.

3.1 *Wall-Bounded Flows*

One of the earliest applications of the POD was by Bakewell & Lumley (1967). They measured two-point correlations of one velocity component in the wall region of a fully developed turbulent pipe flow, and reconstructed the two-point correlation tensor using incompressibility and a closure assumption. The flow is approximately homogeneous in the streamwise and cross stream directions. They computed only one eigenfunction (in addition to the mean) with no streamwise variation. They reconstructed the coherent structure using 0 as a phase relation for the homogeneous cross stream direction, and obtained a pair of counter-rotating rolls. In this sense their work should be considered as a pioneering study.

Herzog performed a complete 3-D POD analysis of the wall layer of a turbulent pipe flow (Herzog 1986). Herzog had relatively low spatial resolution but very well converged statistics which enabled him to compute three significant eigenfunctions for a substantial range of wavenumber pairs in the homogeneous (cross stream and streamwise) directions. This work was a major undertaking and the first full two-point correlation data set to be measured. In his reconstruction Herzog did not apply a rational method for the reconstruction of phase angles. Herzog did measure the two-point correlation with time delay but the processed results are limited to zero time lag.

Moin & Moser (1989), using the channel-flow data base of Kim et al (1987) obtained by direct numerical simulation of a channel flow, performed a comprehensive POD analysis. The computations employ periodic boundary conditions in the spanwise and streamwise directions and the decomposition uses Fourier modes in these directions, as was done by Herzog (see Section 2.4, above). Although it is not certain that the statistics are well converged (less than 200 realizations were used), the spatial resolution is excellent. The main thrust of their work was a systematic application of the shot-noise expansion. Moin & Moser (1989) suggest two additional, ad hoc, methods for the determination of the phase angles. One is based on a "compactness in physical space" condition and the other based on "continuity of eigenfunctions in wavenumber." The reader is referred to Moin & Moser (1989) for details. The results of this work are what they call "characteristic structures" which dominate the production of important statistics.

More recently Sirovich et al (1990a) and Ball et al (1991) have also computed empirical eigenfunctions from a similar direct numerical simulation of channel flow. They took a lower Reynolds number than Moin & Moser and, rather than assuming "compactness" and "continuity" as mentioned above to determine characteristic numbers, they extracted the temporal behavior of the coefficients $a_k(t)$ of the empirical eigenfunction directly. These time series show strong intermittency, as one would expect from the experimental observations of the bursting process (Kline et al 1967, Robinson 1991). In fact Sirovich et al (1990a) investigate the presence of specific structures (oblique plane waves) and their influence as triggers for the bursting events.

In the numerical channel-flow studies mentioned above, when the full channel width is taken as the domain Ω for computation of the eigenvalues, a relatively large number of eigenfunctions (counting Fourier modes) are required to capture, say, 90% of the kinetic energy on average: Ball et al give a figure of about 500 modes in reasonable agreement with Keefe et al's (1987) Liapunov dimension calculations. (Moin & Moser also consider wall region eigenfunctions.) In this respect we stress that in Herzog's study Ω is restricted to the wall layer ($0 < y+ < 40$ wall units), and in this region convergence is considerably faster. However, in all cases, the empirical eigenfunctions representation converges significantly faster than Fourier-Chebyshev representations of flow in the same regions.

3.2 *Free Shear Flows*

One of the first free shear flows to be analyzed was the jet (or annular mixing layer) investigated experimentally by Glauser et al (1987) and Glauser & George (1987a,b). In this work the jet is assumed to be approximately homogeneous in the streamwise direction (the growth of the layer was not accounted for). The main results were: (*a*) demonstration of the effectiveness of the POD in capturing kinetic energy; (*b*) determination of the shape of the POD eigenfunctions (the majority of the energy being in azimuthal invariant modes); (*c*) the proposal of a dynamical mechanism for turbulence production based on the eigenfunctions, and emphasis of the role of nonazimuthal invariant structures in turbulence production, structures which exhibit azimuthal number selection through the POD spectrum.

Sirovich and various coworkers also investigated the jet and mixing layer using numerical simulation and experimental data (Sirovich et al 1990b; Kirby et al 1990a,b). In Sirovich et al (1990b), a conditional form of the POD was applied to the mixing-layer part of the jet, based on a correlation criterion. As a result, the approximate homogeneity in the streamwise direction was broken, resulting in a 2-D eigenfunction problem.

The results of the POD eigenfunctions gave nice descriptions of the lobes responsible for mixing. In Kirby et al (1990b) the POD is applied to a simulated supersonic shear layer. This is again a 2-D problem. The emphasis of this study is on data compression. The time-averaging procedure eliminates interesting large time-scale dynamics. Kirby et al (1990a) is similar in spirit and results to Sirovich et al (1990b), except that Large Eddy Simulation data is used as a basis for the conditional sampling rather than experimental data.

A conditional POD was applied by Glezer et al (1989) to a forced mixing layer. They term their procedure an "extended POD."

3.3 *Convection*

The POD was applied to numerical simulation of Rayleigh-Bénard convection problems by Sirovich & Park (1990), Park & Sirovich (1990), and Sirovich & Deane (1991) (cf Deane & Sirovich 1991). In the former studies, extensive use is made of discrete symmetries of the flow domain (a rectangular box) to simplify the computations by selecting parities (even or odd) for various eigenfunction components. This is especially useful in this problem, since the domain is bounded and there are no homogeneous directions to assist in data reduction. The symmetries are also used to increase the data base over which averages are taken, as discussed in Section 2.4.

The main thrust of the more recent work is to determine scaling properties of POD eigenfunctions as a function of Rayleigh number. Some scaling properties are found. As Sirovich & Deane (1991) point out, if such scaling does hold for asymptotically large Rayleigh numbers, it will be a considerable contribution. Using a definition of $\dim_{KL}$ as the number of modes required to capture 90% of the energy (cf Section 2.5) an empirical relation between the Lyapunov dimension $\dim_L$ and $\dim_{KL}$ is traced out.

3.4 *Mathematical Models*

The POD has been applied to the analysis of the results of several mathematical models—mostly 1-D dissipative partial differential equations. These problems are inherently simpler than real world problems and therefore provide attractive tests for the method.

Chambers et al (1988) applied the POD to simulation of Burger's equation with random forcing and showed that the resulting eigenfunctions exhibit "viscous" boundary layers near the "walls" and have an outer region essentially independent of Reynolds number. Sirovich and coworkers applied the POD to the Ginzburg-Landau equation (Rodriguez & Sirovich 1990, Sirovich 1989) and studied the bifurcation diagram for Galerkin projections and the dependence of the eigenfunctions on an external control parameter such as Reynolds number. They also find that the eigenfunction dependence on parameters is relatively weak. Kirby & Armbruster (1991) applied a conditional POD and a "moving" POD to the study of bifurcation problems in the Kuramoto-Sivashinsky equation. This approach allows the identification of traveling structures. Aubry et al (1991) have also studied the problem via a consideration of the relation of the POD to symmetry groups. Without such a procedure the statistics will be homogeneous and the eigenfunctions will be Fourier modes, as described in Section 2.4 above. The procedure applied in parts of Kirby & Armbruster (1991) is similar to the pattern-recognition techniques described in Section 5.3. Berkooz [1991a] suggested a rigorous procedure that allows the extraction of moving structures without the need to change to a moving reference frame. This is done through the use of third-order statistics. Berkooz (1991a) also provides a procedure that determines whether the signal can be viewed as a small perturbation of a moving structure.

A 2-D model involving the incompressible Navier-Stokes equations with spatially periodic forcing (the Kolmogorov flow) was also studied by Sirovich's group (Platt et al 1991). While they did not use the POD, their Poincaré sections and phase-space reconstructions show evidence of intermittent events somewhat similar to the boundary-layer bursts. Nicolaenko & She (1990) have made similar observations.

4. POD IN DYNAMICAL MODELING

From our (biased) viewpoint we feel that dynamical modeling is perhaps the most innovative recent use of the POD and of the most interest in the context of the dynamics of coherent structures. We make a distinction between two types of applications: a direct or exact simulation, and a model based simulation. In the former, one performs a Galerkin projection and the unresolved modes are neglected with the assumption that they are irrelevant to the dynamics; thus one typically must retain many modes [e.g. $O(10^5)$ even for low Reynolds number turbulence]. In the latter, however, one accounts for the unresolved modes by some model and so the dimension can be much reduced. The latter are exactly the type of low-dimensional models we hope will shed light on the dynamics of coherent structures in turbulent flows.

4.1 *Direct Simulations Using the POD*

The motivation behind such studies is the utilization of the fast energy convergence provided by the POD to get well resolved simulations with a lower number of modes than a simulation using Fourier or other types of standard decompositions. The questions that are normally studied include: (*a*) What is the minimal number of modes that reproduces acceptable dynamics? and (*b*) How do bifurcation sequences differ between a POD simulation and a regular one? Questions relating to dimension are also of interest. Observe that, although the idea of an exact simulation with a smaller (compared to standard) number of modes may seem appealing, the effect of the smaller number may be lost if one is interested in a direct numerical simulation of turbulence since the POD eigenfunctions, in general, are not suited for FFT-type algorithms which offer tremendous savings in computation time.

As we noted in Section 3.4, Sirovich and coworkers (Sirovich 1987, Rodriguez & Sirovich 1990) have used the Ginzburg-Landau equation as a mathematical model. They study the eigenfunctions, Lyapunov exponents, and bifurcations for the simulated systems and show that, for this model at least, the eigenfunctions change rather slowly with external parameters. This implies that an empirical basis computed for a particular Reynolds number will continue to be advantageous in simulations over a range of Reynolds numbers (see also Chamber et al 1988).

Kirby & Armbruster (1991) performed simulations of the Kuramoto-Sivashinsky equation. They commented on the need for increased dissipativity in the system to maintain its stability. This fact is well known in various computational disciplines (Anderson et al 1984) and has recently been studied from a mathematical point of view by Foias et al (1991).

4.2 *Models Based on the POD*

The best studied low-dimensional model is that developed by Aubry et al (1988) for the wall region of a turbulent boundary layer. In this model the neglected modes are modeled through a Smagorinsky-type sub-grid-scale model. In addition there is a model that accounts for the evolution of the local mean velocity profile. An introductory review, with some background material on dynamical-systems theory and coherent structures is given in Holmes (1990). Following the original study, which included five (complex) spanwise Fourier modes, a single ($k_1 = 0$) streamwise mode, and a single family of eigenfunctions in the wall normal direction, several developments have taken place. Stone (1989) showed that smaller truncations, with 3 and 4 spanwise modes—and certain larger ones, with additional nontrivial streamwise modes—continued to exhibit the heteroclinic cycles and bursting behavior characteristic of the unilateral model, which Aubry et al had shown was remarkably similar to the burst-sweep cycle familiar from experimental studies. Recently the more detailed studies of Aubry & Sanghi (1989, 1990) (cf Sanghi & Aubry 1991), which include nontrivial streamwise modes, have added weight to these findings. It is worth noting that

Armbruster et al's (1988) proof of structural stability of such heteroclinic cycles in O(2)-symmetric systems provides a mathematical foundation and does much to explain their robustness and persistence over a range of parameters and with different truncations (cf Armbruster et al 1989, Holmes 1991).

Aubry et al (1990) scaled Herzog's (1986) eigenfunctions in an attempt to study the effect of boundary-layer modification (by polymer addition or riblets, for example) and found a general shift of parameter values but no new dynamical mechanism or structures, in general agreement with the experimental evidence. Berkooz et al (1991b) in a study prompted by comments of Moffatt (1990), showed that if the cross stream and streamwise velocity components are uncoupled and a dynamical model formulated in which they are allowed to evolve independently, then turbulence having no streamwise variation will decay, as expected. The important point is that the decay is not featureless but ghosts of heteroclinic cycles and the resulting bursting behavior are present. In this study, and in Holmes et al (1991), it is pointed out that the coupled vector-valued empirical eigenfunctions that result from direct application of the POD effectively contain streamwise averages of nontrivial structures in the $k_1 = 0$ streamwise components, and so permit sustained extraction of energy from the mean flow, and hence sustained turbulence. From this viewpoint, use of the POD is similar to a closure assumption.

In another development, Stone & Holmes (1990, 1991) showed that small random and (in some cases) deterministic perturbations, characteristic of coupling to the outer region of the flow, produce distributions of bursting events having exponential tails similar in nature to those observed experimentally (cf Bogard & Tiederman 1986, Sreenivasan et al 1983, Holmes & Stone 1991). The thesis of Berkooz (1991b) contained a number of other results related to the original study of Aubry et al (1988). The most relevant in this context is that the low-dimensional models reproduce the energy budget of the full system within acceptable bounds. See also Berkooz et al (1992).

A dynamical model for the jet-annular mixing layer was constructed along the lines of the model for the wall layer by Glauser, George, and coworkers (Glauser, et al 1989, 1990; Zheng & Glauser 1990). In Glauser et al (1989) and Zheng & Glauser (1990) dynamical systems describing the interaction of the POD eigenfunctions are constructed and simulated. The result is a nonrecurring system (unlike the wall-layer system) which corresponds to the physical structures in the jet. Glauser and coworkers see it as a cascade mechanism.

Sirovich et al (1990c) proposed an ad-hoc approximate inertial manifold type of approach (Foias et al 1989, Titi 1990, Constantin et al 1989). The idea is as follows: suppose that $\{\phi_i\}_{i=1}^{\infty}$ are the POD eigenfunctions and we want to resolve only the subset $\{\phi_i\}_{i=1}^{m}$. Denote by P the projection on span $\{\phi_i\}_{i=1}^{m}$ and let $Q = I-P$. Suppose the original equation is $\dot{u} = F(u)$ with $u = p+q$. We can write

$$\dot{p} = PF(p+q),$$
$$\dot{q} = QF(p+q). \tag{4.1}$$

The approximate inertial manifold type model is

$$\dot{p} = PF[p+\Phi(p)], \tag{4.2}$$

where the function $q = \Phi(p)$ is obtained by solution of

$$0 = QF(p+q). \tag{4.3}$$

The rationale is that unresolved, lower energy modes have a faster time scale (van Kampen 1985) and their variations are not felt by the p's. Sirovich et al (1990c) apply this to a weakly chaotic system with a truncation which is not as severe as the boundary layer or jet models. In general the algebraic Equation (4.3) is not solved exactly since it may have multiple solutions and an approximation, or truncation, is used. The results of application of this approach to the Ginzburg-Landau equation are given in Sirovich et al (1990c); also see Sirovich (1990).

Finally, we mention the application of the POD to 2-D flows in complex geometries by Deane et al (1991). They considered flows behind a circular cylinder and in a channel with spanwise, rectangular grooves and used a numerically simulated database to produce empirical eigenfunctions which were in turn used to construct low-dimensional ODE models by Galerkin projection. They found that 4–8 mode models could capture the initial (Hopf) bifurcation and loss of stability of the laminar flow rather well, and that, in the case of the grooved channel, the model performed satisfactorily over a range of Reynolds numbers. The cylinder-wake model was more limited in this respect, and studies of fully developed turbulent flows were not carried out for either problem.

5. RELATION TO OTHER TECHNIQUES

In this section we comment on the connection between the POD and certain other analysis techniques. We start by describing the connection between the POD and linear stochastic estimation, as applied by Adrian and coworkers in Adrian (1979), Adrian & Moin (1988), Adrian et al (1987), and Moin et al (1987). This will lead to relations between the POD and conditional sampling techniques such as described in Blackwelder &

Kaplan (1976), Bogard & Tiederman (1986), Johansson & Alfredsson (1982), Antonia (1981), and references therein. A connection to pattern-analysis techniques such as Stretch et al (1990), Townsend (1979), and Ferre & Girlat (1989) will also be discussed. Recently Aubry et al (1991a) introduced a "new" tool called the bi-orthogonal decomposition. However, a careful examination of their work reveals that their suggestion is in fact a specific case of the rather general POD formulated in Section 2.1. We remark that the formulation of the POD and the results above apply to space, time, or space-time analysis, all depending on the choice of the averaging operator (or equivalently, the measure) as long as the assumptions of Section 2.1 are satisfied. A specific choice of averaging (i.e. a measure concentrated on a finite number of points, as would be encountered in a computer simulation) will produce the decomposition of Aubry et al (1991a).

5.1. *Linear Stochastic Estimation*

Suppose one wanted to find the conditional probability density function (CPDF) of $[u(x)|u(x')]$, where $u(x')$ may be a multipoint event. For the sake of simplicity of the exposition, we limit ourselves to single-point scalar events. There are many good reasons to seek this CPDF, either for closure models (Pope 1985) or for the sake of producing coherent structures. Suppose, moreover, that one seeks an estimate linear in $u(x')$, i.e. instead of the full CPDF we want some representative value which would be our best estimate in a sense which we will define below. This is called a linear stochastic estimate. We outline the method of linear stochastic estimation since it is simple and enlightening. We are seeking $A(x, x')$ such that $A(x, x')u(x')$ will be the estimate for $u(x)$, and we want to find $A(x, x')$ such that

$$\min_{V(x,x')} \langle |u(x) - V(x, x')u(x')|^2 \rangle \tag{5.1}$$

is achieved by $A(x, x')$. We use the calculus of variations [as in the derivation of (2.2) from (2.1), cf Berkooz 1991b]. A necessary condition will be that, for any $V(x, x')$, we have

$$\frac{d}{d\delta} \langle |u(x) - [A(x, x') + \delta V(x, x')] \cdot u(x')|^2 \rangle |_{\delta=0} = 0. \tag{5.2}$$

The expression inside the averaging brackets is equal to

$$\{u(x) - [A(x, x') + \delta V(x, x')] \cdot u(x)\} \cdot \\ \{u^*(x) - [A^*(x, x') + \delta V^*(x, x')] \cdot u^*(x')\}. \tag{5.3}$$

After taking the average of (5.2), differentiating w.r.t. δ and evaluating at $\delta = 0$, and equating to zero we get

$$2\mathrm{Real}\,[V^*(x,x'):\langle u(x)u^*(x')\rangle] = 2\mathrm{Real}\,\{V^*(x,x'):\langle [A(x,x')u(x')]u^*(x')\rangle\}, \quad (5.4)$$

where : denotes the usual tensor contraction. Therefore we require

$$\langle u(x)u^*(x')\rangle = A(x,x')\cdot\langle u(x')u^*(x')\rangle, \quad (5.5)$$

and this implies

$$A(x,x') = \langle u(x)u^*(x')\rangle\cdot\langle u(x')u^*(x')\rangle^{-1}. \quad (5.6)$$

It is natural to ask whether $\langle u(x')u^*(x')\rangle$ is invertible at every point x' (in the vector case). One can convince oneself that this should typically be the case for a turbulent system. Since here we treat u as a scalar, the inverse of $\langle u(x')u^*(x')\rangle$ is just a division.

In (5.6) the average two-point correlation tensor $R(x,x')$ of Section 2.1 appears with some normalization. Results of Adrian (1979) show that the corrections to the CPDF due to higher order nonlinear terms in $u(x')$ are small (recall this is the best linear estimate), at least for homogeneous turbulence. Using our previous results (2.9–2.11) we can write

$$A(x,x') = \frac{\Sigma_{i=1}^{\infty}\lambda_i\phi_i(x)\phi_i^*(x')}{\Sigma_{i=1}^{\infty}\lambda_i|\phi_i(x')|^2} = \sum_{i=1}^{\infty}\phi_i(x)f_i(x'), \quad (5.7)$$

where $f_i(x') = \lambda_i\phi_i^*(x')/\Sigma_{j=1}^{\infty}\lambda_j|\phi_j(x')|^2$. We may interpret $f_i(x')$ as the relative contribution of ϕ_i to $u(x')$ on the average. We conclude that linear stochastic estimation is equivalent to assuming that the estimated value of the POD coefficient of the i-th mode, given the velocity at x', is the average contribution of the i-th mode to the velocity of x' times the given velocity.

The amazing point is that we get exactly the same result from the simplified PDF model based on the POD introduced at the end of Section 2.6. There we assumed $a_i \sim N(0,\lambda_i)$ and that the coefficients a_j were independent. Let us compute the estimator $\langle u(x)|u(x')\rangle$. Since we have an expression for the PDF we can compute this explicitly. Recall from probability theory that if $x_i \sim N(0,\sigma_i^2)$ for $i = 1,\ldots,m$ then

$$\langle x_i \mid \sum_{j=1}^{m} x_j = C\rangle = \frac{\sigma_i^2 C}{\Sigma_{j=1}^{m}\sigma_j^2}. \quad (5.8)$$

[See the formula for the conditional expectation of joint normal variables in Feller (1957).] Using (5.8), we have

$$\{a_i\phi_i(x') \mid \sum_{j=1}^{\infty} a_j\phi_j(x') = u(x')\} = \frac{\lambda_i|\phi_i(x')|^2 u(x')}{\Sigma_{j=1}^{\infty}\lambda_j|\phi_j(x')|^2} \tag{5.9}$$

which gives

$$\{u(x)|u(x')\} = \frac{\Sigma_{i=1}^{\infty}\lambda_i|\phi_i(x')|^2 u(x')\phi_i(x)/\phi_i(x')}{\Sigma_{j=1}^{\infty}\lambda_j|\phi_j(x')|^2}$$

$$= \frac{\Sigma_{i=1}^{\infty}\lambda_i\phi_i^*(x')\phi_i(x)u(x')}{\Sigma_{j=1}^{\infty}\lambda_j|\phi_j(x')|^2}. \tag{5.10}$$

This is exactly the same result obtained from linear stochastic estimation (Equation 5.7).

We conclude that the simple PDF model suggested at the end of Section 2.6 results in the best linear estimator of the conditional PDF of velocity, and that linear stochastic estimation may be viewed as a result of the simple PDF model. This reveals the fundamental connection between the POD and linear stochastic estimation. Aside from this we can make the following technical observations based on our previous results:

1. All fields generated by linear stochastic estimation (LSE) possess any closed linear property that all ensemble members share.
2. Suitable averages of LSE events will produce the POD eigenfunctions.
3. All LSE events are linear combinations of POD eigenfunctions.

Finally we remark that one can apply the geometric result from the POD of Section 2.5 to obtain bounds on the probability of rare LSE events.

5.2 *Conditional Sampling*

In this section we indicate a possible connection between the POD and conditional sampling. This exposition is of a speculative nature which we hope might encourage further work in the area. The general conditional-sampling scheme adopted from Antonia (1981) is given in Figure 4. This scheme may be formulated as

$$R(x, \Delta x, \tau_j) = \frac{1}{N}\sum_{i=1}^{N} c(x, t_i)f(x, \Delta x, t_i + \tau_j), \tag{5.11}$$

where $c(x, t_i)$ is the conditioning function at a point x in space at time t_i which is 1 if a condition is met and 0 otherwise. Once a condition is met, a measurement at a possibly different location (given by $x + \Delta x$) and possibly later in time $(t_i + \tau_j)$ is added to the averaged ensemble. Conditional averaging has been used in the study of turbulent-nonturbulent interfaces, shear layers perturbed by interaction with another turbulent field, and quasi-periodic or periodic flows (such as those behind a turbine),

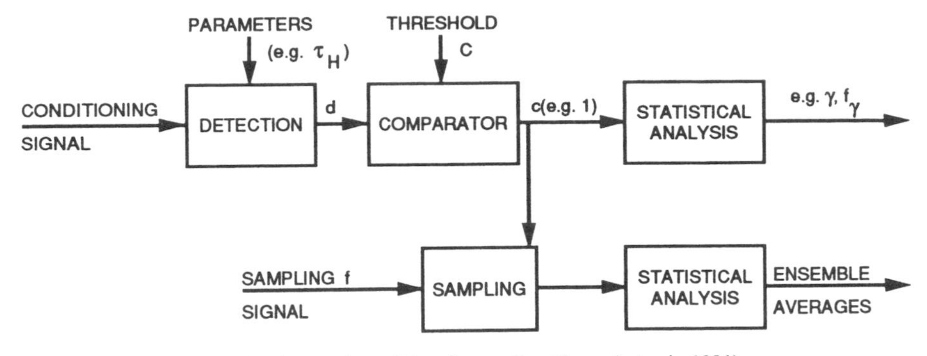

Figure 4 The general scheme of conditional sampling (from Antonia 1981).

and for the study of coherent structures. Our primary interest is in the study of coherent structures. Conditional sampling applied to coherent structures is largely an art: The experimentalist has freedom in defining the threshold and detection criteria and the quantity to be measured. The subjectivity of this procedure may lead to detection of fictitious structures in featureless random fields (Blackwelder & Kaplan 1976) or may yield the wrong structures (Lumley 1981). This subjectivity precludes the possibility of a rigorous analytic connection, since the POD does not offer this kind of freedom. However, one can study the compatibility between the simple PDF model based on the POD and the results of a conditional-sampling study. This is done through the results on linear stochastic estimation presented in Section 5.1 above. Two comments are in order. First, in order to treat conditional sampling in its full generality, one would have to perform a space-time POD—otherwise the comparison will be limited to conditional sampling with no time delays. Second, the two very strong assumptions made in the process of establishing the correspondence through LSE, namely the independent normal distributions for POD random coefficients and the adequacy of stochastic linear estimations, will prevent us from obtaining decisive conclusions in case of a mismatch. On the other hand, showing such a compatibility would be a welcome contribution to the relation between average quantities such as the POD eigenfunctions and average dynamics of coherent structures. References on conditional sampling are Blackwelder & Kaplan (1976), Bogard & Tiederman (1986), Johnansson & Alfredsson (1982), in addition to the survey paper by Antonia (1981) already referred to, and references therein.

Note that the natural combination of the POD and conditional sampling into a conditional POD based on the condition functions currently used might be very illuminating. These conditions also have an interpretation as conditions in phase space and thus are related to the conditional POD mentioned in Section 2.6. Such a combination would give a detailed kinematical description of important processes. If a time-delayed conditional POD were applied, the results might be compared to results of low-dimensional models such as those reviewed in Section 4, thus yielding insight on the dynamics and a comparison for low-dimensional experimental data.

5.3 *Pattern-Recognition Techniques*

In this section we indicate a second possible connection between the POD and pattern-recognition techniques. As above, this section is of a speculative nature. With the advent of digital image processing, pattern recognition has become a vast field (Rosenfeld & Kak 1982). We limit ourselves to the relatively basic procedures used in fluid problems by Stretch et al (1990), Townsend (1979), and Ferre & Girlat (1989). Coherent structures were originally identified in flow visualization. The quest for a quantitative procedure for extracting coherent structures and their dynamics is still a subject of research. Pattern-recognition techniques are designed to mimic the human capability of detecting patterns in a noisy medium and thus the hope for their successful application to the task of identifying structures in a flow. We remark that Sirovich & Kirby (1987) (cf Kirby & Sirovich 1990) have applied the POD procedure directly to reconstruction of images (of human faces); but the feature-extraction method to be discussed below is somewhat different in spirit.

The basic procedure is as follows: One wants to identify a recurrent pattern in a noisy medium. First one picks a template size and fills it with what is conjectured to look like the coherent structures. The template is then moved around in the data set and after each movement a correlation is computed. Every time the correlation attains a local maximum the corresponding pattern is added to the ensemble, which is averaged to produce a modified reference template. This process is repeated until the template undergoes insignificant further change. The final template is the coherent structure. Once the coherent structure is deduced, one attempts to find regions in space well correlated with this structure and to study their contribution to various statistics.

This again, is a subjective procedure, although Stretch et al (1990) suggest it is a robust one, with the final template being practically independent of the initial condition. Our mathematical understanding of the POD may contribute to a better understanding of the results of pattern-recognition applications. Observe the similarity in mission between the pattern-recognition technique and the shot-noise expansion. Both attempt to decompose the flow into building blocks (although in pattern recognition we concentrate on regions of the flow with higher correlation with the template). This suggests caution in the interpretation of the resultant template, since, as we saw, any template with a suitable power spectrum might decompose the flow, with an appropriate sprinkling function. This fear is accentuated by the fact that Stretch et al (1990) show a median correlation of only about 0.3. Based on the shot-noise decomposition one can propose a test for the objectivity of this method to see how well the basic building block is reproduced. Lumley's example (Lumley 1981) is a good starting point for such a quest.

6. DISCUSSION

In this paper we have described the proper orthogonal decomposition technique and illustrated its use in the analysis and modeling of turbulent flows. The POD is already a well-established tool for (statistical) data analysis and in data compression. We have argued that it can also be used to address two further types of questions, typified by the following: 1. Given a complex spatio-temporal signal, what can one determine, via the POD, regarding the system from which it originates? 2. Given a POD-Galerkin projection of a PDE such as the Navier-Stokes equation and the resulting finite-dimensional dynamical system, what can one learn about the behavior of the original PDE? We hope that this paper has convinced the reader that recent progress in these areas promises as much or more than the POD has already delivered in its more conventional roles in signal analysis and data compression.

Given a POD analysis of a black box we can say several things. First of all we can determine whether the system exhibits equipartition of energy between different modes. This is immediately apparent from looking at the POD spectrum. If it is a decaying spectrum the system is not equipartitioned and the notion of "more energetic" and "less energetic" modes is meaningful. We can then use finer results (Proposition 2.7 and 2.8) to examine whether the dynamics could have been generated by a compact attractor in phase space. Looking for symmetries (or invariance properties) and using Proposition 2.6 one can, potentially, detect symmetries in the system which might not be apparent otherwise. If the POD eigenfunctions of a given system all share some linear property, using Corollary 2.2 one can deduce that the system as a whole has that property. For example, if measuring in a turbulent flow, say in air, where it is not obvious whether incompressibility is satisfied, a POD analysis will tell us whether the turbulence is divergence free and, if not, to what extent it is not so.

Given a POD analysis which suggests more energetic and less energetic modes and the equations of motion one can study the question of the interaction of energetic modes or coherent structures. This is done along the lines of the dynamical studies of the models described in Section 4.2. Note that a successful study is liable to require some physical understanding of the system, although there are some promising mathematical approaches too.

ACKNOWLEDGMENTS

Support for this work was provided by the Air Force Office of Scientific Research under grant AFOSR 89-0226.

Literature Cited

Adrian, R. J. 1979. Conditional eddies in isotropic turbulence. *Phys. Fluids* 22(11): 2065–70

Adrian, R. J., Moin P. 1988. Stochastic estimation of organized turbulent structure: homogeneous shear flow. *J. Fluid Mech.* 190: 531–59

Adrian, R. J., Moin, P., Moser, R. D. 1987. Stochastic estimation of conditional eddies in turbulent channel flow. In *CTR, Proc. Summer Prog. 1987,* pp. 7–20

Algazi, V. R., Sakrison, D. J. 1969. On the optimality of the Karhunen-Loève expansion. *IEEE Trans. Inform. Theory* 15: 319–21

Anderson, D. A., Tannehill, J. C., Pletcher, R. H. 1984. *Computational Fluid Mechanics and Heat Transfer*. Washington, DC: Hemisphere

Andrews, C. A., Davies, J. M., Schwartz, G. R. 1967. Adaptive data compression. *Proc. IEEE* 55: 267–77

Antonia, R. A. 1981. Conditional sampling in turbulence measurement. *Annu. Rev. Fluid. Mech.* 13: 131–56

Armbruster, D., Guckenheimer, J., Holmes, P. 1988. Heteroclinic cycles and modulated travelling waves in systems with O(2) symmetry. *Physica D* 29: 257–82

Armbruster, D., Guckenheimer, J., Holmes, P. 1989. Kuramoto-Sivashinksy dynamics on the center unstable manifold. *SIAM J. Appl. Math.* 49: 676–91

Aubry, N., Sanghi, S. 1989. Streamwise and cross-stream dynamics of the turbulent wall layer. In *Chaotic Dynamics in Fluid Mechanics*, Proc. 3rd Joint ASCE-ASME Mech. Conf., UCSD, La Jolla, CA, July 9–12, ed. K. N. Ghia, U. Ghia. New York: ASME

Aubry, N., Sanghi, S. 1990. Bifurcations and bursting of streaks in the turbulent wall layer. In *Turbulence and Coherent Structures*, Proc. Grenoble (France) Conf. on Organized Structures and Turbulence in Fluid Mech., 18–21 Sept., ed. M. Lesieur , O. Métais, pp. 227–51. Dordrecht: Kluwer

Aubry, N., Holmes, P., Lumley, J. L., Stone, E. 1988. The dynamics of coherent structures in the wall region of a turbulent boundary layer. *J. Fluid Mech.* 192: 115–73

Aubry, N., Lumley, J. L., Holmes, P. 1990. The effect of modeled drag reduction on the wall region. *Theoret. Comput. Fluid Dynam.* 1: 229–48

Aubry, N., Guyonnet, R., Lima, R. 1991a. Spatio-temporal analysis of complex signals: theory and applications. *J. Stat. Phys.* 64(3/4): 683–793

Aubry, N., Lian, W.-L., Titi, E. S. 1991b. Preserving symmetries in the proper orthogonal decomposition. *Tech. Rep. 91–35, MSI*

Bakewell, P., Lumley, J. L. 1967. Viscous sublayer and adjacent wall region in turbulent pipe flows. *Phys. Fluids* 10: 1880–89

Ball, K. S., Sirovich, L., Keefe, L. R. 1991. Dynamical eigenfunction decomposition of turbulent channel flow. *Int. J. Numerical Methods Fluids* 12: 585–604

Bartlet, H., Lohmann, A. W., Wirnitzer, B. 1984. Phase and amplitude recovery from bispectra. *Appl. Opt.* 23: 3121–29

Berkooz, G. 1990a. Observations on the proper orthogonal decomposition. In *Studies in Turbulence*,, ed. T. B. Gatski, S. Sarkar, C. G. Speziale, pp. 229–47. New York: Springer-Verlag

Berkooz, G. 1990b. Some examples concerning the POD. Preprint

Berkooz, G. 1991a. Statistical analysis of phase portraits: bifurcations and perturbations of limit cycles. Preprint

Berkooz, G. 1991b. *Turbulence, coherent structures and low dimensional models*. PhD thesis. Cornell Univ.

Berkooz, G., Titi, E. 1992. The POD systems with symmetry and harmonic analysis. In preparation

Berkooz, G., Elezgaray, J., Holmes, P. 1991a. Coherent structures in random media and wavelets. Proc. DARPA/ONERA Workshop on Wavelets and Applications, Princeton, June 1991, Physica D (to appear)

Berkooz, G., Holmes, P., Lumley, J. L. 1991b. Intermittent dynamics in simple models of the wall layer. *J. Fluid Mech.* 230: 75–95

Berkooz, G., Holmes, P., Lumley, J. L. 1992. On the relations between low dimensional models and the dynamics of coherent structures in the wall layer. Preprint

Blackwelder, R. F., Kaplan, R. E. 1976. On the wall structure of the turbulent boundary layer. *J. Fluid Mech.* 76: 89–112

Bogard, D. G., Tiederman, W. G. 1986. Burst detection with single single-point velocity measurements. *J. Fluid Mech.* 162: 389–413

Brillinger, D. R., Rosenblatt, M. 1967. Asymptotic theory of estimates of kth order spectra. In *Spectral Analysis of Time Series*, ed. B. Harris. New York: Wiley

Chambers, D. H., Adrian, R. J., Moin, P. Stewart, D. S., Sung, H. J. 1988. Karhunen-Loève expansion of Burgers' model of turbulence. *Phys. Fluids* 31(9): 2573–82

Constantin, P., Foias, C. 1989. *Navier-Stokes Equations*. Chicago: Chicago Univ. Press

Constantin, P., Foias, C. Temam, R., Nicolaenko, B. 1989. *Integral Manifolds and Inertial Manifolds for Dissipative Partial Differential Equations*. New York: Springer-Verlag

Deane, A. E., Sirovich, L. 1991. A computational study of Rayleigh-Bénard convection. Part I: Rayleigh number scaling. *J. Fluid Mech.* 222: 231–50

Deane, A. E., Kevrekidis, I. G., Karniadakis, G. E., Orszag, S. A. 1991. Low dimensional models for complex geometry flows: application to grooved channels and circular cylinder. *Phys. Fluids A* 3: 2337–54

Farge, M. 1992. The wavelet transform and its applications to fluid mechanics. *Annu. Rev. Fluid Mech.* 24: 395–457

Feller, W. 1957. *An Introduction to Probability Theory and Its Applications*. New York: Wiley

Ferre, J. A., Girlat, F. 1989. Pattern-recognition analysis of the velocity field in plane turbulent wakes. *J. Fluid Mech.* 198: 27–64

Foias, C., Témam, R. 1989. Gevrey class regularity for the solutions of the Navier-Stokes equations. *J. Func. Anal.* 87: 359–69

Foias, C., Sell, G. R., Titi, E. S. 1989. Exponential tracking and approximation of inertial manifolds for dissipative equations. *J. Dyn. Differ. Equ.* 1: 199–224

Foias, C., Manley, O., Sirovich, L. 1990. Empirical and Stokes eigenfunctions and the far dissipative turbulent spectrum. *Phys. Fluids A* 2: 464–67

Foias, C., Jolly, M. S., Kevrekidis, I. G., Titi, E. S. 1991. Dissipativity of numerical schemes. *Nonlinearity* 4: (to appear)

Gay, D. H., Ray, W. H. 1986. Identification and control of linear distributed parameter systems through the use of experimentally determined singular functions. *Proc. IFAC Symp. Control of Distributed Parameter Systems, Los Angeles, 30 June—2 July 1986*,, ed. H. E. Rauch, pp. 173–79. Oxford/New York: Pergamon

Gay, D. H., Ray, W. H. 1988. Application of singular value methods for identification and model based control of distributed parameter systems. *Proc. IFAC Workshop on Model-Based Process Control, Atlanta, GA, June 1988*, ed. J. T. McAvoy, Y. Arkun, E. Zafiriou, pp. 95–102. Oxford/New York: Pergamon

Glauser, M. N., George, W. K. 1987a. Orthogonal decomposition of the axisymmetric jet mixing layer including azimuthal dependence. In *Advances in Turbulence*, ed. G. Comte-Bellot, J. Mathieu. New York: Springer-Verlag

Glauser, M. N., George, W. K. 1987b. Orthogonal decomposition of the axisymmetric jet mixing layer utilizing cross-wire velocity measurements. In *6th Symp. Turbulent Shear Flows, 1987*

Glauser, M. N., Leib, S. J., George, W. K. 1987. Coherent structures in the axisymmetric turbulent jet mixing layer. In *Turbulent Shear Flows 5*, ed. F. Durst, B. E. Launder, J. L. Lumley, F. W. Schmidt, J. H. Whitelaw. New York: Springer-Verlag

Glauser, M. N., Zheng, X., Doering, C. R. 1989. The dynamics of organized structures in the axisymmetric jet mixing layer. In *Studies in Turbulence*, ed. T. Gatski, S. Sarkar, C. Speziale, pp. 207–22. New York: Springer-Verlag

Glauser, M. N., Zheng, X., George W. K. 1990. The streamwise evolution of coherent structures in the axisymmetric jet mixing layer. In *The Lumley Symposium: Recent Developments in Turbulence*, Newport News, VA, Nov. 1990, ed. T. Gatski et al., pp. 207–22. New York: Springer-Verlag

Glezer, A., Kadioglu, A. J., Pearlstein, A. J. 1989. Development of an extended proper orthogonal decomposition and its application to a time periodically forced plane mixing layer. *Phys. Fluids A* 1: 1363–73

Herzog, S. 1986. *The large scale structure in the near wall region of a turbulent pipe flow*. PhD thesis. Cornell Univ.

Holmes, P. J. 1990. Can dynamical systems approach turbulence? In *Whither Turbulence? Turbulence at the Crossroads*, ed. J. L. Lumley, Lect. Notes Phys. 357: 195–249

Holmes, P. J. 1991. Symmetries, heteroclinic cycles and intermittency in fluid flow. Proc. I. M. A. Workshop on Dynamical Theories of Turbulence (Minnesota, May 29—June 2, 1990) New York: Springer-Verlag. In press

Holmes, P. J., Stone, E. 1991. Heteroclinic cycles, exponential tails and intermittency in turbulence production. In *Studies in Turbulence*, ed. T. Gatski, S. Sarkar, C. Speziale, pp. 179–89. New York: Springer-Verlag

Holmes, P. J., Berkooz, G., Lumley, J. L. 1991. Turbulence, dynamical systems and the unreasonable effectiveness of empirical eigenfunctions. In *Proc. ICM—90*, Kyoto, Japan. New York: Springer-Verlag

Hopf, E. 1957. On the application of functional calculus to the statistical theory of turbulence. In *Proc. Symp. Appl. Math.* Am. Math. Soc.

Jimenez, J., Moin, P. 1991. The minimal flow unit in near-wall turbulence. *J. Fluid Mech.* 225: 213–20

Johansson, A. V., Alfredsson, P. H. 1982. On the structure of turbulent channel flow. *J. Fluid Mech.* 122: 295–314

Karhunen, K. 1946. Zur spektral theorie stochastischer prozesse. *Ann. Acad. Sci. Fennicae* Ser. A1: 34

Keefe, L., Moin, P., Kim, J. 1987. The dimension of an attractor in turbulent Poiseuille flow. *Bull. Am. Phys. Soc.* 32: 2026

Kim, J., Moin, P., Moser, R. J. 1987. Turbulence statistics in a fully developed channel flow at low Reynolds number. *J. Fluid Mech.* 177: 133–66

Kirby, M., Armbruster, D. 1991. Reconstructing phase space for PDE simulations. Preprint

Kirby, M., Sirovich, L. 1990. Application of the Karhunen-Loève procedure for the characterization of human faces. *IEEE Trans. Pattern Anal. and Mach. Intell.* 12: 103–8

Kirby, M., Boris, J., Sirovich, L. 1990a. An eigenfunction analysis of axisymmetric jet flow. *J. Comput. Phys.* 90(1): 98–122

Kirby, M., Boris, J., Sirovich, L. 1990b. A proper orthogonal decomposition of a simulated supersonic shear layer. *Int. J. Numerical Methods Fluids* 10: 411–28

Kline, S. J., Reynolds, W. C., Schraub, F. A., Rundstadler, P. W. 1967. The structure of turbulent boundary layers. *J. Fluid Mech.* 30: 741–73

Kosambi, D. D. 1943. Statistics in function space. *J. Indian Math. Soc.* 7: 76–88

Liepmann, H. W. 1952. Aspects of the turbulence problem. Part II *Z. Angew. Math. Phys.* 3: 407–26

Lii, K. S., Rosenblatt, M., Van Atta, C. 1976. Bispectral measurements in turbulence. *J. Fluid Mech.* 77: 45–62

Liu, J. T. C. 1988. Contributions to the understanding of large scale coherent structures in developing free turbulent shear flows. *Adv. Appl. Mech.* 26: 183–209

Loève, M. 1945. Functions aleatoire de second ordre. *C. R. Acad. Sci. Paris* 220

Lumley, J. L. 1967. The structure of inhomogeneous turbulence. In *Atmospheric Turbulence and Wave Propagation*, ed. A. M. Yaglom, V. I. Tatarski, pp. 166–78. Moscow: Nauka

Lumley, J. L. 1971. *Stochastic Tools in Turbulence*. New York: Academic

Lumley, J. L. 1981. Coherent structures in turbulence. In *Transition and Turbulence*, ed. R. E. Meyer, pp. 215–42. New York: Academic

Lumley, J. L. 1989. The state of turbulence research. In *Advances in Turbulence*, ed. W. K. George, R. Arndt, pp. 1–10. Washington, DC: Hemisphere

Matsuoka, T., Ulrych, T. L. 1984. Phase estimation using the bispectrum. *Proc. IEEE* 72: 1403–22

McComb, W. D. 1990. *The Physics of Turbulence*. Oxford: Clarendon

Meyer, Y. 1987. Ondelettes, fonctions splines at analyses graduées. *Rapp. CEREMADE 8703*, Univ. Paris-Dauphane

Moffatt, H. K. 1990. Fixed points of turbulent dynamical systems and suppression of nonlinearity. In *Whither Turbulence? Turbulence at the Crossroads*, ed. J. L. Lumley, Lect. Notes Phys. 357: 250–57

Moin, P., Moser, R. D. 1989. Characteristic-eddy decomposition of turbulence in a channel. *J. Fluid Mech.* 200: 471–509

Moin, P., Adrian, R. J., Kim, J. 1987. Stochastic estimation of organized structures in turbulent channel flow. In *6th Turbulence Shear Flow Symp.*, Toulouse, France, pap. 16–9

Monin, A. S., Yaglom, A. M. 1987. *Statistical Fluid Mechanics*, ed. J. L. Lumley. Cambridge, MA: MIT Press

Nicolaenko, B., She, Z.-S. 1990. Temporal

intermittency and turbulence production in the Kolmogorov flow. In *Topological Fluid Mechanics*, ed. H. K. Moffatt, p. 256. Cambridge: Cambridge Univ. Press

Novikov, E. A. 1963. Variability of energy dissipation rate in a turbulent flow and the energy distribution over the spectrum. *Prikl. Mat. Mekh.* 27(5): 944–46

Obukhov, A. M. 1954. Statistical description of continuous fields. *Tr. Geophys. Int. Akad. Nauk. SSSR* 24: 3–42

Papoulis, A. 1965. *Probability, Random Variables, and Stochastic Processes*. New York: McGraw-Hill

Park, H., Sirovich, L. 1990. Turbulent thermal convection in a finite domain: Part II numerical results. *Phys. Fluids A* 2(9): 1659–68

Perrier, V., Basdevant, C. 1989. Periodic wavelet analysis: a tool for inhomogeneous field investigation. Theory and Algorithms. *Rech. Aérosp. No.* 1989–3, pp. 54–67

Platt, N., Sirovich, L., Fitzmaurice, N. 1991. An investigation of chaotic Kolmogorov flows. *Phys. Fluids* 3(4): 681–96

Pope, S. B. 1985. Pdf methods for turbulent reactive flows. *Prog. Energy Combust. Sci.* 11: 119–92

Pougachev, V. S. 1953. General theory of the correlations of random functions. *Izv. Akad. Nauk. SSSR, Ser. Mat.* 17: 1401–2

Preisendorfer, R. W. 1988. *Principal Component Analysis in Meteorology and Oceanography*. Amsterdam: Elsevier

Promislow, K. 1991. Time analyticity and Gevrey regularity for solutions of a class of dissipative pde's. *Nonlinear. Anal.* 16(11): 959–80

Reynolds, O. 1894. On the dynamical theory of incompressible viscous fluids and the determination of the criterion. *Phil. Trans. R. Soc. London* 186: 123–61

Rice, S. O. 1944. Mathematical analysis of random noise. *Bell Sys. Tech. J.* 23: 282–332

Riesz, F., Nagy, B. 1955. *Functional Analysis*. New York: Ungar

Robinson, S. K. 1991. Coherent motions in the turbulent boundary layer. *Annu. Rev. Fluid Mech.* 23: 601–39

Rodriguez, J. D., Sirovich, L. 1990. Low-dimensional dynamics for the complex Ginsburg-Landau equation. *Physica D* 43: 77–86

Rosenfeld, A., Kak, A. C. 1982. *Digital Picture Processing*. New York: Academic

Ruelle, D., Takens, F. 1971. On the nature of turbulence. *Comm. Math. Phys.* 20: 167–92; 23: 343–44

Sanghi, S., Aubry, N. 1991. Mode interaction models for near wall turbulence. Levich Inst. Preprint

Sirovich, L. 1987. Turbulence and the dynamics of coherent structures, parts i–iii. *Quart. J. Appl. Math.* 45(3): 561–90

Sirovich, L. 1989. Chaotic dynamics of coherent structures. *Physica D* 37: 126–45

Sirovich, L. 1990. Empirical eigenfunctions and low dimensional systems. *Cent. Fluid Mech. Rep.* 90–202, Brown Univ.

Sirovich, L., Deane, A. E. 1991. A computational study of Rayleigh-Bénard convection. Part II. Dimension considerations. *J. Fluid Mech.* 222: 251–65

Sirovich, L., Kirby, M. 1987. Low-dimensional procedure for the characterization of human faces. *J. Opt. Soc. Am. A* 4: 519–24

Sirovich, L., Knight, B. W. 1985. The eigenfunctions problem in higher dimensions: asymptotic theory. *Proc. Natl. Acad. Sci.* 82: 8275–78

Sirovich, L., Park, H. 1990. Turbulent thermal convection in a finite domain: Part I. Theory. *Phys. Fluids A* 2(9): 1649–58

Sirovich, L., Ball, K. S., Keefe, L. R. 1990a. Plane waves and structures in turbulent channel flow. *Phys. Fluids A* 2(12): 2217–26

Sirovich, L., Kirby, M., Winter, M. 1990b. An eigenfunction approach to large scale transitional structures in jet flow. *Phys. Fluids A* 2 (2): 127–136

Sirovich, L., Knight, B. W., Rodriguez, J. D. 1990c. Optimal low-dimensional dynamical approximations. *Quart. J. Appl. Math.* 48(3) : 535–48

Sreenivasan, K. R., Narashima, R., Prabhu, A. 1983. Zero-crossings in turbulent signals. *J. Fluid Mech.* 137: 251–72

Stone, E. 1989. A study of low dimensional models for the wall region of a turbulent boundary layer. PhD thesis. Cornell Univ.

Stone, E., Holmes, P. J. 1990. Random perturbations of heteroclinic cycles. *SIAM J. Appl. Math.* 50(3) : 726–43

Stone, E., Holmes, P. J. 1991. Unstable fixed points, heteroclinc cycles and exponential tails in turbulence production. *Phys. Lett. A* 155: 29–42

Stretch, D., Kim, J., Britter, R. 1990. A conceptual model for the structure of turbulent channel flow. In *Notes for Boundary Layer Structure Workshop*, ed. S. Robinson, NASA Langley, Aug 28–30. See also Stretch, D. 1991. Automated pattern eduction from turbulent flow diagnostics. *1991 Annu. Res. Briefs*, Cent. for Turb. Res., Stanford Univ., pp. 145–57

Témam, R. 1988. *Infinite-Dimensional Dynamical Systems in Mechanics and Physics*. New York: Springer-Verlag

Tennekes, H., Lumley, J. L. 1972. *A First*

Course in Turbulence. Cambridge, MA: MIT Press
Titi, E. S. 1990. On approximate inertial manifolds to the Navier-Stokes equations. *J. Math. Anal. Appl.* 149: 540–57
Townsend, A. A. 1956. *The Structure of Turbulent Shear Flow*. Cambridge, UK: Cambridge Univ. Press
Townsend, A. A. 1979. Flow patterns of large eddies in a wake and in a boundary layer. *J. Fluid Mech.* 95: 551
van Kampen, N. G. 1985. Elimination of fast variables. *Phys. Rep.* 123: 69–160
Zheng, X., Glauser, M. N. 1990. A low dimensional description of the axisymmetric jet mixing layer. *ASME Comput. Eng.* 2: 121–27

Annu. Rev. Fluid Mech. 1993. 25 : 577–602

THE IMPACT OF DROPS ON LIQUID SURFACES AND THE UNDERWATER NOISE OF RAIN

Andrea Prosperetti and Hasan N. Oğuz

Department of Mechanical Engineering, The Johns Hopkins University, Baltimore, Maryland 21218

KEY WORDS: bubble entrainment, bubble noise, liquid impact, splash

INTRODUCTION

> There will be but few of my readers who have not, in some heavy shower of rain, beguiled the tedium of enforced waiting by watching, perhaps half-unconsciously, the thousand little crystal fountains that start from the surface of pool or river; noting now and then a surrounding coronet of lesser jets, or here and there a bubble that floats for a moment and then vanishes.
>
> It is to this apparently insignificant transaction, which always has been and always will be so familiar, and to others of a like nature, that I desire to call the attention of those who are interested in natural phenomena; hoping to share with them some of the delight that I have myself felt, in contemplating the exquisite forms that the camera has revealed, and in watching the progress of a multitude of events, compressed indeed within the limits of a few hundredths of a second, but none the less orderly and inevitable, and of which the sequence is in part easy to anticipate and understand, while in part it taxes the highest mathematical powers to elucidate.

Thus begins the book *A Study of Splashes* (1908) in which A. M. Worthington (1852–1916) presents "in a form acceptable to the general reader the outcome of an inquiry conducted by the aid of instantaneous photography, which was begun about fourteen years ago." Worthington's fascination with these phenomena actually went all the way back to 1875 (he was then 23 years old), when H. F. Newall, a student at the famous Rugby boys' public school, gave a report at the Rugby Natural History

0066–4189/93/0115–0577$02.00

Society about the curious "marks of accidental splashes of ink drops that had fallen on some smoked glasses."[1]

After experimenting with drops splashing on solid surfaces from 1876 to 1894, Worthington turned to studying the impact of drops on liquid surfaces—which is the object of the present review. With considerable ingenuity he was able to take remarkably sharp photographs of which he reproduced abundant examples in his book (Figure 1). As a light source he used electric sparks and was able to achieve exposure times of a few μsec. Thirty years later Edgerton improved on this method by using his newly invented electronic flash and published some very high-quality photographs of drops falling on a liquid surface in his book illustrating "the unseen by ultra high-speed photography" (Figure 2).

While Worthington was motivated purely by intellectual curiosity,[2] more recent investigations of this problem have been prompted by the very specific puzzle posed by the underwater noise of rain. This and further applications will be described in what follows.

Occasional references to the falling of drops and solid objects on liquid surfaces can be found scattered in the scientific literature in the following fifty years. Mallock (1919), in a paper titled *Sounds Produced by Drops Falling on Water*, states on the first page that "the same class of sounds were produced whether the falling body was a liquid drop or a solid sphere. The experiments, therefore, were made with solid spheres." He never mentions drops again and proceeds in fact to propose an incorrect explanation of the sounds in question. Several studies of impacting solid spheres followed Mallock's, notably by the celebrated Indian physicist C. V. Raman (Raman & Dey 1920), and E. G. Richardson (1948, 1955). In a brief note, Jones (1920) reported on some earlier observations of airborne drop sounds. Minnaert, with remarkable insight, concluded his famous 1933 paper (in which he found for the natural frequency of oscillation f_0 of a bubble of radius R the formula

$$f_0 = \frac{1}{2\pi R}\sqrt{\frac{3\gamma P}{\rho}}, \tag{1}$$

[1] Worthington's book is "Dedicated to the Natural History Society of the Rugby School and its Former President Arthur Sidgwick in Remembrance of the Encouragement Given to the Early Observations Made in Boyhood by my Old School-Friend H. F. Newall from which this Study Sprang." Later (1885) Newall and J. J. Thomson published the paper "On the formation of vortex rings by drops falling into liquids, and some allied phenomena."

[2] In 1894, opening a lecture delivered in front of the Royal Society—of which he was later to become a member—he said "... it may seem to some that a man who proposes to discourse on the matter for an hour must have lost all sense of proportion."

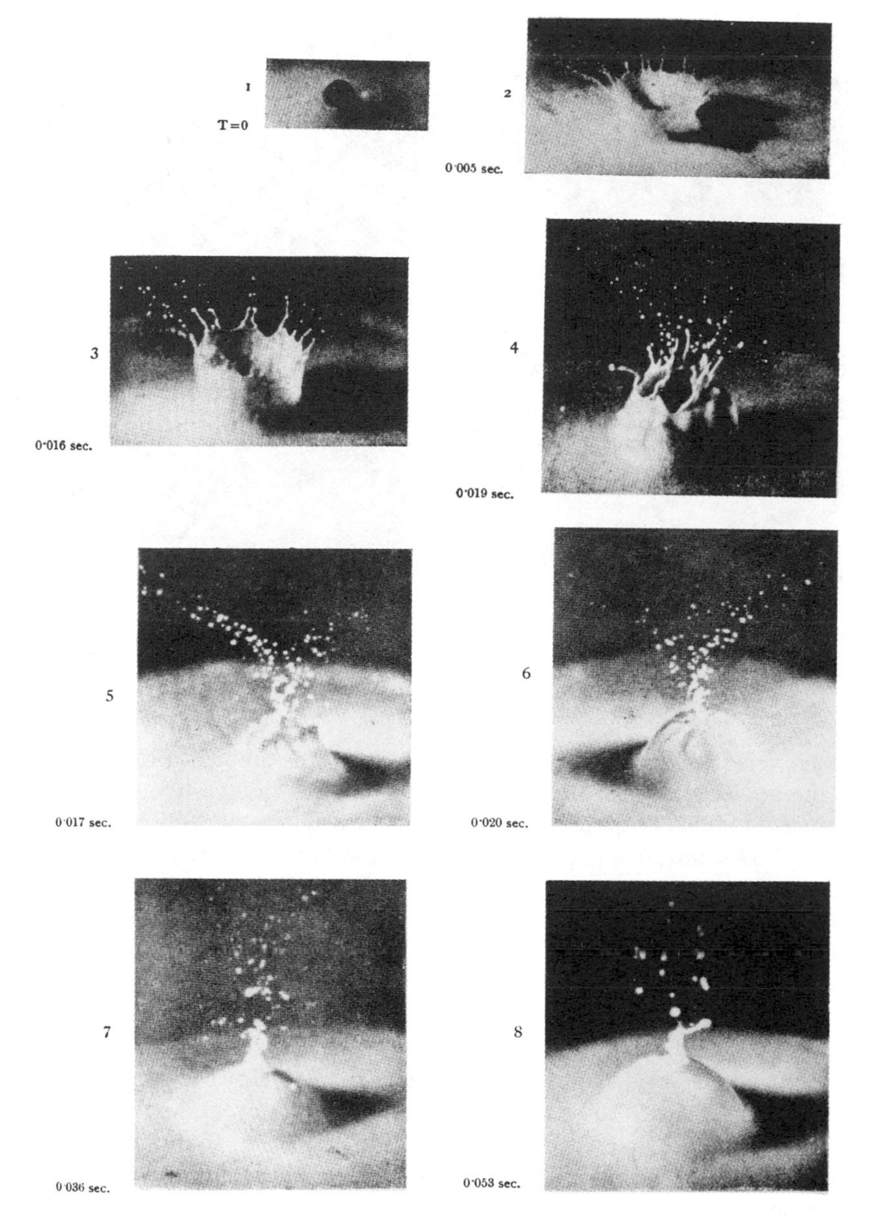

Figure 1 An example of Worthington's (1908) photos for a 9.1 mm diameter water drop impacting with a speed of approximately 5 m/sec (height of fall = 1.37 m). Each frame was obtained with a different drop released under nominally identical conditions. Note the thin splash that eventually closes with an upward jet originating at the point of closure. The photos are 5, 16, 16, 19, 17, 20, 36, and 53 msec after the first contact shown in the first image. [Reproduced with permission from Worthington (1963).]

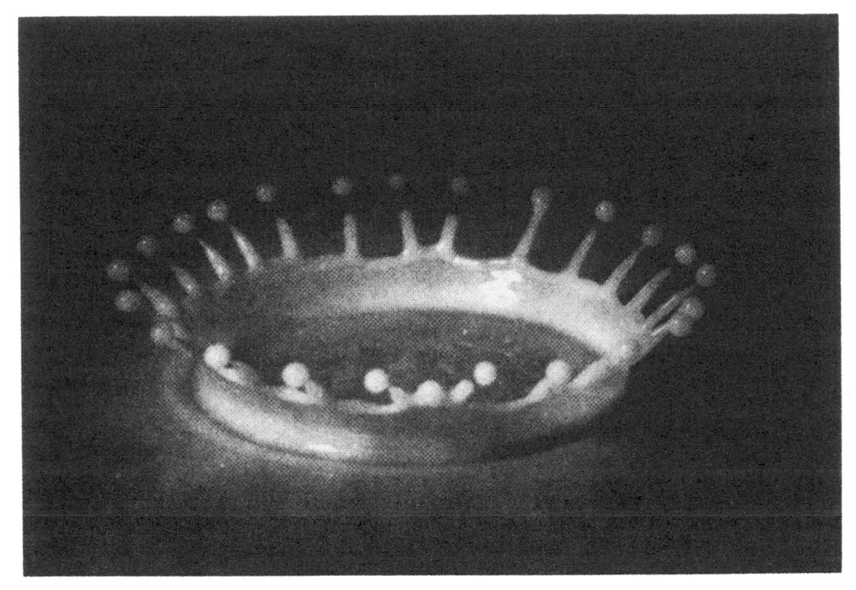

Figure 2 The coronet produced soon after the first contact of a drop of milk falling upon a plate covered with a thin layer of milk. [Reproduced with permission Edgerton & Killian (1987).]

where γ is the ratio of the gas specific heats, p the ambient pressure, and ρ the liquid density), with the words "It remains to investigate ... if the sounds of falling drops cannot have the same origin as the bubble sounds."

In 1959 Franz, in a paper titled *Splashes as Sources of Sound in Liquids*, discussed several mechanisms of noise production by impacting drops and solids and then describes an experiment devised "to substantiate some of the speculations about how splashes produce sound." He seems to have been the first one to make the crucial discovery that "Under some conditions, an air bubble can be entrained in the water ... The total sound energy radiated by an individual bubble was often greater than the sound energy radiated by the impact of the water droplet." However, since bubble entrainment seemed to be a random and infrequent event, he felt that, on average, bubbles had a relatively minor importance in sound generation. As we shall see, recent research has instead proven quite the opposite to be true.

The complexity of the fluid mechanic process uncovered by the photographs of Worthington and Edgerton was such as to discourage attempts at a theoretical interpretation of the observed flow. The first contribution of this nature was an early application of the MAC computational method of Harlow and Amsden (Harlow & Shannon 1967a,b). Later research, both theoretical and experimental, has proven however that the numerical method was too dissipative to give precise results.

THE RAIN NOISE PARADOX ...

World War II research had shown that rain produces a substantial amount of underwater noise (Knudsen et al 1948). In a short note, Heindsman et al (1955) published the first rain noise spectra, but their inability to obtain reliable data beyond 10 kHz prevented them from discovering the unique acoustic signature of rain and they concluded that the spectral "characteristic was that of wide band white noise," a misconception reinforced by some aspects of Franz's work (1959) and reiterated in the well-known review paper of Wenz (1962). A few years later Bom (1969) took better data in an Italian lake but, due to his study of the same frequency range, he was only able to confirm that the noise level increases with rainfall rate and exhibits a slight decline with frequency.

In the mid-1970s, two factors were coming together to rekindle specific interest in the underwater noise of rain. One—as always with underwater noise studies—was the military need for a better understanding of the natural mechanisms of underwater noise generation brought about by improvements in sound detection, signal processing, and the silencing of vessels. The other factor was the beginning of the use of acoustical techniques to probe phenomena of geophysical significance. In a paper devoted to the acoustical measurement of wind speed and stress at the ocean surface, Shaw et al (1978) realized the possibility that the results could be contaminated by rain noise. By using acoustic monitoring at three different frequencies, Lemon et al (1984) were able to identify the rain periods of their time traces, but their attempt to infer the amount of rainfall was not successful.

Some time before the study of Lemon et al, Walter Munk had suggested to J. A. Nystuen—then a graduate student at Scripps—that he study the possibility of using underwater noise to infer rainfall rates. This is an important quantity because rain is an essential component of the atmospheric heat balance, but about 80% of it falls on the oceans where measurements with good spatial resolution and over long periods of time are next to impossible. In the fall of 1982 Nystuen obtained the first data at Clinton Lake in Illinois over a frequency range much broader than used until then and found an astonishingly pronounced peak around 15 kHz. Publication of this result was however delayed until 1986. In 1984 Scrimger and coworkers (who had been in contact with both Nystuen and Farmer) made similar measurements in Canada with analogous results and were the first to report the discovery in the open literature (Scrimger 1985; Scrimger et al 1987, 1989).

The remarkable results obtained by these investigators can be illustrated with reference to Figure 3 (from Scrimger et al 1987). The left-hand side of the figure shows drop size spectra obtained with a distrometer during two rain events, labeled (*a*) and (*b*), while the right-hand side shows the corresponding underwater acoustic spectra. While it is obvious from the distrometer records that the data correspond to different rainfall rates—0.4 mm/hr for (*a*) and 0.3 for (*b*)—and drop sizes, the noise spectra are remarkably similar above about 10 kHz. This similarity between the spectra of different rain events is quite common at light rainfall rates and low wind speeds and is so striking that one may refer to a *universality* of the underwater noise of rain above 10 kHz or so.

. . . AND ITS RESOLUTION

In a footnote to their paper mentioned before, Raman & Dey (1920) state that "the splash of a liquid droplet is practically soundless unless the height of fall exceeds a certain minimum." This remark prompted the publication in the same year of a one-page note in *Science* in which A. T. Jones mentioned some preliminary observations he had conducted in 1915 where he found "not only a single minimum height . . . , but also other greater heights of fall for which the drops enter the water without sound." Curiously, this absolutely key observation was overlooked and Jones's note completely forgotten.

For his undergraduate thesis at Cambridge University, under the guidance of A. J. Walton, H. C. Pumphrey in 1984 had started a systematic investigation of the impact of a single drop on a liquid surface (Pumphrey & Walton 1988). Upon graduation, he entered the physics PhD program at the University of Mississippi where L. A. Crum had recently become involved in oceanic ambient noise research. Pumphrey conducted a very careful and exhaustive series of experiments releasing drops of different sizes from various heights, measuring the sound and, at the same time, taking high-speed movies (Pumphrey 1989, Pumphrey et al 1989, Pumphrey & Crum 1988,1990). Examples of sequences of frames from these movies are shown in Figures 4 and 5. In the first case (Figure 4), as a result of the impact, a small bubble is entrapped in the liquid. Figure 6 shows a detailed view of the formation and detachment of such a bubble. In the case shown in Figure 5, on the other hand, the dynamics of the cavity is quite different and no bubble is produced. A decisive innovation with respect to earlier research was to photograph *on the same frame* both the physical event and the trace of an oscilloscope driven by a hydrophone

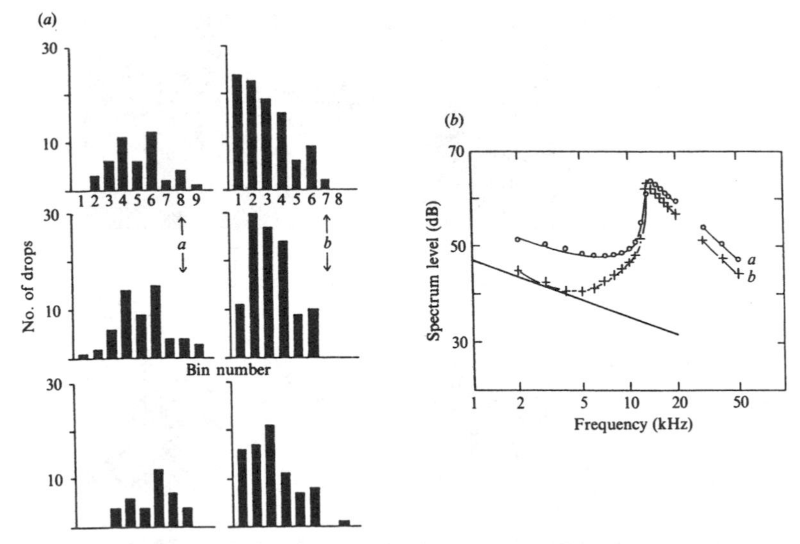

Figure 3 The left part of the figure shows three sets of 30 sec distrometer data corresponding to two different rain events (*a*) and (*b*). The right part shows the corresponding underwater sound spectra with the solid line indicating the background noise level. The wind speed was 0.9 m/sec and the rainfall rate was 0.4 mm/hr for event (*a*) and 0.3 mm/hr for event (*b*). Bin numbers 1 to 5 correspond to drop diameters between 0.3 and 0.8 mm in 0.1-mm intervals, larger bin numbers correspond to 0.2-mm intervals. [Reproduced with permission from Scrimger et al (1987).]

located near the point of impact. In this way it became apparent that, with the level of amplification used, the impact itself produced an acoustic signal indistinguishable from the electronic noise, while the bubble noise was very strong. While these observations confirmed earlier work, Pumphrey was able to determine that there was a rather limited and well defined region of impact velocities and drop diameters that resulted in bubble entrainment and, therefore, in the production of significant noise (Figure 7). At this point the resolution of the rain noise paradox was at hand (Prosperetti et al 1989). Although in the laboratory drop size and impact velocity are independent, in natural rain they are connected by the terminal velocity curve, which is the steeply rising line plotted on the left of Figure 7. If bubble-entraining drops are responsible for the underwater noise of rain above 10 kHz, it is clear that the only relevant size range is that where the terminal velocity curve crosses the bubble-entrainment region. Due to the steepness of the terminal velocity line, this range is quite narrow, extending from a diameter of about 0.82 mm to 1.1 mm, which happens to correspond

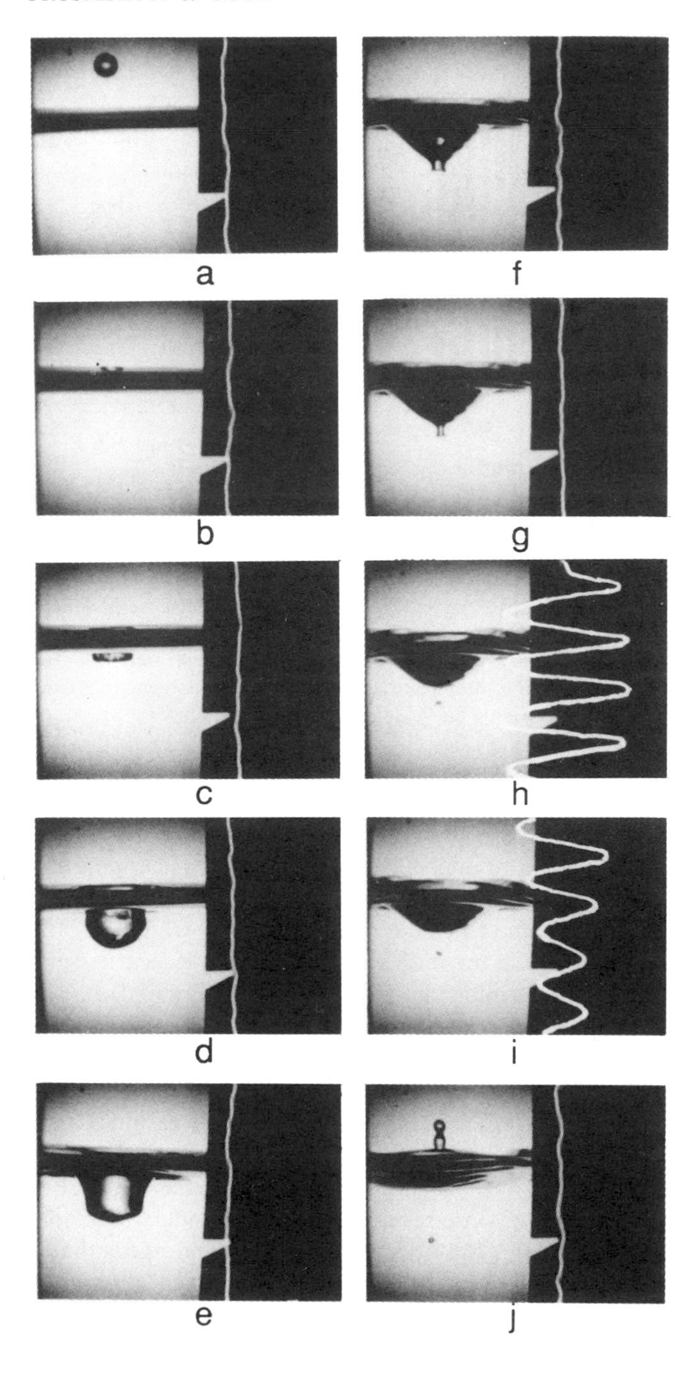
a
f
b
g
c
h
d
i
e
j

to sizes quite common in natural rain. To the extent that all raindrop spectra contain drops of this size, the acoustic signature of rain is therefore nearly universal.

In the presence of wind and waves both the velocity and the angle of impact of the drops on the water surface are affected and the bubble entrainment probability decreases (Medwin et al 1990). Furthermore, in these conditions, the bubble layer that forms at the ocean surface strongly attenuates the surface-generated sound (Farmer & Lemon 1984). Nevertheless, the 14 kHz peak remains a clearly recognizable feature of the acoustic spectra for wind velocities at least up to several m/sec (Medwin et al 1992).

This explanation of the nature of rain noise has received several confirmations. For example, a very simple and striking observation is that rain noise in an outdoor tank can be nearly completely shut off by the addition of a small amount of a surfactant that also inhibits bubble entrapment (Pumphrey & Crum 1988).

HIGHER-ENERGY IMPACTS

In the paper cited earlier, Jones (1920) mentions not one, but two ranges of impact velocity in which noise is associated with the drop impact. An investigation of high-kinetic-energy impacts along the terminal velocity curve has recently been carried out by Medwin et al (1992). The results are in agreement with Jones's in that bubble entrainment for impact velocities higher than those considered in the previous section is also encountered. By high-speed cinematography it was found that the mechanism of air entrainment is however completely different from the one described earlier. A reproduction of tracings from the high-speed film is shown in Figure 8, which may be compared with Figure 1.

The early stages of the process are qualitatively identical to those found upon the impact of a solid object and already noted and illustrated by Worthington (1908).[3] At first the drop creates a nearly vertical thin splash

←

Figure 4 Successive frames of a single 3 mm diameter water drop impacting on water at 2 m/sec. The right half of each frame shows the screen of an oscilloscope driven by a hydrophone placed in the vicinity of the impact point. Note the strong damped sinusoid that appears in correspondence with the detachment of a bubble from the nipple formed at the bottom of the crater. The times, in msec, are: (*a*) –3; (*b*) 0; (*c*) 2; (*d*) 6.5; (*e*) 14.5; (*f*) 20; (*g*) 20.5; (*h*) 21; (*i*) 21.5; (*j*) 35.5. (Courtesy of H.C. Pumphrey.)

[3] who, in this connection, observes: "I can recommend any reader who is not afraid of being late for breakfast to keep a bag of marbles in his bath-room."

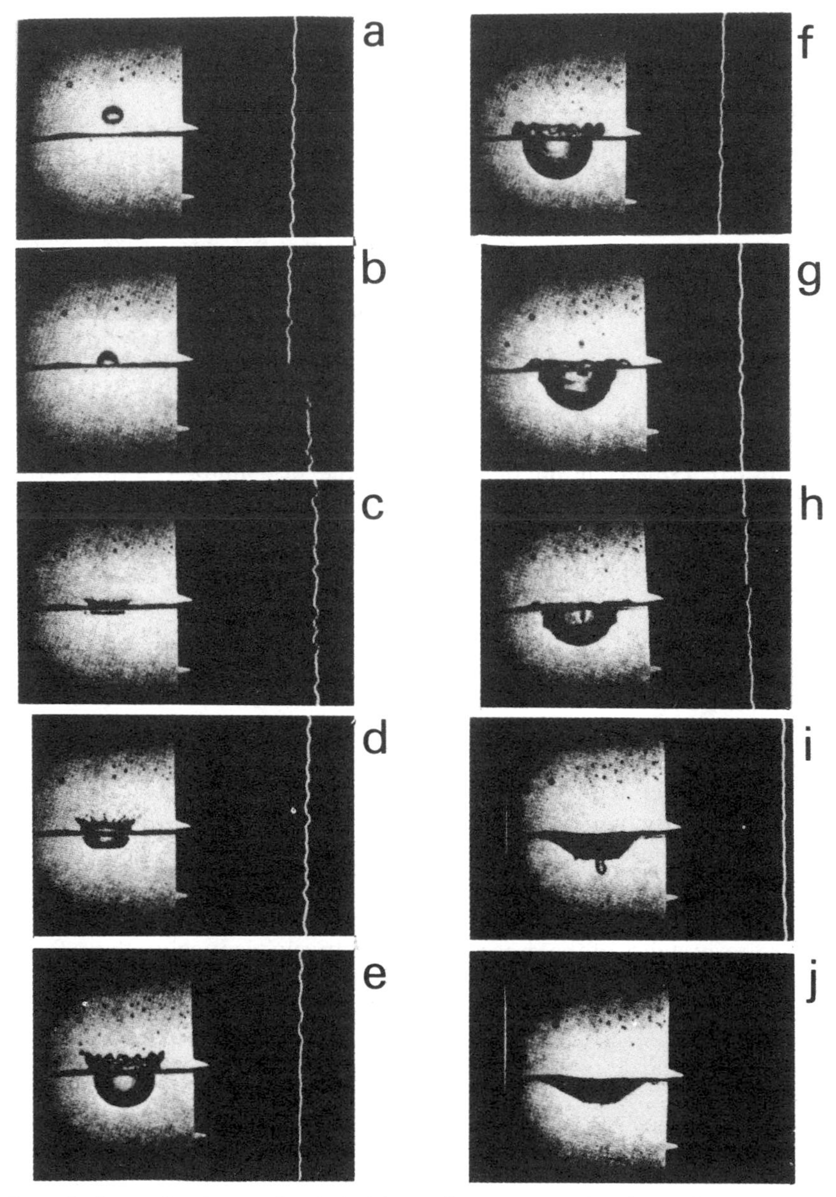

Figure 5 Successive frames of a single 2.9 mm diameter water drop impacting on water at 2.4 m/sec. The right half of each frame shows the screen of an oscilloscope driven by a hydrophone placed in the vicinity of the impact point. Although in these conditions no bubble is entrained by the primary drop, a tiny bubble is seen to be entrained in frame (*i*) by a Plateau's spherule [visible in frame (*g*)] that follows the main drop. The times, in msec, are: (*a*) –3; (*b*) 0; (*c*) 2; (*d*) 5; (*e*) 13; (*f*) 22; (*g*) 30 ; (*h*) 35; (*i*) 44; (*j*) 46. [Reproduced with permission from Pumphrey & Crum (1988).]

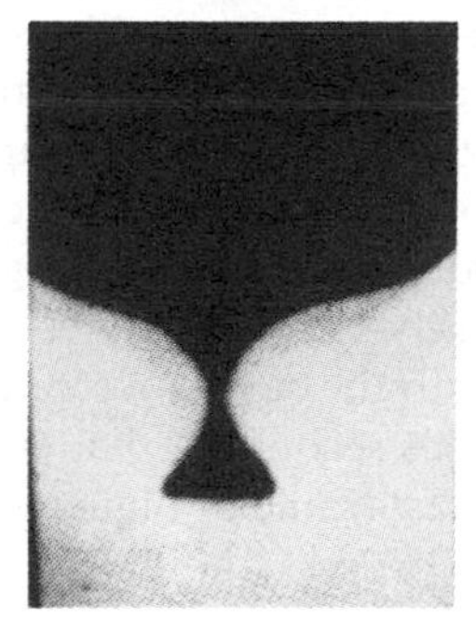

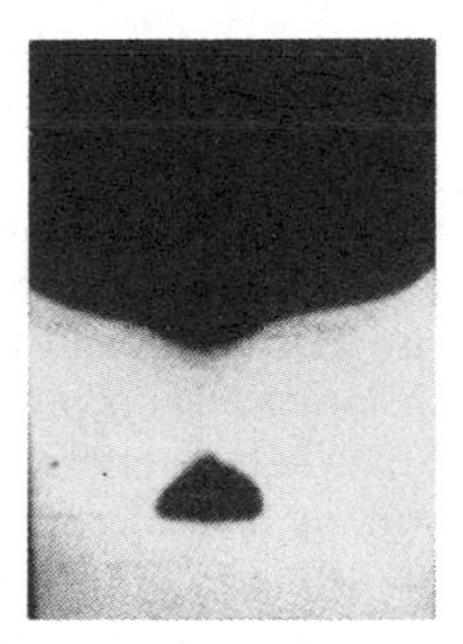

Figure 6 Detail of bubble formation and detachment for a 2.5 mm diameter water drop impacting at 2 m/sec. [Reproduced with permission from Chahine et al (1991).]

which has been shown by Engel (1967) to consist of receiving liquid rather than drop liquid. At a certain point, if the drop's energy is sufficiently large, the thin-walled liquid cylinder thus created closes at the top and forms "a bubble that floats for a moment and then vanishes." Worthington

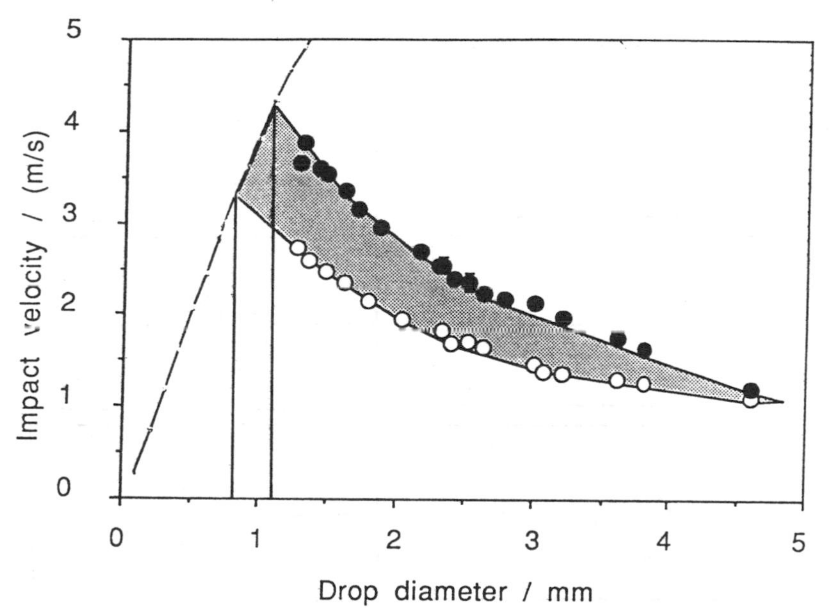

Figure 7 The shaded area corresponds to the parameter range where a bubble is entrained by every water drop. The dashed line on the left is the terminal velocity curve, and the two vertical parallel lines indicate the size range where drops impacting at their terminal velocity entrain bubbles. [Reproduced with permission from Pumphrey et al (1989).]

hypothesized that this process is governed by surface tension forces, and indeed this must be so since the duration of this phase is too short for gravity to have any effect. The precise mechanism, however, cannot be said to be clearly understood. As the canopy closes, the motion must stop abruptly and the inward-directed momentum generates a sharp pressure rise that results in two nearly symmetrical jets, one directed upward and the other one downward into the cavity. According to the observations of Medwin et al (1992) it is this second jet that entrains an air bubble upon striking the water surface (last frame of Figure 8). Understandably, in view of the higher energy of this process and the precise timing required, about half the drops entrain bubbles in this way as opposed to the previous mechanism for which the entrainment probability was essentially 100%. The size of the entrained bubble also exhibits a much greater variation from drop to drop.

These results have a substantial bearing on rain noise at frequencies below the peak discussed in the previous sections. As already noted by

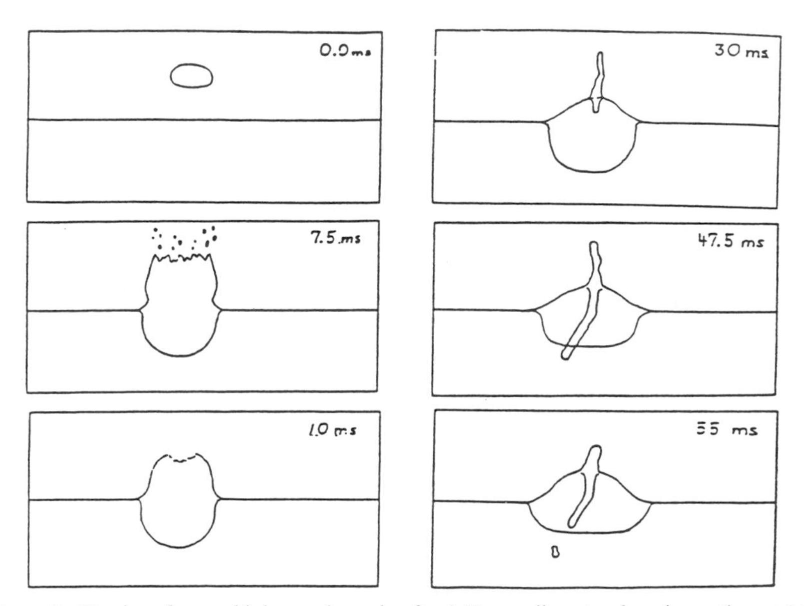

Figure 8 Tracings from a high-speed movie of a 4.57 mm diameter drop impacting at 9.3 m/sec; times in msec. This sequence is comparable to that shown in Figure 1. Note the bubble entrained by the impact of the downward jet formed by the closing of the canopy. [Reproduced with permission from Medwin et al (1992).]

Knudsen et al (1948), and later confirmed by the work of Heindsman et al (1955) and Bom (1969), there is a clear correlation between rainfall rate and underwater noise below 10 kHz. Medwin et al (1992) present convincing arguments that the noise in this range is the product of the alternative entrainment mechanism they describe, which becomes more and more important with increasing rainfall rates due to the presence of greater numbers of larger drops.

A further mechanism, first reported by Franz (1959) and confirmed by Pumphrey & Elmore (1990), can entrain bubbles in the case of fairly large and energetic drops. In this mechanism, the liquid column collapses and fills the crater produced by the impact.

An extreme case is that of huge "drops" (in fact, several liters of water) released above a liquid pool (Kolaini et al 1991). Here again large bubbles are entrained with acoustic emissions at frequencies of the order of a few tens of Hz.

NUMERICAL SIMULATIONS

Harlow & Shannon's (1967a,b) work already mentioned, although remarkable at the time, employed too crude a numerical method to disentangle the subtleties of the physical process found by Pumphrey in his experiments. This was confirmed by calculations by Nystuen (1986) who used the same MAC code without being able to improve on those results. We decided therefore to use a potential-flow boundary-integral method that had been found to work remarkably well in other simulations of free-surface flows. Of course this implied doing away with viscous effects and therefore could only be expected to be useful for times t much shorter than a^2/ν, where a is the drop radius and ν the liquid's kinematic viscosity. For water drops of $a = 1$ mm, however, $a^2/\nu \simeq 1$ sec, so that the restriction is not too stringent.

The early attempts were plagued by severe numerical instabilities. It was necessary to develop a new boundary-integral formulation resulting in a Fredholm integral equation of the second, rather than the first, kind, and novel time stepping techniques and surface parametrizations which are described in detail in the original paper (Oğuz & Prosperetti 1990). We were finally successful and some typical results of the computations are shown in Figure 9, to be compared with the experiment of Figure 4. Other examples are given in Oğuz & Prosperetti (1990). In spite of some differences, the important features of both bubble-entrapping and non-entrapping events are well reproduced numerically.

Given this early success, we decided to attempt a completely synthetic—i.e. numerical—calculation of underwater rain noise. The only exper-

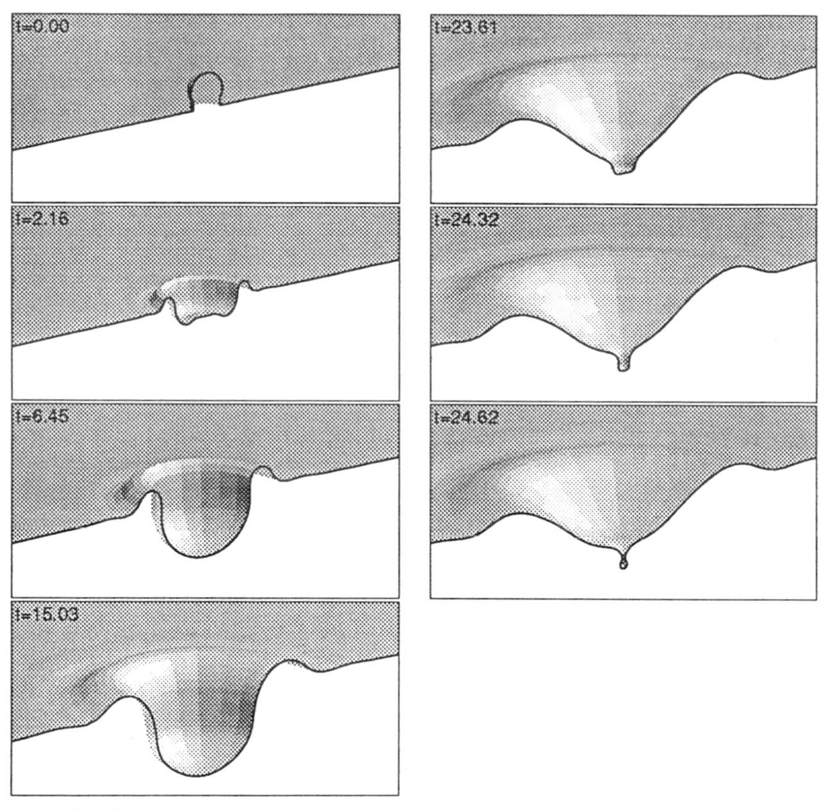

Figure 9 Numerical simulation of a 2.86 mm diameter drop impacting at 2 m/sec (Fr = 285.4, We = 79.4). The times are in msec.

imental input was to be the background noise and the total drop count in the bubble-entrapping size range. The procedure was as follows. A number of drop radii were chosen along the terminal velocity line and the corresponding simulations were run to determine the size of the entrained bubble and its initial energy. After closure, the bubble was allowed to oscillate with the natural frequency and damping rate corresponding to its size, and the acoustic emissions from all the bubbles were added incoherently with the assumption of dipole radiation due to the presence of the neighboring free surface.

The plan was rather ambitious. It was known from Pumphrey's laboratory study that the reproducibility of the bubble is not very good in the neighborhood of the boundaries. For instance, in the case of a 1 mm-diameter drop, the experimental standard deviation of the frequency of bubble emission is of the order of 0.3% near the center of the entrainment

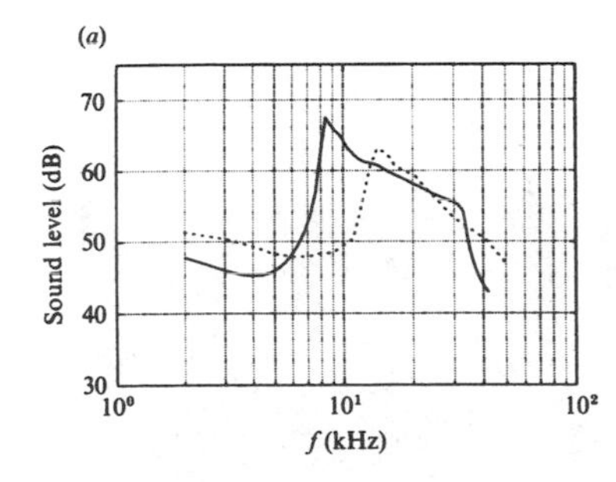

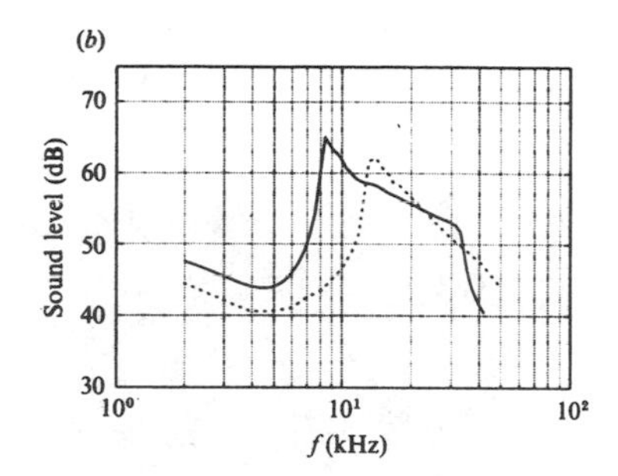

Figure 10 Comparison between measured (*dotted lines*) and computed (*solid lines*) underwater noise spectra for the two rain events (*a*) and (*b*) of Figure 3. [Reproduced from Oğuz & Prosperetti (1991).]

region, but becomes of the order of 100% near the upper and lower velocity boundaries. Furthermore, it can be appreciated from Figure 9 that the bubble region is only a small fraction of the flow domain, so that its good numerical resolution is not trivial. Nevertheless our attempt was moderately successful, as shown by the comparison between measured and calculated results given in Figure 10. The general shape of the peak is in good agreement with the measured one, and the noise level is also very close. The biggest discrepancy is in the position of the peak, which the calculations put at around 9 kHz rather than 14, thus indicating—from Equation 1—that the size of the "numerical" bubbles is somewhat greater than that of the physical ones. We have carried out a detailed analysis of the origin of this discrepancy (Oğuz & Prosperetti 1992) but the results are rather ambiguous. The indications are that it is probably a consequence of the neglect of viscous effects in the air as well as in the water.

The computations do however afford an interesting insight into the fluid dynamics of the process.[4] For example, we show in Figure 11 the last computed configuration of the very bottom of the drop crater for several drop diameters and impact velocities along the terminal velocity curve. It is seen that, for small drops, the crater is not deep enough to "pinch" an air bubble. As the drop's energy increases, the crater becomes deeper and

[4]This would have been of some solace to Worthington who ruefully observes: "But even were the photographic record complete, what does it amount to? All that we have done has been merely to follow the rapid changes of form that take place in the bounding surface of the liquid. The interior particles of the liquid itself have remained invisible to us. But it is precisely the motion of these interior particles that the student of hydrodynamics desires to be able to trace."

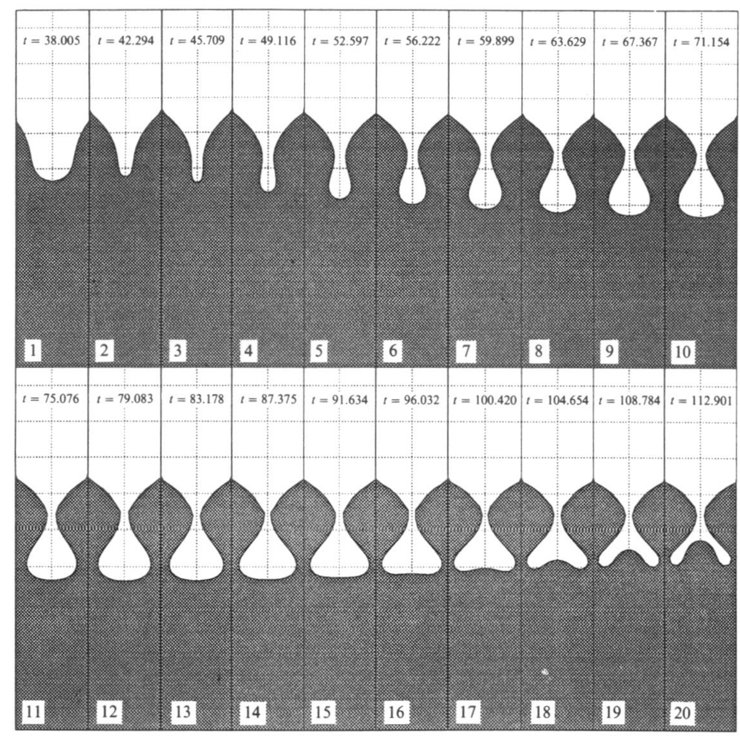

Figure 11 Last computed frame for the impact of water drops along the terminal velocity curve and corresponding dimensionless times Vt/a. The first panel is for a drop diameter of 0.82 mm, the second one for 0.84 mm, and so on; the last one is for 1.2 mm. The terminal velocity for a 0.80 mm drop is 3.28 m/sec, for 0.90 mm—3.66 m/sec, for 1 mm—4.01 m/sec, and for 1.2 mm—4.65 m/sec. The spacing of the dashed lines equals one drop radius. [Reproduced from Oğuz & Prosperetti (1991).]

the sideways motion reverses faster than the downward one, so that a bubble remains entrapped. At still higher energies, the directionality of the drop's momentum becomes less important and the crater grows more nearly spherically (compare Figures 4 and 5). Now the bottom of the crater is the highest-energy point and the downward motion tends to reverse earlier and earlier, until eventually the inward lateral velocity is not sufficient to close the cavity before the bottom jet shoots out of it and prevents the formation of the bubble.

THEORETICAL ASPECTS

The sequence of events uncovered by experiment and computation is indeed so complex as to "tax the highest mathematical powers to elucidate." So much so, as a matter of fact, that not much progress has been made along this line.

If the drops are assumed to strike as spheres (a point to which we shall return below), there are three basic dimensionless groups on which the process depends, the Froude Fr, Weber We, and Reynolds Re numbers defined by

$$Fr = V^2/ga, \qquad We = \rho V^2 a/\sigma, \qquad Re = aV/\nu. \tag{2}$$

Here V is the impact velocity, σ the surface tension coefficient, and g the acceleration of gravity. With the neglect of viscous effects, the controlling

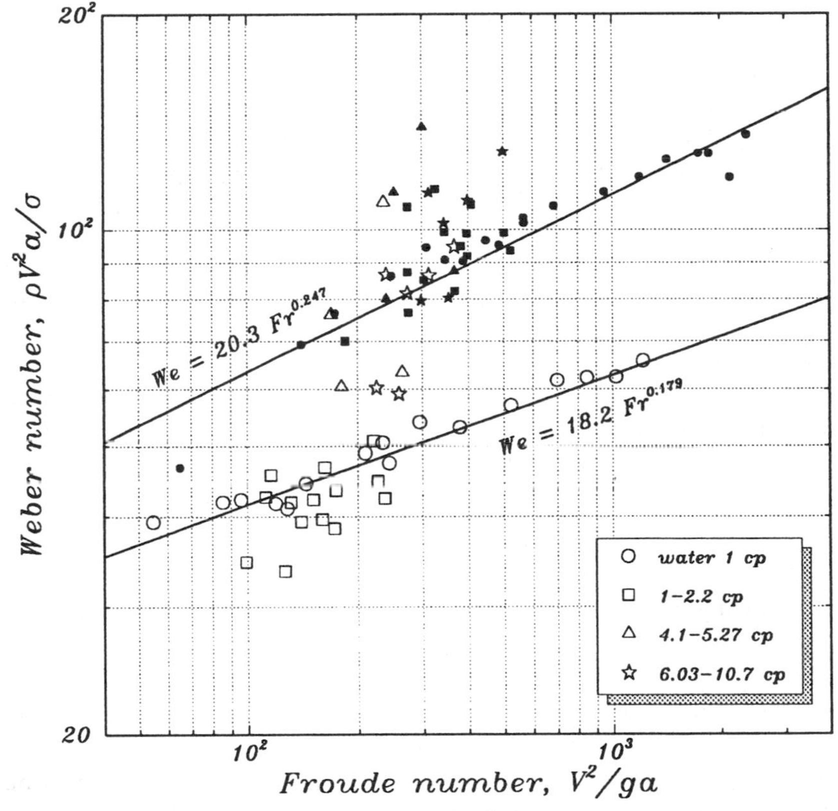

Figure 12 Nondimensional plot of the regular entrainment region. The circles are the water data of Figure 7. The other data points are for different liquids of various viscosities as specified in the legend. The straight lines are fits to the curves of Figure 7.

parameters are *Fr* and *We* and one can replot the boundaries of the entrainment region of Figure 7 in the (*Fr*, *We*) plane. We show such a plot in Figure 12 where the data of Figure 7 are supplemented by others (indicated by squares, triangles, and stars), due to Detsch & Harris (1992), and taken with other liquids. The black symbols are for the upper entrainment boundary and the open ones for the lower boundary. Circles are for water, squares for other low-viscosity liquids ($\mu = 1$ to 2.2 cp), triangles for intermediate viscosities ($\mu = 4.1$ to 5.27 cp), and stars for high-viscosity liquids ($\mu = 6.03$ to 10.7 cp). A general consistency of the data for water and other low-viscosity liquids is present, but the amount of scatter is significant. This is not surprising in view of the results previously mentioned concerning the large variance of the data near the entrainment boundaries. The lines shown in the figure are the same as those shown in Figure 7. They are power laws of the form $We = kFr^{\alpha}$, with k a constant and $\alpha = 0.247$ for the upper line and $\alpha = 0.179$ for the lower line. Simple arguments given in Oğuz & Prosperetti (1990) lead to the values $\frac{1}{4}$ and $\frac{1}{5}$, respectively.

It may be noted that if, rather than by using a different liquid, surface tension is varied by the addition of a surfactant, the previous scaling is found to be incorrect (Pumphrey & Elmore 1990). The spreading of the surfactant on the newly formed liquid surface is evidently an important phenomenon under these conditions.

The maximum size of the crater can be estimated by equating the potential energy stored into it to the initial energy of the drop. For this argument to work, it is necessary that the velocity field vanish nearly simultaneously in the liquid, which is only approximately true. If the additional assumption is made that the crater is a hemisphere (which could be corrected by the use of a dimensionless shape factor of order 1), one finds the crater radius R_c to be

$$\frac{R_c}{a} = \left[2\left(\tfrac{2}{3}Fr + 4\frac{Fr}{We} + \frac{Fr^2}{We^2}\right)^{1/2} - 2\frac{Fr}{We}\right]^{1/2}. \tag{3}$$

Numerically Fr/We is much smaller than Fr and the equation can be simplified to

$$\frac{R_c}{a} = (\tfrac{8}{3}Fr)^{1/4} \simeq 1.278\,Fr^{1/4}. \tag{4}$$

This expression is found to be in very good agreement with the data of Pumphrey & Elmore. At the higher energies of Engel's (1967) experiments, the numerical constant is somewhat lower—about 1.05.

The dynamics of the crater after it reaches the conical shape visible e.g.

in the 6th and 7th frames of Figures 4 and 9 has been investigated by Longuet-Higgins (1990). By assuming a self-similar potential flow solution of the form

$$\phi = \tfrac{1}{2}A(t)r^2(2\cos^2\theta - 1), \tag{5}$$

and neglecting the effect of gravity and surface tension, he found that a limit angle exists past which the pressure gradient is unable to further increase the aperture of the cone. This limit angle is 109.5°, and compares very well with the numerical calculations of Oğuz & Prosperetti (1990). On the basis of his solution, Longuet-Higgins was able to estimate the acoustic dipole moment of the entrained bubble in good agreement with experiment.

ACOUSTICS OF IMPACT

Insofar as they originate from the entrapped bubble, the acoustic emissions considered so far may be regarded as a byproduct of the fluid flow generated by the impacting drop. The impact itself, however, is also a source of sound. The two acoustic emissions can clearly be differentiated as seen in Figure 13, where the initial signal is the impact and the subsequent damped sinusoid is the bubble noise. Although it is now known that the acoustic energy directly due to the impact represents an insignificant fraction of the total radiated energy, it is interesting to briefly consider its mechanics.

As described by Lesser & Field (1983), at the very early stages after contact, a compression wave propagates supersonically in the two liquids following the progress of the geometric circle of contact between them. Guo & Ffowcs Williams (1991) find that this supersonic stage lasts until a time t_c given approximately by

$$t_c \simeq \tfrac{1}{2}M^2\frac{a}{V}, \tag{6}$$

where $M = V/c$ is the Mach number of the impact. For $V = 2$ m/sec and $a = 1$ mm, this gives $t_c \simeq 0.4$ nsec. After this time the liquid can expand and the initial thin splash is formed. A calculation of the acoustic energy E radiated during this initial stage gives

$$E \simeq \frac{3}{16}M^3T, \tag{7}$$

where $T = 2/3\pi\rho a^3V^2$ is the drop's kinetic energy.

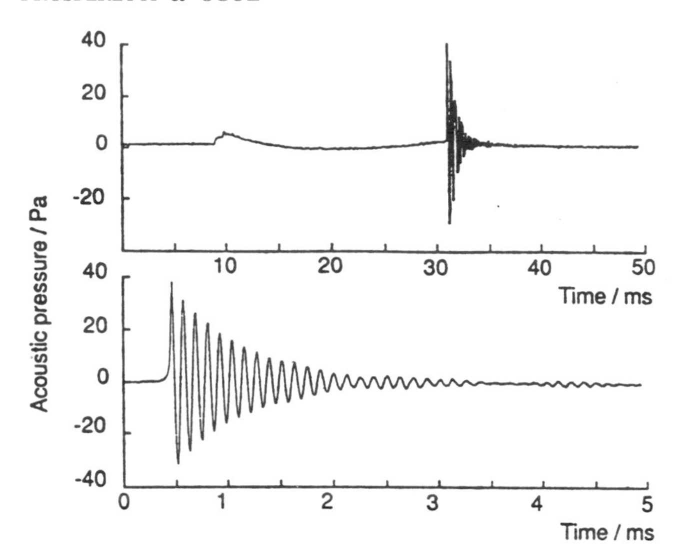

Figure 13 Near-field pressure disturbance produced by an impacting drop. The upper trace shows the initial impact and the bubble emission about 20 msec later. The lower part of the figure is a magnification of the bubble noise. (Note: Origin has been shifted for clarity.) [Reproduced with permission from Pumphrey et al (1989).]

The analysis beyond t_c is far from simple, but a dimensional argument can be constructed as follows (Oğuz & Prosperetti 1991). As expected, and as confirmed by Pumphrey & Elmore (1990), the radiated pressure p_r has the character of a dipole. The source/sink pair that constitutes it is separated by a distance of order a and must each have a strength of the order of the mass injection rate $\rho V a^2$ times the inverse characteristic time a/V. Thus

$$p_r \sim \frac{\cos\theta}{r} \frac{a}{ac/V} \frac{V}{a} \rho V a^2 \Phi(Vt/a, M). \tag{8}$$

Here θ and r are polar coordinates measured from the point of impact and the factor $a/(ac/V) \sim a/\lambda$, with λ the acoustic wavelength, converts the monopole to dipole radiation. The dimensionless function Φ, taken to depend on the only available dimensionless variables, accounts for the details of the process. Since $M \ll 1$, for times that are not too small, we can approximately take $\Phi(Vt/a,M) \simeq \Phi(Vt/a,0) \equiv \Phi(Vt/a)$. The total energy radiated can readily be calculated from the form (8) of the pressure field and is

$$E = \frac{2\pi a}{3\rho Vc}\left(\frac{\rho a V^3}{c}\right)^2 \int_0^\infty \Phi^2(t_*)\, dt_*, \tag{9}$$

where $t_* = Vt/a$. By using Parseval's equality, the integral over time can be converted into an integral over the dimensionless frequency $f_* = fa/V$, from which the radiated energy in the frequency band between f and $f + df$ can be read off directly to find

$$\hat{E}(f) = \frac{4\pi a}{V} TM^3 |\hat{\Phi}(f_*)|^2, \tag{10}$$

where $\hat{\Phi}(f_*)$ is the Fourier transform of $\Phi(t_*)$. This argument leads to the expectation that the dimensionless function

$$\hat{E}_*(f_*) \equiv 4\pi |\hat{\Phi}|^2 = \frac{V\hat{E}(f)}{aTM^3} \tag{11}$$

is a universal function, as indeed was found by Franz (1959), whose data are plotted in the form suggested by this equation in Figure 14.

ALLIED PHENOMENA

However complex and intriguing the fluid mechanics processes touched upon in the previous sections, by far they do not exhaust the astonishingly rich range of phenomena related to the impact of drops on liquid surfaces.

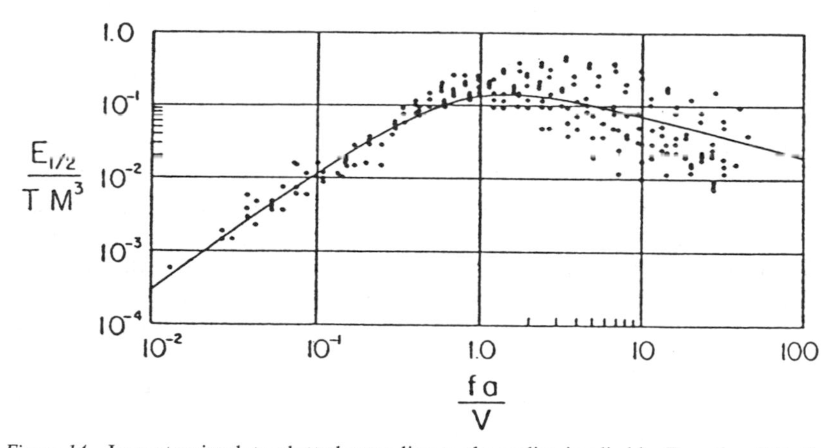

Figure 14 Impact noise data plotted according to the scaling implied by Equation (11). $E_{1/2}$ is the energy per half-octave band, T the drop's kinetic energy, M the Mach number at impact, f the frequency, a the drop radius, and V the impact velocity. [Reproduced with permission from Franz (1959).]

Still very little understood, for instance, is the formation and penetration of vortex rings into the receiving liquid. This process was first reported in 1858 by William Barton Rogers,[5] and later (1885) studied by J. J. Thomson & H. F. Newall. They observed that certain heights produce the most penetrating vortex rings, and that these heights are spaced so as to permit one drop oscillation between them. A beautiful photo by Peck and Sigurdson of the very early stages of vortex ring formation has recently been published by Reed (1991), and one corresponding to a few instants later, taken by Okabe & Inoue (1961), is reproduced in Batchelor's (1967) *An Introduction to Fluid Mechanics*. According to a study by Rodriguez & Mesler (1988), there is an effect of the drop shape upon impact which, rather than being direct as hypothesized by Chapman & Critchlow (1967), is mediated by the crater dynamics. After its formation, the vortex ring expands and eventually becomes unstable—a regime studied by Kojima et al (1984) in the low Reynolds number limit.

Vortex ring formation is a phenomenon typical of relatively low-velocity impacts (Hsiao et al 1988). Most characteristically, at higher velocities the drop fluid mixes only negligibly with the receiving fluid and collects at the tip of the jet that ultimately results from the filling up of the crater. This fact was known to Worthington, who observes:

> The reappearance of the original drop at the head of the rebounding column ... is easily verified by naked-eye observation. Let the reader when he next receives a cup of tea or coffee to which no milk has yet been added, make the simple experiment of dropping into it from a spoon at the height of fifteen or sixteen inches above the surface, a single drop of milk. He will have no difficulty in recognizing that the column which emerges carries the white-milk drop at the top only slightly stained by the liquid into which it has fallen.

An intriguing phenomenon is found when the depth of the receiving liquid is so small that bottom effects become important. As first reported by Hobbs & Osheroff (1967), and recently studied by Shin & McMahon (1990), the height of the rebounding jet reaches a maximum for a certain pool depth. This is the effect of two competing mechanisms. For very shallow pools, the jet's height is limited by the fact that not enough liquid is readily available to feed it. For deep pools, on the other hand, a substantial amount of liquid flows into the jet which is therefore thicker and limited by gravity in its ascent. The jet velocity can be as large as four times the velocity of the drop that originates it. Mori et al (1987) have also studied this phenomenon and reported that the size of the bubble entrained grows at the same time as the pool depth and the jet height.

[5] Although at the time he was actively engaged in efforts that ultimately led to the establishment of M.I.T. seven year later, evidently he still had time for science.

For high-velocity impacts, the drops that form the coronet along the rim of the initial splash (Figure 2) detach. Hobbs & Kezweeny (1967) studied their number and the process of electric charge separation that takes place in these circumstances. Hashimoto & Sudo (1980) gave a detailed account of bubble entrainment by splashing drops in a vertically vibrated liquid column, and Sudo et al (1991) studied the impact of drops of a magnetic liquid with and without a magnetic field.

A point of fundamental interest that has not been addressed yet is the behavior of the two fluids at the very initial instant when contact is established. From a macroscopic point of view, geometry forces the free surface to possess a cusp, which is incompatible with the standard requirement of classical fluid mechanics that the jump in the normal stresses be compensated by surface tension times the local curvature. This boundary condition is a consequence of the assumption of local thermodynamic equilibrium, and therefore the presence of a cusp would indicate the prevalence of nonequilibrium conditions. Such a situation, therefore, must relax over molecular time scales to an equilibrium one with the sharp edge of the cusp softened into a region of very high local curvature. The subsequent behavior of the free surface has been studied in Oğuz & Prosperetti (1989) with the striking conclusion that further closing of the surface separating the two contacting liquid masses occurs in jumps that incorporate tiny air bubbles. Striking photos of small bubbles (diameter of the order of a few tens of μm) along the surface of the entering drop have been published by Sigler & Mesler (1990) at low impact velocities. It is difficult to say whether these are the bubbles predicted in Oğuz & Prosperetti, the effect of the breakup of a lubrication-type air film that remains entrapped between the two approaching liquid surfaces, or a combination of the two.

CONCLUSIONS

"I have some hope that ... I may have succeeded in producing in the mind of my reader some sympathy with the state of perplexity of Mr. Cole and myself," says Worthington. While nearly a century of progress has helped to dissipate some of that "perplexity," it is clear that we are still a long way from an understanding of drop impact phenomena in all their subtleties. The range of "very difficult hydro-dynamical questions involved" encompasses some fundamental problems, such as the role of surface tension when two liquid surfaces come together. A host of other absolutely nontrivial fluid mechanical phenomena has been mentioned in the previous pages.

While "One can almost regret that so beautiful a process should have been so long unwatched," in addition to aesthetic, scientific reasons for its investigation are readily found. A case in point is rain noise, the explanation of which has been the most notable success of recent progress in this area. In addition, bubbles in the upper layers of water bodies exert a strong influence on the exchange of atmospheric gases, with implications that range from the trophism of planktonic organisms at the beginning of the food chain to the absorption of CO_2 by the ocean. The tiny drops produced by a variety of mechanisms by the impact of drops and splashes are responsible for seeding the atmosphere with the salt grains that act as condensation nuclei for rain (Blanchard & Woodcock 1957). In plant physiology one talks of the splash-cup dispersal mechanism (Brodie 1951), according to which the tiny drops produced by raindrops hitting the splash cup are responsible for the dispersal of spores and gemmae of certain plants.

ACKNOWLEDGMENTS

The authors wish to express their gratitude and appreciation to the many colleagues who have in different ways contributed to this paper: G. Chahine, L. A. Crum, R. M. Detsch, P. A. Elmore, H. Medwin, J. A. Nystuen, H. C. Pumphrey, and J. A. Scrimger.

This paper and the work on which it is based have been sponsored by the Underwater Acoustics Program of the Office of Naval Research under the supervision of Dr. Marshall Orr, whose support is gratefully acknowledged.

Literature Cited

Batchelor, G. K. 1967. *An Introduction to Fluid Dynamics*. Cambridge: Cambridge Univ. Press

Blanchard, D. C., Woodcock, A. H. 1957. Bubble formation and modification in the sea and its meteorological significance. *Tellus* 9: 145–58

Bom, N. 1969. Effect of rain on underwater noise level. *J. Acoust. Soc. Am.* 45: 150–56

Brodie, H. J. 1951. The splash-cup dispersal mechanism in plants. *Can. J. Bot.* 29: 224–34

Chahine, G. L., Wenk, K., Gupta, S., Elmore, P. A. 1991. Bubble formation following drop impact at a free surface. In *Cavitation and Multiphase Forum*, ed. O. Furuya, H. Kato, FED Vol. 109, pp. 63–68. New York: Am. Soc. Mech. Eng.

Chapman, D. S., Critchlow, P. R. 1967. Formation of vortex rings from falling drops. *J. Fluid Mech.* 29: 177–85

Detsch, R. M., Harris, I. A. 1992. Bubble entrainment by impacting drops in various liquid solutions. *J. Fluid Mech.* Submitted

Edgerton, H. E., Killian, J. R. 1939. *Flash*. Boston: Branford. See also 2nd ed.(1954) and Edgerton & Killian (1987)

Edgerton, H. E., Killian, J. R. 1987. *Moments of Vision*. Boston: MIT Press

Engel, O. G. 1967. Crater depth in fluid impacts. *J. Appl. Phys.* 37: 1798–808

Farmer, D. M., Lemon, D. D. 1984. The influence of bubbles on ambient noise in the ocean at high wind speeds. *J. Phys. Oceanogr.* 14: 1762–778

Franz, G. J. 1959. Splashes as sources of sound in liquids. *J. Acoust. Soc. Am.* 31: 1080–96

Guo, Y. P., Ffowcs Williams, J. E. 1991. A theoretical study on drop impact sound and rain noise. *J. Fluid Mech.* 227: 345–55

Harlow, F. H., Shannon, J. P. 1967a. Distortion of a splashing liquid drop. *Science* 157: 547–50

Harlow, F. H., Shannon, J. P. 1967b. The splash of a liquid drop. *J. Appl. Phys.* 38: 3855–66

Hashimoto, H., Sudo, S. 1980. Surface disintegration and bubble formation in vertically vibrated liquid column. *AIAA J.* 18: 442–49

Heindsmann, T. E., Smith, R. H., Arneson, A. D. 1955. Effect of rain upon underwater noise levels. *J. Acoust. Soc. Am.* 27: 378–79

Hobbs, P. V., Kezweeny, A. J. 1967. Splashing of a water drop. *Science* 155: 1112–14

Hobbs, P. V., Osheroff, T. 1967. Splashing of drops on shallow liquids. *Science* 158: 1184–86

Hsiao, M., Lichter, S., Quintero, L. G. 1988. The critical Weber number for vortex and jet formation for drops impinging on a liquid pool. *Phys. Fluids* 31: 3560–62

Jones, A. T. 1920. The sound of splashes. *Science* 52: 295–96

Knudsen, V. O., Alford, R. S., Emling, J. W. 1948. Underwater ambient noise. *J. Mar. Res.* 7: 410–29

Kojima, M., Hinch, E. J., Acrivos, A. 1984. The formation and expansion of a toroidal drop moving in a viscous fluid. *Phys. Fluids* 27: 19–32

Kolaini, A., Roy, R. A., Crum, L. A. 1991. An investigation of the acoustic emissions from a bubble plume. *J. Acoust. Soc. Am.* 89: 2452–55

Lemon, D. D., Farmer, D. M., Watts, D. R. 1984. Acoustic measurements of wind speed and precipitation over a continental shelf. *J. Geophys. Res.* 89: 3462–72

Lesser, M. B., Field, J. E. 1983. The impact of compressible liquids. *Annu. Rev. Fluid Mech.* 15: 97–122

Longuet-Higgins, M. S. 1990. An analytic model of sound production by rain-drops. *J. Fluid Mech.* 214: 395–410

Mallock, A. 1919. Sounds produced by drops falling on water. *Proc. R. Soc. London Ser. A* 95: 138–43

Medwin, H., Nystuen, J. A., Jacobus, P. W., Ostwald, L. H., Snyder, D. E. 1992. The anatomy of underwater rain noise. *J. Acoust. Soc. Am.* 92: 1613–23

Medwin, H., Kurgan, A., Nystuen, J. A. 1990. Impact and bubble sound from raindrops at normal and oblique incidence. *J. Acoust. Soc. Am.* 88: 413–18

Minnaert, M. 1933. On musical air bubbles and the sounds of running water. *Phil. Mag.* 16: 235–48

Mori, Y., Mizukami, M., Bushimata, T. 1987. Study of sound generation by rigid spheres and droplets falling on a water surface and by water flow in tubes (1st report, sound generated by falling sphere and droplet). *Trans. Jpn. Soc. Mech. Eng.* 53B: 894–902. (Abstract in English, article in Japanese)

Nystuen, J. A. 1986. Rainfall measurements using underwater ambient noise. *J. Acoust. Soc. Am.* 79: 972–82

Oğuz, H. N., Prosperetti, A. 1989. Surface-tension effects in the contact of liquid surfaces. *J. Fluid Mech.* 203: 149–71

Oğuz, H. N., Prosperetti, A. 1990. Bubble entrainment by the impact of drops on liquid surfaces. *J. Fluid Mech.* 219: 143–79

Oğuz, H. N., Prosperetti, A. 1991. Numerical calculation of the underwater noise of rain. *J. Fluid Mech.* 228: 417–42

Oğuz, H. N., Prosperetti, A. 1992. Drop impact and the underwater noise of rain. In *Natural Physical Sources of Underwater Sound*, ed. B. R. Kerman. Dordrecht: Reidel

Okabe, J., Inoue, S. 1961. The generation of vortex rings, II. *Rep. Res. Inst. Appl. Mech., Kyushu Univ.* 9: 147–52

Prosperetti, A., Pumphrey, H. C., Crum, L. A. 1989. The underwater noise of rain. *J. Geophys. Res.* 94: 3255–59

Pumphrey, H. C. 1989. *Sources of ambient noise in the ocean: an experimental investigation.* PhD thesis. Univ. Mississippi

Pumphrey, H. C., Crum, L. A. 1988. Acoustic emissions associated with drop impacts. In *Natural Mechanisms of Surface-Generated Noise in the Ocean*, ed. B. R. Kerman, pp. 463–83. Dordrecht: Reidel

Pumphrey, H. C., Crum, L. A. 1990. Free oscillations of near-surface bubbles as a source of the underwater noise of rain. *J. Acoust. Soc. Am.* 87: 142–48

Pumphrey, H. C., Crum, L. A., Bjørnø, L. 1989. Underwater sound produced by individual drop impacts and rainfall. *J. Acoust. Soc. Am.* 85: 1518–26

Pumphrey, H. C., Elmore, P. A. 1990. The entrainment of bubbles by drop impacts. *J. Fluid Mech.* 220: 539–67

Pumphrey, H. C., Walton, A. J. 1988. Experimental study of the sound emitted by water drops impacting on a water surface. *Eur. J. Phys.* 9: 225–31

Raman, C. V., Dey, A. 1920. On the sound of splashes. *Phil. Mag.* 39: 145–47

Reed, H. L. 1991. Gallery of fluid motion. *Phys. Fluids A* 3: 2027–37

Richardson, E. G. 1948. The impact of a solid on a liquid surface. *Proc. Phys. Soc.* 61: 352–67

Richardson, E. G. 1955. The sounds of impact of a solid on a liquid surface. *Proc. Phys. Soc.* 68: 541–47

Rodriguez, F., Mesler, R. 1988. The penetration of drop-formed vortex rings into pools of liquid. *J. Colloid Sci.* 121: 121–29

Rogers, W. B. 1858. On the formation of rotating rings by air and liquids under certain conditions of discharge. *Am. J. Sci. Arts, 2nd Ser.* 26: 246–58

Scrimger, J. A. 1985. Underwater niose caused by precipitation. *Nature* 318: 647–49

Scrimger, J. A., Evans, D. J., McBean, G. A., Farmer, D. M., Kerman, B. R. 1987. Underwater noise due to rain, hail, and snow. *J. Acoust. Soc. Am.* 81: 79–86

Scrimger, J. A., Evans, D. J., Yee, W. 1989. Underwater noise due to rain—open ocean measurements. *J. Acoust. Soc. Am.* 85: 726–31

Shaw, P. T., Watts, D. R., Rossby, H. T. 1978. On the estimation of oceanic wind speed and stress from ambient noise measurements. *Deep-Sea Res.* 25: 1225–33

Shin, J., McMahon, T. 1990. The tuning of a splash. *Phys. Fluids A* 2: 1312–17

Sigler, J., Mesler, R. 1990. The behavior of the gas film formed upon drop impact with a liquid surface. *J. Colloid Sci.* 134: 459–74

Sudo, S., Hashimoto, H., Katagiri, K. 1991. Impact between a magnetic fluid drop and a liquid free surface. *Trans. Jpn. Soc. Mech. Eng.* 57B: 75–81 (In Japanese)

Thomson, J. J., Newall, H. F. 1885. On the formation of vortex rings by drops falling into liquids, and some allied phenomena. *Proc. R. Soc. London* 29: 417–36

Wenz, G. M. 1962. Acoustic ambient noise in the ocean: spectra and sources. *J. Acoust. Soc. Am.* 34: 1936–56

Worthington, A. M. 1908. *A Study of Splashes*. London: Longman & Green; reprinted 1963, New York: Macmillan

SUBJECT INDEX

C

T

CUMULATIVE INDEXES

CONTRIBUTING AUTHORS, VOLUMES 1–25

CHAPTER TITLES, VOLUMES 1–25

ANNUAL REVIEWS INC.

a nonprofit scientific publisher
4139 El Camino Way
P. O. Box 10139
Palo Alto, CA 94303-0897 • USA

ORDER FORM

ORDER TOLL FREE
1-800-523-8635
from USA and Canada

FAX: 415-855-9815

Annual Reviews Inc. publications may be ordered directly from our office; through booksellers and subscription agents, worldwide; and through participating professional societies. **Prices are subject to change without notice.** California Corp. #161041 • ARI Federal I.D. #94-1156476

- **Individual Buyers:** Prepayment required on new accounts by check or money order (in U.S. dollars, check drawn on U.S. bank) or charge to MasterCard, VISA, or American Express.
- **Institutional Buyers:** Please include purchase order.
- **Students/Recent Graduates:** $10.00 discount from retail price, per volume. Discount does not apply to Special Publications, standing orders, or institutional buyers. **Requirements:** [1] be a degree candidate at, or a graduate within the past three years from, an accredited institution; [2] present proof of status (photocopy of your student I.D. or proof of date of graduation); [3] Order direct from Annual Reviews; [4] prepay.
- **Professional Society Members:** Societies that have a contractual arrangement with Annual Reviews offer our books to members at reduced rates. Check your society for information.
- **California orders** must add applicable sales tax.
- **Canadian orders** must add 7% General Sales Tax. GST Registration #R 121 449-029. Now you can also telephone orders Toll Free from anywhere in Canada (see below).
- **Telephone orders,** paid by credit card, welcomed. **Call Toll Free 1-800-523-8635** from anywhere in USA or Canada. From elsewhere call 415-493-4400, Ext. 1 (not toll free). Monday – Friday, 8:00 am – 4:00 pm, Pacific Time. Students or recent graduates ordering by telephone must supply (by FAX or mail) proof of status if current proof is not on file at Annual Reviews. Written confirmation required on purchase orders from universities before shipment.
- **FAX: 415-855-9815 – 24 hours a day.**
- **Postage paid** by Annual Reviews (4th class bookrate). UPS ground service (within continental U.S.) available at $2.00 extra per book. UPS air service or Airmail also available at cost. UPS requires a street address. P.O. Box, APO, FPO, not acceptable.
- **Regular Orders:** Please list below the volumes you wish to order by volume number.
- **Standing Orders:** New volume in series is sent automatically each year upon publication. Please indicate volume number to begin the standing order. Each year you can save 10% by prepayment of standing-order invoices sent 90 days prior to the publication date. Cancellation may be made at any time.
- **Prepublication Orders:** Volumes not yet published will be shipped in month and year indicated
- **We do not ship on approval.**

ANNUAL REVIEWS SERIES *Volumes not listed are no longer in print*			**Prices, postpaid, per volume.** USA / other countries (incl. Canada)	**Regular Order Please send Volume(s):**	**Standing Order Begin with Volume:**
Annual Review of		**ANTHROPOLOGY**			
Vols.	1-20	(1972-1991)	$41.00/$46.00		
Vol.	21	(1992)	$44.00/$49.00		
Vol.	22	(avail. Oct. 1993)	$44.00/$49.00	Vol(s). ______	Vol.______
Annual Review of		**ASTRONOMY AND ASTROPHYSICS**			
Vols.	1, 5-14	(1963, 1967-1976)			
	16-29	(1978-1991)	$53.00/$58.00		
Vol.	30	(1992)	$57.00/$62.00		
Vol.	31	(avail. Sept. 1993)	$57.00/$62.00	Vol(s). ______	Vol.______
Annual Review of		**BIOCHEMISTRY**			
Vols.	30-34, 36-60	(1961-1965, 1967-1991)	$41.00/$47.00		
Vol.	61	(1992)	$46.00/$52.00		
Vol.	62	(avail. July 1993)	$46.00/$52.00	Vol(s). ______	Vol.______